MINERAL PROCESSING AND EXTRACTIVE METALLURGY

100 YEARS OF INNOVATION

EDITED BY CORBY G. ANDERSON, ROBERT C. DUNNE, JOHN L. UHRIE

PUBLISHED BY THE
SOCIETY FOR MINING, METALLURGY & EXPLORATION

Society for Mining, Metallurgy & Exploration (SME)
12999 E. Adam Aircraft Circle
Englewood, Colorado, USA 80112
(303) 948-4200 / (800) 763-3132
www.smenet.org

SME advances the worldwide mining and minerals community through information exchange and professional development. SME is the world's largest association of mining and minerals professionals.

ISBN 978-0-87335-371-7
Ebook 978-0-87335-372-4

Library of Congress Cataloging-in-Publication Data

Mineral processing and extractive metallurgy : 100 years of innovation / edited by Corby G. Anderson, Robert C. Dunne, John L. Uhrie.
pages cm
Includes bibliographical references and index.
ISBN 978-0-87335-371-7 (print) -- ISBN 978-0-87335-372-4 (ebook)
1. Minerals--Biotechnology. 2. Extraction (Chemistry) 3. Mineral industries--Technological innovations. I. Anderson, Corby G., editor of compilation. II. Dunne, Robert C., editor of compilation. III. Uhrie, John L., editor of compilation.
TN688.3.B33M564 2014
669--dc23

2013047622

*This book is dedicated to **Robert C. Dunne**,*
the originator and driving force for the symposium and this book.

Contents

Preface

"We stand on the shoulders of giants" is possibly an overused phrase; however, in the production of the raw materials necessary for our ever improving standard of living and the advancement of society, it holds true time and again. With the eye of a historian, it is easy to see the advancement of metallurgy, starting with such valuable historical texts as Georgius Agricola's *De Re Metallica*, widely accepted to be the first text on minerals processing, to other groundbreaking and still invaluable books such as Arthur F. Taggart's *Handbook of Mineral Dressing* and Norman L. Weiss's *SME Mineral Processing Handbook*, to the plethora of high-quality journals and books of today. Through these texts, time reveals how innovation continues to push our industry forward.

And as such, the Minerals and Metallurgical Processing Division of the Society for Mining, Metallurgy & Exploration is very proud to publish these proceedings of the successful symposium of the same name held in 2013: *Mineral Processing and Extractive Metallurgy: 100 Years of Innovation*.

Innovation is different than invention or improvement. Innovation represents creative thinking that provides value and the capacity for a business to achieve the future it desires and hence allows a new avenue or path for growth. Other books are available on how innovation has worked in the commercial consumer sector, but very little information exists for the mining industry. This volume presents a century of the innovation drivers that have changed the U.S. mining industry with topics from experts in pyro-, hydro-, and electrometallurgy. In addition, current and future trends, developments, and improvements in these areas are discussed in depth, and likely areas for future innovations are explored.

The labors of many have gone into the preparation of this volume. We are indebted to the individual authors for contributing their life's knowledge, experience, and passion to this text, but also to their respective employers for allowing them the time and resources necessary to produce and share their work. Thanks must also be given to the employers of the organizing committee, Newmont Mining Corporation, and Colorado School of Mines for allowing the time to develop the idea of the symposium and to see it through to fruition.

We thank Henry Kasaini of RER for graciously and expertly stepping in on short notice to give the rare earths presentation. Also, we thank Barney Guarnera of Behre Dolbear for his opening keynote presentation.

The symposium could not have been possible without the direct support and sponsorship of Tetra Tech Engineering, Cytec Industries, Weir International, Hatch Engineering, and, most importantly, Air Products.

We are further indebted to Jane Olivier, Diane Serafin, Tara Davis, and Tessa Baxter for their tireless support. We also acknowledge Christie Kelsay, Yessica O'Collor, Katherine Moyle, and Evody Tshijik Karumb for their tireless efforts typing and formatting the manuscripts.

Most importantly, we extend our gratitude to those who are directly responsible for the innovation that advances the mining industry: from the highest-level senior executive responsible for corporate metallurgy to the metallurgical technician standing on the mill floor. These individuals are truly our giants, the ones with the broad shoulders, who are responsible for the advancements of yesterday. These are the innovators who will lead us to the improvements of tomorrow. It is the metallurgical community to whom we give our deepest, most sincere thanks.

Time and conditions change so rapidly that we must keep our aim constantly focused on the future. —Walt Disney

Change is the law of life. And those who look only to the past or present are certain to miss the future. —John F. Kennedy

MANAGING EDITORS

Corby G. Anderson
Robert C. Dunne
John L. Uhrie

CONTRIBUTING AUTHORS

Peter Amelunxen
Corby G. Anderson
Joseph L. Anjier
Bo R. Arvidson
Wolfgang Baum
David A. Beattie
Michael D.S. Blois
Jeremy Bosman
James A. Brierley
Nikhil Dhawan
Meg Dietrich
Robert C. Dunne
Hassan El-Shall
Magnus Ericcson
Mark T. Erickson
Raymond S. Farinato
Linda A. Figueroa
Brian Flintoff
Kevin Galvin
Andrea R. Gerson
Stephen Grano
Karl A. Gschneidner Jr.
Olivier Guyot
Fathi Habashi
Len Harris
Brian R. Hart
Steve Hearn
J. Brent Hiskey
Glenn E. Hoffman
Peter Holtham
Rick Honaker
Sarma S. Kanchibotla
Martin Kuhn
Jens Lichter
Larry A. Lien
Gerald H. Luttrell
Catherine "Carrie" MacDougall
John O. Marsden
Mick L. McCaslin
Jay McCloskey
Jeff McKay
Stephen Morrell
Brij Moudgil
Keith J. Mounteer
D.R. Nagaraj
Tim Napier-Munn
Michael G. Nelson
C. Alex Norgren
Daniel Norrgran
Brian Park
Eric H. Partelpoeg
Joe Pease
Michael Phillips
Neville Plint
Raj K. Rajamani
Samira Rashidi
Robert W. Reisinger
Tim Robinson
Kym Runge
S. Andrew Sass
Henry Schnell
D. Erik Spiller
Patrick R. Taylor
Larry Twidwell
Roger St.C. Smart
Regis Stana
Thomas Wildeman
Stephen Woollam
Hermann Wotruba
Dirk van Zyl
Katherine Vatterrodt
Andre Vien
Courtney Young
Massimiliano (Max) Zanin
Patrick Zhang

EDITORIAL REVIEW COMMITTEE

Corby G. Anderson
Dennis Appelhans
Sarah Appleby
David Brosnahan
Charles Bucknam
Robert C. Dunne
Aidan Giblett
Steve Hearn
Ronel Kappes
John O. Marsden
Mike Nees
Patrick R. Taylor
John L. Uhrie

SESSION CHAIRS

Corby G. Anderson
Michael D.S. Blois
Robert C. Dunne
Rick Gilbert
Rich Johnson
Mark Jorgensen
Mike Moats
Holgar Plath
Jayson Ripke
Anthony Staley
Patrick R. Taylor
Phil Thompson
John L. Uhrie
Courtney Young

Introduction

Supply and Demand: Innovation Drivers in the Minerals Industry

Magnus Ericcson
Luleå University of Technology, Sweden

ABSTRACT: Reports of economic crises, investor anxiety and the impact of climate change all help paint a gloomy picture of the future. More optimistic predictions may be met with scorn and accused of lack of 'reality.' Nonetheless, one of the many lessons of history is that accurate factual data provides by far the best basis for discussion of future alternative paths of development, optimistic or otherwise (Raw Materials Database 2012).

The objectives of this paper are to show: (1) that positive future trends related to mining and metals are evident (Ericsson and Hodge 2012); and (2) that a number of gloom-laden myths should be refuted.

MINERALS AND MINING—A POWER BASE

Economic development generates demand for metals, and production of metals has created powerful nations. Over the past one hundred and fifty years a succession of countries has built their strength on metals. Europe dominated metals production until the mid-19th century, when the rise of the USA was closely linked to expansion of its production of steel, copper and other metals. The Soviet Union chose a different, but nevertheless metals based industrialisation strategy exploiting the minerals of the Urals and Siberia. All through the 20th Century the centre of gravity of mining has shifted south of the equator: mineral production in Latin America, Oceania and Asia has grown consistently (Carl-Wolfgang 1974). African mine production grew until a peak was reached in the late 1970s to early 1980s. In the early 21st Century, emerging economies account for much of the metals mined globally.

The period from 1850 coincided with a decline in the traditional Asian power houses; India was colonised by Britain and ancient China collapsed. When these populous and historically dominant global economies together with other fast growing Asian countries began to re-emerge on the world scene in the early 21st Century, the global mining industry was shaken by its roots. Demand grew by double digit figures annually and production of metals and minerals in Asia and around the world soared.

During the 20th Century metal production increased manyfold, but during this period there have also been several peaks and troughs. Mining is a cyclical industry and historical peaks in production have always been followed by periods of continued growth. Metal prices have also shown cyclical swings even though over the entire century the trend has been downwards. In spite of this, the mining industry has been able to meet the continuously increasing demand.

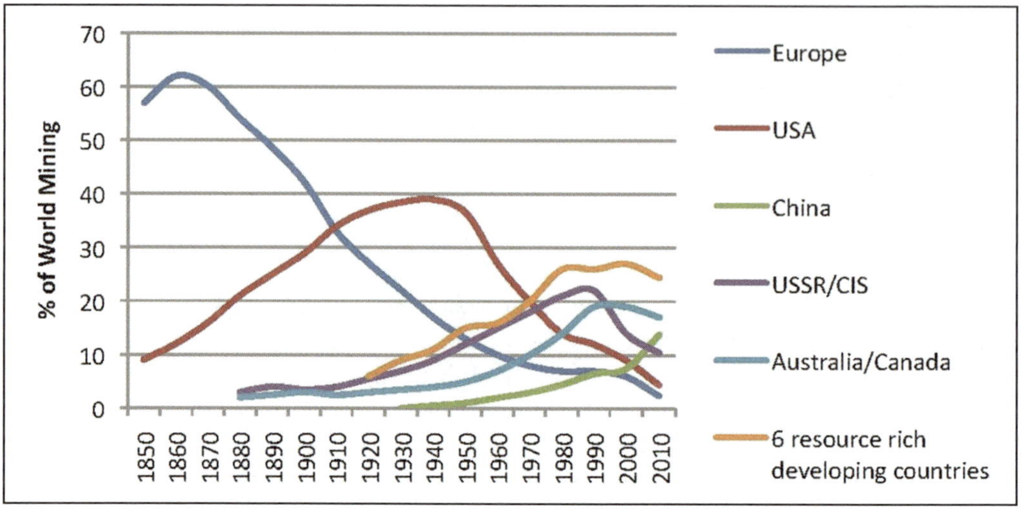

Figure 1. Location of world mining 1850 to present (Raw Materials Database 2012)

In the 2010s, China is one of the largest mining countries of the world in a neck to neck race with Australia for the top position. Metal production has stagnated in Europe including Russia and the former Soviet republics. African mining has also declined but a revival is under way. There is still an important mining sector in the US, contrary to popular belief. If also coal is included China accounts for roughly a third of all mine production in the world.

CONTINUED GROWTH IN SOCIETAL DEMAND FOR METALS

The following three factors drive the inevitable growth in societal demand for metals (Graedel 2005):

1. Continuing population growth across the world
2. Increasing urbanization and industrialisation, particularly in emerging economies
3. The special characteristics that make metals essential for the provision of goods and services needed in modern societies

At base, the continuing growth in demand for metals comes from the millions of individuals in the emerging economies who are striving for a better material standard of life at a time when demand in developed countries is still strong, though stable. Even if recycling rates are increased, a large number of new mines are needed to support individuals, communities and countries emerging from poverty. Nothing short of a global environmental or social disaster—whether a plague or total climate collapse—can more than temporarily stop the growing demand for metals and minerals.

Metals and minerals are key to all services and infrastructure that are used in today's society: food and water supply, sewage treatment, heat and light, shelter, transportation, construction, manufacturing, education, health, communication, entertainment, the arts, tourism, and the vast range of associated consumer goods and services. As the material standard of living increases, metal and mineral demand also grows. This pattern has been followed by all developed countries. When per capita income in a country reaches US$5–10,000 per year, metal demand grows particularly quickly. When countries with huge populations such as China and India go through phase of development, the effects on metal demand is dramatic.

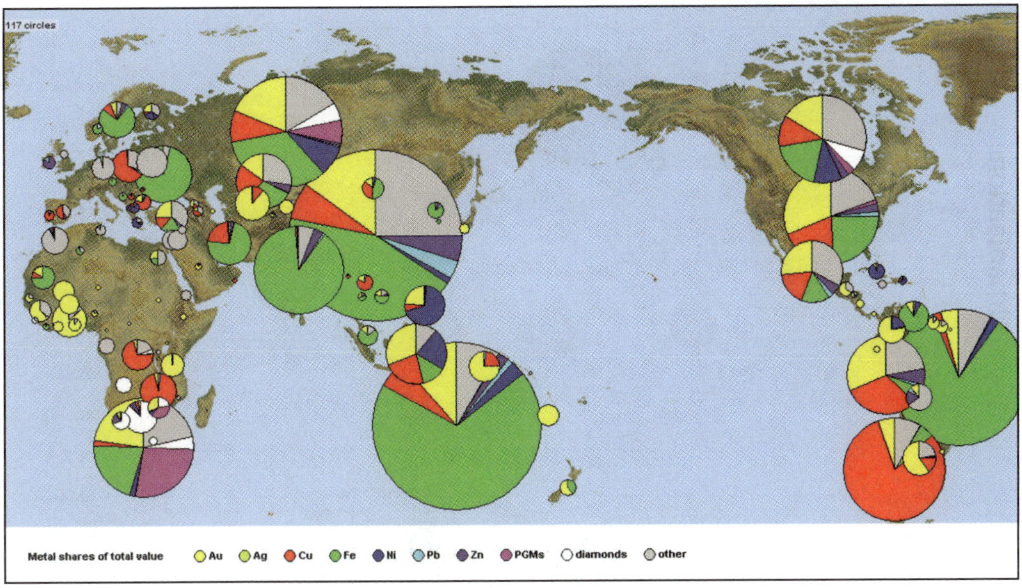

Figure 2. Value of mine production by country in 2011 (% of world total), distributed by metal and excluding coal (Raw Materials Database 2012)

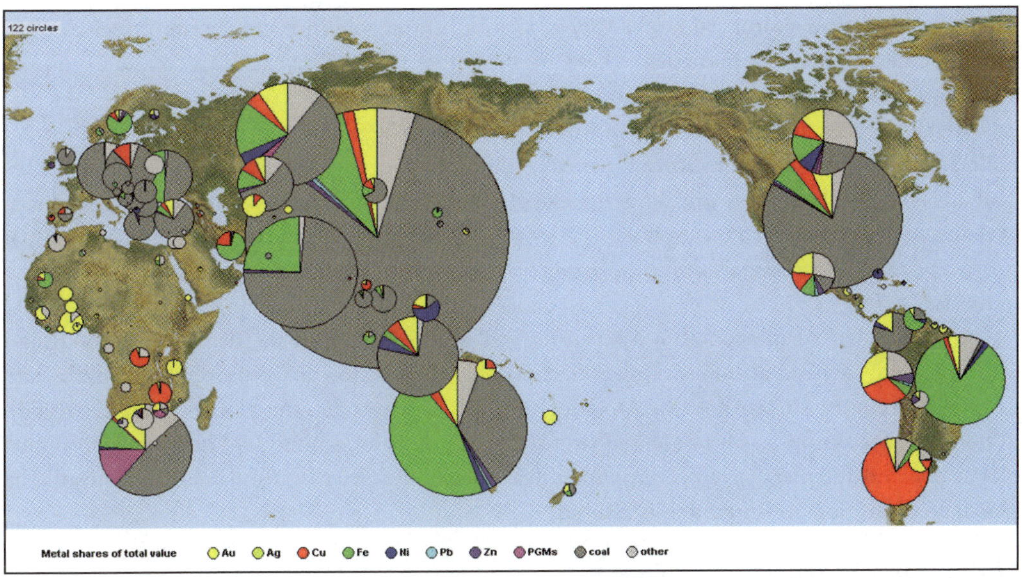

Figure 3. Value of mine production by country in 2011 (% of world total), distributed by metal and including coal (Raw Materials Database 2012)

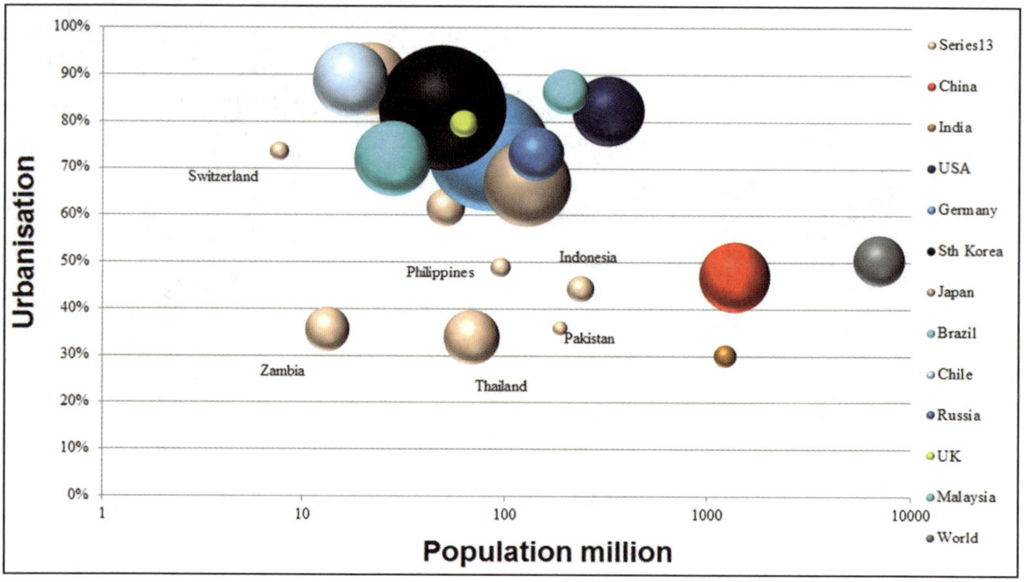

Figure 4. Copper use and urbanization (Raw Materials Database 2012)

SPECIAL CHARACTERISTICS OF METALS

Metals have particular properties which give them a central role in everyday life in general and economic development in particular. These are their high strength, capacity to conduct heat and electricity, reasonable cost, and the aesthetic appeal of their appearance.

Metals are elements and therefore have the potential to be indefinitely recyclable. While other materials can replace and substitute for metals, the scale and cost generally give metals a significant advantage. Substitution will certainly take place but with increasingly complex final products, exchange of one metal by another is getting more and more difficult in the short term and only if relative prices between metals shift in a lasting way will substitution have strong effects of demand patterns.

It is true that when economic downturns lead to a reduction in the demand for goods and services, metal demand also falls. However, inevitably these swings have proven to be cyclic and as growth recovers, so too does demand for minerals and metals. The global financial crisis in 2008/2009, for example, was not felt equally across all countries, smoothing the overall continued global demand for metals. However, with metals demand driven by fundamentals beneath the global economy, long-term growth is assured.

SUPPLY OF METALS AND MINERALS

Mining is dominated by production of three metals: iron ore, copper and gold which together account for 66 percent of the total value at the mine stage of all metals produced globally in 2012. These three are by far the most economically important metals.

The scale of production varies from almost 2 billion tonnes of iron ore products every year, over a little less than 20 million tonnes of copper, 2,000 tonnes of gold and only 200 tonnes of PGMs (Platinum Group Metals, mainly platinum and palladium). While the remaining share of the value of mine production may not be as economically significant from a global perspective,

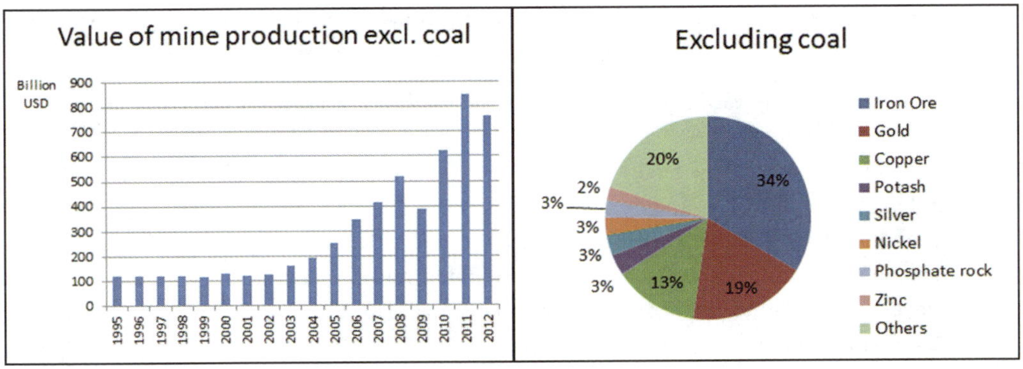

Figure 5. Value of mine production by country (2011), metals and industrial minerals (Raw Materials Database 2012)

these other minerals and metals also play a vital role in society. For example, nickel, vanadium, chrome, and platinum group minerals (PGMs) are all critical in today's economy, but the fact that they are produced in much smaller volumes makes them much less critical from an absolute supply aspect. Rare earths illustrate an extreme case. Total production of these globally is a few hundred thousand tonnes annually only. They are presently produced almost exclusively in China, but a handful of relatively small mines could cover most of global demand at a moderate investment. Their true strategic importance is hence much lower than generally perceived. The sensitivity to price increases for most of these products is further low as their contribution to total costs for the consumer is miniscule, i.e., the price elasticity is low.

The increased output of metals and the increased value of most metals have resulted in a rise in value of the global metal mining industry from US$128 billion in 2000 to US$849 billion by 2011. This has opened a new opportunity for emerging economies to benefit from the mining industry as a lever of economic development.

Even if the volume of final product differs substantially, the amount of ore that has to be moved to produce the iron ore, the copper and the gold demanded annually is of the same order of magnitude, 1,500–3,000 Mt each. This depends on the grade of the ores: gold grades are around 1 g gold/tonne of ore, copper ore contain around 1% metal, while iron ore with a grade of around 60%. The production of ores containing PGMs is however much smaller, only 100 Mt. Most ores are produced in open pit operations, as much as 85% varying from metal to metal. Zinc is the metal being most widely produced by underground methods, over 50% of total world production outside China. There was a fast increase in open pit mining during the end of the 20th Century but during the past decade, the share has been more or less constant. In spite of new exploration technologies penetrating deeper into the earth's crust, it will still be easier to locate a huge low grade surface ore body than a richer one at depth. With lower commodity prices and a growing focus on costs, advantages of scale will be crucial and large scale open pit operations which undoubtedly give much lower costs than underground operations will continue to dominate.

The mining industry has already been trading in high value metals such as copper, gold, and tin for centuries. With the development of huge ocean going vessels in the late 20th Century, international trade of low value bulk mineral commodities such as iron ore, coal and bauxite became possible. The industry turned truly global before most other branches of the world economy.

A few structural trends are identifiable which contribute to increased production cost in the mining sector and in the mid-term—5 to 10 years—sets a floor for metal prices:

- New mines are increasingly developed in remote locations, far from markets and necessary infrastructure.
- New mining areas are generally found in areas with extreme temperatures, arid conditions and high altitudes.
- More complex, lower grade ores have to be mined when the richer deposits are depleted making extraction technically more difficult and lowering yields.
- Ore bodies are found at greater depth whether suitable for open pit or underground mining.
- Longer permitting processes are now typical.

It has always been a long process from the exploration find to mine construction and startup. To do this in less than 10 years is virtually impossible today, and in many cases, major mines will take even longer than that. Hence the window of opportunity for developing countries will remain open for a long period of time.

R&D AND EXPLORATION

In recent decades when metal prices have dropped, mining companies have made cost cuts. These have often been to research and development budgets, including exploration. Yet these areas are essential for a successful bounce-back when the market turns up again.

Given the difficulties and risk to introduce new technologies in existing mines, the mining industry has traditionally been conservative and slow in adopting new technology. The industry has further relied on outside technology providers and there has only been limited proprietary

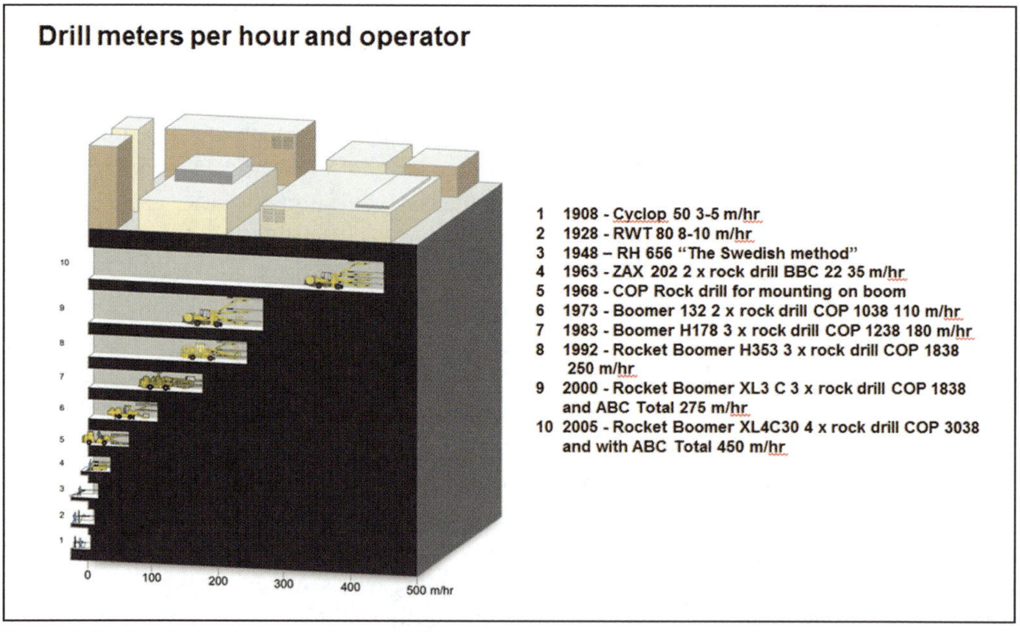

Figure 6. Rock drill development 1905–2010 (courtesy of Atlas Copco)

technologies developed in-house. Nevertheless technological developments have increased productivity over the past century by as much as 100%.

In the coming years the mining companies and the equipment suppliers which set aside more funds to R&D will likely become gradually more competitive. However, such changes will take time, and in the near future, the focus will most probably continue to be scaling up rather than introducing completely new concepts. The move away from the dominating, century old batch-wise drilling, blasting, mucking concept will however be slow. Initially new concepts such as in-situ leaching, continuous mining and blast-free mining will be introduced in special applications. Hard-rock cutting, for example, will most likely first get introduced in narrow vein and reef mining. These shifts, whenever they get going, will gradually lead to hydrometallurgical processes being employed, applied to ores with lower grades, but still being more efficient than the traditional processes.

With lower commodity prices and a growing focus on costs, advantages of scale will be crucial and large scale open pit operations, which undoubtedly give much lower costs than underground operations will continue to dominate. In spite of new exploration technologies penetrating deeper into the earth's crust, it will still be easier to locate a huge low grade surface ore body than a richer one at depth.

From a societal point of view the strong, short term variations up and down in exploration expenditure are increasing the uncertainties about future metal production levels. These fluctuations introduce inefficiencies with effects reaching far into the future. Many projects get started only to be closed down before they have come to an optimal stage and can be properly evaluated. There are considerable costs associated with starting and stopping projects, which result in a waste of funds and human resources when projects are operated in an on/off manner rather than extended until they are completed according to original plan. The industry, and society, would be better served by a steady flow of exploration and development projects.

If metal production grows, which it has done in recent years, there should be an increase in exploration expenditure to deal with the increased speed of depletion. There are also some doubts over the success rate of each exploration dollar spent and there are indications that the success rate of exploration is dropping. This might partly be explained by the fact that exploration is getting more difficult because of several factors including the five discussed above. There could also be a decrease in exploration efficiency for other reasons, such as difficulties in attracting the best students to study geology (Schodde 2012).

To break this recurring pattern, research and development and exploration expenditures need considerable expansion in good times to enable it to continue during less good times. Whether such expansion should be on an individual company or on an industry basis or in cooperation with governments is a moot point.

CORPORATE STRUCTURE

In spite of M&A activities during recent years, the globally active mining companies are relatively small, compared with many other branches of industry. A slow increase only of corporate concentration is the result of these corporate machinations, but there are nevertheless a number of large transnational companies active all around the world.

Table 1. Top metal mining companies 2011

Rank world 2011	Company name	Country	Share of value of world mine prod of all minerals 2011 (%)	Cumul share of world 2011 (%)	*Main metal value share (%)
1	Vale SA	Brazil	7.725	7.72	Fe 87
2	BHP Billiton Group	Australia	5.374	13.10	Fe 67
3	Rio Tinto Group	UK	4.929	18.03	Fe 80
4	Anglo American Plc	UK	2.421	20.45	Fe 38
5	Barrick Gold Corp	Canada	1.685	22.13	Au 87
6	Corporacion Nacional del Cobre (Codelco)	Chile	1.600	23.73	Cu 90
7	Freeport-McMoran Copper & Gold Inc	USA	1.591	25.33	Cu 73
8	Xstrata Plc	Switzerland	1.384	26.71	Cu 53
9	Norilsk Nickel Mining & Metallurgical Company	Russia	1.234	27.94	Ni 39
10	ArcelorMittal	UK	1.137	29.08	Fe 100
11	Newmont Mining Corp	USA	1.098	30.18	Au 91
12	Fortescue Metals Group Ltd	Australia	0.991	31.17	Fe 100
13	Cliffs Natural Resources Inc	USA	0.852	32.02	Fe 100
14	Metalloinvest Holding Company OJSC	Russia	0.833	32.85	Fe 100
15	Anglogold Ashanti Ltd	South Africa	0.826	33.68	Au 99
16	State of India	India	0.802	34.48	Fe 91
17	Grupo Mexico SA de CV	Mexico	0.801	35.28	Cu 79
18	Vedanta Resources Plc	UK	0.775	36.06	Fe 56
19	Eurasian Natural Resources Corp Plc	UK	0.731	36.79	Fe 60
20	Goldcorp Inc	Canada	0.687	37.48	Au 70

Three trends among these major companies are visible:

1. Diversification in terms of metals produced
2. Vertical integration towards smelting and refining
3. More companies are headquartered in emerging economies

By producing several metals and minerals, the risk of a declining price of one commodity is mitigated. The number of one-metal companies is decreasing. To become one of the majors, you have to be involved in at least two of the three dominating metals: gold, copper and iron ore. The global outreach of the mining industry further reduces the geographical risk. Vertical integration ensures that variations in metal prices are fully reflected in the value of the concentrates produced at the mine. As production moves towards emerging economies countries, so is also control over these companies. The only major mining country which does not host any of the majors is really China. There is however no doubt that there will be also Chinese ownership of one of the majors. The question is only when this will happen, probably sooner than later.

POLITICS OF MINING

During the last decades of the 20th Century, mining was considered a "sun-set" industry, particularly in West Europe, where many mines received state subsidies and were gradually shut down leaving huge environmental debts behind and often socially run-down areas. Mining became a political and economic burden. Political focus switched to growth sectors such as IT and biotechnology. With the boom beginning around 2005, this has partially changed. The mining sector is certainly still accused of lacking sustainability and, at least in Europe, recycling is seen as a panacea to many supply problems. Security of supply of the so called critical minerals/metals is attracting growing interest from politicians. The reasons for this renewed interest are mainly the extended boom with high raw material prices in general and the fear that China will, in various ways, affect the global demand/supply balances to the detriment of European metal users. Linked to the growth in demand, a wide spread fear of running out of resources as expressed for example in the various peak theories has gathered followers.

It is however important to recognise that the amount of metals available is not generally a geological problem but mostly an economic issue. When prices rise due to increasing demand, this is not a sign of metals running out but of an imbalance between demand and supply. Further, when prices increase, the resources increase as with higher prices more deposits can be extracted profitably than at lower prices. In the period from 1950–2000, resources available for mining have also increased manyfold in spite of total production in this period being higher than in all previous history. Furthermore, metals are elements and hence all copper, which has been mined over the centuries, is still around and in principle available for recycling. Hence there is always need for additional volumes of virgin mined material in growing economies such as in most of the emerging world today.

Another important corollary of the peak theories (Hubbert 1962; Laherrere 2005) is that the emerging economies of the world will not be given the opportunity to develop economically and socially using metals and minerals as key inputs. This route has been successfully used by the industrialised countries of today during the 19th and 20th Centuries. Recent developments in China, India and other developing countries show that this route is followed also today, and that metals and minerals have not lost their importance in spite of technological progress bringing new materials and processes. This does not preclude that emerging economies can develop more efficiently than has been the case in the past. As long as the peak theories are only discussed among academics, they will not cause much harm, but if peak theorists manage to gain support from politicians and political action is taken based on the unproven theories, damage could be done. Today's emerging economies would perhaps be forced to try to find new, so far unknown routes for their future economic development. They would not be able to use metals and minerals for their quick and efficient economic development. Exactly what political and social implications this will have is difficult to imagine, but when it becomes clear that the peak theory was not sufficient to predict *when* metals are becoming seriously scarce, it is certain that the result could be serious. The mistrust between developed and emerging countries, between the haves and have-nots could deepen further and to rebuild it will be take long (Ostrowski and Dannreuter 2013).

Metals have often been called non-renewable resources implicating a poor sustainability. It is certainly true that a specific mine can be depleted and there are numerous examples of mining districts turning to ghost towns. But on the other hand, a mine cannot be removed, it stays where the deposit is, and there are equally many examples of mining towns that thrive after more than 100 years in operation. How many other industries remain in the same place for centuries? Mining can also act as the catalyst for cross sectoral linkages. The Nordic countries, Canada and Australia are examples of countries successfully using mining as a lever for economic and social development during the past century. Gauteng and the Antofagasta province in northern Chile are examples of regions within emerging economies which have developed a sustainable economy out of a mining boom.

Political worries over China taking control over mineral and metal resources of the world are grossly overstated. High import dependence is a serious economic and political problem mainly for the Chinese themselves but not so much for other nations. As discussed earlier, the supply of rare earths, the only metal which China controls tightly, is a small problem in spite of their critical importance compared to the supply of iron ore, copper, phosphate and potash (the two latter ones are key components of fertilizers). The commodities are demanded in high volumes requiring huge annual investments to develop. Chinese import of iron ore alone is costing more than US$50 billion per year. The total value of all rare earths production is more difficult to determine, but probably only around a few hundreds of million dollars per year.

The Chinese quest to control a larger share of its imports has not thus far been very successful. In Africa, Chinese taking control over large chunks of the mineral resources has been front page news for a long time. The worries seem, however, to be mostly unfounded. In total, the Chinese control around 1% of total African metal production (Ericsson 2011). The largest controlling entity is the major mining company Anglo American with 28% of the region's production under its control. Other transnational companies such as Rio Tinto, Barrick and others are also each in control of a larger share of African production than the Chinese.

It is true that the Chinese have been trying to their share of African mining, but most often with only limited success. There are several reasons behind this failure. Most important is the fact that the Chinese in general lack large scale mining experiences, as most of China's domestic mines are small with low productivity. This does not mean that China will not acquire control over a larger share of African mining nor of the world in the future, but it is important to realise that it has not happened yet. There is still time to try to find solutions including the Chinese rather than excluding them. This opens opportunities both for transnational companies and African governments. A healthy competition could replace the previous colonial quest, provided African countries find the human resources to manage their own mineral resources (UNECA 2011).

FUTURE

Overall, although there are always uncertainties about the future, this review indicates that future trends for the mining industry remain positive. It is acknowledged that no predictions can be infallible—even ones based on extensive datasets–but nonetheless an optimistic view remains.

Population growth and economic development in the emerging economies are still positive and provide a strong base for growth in metal demand and hence the need for increased mine production and new investment projects. The mining industry is set, if properly managed, to provide both opportunities for economic and social development and metals necessary for an improved standard of living.

New mineral resources remain to be discovered and developed. Metals are not running out. New technologies will help find and develop these new deposits. These technologies are only developed if considerable efforts are set aside for R&D. The decision to do so is one which governments and companies can make together. If new technologies are not introduced, metal prices will remain on much too high a level for a long period. Metal prices will however remain high also because of lower grades and larger depths etc of the deposits exploited. These trends will also mean higher environmental and social costs of mining which must be covered. Furthermore, in the short term there is a lack of trained and experienced staff. Although this is not a structural issue, it is nevertheless a perennial problem, which has continuously worsened over the past 20 years. It will take at least another 10–15 years to solve.

The trend of metal production moving to the emerging economies will continue. Even if mid-term aspects of political stability will steer investments away from some emerging but resource rich countries. There are two major land areas in the world (excluding Antarctica) which are not fully explored: Africa and the circumpolar Arctic region.

While the industry is facing many difficult challenges, for many remote regions of industrialised countries and many entire emerging economies mining, remains the only way to kick-start economic growth and hence bring many inspiring opportunities.

ACKNOWLEDGMENTS

I am indebted to Anthony Hodge of ICMM and Gill Burke of Raw Materials Group for providing some of the present text. Responsibility for all conclusions remains my own.

REFERENCES

Carl-Wolfgang, S., 1974. Die Zukunft der Metalle, Suhrkamp Verlag, Frankfurt am Main.

Ericsson, M., and Hodge, A. 2012. Trends in the mining and metals industry. ICMM, Brief 3.

Ericsson, M. Mineral supply from Africa: China's investment inroads to the African mineral resource sector. SAIMM, 111, pp. 1–4.

Graedel, T. 2005. Introductory presentation, Second Annual Meeting Steel Advisory Committee, Yale University, March.

Hubbert, M.K. 1962. Energy Resources, National Academy of Sciences, Publication 1000D. pp. 81–83.

Laherrere, J. 2005. Forecasting production from discovery, ASPO, Lisbon.

Ostrowski, W., and Dannreuter, W. 2013. Global resources: Cooperation and conflict, Palgrave McMillan, London.

Raw Materials Database. 2013. www.IntierraRMG.com.

Schodde, R. 2012. Recent trends in mineral exploration—Are we finding enough? Exploration Seminar, Luleå University of Technology, September.

UNECA. 2011. Minerals and Africa's development, International Study Group on Africa's Mineral Regimes.

ABOUT THE AUTHOR

Magnus Ericsson's key qualifications lie within the areas where mineral economics, mineral policy/strategy, mineral legislation and taxation, institutional organization, and investment promotion meet. He has extensive experience in the field of mineral policy, including taxation and legislative issues, and investment promotion strategies and their implementation and has performed several analyses of mineral industries and their competitiveness. He is an advisor to the African Union Commission and the United Nation Economic Commission for Africa and is at present deeply involved in the setting up of the African Mineral Development Centre in Addis Ababa. He has been appointed rapporteur by the European Commission of the European Innovation Partnership on Raw Materials for international issues. Since the mid-2000s, Ericsson is one of the driving forces behind the Nordic Mining for Development Initiative. He has a wide network of contacts in China, India, and Japan. Since the early 2000s, he has had special focus on the iron ore industry and has been responsible for publication of the annual iron ore market review on behalf of UNCTAD, Geneva. In the mid-1980s, he was one of the founders of Raw Materials Group and developed the global supply side database Raw Materials Data. In 2009, Ericsson was appointed professor of mineral economics at the Luleå University of Technology.

Rules of Engagement for the Innovator in the Minerals Industry

Martin Kuhn
Minerals Advisory Group, Tucson, Arizona, United States

Robert C. Dunne
Newmont Mining Corporation, Denver, Colorado, United States

ABSTRACT: The word *innovation* derives from the Latin word *innovates*, "to renew or change." Innovation differs from invention and improvement and possibly the more useful of its many meanings is *significant positive change*. Innovations theory that seeks to explain how, why, and at what rate new ideas and technology spread through cultures was first started in 1903 by seminal researcher Gabriel Tarde. Clayton Christensen (2003) classifies innovation into three categories: efficiency innovations, which produce the same product more cheaply; sustaining innovation, which turns good products into better ones; and disruptive innovations, which transform expensive, complex products into affordable, simple ones. A company's biggest potential for growth lies in disruptive innovation. A number of books and survey information are available on how innovation has worked (or not worked) in the commercial consumer sector but very little information exists for the mining industry. In order to get a better understanding of how innovation has worked in the mining industry a survey was undertaken using a selected group of people that have been involved in this industry over the last 50 years. A review of this survey is provided.

DEFINITION OF INNOVATION

The word *innovation* derives from the Latin word *innovates*, which is the noun form of *innovare* "to renew or change," stemming from *in*, "into" plus *novas*, "new." Innovation differs from invention in that innovation refers to the use of a better and, as a result, novel idea or method, whereas invention refers more directly to the creation of the idea or method itself. Innovation differs from improvement in that innovation refers to the notion of doing something different rather than doing the same thing better. Possibly more useful of its many meanings to be found in dictionaries, the most potent is *significant positive change* (Berkun 2010). If the offering represents a significant positive change for whomever it is offered to, by definition, it's an innovation. This calls into question statements such as "we innovate every day" or "we are in the innovation business" (Berkun 2010) because if something is done regularly, how can it represent significant change?

It would appear that innovation in its widest sense represents fresh thinking that provides value and the capacity for a business to achieve the future it desires and hence allows a new avenue or path for growth.

THE SCIENCE, METHODOLOGY, AND DEVELOPMENT OF INNOVATION

The history of innovation is large enough that all the sayings, from Plato's famous "Necessity is the mother of invention" to Emerson's "Build a better mouse trap and the world will beat a path to your door" hold a lot of truth.

Diffusion of innovation is a theory that seeks to explain how, why, and at what rate new ideas and technology spread through cultures. It would appear research was first started in this field in 1903 by seminal researcher Gabriel Tarde, who first plotted the S-shaped diffusion curve (Tarde 2010). Everett Rogers, a professor of rural sociology, popularized the theory in his book *Diffusions of Innovations* (1962). He explains that diffusion is the process by which an innovation is communicated through certain channels over time among the members of a social system. The innovation adoption curve of Rogers is a model that classifies adopters of innovations into various categories, based on the idea that certain individuals are inevitably more open to adaptation than others. Innovators and early adopters make up only a small proportion of any population. Another important concept described by Rogers (1962) is the S-shaped adoption curve where successful innovation goes through a period of slow adoption before experiencing a sudden period of rapid adoption and then a gradual leveling off (forms an S-shaped curve). Rapid expansion of most successful innovations will occur when social and technical factors combine to permit the innovation to experience dramatic growth.

Both Csikszentmihalyi (1996) and Medina (2008) studied the areas of creative thinking and problem solving. They looked at how creative people worked and lived to understand their perceptions of innovation, unfiltered by the often stifling and occasionally self-defeating rigors of hard science. One conclusion was that creativity arises from the synergy of many sources and not only from the mind of a single person. Simonton (1999) on the other hand examined the common traits that produce exceptionally creative people. He proposes that creativity can best be understood as a Darwinian progression of variation and selection. Koestler (1964) concludes from his studies that all humans have the capacity for creative activity. However, this is usually suppressed by the automatic routines of thought and behavior that dominate ones daily lives and routine. Furthermore, he suggests that people are at their most creative when rational thought is suspended, for example, in dreams and trance-like states. Simonton (1999) investigated the personality traits of exceptionally creative people (e.g., Darwin, Shakespeare, Einstein). Johnson (2010) explores and explains how particular environments will generate new ideas continually while other environments appear to suppress idea generation.

Basalla (1989) and Utterback (1994) provide details of innovations that have dominated history, examples are the wheel, the hammer, the automobile, the truck and the typewriter. Utterback goes on to show that the emergence of a dominant invention is not necessarily predetermined, but is the result of the interplay between technical and market choices at any particular time. Drucker (1986) comments that entrepreneurs who start out with the idea that they will make it "big-and in a hurry-can be guaranteed to fail." He concludes they are bound to do the wrong things and the outcome is more than likely to be nothing more than a technical 'virtuosity.' Csikszentmihalyi (1996) points out that the transformation of an idea into reality takes up the most time, resources and involves the hardest work. Hargadon (2003) and Berkun (2010) both provide insights to how innovations take place. They claim that revolutionary innovations do not result from "flashes of brilliance" or "eureka moments" but are really about creatively recombining ideas and objects from past and present technologies in ways that spark new technological revolutions. Furthermore it is argued that romantic notions and trivializing innovation undermines the ability to pursue breakthrough/disruptive innovations.

Burns and Stalker (1961) provide insights on the different management styles, structures (organization, team, culture) and systems that have developed over time within businesses to cater for Research & Development and innovation departments. On the basis of the findings recommendations are provided for managing organizations that operate under a stable/predictable business model, those that have a constant predicable rate of change of novelty in their business and a business that experiences ongoing change to cater for continuous innovation. Raynor (2007) suggests that the only way for a company to successfully plan for the future if it chooses to pursue innovation is to develop practical strategies based on multiple choices that respond to the different requirements of several possible futures (i.e., hedge one's bets). He suggests that 'strategic uncertainty' be managed by having a portfolio of options that incorporate scenarios of the future with options and strategies for each scenario. Interestingly, Steve Jobs, founder of Apple and Pixar, was once asked how does one systematize innovation? His answer was, "You don't." The conclusion from this is that one might have a system to try to manage innovation, or a strategy for managing the risks of new ideas, but this is very different to systematizing innovation.

Christensen (1997) classifies innovation into three type categories, efficiency innovations, which produce the same product more cheaply, sustaining innovation, which turns good products into better ones, and disruptive innovations, which transform expensive, complex products into affordable, simple ones. According to Christensen a company's biggest potential for growth lies in disruptive innovation, the other types could just as well be called ordinary progress and usually does not create more jobs or business. The problem with disruptive innovation is that it can take five to eight years to bear fruit and so companies lose patience.

How can one make innovation truly happen, for this to occur innovation must become part of a company's culture (Skarzynski and Gibson 2008). One of the most difficult objectives for any business is to achieve continuous growth. It seems that this is not due to a lack of great ideas or capable managers. Growth fails because organizations unwittingly strip the disruptive potential from new ideas before they ever see the light of day (Christensen and Raynor 2003).

Chesbrough (2003) points out that while innovation is important to a company's growth the usual "old process" of managing innovation does not seem to work anymore. Furthermore, while ideas may abound in an organization the internal research organization is less effective. He promotes the concept of Open Innovation that he says is a paradigm shift from what he refers to as the old Closed Innovation model. This model he says promotes the concept that successful innovation requires total control of all aspects of innovation (ideas, development, servicing and financing) within a company. Open Innovation promotes the concept that firms can and should use internal and external ideas as well as external partnerships to advance technologies to market. Crowdsourcing (Howe 2006) is a tool that can be applied in conjunction with the Open Innovation model to enhance problem solving by employing a large external network of diversified people. Ogle (2007) points out that creative breakthroughs are born when individuals and groups access new *idea-spaces* and exploit the principles that govern them. These include "hotspots and small-world networks."

THE MINING INDUSTRY IS DIFFERENT FROM THE MANUFACTURING INDUSTRIES

Mines are built around mineral ore deposits that are today to be found in remote locations that usually have limited or no infrastructure (i.e., water, energy, skilled labor). Given this situation it is obvious that mining company do not "control" the operating environment in which they do business. The situation is different for a manufacturing company that can choose the most appropriate location to build its facility to optimize the operating cost of the product or products it produces. Over and above this, each mineral ore deposit has its own unique geology and mineralization requiring its own specific mining and metallurgical solutions. As mining companies do not as a rule manufacture their own equipment (e.g., drills, trucks, crushers, grinding mills, flotation cells, thickeners, filters, pumps) and consumables (e.g., drill bits, explosives, grinding balls, chemical reagents) they are constrained by the available products and technologies to find solutions to their problems. Lastly and most importantly mining companies are "price takers" for their products, and market prices can fluctuate dramatically with limited scope to substantially change operating costs.

DRIVERS THAT LED TO INNOVATIVE CHANGES IN THE MINERALS INDUSTRY

The following innovation drivers have changed the USA mining industry and have also had ramifications worldwide.

- Russian launch of Sputnik in 1957 created the Space Act of 1958 and reauthorizations of the National Defense Education Act in 1963 provided substantial funds for education and research in the sciences.
- The Clean Air Act of 1963 and regulatory amendments in 1970 changed copper, lead, zinc and iron smelting forever. This caused the birth of hydrometallurgical processes to replace smelters.
- The Clean Water Act of 1948 and as amended in 1972 focused industry on effluent problems and developing new analytical systems and techniques.
- Nationalization of foreign resources have required domestic operations to become profitable.
- Lower grade resources and more difficult mineralogies.

HOW INNOVATION HAS WORKED IN THE MINING INDUSTRY

Surveys are recognized as a means to evaluate processes, outcomes and impacts of programs and policies within a company or industry. A number of surveys have been carried out over the last number of years to ascertain the progress and impact of innovation in the following sectors, manufacturing, energy, high tech, IT, retail, health care and business/professional/legal services (McKinssey 2010; Forbes 2011; Groonstone and Looney 2011; Jaruzelski et al. 2012).

For the mineral industry there have been two publications in recent years that provide an excellent overview of the historical developments in comminution (Lynch and Rowland 2005) and flotation (Lynch et al. 2010) in the industry. However, a literature search spanning a 20-year period showed no details of any surveys been undertaken to determine the impact and progress of innovation in the mining industry. In order to get a better understanding of the current progress of innovation and the status of generating new ideas in the industry it was decided to undertake a survey of key figures in the USA minerals industry to get their view on these subjects.

PARTICIPANTS OF THE MINING INNOVATION SURVEY

Study methodology involved the participation of accomplished-senior engineers and/or scientists engaged in research, development and engineering activities with careers spanning 40–60+ years. Using estimates only, the group represents a collective experience of about 550 relevant years. A list of study questions was developed to provide guidance and direction in order to allow continuity in the developed data. Data collection was by personal one-on-one interviewing and refining of comments and observations. Participant input was typed into a generic questionnaire during the interview, refined and corrected, and sent to the participant for correction and approval for presentation in the study. A list of study participants follows:

- James W. White, retired. CEO and chairman at Modular Mining Corporation; founded with others Modular Mining to develop jointly with Phelps Dodge the first computer-driven truck dispatch system and now autonomous haul trucks. Former processor of chemical engineering at University of Arizona and former director of computer services at Rhom and Haas (50 years).
- T.P. McNulty, past CEO and chairman at Hazen Research, and principal at McNulty and Associates. Manager of research and development for The Anaconda Company, vice president–technology for Kerr-McGee Chemical Company (50 years).
- Roshan B. Bhappu, owner, CEO, and chairman at Mountain States Research and Development International; metallurgist at Miami Copper Corporation; professor and chairman of the Metallurgy Department at New Mexico Tech (60 years).
- Gene McClelland, owner, CEO, and chairman at McClelland Laboratories. Developed agglomeration heap leach technology and carbon adsorption for gold with the U.S. Bureau of Mines (40 years).
- Robert Bartlett, former VP of research and development at The Anaconda Company and former dean of Idaho School of Mines; professor of metallurgy at Stanford; and senior scientist at Kennecott Copper (55 years).
- James Reeves, retired. Senior scientist at DuPont. Developed chlorination processes for titanium dioxide production. Developed smelting technologies for rutile/ilmenite heavy mineral sands (65 years).
- Patrick Taylor, professor and director of Kroll Institute at the Colorado School of Mines; professor at the University of Idaho (40 years).
- Nick Hazen, president and chairman at Hazen Research. Hazen Research is a leading hydrometallurgical, smelting, and mineral processing laboratory, specializing in autoclave leaching and metal recovery (40 years).
- Micheal Virnig, senior scientist at COGNIS Corporation. Involved with the development of solvent extraction reagents and technology for General Mills, Henkel, and COGNIC (40 years).
- Brent Hiskey, former dean of research for the College of Engineering at University of Arizona; professor in the Mining and Geological Engineering Department at University of Arizona; senior research engineer for Kennecott, specializing in hydrometallurgy (45 years).
- Martin C. Kuhn, former president of Western States Engineering; owner and principal at Minerals Advisory Group and Minerals Advisory Group Research and Development. Formerly vice president and director of Mountain States Mineral Enterprises; manager of

UOP Technical Development Center; and manager of process metallurgy for The Anaconda Company (45 years).

- P.J. Crescenzo, former VP of engineering at Newmont Mining Corporation; owner and principal, Minerals Advisory Group and Minerals Advisory Group Research and Development. Area manager of six processing plants for Pickens Mather (60 years).

SURVEY QUESTIONS AND SUMMARY OF RESPONSES

The Organization

Question 1: What organizational structure, groups of people, and/or organizations or individuals have successfully innovated and resulted in a lasting technology, process or practice?

Answers to Question 1

Historically, innovation has come from academia, industrial R&D groups, manufacturers, governmental research laboratories, commercial laboratories, entrepreneurial start-up companies and collaborative efforts.

Currently, collaborative efforts appear to offer the best opportunities for innovation, assuming the right problems or opportunities are identified.

The organization needs to foster and support a culture of innovation.

Question 2: How best should in-company and non-company resources be organized to maximize a results oriented corporate program?

Answers to Question 2

Real problems and opportunities must be identified and the execution of the work must not be susceptible to whims of management or the price of gold, copper or other commodity.

The effort should be funded from central corporate sources with support from operating groups and be seen as a long term commitment by the company.

The innovative efforts must be sanctioned and championed at the highest levels within the corporation otherwise little will be accomplished.

The Team

Question 1: What are the appropriate skills and/or experience that are most critical to successful innovation?

Answers to Question 1

Developments in innovation must recognize the chemical and physical constraints. Therefore, a basic understanding of engineering, scientific principals and operating realities is required. Channeled curiosity, energy and ability to think outside the box in an accepting environment are more important than academic degree level.

Question 2: Who or what groups or companies are best suited to identify potential projects that may result in innovative technologies, processes, practices and/or products?

Answers to Question 2
The personnel charged with operating an enterprise and the responsibility for making a profit are in the best position to recognize the improved profit potential by defining the goals. These are the same individuals who can provide financial and manpower resources in pursuit of innovation.

The Concept of Competitive Edge and Focus

Question 1: How should a company focus its resources to maximize its research and development efforts in support of innovation necessary not only to compete but gain a competitive edge on its competition?

Answers to Question 1
It is not possible to gain a lasting competitive edge in the mining industry. Once meaningful innovations become operating realities, the innovator often loses control of the technology.

Assume a leadership role in organizing and leading a cooperative industry/academic/consultant/commercial laboratory consortium to attack problems and opportunities specific to corporate agenda's.

Consolidate highly qualified, diverse (expertise) staff and consultants to work on the more innovative problems. They must be well informed on objectives and operations and have access to the highest levels of management.

Technology is a target rich environment. Innovation requires a directed approach.

Impediments or Road Blocks

Question 1: Innovation is expensive and risky. Who benefits most from the innovation in the industry?

Answers to Question 1
Normally the initial innovator does not derive the most benefit from innovative technology. Heap leach/agglomeration/carbon/EW for low grade oxide gold ores was significantly advanced through the US Bureau of Mines. Herald Heinen, Gene McClelland and George Potter the main innovators but other than name recognition received little to no remuneration.

In mining, most companies adopted a "wait & see" attitude before jumping on the innovation band wagon. A good example is the old Phelps Dodge Company waiting for 17 years before committing to SX/EW for copper recovery from heap and dump copper leach solutions.

Question 2: Does the cyclical nature of the mineral business interrupt progress and decrease the probability of future success?

Answers to Question 2
Absolutely! Long-term technical commitments are terminated due to short-term cash flow problems.

A very big interruption in the mid-80s, for example. Research departments vanished and many technical people changed careers.

Time value of money-management philosophies caused a myopic view of the long-term plans for technology development.

The longevity of senior management can have a negative impact on long-term innovation, due to changes in focus by new management.

Developing a Portfolio

Question 1: What are the drivers?

Answer to Question 1

Generally speaking, 'necessity is the mother of invention.'

Question 2: List motivating factors that lead to innovation.

Answers to Question 2.

Within a for-profit enterprise, three principal reasons exist.

- A real problem needing an economical solution is identified (most often by operations).
- An innovative solution is needed to exploit an economic opportunity.
- Governmental regulation or national defense requires industry to innovate.

Within people, motivating factors include panic and/or fear of failure (creative stress) to innovate.

Avoid the impulse to innovate for the sake of innovation and do not waste resources on perceived rather than real problems.

Question 3: Who or how should identified projects be directed and/or administrated?

Answers to Question 3

The direction and guidance needs to be rooted at the President or CEO level with good internal communications to the operating research or technical Vice President level.

It is not the organizations chart but who has the communication and technical skills in order to effectively keep senior management informed and maintain the trust of upper management.

At the operational level, the identified projects must have a champion willing to sell and protect the selected projects.

Question 4: What are the most important components of a program to produce economically viable innovations?

The main component of any program is a clear statement of need, goal, cost and benefit.

The people involved in the program. The leader must have a clear view of the economic drivers; understand the technology, have good communication skills and have the ear of the CEO. The developed organization trying to innovate must not be looking over their shoulder wondering when the price of metal drops where their next job may be.

Implementation

Question 1: Over the years numerous companies have worked on projects thought to be innovative with little to nothing to show for their efforts at the end of the day. What are the most apparent reasons success was not achieved?

Answers to Question 1

Goals were not practical or clearly identified as having an economic benefit of sufficient magnitude.

Program continuity was lacking due to changing market conditions.

Programs were underfunded.

Management changed and research efforts were re-directed.

Politics got in the way and the champion of the technology was undercut by corporate competitors.

The corporation lacked sufficient commitment.

Question 2: Most corporate research and development departments or groups are centers for technical support and testing. List and discuss research and development groups that over the years have consumed significant budgets in search of innovative technologies and/or processes. Were significant advances realized?

Answers to Question 2

Yes.

Historically, the following partial list of company research and development organizations contributed: INCO, Sherritt Gordon, Anaconda, Kennecott, Cominco, Tech/Cominco, Phelps Dodge, BHP, CVRD/VALE, HC Stark, Outokumpu, Boliden, Newmont, Anglo Gold, ASARCO South Plainfield, Exxon Florham Park, Noranda, Falconbridge, Dow, DuPont, Universal Oil Products, 3M, Inspiration, St. Joe Lead, Amax, Climax Molybdenum, Chevron Minerals, Conoco, Rio Tinto, Texas Gulf Sulfur, Freeport, U.S. Steel, Bethlehem Steel, Alcoa, Kaiser Aluminum and others.

Currently, few of the R&D organizations listed above are staffed for or involved in innovative mineral processing/chemical metallurgy/pyrometallurgical research. Surviving R&D organizations are more involved in technical service and standard testing.

Question 3: Have you come across "crowd sourcing" and as a way to increase innovative idea generation and what are your thoughts and experience with this concept? Crowd sourcing-ideas from "all and sundry" via internet and companies that do this for a living.

Generally the respondents had little experience with crowd sourcing.

Question 4: Does academic research provide cutting edge technology or science that benefits industry? If yes, list examples of the work and the schools involved.

Answers to Question 4

Historically, yes. Today, no.

The possibility exists for academia to contribute; however, academic research is normally not directed to the applied side of technology. The role of academia is to train and educate students not provide cutting edge technology.

Question 4b: If yes, is it possible for industry to benefit from the work?

Yes, specifically to understand the fundamental behavior of an innovation.

Question 4c: If no, what can be done to make academic research more relevant to industrial needs?

Bring professors and students to industry to understand real world problems.

Table 1. Innovations that have changed the mining industry

Mining	Comminution	Separation	Hydrometallurgy	Smelting
In-pit crushing and conveying	Bond theory of comminution	Froth flotation	Cyanidation	Fluidized bed roasting
Mechanized mining (coal)	SAG/AG grinding	Flash flotation	Merrill-Crowe	Flash smelting
Block caving	High pressure grinding rolls	Column flotation	CIP and CIL	
Modular mining despatch	Ultra-fine grinding	Jameson flotation cell	Chlorination of preg-robbing gold ores	
		Spirals	Heap leaching copper and gold ores	
		Reichert cones	Autoclave leaching	
		Centrifugal gravity separators	Bioleaching	
		Rare earth magnetic separators	SX/EX (copper)	

EXAMPLES OF INNOVATIONS IN THE MINERALS INDUSTRY

Table 1 provides some examples of innovations that have changed the mining industry.

OBSERVATIONS AND COMMENTS

In the interdependent and global economy the mining industry must find competitive advantages based on new ideas and capabilities. Delivering a solution that is unique to each mining project is becoming more important than delivering a standard solution with good implementation. Mining companies will need to manage growth by offering innovative solutions to problems that will lower both capital and operating costs given the ever deceasing ore grades that face all sectors of the mining industry. The question that each mining company will face is how well it is equipped to innovate and offer appropriate solutions to sustain the business and the more difficult task to grow the business.

There are three ascending tiers of innovation that a mining company can pursue. The level that is chosen will dictate the simplicity or complexity of the journey to be undertaken. Unfortunately, many companies expect or will expect to achieve the highest levels of innovation while only providing the strategy, tools, and support for lower-level success, at best. Importantly radical Innovation does not happen when you bring people an incremental improvement of what they already know. Furthermore the premise is simply you cannot lead by following the leader.

When it comes to innovation, bigger appears not always to be better. Smaller companies appear to be more agile in generating and developing innovative ideas. However, larger companies with adequate financial resources are better at executing them. Turning an idea into a product or process requires a certain level of scale and internal resources. But at some point size begets bureaucracy, which can interfere with successful execution.

Let's now consider some of the internal structural issues that need to be considered when adding an innovation function to a company. Firstly and most importantly leadership behaviors and company cultures greatly impact the effectiveness of innovation (Schein 2010; Jaruzelski et al. 2011). When choosing the innovation team it should be recognized that most people have the capacity to be innovative, the problem is that each person approaches innovation and change in different ways. Each individual has different habits, talents, knowledge, values, interests, and ways of expressing themselves. Recognizing the different ways that an individual likes to innovate is a key to working together successfully—in a team or in an organization. Sometimes those who may be the least creative during idea generation can offer the most help during implementation. Innovation and the enterprise therefore benefit strongly from a diverse set of workforce skills.

Innovation is also about change and change comes from many sources. The small changes from incremental innovation often come from the realm of implementation, so the organization, customers, and other stakeholders can generally adapt. Large changes generated by disruptive innovation, often come from the imagination, and so these often disrupt not only the market but the internal workings of an organization and they also require a lot of explanation. The organization may have to adapt to disruptive innovations by hiring different types of employees, re-training existing employees, accounting for revenue in a different way, or going about production in a new way.

Arguably, the corporate Research & Development (R&D) department, if one exists, is the unit most capable of generating new ideas. But engineers and scientists may lack the abilities to persuade, communicate and politically maneuver to see their ideas through to commercialization. Building alliances and relationships between R&D and others internally and externally is an important consideration of developing an innovation portfolio strategy. Finding effective influential sponsors early on can greatly increase chances of success. Ideas from those without influence can lead to creativity without results. Innovation is not cheap, let alone free; and it is more often firms themselves, rather than external finance providers, that fail to allocate resources to it. If strategic plans do not allow for the resources to pursue serendipitous innovation, then they can stifle it altogether.

In today's world, Open Innovation has the ability to identify solutions to complex problems, capitalize on opportunities and develop new technologies which would not have been possible if the organization had merely looked internally for ideas (Lindegaard 2010). Crowdsourcing as a strategy to enhance open innovation has huge potential, including the ability to provide organizations with an Open Innovation model to access a wider variety of expertise, resources and technology. A number of online internet sites are available that specialize in bringing together different parties or groups to work on a particular project or solve a specific problem.

Execution of innovative ideas and products, however, is a very different problem. Successful implementation is where many companies 'fall through the cracks" (Glatstein 2008). The corporation and its bureaucracy in many cases seems to get in the way.

REFERENCES

Basalla, G. 1989. The Evolution of Technology. Cambridge University Press.

Berkun, S. 2010. The Myths of Innovation. O'Reilly Media Inc.

Burns, T., and Stalker, G.M. 1961. The Management of Innovation. Oxford University Press Inc. New York.

Chesbrough, H. 2003. Open Innovation: The new imperative for creating and profiting from technology. Harvard Business School Press.

Christensen, C.M. 1997. The Innovator's Dilemma. Harvard Business School Press.

Christensen, C.M., and Raynor, M.E. 2003. The Innovators Solution: Creating and sustaining successful growth. Harvard Business School Press.

Csikszentmihalyi, M. 1996. Creativity: Flow and Psychology of Discovery and Invention. New York, Harper Perennial.

Drucker, P.F. 1986. Innovation and Entrepreneurship. Harper Collins Publishers, Inc.
Forbes Insights. 2011. How entrepreneurial executives mobilize organizations to innovate. Nurturing Europe's Spirit of Enterprise.
Glatstein, S. 2008. Strategy Activation. Business Leaders Press.
Groonstone, M., and Looney, A. 2011. A Dozen Economic Facts About Innovation. The Hamilton Project, Brookings.
Hargadon, A. 2003. How Breakthroughs Happen: The surprising truth about how companies innovate. Harvard Business School Press.
Howe, J. 2006. The Rise of Crowdsourcing. Wired Magazine, Issue 14.06.
Jaruzelski, B., Loehr, J., and Holman, R. 2011. The Global Innovation 1000: Making Ideas Work. Strategy and Business Magazine, Brook and Company.
Jaruzelski, B., Loehr, J., and Holman, R. 2012. The Global Innovation 1000: Why Culture is Key. Strategy and Business Magazine, Brook and Company.
Johnson, S. 2010. Where Good Ideas Come From: The natural history of innovation. Penguin Group.
Koestler, A. 1964. The Act of Creation. Hutchinson & Co.
Lindegaard, S. 2010. The Open Innovation Revolution. Essentials, Roadblocks, and Leadership Skills. John Wiley and Sons, New Jersey.
Lynch, A.J., and Rowland, C.A. 2005. The History of Grinding. SME, Littleton, Colorado.
Lynch, A.J., Harbort, G.J., and Nelson, M.G. 2010. History of Flotation. AusIMM, Carlton, Australia.
McKinssey. 2010. Innovation and commercialization, 2010. McKinsey Global Survey.
Medina, J. 2008. Brain Rules. Pear Press.
Ogle, R. 2007. Smart World: Breakthrough Creativity and the New Science of Ideas. Harvard Business School Press, New York.
Raynor, M.E. 2007. The Strategy Paradox. Doubleday. New York.
Rogers, E.M. 1962. Diffusion of Innovations. Free Press, New York.
Schein, E.H. 2010. Organization Culture and Leadership, Jossey Bass Business and Management Series.
Simonton, D.K. 1999. Origins of Genius, Oxford University Press, New York.
Skarzynski, P., and Gibson, R. 2008. Innovation to the core: A blueprint for transforming the way your company innovates. Harvard Business Press, Massachusetts.
Tarde, G. 2010. On Communications and Social Influence. University of Chicago Press.
Utterback, J.M. 1994. Mastering the Dynamics of Innovation. Harvard Business School Press, Boston.

ABOUT THE AUTHORS

Martin Kuhn was born in Tucson, Arizona, and is the son of Truman Kuhn, a geologist for Magma Copper Company, and was quickly moved to Superior, Arizona. Two years later, he moved to Golden, Colorado, and eventually attended the Colorado School of Mines, majoring in metallurgy. After two years in the army as OIC of the 1st Chemical Detachment, Kuhn returned to CSM for his MS and PhD degrees in metallurgy and thereafter joined The Anaconda Company as a senior process engineer and advanced to manager of process technology before joining Universal Oil Products as manager of UOP's Tucson Development Center. Kuhn joined Mountain States Mineral Enterprises and was VP and GM of Minerals Separation Corporation, a director of MSME, and a VP of MSRDI. In 1992, he founded and helped build Western States Engineering as its president and formed Minerals Advisory Group in 1993. In 2010, MAG incorporated MAG R&D in Tucson. He has written and co-authored numerous papers and book chapters, edited books, and has been issued 10 U.S. patents. In 1998, he organized and edited *Managing Innovation in the Minerals Industry*, an SME publication.

Robert C. Dunne is the group executive–metallurgical fellow at Newmont Mining Corporation, Denver, Colorado, USA. He is a graduate of the University of the Witwatersrand, South Africa, and is an adjunct professor at the Julius Kruttschnitt Mineral Research Centre, Brisbane, Australia. Over the last 30 years, Dunne has worked for a number of mining companies, including Anglo American, Anglovaal, and Newcrest Mining. He also spent time at the Western Australian School of Mines lecturing and at Mintek, South Africa, where he was involved in introducing new technologies into the mining industry. He has authored and co-authored more than 50 papers in a variety of fields, including mine-to-mill, comminution, gravity separation, flotation, gold hydrometallurgy, water treatment, and metallurgical management.

History of Innovations in Extractive Metallurgy

Fathi Habashi
Laval University, Quebec City, Canada

ABSTRACT: People learned melting of ores to get metals since ancient times but the use of aqueous solutions for this purpose was introduced relatively recently in the Middle Ages. The most recent technology, however, came when electric current became available on commercial scale towards the middle of the nineteenth century. The present article records in general terms the development in pyro-, hydro-, and electrometallurgy.

INTRODUCTION

For the sake of convenience in presentation, extractive metallurgy is divided into pyro-, hydro-, and electrometallurgy. The use of fire made it possible for ancient people to produce gold articles by melting and casting tiny particles loose in the ground or found in rocks (Figure 1). It must have been noticed in an early date that blowing the fire with breath or exposure to powerful wind resulted in its intensification. This observation might have been the reason for improving furnace operation by using blowpipes (Figure 2). Gradually, furnaces enlarged and work became more complicated; the smith found the blowpipe not enough for his purpose. This led to the invention of bellows made of goat skin with a tube leading to the fire (Figure 3).

Figure 1. Ancient Egyptian wall painting showing workers melting and casting gold

Figure 2. Ancient Egyptian wall painting showing the use of blowpipes

Figure 3. Ancient Egyptian wall painting showing the use of bellows to blow air in a furnace

Figure 4. A primitive furnace for smelting ores. In the background, a kiln used to make charcoal from timber.

PYROMETALLURGY

Pyrometallurgy is most suitable for the reduction of high grade oxide ores (Figure 4). However, due to the absence of other technology, thermal methods were also applied for the treatment of massive sufides. Even when the sulfides did not yield a metal directly on smelting as in the case of copper and lead sulfides, roasting was invented to transform the sulfide to oxide which was then reduced to metal.

Iron production has undergone tremendous changes: from a primitive furnace producing a kilogram of metal per day to a complex structure producing 10,000 tonnes/day. The early iron lumps produced in small furnaces were contaminated with slag because the temperature was not high enough to melt the charge. Steel, once a rare and expensive alloy became a material of construction for ships, bridges, and sky scrapers. Together with the steam engine, pyrometallurgy was the origin of the Industrial Revolution which took place in England between the 17th and 19th centuries and made England the master of the world. It also changed our civilization. On the other hand, the introduction of new hydrometallurgical techniques at the end of the 19th century has gradually displaced some of the polluting or energy intensive pyrometallurgical processes. The use of electric current in metallurgy became possible when the dynamo was invented in the 19th century and replaced the bulky and expensive batteries. Hence it was used immediately in the refining of copper needed for the growing electric industry.

The Vertical Furnace

In the primitive metallurgical operations, only high grade ore in form of lumps could be treated in small vertical furnaces. As a reducing agent charcoal made by partial burning of timber was exclusively used. Larger bellows were later used and these were mechanically driven by horses, later water wheels which became an important part of a smelter. Waterpower became more important than the availability of ore when choosing the location for smelters. The blast produced by large water-driven bellows (Figure 5) resulted in the following improvements:

Figure 5. Large water-driven bellows used to blow air in a shaft furnace

- The intense heat generated due to these large bellows resulted in the melting of the charge. Once iron was molten it dissolved appreciable amounts of carbon and the solidified product became fragile and necessitated further treatment to remove the carbon.
- The dissolved carbon also resulted in lowering the melting point of iron. The liquid iron was directly cast into finished products such as furnace plates and cannon balls, or bars for sale or for further processing in the finery.
- The furnaces could be charged without interrupting the process (Figure 6).
- This in turn enabled relatively larger furnaces to be operated hence increased productivity.

Figure 6. A shaft furnace from the 18th century showing continuous charging at the top

The Use of Coke

In the early iron ore smelting processes, charcoal made from forest timber was used. Later on, mined coal was introduced, but it was soon realized that it softened in the furnace and spoiled the metallurgical processes. This factor, together with the mining difficulties such as coal-gas explosions in mines, ventilation and drainage, as well as the laws prohibiting the destruction of forest contributed to the increased price of timber. This problem was solved when in 1607 coal was first converted into coke. This simple step revolutionized the iron industry, for coke being hard and porous is able to withstand a far greater burden without crushing, thereby making possible the construction of much larger furnaces with resultant increase in output.

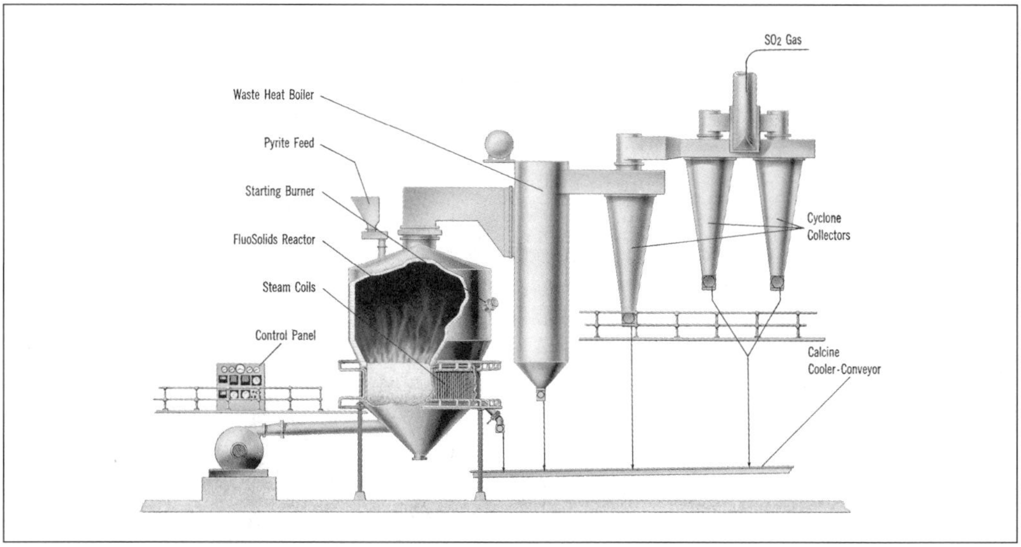

Roasting

The early smelter operator was mainly concerned with the treatment of oxide ores. When he attempted to treat sulfides the material melted but no metal was produced. Gradually, the smelter operator learned that the sulfide ores could be heated under certain conditions to get a calcined product (an oxide) that can be smelted to metal in a second operation. The transformation of sulfide to a calcined product without melting underwent numerous stages of development: roasting in heaps, in small vertical kilns made of bricks known as "stalls," in rotary kilns, in reverberatory furnaces, in moving grate furnace, in multiple hearth furnace, and finally in fluidized bed reactors.

The Horizontal Furnace

When high grade sulfide ores were exhausted and the metallurgist was obliged to beneficiate low grade ores by crushing, grinding, and later by flotation, he obtained a fine material that was not suitable for charging to a vertical furnace. Consequently, the horizontal furnace that was extensively used for glass manufacture was adopted (Figure 7). While the vertical furnace is an excellent reactor: it is a heat exchanger as well as counter-current mass transfer reactor, the horizontal furnace suffers from inefficient heat transfer, excessive dust formation necessitating installing large dust recovery system, and gases leaving the furnace are at high temperature necessitating installing bulky and expensive heat recovery systems.

Flash Smelting

Finally, the horizontal furnace was gradually displaced by flash smelting furnace starting in the 1945. In this furnace preheated air or oxygen were used to ignite the finely ground concentrate and the heat generated was enough to melt the bulk of the charge thus greatly economizing the energy (Figures 8 and 9).

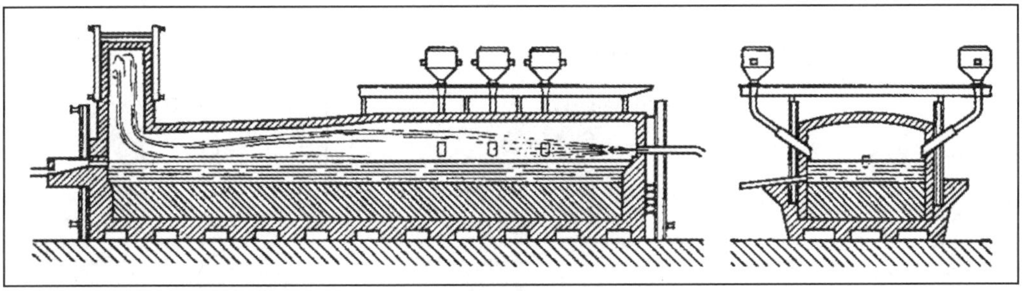

Figure 7. Horizontal furnace

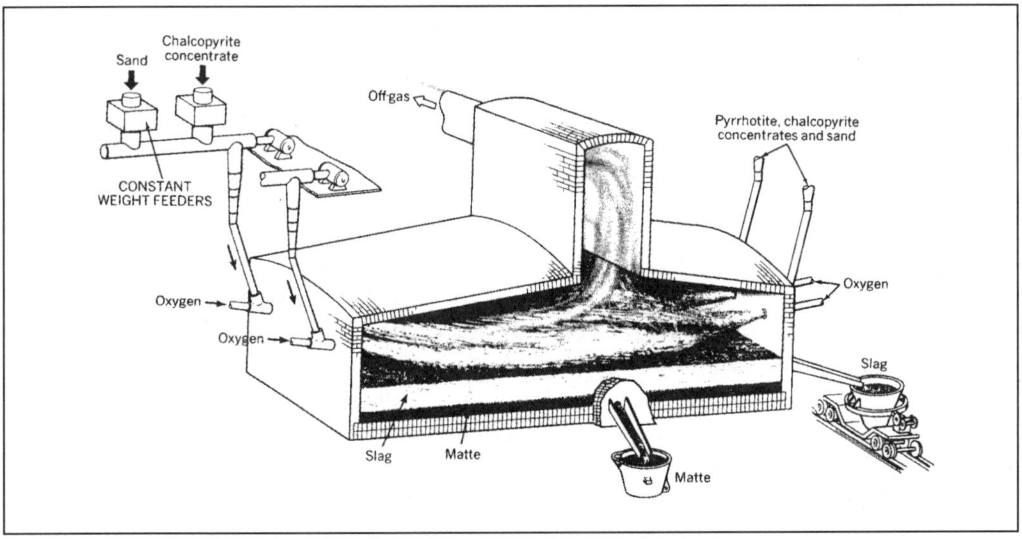

Figure 8. Flash smelting furnace using oxygen (INCO)

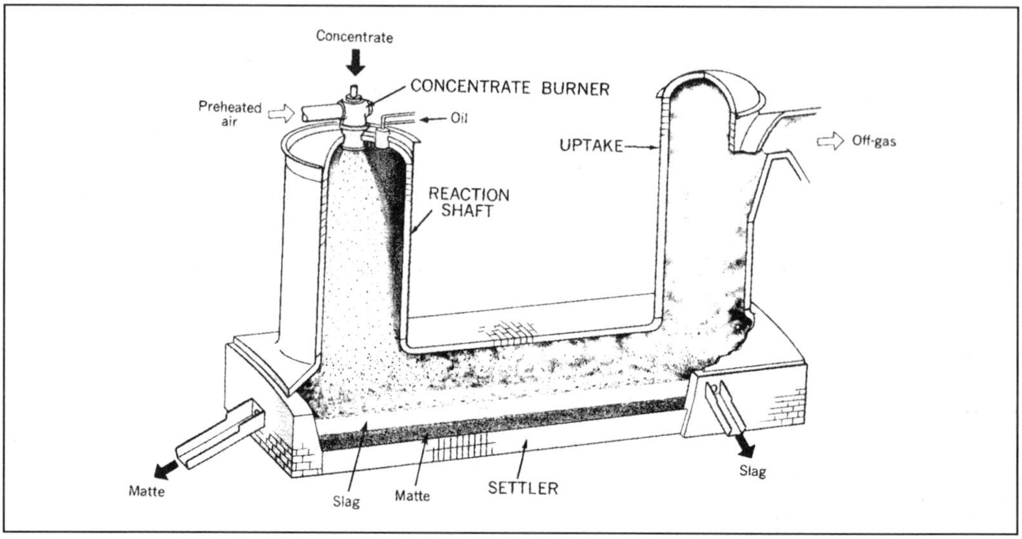

Figure 9. Flash smelting furnace using preheated air (Outokumpo)

Figure 10. Henry Bessmer (1813–1898)

Figure 11. William Kelly (1811–1888)

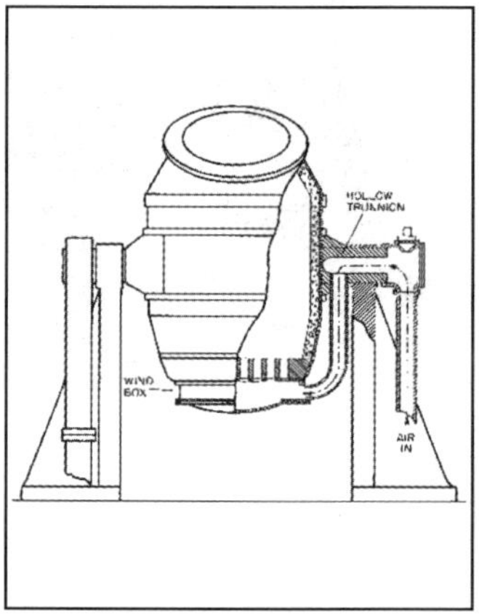

Figure 12. Bottom-blown converter using air

The Converter

Henry Bessemer (1813–1898) (Figure 10) in England and independently William Kelly (1811–1888) (Figure 11) in USA invented the revolutionary process in 1856 to produce steel from iron by blowing air through the molten material in a converter (Figure 12) thus replacing the old process of exposing the molten material to an oxidizing draft in a horizontal furnace known as the "Puddling process." In this process, the time to produce a batch of steel was reduced from few days to few minutes and at the same time the need to use fuel for making the transformation was eliminated. This technique resulted in the tremendous increase in steel production. In 1880, Pierre Manhès (1841–?) (Figure 13) in France adapted Bessemer's steelmaking process to the copper industry.

To enhance the oxidation process, oxygen was used instead of air. However, to avoid the melting of the refractory nozzles because of the intense heat of reaction as a result of using oxygen,

Figure 13. Pierre Manhès (1841–?)

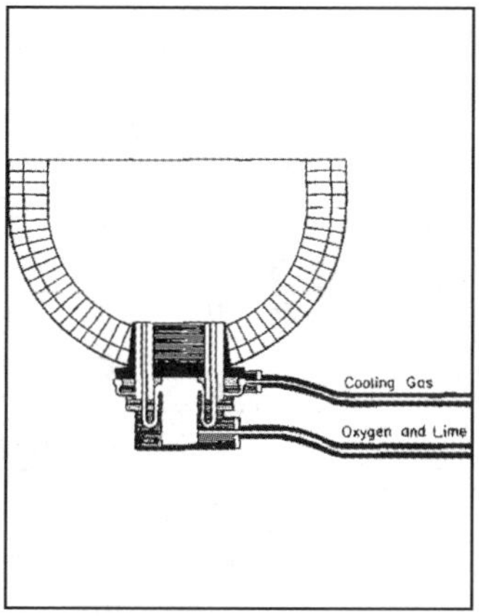

Figure 14. Bottom-blown converter using oxygen

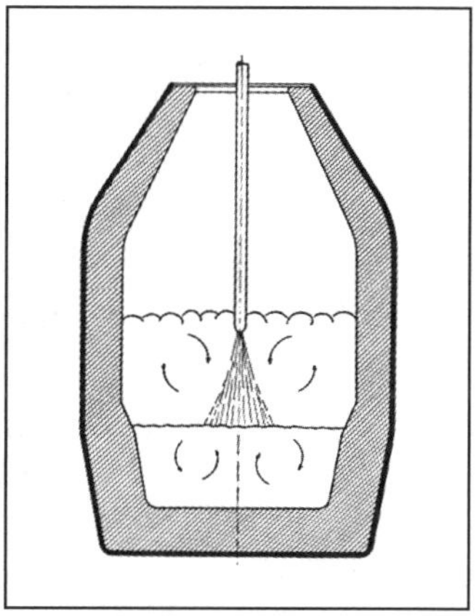

Figure 15. Top-blown converter using oxygen

cooled nozzles were introduced. Cooling with a water jacket is possible but too risky because of the explosion hazard if water gets in contact with the molten metal. The method successfully adopted is by direct cooling with natural gas or fuel oil. The endothermic cracking of these fluids at the tip of the nozzles avoided the melting of the refractories. The amount of these materials used is only about 4% of the amount of oxygen (Figure 14).

Top-Blown

The top-blown converter (Figure 15) was developed in Austria in 1951 by Vereinigte Österreichischen Eisen- und Stahlwerke (VÖEST) in Linz, and the Österreichische Alpin-Montangesellschaft in Donawitz. It became known as the Linzer Düsenverfahren, i.e., the Linz lance process and abbreviated L-D process. Oxygen at high speed is introduced to the molten bath by means of a water-cooled lance situated at some distance from the surface of the metal.

The oxidizing action takes place in the bath in the immediate vicinity of the lance but, due to convection currents, it spreads rapidly through the whole batch. When the reaction is complete, which usually takes place in about 15 minutes, the flow of oxygen is stopped, the lance is raised, and the converter is tilted to empty its contents. Historically, after the invention of the Bessemer process, the top-blown converter was used which was later replaced by the bottom-blown converter using oxygen.

Rotating Converters

These are in the form of a short cylinder laid horizontally on a slight inclination and rotate at slow speed around their horizontal axis (Figure 16). Oxygen is introduced by means of a lance from the top opening. The combined action of oxygen and rotation enhances the oxidation reaction in the molten bath. These reactors are used to some extent in steelmaking (Kaldo and Rotor

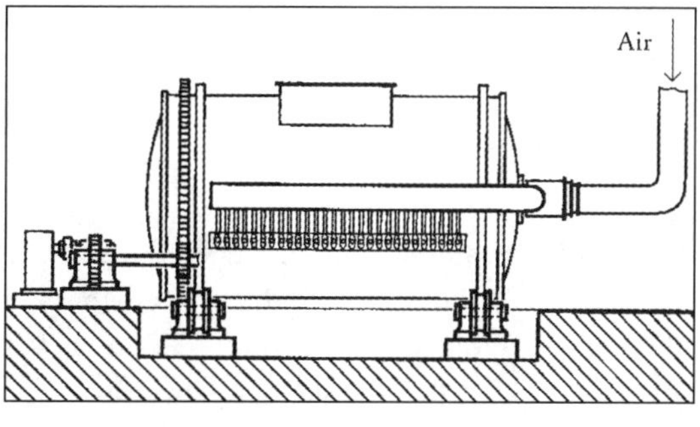

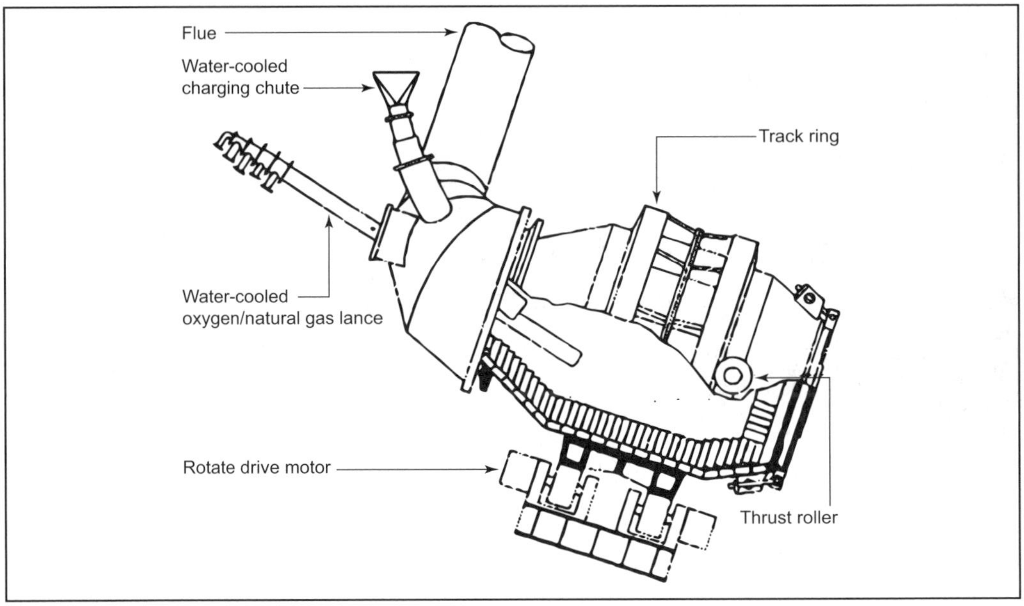

Figure 16. Top-blown rotary converter

Processes) as well as for copper and nickel production (TBRC Process, abbreviation for Top Blown Rotary Converter).

Metallothermic Reactions

Chemists of the early 19th century used the alkali metals to liberate metals from their compounds—a reaction that became known as metallothermic reaction. The Swedish chemist Jöns Jakob Berzelius (Figure 17) isolated zirconium and titanium in 1824 for the first time by this method followed by the Danish scientist Hans Christian Oersted (1777–1871) (Figure 18) who isolated aluminum a year later.

The method was developed on industrial scale in 1850s by the French chemist Henri Saint-Claire Deville (1818–1881) (Figure 19) who produced the first aluminum by heating $AlCl_3$-NaCl with metallic sodium. Once aluminum became available in large quantities, it was also used to liberate metals from their compounds. During the Manhattan Project to produce an atomic bomb in USA in World War II the method was extensively used to prepare metallic uranium by reducing UF_4 by magnesium. In the 1950s the method was used by the Luxemburg metallurgist Wilhelm Kroll (1889–1973) to produce the first metallic titanium on a pilot then on a commercial scale by reducing $TiCl_4$ by magnesium.

HYDROMETALLURGY

The Origins

The roots of hydrometallurgy may be traced back to the period of alchemists when the transmutation of base metals into gold was their most important occupation. Some of these operations involved wet methods. For example, when an alchemist dipped a piece of iron into a solution of copper vitriol, i.e., copper sulphate, the iron was immediately covered by a layer of metallic copper.

Figure 17. Jöns Jakob Berzelius (1779–1848)

Figure 18. Hans Christian Oersted (1777–1871)

Figure 19. Henri Saint-Claire Deville (1818–1881)

The major question, however, that remained unanswered was: how can the transmutation of iron or copper into gold be effected? Gold, the most noble of all metals was insoluble in all acids or alkalis known at that time.

The discovery of *aqua regia* by Jabir Ibn Hayyan (720–813 AD) (Figure 20), the Arab alchemist, may be considered as the first milestone marking the beginning of hydrometallurgy. *Aqua regia*, i.e., royal water is a mixture of HCI and HNO_3 that dissolves gold; neither of the acids alone has any dissolving action on gold. In the Middle Ages, certain soils containing putrefied organic matter were leached to extract saltpetre (salt of stone, potassium nitrate), a necessary ingredient for the manufacture of gunpowder. The process was fully described by Vannoccio Biringuccio (1480–1539) in his *Pirotechnia* published in 1540.

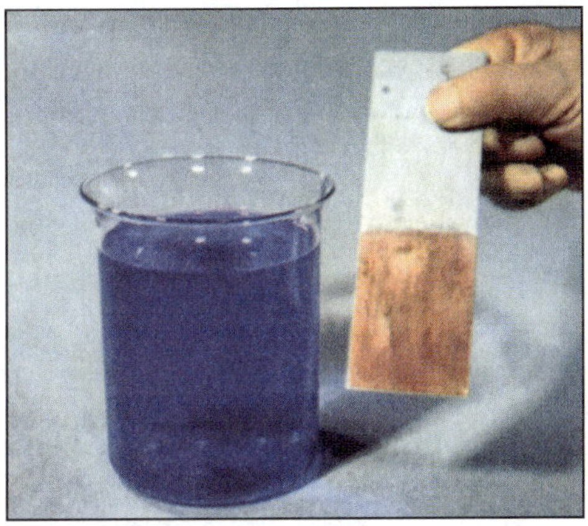

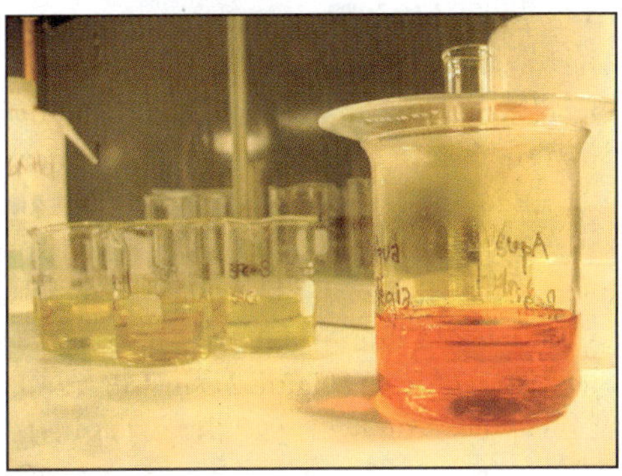

In the 16th Century, the extraction of copper by wet methods received some attention. Heap leaching was practised in the Harz Mountains area in Germany and in Rio Tinto mines in Spain. In this operation, pyrite containing some copper sulphide minerals was piled in the open air and left for months to the action of rain and air, whereby oxidation and dissolution of copper took place. A solution containing copper sulphate was drained from the heap and collected in a basin. Metallic copper was then precipitated from this solution by scrap iron, a process that became known as 'cementation process.' This is

Figure 20. The Arab alchemist Jabir Ibn Hayyan (720–813 AD), discoverer of *aqua regia* (royal water) for dissolving gold, the king of metals

the same process that was already known to the alchemists and was in operation to an appreciable extent until recently when it was replaced by solvent extraction-electrowinning.

The Modern Era

The birth of modern hydrometallurgy dates back to 1887 when two important processes were invented. The first, the cyanidation process for treating gold ores, and the second, the Bayer process for treating bauxite.

Cyanidation Process

The dissolving action of cyanide solution on metallic gold was known as early as 1783 by the Swedish chemist Scheele. Elsner in Germany in 1846 studied this reaction and noted that atmospheric oxygen played an important role during dissolution. The application of this knowledge to extract gold from its ores was proposed and patented much later by the Scottish chemist John Stewart MacArthur (Figure 21) in 1887, and became known as the cyanidation process. The impact of this process on hydrometallurgy has been tremendous. Extremely large reactors known as Dorr agitators equipped with compressed air injection in the pulp in which the finely ground ore was agitated with the cyanide leaching agent had been designed and built by the metallurgical engineer John Dorr. Huge sedimentation equipment, which became known as Dorr thickeners, was constructed followed by large filtration plants designed to obtain clear leach solutions for metal

Figure 21. John S. MacArthur (1856–1920)

Figure 22. Karl Josef Bayer (1847–1904)

Figure 23. Vladimir N. Ipatieff (1867–1952)

recovery. The ancient process known as cementation, which was applied for precipitating copper from solution by scrap iron, was applied to gold solutions, iron being replaced by zinc.

Bayer Process

The second major hydrometallurgical process of this era was the process invented by Karl Josef Bayer (Figure 22) for the preparation of pure Al_2O_3 and is known as the Bayer process. This process was concerned with leaching bauxite with sodium hydroxide solution above its boiling point in a pressure reactor. After separating the insoluble material, the pure solution was then seeded to precipitate pure crystalline aluminum hydroxide, which was filtered, washed, dried, and calcined to pure Al_2O_3 suitable for charging to the electrolytic reduction cell invented two years earlier.

Other Processes

At the beginning of the 20th century numerous leaching and recovery processes were proposed. Some of them were put into practice, and others had to wait for about half a century until they were applied, while others never developed beyond a pilot plant.

- Ipatieff's (Figure 23) work in Russia in 1904–1920 using hydrogen under pressure to precipitate metals from solution had to wait until 1953 when it was applied commercially because hydrogen before that time was an expensive gas. It became only cheap after the discovery of reforming processes for petroleum gases.
- World War I created a demand for zinc for the manufacture of cartridge brass. Zinc, for this purpose, used to be obtained by the distillation of commercially available metal in Belgium and Germany from ore supplied by Australia. This situation inspired industry in North America to supply additional metal from ores that were not amenable to standard methods. The metallurgists at Trail and Anaconda developed the process for electrolytic zinc, and as a result, the leaching of a large tonnage of ZnO by H_2SO_4 was introduced. The process is mainly based on a patent by Letrange issued in 1881. Cadmium gradually emerged as an important by-product of this process.

- In the 1940s, the technology of uranium production was introduced in connection with the Manhattan Project, a US project aimed at producing an atomic bomb. Numerous new techniques became suddenly used on large scale. Some of these are, for example, the use of Na_2CO_3 as a leaching agent, ion exchange, solvent extraction, and many processes for precipitation from aqueous solution. A large number of synthetic resins for use as ion exchangers, and similarly a large number of organic solvents were specially synthesised for use as extractants for uranium.
- In the 1950s it became apparent that hydrometallurgy can solve the pollution problems in smelters by producing elemental sulfur (**Figure 24**). This was materialized in the 1970s for the treatment of zinc sulfide concentrates in Canada (**Figure 25**). The new process rendered the production of zinc a fully hydrometallurgical process and has also the advantage of liberating the zinc industry from the necessity of producing sulfuric acid.
- Separation of the rare earths by ion exchange replaced the tedious fractional crystallisation from solution. Later on, solvent extraction replaced ion exchange and an industrial plant by Molycorp went into operation at Mountain Pass in USA producing high purity rare earths using about one thousand mixer-settlers.
- In the 1950s [Korean War], pressure hydrometallurgy was introduced for leaching laterites (Moa process), for sulfide concentrates (Sherritt-Gordon process), and for tungsten ores.
- Work at the Mines Branch in Ottawa (known today as CANMET) demonstrated that pyrrhotite-penthlandite concentrate could be treated in autoclaves at 120°C under oxygen pressure to get nickel in solution while Fe_2O_3 and elemental sulphur remain in the

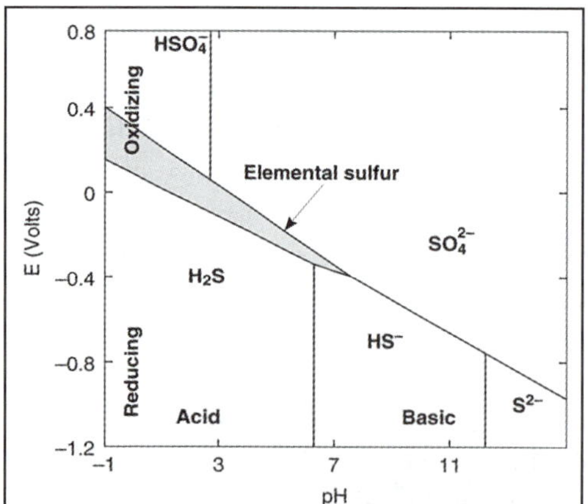

Figure 24. Conditions for the formation of elemental sulfur during leaching sulfide ores

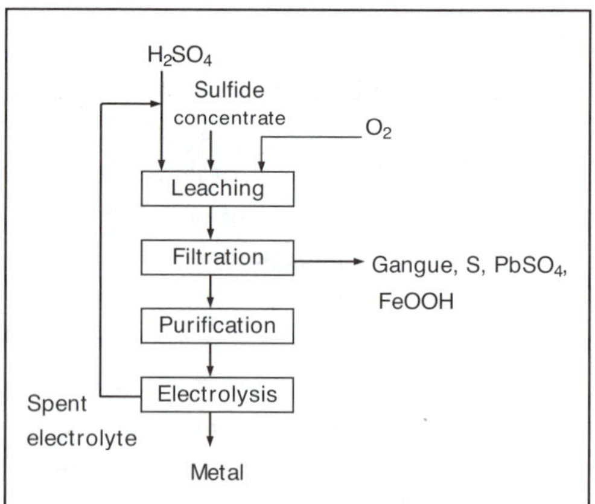

Figure 25. Flowsheet showing the formation of elemental sulfur during the aqueous oxidation of sulfide concentrates

residue. The process was later applied by the Russians at Norilsk plant for nickel recovery and at INCO's Voisey Bay recent operations.

- In the 1960s, the role of bacteria in leaching became known and the wide-spread use of heap and in situ leaching for extracting copper was practiced on large scale. The same technique was later adopted for leaching low-grade uranium and gold ores.
- In the same period the application of organic solvents, mainly oximes, for extracting copper from solution was done.
- In the 1980s, the hydrometallurgy of gold dominated the scene: widespread application of activated charcoal for gold adsorption and the aqueous oxidation of gold refractory ores were industrialised. Large autoclaves are now used for this purpose.

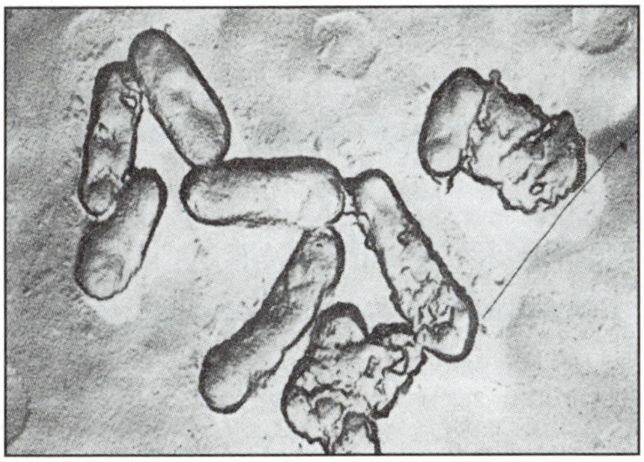

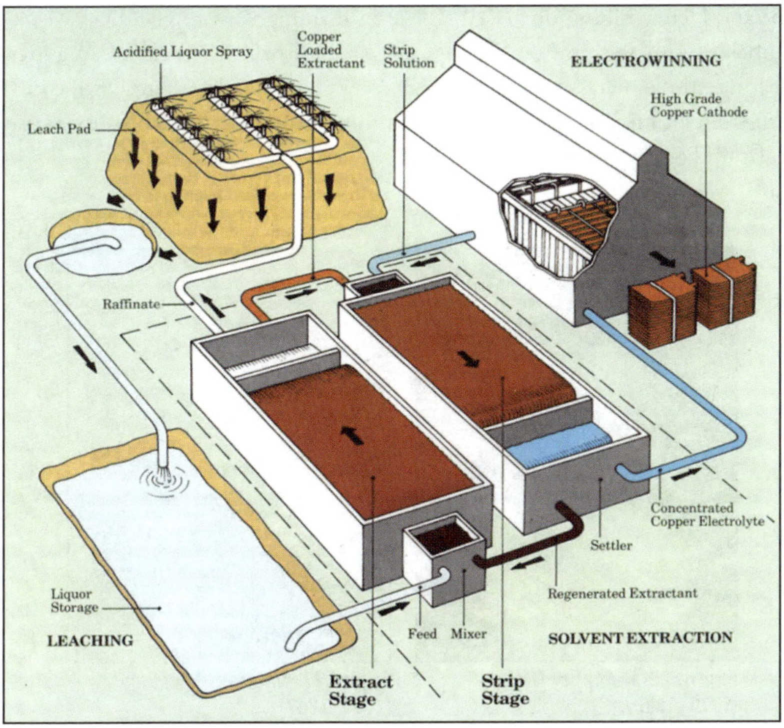

- In 1980 the metallurgy of zinc became fully hydrometallurgical with recovery of elemental sulfur:

  $ZnS + 2H^+ + ½O_2 \rightarrow Zn^{2+} + S + H_2O$

- In 2007 Phelps Dodge (now Freeport-McMoran) at Morenci, Arizona commissioned a pressure leaching plant for chalcopyrite concentrate at 160°C and 2,100 kPa to extract copper and recover elemental sulfur.

  $2CuFeS_2 + 4H^+ + \frac{5}{2}O_2 \rightarrow 2Cu^{2+} + Fe_2O_3 + 4S + 2H_2O$

ELECTROMETALLURGY

The Pioneers

The first application of Volta's pile in metallurgy was in 1807 when Humphry Davy (1778–1829) (Figure 26) at the Royal Institution in London built a large battery and was able to decompose caustic potash and caustic soda and identify potassium and sodium as metals for the first time. This was followed in the next year, using the same technique, to identify barium, calcium, magnesium, and strontium.

Michael Faraday (1791–1867) (Figure 27) who was Davy's assistant, continued this type of research further and in the early 1830s found the relationships between current used and quantity of metal deposited. He was the first to have clear ideas concerning the quantity of metal deposited and intensity of electricity, i.e., the quantities now measured in terms of amperes and volts. Faraday introduced the terms *ion, cation, anion, electrode, electrolyte,* etc., which are in common use today. During the same period, electroplating of silver was commercialized by George Richard Elkington (1801–1865) and Henry Elkington (1810–1852) in Birmingham, England.

Davy's liberation of sodium, potassium, and calcium from their salts had a great impact on metallurgy because these metals were used in 1841 by Henri Sainte Claire Deville in Paris to liberate aluminum from aluminum chloride. Once aluminum became available, it was used to liberate other metals from their oxides.

Figure 26. Humphry Davy (1778–1829)

Figure 27. Michael Faraday (1791–1867)

Figure 28. Robert Bunsen (1811–1899)

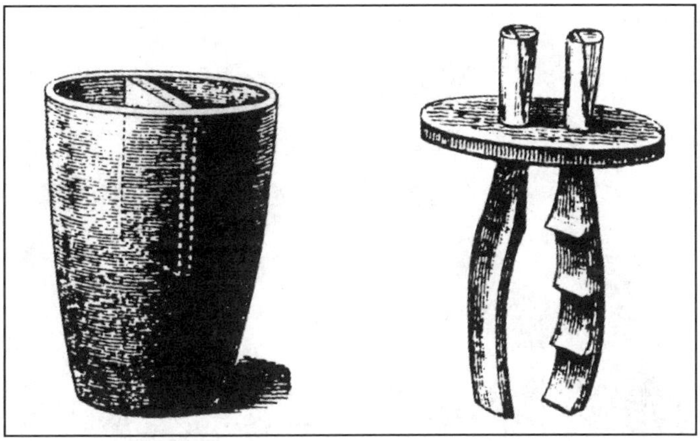

Figure 29. The porcelain crucible and the carbon cathode used by Bunsen to electrowinn light metals from fused salts

Fused Salt Electrolysis

The first use of electric current to deposit a metal from its fused salt was conducted by Robert Bunsen (**Figure 28**) in Heidelberg in 1852 to prepare magnesium from $MgCl_2$ and in 1858 to prepare lithium from LiCl using the porcelain crucible shown in **Figure 29**. The carbon cathode was cut so as to form pockets inside which the metal was deposited. Unless this precaution was taken, the metal, being lighter than the fused chloride, rose to the surface and burned. The fused salt method was adopted later for preparation of other reactive metals such as beryllium, niobium, tantalum, and others.

In 1886 Paul Héroult (1863–1914) (**Figure 30**) in France and Charles Hall (1863–1914) (**Figure 31**) in USA discovered a new process for producing aluminum by electrolysis of fused bath containing Al_2O_3 dissolved in cryolite. The process was adopted on commercial scale in 1887 and gradually replaced Sainte Claire Deville's chemical process by heating metallic sodium with $AlCl_3$-NaCl mixture. It is adopted today practically unchanged but on a very large scale. Carl Wilhelm Søderberg (1876–1955) (**Figure 32**) introduced the self-baking electrode in the early 1930s which was considered an important improvement in the technology. But, due to pollution problems associated with this type of electrode, the industry is now going back to the pre-baked electrodes.

Refining of Metals

In 1876, copper was refined electrolytically in Germany by the Norddeutsche Affinerie in Hamburg (**Figure 33**). This was followed by gold refining in 1878 using the Wohlwill process, named after the German chemist Emil Wohlwill (1835–1912) (**Figures** 34 and 35). The first copper refinery built in America was at Newark by the Balbach Smelting and Refining Company in 1883. In 1884, Bernhard Möbius (1852–1898) invented in USA a process for silver refining (**Figures** 36 and 37). During the same period silver was produced with the development of the Balbach–Thum cells. The first Betts refinery for the production of lead was established at Trail in Canada in 1902.

In 1905, Viktor Hybinette (1867–1937) invented a method of refining nickel which used a divided cell (**Figure 38**). Electrolyte from the anode compartment overflows from the cell and passes through a purification system where the impurities are removed (**Figure 39**). The clean solution is

Figure 30. Paul Héroult (1863–1914)

Figure 31. Charles Hall (1863–1914)

Figure 32. Carl Wilhelm Søderberg (1876–1955)

Figure 33. The first copper electrolysis tanks in which copper was refined electrolytically in the old works of the Norddeutsche Affinerie in Hamburg in 1876

Figure 34. Emil Wohlwill (1835–1912)

Figure 35. Wohlwill cell for electrolytic refining of gold (1878)

Figure 36. Bernhard Möbius (1852–1898)

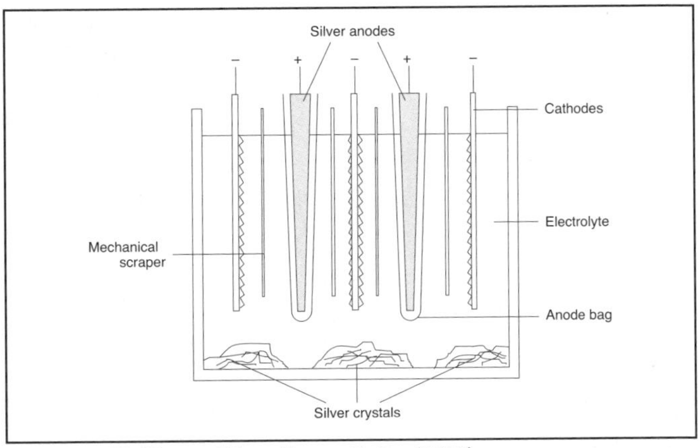

Figure 37. Möbius cell for silver refining (1883)

Figure 38. Viktor Hybinette (1867–1937)

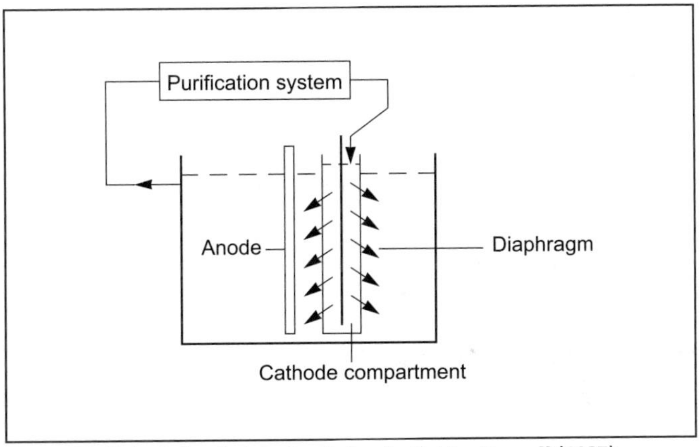

Figure 39. Principle of nickel refining in Hybinette cell (1905)

then fed back to the cathode compartment, where pure metal is deposited. A slight head of solution is maintained in the cathode compartment and the circuit is completed by electrolyte diffusing through the diaphragm to the anode compartment. This solves the problem of co-deposition of impurities on the cathode. This development enabled nickel to be refined electrolytically for the first time and one of the first refineries was set up in Kristiansand, Norway, in 1910.

Electrowinning

During 1912–1915, two important applications in electrometallurgy took place:

1. The large scale copper electrowinning in Chile.
2. The first electrowinning of zinc on commercial scale in USA and Canada.

In 1965 the first plant for recovering copper from low-grade ores by solvent extraction-electrowinning went into operation in Arizona, USA. This marked the beginning of a new technology that gradually displaced the cementation of copper by scrap iron.

Anodic Dissolution of Sulfides

In 1957 Louis Renzoni (1915–1994) (**Figure 40**) and his coworkers at INCO developed the electrorefining of nickel sulfide anodes and producing elemental sulfur. With engineering developments the process was adapted at INCO's nickel refinery at Thompson, Manitoba then used in Russia and Japan.

Figure 40. Louis Renzoni (1915–1994)

SUMMARY

Pyrometallurgy was the oldest art of metal extraction and expanded greatly after the Industrial Revolution. Hydrometallurgy was introduced after the work of alchemists in the Middle Ages. Now it is dominated by the cyanidation process for gold extraction and Bayer process for alumina production. Electrometallurgy came into existence after the discovery of electric current at the beginning of the 19th century followed by the invention of the dynamo. Davy, Faraday, and Bunsen laid its foundation and it had a great impact on other areas of metallurgy. Electrometallurgy is presently dominated by the production of aluminum, the electrorefining of copper, and the electrowinning of zinc. It has also been successful in the electrowinning of copper from solutions obtained by leaching-solvent extraction technology.

SUGGESTED READINGS

F. Habashi, editor, *A History of Metallurgy*, Métallurgie Extractive Québec, Québec City, Canada 1994. Distributed by Laval University Bookstore "Zone." www.zone.ul.ca.

F. Habashi, *The Future of Extractive Metallurgy*, booklet published by CETEM, Rio de Janeiro, Brazil 1996. Condensed version in *Bull. Can. Inst. Min. & Met.* 92 (1028), 161–165 (1999).

F. Habashi, editor, *Handbook of Extractive Metallurgy*, 4 volumes, WILEY-VCH, Weinheim, Germany 1997.

F. Habashi, *A Textbook of Hydrometallurgy*, 2nd edition, Métallurgie Extractive Québec, Quebec City, Canada 1997. Distributed by Laval University Bookstore, www.zone.ul.ca.

F. Habashi, *Textbook of Pyrometallurgy*, Métallurgie Extractive Québec, Quebec City, Canada 2006. Distributed by Laval University Bookstore, www.zone.ul.ca.

F. Habashi, *Principles of Extractive Metallurgy*, Volume 4. Amalgam & Electrometallurgy, Métallurgie Extractive Québec, Quebec City, Canada 1998. Distributed by Laval University Bookstore, www.zone.ul.ca.

F. Habashi, "Metallurgy, History of," pp. 5537–5541 in *Encyclopedia of Materials: Science & Technology*, Elsevier 2001.

F. Habashi, *Readings in Historical Metallurgy*, *i*, Métallurgie Extractive Québec, Quebec City, Canada 2006. Distributed by Laval University Bookstore, www.zone.ul.ca.

F. Habashi, *Extractive Metallurgy Today. Progress and Problems*, Métallurgie Extractive Québec, Quebec City, Canada 2006. Distributed by Laval University Bookstore, www.zone.ul.ca.

J. Cole and J. Wilmot, "Morenci Concentrate Leach Plant First Year Review," Preprint 09–028, SME Annual Meeting, Denver, CO 2009.

ABOUT THE AUTHOR

Fathi Habashi, professor emeritus of extractive metallurgy at Laval University in Quebec City, Canada. He was born in Al Minia, Egypt, on October 9, 1928. He holds a B.Sc. degree in chemical engineering from Fouad Al Awwal University (now known as University of Cairo), and a Dr. techn. degree in inorganic chemical technology from the University of Technology in Vienna. He is an honorary professor at the Technical University of Oruro in Bolivia and was awarded honorary doctorates from the Saint Petersburg Mining Institute in Russia and the National Technical University in Lima, Peru. He is a Fellow of the Canadian Institute of Mining, Metallurgy, and Petroleum, and was awarded its Silver Medal. He was a holder of the Canadian Government Scholarship in Ottawa (1960–62), taught at Montana School of Mines [now Montana College of Mineral Science & Technology] (1964–1967), worked at the Extractive Metallurgical Research Department of Anaconda Company in Tucson, Arizona, before joining Laval University in 1970. Habashi was guest professor at a number of foreign universities, a guest lecturer at a number of institutions, and consultant to the United Nations Development Program for the Cuban Nickel Laterite Project. He authored *Principles of Extractive Metallurgy* in four volumes as well as a number of textbooks, and edited *Handbook of Extractive Metallurgy* in four volumes.

Comminution

Mine to Mill Value Chain Optimization—Role of Blasting

Sarma S. Kanchibotla
Julius Kruttschnitt Mineral Research Centre, University of Queensland, Australia

ABSTRACT: In metal mining operations the in-situ ore is separated from the waste rock and is subjected to a series of breakage and separation processes to convert it into a valuable product. Profitability in this industry depends on how efficiently the in-situ rock is converted into the final product.

Despite the fact that the breakage and separation processes that take place at the mine and mill are inter-dependent, traditionally, mining and milling processes are managed as separate cost centers and optimized with little understanding of the impact of one over the other. During the past two decades, several researchers have demonstrated that all the processes in the mine to mill production value chain are inter-dependent and understanding the impact of blasting outcomes on the overall process efficiency is critical to improve the overall profitability. This paper explains the mine to mill optimization methodology and the critical role of blast outcomes in improving overall efficiency. The value of this approach is demonstrated by using data from field studies.

INTRODUCTION

In most metal mining operations in-situ ore is separated from the waste rock and is subjected to a series of breakage and separation processes to convert it into a valuable product. Profitability of any operation in this industry depends on how efficiently the in-situ rock (ore + waste) is converted into the final product (operating cost) and how much value (price) it commands in the market.

Breakage of in-situ rock generally takes place in three stages:

1. Blasting to prepare the ore for excavation, transportation and subsequent crushing and grinding
2. Primary crushing to improve its handling characteristics and to prepare the ore for subsequent size reduction
3. Grinding in semi-autogenous (SAG) or autogenous (AG) mills followed by ball mills and fine grinding mills or further crushing in either crushing and screening circuits or high pressure grinding rolls (HPGR) followed by ball mills and fine grinding mills

In large open pit metal mines, the energy requirements and cost of breakage in crushing and grinding are generally an order of magnitude higher than that of blasting (Table 1).

Separation of waste and ore also takes place at the mine and at the mill. It starts at the mine, where strip ratio (waste:ore) determines the amount of waste that needs to be removed in order to expose the ore. Generally good stable walls resulting from careful blasting can permit steeper walls compared to unstable walls.

In many mines, especially in narrow vein ore bodies, ore and waste are blasted together and during blasting the rock mass within the blast volume is fractured and displaced relative to one

Table 1. Relative energy and costs of breakage (Scott et al. 2000)

	Energy, kWh/t	Cost, $/t	Energy Factor	Cost Factor
Drill and blast	0.2	0.15	1	1
Crushing	2	0.75	10	5
Grinding	20	3.75	100	20

another. If ore and waste are mixed or if the waste is mis-classified as ore and sent to the mill for further processing it can lead to dilution. Similarly, if the ore is mis-classfied as waste and sent to waste dumps, it can lead to ore loss.

At the mill, separation of ore and waste takes place in different stages depending on the mineralogical characteristics of the ore. A number of factors determine the recovery of the valuable minerals in the subsequent processes following the size reduction stage. One of the key factors is the final grind size of the product that is fed to the flotation and/or leaching plants.

MINE TO MILL VALUE CHAIN OPTIMISATION

Despite the fact that mining and milling processes are interdependent, traditionally, mining engineers optimize blasts to minimize the overall mining costs ie. drilling, blasting, secondary blasting, loading and hauling (McKenzie 1965, Dinis da Gama 1990). Such an approach is true for operations where the sole objective of mining is to shift the dirt from place A to place B, but not for operations where the Run of Mine (ROM) ore is subjected to further comminution and separation processes. This approach is still a common practice in many operations even now because mining and milling are managed as separate cost centers and each unit tries to minimize their overall unit costs.

Eloranta (1995) and Moody et al. (1996) have extended the blast optimisation concept to primary crusher and shown that crusher throughput and power consumption are strongly influenced by the ROM fragmentation distribution which in turn is influenced by blasting.

McKee et al. (1995) further extended the concept to grinding through modelling and simulation showed that by tailoring the blast fragment size distribution, autogenous mill throughput could be increased by as much as 20%. This work has resulted in AMIRA/JKMRC Mine to Mill research project P483A, Optimisation of Mine Fragmentation for Downstream Processing. The aim of the project was *"To aid in the development of a mining and milling strategy which minimises the overall cost per tonne treated in the mining and milling stages and maximises company profit"(JKMRC 1998).* This research project clearly demonstrated the impact of blast fragmentation on the mine to mill value chain through modeling, simulations and case studies (Kanchibotla et al. 1998, Kojovic et al. 1998, Valery et al. 1999, and Scott et al. 1999). During the same time, a number of other investigations (Nielsen and Kristiansen 1995, Nielsen 1998, and Simkus and Dance 1998) have also confirmed the impact of blast results and feed size effect on crushing and grinding circuit efficiency.

Another aspect of blast optimization focused on reducing ore loss and dilution resulting from blasting. In this approach, the blast movement dynamics are monitored and understood and ore/waste boundaries are adjusted to account for that movement (Taylor et al. 1996, Thornton et al. 2005, Thornton 2009, Rogers et al. 2012, Engmann et al. 2012). By doing so, the ore and loss and dilution from the blast movement are reduced.

Some sections of the mining industry think that mine to mill optimization is just increasing blasting energy to produce finer fragmentation. It may be true in some cases but in most cases producing a finer fragmentation from blasting may not result in improved efficiencies in the mine to mill production value chain because the crushing and grinding circuit is not optimized for the finer feed size. Sometimes, an indiscriminate increase of blast energy can also result in increased ore loss, dilution and damage. *Therefore, proper optimization of mine to mill value chain requires an understanding of the leverage of each process on the overall value chain efficiency and then using that leverage to maximize the overall profitability of the operation rather than just the individual process.*

ROLE OF BLASTING ON MINE TO MILL PRODUCTION VALUE CHAIN

Blasting is the first step in the breakage and separation process and plays an important role in optimising the mine to mill value chain. Hence, a good understanding of the impact of blasting outcomes such as fragmentation, movement, dilution, damage and rock softening on the overall process efficiency is essential to the success of this approach.

Blast Fragmentation

Fragmentation has a direct influence on the performance of mining production equipment, crushers and grinding mills. The definition of optimum particle size distribution (PSD) depends on the site specific conditions but traditionally blasting engineers are interested in +250 mm size rocks because these size particles are the ones that typically affect the efficiency of the loading and hauling operations which are a part of the mining cost centre. However, fragmentation from blasts also affect the efficiency of crushing and grinding operations and in some cases the recovery and the final product value.

The PSD from blasting must not only physically fit into the bucket of the excavator but must do so without unduly reducing the bucket fill time. The oversize fragments not only reduce the productivity of the excavator but also increase the secondary blasting costs and equipment maintenance costs. The definition of oversize depends on the size of excavating equipment and large excavators are more tolerant to large fragments than a small backhoe or front end loader. In addition to over size, the amount of fines and the overall PSD also affect the excavator productivity. Generally finer fragmentation created by the high energy mine to mill blasts improves excavator productivities. Figure 1 shows a comparison of loading times for hundred trucks from muckpiles with higher energy blasts and conventional blasts. The rock mass characteristics in both blasts are similar but high energy blasts produced much finer PSD.

The results clearly show that finer fragmentation from high energy blasts can reduce truck loading times by up to 30%. Even though it is obvious that finer PSD improves the diggability of an excavator it may be very difficult to correlate it to the overall productivity because of the following factors:

- The digging time is only a minor fraction of the overall truck cycle times
- In under-trucked operations any reduction in digging time will be lost in truck waiting times
- Diggability is also dependent on muckpile shape, looseness and the operator

Michaud and Blanchet (1995) have shown that the payload of a truck loaded with finely fragmented rock is significantly higher than a truck loaded with coarsely fragmented rock (Figure 2). It is important to note that the maximum truck pay load is also determined by the specifications of the truck and mine operating practices (i.e., number of passes an excavator takes to fill a truck). In

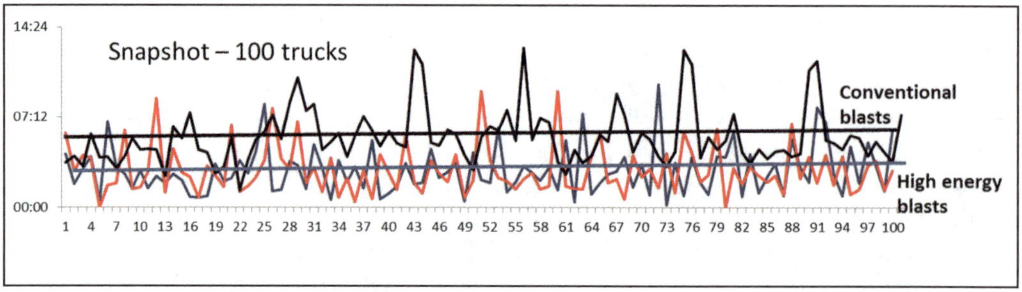

Figure 1. Impact of fragmentation on truck loading times

Figure 2. Impact of fragmentation on truck fill factors

addition to improved loading times and pay loads with finer PSD will also impose less stresses on load and haul equipment and can potentially reduce maintenance costs.

Primary crusher throughput and power consumption are strongly influenced by the ROM PSD distribution especially by the coarse particles. Eloranta1995 and Eliiot et al. 1999 have demonstrated that by reducing the coarse particles in ROM can increase the crusher efficiency. Particle shape also plays an important role in crusher blockages. Flakey and platy particles can block the crusher opening and bridge very quickly. Similarly, clayey material with fine PSD coupled with moisture can bog the crusher pocket and affect its performance.

SAG mills operate most efficiently when they are fed with a bi-modal feed size distribution shown as the continuous line in 3. The dotted line is a typical sizing curve. Ore fractions smaller than the mill discharge grates and trommel screen apertures are considered as 'free grind' material as they pass through the primary mill and are easily broken down further. Fractions of competent ore between 100–200 mm act as grinding media and potentially reduce the required load of steel balls. Fractions between 20 mm and 80 mm are generally called "critical size" breaks slowly in SAG mills and hence should be minimized in the mill feed. Ideally a SAG mill runs efficiently when it is fed with lots of fines (–10 mm), very little critical size and some coarse rock to work as grinding media (Napier-Munn et al. 1996). Impact of critical size on mill efficiency can be minimized by installing large holes in the grate (pebble ports) and crushing them with pebble crushers. However, removing critical size from the mill feed before it reaches the mill is even more beneficial.

SAG mill feed size distribution is influenced by the primary crusher and blasting processes. Generally primary crushers do not produce much fines (–10 mm) and most of the fines in the SAG

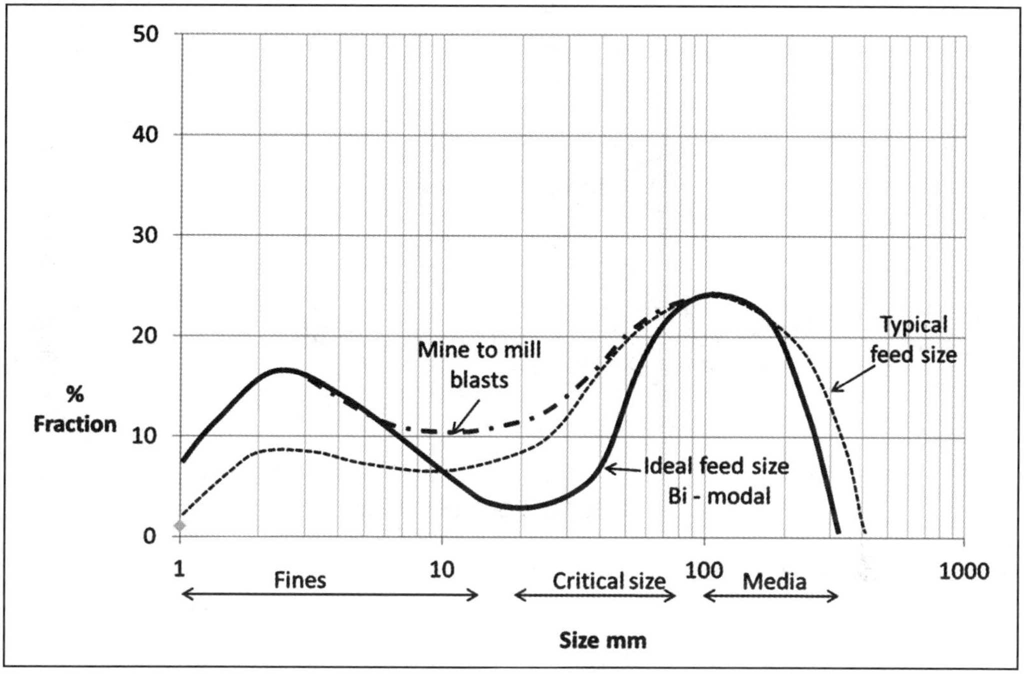

Figure 3. Feed size distribution for SAG mills (Kanchibotla et al. 2011)

feed are produced during blasting (Kanchibotla et al. 1999). Primary crushers are mostly used to reduce the top size from the ROM. An inappropriate feed size distribution shown (shown as dashed line in Figure 3) to the SAG mill reduces its throughput and make them unstable. This is because the mill charge and grinding media are obtained from the coarse fraction of the feed. Changes in feed size distribution alter the size distribution of the charge, which is the grinding media which in turn affects the breakage rates of different size fractions in the SAG mill.

The ideal size distribution for a SAG mill depends on the breakage characteristics of the ore (rock strength) as well as the operating conditions of the mill (lifter design, grate design, mill speed and rock charge). A well designed mine to mill blast combined with appropriate primary crusher setting can produce a feed size with lots of fines and controlled top size (media) but it has little control of the critical size (shown as dashed and dotted lines in Figure 3) and can sometimes negate the benefits of additional fines. However, if the grinding circuit has enough pebble crushing capacity to crush the critical size, the negative impact of additional critical size material from the feed size can be negated.

There are considerable advantages in operating crushing and grinding circuit with a finer and more consistent feed. Figure 4 shows the impact of feed size (F80) on SAG mill throughput and specific energy consumption. Optimising a mineral processing plant with a wide range of feed sizes is considerably more complicated and challenging compared to when it is fed with consistent feed size distribution.

Figure 5 shows the impact of fines on SAG mill throughput per operating hour a large open pit metal mine and it can be clearly seen that SAG mill throughput increases with increased fines.

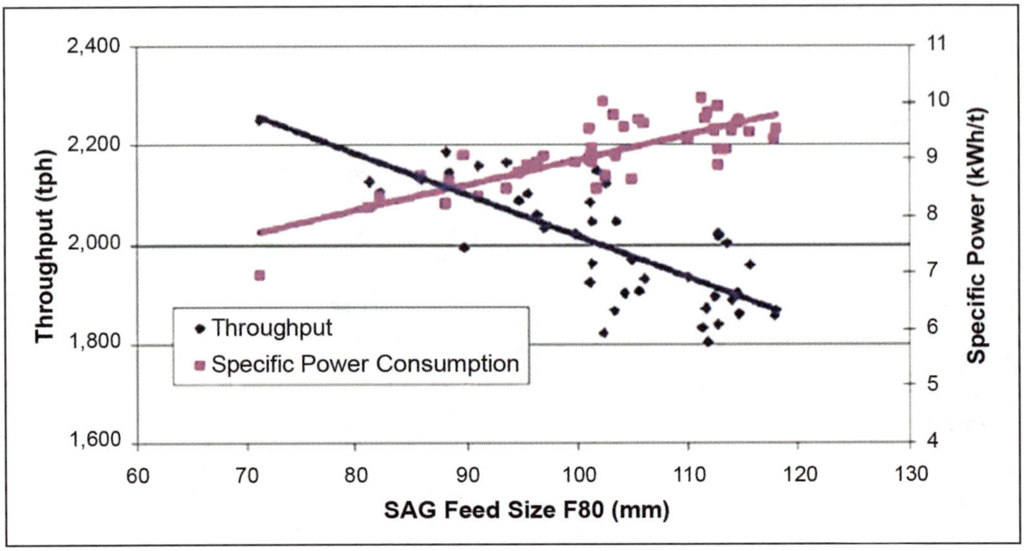

Figure 4. Effect of SAG feed size F80 on throughput and specific power (Hart et al. 2001)

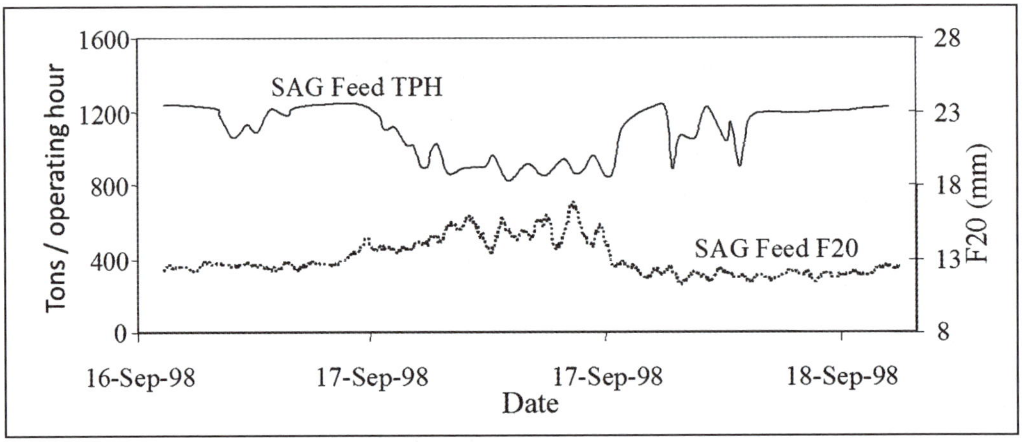

Figure 5. Effect of SAG feed fines on throughput

In some operations the contrary is true. In heap leaching operations excessive fines can reduce leach solution percolation rates and thus can impact the final recoveries (Scott et al. 1998). Iron ore and some quarry products are sold on the basis of size and shape and are particularly prone to a reduction in value due to excess fines.

Blast Movement and Its Impact on Ore Loss and Dilution

In standard ore control procedures various ore zones and waste are excavated based on the pre-blast markings. During a blast the rock mass within the blast volume is fractured and displaced. The direction and magnitude of movement is a function of ore body shape, free face conditions and blast design parameters. Excavation of ore zones without understanding the blast movement can cause significant dilution and ore loss as shown in the Figure 6.

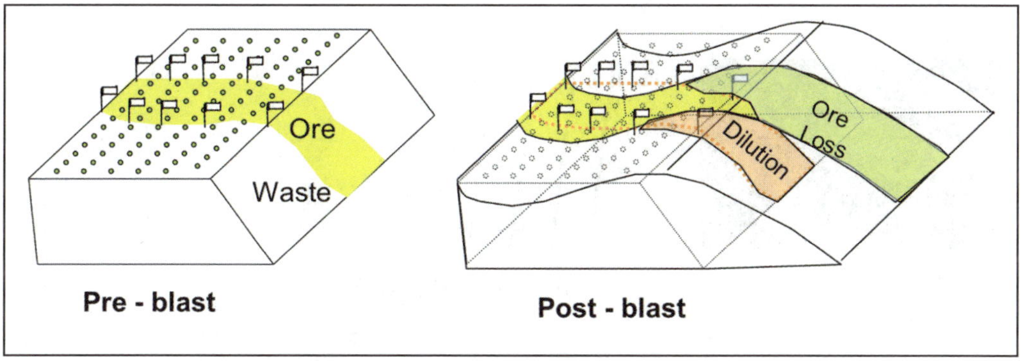

Figure 6. Impact of blast movement on ore loss and dilution

Figure 7. Uncontrolled movement in a blast

Improper blast design and implementation can result in uncontrolled movement as shown in the Figure 7. If such blasts contain both ore and waste rock, it can lead to significant ore loss and dilution.

In operations where the dilution is a critical issue, blasts are usually designed to minimise any lateral movement of the ore. The success of such practices depends on a trade-off between production efficiency and any ore loss or dilution. However, proper blasting and grade control procedures coupled with ore boundary adjustments to account for blast movements can reduce ore loss and dilution (Taylor et al. 1996, Thornton et al. 2005, Thornton 2009, Rogers et al. 2012, Engmann et al. 2012).

Blast Damage

Some damage to the rock mass is inevitable during the blasting process but the incentive to limit this damage is significant. Often the back break caused by the blast loosens and partially dislodges in situ blocks (Figure 8). If there is excessive back break, drills from the next blast cannot drill holes near the crest resulting in excessive burdens and insufficient explosive energy in the burden rock. When the next blast is fired, it will push the in-situ blocks formed by the joints and back break without breaking them resulting in coarse fragmentation (Figure 8a).

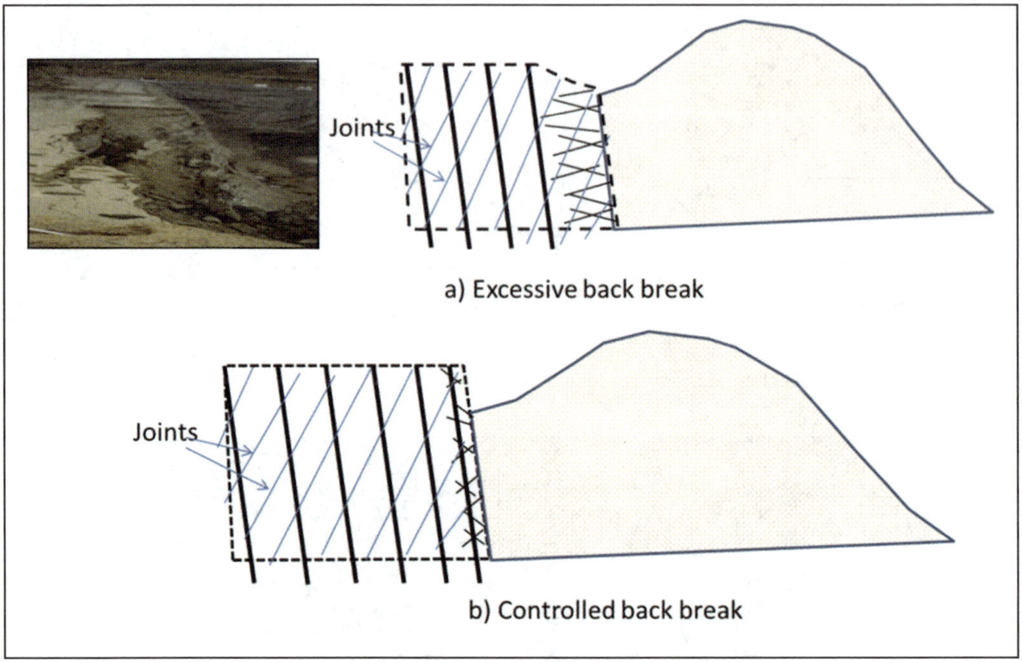

Figure 8. Impact of blast damage on fragmentation

A controlled blast can limit the back break so that holes can be drilled close to the crest to minimize coarse fragmentation from the edges (Figure 8b). Blast damage can also lead to sterilization of ore and pose safety risk to both machinery and life.

ORE SOFTENING AND GRADE CONCENTRATION

During blasting, the sudden impact of the high pressure explosion gases on the borehole wall transmits a stress wave into the surrounding rock. Interaction of this stress wave with free faces and open joints generates tensile and shear waves. All these waves extend pre-existing cracks. Because of the short duration of this stress transient, these cracks may not propagate very far but their total surface area in the rock mass may be 10–100 times larger than the total surface of the primary fragments in a blast (Revnivstev 1988). Therefore blasting action not only produces fractures dividing the rock mass into fragments but also creates a network of micro cracks within the fragments.

Laboratory experiments conducted by several researchers suggest that these micro cracks reduce the crushing and grinding resistance of post-blast fragments (Revnivstev 1988, Chi et al. 1996, Nielsen and Kristiansen 1995). Kojovic and Wedmair (1995) found that the breakage strength of the post-blast fragments is less than the strength of fragments not subjected to blast loading. Nielsen (1998) through his industrial and laboratory experiments demonstrated that the degree of reduction in strength is a function of explosive type and powder factor. Microscopic examination of some of these blast fragments showed that the blast induced micro cracks occurred mainly along the grain boundaries. This preferred formation of micro cracks along the grain boundaries can improve the liberation characteristics.

In ore bodies where mineralization is concentrated along the veins and fracture planes, increased fracturing and fragmentation along these fracture planes can concentrate the mineralized

Figure 9. Comparison of fragmentation from trial blasts

rock or grade in fines. Blast induced stress concentrations near the veins and fracture planes can be increased with advanced timing techniques. Preliminary results from one of the recent studies conducted by the author at a gold mine in Australia indicated significant ore softening and grade increase in finer size fractions with high intensity blasts with advanced priming and timing designs. Both the mechanisms will have significant positive impact on mill throughput and metal recovery, hence further studies are being conducted to confirm the mechanisms.

CASE STUDY 1: IMPACT OF BLAST FRAGMENTATION ON CRUSHING AND GRINDING EFFICIENCY

The first case study was conducted at a copper–zinc mine with an objective to increase the SAG mill throughput for harder ores by optimizing the feed size distribution (Kanchibotla et al. 2008 and Kanchibotla and Valery 2010). During this study baseline blasts and high energy blasts with tighter patterns and higher powder factors were conducted to quantify the impact of ROM PSD on SAG mill throughput. The high energy blasts had about 30% more energy, better distribution of energy and better control of energy due to good stemming. In addition, high energy blasts were also primed and fired with electronic detonators and advanced timing designs to increase shock wave interactions.

The high energy blasts resulted in much finer ROMPSD compared to the baseline blasts (Figure 9). Ore from the baseline and high energy blasts was campaigned through the mill and feed size distribution to the mills was measured by taking belt cuts of primary crusher product. The

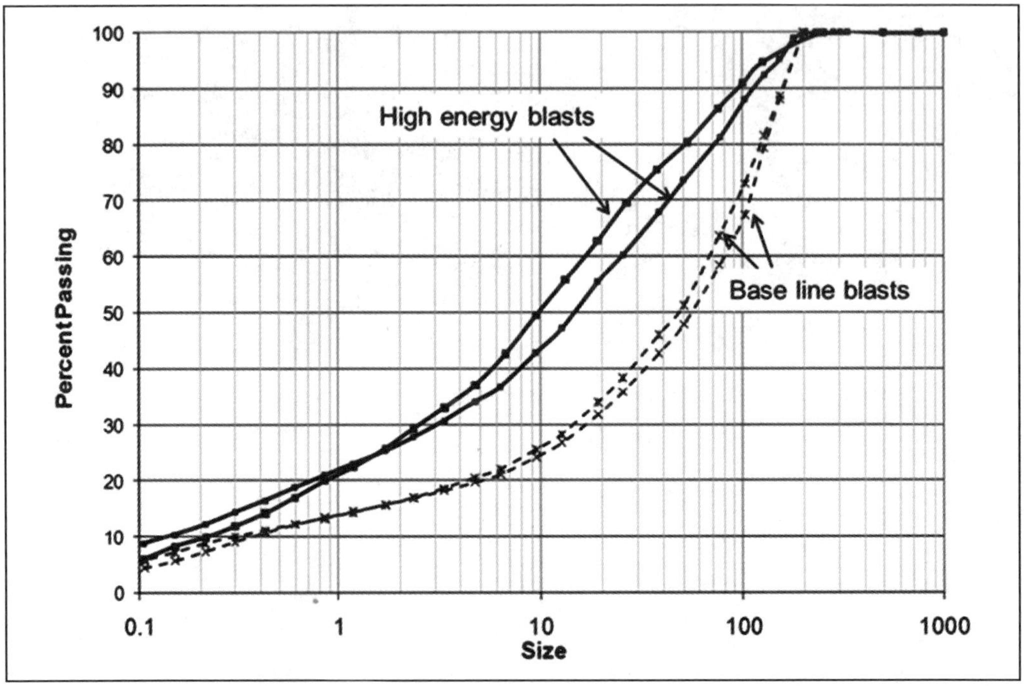

Figure 10. SAG mill feed size from trial blasts

results show that mill feed from the high energy blasts are much finer compared to the baseline blasts (Figure 10).

During this campaigning, crushing and grinding circuits were surveyed and all the operating parameters were carefully monitored for the baseline feed and high energy blasts feed (Table 2). In the case of the primary crusher, the peak power draw was reduced from around 800 kW to less than 250 kW, lowering the thermal capacity which is very important for smooth running of primary crushers. The fine feed from high energy blasts increased the SAG mill throughput by 15% and reduced the power drawn by the mill from 14 MW to 11.7 MW. Overall the fine feed from the high energy blasts reduced the total power consumption of crushing and grinding circuit by approximately 30%. By continued application of the mine to mill integration and optimization philosophy, the concentrator exceeded its target of 4,400 tph for harder CuZn ores by 2010 (Rybinski et al. 2011).

CASE STUDY 2: IMPACT OF FRAGMENTATION AND BLAST MOVEMENT

The second case study was conducted at Newmont Ahafo Gold Mine in Ghana. At Ahafo mine the mine to mill process optimization philosophy was implemented in two phases. The objective of first phase was to improve the grinding circuit performance by optimizing ROM PSD from blasting and the objective of second phase was to control dilution and ore loss from blast movements.

The first phase blast fragmentation optimization project was implemented during 2009 and 2010. During this period changes were made to blast designs to achieve a finer ROMPSD. Ahafo have since standardised on this blast designs for all Apensu ore zones and have been able to sustain higher throughput compared to historical performance on similar material. Modified blasting

Table 2. Crushing and grinding performance for trial blasts

	Baseline Blasts	High Energy Blasts
Primary crusher		
Throughput (t/hr)	4,847	5,133
Average power (kWh)	190.5	124
Peak power (kWh)	816	237
SAG mill		
Throughput (t/hr)	3,467	4,100
Power (kWh)	14	11.7
Specific power (kWh/t)	4.64	3.24

practices resulted in a reduction of SAG mill specific energy by over 20% for similar hardness feed material (Dance et al. 2011).

In the second phase a comprehensive blast movement monitoring program to minimize ore loss and dilution was conducted. A detailed description of this study is given by Rogers et al. (2012) and Engmann et al. (2012). Based on the understanding of blast movement dynamics from the study, alternative strategies were developed implemented to reduce blast induced ore loss and dilution:

- Implementation of blast designs to promote consistent movement along the strike of the ore body and minimise inconsistent movement from edge effects, uneven free faces and cratering especially along the ore/waste boundaries.
- Adjusting the post-blast ore boundaries to account for expected blast movement and to ensure that excavation follows the adjusted polygons.

The alternative strategies developed from this study have been incorporated in the standard site operating procedures. Reconciliation data shows improved agreement between the mine and mill grade and reduction in diluted tons since the implementation of the blast movement study in April 2011 (Engmann et al. 2012) (Figure 11).

In addition to blast movement adjustments, a number of other improvements such as increased blast hole sampling and pit supervision have already been implemented. The geologists at Ahafo mine consider that the biggest gains have been made from post-blast movement adjustments. Post-blast polygon adjustments have enabled ore to be recovered correctly. This project has also improved focus on heave mining and excavation control to ensure that grade blocks are captured correctly in dispatch and the correct sampling techniques are being followed.

CHALLENGES IN IMPLEMENTATION

Even though the benefits of mine to mill process optimization methodology are obvious, based on author's experience, sustenance of the benefits in long term is challenging. Management commitment, continuous cooperation and communication between the personnel from geology, mining and milling is essential to the successful adaptation of this approach. Establishment of appropriate KPIs and awareness of their impact on other downstream process are critical to drive the behavioral change and for long term sustenance of the process. Reliable systems are required to monitor the KPIs accurately and report timely through the production chain on a daily basis. It is important that the new methodology, designs and operating procedures are embedded in the day to day standard operating procedures and all key site personnel are well trained to implement them. Regular

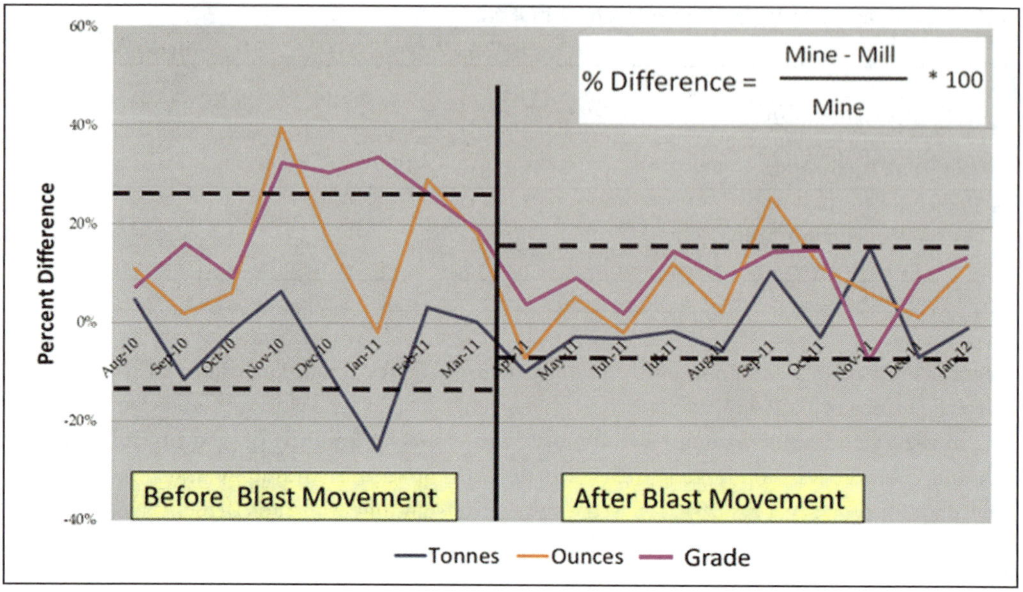

Figure 11. Grade reconciliation between mine and mill

process audits can help to monitor the compliance and also to many necessary changes as per the operational and market requirements.

CONCLUSIONS

Traditional approach of optimizing mining and milling operations separately can fail to recognise the potential benefits that can otherwise be achieved by optimizing the entire process. All the processes in the mine to mill production value chain are inter-dependent and the results of the upstream mining processes can have a significant impact on the efficiency of downstream processes. Blasting plays an important role in this approach, hence, understanding the impact of blasting outcomes on the overall process efficiency is critical to the success of this approach.

The successful mine to mill process optimization case studies demonstrate the value of mining and concentrator personnel working together to reduce the overall cost of an operation rather than reducing their individual operation costs. The benefits derived from designing blasts to improve the efficiency of crushing and grinding stages far outweigh the additional blasting costs. Ore softening and grade increase in finer fractions with advanced blasting practices can further increase the leverage of blasting on the production value chain. Possible negative side-effects such as ore loss, dilution and blast damage can be controlled with improved understanding of blasting process based on systematic monitoring.

REFERENCES

Chi, G., Fuerstenau, M.C., Bradt, R.C., and Ghosh, A. 1996. Improved comminution efficiency through controlled blasting during mining, *Int. J. Miner. Process.* 47:93–101.

Dance, A., Mwansa, S., Valery, W., Amonoo, G., and Bisiaux, B. 2011. Improvement in SAG Mill Throughput from Finer Feed Size at the Newmont Ahafo Operation, SAG2011, Vancouver, Canada, 25–28 September.

Dinis da Gama, C. 1990. Reduction of cost and environmental impacts in quarry rock blasting. Proc. 3rd Intl. Symp. on Rock Fragmentation and Blasting. Brisbane.

Eliiot, R., Either, R., and Levaque, J. 1999. Lafarge Exshaw finer fragmentation study. Proc. 25th ISEE Conf., pp. 333–354, Nashville, TN, USA.

Eloranta, J. 1995. Selection of powder factor in large diameter blastholes. EXPLO '95 Conference, AusIMM, Brisbane, 25–28 September.

Engmann, E., Ako, S., Bisiaux, B., Rogers, W., and Kanchibotla, S. 2012. Measurement and Modelling of blast movement to reduce ore loss & dilution at Ahafo Gold Mine in Ghana. Proc. 2nd UMaT Biennial International Mining & Mineral Conference–UBIMMC, Tarkwa, Ghana.

Hart, S., Valery, W., Clements, B., Reed, M., Song, M., and Dunne, R. 2001. SAG2001, Vancouver, Canada, 25–28 September.

JKMRC. 1998. Final Report on P 483, Optimization of Mine Fragmentation for Downstream Processing. Submitted to Australian Mineral Industries Research Association Limited.

Kanchibotla, S.S., Morrell, S., Valery W., and O'Loughlin, P. 1998. Exploring the effect of blast design on sag mill throughput at KCGM. Proc. Mine to Mill Conf., Brisbane.

Kanchibotla, S.S., Valery, W., and Morrell, S. 1999. Modelling fines in blast fragmentation and its impact on crushing and grinding. Explo-99, Kalgoorlie.

Kanchibotla, S.S., Valle, R., Corsini, J., Colacioppo, J., Serkan, D., Valery, W., and Baguley, P. 2008. Final Report on Blast Optimisation Projects at Compañía Minera Antamina. Internal report.

Kanchibotla, S., and Valery, W. 2010. Mine to mill process integration and optimisation—Benefits and challenges. Proc. 36th ISEE conference, Orlando.

Kanchibotla, S.S., Tello, S., Morrisson, R., and Adair, B. 2011. Energy Footprint Reduction at KUC. JKMRC interal report.

Kojovic, T., and Wedmair, R. 1995. Prediction of fragmentation and the lump/fines ratio from drill core samples at Yandicoogina. JKMRC Internal report.

Kojovic, T., Kanchibotla, S., Poetschka, N., and Chapman, J. 1998. The effect of blast design on the lump:fines ratio at Marandoo iron ore operations. Proc. Mine-Mill Conf., Brisbane.

McKenzie, A.S. 1965. Cost of explosives—Do you evaluate it properly? Proc. American Mining Congress, Las Vegas, NV.

McKee, D.J., Chitombo, G.P., and Morell, S. 1995. The relationship between fragmentation in mining and comminution circuit throughput. *Minerals Engineering* 8(11):1265–1274.

Michaud, P.R., and Blanchet, J.Y. 1995. Establishing a quantitative relation between post blast fragmentation and mine productivity: A case study. Proc. of the Fragblast 5, pp. 389–396, Montreal, Canada.

Moody, L., Cunningham, C., and Lourens, H. 1996. Measuring the effect of blasting fragmentation on hard rock quarrying operations. Proceedings of Fragblast 5, Fragmentation by Blasting. pp. 353–359, Montreal, Quebec, Canada. August 25–29.

Napier-Munn, et al. 1996. Mineral comminution circuits. JKMRC Monograph Series in Mining and Mineral Processing 2, Ed. T.J. Napier-Munn.

Nielsen, K., and Kristiansen, J. 1995. Blasting and grinding—An integrated comminution system. EXPLO '95 Conference, AusIMM, Brisbane, September, pp. 427–436.

Nielsen, K. 1998. Economic optimisation of the blasting-crushing-comminution process in a low grade iron ore (Taconite) operation. Proc. Mine to Mill Conf., Brisbane.

Revnivstev, V.I. 1988. We really need revolution in comminution. Proc. 16th International Mineral Processing Congress.

Rogers, W., Kanchibotla, S., Tordoir, A., Ako, S., Engmann, E., and Bisiaux, B. 2012. Solutions to reduce blast-induced ore loss and dilution at the Ahafo open pit gold mine in Ghana. 2012 *Transactions SME*, Vol. 332.

Rybinski, E., Ghersi, J., Davila, F., Linares, J., Valery, W., Jankovic, A., Valle, R., and Dikmen, S. 2011. Optimisation and Continuous Improvement of Antamina Comminution Circuit. SAG2011, Vancouver, Canada, 25–28 September.

Scott, A., David, D., Alvarez, O., and Veloso, L. 1998. Managing fines generation in the blasting and crushing operations at Cerro-Colorado mine. Proc. Mine to Mill Conf., Brisbane.

Scott, A., Segui J., and Kanchibotla, S.S. 2000. Ore characterization for mine to mill fragmentation. Proc. 4th International Mining Geology Conference, Coolum.

Scott, A., Kanchibotla, S.S., and Morrell, S. 1999. Blasting for Mine to Mill Optimisation. Explo-99, Kalgoorlie.

Simkus, R., and Dance, A. 1998. Measuring and monitoring the ROM properties at Highland Valley copper. Proc. Mine-Mill Conf., Brisbane.

Taylor, S.L., Gilbride, L.J., Daemen, J.J.K., and Mousset-Jones, P. 1996. The impact of blast induced movement on grade dilution in Nevada's precious metal mines. In: B. Mohanty (ed.), Proc. Fifth Intl. Symp. on Rock Fragmentation by Blasting—Fragblast 5. A.A. Balkema, Montreal, Canada, pp. 407–413.

Thornton, D., Sprott, D., and Brunton, I. 2005. Measuring blast movement to reduce ore loss and dilution. Proc. 31st Annual Conference on Explosives and Blasting Technique. International Society of Explosives Engineers, Orlando, FL, pp. 189–200.

Thornton, D. 2009. The implications of blast-induced movement to grade control. In Seventh International Mining Geology Conference, Perth, WA, Australia.

Valery, J.W., Kojovic, T., Tapia-Vergara, F., and Morrell, S. 1999. Optimisation of blasting and sag mill feed size by application of online size analysis. IRR Crushing and Grinding Conference, Perth, WA, 29–31 March.

ABOUT THE AUTHOR

Sarma Kanchibotla has over 30 years of industrial and research experience in the fields of drilling, blasting and mine to mill process optimisation. He implemented several mine to mill optimisation projects around the world and also managed several mine optimisation projects in open-cast coal mines in Australia and Indonesia. Kanchibotla is currently a professor of energy efficient mining at the Julius Kruttschnitt Mineral Research Centre at the University of Queensland and also manages the mining consulting division of JKTech. Before joining the University of Queenlsand, Kanchibotla worked at Metso, Thiess mining contractors, Dyno Nobel, and IDL Chemical Ltd.

Innovations in Comminution Equipment: Crushers, High Pressure Grinding Rolls, Semi-Autogenous Grinding, Ball Mills, and Regrind Mills

Mark T. Erickson
Newmont Mining Corporation, Denver, Colorado, United States

ABSTRACT: There have been numerous innovations and advancements in the design, size, serviceability, maintainability and application of comminution equipment over the last 10 years. Specific innovations/developments for gyratory and cone crushers, HPGRs, SAG and ball mills, and regrind mills will be discussed.

INTRODUCTION

Over the past 10 years or so, there have been a number of innovations and advancements in the design, size, serviceability/maintainability and application of comminution equipment. In general, the comminution equipment has become larger with higher installed power, resulting in greater throughput and fewer pieces of equipment.

In addition to the sizes and installed power increases, innovations in measurement and analysis of wear in the equipment has also been developed and incorporated into the operator's tools to improve maintenance predictions and increase operating time.

COMMINUTION EQUIPMENT TECHNOLOGY

Crushing—Gyratory Crushers

There have been at least three areas of innovation and development in the design and operation of gyratory crushers. These innovation areas include:

1. Gyratory crushers with greater installed power
2. Serviceability/maintainability of bushings and crusher drive components
3. Measurement and analysis of wear components in the crusher, i.e., mantles and concaves

Sizes and Installed Power

The general sizes of gyratory crushers have remained largely unchanged for the past 20 plus years. However, the throughput capabilities of these gyratory crushers have been increased due to the increases in installed power of these crushers.

FLSmidth (FLS) developed their UD (Ultra Duty) design which incorporates improved drive, structural and operating components. The largest FLS UD design is the 60" × 113" with 1MW (1,000 kW) installed power.

Serviceability/Maintainability of Bushing and Crusher Drive Components

In the past, servicing and maintaining of gyratory crusher bushings, drive components and other components is typically performed from below the crusher, in the crusher discharge or eccentric area. This maintenance required the use of eccentric cart and maintenance access into the crusher discharge bin in order to maintain the crusher.

The development of the FLSmidth top service gyratory crusher (TS) allows service to be performed on the crusher by removing parts from the top of the crusher, thereby eliminating the need to access the crusher systems from crusher discharge area (**Figure 1**; FLSmidth 2012). Users have selected the top service gyratory crushers due to

- Increased safety of maintenance employees—no maintenance access required under the crusher;
- Faster maintenance turn-around time; and
- Decreased construction costs of the gyratory crusher installation, i.e., the height of the crusher station can be reduced since there is no crusher maintenance access required under the crusher.

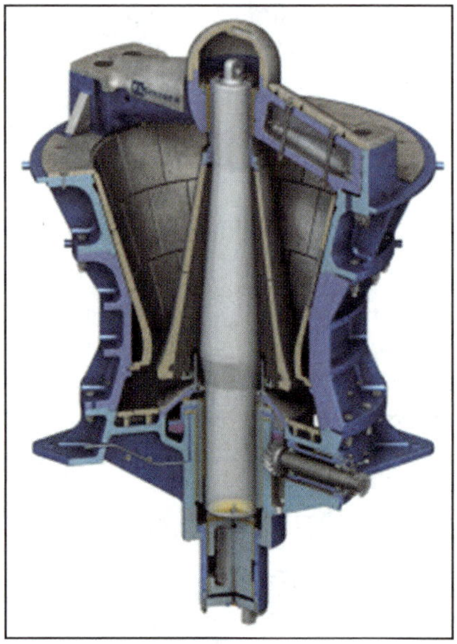

Figure 1. FLS top service crusher

The use of the top service crusher improves the safety of the crusher maintenance by eliminating fall hazards, overhead hazards, and work in confined spaces (FLSmidth 2010).

Measurement and Analysis of Wear Components in the Crusher

Maintenance of the wear components in a gyratory crusher is one of the major operating costs of the crusher. Predicting the wear and wear life of the crusher wear components, i.e., mantles and concaves, is one of the prime activities for the crusher maintenance staff. Recently, techniques to measure the profile of the mantle and concaves, in situ, have been developed.

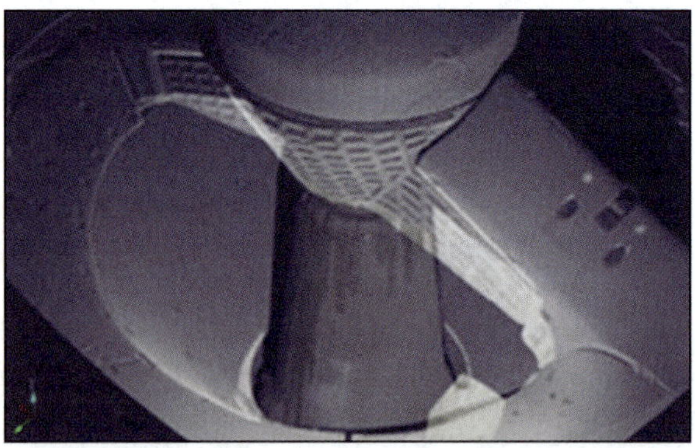

Figure 2. 3D Model of crusher chamber using CrusherMapper

Scanalyse (Scanalyse 2012) has developed a laser scanner tool, CrusherMapper™, which surveys the physical dimensions of the crushing chamber producing a 3D model of the crushing chamber and mantle (Figure 2).

The CrusherMapper™ is a unique service which allows plant operations

Table 1. Large cone crushers

Metso Minerals	FLSmidth
MP800 (800 hp (597 kW))	XL900 (900 hp (671 kW))
MP1000 (1,000 hp (746 kW))	XL1100 (1,000 hp (746 kW))
MP1250 (1,250 hp (932 kW))	XL2000 (2,000 hp (1,491 kW))

and maintenance personnel to make informed decisions on reline scheduling, mantle and concave design and crusher operation.

CrusherMapper™ can be used to provide a time based view of crusher component wear (from new profile to totally worn profile), to identify asymmetrical wear patterns, measure CSS (closed side setting) and OSS (open side setting) and optimize mantle and concave liner designs. Furthermore, the use of CrusherMapper™ improves the safety of crusher maintenance by reducing or eliminating the need for confined space entry (into the crushing chamber) when compared to traditional ultrasonic testing programs.

Cone Crushers

The innovations in cone crushers have been in the size and installed power of the units. Two manufacturers have been the major drivers in the development of cone crushers of increased size and power, Metso Minerals and FLSmidth (see Table 1).

The primary objective in the development and operation of these larger cone crushers is increasing capacity without significantly increasing operating costs (Metso 2010), reducing maintenance and increasing reliability.

Both manufacturers have utilized state-of-the-art design tools, manufacturing materials and techniques, and high quality components to accomplish this increased reliability and performance.

Grinding

High Pressure Grinding Rolls (HPGR)

High pressure grinding rolls (HPGR) were initially introduced to the cement industry by Klaus Schönert in 1979 (Schönert 1982, 1987). In those experiments, Schönert loaded (compressed) a bed of particles utilizing extremely high pressures (50 to 300 N/mm^2) resulting in the formation of a compacted flake of product material (McIvor 1997). Schönert indicated that the total energy needed for compressing and disagglomeration of the limestone and cement clinker was not only significantly less than for ball milling alone, but that the particle fracturing induced at the high pressures resulted in substantial savings (approx. 25%) in subsequent ball mill energy requirements (Schönert 1987).

Since the original development by Schönert, several equipment manufacturers have developed HPGR equipment/systems for use in the cement and mineral industry. These equipment companies include but are not limited to the following:

- KHD Humboldt Wedag AG (Germany)
- Krupp Polysius AG (Germany)
- FLSmidth (Denmark and USA)
- Maschinenfabrik Koppern GmBH (Germany)
- Alpine AG (Germany)

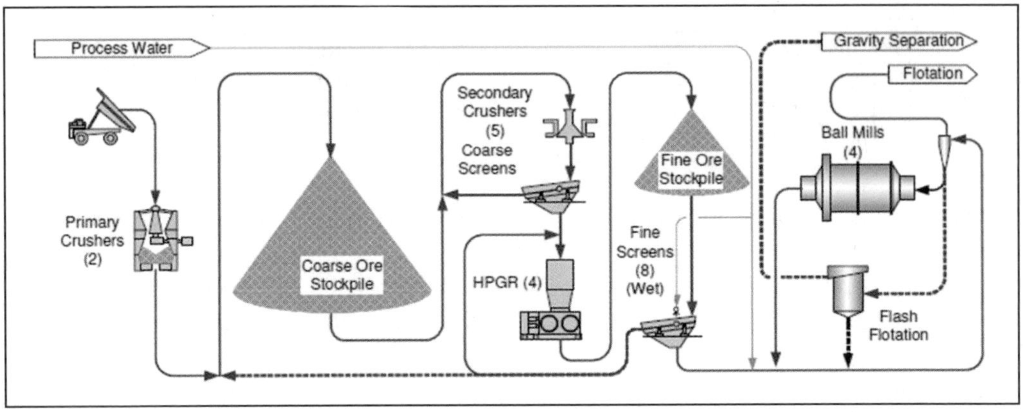

Figure 3. Cerro Verde flowsheet (Vanderbeek et al. 2006)

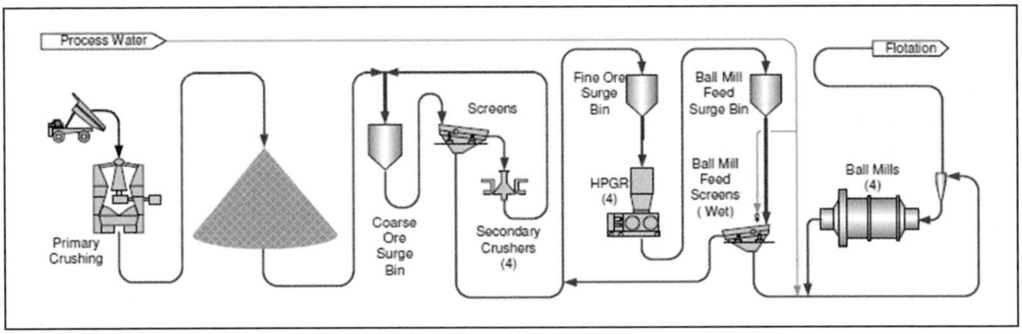

Figure 4. Boddington flowsheet (Dunne et al. 2007)

In addition to the initial use in the cement industry (for both raw and clinker grinding), high pressure grinding rolls (HPGR) have been applied in the following mineral applications:

- Diamonds
- Iron ore
- Tin
- Copper
- Chrome
- Gold

Over the past 30 years, there have been numerous applications and installations that have both evaluated and installed HPGRs on projects to take advantage of either the breakage characteristics of the rock, the power savings, or both (Klymowsky et al. 2006; Patzelt et al. 2006; Rosario et al. 2010; Burchardt et al. 2011; Koski et al. 2011). At present, the HPGR sizes range from the small units with 2 × 220 kW maximum power (total power 440 kW) and 2,700 kN grinding force up to the largest sizes with 2 × 3,400 kW maximum power (total power 6,800 kW) and 20,000 kN grinding force (ThyssenKrupp Polysius 2010). Various circuit configurations are possible in combination with SAG mills, ball mills, or HPGRs as the main comminution device.

Probably the most significant application and installations of high pressure grinding rolls is Freeport Indonesia's Cerro Verde operating in Peru (Vanderbeek et al. 2006; Koski et al. 2011) and at Newmont Mining's Boddington Gold operation in Western Australia (Dunne et al. 2007; Hart et al. 2011). Based on crushing and grinding tests performed in the design phase of these projects, both Cerro Verde and Boddington Gold determined that the most cost effective design was to have two stage coarse crushing (gyratory and cone crushers) followed by a third stage of crushing utilizing HPGRs, followed by single-stage ball mills (**Figure 3** and **Figure 4**).

As with any project, the selection of the comminution devices can be dependent on a number of factors, including the toughness and grindability of the rock, the energy costs, metallurgical implications, etc. The same is true for HPGR systems. HPGR might not be the best unit for all rock types, but when applied correctly the can be a real benefit to the operation and economics of the project development.

SAG and Ball Mills

SAG (Semi Autogenous Grinding) and ball mills have seen the same size and installed power escalation as many of the other crushing and grinding devices. Some of the size/power increases in SAG mills is illustrated in Table 2.

In conjunction with the "first of their kind" mill developments in table above, there have been a number of mill size, design and installed power development milestones over the past decade:

- 2010—Metso sold the largest SAG mill to Newmont Mining for their Conga Project in Peru. Mill is 12.8 m diameter by 7.62 m long (6.86 m effective grinding length, EGL) 42 ft. diameter by 25 ft. long (22.5 ft. EGL), 28 MW (37,520 hp) gearless drive.
- 2010—FLSmidth sold the largest dual pinion drive to a project in Canada. Mill is 11.0 m diameter, 2 × 7.5 MW (2 × 10,055 hp) dual pinion drive.
- 2007—FLSmidth sold the largest ball mills to follow a SAG mill to Antofagasta Minerals for their Esperanza Project in Chile. Mills are two (2) 8.53 m diameter by 13.6 m long (28 ft. diameter by 44.5 ft. long), 22 MW (29,480 hp) gearless drive.

Table 2. First of a kind AG/SAG mills (Jones and Fresko 2010)

Year	D (ft)	D (m)	Type	kW	HP
2010	42	12.8	SAG	28,000	37,520
1996	40	12.2	SAG	20,000	26,800
1996	38	11.6	SAG	20,000	26,800
1986	32	9.75	1st Gearless SAG	8,200	11,000
1979	34	10.4	AG	6,560	8,800
1973	36	11.0	AG	8,950	12,000
1970	30	9.14	AG	5,220	7,000
1970	26	7.93	AG	2,240	3,000
1965	32	9.75	AG	4,470	6,000
1965	20	6.10	AG	370	500
1962	28	8.53	AG	2,610	3,500
1962	24	7.32	AG	1,300	1,750
1959	22	6.71	AG	930	1,250
1959	18	5.49	AG	450	600

Figure 5. CP Mining, Sino Iron ball mills

- 2007—CITIC sold the largest value AG (Autogenous Grinding) mill order to the CP Mining, Sino Iron Project in Australia. Order was ultimately for six (6) 12.2 m diameter by 11.0 m long (40 ft. diameter by 36 ft. long), 28 MW (37,500 hp) gearless drive. Ball mills shown in Figure 5.
- 2006—Metso sold the world's largest AG and pebble mills for an expansion at the Boliden Aitik mine in northern Sweden. Mills are two (2) 11.6 m diameter by 13.7 m long (38 ft. diameter by 45 ft. long) AG mills, 22.6 MW (30,300 hp) gearless drives, and two (2) 9.14 m diameter by 11.6 m long (30 ft. diameter by 38 ft. long) pebble mills, 10 MW (13,400 hp).

Roller Mills

Roller Mills, from manufacturers such as Loesche (Figure 6; Loesche 2012), Polysius, FLSmidth, Gebr. Pfeiffer, Raymond and others have seen wide use in the cement and coal industries. In a roller mill, the material to be ground is essentially crushed (compressed) between the horizontally rotating grinding table and the individual grinding rollers. Depending on the manufacturer, application and equipment size, the roller mill might have 2, 3, 4 or 6 grinding rolls.

To date their application in the comminution of minerals has been very limited, although testwork has been performed by several mining companies.

When applying the roller mill to mineral applications, there are several issues that should be considered:

- Both milling and classification can be performed in a single unit
- Higher efficiency in the size reduction of minerals
- Depending on the manufacturer and roll configuration, it may be possible to perform maintenance on individual rolls while the roller mill is in operation
- With units with 4 or 6 rolls, it may be possible to perform maintenance on 2 rolls on a 4-roll unit or 2 or 3 rolls on a 6-roll unit, while the mill remains in operation.

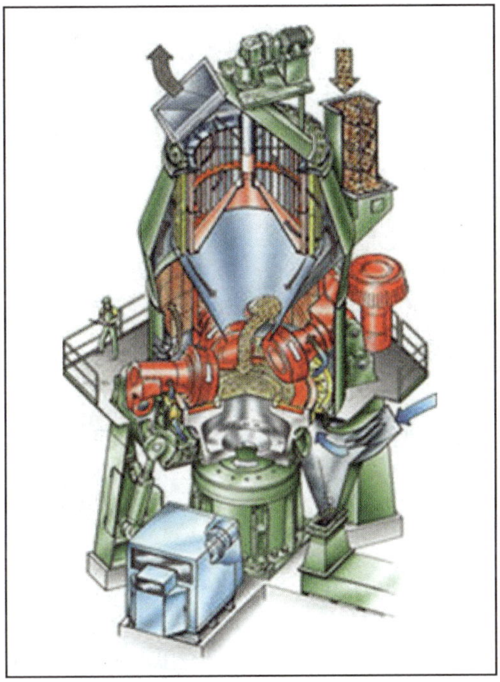

Figure 6. Loesche roller mill

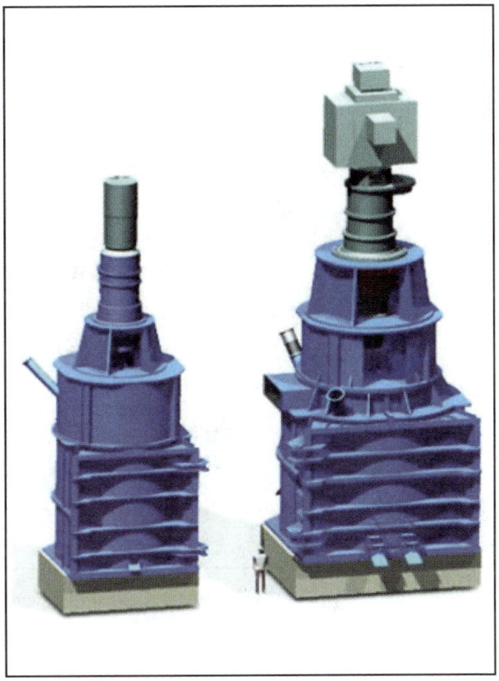

Figure 7. Metso VertiMill

Regrind Mills

Regrind mills have also seen a number of innovations. As with many of the other comminution devices, the major innovations have been in larger sizes and increased installed power. The two main players in the large regrind mills are the Metso VertiMill™ and the Xstrata IsaMill.

VertiMills™

Originally developed in the 1950s for industrial applications, VertiMills (Figure 7; Lichter and Davey 2006) have emerged as the preferred regrind mill application for many operations and applications (Lichter and Davey 2006). VertiMills are a very simple machine with an agitating screw suspended into the grinding chamber. VertiMills have been proven to grind more efficiently than conventional ball bills with feeds as coarse as 6 mm to products finer than 20 microns. Beyond the grinding efficiency, VertiMills offer reduced media consumption, lower installation costs, minimal maintenance and minimal liner wear.

Recently, Metso developed and installed their largest unit, VTM-3000-WB (3,000 hp (2,240 kW)) at Newcrest's Cadia Valley, Ridgeway Concentrator in New South Wales, Australia.

The Ridgeway Concentrator had been continuously modified to increase throughput and concentrate grade (Allen et al. 2011) by

- Converting AG to SAG mill,
- Adding pebble crushing,
- Adding SMD (Stirred Media Detritor) and VTM-1250-WB VertiMill in regrind (Davey 2002),

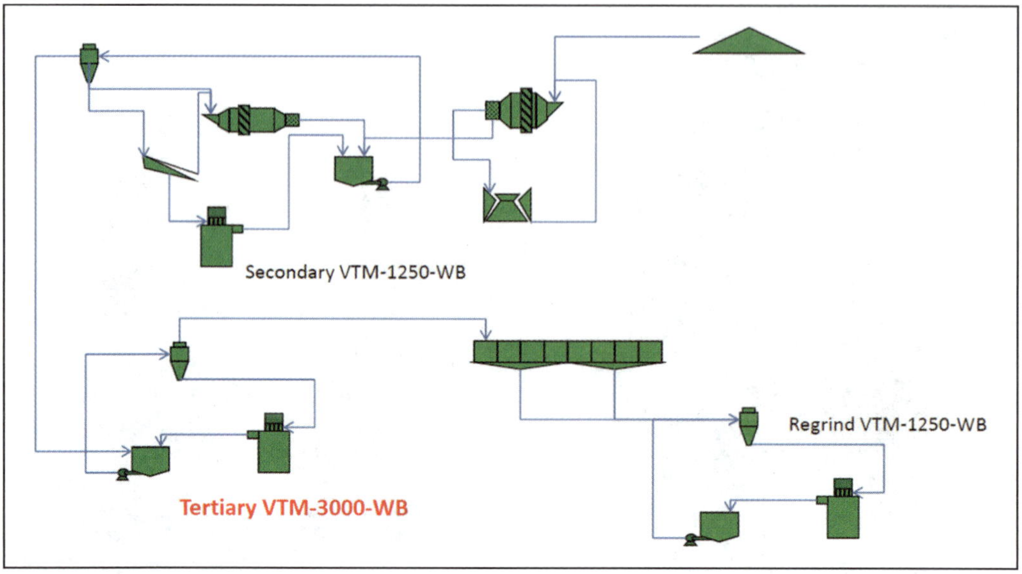

Figure 8. Newcrest Cadia/Ridgeway flowsheet (Allen and Knorr 2011)

- Adding the VTM-1250-WB VertiMill for secondary grinding (parallel to existing ball mill), and
- Adding the VTM-3000-WB VertiMill for tertiary grinding.

Figure 8 illustrates the simplified comminution flowsheet currently being used at Newcrest's Cadia/Ridgeway facility. The Cadia/Ridgeway flowsheet illustrates the use of the secondary VTM-1250-WB in parallel with the horizontal ball mill and the new VTM-3000-WB VertiMill as a tertiary grinding mill feeding the roughter-scavenger flotation circuit.

IsaMills

For grinding applications with product size requirements finer than 25 microns, the use of conventional horizontal and vertical ball milling has been found to be non-economical due to low energy efficiency, low power intensity and high grinding media consumption (Figure 9).

Investigations into fine grinding started at Mount Isa in the 1970s using conventional grinding technologies to increase mineral liberation by grinding to finer sizes. These conventional technology investigations yielded unacceptable results with respect to high power consumption and poor flotation performance due to pulp chemistry and iron contamination of the mineral surfaces from the steel grinding media.

Testing of various grinding technologies from other non-mineral industries resulted in a number of potential stirred media comminution devices, but none that could address the continuous, high tonnage and low value streams of the minerals industry.

Over a period of several years of development and prototype testing, the IsaMill was developed by the Mount Isa operations team and the first full scale IsaMill™, 1.1 MW, was put into operation in 1994 at the Mount Isa concentrator (Johnson et al. 1998; Nihill et al. 1998; Gao et al. 2002). Since the original installation at Mount Isa, IsaMills totaling 130 to 140 MW of installed power have been installed in various facilities around the world (Ellis and Gao 2002; Rule 2008, 2011).

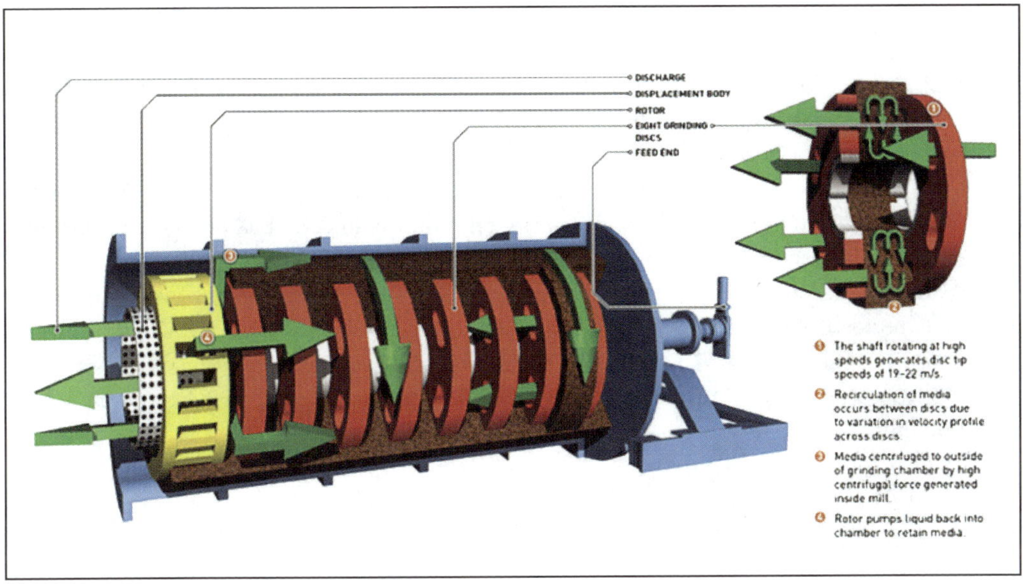

Figure 9. IsaMill grinding mechanism (Xstrata 2012)

Various sizes are available but the M10,000 IsaMill is the most commonly installed unit. As of 2009, the larger 8 MW M50,000 IsaMill has also been in development.

IsaMill advantages include the following:

- Energy efficiency
- Improved recovery
- High intensity
- Open circuit configuration
- Simplicity in maintenance
- Large scale
- Horizontal layout
- Low installation cost
- Accurate scale-up

SUMMARY

Innovations and advancements in the design and use of comminution equipment have been in three major areas:

1. Increase in equipment sizes and installed power
2. Serviceability and maintainability of equipment
3. Measurement and analysis of wear components

By far the greatest contribution is in the area of increased equipment sizes and installed power. Crushers, both gyratory and cone crushers, have seen increases in installed power and throughput with the development of FLSmidth Ultra Duty gyratory crusher (1,000 kW installed power), the

FLSmidth XL2000 cone crusher (2,000 hp installed power) and the Metso MP1250 cone crusher (1,250 hp).

High pressure grinding rolls (HPGR) have been installed in a number of mineral applications, the most notable installations being the FMI Cerro Verde project in Peru and the Newmont Boddington project in Australia.

SAG mills and ball mills have seen increases in size and installed power with the Metso 42 ft. SAG mill (28 MW) to be installed at Newmont's Conga project in Peru. Ball mills and pebble mills in the 28 to 30 ft. diameter range have also been installed in a number of projects.

Regrind mills have also seen an increased in size and installed power. Two manufacturers are at the forefront of development of large regrind mills, i.e., Metso VertiMill and Xstrata Technology IsaMill. Metso VTM-3000-WB VertiMill (3,000 hp) has been installed in Newcrest's Cadia Ridgeway property in Australia. Xstrata Technology has developed IsaMills up to the 8 MW M50,000 unit.

REFERENCES

Allen, J. and Knorr, B., 2011. Development and implementation of the new 2240 kW (3000 hp) VertiMill grinding mill for Newcrest Cadia Valley Operations, *8th International Mineral Processing Seminar*, Procemin.

Burchardt, E., Patzelt, N., Knecht, J., and Klymowsky, R., 2011. HPGR's in minerals: What do existing operations tell us for the future? *SAG 2011.*

Davey, G., 2002. Ultrafine and fine grinding using the Metso Stirred Media.

Detritor (SMD); *Proc 34th CMP*. pp. 153–169.

Dunne, R., Hart, S., Parker, B. and Veilette, G., 2007. Boddington gold mine—an example of sustaining gold production for 30 years, *World Gold 2007 Conference*, AusIMM, pp. 213–220.

Ellis. S. and Gao, M., 2002. The development of ultrafine grinding at KCGM, SME Annual Meeting, Preprint 2-072.

FLSmidth, 2010. Latest generation crusher upgrades quarry production, *http://www.flsmidth.com*, April.

Gao, M., Young, M.F. and Allum, P., 2002. IsaMill fine grinding technology and its industrial applications at Mount Isa Mines, Proc 35th CMP, pp. 171–188.

Hart, S., Parker, B., Rees, T., Manesh, A. and Mcgaffin, I., 2011. Commissioning and ramp up of the HPGR circuit at Newmont Boddington Gold, *SAG2011.*

Johnson N.W., Gao, M., Young M F and Cronin, B., 1998. Application of the IsaMill (a horizontal stirred mill) to the lead–zinc concentrator (Mount Isa Mines Limited) and the mining cycle, AusIMM '98—The Mining Cycle, Mount Isa, AusIMM, p. 291–297.

Jones, S. and Fresko, M., 2011. Autogenous and semi-autogenous mills, 2010 Update, *SAG 2011.*

Klymowsky, R., Ratzelt, N., Knecht, J., and Burchardt, E., 2006. An overview of HPGR techonology. *SAG2006,* 4. pp. 11–26.

Koski, S., Vanderbeek, J.L. and Enriquez, J., 2011. Cerro verde concentrator—four year operating HPGRs. *SAG 2011.*

Lichter, J.K. and Davey, G., 2006. Selection and sizing of ultrafine and stirred grinding mills. *Advances in comminution* (ed. S.K. Kawatra), SME. pp. 69–85.

Loesche, 2012. http://www.Loesche.com.

McIvor, R., 1997. High Pressure Grinding Rolls—A Review, *Comminution Practices* (ed. S.K. Kawatra), SME, pp. 95–98.

Metso, 2010. Nordberg Cone Crushers–MP Series Cone Crushers Product, *Brochure.*

Nihill, D.H., Stewart, C.M. and Bowen, P., 1998. The McArthur River mine—the first years of operation, *AusIMM '98—The Mining Cycle, Mount Isa,* AusIMM, pp. 73–82.

Patzelt, N., Klymowsky, R., Knecht, J., and Burchardt, E., 2006. High pressure grinding rolls for gold/copper applications, *Advances in comminution* (ed. S.K. Kawatra), SME, pp. 51–67.

Rosario, P., Grundy, M. and Hall, R., 2010. Some approaches to high tonnage comminution, *SME Annual Conference*, Preprint 10–114.

Rule, C.M., 2008. Energy considerations in the current PGM processing flowsheet utilizing new technologies, *3rd International Platinum Conference*, SAIMM pp. 45–52.

Rule, C.M., 2011. Stirred Milling—New Comminution Technology in the PGM Industry. SAIMM. Feb, pp. 101–107.

Scanalyse, 2012. CrusherMapper™—Introducing CrusherMapper™ for gyratory crushers, http://www.scanlyse.com, September.

Schönert, K., 1982. Method of fine and very fine comminution of materials having brittle behaviour, *US Patent 4,357,287*.

Schönert, K., 1987. A first survey of grinding with high-compression roller mills, *Int. J. Min. Proc.*, 22, pp. 401–412.

ThyssenKrupp Polysius, 2012. POLYCOM high pressure grinding roll for the mineral industry, *Brochure*.

Vanderbeek, J.L., Linde, T.B., Brack, W.S. and Marsden, J.O., 2006. HPGR implementation at Cerro Verde. *SAG2006, 4.* pp. 45–61.

Xstrata Technology, 2012. IsaMill Advantages, http://www.isamill.com.

ABOUT THE AUTHOR

Mark T. Erickson has worked for 36 years in the mining industry focusing on comminution equipment application, process design, and project engineering. His career includes assignments with U.S. Steel Minntac, Allis Chalmers, Boliden Allis, Svedala, Ore Sorters North America, Bateman Engineering, Fluor Daniel Mining and Minerals, Bechtel Mining and Metals, FLSmidth, and Newmont Mining. At Bechtel, he was the director of the Copper Center of Excellence in Santiago, Chile. He joined Newmont Mining in 2008 as the manager of project engineering. He is currently a technical advisor/subject matter expert and member of the Value Assurance team.

Innovations in Comminution Modelling and Ore Characterisation

Stephen Morrell
SMCC Pty Ltd, Queensland, Australia

ABSTRACT: If one reads the literature from the last 75 years or so, it is full of many good ideas and theories concerning the modeling of comminution circuits and the associated field of rock hardness testing. Very few, however, have stood the test of time and become established practice and hence can claim to be recognized as being "state-of-the-art." An attempted is made here to describe what the current "state-of-the-art" is in the fields of comminution modeling and ore characterization. In fact in the author's view first prize for state-of-the-art in comminution is still the approach first published by Fred Bond over 60 years ago. This is not to say that our industry has stood still during this time. One of the major advances in technology has been the development of simulation modelling, at the forefront of which has been the JKMRC. The result of their research of over 30 years has been the introduction and acceptance of JKSimMet and the Drop-weight Test as being a commonplace, reliable design and optimization tool for comminution circuits. Geometallurgy, unheard of in Bond's day, has seen a relatively rapid introduction to our industry with major benefits for circuit design and accurate forecasting. What has helped its introduction and success has been the development and commercialization of bench-scale ore characterization tests such as the SPI® and SMC Test® which are relevant to AG/SAG mill grinding circuits which have come to dominate circuit design.

INTRODUCTION

If one reads the literature from the last 75 years or so, it is full of many good ideas and theories concerning the modelling of comminution circuits and the associated field of rock hardness testing. Very few, however, have stood the test of time and become established practice and hence can claim to be recognised as being "state-of-the-art." One might ask the question in the first place as to what constitutes "state-of-the-art" and it is appropriate to provide a definition, at least as it has been applied in choosing the subject matter that will be discussed in this paper. For the purposes of this paper a comminution model or ore characterisation test has been considered as being state-of-the-art if it meets the criterion "Is it currently a major contributor in the execution of comminution circuit design and optimisation studies?" Having worked on numerous comminution studies over the last 30 years the author's choices from answering this question are as follows.

Modelling:

- Bond
- JKSimMet

Ore Characterisation:

- Bond
- JK Drop-weight/SMC Test®
- SPI®

MODELLING

Bond

At first sight it might come as a surprise to see Bond in this section as often modelling is seen as the domain of relatively complex equations linked together in a simulator framework. However, models need not be like this. Any equation, however simple, that predicts a real-world variable can be classed as a model and Bond's equation (Bond 1952, 1961) for predicting specific comminution energy has become the most widely used and successful model of all. It is used in its simplest form (Equation 1) or with various factors applied to it, these factors being either published (Rowland 1982) or in some cases proprietary.

$$W = W_i\left(\frac{10}{\sqrt{P}} - \frac{10}{\sqrt{F}}\right) \quad \textbf{(1)}$$

where

W = specific energy
W_i = work index
P = 80% passing size for the product
F = 80% passing size for the feed

Equation 1 is one form of a so-called energy-size relationship and Bond was by no means the originator of the idea. As far back as the mid/late 19th century von Rittinger (1867) and Kick (1885) published theirs. As pointed out by Hukki (1961) all of the equations that these researchers developed are special forms of the same differential equation as proposed by Walker et al. (1937). This equation can be written as

$$dE = -C\frac{dx}{x^n} \quad \textbf{(2)}$$

where

E = net energy required per unit weight (specific energy)
x = index describing the size distribution, e.g., p80
n = exponent indicating the order of the process
C = constant related to material properties and the units chosen to balance the equation

If the exponent in Equation 1 is set to values 2, 1.5, and 1, then integrated, the equations proposed by von Rittinger, Bond, and Kick, respectively, are obtained:

$$E_{1-2} = C\left(\frac{1}{x_2} - \frac{1}{x_1}\right) \quad \textbf{(3)}$$

$$E_{1-2} = C\left(\frac{1}{\sqrt{x_2}} - \frac{1}{\sqrt{x_1}}\right) \tag{4}$$

$$E_{1-2} = C\left(\ln(x_1) - \ln(x_2)\right) \tag{5}$$

where

E_{1-2} = net specific energy to reduce size distribution 1 to distribution 2

x_1, x_2 = stize distribution indices where $x_1 > x_2$

Bond's contribution and what made his equation so successful is that he combined it with a series of associated laboratory tests for crushers, rod mills and ball mills that, when applied to practical design situations, provided reasonably life-like responses, i.e., it worked. Interestingly Bond never published any data to support the validity of his approach, but it is known he did conduct experiments at the larger scale to collect data with which he could test and tune his equations. The fact that 60 years after his 1952 paper his methodology is still widely used is testament to the fact that he must have got it largely right. No industry is kind to failures and over the years practical experience has shown, that as applied to crushers, rod and ball mills, his equations provide acceptable accuracy in design applications. In his day Autogenous (AG) and Semi-autogenous (SAG) mills were not common and hence he never expanded his approach to encompass these machines. Proprietary techniques have been developed which have adapted his equations with varying degrees of success. By and large, however, experience has shown that caution needs to be exercised when employing methodologies based on Bond's approach to AG/SAG mill situations.

JKSimMet

The advent and rapid development of high speed computers provided the opportunity to develop complex models of comminution machines that could execute their calculations very rapidly and which could be linked together such that entire circuits could be simulated. Such models contrast significantly with the energy-size relationship of Bond and are aimed at reproducing the overall response of comminution machines in terms of throughput, power draw and product size distribution. By their very nature such models are relatively complex and may have many interactions between the various sub-processes that describe the machine in question, and which without computers would be almost impossible to apply in a reasonable time-frame. In parallel with the development of modern computers, researchers started developing mathematical simulation models of AG/SAG mills, ball mills, crushers and classifiers. Some of this early research was purely academic and had little relevance to practicing Minerals Processors. Lynch (1977) and the team of researchers he put together at the Julius Kruttschnitt Mineral Research Centre (JKMRC), however, were at the forefront of developing models that were practically relevant. This was brought about in two ways. Firstly they developed models with the aid of data collected from industrial comminution circuits, which by definition made them applicable in this arena. This contrasted with what was done by many other research teams, which was to provide either theoretical equations or equations based on the response of simple batch laboratory systems that bore little relationship to full-scale continuous operations. The second development was to create a user-friendly interface with a simulation framework which efficiently linked the various comminution models such that the response of an entire comminution circuit could be mimicked—JKSimMet.

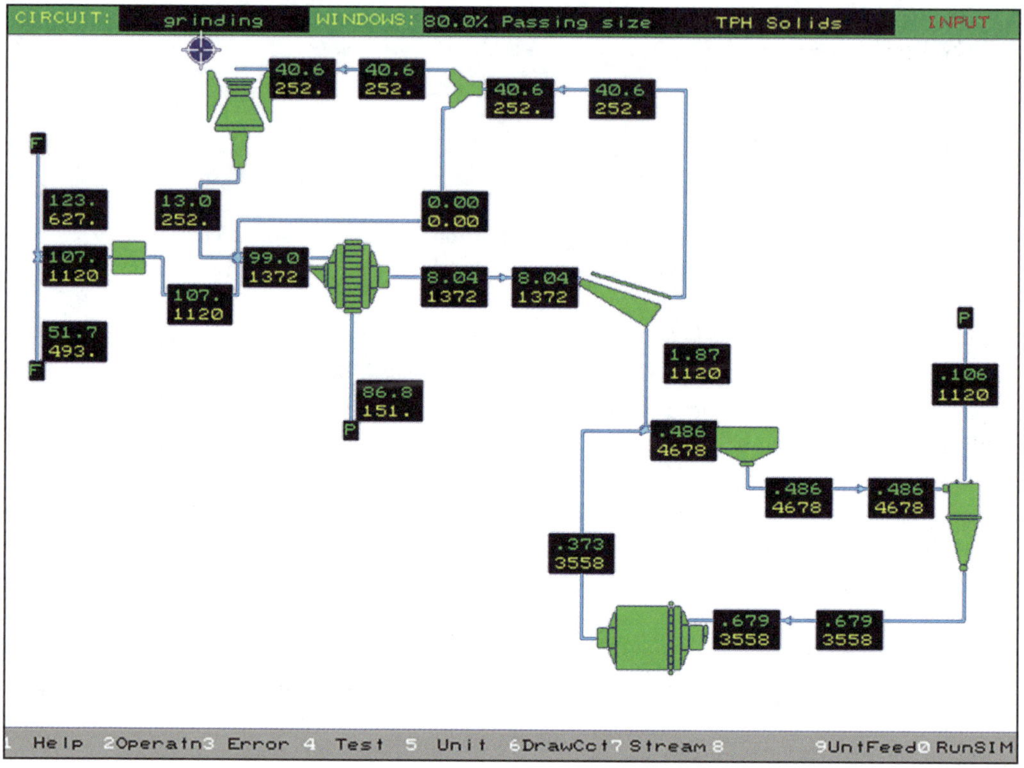

Figure 1. Typical screen view of the original DOS version of JKSimMet

JKSimMet was commercialised in 1986 and to date 480 copies have been sold. Its original purpose was to give practicing Minerals Processors access to an easy-to-use, life-like comminution circuit simulator with which to run and assess optimisation strategies to improve plant performance (Figure 1 shows a typical screen view of the original DOS version). This cut down significantly the time consuming and potentially costly trial-and-error field experimentation that was the usual methodology up until then. The models in the original version were all largely applied by initially fitting key parameters from survey data obtained from the plants that were required to be optimised. As such it had very limited use in design situations, as there were no existing plants which could be used to fit the necessary parameters.

The rapid rise in popularity of AG/SAG mills from the early 1980s and the observation of many comminution practitioners that Bond's equations were not entirely suitable for designing AG/SAG mill circuits, prompted a number of mining companies to look to simulation models as a potential alternative in design projects. This was reflected in the focus of the AMIRA P9 research programme which at that time was driven by these mining companies to develop an AG/SAG mill model for JKSimMet that could be used for design purposes, rather than only for optimisation. The result was what has become known as the "Variable Rates" model (Morrell and Morrison 1996). This model was built using Leung's framework (Leung 1987) and a series of equations which related the model's breakage rates to mill dimensions and operating conditions such as ball load and rotational speed (see Figure 2). These equations were developed using a data base of pilot and full scale mill AG/SAG mill circuits. The equations hence encapsulated the performance trends of the mills

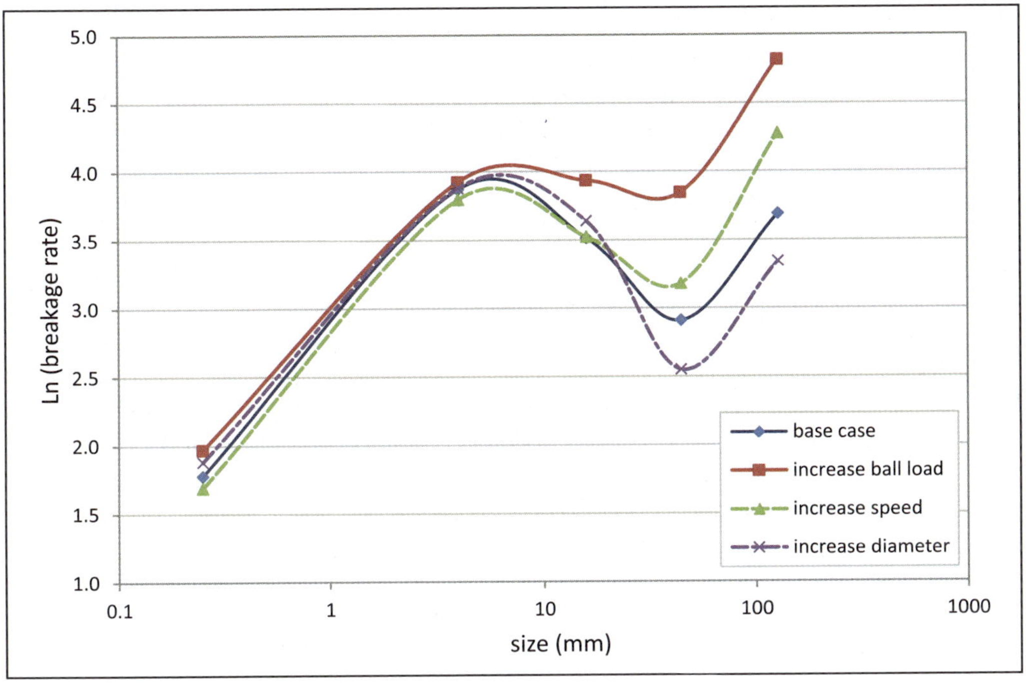

Figure 2. Breakage rates dependencies

in the data base, which effectively took the place of an existing circuit in the optimisation scenario. Since the inclusion of this model in JKSimMet its use in circuit design has become established practice and has greatly enhanced our ability to correctly size comminution circuits.

ORE CHARACTERISATION

When it comes to predicting comminution circuit performance there are two principle requirements. The first is that the equations/models which describe the crushing and grinding machines' operational characteristics are life-like and the second is that the ore hardness has been appropriately characterised. The necessity of characterisation cannot be stressed enough, both in terms of being able to obtain the overall hardness of an ore body but also how it varies spatially. This latter aspect has only been taken seriously by most mining companies over the last 5–10 years and is embodied in so-called Geometallurgy. Geometallurgy relates to the practice of combining minerals processing (metallurgy) with geology and geostatistics to create a spatially or geologically based predictive model for, in this case, crushing and grinding plants. Effectively this entails being able to populate a block model with comminution parameters which allow the accurate forecasting of crushing and grinding circuit throughput. At the design stage this enables the most appropriate circuit to be chosen so that the LOM plan can be successfully executed and is far superior to a decision based on a global ore hardness such as a mean or percentile value (e.g., 80th). Whichever route is taken, samples of the ore body must be obtained and this usually entails core drilling. The ore characterisation test(s) chosen must therefore be able to utilise drill core samples.

Bond Tests

To go with his general equation for predicting specific energy (Equation 1), Bond developed three rock characterisation test procedures, each one being associated with a particular comminution machine (he also developed an additional procedure associated with wear but this is not within the scope of this paper). In his day crushers, rod mills and ball mills dominated circuit design and hence the tests that he developed related to these machines. All of the tests he developed require relatively small amounts of material and hence are readily suited for use with drill core samples. However, ideally the crushing work index test needs whole PQ core. This tends to be less common (and more expensive) than NQ and HQ core, which are suitable for use in Bond's rod and ball mills tests.

Crushing Test

For crushing applications he developed an apparatus comprising two opposing 30-lb hammers which came together through the action of two counter-rotating wheels (see Figure 3). The left hand wheel was initially rotated clockwise and, through a connecting device, the right-hand wheel was synchronised and rotated counter-clockwise by a similar amount. As the hammers were attached to the wheels the hammers were lifted up by this rotation and when the wheels were released the hammers would collide with one another at the 6 o'clock position. By measuring how high the hammers were raised in relation to their final rest position the potential energy could be estimated. A rock specimen was mounted on a plinth at the point of collision of the hammers and hence was broken with an amount of energy equivalent to the estimated potential energy. Figure 3 is a copy of a picture of the device from very old source and hence is not of ideal quality. Where the hammers collide (centre of the v-frame toward the bottom of the picture) is therefore not clear. Figure 4 shows a clearer view of this region and has been taken from a modern machine which is faithful to Bond's original design.

Only particles of a certain size should be tested in this machine and this was specified by Bond to be in the range –3"+ 2.5". The procedure that must be followed, according to Bond, is

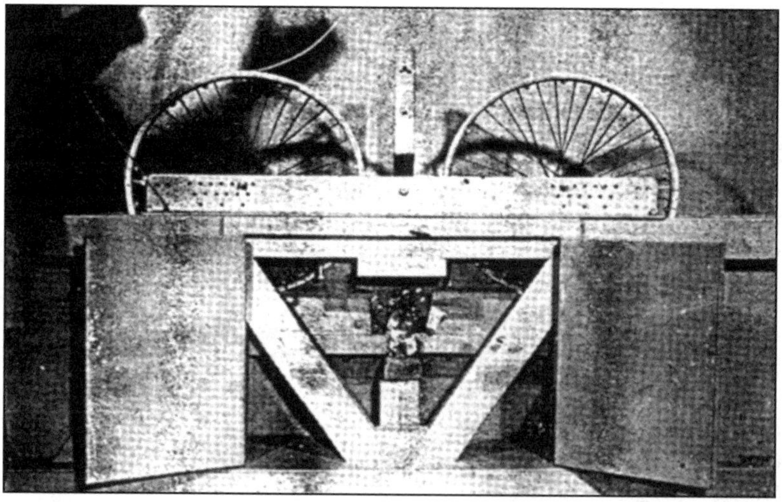

Figure 3. Bond's original crushing work index test machine

Figure 4. Region where breakage takes place in Bond's crushing work index machine

to successively raise the height of the hammers until the rock breaks. It hence is aimed at trying to measure the energy needed to just break the rock. Bond recommended 10 rocks should be broken. The so-called crushing work index (W_{ic}) was then found using Equation 6, based on the average result from the 10 rocks. By using this value in Equation 1 the crusher specific energy is then estimated.

$$W_{ic} = 2.59C/sg \tag{6}$$

where

W_{ic} = crushing work index (kWh/short ton)
$C = E/D$
E = breakage energy (foot-pounds)
D = rock thickness (inch)

Rod Milling Test

For rod milling Bond developed a dry locked-cycle test which uses a 12" diameter, 24" long batch mill with wave liners and running at 46 rpm (see **Figure** 5). The mill is charged with a specified quantity of rods of given sizes. The rock being characterised is crushed to 100% passing 12.7mm and ground, dry, such that the recycle load is 100%. The net grams of final product produced when the recycle load is 100% is measured and inserted in to Equation 7 to obtain the rod mill crushing work index. By inserting this value into Equation 1, Bond claimed that the specific energy of a wet 8-ft overflow rod mill could be obtained.

$$W_{ir} = \frac{68}{P_1^{0.23}(Gbp)^{0.625}\,10\left(\frac{1}{\sqrt{P}} - \frac{1}{\sqrt{F}}\right)} \tag{7}$$

Figure 5. Bond rod mill

where

W_{ir} = Bond laboratory rod work index (kWh/tonne)
P_1 = closing screen size in microns
Gbp = net grams of screen undersize per mill revolution
P = 80% passing size of the product in microns
F = 80% passing size of the feed in microns

Ball Milling Test

For ball milling Bond developed a dry locked-cycle test which uses a 12" diameter, 12" long batch mill with rounded corners and smooth liners and running at 70 rpm (see **Figure 6**). The mill is charged with a specified quantity of balls of given sizes. The rock being characterised is crushed to 100% passing 3.35mm and ground, dry, such that the recycle load is 250%. The net grams of final product produced when the recycle load is 250% is measured and inserted in to Equation 8 to obtain the rod mill crushing work index. By inserting this value into Equation 1, Bond claimed that the specific energy of a wet 8ft overflow ball mill could be obtained.

$$W_{ib} = \frac{49}{P_1^{0.23}(Gbp)^{0.82}\,10\left(\frac{1}{\sqrt{P}} - \frac{1}{\sqrt{F}}\right)} \qquad \textbf{(8)}$$

where

W_{ib} = Bond laboratory ball work index (kWh/tonne)
P_1 = closing screen size in microns
Gbp = net grams of screen undersize per mill revolution
P = 80% passing size of the product in microns
F = 80% passing size of the feed in microns

Figure 6. Bond ball mill

JK DROP-WEIGHT TEST

The original idea for what became known as the JK Drop-weight Test came from Douglas Brown in 1992 and was precipitated by the need for a machine that could break larger rocks than the twin-pendulum device that was in use at the time, and also for a machine that was simple to use, easy to maintain, and was relatively precise (i.e., had good repeatability). A picture of the device is shown in Figure 7. Unlike Bond's laboratory characterisation machines which produce a single work index value, the objective of the drop-weight tester is to produce much more detailed information concerning the size distributions of the progeny of broken rocks and their relationship to input energy. This is required to satisfy the needs of comminution simulation models whose aim among other things is to predict the product size distributions of crushers and mills. The JK Drop-weight Test is thus intimately linked to JK comminution models and it is no coincidence that the Drop-weight Test was developed at about the same time as the Variable Rates AG/SAG mill model.

The test itself involves breaking rocks from five different size fractions: –63+53mm, –45+37.5mm, –31.5+26.5mm, –22.4+19mm, –16+13.2mm. Each rock size fraction is broken with a range of three input energies and the resultant broken particles are sized. In total, therefore fifteen sets of data are generated (5 size fractions × 3 energy levels). The so-called t_{10} is then determined from each set of data and the relationship between the resultant t_{10} values and the associated input specific energies (*Ecs*) is fitted using the following Equation (Leung 1987):

$$t_{10} = A\,(1-e^{-b \cdot Ecs}) \quad \textbf{(9)}$$

where

A and *b* = ore-specific parameters

Ecs = specific comminution energy (kWh/t)

(a) (b)

Figure 7. (a) Drop-weight test in operation; (b) close-up of impact head and anvil

As the impact energy is varied, so does the t_{10}, higher impact energies producing higher values of t_{10}, which are reflected in products with finer size distributions (**Figure 8**).

The t_{10} parameter and its associated use in modelling for reproducing size distributions was originally developed by Narayanan (Narayanan and Whiten 1988). As applied to the breakage of a single particle, the t_{10} is the percentage of broken particles that pass a sieve that is 1/10th of the original particles size. **Figure 8** illustrates how the t_{10} is estimated from breaking 1 inch rocks at 2 different energy levels. **Figure 9** shows how for a given rock size the t_{10} varies with input energy following the shape of curve described by Equation 9. Hence, when modelling comminution machines, if the energies applied in breaking rocks can be estimated, Equation 9 can be used to give the associated t_{10} values. As simulation models are required to predict product size distributions this still leaves a link from the t_{10} to a size distribution to be made. This is done through the use of the relationship given in **Figure 10**. Just as the t_{10} is the percentage of broken particles that pass a sieve that is 1/10th of the original particles size, the t_2 is the percentage of broken particles that pass a sieve that is ½ of the original particles size and the t_{75} is the percentage of broken particles that pass a sieve that is 1/75th of the original particles size etc. With reference to **Figure 8** it is therefore possible to determine not only a t_{10} from a particular product size distribution but a t_2, t_4 etc.—in fact, a whole family of tn values can be determined. Narayanan found that if the t_{10} values from breaking rocks at different

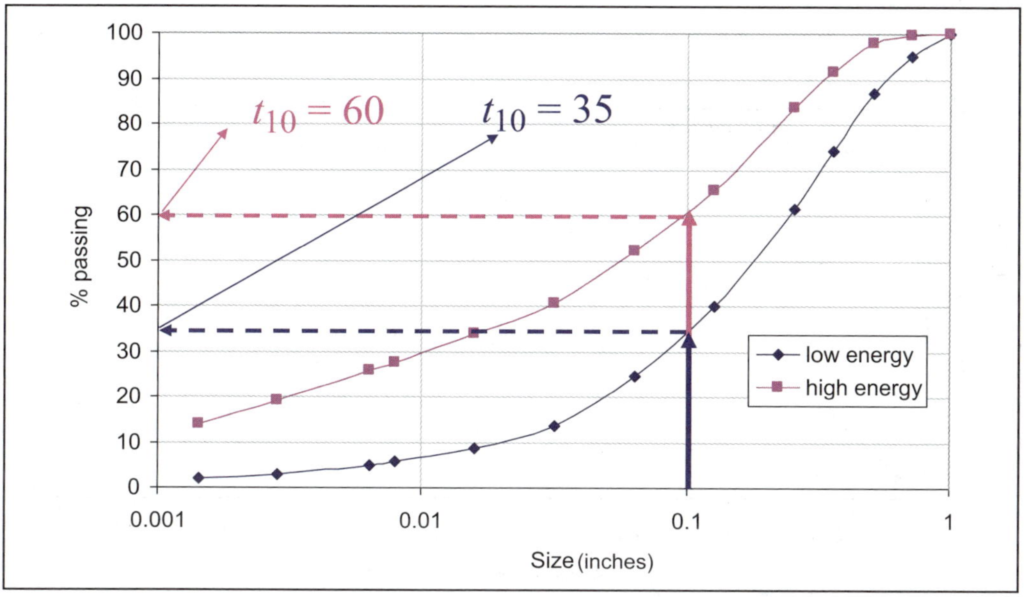

Figure 8. Estimating the t_{10} from the breakage of 1-inch particles

energies are plotted against the associated t_2, t_4, etc., values, the graph in Figure 10 was obtained. Narayanan found that regardless of the rock being broken, the trends seen in Figure 10 remained largely the same, i.e., it was a universal "map" of breakage distributions. Hence if the t_{10} is known (e.g., from Equation 9) then the associated t_n values can be read-off Figure 10. These t_n values are different points on the product size distribution curve which can thus be fully reproduced. In JK comminution modelling Figure 10 and its associated Equation 9 have become central to its modelling approach and has been a major contributor, in conjunction with the drop-weight tester to the success of its AG/SAG mill model.

One of the big drawbacks of the drop-weight test, at least in terms of design and geometallurgical modelling, is that usually over 60 kg of rock are required for a single test and that ideally this rock needs to be sourced from whole PQ core. This therefore precludes the use of smaller diameter NQ and HQ core. This problem was overcome in 2004 with the introduction of the SMC Test® (Morrell 2004), which can be used with samples as small as quartered NQ core. The test gives the *A* and *b* values for Equation 9 but also provides parameters that can be used to predict the specific energy of AG/SAG mills crushers and High Pressure Grinding Rolls (Morrell 2009). Since 2004 over 25,000 tests have been conducted, making it one of the most commonly used rock breakage tests today.

SPI® (SAG Power Index)

In line with the increasing requirement for rock characterisation tests that require relatively small amounts of sample, the SPI was targeted at producing a parameter that could be used to predict SAG mill specific energy from small diameter drill core. It therefore sought to fill the gap in Bond's suite of tests, which does not have a SAG mill-specific test.

The fact that the SPI test requires relatively small amounts of sample and targets SAG mills, which have come to dominate comminution circuit design, makes it well suited for use in

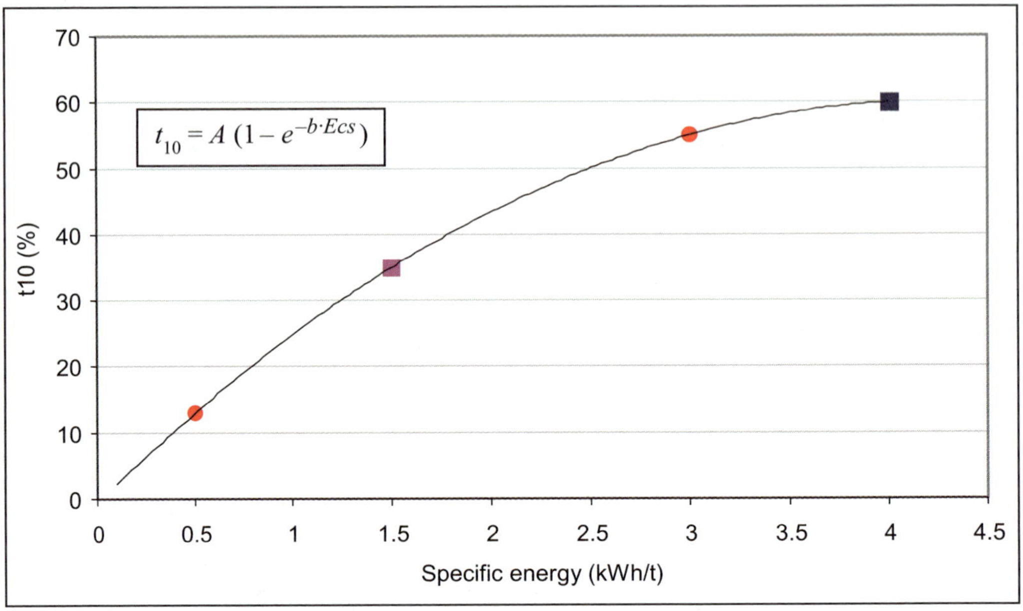

Figure 9. Variation of t_{10} with input energy for breakage of rocks in a fixed narrow size class

geometallurgical projects and is used extensively for this purpose. Indeed, its development and commercialisation has been a major influence on the increased use and acceptance of geometallurgy.

The test was developed in the 1990s (Starkey et al. 1994) and uses a 12" × 4" (D × L) batch laboratory mill, loaded with 15% by volume of 1" balls. A picture of the mill is shown in Figure 11. The mill is loaded with 2kg of –19mm (P80 = 12.7mm) of sample material and run until it is ground to 80% passing 1.7mm. The time taken to reach this grind size is designated the SPI value.

The SPI is subsequently used in the following equation (Dobby et al. 2001):

$$\text{SAG kWh/t} = K\,(\text{SPI} \times \text{T80}^{-0.5})^{n} f_{sag} \quad \textbf{(10)}$$

where

T80 = transfer size which needs to be specified

K, n and f_{sag} = proprietary parameters which are related to factors such as feed size and whether there is a pebble crusher in-circuit

CONCLUSIONS

This paper has attempted to describe what the current "state-of-the-art" is in the fields of comminution modelling and ore characterisation. To do so it adopted a definition for state-of-the-art such that to qualify the model/ore characterisation technique had to be a proven major contributor to the current practice of comminution circuit design and optimisation. This does not necessarily mean "cutting-edge" as past literature is full of, what at the time were, cutting-edge ideas that time has subsequently shown have fallen by the wayside and have not found favour with our industry. In fact in the author's view first prize for state-of-the-art in comminution is still the approach first published by Fred Bond over 60 years ago. This is not to say that our industry has stood still

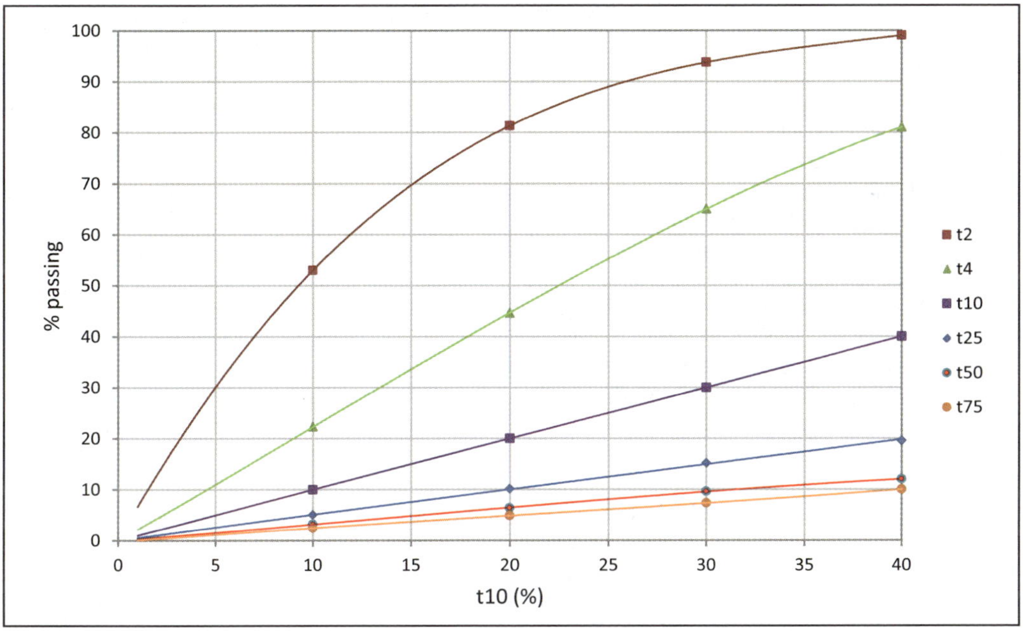

Figure 10. Relationship between t_{10} and the t_n "family" of curves

during this time. One of the major advances in technology has been the development of simulation modelling, at the forefront of which has been the JKMRC. The result of their research of over 30 years has been the introduction and acceptance of JKSimMet and the Drop-weight Test as being a commonplace, reliable design and optimisation tool for comminution circuits. Geometallurgy, unheard of in Bond's day, has seen a relatively rapid introduction to our industry with major benefits for circuit design and accurate forecasting. What has helped its introduction and success has been the development and commercialisation of bench-scale ore characterisation tests such as the SPI® and SMC Test® which are relevant to AG/SAG mill grinding circuits which have come to dominate circuit design.

Figure 11. SPI® mill

REFERENCES

Bond, F.C. 1952. The third theory of comminution. Trans SME/AIME, 193, 484–494.

Bond, F.C. 1961. Crushing and Grinding Calculations Part I and II. British Chem Eng 6(6) 378–385.

Dobby, G., Bennett, C., and Kosick, G. 2001. "Advances in SAG Circuit Design and Simulation Applied to the Mine Block Model," Proceedings of the Conference on International Autogenous and Semi-autogenous Grinding Technology 2001 (SAG 2001), vol. 4, Vancouver, British Columbia.

Hukki, R.T. 1962. Proposal for a solomnic settlement between the theories of von Rittinger, Kick and Bond, Trans AIME, vol. 223, pp. 403–408.

Kick, F. 1885. Das Gesetz der proportionalen Widerstande und seine anwendung felix, Leipzig.

Leung, K. 1987. An energy-based ore specific model for mautogenous and semi-autogenous grinding. PhD thesis, University of Queensland (JKMRC).

Lynch, A.J. 1977. Mineral crushing and grinding circuits: their simulation, optimaisation, design and control. Elsevier, 340 pp.

Morrell, S. 2004. Predicting the Specific Energy of Autogenous and Semi-autogenous Mills from Small Diameter Drill Core Samples. Minerals Engineering, Volume 17, Issue 3, March, pp. 447–451.

Morrell, S. 2009. Predicting the overall specific energy requirement of crushing, high pressure grinding roll and tumbling mill circuits, Minerals Engineering, Volume 22, Issue 6, May, pp. 544–549.

Morrell, S., and Morrison, R. 1996. AG and SAG mill circuit selection and design by simulation. Proc SAG '96, Vancouver, pp. 769–790.

Narayanan, S.S., and Whiten, W.J. 1988. Determination of comminution charcteristics from single particle breakage tests and its application to ball mill scale-up. Trans Inst Min Metall, 97, C115–124.

Rowland Jr., C.A. 1982. Selection of rod mills, ball mills, pebble mills and regrind mills. Design and installation of comminution circuits (Eds Mular and Jergensen), SME/AIME, New York, pp. 393–438.

Starkey, J., Dobby, G., and Kosick, G. 1994. "A New Tool for SAG Hardness Testing," Proc. Canadian Mineral Processor's Conference, Ottawa.

von Rittinger, P.R. 1867. Lehrbuch der Aufbereitungskunde, Ernst and Korn, Berlin.

Walker, W.H., Lewis, W.K., McAdams, W.H., and Gilliland, E.K. 1937. Principles of Chemical Engineering, New York.

ABOUT THE AUTHOR

Stephen Morrell graduated from the Royal School of Mines in 1980 after which he joined DeBeers and worked in Southern Africa at a number of mines until 1987. In 1987 he joined the JKMRC as a research scholar and was awarded his masters of engineering degree in 1987 in the field of AG/SAG mill modelling. He subsequently joined the staff of the JKMRC, going on to lead their comminution research program for 10 years. Part-time study at the JKMRC also gained him a PhD in 1993 for his work on mill power modelling. In 2000 he started as an independent consultant in the fields of comminution circuit design and optimisation via his company, SMCC Pty Ltd. In 2003 he developed the SMC Test® for rock hardness characterisation and commercialised it via a new company, SMC Testing Pty Ltd. In late 2012 SMCC was acquired by CITIC Heavy Industries, which has subsequently formed CITIC-SMCC Process Technology Pty Ltd, which Morrell is currently helping to develop under a strategic partnership agreement.

Innovations in Comminution Instrumentation and Control

Brian Flintoff
Metso Technology Development, Kelowna, British Columbia, Canada

Olivier Guyot
Metso Cisa, Orleans, France

Jeff McKay
Metso Cisa, Salt Lake City, Utah, United States

Andre Vien
Metso Technology Development, Kelowna, British Columbia, Canada

ABSTRACT: In one form or another process control in comminution has been around since man first started breaking rock. However, as we usually associate the term 'process control' with automatic process control, this history dates back to the 1960s when analog controllers were first introduced to mineral processing plants. The past five decades have seen significant change in the practice of process control, and innovations in this area have occurred much more quickly than, say, innovations in the associated process equipment portfolio. In this evolution, automation has gone from nice-to-have to need-to-have, and plant-wide process control systems are now an essential element of every operation. Innovation continues in mainstream process control (control hardware, field instrumentation, control algorithms, interfaces, communications, etc.), but we are also witnessing the beginning of innovation in several related areas, such as condition monitoring, data/image mining, simulation, as well as a future that will include autonomous operation.

INTRODUCTION—A PRIMER ON PROCESS CONTROL

Figure 1 provides a simplified view of the IT space in a modern mineral processing plant. Traditionally, the process, the analyzers/instruments/final control elements as well as other networks like Laboratory Information Management Systems and the process control hardware itself comprise the core of industrial Process Control. Having said that, Unit Automation (vendor supplied packages), Advanced Process Control (APC) and Plant Information Management Systems (PIMS) are now well integrated as important elements of the bigger process control picture. Recently, real time condition monitoring (here included in Predictive Maintenance) has generated considerable interest among operators, and these systems, along with the DCS and/or PLC systems feed valuable information on machine health to the Asset Management System. These later linkages are still in an immature state, as is the connection with the business system, here designated as the ERP. Finally, a topic of increasing interest is Remote Access/Monitoring (and Control), which essentially allows designated experts to provide on-line diagnostic and remedial actions across a range of applications.

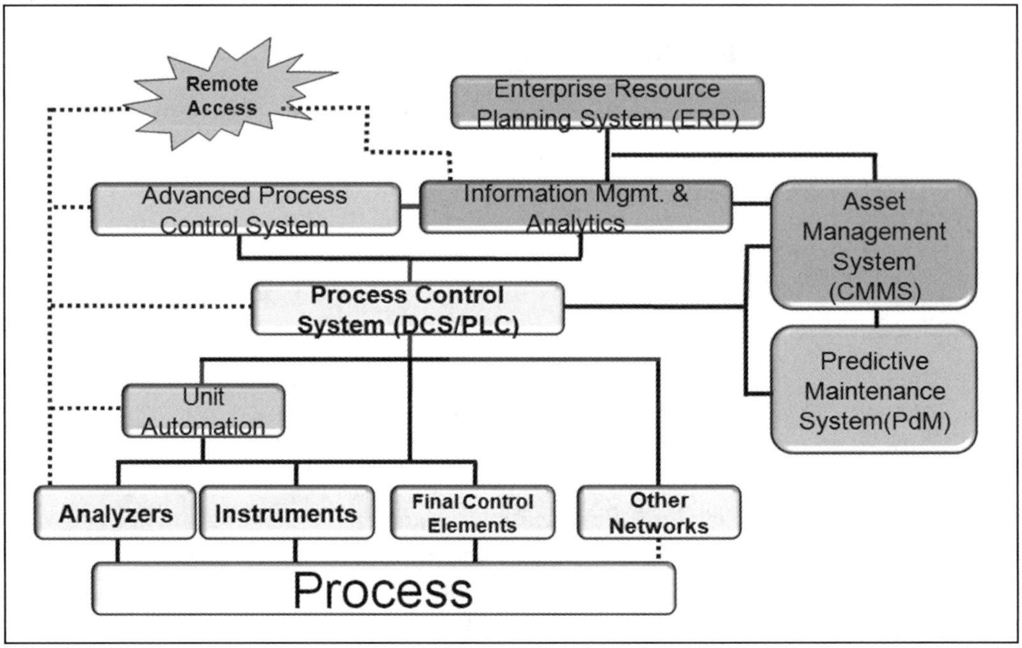

Figure 1. Process control in the IT space of a modern plant

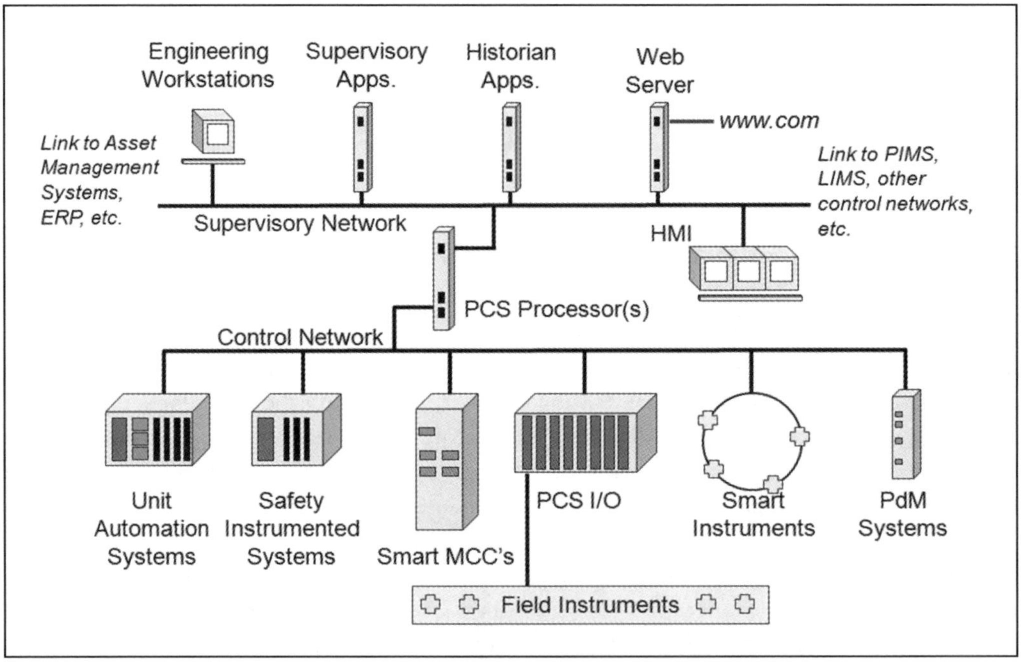

Figure 2. Typical modern process control system (PCS) architecture

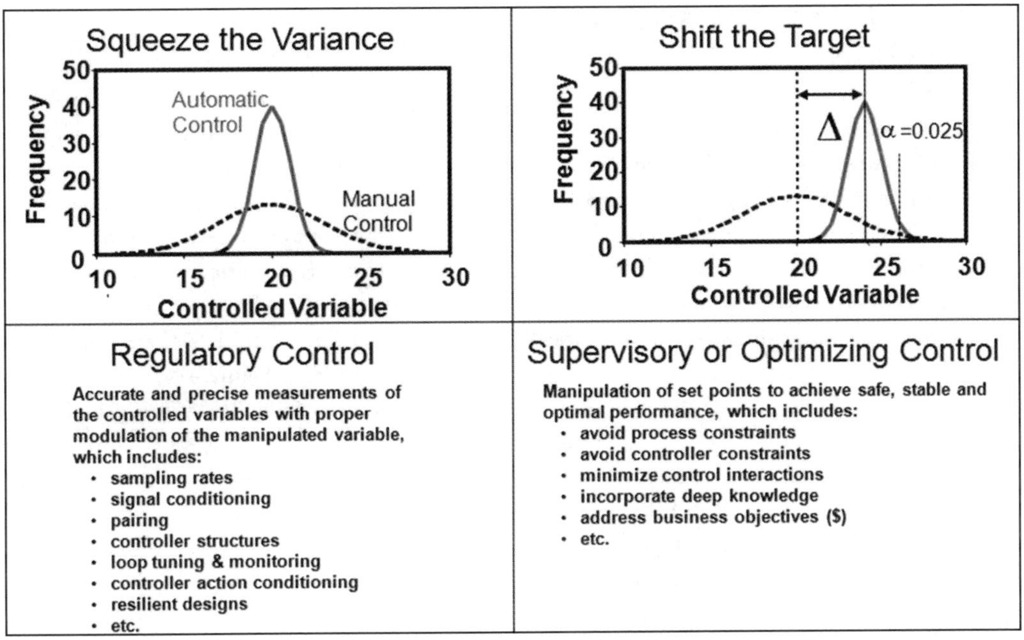

Figure 3. The fundamental goals of process control

The usual elements of the process control system are illustrated in the 'typical architecture' shown in Figure 2.

Finally, Figure 3 is a simplified attempt to partition the two principal goals of process control—take control by reducing the variance of the process variables, and then shift the operating target to optimize process performance.

These figures are important as they provide context for the discussion that follows, and they also help to map the evolution of innovation in this field of mineral processing. From Figure 3, developments started with regulatory control and moved to the supervisory or optimizing level, which require an effective regulatory control foundation. From Figure 1, there were simultaneous developments in all of those lightly shaded areas, which served to fuel rapid innovation in the early years, extending progressively to the darker shaded areas. Finally, Figure 2 highlights what might be considered today's best practice in terms of overall process control functionality.

A BRIEF HISTORY OF INNOVATION

The history of innovation in the control of mineral processing systems is really the subject for a book, so in this paper we are constrained to present only some of the highlights. In addition, some interesting topics such as developments in control room design, modern HMI design, intelligent alarm management, etc., have been deliberately excluded.

Like most journeys of innovation, this one had many players, some from operations, some from automation and instrumentation suppliers, some from engineering companies, some consultants, some specialty companies, some academics, etc. While there was some synergy in these processes, the broad community of participants was necessary to ensure the simultaneous development on multiple fronts, as described in the previous section.

Control Systems Hardware and Related Systems/Software

This story starts in the early 1960s when the best practice of the day meant centralized control rooms with push buttons for equipment stop/starts, and alarm annunciation (usually visual and sound), all based on hard-wired relay logic. Continuous or regulatory control was achieved with single input–single output (SISO) analog controllers, and in some cases cascade or ratio control was implemented by nesting these controllers. Trends were 'visualized' with strip or circular chart recorders, and operating strategy and set points were dictated by the shift foreman. In many cases, different foreman had different levels of process and business understanding, so the operational objectives changed every 8 hours or so. We've come a long way.

From a hardware point of view, the first major breakthrough of the early 1970s was the confluence of newly introduced minicomputers for direct digital control (e.g., Figure 4). They arrived on the plant floor about the same time as freshly minted process engineers, who had some formal training in applying computer technology. By today's standards the usual RAM memory complement of ≤12k of the minicomputer was seemingly very small, nevertheless, as these systems programmed in some flavor of BASIC or FORTRAN, it was relatively easy for the process engineers to conceive, test and implement new control algorithms. Especially useful were the IF-THEN constructs which allowed creative souls to explore the precursor to crisp rule-based Expert Systems (see for example Bradburn et al. 1976). The real benefits of these systems lay in the fact that a consistent (best operator) control strategy was implemented 24×7, and logic improvements could be easily made by plant staff.

Minicomputers were a great boon to the process engineers and instrumentation technicians who focused more on regulatory control applications, but the Programmable Logic Controller (PLC) was soon to supplant the old hard-wired relay systems, and had a similar and profound effect on electrical engineers and electricians working mainly in the discrete control world. PLCs made their way into process plant design in the mid 1970s, followed shortly thereafter by the newer version of the regulatory control solution, the Distributed Process Control System (DCS). Rather remarkably, these two 'solitudes' would co-exist for some 30+ years, which has led to several decades

The Fisher Controls DC² minicomputer system – a key technology enabler in early Canadian process control development (*Photo courtesy of Emerson Process Management*)

1970's era process control retrofit with DDC for analog control and HMI/reporting and hard wired relay logic for discrete control and alarms.

Figure 4. Plant-wide process control in the mid-1970s

of (heated) debate on whether the PLC or DCS should be the preferred hardware option. (Over those years the functionality did converge, although even today the systems retain some differentiation, e.g., Montague 2006 and Nelson & Stauffer 2008.) A common solution was a hybrid system with the PLC serving the safety and process interlocks, stop/starts, alarms, and other electrical duties requiring fast response times (milliseconds). The DCS then provided the regulatory control, the HMI, alarm management, the reporting functions, etc., all of which were far less time sensitive (seconds). Today, a single system is usually selected.

One of the biggest changes over the past four decades is the I/O count which has grown considerably as automation has been proven to be one of the most cost effective investments an operator (or a supplier) can make. In the 1980s plants with total I/O counts of 5000 were considered to be very large control installations. These days I/O counts approaching 20,000 are not uncommon, and one (admittedly complex) heap leaching installation in Finland has an I/O count of over 30,000 (Kytokari 2012).

Another big change has come in communications as pneumatics (3–15 psig) have given way to electronic (0–24 V or 4–20 mA) signals, which in turn have given way to digital bus technology. HART, Foundation Fieldbus, ProfiBus, ModBus+, DeviceNet, CAN, Ethernet, OPC, and a host of other protocols are now commonly used in industrial process control (Verhappen 2011). However, bus design can be complex and sometimes installation is costly, so there is still a significant reliance on twisted pair cables routed to the field instrumentation on cable trays. There is of course a good deal of interest in wireless technology these days, but deployment thus far has been mainly limited to remote (telemetric type) applications.

Many mining companies have looked at setting standards for process control hardware, e.g., Noranda settled on Fisher Provox for its plants in the 1980s. The benefits were expected to flow in the form of consistent and fair pricing, good vendor technical support, an ability to migrate good control practices and people more easily, and simplified central technical support. There was some resistance to this from the various operating sites, but those who went through the process did recognize the expected benefits in the longer term. In general, this practice failed to get much traction in the industry, in part because of the autonomy of the operations themselves, and in part because engineers are generally looking for what they regard as best available technology when they make a new purchase. It is somewhat ironic then to look back and see the similarity in buying practices in various regions over time—presumably a risk mitigation step. Crudely speaking, looking at the DCS market in North America, Fisher Provox and Bailey Network 90 were sort of defacto standards in the 1980s, Foxboro I/A in the 1990s and Emerson Delta V in the first decade of the new millennium. These companies have enjoyed a long and mutually beneficial relationship with the mining industry dating back to their minicomputer products in the early 1970s. A similar case can be made for Schneider Modicon and Rockwell Allen-Bradley in the PLC market.

To conclude this discussion it is interesting to note that with smart instrumentation, an increase in unit automation, and specialty suppliers of APC and Data Historian applications, the role of the DCS/PLS systems have been changing—and process control is now more often referred to as production management in this sector. In addition, the value proposition of these automation houses (including ABB, Siemens, GE, Honeywell, Yokogawa, etc.) is also changing as the larger supply companies are now bundling process control hardware and related services with other items in their portfolio. This could include field instrumentation, APC, analytics, predictive maintenance, ACM, drives, motor controls and switchgear, and even power distribution—a new twist on corporate standards. As one junior mining company executive put it: "*I want to buy everything dealing with electrons from one company.*" So far this idea is getting some interest in the upper echelons of the mining companies, but it is difficult to demonstrate best available technology across a broad portfolio of products and services.

Field Instrumentation

Of all the components of process control, this is one that arguably hasn't changed as quickly as is the case with the other areas. We still measure pressure, level, flow, temperature, density, etc. in more or less the same way, albeit with newer, more robust and smarter instruments.

A good example is slurry density, where the standard is still the ubiquitous γ gauge, which has been used in this application since the 1960s. This instrument has the advantage that it is non-contact and can be attached directly to slurry pipelines. It has the disadvantage that it employs a radioactive source, which complicates maintenance, and in some countries makes it practically impossible to use. Of course, with a microprocessors on board the γ gauge has become 'smarter' and the source strengths have been reduced. Nevertheless a viable optional technology has apparently not emerged and there is seemingly little development effort in this area. Contrast this with Cidra's SONARtrac® flow measurement system (Maron & O'Keefe 2008) is now seen as a viable, and sometimes superior option to the magnetic flow meter, which has been around practically as long as the γ gauge. Since instrument suppliers need to look across the process industries to develop sales volumes that are interesting from a business point of view, without proactive engagement, mineral processing must apparently wait for others to share an interest in measurement problems.

What has changed over the 15 years or so is that industrially hardened microprocessors have made their way on board instruments (and final control elements) to make them 'smart.' The result is that these smart instruments can perform not only their basic function(s), but they can implement diagnostics (e.g., valve stiction or backlash), perform basic control (e.g., PID loops), be remotely accessed/configured, simplify asset management, and of course communicate via digital bus technology. The predictive maintenance capabilities alone can lead to significant savings in maintenance and lost production. Having said that, in a recent publication (Livelli 2012) it was estimated that only 10% of the 30 million or so HART devices installed since 1989 have a digital pathway back to the host.

Given the cost savings for installation of digital bus technology (wired or wireless), the advantages of harvesting the additional information available from these devices, the benefits of distributing regulatory control functionality in large and complex automation architectures, one would expect to see greater use of this technology in the near term.

Control Strategies

This is probably the area most engineers immediately think of when the subject of process control is raised. There is a relatively large body of literature devoted to this topic, as some conferences are dedicated to automation in the mining industry (e.g., AutoMining or the IFAC symposia), books have been written on the subject (e.g., Herbst 1984, Rajamani and Herbst 1990, Flintoff and Mular 1992, Herbst 2000, and Sbarbaro and Del Villar 2010) and almost every conference has a least one session on the subject of control, mainly focused on algorithms and applications.

The PID controller was and still is the foundation of regulatory control in the process industries. It is used to regulate solids flow, liquid and slurry flow, tank levels, temperature and pressure, etc. Despite its importance to the process industries in general, a seminal study (Bialkowski 1992) showed that only 20% of these loops decrease variance over the manual mode of operation. A more recent study in our industry (Ruel 2011) shows that things are still not as good as they should be, with the value now only up to 30%. (Obviously the critical loops in a plant are among the 30%.) Control system suppliers and specialist firms have introduced "asset management" software to help identify problematic control loops and facilitate repairs (e.g., Process Doctor or Expertune Plant Triage). Loop tuning remains one of the major contributors, and some loops can still be found operating with the default parameters set by the hardware manufacturer on shipment. In the past, loop tuning was part art and part science, and usually practiced only by skilled instrument technicians. This was largely solved in the 1990s as model based tuning strategies were introduced, which required no specific art. Among these 'lambda tuning' has probably received the most attention from the process industries (Olsen and Bialkowski 2002).

Despite their utility, PID loop performance will degrade in the presence of dead time, process non-linearities, non-minimum phase behavior, because of loop interactions, etc. For this reason there was some early work employing self-tuning regulators and dead-time compensation (e.g., model-based approaches, as discussed in Edwards and Flintoff 1992). In fact, these ideas led to some very interesting developments in what might best be termed advanced regulatory control—e.g., multiple feeder control (Vien et al. 2000). More recently, operators have made more use of the built in functionality of the DCS systems by employing feed forward controllers, the Smith Predictor, and even Model Predictive Control (MPC) on some of the difficult SISO and MISO control loops. To conclude, the concept of resilient process designs (i.e., make control as simple as possible) emerged from efforts to minimize these and other performance issues (e.g., Morari 1983), although it is probably fair to say this isn't widely practiced, yet.

In an earlier section it was mentioned that the first efforts to tackle the constrained multivariable problem that is the dynamic optimization of comminution circuits was to employ 'simple IF-THEN' logic-based strategies hard coded in the minicomputer programming language. Although primitive by the standard of today's tools, these systems were very effective and thus raised the visibility and importance of process control in the industry, which then fueled innovation. The DCS systems had essentially transformed to configurable, as opposed to programmable minicomputers. Couple that with the notion that process control required deep process knowledge, it is not surprising that the early days of Advanced Process Control (APC) generally featured small specialty companies working on PC-based applications that would link with the DCS or PLC. In fact, this has been the dominant model through the middle part of this past decade.

The use of heuristic strategies in crushing and grinding control lent themselves self very well to the application of real-time expert systems and the early innovators in this area included Hales and Herbst 1987 (then Control International) and Harris and Kosick 1988 (then Comdale Technologies). Much of the work that followed was based on the pioneering efforts of these individuals. It was realized early on that the expert systems could benefit from deeper process knowledge embedded in the control strategy and model-based expert control was born. In this instance there were two approaches, one that favored the use of mathematical models (e.g., Herbst et al. 1989) and another that favored the use of statistical models (e.g., Hales et al. 1996). Both showed improved control performance over expert systems alone, although neither one was especially easily maintained by the site owners of APC. (It is worth noting that adaptive model-based control approaches, when applied on their own, were generally unstable longer term because of the nature

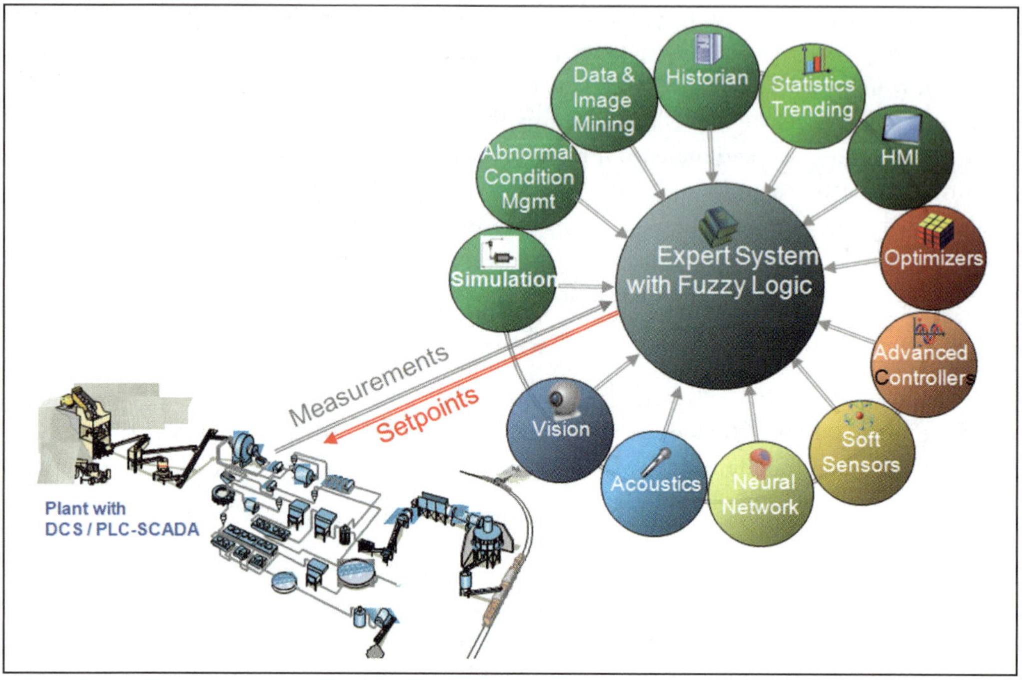

Figure 5. A modern (OCS©) advanced process control platform

of the disturbances and operational upsets in comminution. As Astrom et al. 1986 suggested, all of these approaches require some kind of expert shell or jacket software to ensure longer term stability.)

For the decade or so starting in the mid-1990s, the focus was not so much on innovation in process control strategies and tools per se, as it was on other aspects and especially sensing. Perhaps the biggest exceptions included the incorporation of Fuzzy Logic into the Expert System framework, which provided the system integrator with another important degree of freedom. This was followed a few years later by MPC, which started to appear in plant process control. Model Predictive Control had its roots in the chemical industry of the mid 1980s with DMC, and to our knowledge the first industrial demonstration of MPC in mineral processing was in the late 1980s (Vien et al. 1991). While it performed well, some lessons were learned regarding the need for the shell software mentioned above and/or the need for local expert support. MPC has been successfully applied in many other process industries, and as the large automation houses had developed expert system functionality to improve control strategy robustness, operators began to show interest. Small specialty firms (e.g., Andritz's Brainwave) and large companies (e.g., Honeywell's Profit Controller) have been among the leaders in successfully moving MPC into operating plants, in some case displacing older APC applications.

Today, the tools available to an APC integrator are plentiful and proven. Figure 5 is an attempt to illustrate this using the Metso Cisa's OCS© architecture as an example. Clearly, this functionality goes beyond the topics is this section, but most of these will be covered in the remaining sections of the paper.

Figure 6. Installation of a rock vision system (VisioRock™) on a conveyor

Analyzers (Sensors)

Here we draw a distinction between instrumentation, as the kind of device that can be used across many industries, and analyzers and/or sensors, which are more or less dedicated to mineral processing applications. Mineral processors have long believed in the message later coined by the SPC guru W. Edwards Deming: "*You have to measure it to manage it.*" In comminution, the principal variables of interest are particle size and the operating state of the mills.

Particle Size in Crushing and Coarse Grinding

While there may have been other techniques explored, the standard method for measuring particle size distribution of essentially dry solids on a moving conveyor employs image analysis. The first such device of which the authors are aware was the Foxboro 95PSD (Vignos & Bruno 1979). This device used a lens/detector assembly to deduce the chord length of particles, which was then transformed to an equivalent particle size distribution. Although some of the test results were 'intriguing,' the device did not get a lot of traction in the market. Over the next decade researchers (e.g., Waterloo University's Maerz et al. 1987 and University of Utah's Lin & Miller 1992) demonstrated signal processing techniques could be used to process images obtained with a camera. WipWare's Wipfrag system, which is based on the Waterloo research, claims to be the pioneer in photoanalysis solutions for automation. In the past decade or so Split Engineering's Split OnLine and Metso Cisa's VisioRock™ have also introduced rock vision systems that have gained significant market share. Figure 6 shows a typical installation of such a system. These companies also have off-line versions for analyzing still photos, and on-line versions for analyzing the payload of a haul truck dumping into a crusher pocket.

Particle Size in Fine Grinding

In the very early 1970s, Autometrics launched their Particle Size Monitor (PSM) (reviewed by Anderson 1979) and a number of companies installed these systems to monitor the particle size produced by the grinding circuit. It was pretty clear that to fully realize the value of this analyzer, it would need to be embedded in the control strategy, either as a constraint or a controlled variable. The control solution was a little more difficult to achieve as grinding is a multivariable process and the control techniques to deal with this (loop pairing, decoupling, etc.) were not all that well understood on the plant floor at the time. Still, the PSM was quite broadly applied and has seen several generations of development to its current form.

Over the following years others particle size analyzers were developed and tested, all with the goals of improving accuracy and reliability, while reducing capital and operating costs. For example, Rosenblum et al. 1984, automated the laboratory sedimentation test. CSIRO used advanced acoustic and nuclear methods to infer particle size distribution, especially for finer mineral slurries (Coghill et al. 1997). In the interval, Outotec introduced the PSI-200 (Miettunen et al. 1993), which along with the PSM has undergone several generations of development, and these two systems have essentially become the standard technologies in the industry.

The direct sensing technologies above have also been complemented with considerable work around soft sensors (e.g., Gonzalez et al. 1986, Gonzalez & Meyer 1990, Del Villar et al. 1996, and Sbarbaro et al. 2008). Undoubtedly there were other published and unpublished works in this area, but the point to be made here is that innovation requires tenacity in the time varying, non-linear and rugged world of mineral processing.

Mill State of Operation (Load)

As we'll attempt to illustrate, this has been particularly fertile ground for innovation, especially as it relates to the measurement/inference of SAG mill load, since this unit is often the rate limiting step grinding.

Efforts in this area began in the early 1970s when rod mill–ball mill grinding circuits were still quite common. Experienced operators used mill power and mill density to infer something about the state of operation, and they could tell a lot about the mill from the sound. Not surprisingly then, some of the earliest work was around rod mill sound, and establishing the maximum throughput which could be achieved before mill performance deteriorated and rods were broken prematurely, and the risk of a rod tangle increased (see, for example, Bradburn et al. 1976 and de Mayo et al. 1985)

Non-contact acoustic systems on AG/SAG mills (e.g., Jaspan et al. 1986 and Pax 2001) were also under investigation, but it seemed that success was fleeting, and control strategies involving acoustic signals often required retuning as conditions inevitably changed. On-the-shell systems (e.g., Barrientos and Telias 1997) have been explored and while performance is seemingly reasonable, market uptake has been slow. However, the research tenacity described earlier has prevailed and there are a few acoustic (or vibration sensors) that are giving good and consistent results in control applications, e.g., non-contact: AudioMill™ (Broussaud et al. 2011) and on-the-shell: CSIRO monitor (Davey et al. 2012).

Soft sensors for SAG mill load have also been common, the earliest employing bearing pressure and/or mill load along with mill power to infer the magnitude and trending of the SAG charge. Like acoustics, this approach was also affected by changes (new liners, lube filter conditions, etc.), however, these signals are still employed in virtually all SAG mill control strategies. The most robust implementation is the incorporation of these signals into a soft sensor framework (e.g., Samskog et al. 1996) where non-stationary behavior is accommodated by adapting the process model.

A number of efforts have been made with on-the-shell devices to infer the load in tumbling mills, and while all showed promise, the authors are unaware of any longer term installations. Included in this class of analyzer would be those based on strain, e.g., for steel liners, Herbst and Pate 1996, and for rubber liners, Dupont and Vien 2001. Van Nierop and Moys (1995), developed a conductivity probe that could be embedded in a liner. As with the on-the-shell acoustic devices, these systems were primarily to measure toe and shoulder positions of the charge in the mill, which permitted the back calculation of the charge volume.

Systems using special signal processing algorithms on power have also been developed and employed on a limited basis, for example Outotec's MillSense® (Koivistonen and Miettunen 1989) and MONSAG (Pontt et al. 1997).

Other Efforts

Over the years there have been efforts to measure other important process variables in comminution, one of which is mill density or viscosity. The soft sensor approach (a dynamic circuit mass balance) is probably the most common solution here, yet the number of installations is still quite small. Other efforts have involved attempts to adapt lab measurement devices to on line systems, or alternative forms of soft sensing, such as the use of temperature in an energy balance approach (van Drunick and Moys 2001).

Analyzer development/refinement remains a very high priority with the industry, so this is expected to be an area of innovation investment. Looking ahead, on-line liberation analysis is the next big challenge, and one that is becoming increasingly critical as mining companies focus on energy costs and related issues in sustainability.

OBSTACLES AND CHALLENGES

The journey of innovation in comminution automation has been one of considerable success, but there have been obstacles and challenges along the way. Some of the more important factors are outlined below.

- People—The mining industry has been increasingly challenged to recruit and retain people with skills in the design, maintenance and development of effective industrial automation solutions. Moreover, the tumultuous economic times of the late 1980s and 1990s meant fewer people were hired and as a consequence, 'bench strength' was depleted and traditional on-the-job mentoring broke down. As systems get larger and more complex, the lack of on-site expertise will likely create issues with system reliability and efficacy.
- Risk Aversion—Mining operations are usually remote and self sustainability is an issue. Couple this with the fact that the automation system is, in effect, the 'central nervous system' for the mineral processing plant, and it is easy to see how the old adage—*the first to be second*—applies to risk taking around key components of the plant automation system. This has slowed progress.
- Market Size & Stability—There is currently much activity and optimism in the industry, but it is a relatively small market and one that has traditionally exhibited business cycles. These factors have tended to limit supplier investment in specialty devices for mineral processing.
- Succession—Arguably this is also a people problem, although here the issue is the continuity of successful control installations. The lack of documentation, understanding, education, etc., has often led to new engineers or technicians systematically crippling effective controls, in the name of continuous improvement.
- Complexity—We hope that this paper makes it clear that mineral processing plants are increasingly dependent on automation, and that the field is rapidly branching into a host of

new areas. This complexity presents real challenges for the site staff charged with maintaining and developing the systems and is forcing operations to look at new business models such as remote support, and even vendor managed assets.

- Project Models—It is somewhat perplexing to look at the project model that is used in the usual EPC/M work processes for new plants; there is a lot of rework on every project. In part this is because there has been a lack of tools that can be used to simplify the automation design process, and to efficiently transfer data to and from related design processes. Imagine the widespread use of process simulation, DCS emulation, and APC as a means of holistic design and delivery of process control systems ready for start-up.
- Training and Education—As a wise man (Bill Bialkowksi) once said, the half life of detailed technical knowledge is ~6 weeks; truly a case of use it or lose it. In the same way that we develop maintenance programs for physical assets, it would make sense to apply this same logic to the intellectual assets. However, there are challenges here as topical short or on-line courses are not that common.

We have our work cut out for us in trying to ensure that we continue the record of innovation of our predecessors, and to even find ways to accelerate the pace. For some of our more business oriented colleagues, in these times they are at the point where instant gratification isn't fast enough—the pressure is on, and doing nothing is no longer an acceptable option.

CURRENT AND FUTURE TRENDS

There is much to discuss in terms of the current and future areas of automation research, development and implementation, but time and space constraints prohibit more than a few introductory remarks on selected topics. What is perhaps most interesting here is that, strictly speaking, most of the new work is arguably peripheral to the discipline of process control, but builds on the real-time platform and support functions that comprise an effective plant-wide process control system.

Condition (Based) Monitoring (CM or CBM)

Essentially all of the major automation houses (e.g., Rockwell, GE, Emerson, Metso) now provide CM sensors and analytics software that are well integrated in their process control offering. Some bearing suppliers (e.g., SKF, CBC/AzimaDLI) offer stand-alone on-line systems. The result is that manual inspections associated with plant Predictive Maintenance (PdM) programs are giving way to real-time CM systems employing these tools, especially in 'mission critical' applications.

Much has been written on CM using sensors that monitor vibration and/or temperature and/or acoustic emissions and/or lubricant analysis, etc. The diagnostics for, say, bearing faults on higher speed rotating equipment as detected by vibration are well established. However, in mineral processing there are some interesting challenges posed by slowly rotating equipment and in a number of applications where direct sensing isn't readily applicable.

In an earlier paper, Herbst and Flintoff (2011), presented an example of the use of vibration analysis with Synchronized Time Averaging was presented to show how CM can be adapted to examine the mechanical condition of large gear driven equipment, like grinding mills. Here, we look briefly at the use of an alternative to direct sensing—model-based CM applications.

Perhaps the simplest predictive model is the common practice of using cumulative tonnage or energy consumption to infer the wear state of liner components. However, with a change in

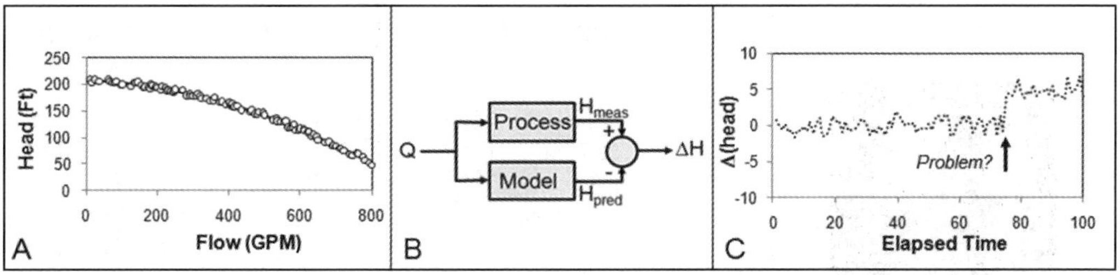

Figure 7. An example of a model-based CM application in slurry pumping

mindset, one can see many other possibilities when considering the use of process data for mechanical diagnostic purposes. For example, in the recent IMPC conference an example of detecting changes in pump performance was presented, and this is illustrated in Figure 7. In block A, a simple model is fit to historical flow/head data for the pump. Thus, with a flow measurement, head can be predicted and compared to the measured variable, as shown in block B. Normally the prediction error will be within established bounds, but trends, spikes or step changes (block C) give an early indication of a problem.

Real-time CM, involves sensed data, signal processing with advanced mathematical and statistical tools, the usual reporting and alarming, and where practical, compensating adjustments to the equipment or process—just like process control. The confluence of process and mechanical diagnostics and/or prognostics in a single platform offers operators a holistic solution to the asset management problem. One should anticipate a good deal of development in this area in the near future.

Sensor Fusion

Sensor fusion, multisensors, multi-sensor data fusion, etc. are among the terminology employed to describe a variety of techniques that generally use data derived from multiple information sources. The whole area of "fusion" is emerging as a hotbed of research. Definitions abound, but for the purposes of this paper that of Shah et al. (2012) will be employed: "*Sensor fusion is the combining of sensory data or data derived from sensory data from disparate sources such that the resulting information is in some sense better than would be possible when these sources were used individually.*"

Perhaps a good illustrative example is the possible use of a laser and a camera over a conveyor belt, where the rock profile as defined by the laser can be used to infer the volume of rock on the belt, giving an estimate of the mass flow. Some recent research shows that it is then possible to analyze the profile to deduce the particle size distribution on the belt.

This is a relatively new area in the control of mineral processing systems, but one that would appear to hold great promise. Consider the example presented in Figure 8. At various times and in various ways, all of these signals have been used or considered (all as standalone applications) to infer elements of the operating state of a SAG Mill: total load, ball load, rock to ball ratio, pooling, liner wear, broken liners, ball on liner collisions, alignment, lube system issues, etc. What if we could combine some or all of them, perhaps within a model, that would provide a fairly clear and robust estimation of the state of SAG mill operation and repair.

Progress in this area will most probably occur through co-creation efforts involving operators, equipment suppliers, and researchers, but there do seem to be easily achievable goals.

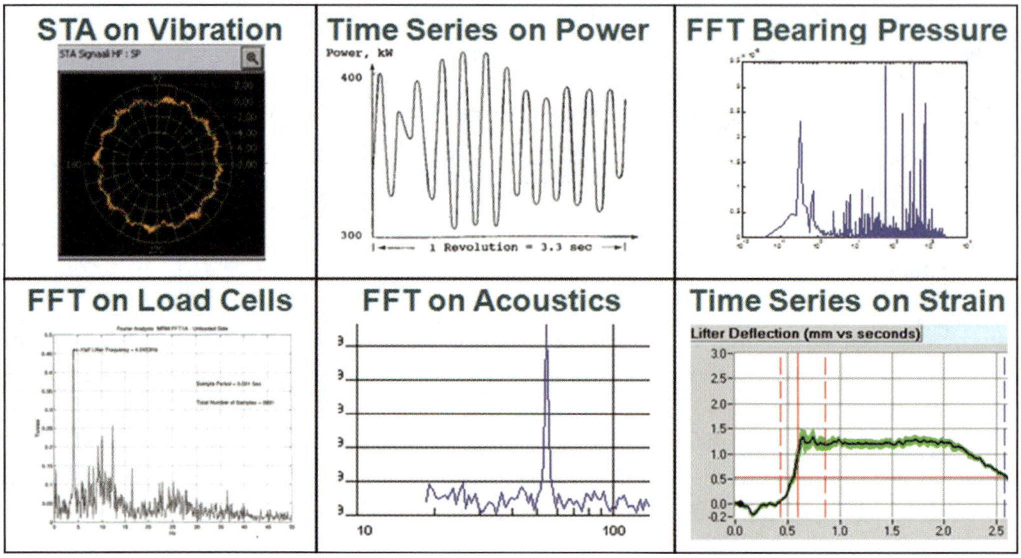

Figure 8. Candidates for sensor fusion of SAG mill process and mechanical health

Abnormal Condition/Situation Management (aka Critical Condition Management)

This topic has been discussed for more than a decade, and appears to have its origins in manufacturing. A working definition (ARC 2002) is "*Abnormal Conditions are usually caused by a combination of events that are not normally expected to happen at the same time, which the safety interlocks and exception logic in a conventional control system cannot handle, and the operators have little experience in dealing with.*" In this same article, the authors attribute typical gains to advanced process optimization to ~3%, whereas ACM can yield 5% or more to profits. Some of the factors that are emphasizing the need for ACM are automation system complexity, maximum utilization, environmental and regulatory concerns, safety, and staff turnover/training. While ACM is not a phrase heard often in mineral processing, it does offer a convenient framework to think about some nontraditional process control applications, a few examples of which are outlined below.

Frozen charges in the start-up of large tumbling mills can be problematic over time, and sometime with catastrophic consequences, as one or two operations have discovered. Both Siemens and ABB have developed solutions into their large electric motor technology which will detect a frozen charge from the power signature on startup, take protective measures, and then institute a sequence intended to break the charge loose. This can save considerable damage and the downtime incurred after a mechanical failure, as well as time lost to operators inching the mill as a normal part of a safe startup.

While it might be something of a stretch to include it here, automated startup (and shutdown sequences) can also be viewed as ACM strategies. Modern advanced process control systems can apply logic to this process, but work at a much higher speed in terms of safely executing the steps as quickly as possible, always in the right sequence, and properly filtering the inevitable alarms. This logic has been implemented in a large concentrator and the startup time for a grinding circuit was reduced by ~10 minutes, leading to pretty significant savings in lost production.

Table 1. Functionality of process control and ACM systems

Function	Control System	ACM
Alarm detection	Reactive	Predictive
Alarm management	Non-deductive	Deductive notification
Personnel guidance	Little	Significant
Inputs	Largely from sensors	Sensors and expert knowledge
Corrective actions	Automatic	Operator initiated
Used by	Operators	Operators and safety personnel

Source: ARC 2002.

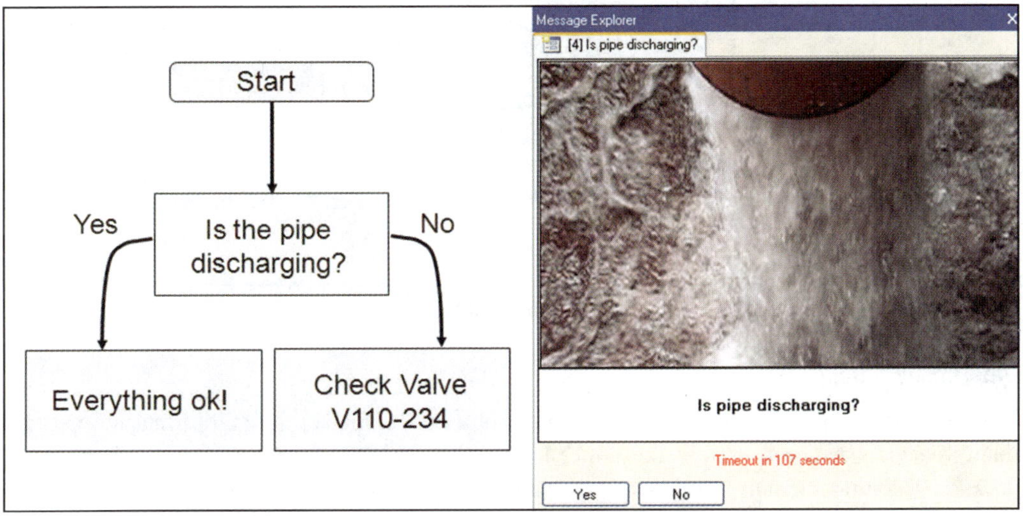

Figure 9. A simple illustration of an ACM application using OCS-4D©

For the most part, mineral processing operations don't have the inherent safety risks of nuclear, petrochemical or chemical processes. However, with the industry focus on safety first, and the desire to maximize asset utilization, all in the face of employee turnover or job rotations, there is more thought being given to operator decision support tools. Table 1 highlights the different functionality between the usual DCS/PLC control system and ACM.

Metso Cisa's OCS-4D© is an example of an intelligent software package that can detect/infer abnormal conditions, identify these to the operator and control system, and then step through the process of resolution, prompting real-time sensor data when necessary, and requesting input from, and providing instruction to the operator as and when required. Figure 9 provides a very simple example of this hybrid activity. In this instance something has triggered a branch to a query relating to the discharge from a pipe. The system prompts a live video feed from the pipe discharge and requests the operator indicate the status. In the event the operator is away from his station, fail safe strategies will be employed.

This is an area with some good potential in mineral processing, mainly because our industry is focused on safety and productivity, and there is good scope to quickly adopt best practices in this area.

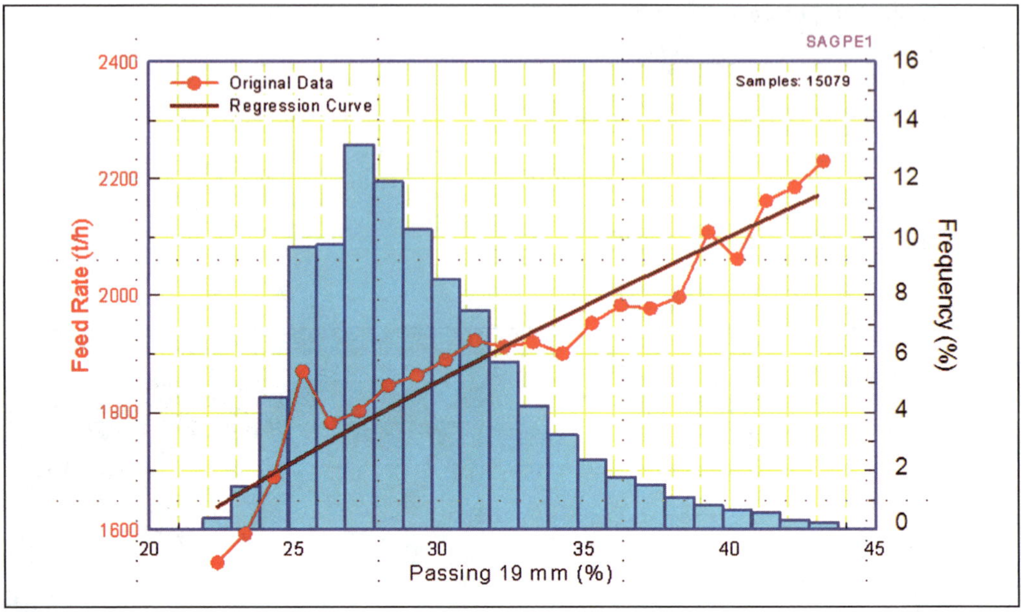

Figure 10. Data visualization for SAG mill feed size and throughput

Data/Image Mining

Data historians (e.g., OSIsoft's PI®, AspenTech's InfoPlus.21®), aka process information management systems, have been used in the mineral processing industry for most of the past couple of decades. The fundamental appeal is that the real-time data generated through the process control system provides valuable insights into process performance, issues and opportunities to optimize, so it should be saved. Tools exist to analyze and visualize process data from these systems, although in many cases users construct their own custom reports or process dashboards, and the analytics work is often performed on data exported to Excel. It is the authors' opinion that this is still an area with tremendous potential for operators, and one that now seems to be getting some renewed attention.

Much has been written on data mining, albeit usually in the transactional world of companies like Wal-Mart and Costco, and little of which is directly applicable to real-time process data. There is fertile ground for development here. To illustrate by example, Figure 10 is a visualization of process data from a SAG mill. In this case, only simple statistical tools have been used to examine about a half year of data (15-minute averages, so more than 15,000 points), which has not been classified in any other way. In this figure the histogram shows the distribution of SAG feed size (% –19 mm from a vision system) and the throughput bullets are the simply averages for the corresponding histogram "bin." The straight line is the linear regression of throughput on size. Simply put, size does matter in SAG milling.

SAG bearing pressure and noise are both known to correlate to the load in the mill. Figure 11 is drawn from the same process data set used above, and it clearly shows the correlation is nonlinear between these two measured variables. (Although not shown here, the pressure data had to be de-trended to account for liner wear—a simple step with these tools.) Of course, this sort of

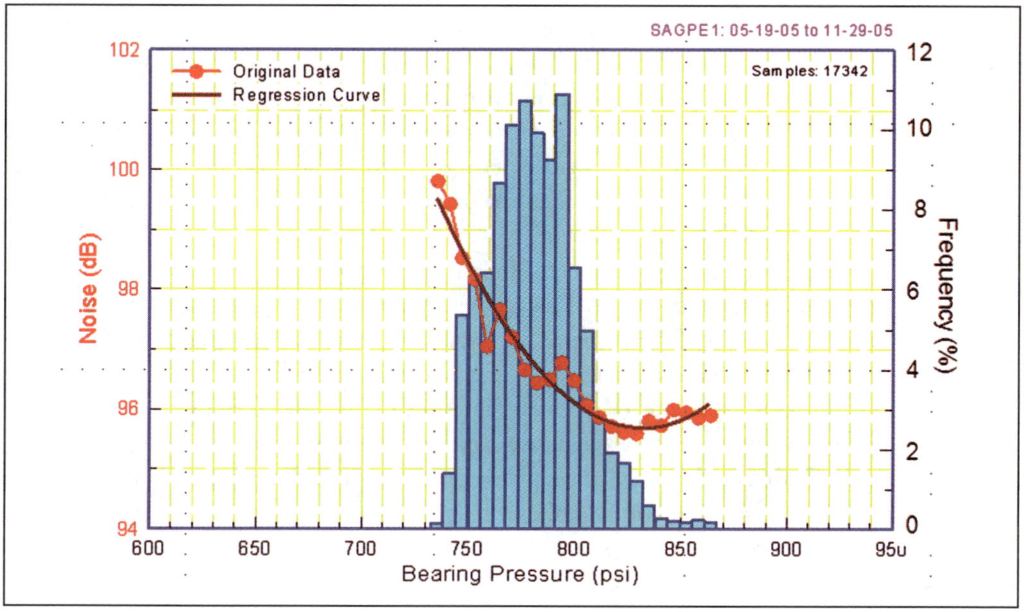

Figure 11. Data visualization for SAG bearing pressure (de-trended) and noise

quantitative graphical data can be useful in structuring both data validation and expert control rules, including choosing appropriate parameter values.

While most of the statistical data mining has been done off-line with desktop software exploiting databases, the advanced control suppliers, like Metso Cisa (DataMiner™) are now moving into this area. Given their interest and experience with real-time data acquisition and filtering, as well as with statistical and mathematical modeling, real-time data mining is the next logical step. In Metso Cisa's case, they have embedded vision systems, which means it is possible to capture and store images either on some schedule (e.g., every 10 seconds or every minute) or in response to a particular event (ImageMiner™). In combination with DataMiner™ it is also possible to display the process variable trends, and for particular windows of interest, perform the kind of analyses and visualizations shown above. These are powerful tools to playback specific operating conditions, and use the information to diagnose, repair and improve upon both operations and controls knowledge and strategies. Since it is available in real time, the data-mining statistical outputs, calculated on a moving horizon, can be used in control strategies. Finally, and this is the latest development, SoundMiner™ has the exciting ability to store sound samples (e.g., from SAG mill microphones) and perform sound-related data mining queries.

As has been said by others (e.g., Hales and Hales 2005), the process control development at the mine site will be driven more quickly and more accurately with these kinds of tools.

Simulation

This section is included for completeness and has been reproduced from Herbst and Flintoff (2011) with the kind permission of SME.

Dynamic simulation of mineral processing systems (e.g., Kenwalt's SysCAD, Andritz's Ideas™, Metso's ProSim©) has also been around for many years, but like many innovations in our industry, is only now emerging as a useful tool for operator training, for process design, and especially for the development and evaluation of process control strategies. Visionaries see a day in the not too distant future when the entire control package for a concentrator will be developed and tested using a variety of simulation tools, and then ported to the plant to help crash ramp-up schedules. This is a reality in some of the other process industries, e.g., nuclear.

Dynamic simulators are usually constructed from dynamic process models, such as Equation 2. In some instances, for example in the case of cyclones, static models are usually employed as the dynamics of these units are very fast compared to, say, mixing in a sump, or changes in the holdup of a large grinding mill. To accurately model dynamics with the highest fidelity, it is necessary to include pipelines, conveyors and other dynamic elements that significantly impact transient responses. These are usually what are referred to as continuous time simulations. Incorporating discrete time capabilities to model maintenance (planned and unplanned outages) and other random events that can affect production will complicate matters, but some of the commercial software (e.g., Metso's ProSim©) provides for this capability. Despite the impressive offering, dynamic simulation hasn't really had a major impact on process control design per se.

More recently the advanced control suppliers have begun to offer simulation tools, and the Metso Cisa approach to simulation in control is presented in the summary graphic in Figure 12. This purpose driven integration of simulation and control seems to have been the catalyst to drive industry interest. The main difference with previous simulation tools is the full emulation of the controls layer interacting with the dynamic models. It allows for a 'real-world' experience when using the simulator whether for training purposes or for testing control loops. Provided the dynamic models are a good representation of the process, then one can use the simulation to facilitate such process control related tasks:

- Instrument selection—What is the minimum instrument complement needed to get adequate control performance and where should it be placed for maximum control "resilience?" Example: should a weightometer be closer to the stockpile or to the mill.
- Fault tolerance testing–What if instruments fail, what if part of the process fails? Example: loss of a level sensor on a cyclone feed sump.
- Control loop design and tuning—Where are the PID loops and what are the tuning parameters, are more complex controllers needed, is decoupling an issue, etc.? Example: a bin-conveyor variable speed feeding system for crushers or HPGRs will highly benefit from control loop simulation.
- HMI design—How should the process info best be displayed to the operators, what are the navigation tools, etc.? Example: control room HMIs may only be ready very late in a project development phase. With the simulator HMIs being ready before, operators can help finalizing the HMI design and get some real end-user experience before startup.
- Operator training—Stopping/starting and running the process in the face of expected disturbances, before it gets built, and then refreshing as needed during the operations phase. Example: running the mill with new mill liners vs. running with 6-month-old liners (latest dynamic models will account for running-time/wear, reflecting the drastic operational changes observed in the real life).

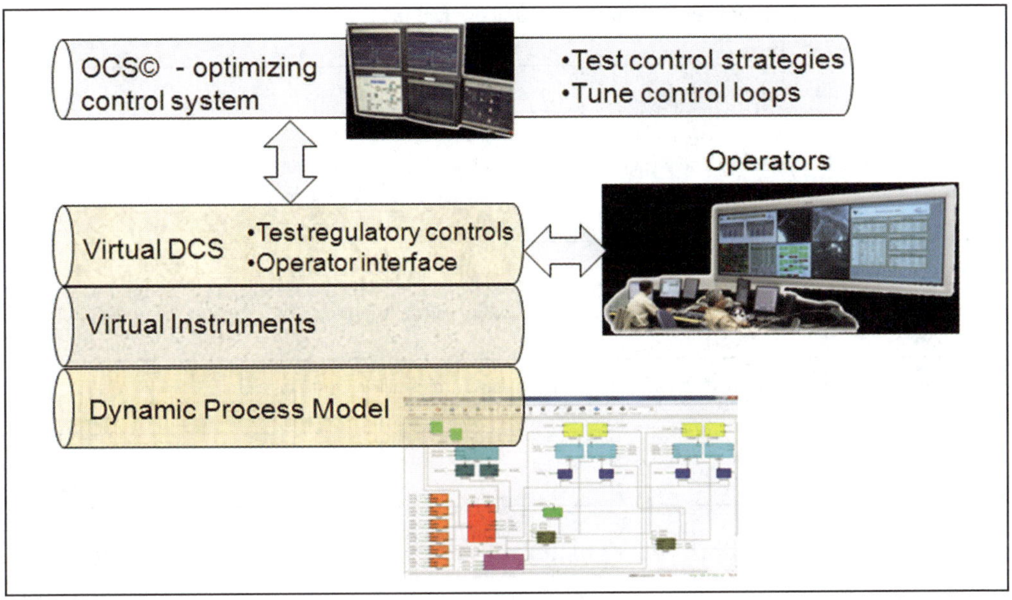

Figure 12. Use of simulation for process control design in OCS©

Ongoing operator training is taking on increased importance, as the extent and complexity of plant automation increase. For example, start-ups, shutdowns or recovery from major upsets are highly non-linear and batch-like processes, which cannot always be handled by ACM applications. As is the case with airline pilots, simulation can be a very useful means of keeping the operators "tuned" to these matters, although the simulators themselves tend to lack fidelity in these areas—a development need.

- Designing and evaluating advanced regulatory control strategies, e.g., cyclone and pump speed logic to maintain pressure and sump level or MPC or MBC techniques.
- Designing, troubleshooting, pre-tuning and evaluating advanced control strategies e.g., let the expert system run the simulator alone. And watch.
- Evaluating the implications of process design changes, e.g., what happens to control and operability if a pebble crusher is added?

Clearly, this is an exciting area of convergence between process control and dynamic process simulation and one that end users now seem very keen to pursue. It will be very interesting to see how this evolves in the next 2–4 years, as the benefits are documented and communicated.

Remote Support and Control

This section isn't so much about recent advances in technology as it is about recent advances in strategic thinking. Remote control is not a new concept in the mining industry and the remote operation of mining equipment (e.g., haul trucks, LHDs, drills, etc.) was first demonstrated many years ago, and much development has occurred since. In mineral processing, the concept of a central control room or even a hand held mobile interface that can be used anywhere on the plant floor are also arguably examples of remote control. However, an emerging trend today is the development of remote operations (support and control) centers that are located far from the mine site. This

Table 2. Current pros and cons for remote support and control in milling

Pros	Cons
Allows a way to deal with increasingly complex systems	Security issues
Provides greater stability and leverage for top technical talent	Never been done before (cultural issues). May face unexpected obstacles, like time zone difference: having experts work on night shifts?
Encourages team-based solutions	Requires intelligent devices
Effective way to capture and quickly disseminate best practices	Bandwidth/integrity
Identify need and deliver context sensitive training at the right time	Instrumentation must be right
Population of a decision support database (KM development)	Requires new analytics
Virtual teaming is easier	Requires development of diagnostics
Opens an opportunity to outsource	Documentation authenticity critical
Cloud computing is already here, virtualization of resources too	Removes skills and knowledge from sites
Remote support and assistance already routinely practiced by vendors	
Advanced Remote Process Control a good candidate to start with: higher level controls are less real-time-critical and run separately from the embarked unit automation which has to be on site	
Vision systems and CCTV allow for remote inspection	Operations still highly depend on visual inspection/feedback
Protection against turnover of staff at sites	Turnover becomes critical at the remote control center location

is motivated in part by improvements in bandwidth and communications reliability, by developments in artificial intelligence and sensors, by the convergence of on-line mechanical and process diagnostics (see real-time condition monitoring below), but perhaps the biggest driver is the desire to recruit, retain and better leverage the best people by offering them an attractive and stimulating place to live and work. Rio Tinto's Mine of the Future™ concept is an example of such an initiative, although, like others (Newcrest, Vale, Stracon, Codelco, etc.), this is currently more focused on mining and transportation processes.

Mining equipment is generally mobile and therefore more autonomous than the stationary equipment in a mill, there are generally larger populations, and there is usually only one or two OEMs in the supply chain. Consequently, there has been a strong supplier focus on embedding intelligence in mining equipment to monitor mechanical health and to enable operational productivity and safety improvements. Speaking metaphorically, this "fly by wire" concept is one of factors that have allowed operators to move out of the cab and into control centers many miles away. Of course, there are also some significant advantages when an OEM can choose a standard hardware platform and then focus on developing value-adding content in the form of diagnostics, specialized controllers, KPI dashboards, etc., and to develop these tools with data and information across a very wide range of applications in different mining environments. For these reasons, it is easier to imagine mining processes moving to multi-agent or holonic controls, i.e., toward fully autonomous operation, more quickly than will likely be the case for mineral processing.

In mineral processing, suppliers are moving to embedding intelligence on their equipment (e.g., PLC packages for (i) lube systems; (ii) filtration sequencing; (iii) crusher controls), although complying with site hardware standards has tended to put the focus more on the hardware platform and less on the value adding step of technical content development. Security and data ownership are still important factors in operational thinking, but there are some positive examples (e.g., remote support from software vendors, such as DCS/PLC, analyzers, advanced control packages), and the doors to greater collaborative support and development appear to be opening.

To briefly conclude, it would certainly appear that in mineral processing we will be moving more and more toward remote support and control, driven in part by the trends in the mining operations. (This concept appears in mining companies technology roadmaps.) Unfortunately, our history of collaboration between the equipment suppliers and the operators isn't as rich as is the case for mining, so we have a good deal of work in front of us. Table 2 summarizes some of the pros and cons the authors can envisage as we embark on this probably bumpy, but ultimately rewarding journey.

Unit Automation and Embedded Intelligence

Unit automation is usually delivered on a PLC and contains vendor content for monitoring and managing specific tasks. It has been around for most of the past two decades. Machines these days often have a very high level of automation, for example the Metso Hydraulic Roll Crusher (HRC®) has some 400 I/O tags associated with the unit automation package. Of course the OEMs would prefer to deliver these unit automation applications on their standard PLC platform, and spend most of their time working on enhancing the value-adding content. However, each operation usually has hardware standards, and more often than not, the vendor's choice is not compliant. Moreover, the operator wants easy access to the source code, should the need for some sort of maintenance or development change ever occur. These two factors mean that there is a higher risk for errors and a lot of re-work is performed porting the applications from one platform to another. Consequently, the incentive for the OEM to innovate around the on-board prognostics is limited, as there is no real protection of the resulting IP. At the risk of overstating the situation, while the world around us has become "smart" through ubiquitous computing, embedded intelligence, wireless and the internet, we have been slow to evolve in mineral processing. Things will change.

As an example, consider a cone crusher, which these days is almost always shipped with a unit automation package, although, at the buyer's discretion, not all of the functionality is enabled. **Figure 13** provides a glimpse of the current capabilities, and how such a unit might evolve in a 2–3 year and longer time frame. Such systems would likely be virtually invisible to the end user, and delivered on very robust, easily networked platforms, with architectures that can accommodate the diversity of applications and needs.

An interesting area that could soon impact the design of smart equipment is micro-mechatronics, specifically micro electrical mechanical systems (MEMS)—think ubiquitous sensing (and

3⁺ Yrs	• context sensitive help with augmented reality for difficult procedures • CBT for maintenance systems on board • data historian and maintenance analytics package	• remote support • multisensoring on power, acoustics and vibration • CBT for operations on board • embedded ACM support • performance dashboard with KPI's
2 – 3 Yrs	• tramp metal vs. packing discrimination • CM on bearings and bushings • parts list on board • manuals on board • FFT on vibration and acoustics	• internet compatible HMI • product particle size control (with vision) • variable speed drive for choke feed control • adaptive on-line model for training, scenario evaluation and control
Current	• lubrication monitoring, alarming and interlocks • ring bounce detection , pressure release and interlocks • bearing and busing RTD's • system tarts/stops	• PID controls and some simple strategies (e.g. bin level and feeder speed) • Override control (level & power) • Automatic gap adjustment

Figure 13. Potential evolution path for a "smart" cone crusher

sensor fusion). These miniature sensors, which will measure flow, pressure, velocity, acceleration/vibration/tilt, are seen to have a strong future in machine health applications.

To conclude, on the topic of embedded intelligence and diagnostics, automation futurist Jim Pinto (2008) has said: *"Industrial equipment will increasingly contain operating information and self-diagnostics to minimize the need for scarce and expensive factory-service personnel. Beyond just showing cause of failure (after the failure has occurred), integrated diagnostics will increasingly be predictive (before failure), preventive (precautionary and deterrent) and advisory (providing maintenance and repair instructions)."* He goes on to comment on machine-to-machine (M2M) communications, which "*provides major benefits to suppliers and end users, leading to improved asset management, dramatic cost reductions and enhanced service initiatives. Growth in M2M is still emerging, and will start generating significant productivity results in the coming years.*" These trends suggest something of a paradigm shift from the centralized control concept now practiced, to distributed multi-agent or holonic systems, and greater autonomy in operations.

CONCLUSIONS

The mineral processing industry has a rich history of achievement in instrumentation and control. This has come in the face of many unique challenges, and has required contributions from many different players in the larger community. The benefits have accrued in significant productivity and availability increases, reduced staffing, safer operations, and the systems have provided a platform for delivering new solutions.

The early efforts were focused of regulatory control, driven mainly by the introduction of minicomputers and computer savvy process engineers. These successes helped drive development in new analysis techniques and the early adoption of new generations of control hardware. Regulatory control improvements led to an interest in advanced process control, and this success sparked the exploration of advanced instrumentation. Innovation continues in all fronts.

Over the past few years, and based largely on the process control platform (infrastructure and needs), a new innovation path has emerged—peripheral applications that harmonize well with the goals of process control. The journey continues.

ACKNOWLEDGMENTS

The authors acknowledge Ming Song of Metso Technology Development for his assistance with the data mining examples.

REFERENCES

Anderson, A.E., 1979, Armco Autometrics Particle Size Measurement and Control. SME Preprint 79-375.

ARC, 2002, Use Critical Condition Management to Improve Your Bottom Line, ARC Strategies—ARC Advisory Group, April.

Asrom, K., Anton, J.J., and Arzen, K.E., 1986, Expert Control, Automatica, Vol. 22, No. 3, pp. 277–286.

Barrientos, R., and Telias, M. 1997, Nuevos Sonsores en el Circuit SAG, Workshop SAG, '97, Chile.

Bialkowski, W.L., 1992, Dreams vs Reality: A View from Both Sides of the Gap, Control Systems '92, p. 283.

Bradburn, R.G., Flintoff, B.C., and Walker, R.A., 1976, A Practical Approach to Digital Control of a Grinding Circuit at Brenda Mines Ltd., SME Preprint 76-106.

Broussaud, A., Legrand, G., Kok, D., Roux, E., Guyot, O., and Revalor, M., 2011, Integrated Advanced Grinding Control System and Newmont Ahafo, Proc SAG 2011, Vancouver, Paper No. 162.

Coghill, P.J., Mille, M.J., Rainey, S., and Sowerby B.D., 1997, On-Line Particle Size Analysis of Fine Mineral Slurries, Proc. 6th Mill Ops Conf., Madang, PNG, pp. 193–197.

Davey, K.J., Spencer, S.J., Phillips, P.L., Barker, D.G., and Holmes, R.J., 2012, Response of a Primary Grinding Mill Performance To Changes in Operating Conditions Using an On-Line Surface Vibration Monitor, Proc. XXVI IMPC, Delhi, Sept. 24–28, Paper no. 776.

De Mayo, F., Soto H., Hernandez, R., Gonzalez, G., Gonzalez, C., and Zamora, C., 1985, Audiometric Study to Detect Overloads in Rod Mills, Proc. Auto,. For Mineral Resource Development, Brisbane, pp. 253–257.

Del Villar, R.G., Thibault, J., and Del Villar, R., 1996, Development of a Soft Sensor for Particle Size Monitoring, Min. Engng., Vol. 9, No. 1, pp. 55–72.

Dupont, J.F., and Vien, A., 2001, Continuous SAG Volumetric Charge Measurement, Proc 33rd Ann. CMP. Ottawa.

Edwards, R.P., and Flintoff, B.C., 1992, Process Control in Crushing, In Comminution, Theory and Practice, Ed S.K. Kawatra, SME.

Flintoff, B.C., and Mular, A.L., 1992, A Practical Guide to Process Controls in the Minerals Industry, UBC Press.

Gonzalez, G., and Meyer, W., 1990, Replacement of a Particle Size Distribution Sensor in a Grinding Plant by an Estimator, Proc. 4th Conf. on Control Engng., Gold Coast, Aus., pp. 65–69.

Gonzalez, G., Meyer, W., and de Mayo, F., 1986, A Dynamic Compensation for Static Particle Size Distribution Estimators, ISA Trans. Vol. 25, No. 1, pp. 47–51.

Hales, L.B., and Hales, M.L., 2005, Improving Expert Control Systems Trough Deep Data Mining of Plant Data (Available from the KSX website: www.kscape.com/papers).

Hales, L.B., and Herbst, J.A. 1987, The Use of Expert Systems in Concentrator Control, SME Preprint 87–159.

Hales, L.B., Colby, R.W., and Ynchausti, R.A. 1996, Optimization of AG ans AG Mill Circuits Using Intelligent Process Control Software, Proc. SAG 1996, Vancouver, Vol. 2, p. 632–641.

Harris C., and Kosick G., 1988, Expert System Technology at the Polaris Mine, Proc. 20th AGM CMP, pp. 149–174.

Herbst, J.A., 1984, Ed. of Control '84, AIME.
Herbst, J.A. 1990, Ed of Control '90—Mineral and Metallurgical Processing, SME.
Herbst, J.A. 2000, Ed of Control 2000—Mineral and Metallurgical Processing, SME.
Herbst, J.A., and Flintoff, B.C., 2011, Recent Advances in Modeling Simulation and Control of Mineral Processing Operations, in Separation Technologies for Minerals, Coal, and Earth Resources," Ed. Courtney A. Young and Gerald H. Luttrell: SME, pp. 667–698.
Herbst, J.A., and Pate, W.T., 1996, On-Line Estimation of Charge Volumes in SemiAutogenous and Autogenous Grinding Mills, Proc SAG 1996, Vancouver, Vol. II, pp. 817–827.
Herbst, J.A., Pate, W.T., and Oblad, A.E., 1989, Experiences in the Use of Model-Based Expert Control Systems in Autogenous and SemiAutogenous Grinding Circuits, Proc. SAG 1989, Vancouver, Vol. II, pp. 669–686.
Jaspan, R.K., Young, G.J.C., and Mellor, M.S., 1986, ROM Mill Poer Draft Control Using Multiple Microphones to Determine Mill Load, Proc. Gold 100, Extractive Metallurgy of Gold, Ed. C.E. Fivaz and R.P. King, Vol. 2, pp. 483–492.
Koivistoinen, P., and Miettunen, 1989, The Effect of Mill Lining on the Power Draw of a Grinding Mill and Its Utilization in Grinding Control, Proc. SAG 1989, Vancouver, pp. 687–695.
Kyotokari, J., 2012, Personal Communication (Metso Automation, Tampere, Finland).
Lin, C.L. and Miller, J.D., 1992, The Development of a PC Inage Based On-Line Particle Size Analyzer, SME Preprint 92–10.
Livelli, G., 2012, Improving Productivity and Accuracies While Reducing Downtime, Power Engineering, Vol. 116, Issue 7.
Maerz, N.H., Franklin J.A., Rothenburg, L., and Coursen, D.L., 1987, Measurement of Rock Fragmentation by Digital Photoanalysis, ISRM 6th Int. Cong. On Rock Mechanics, Montreal, Canada, Vol. 1, pp. 687–692.
Maron, R.J., and O'Keefe, C., 2008, Application of Non Intrusive Sonar Technology to Flow Measurement and Control in Mineral Processing Plants, Proc Automining, Santiago, April.
Miettunen, J., Lahteenaki, S., and Jamsa-Jounela, S.L., 1993, New Methods in the Control of Autogenous Grinding Circuit, Proc. XVIII IMPC, Melbourne, pp. 301–317.
Montague, J., 2006, The DCS Spirit Lives on Everywhere, Control Magazine, Dec.
Morari, M., 1983, Design of Resilient Processing Plants—III. A General Framework for Assessing Dynamic Resilience, Chem. Engng. Sci., Vol. 38, No. 11.
Nelson, B., and Stauffer, T., 2008, PLCs and DCSs Converge, Control Engineering, Vol. 55, No. 5.
Olsen, T., and Bialkowski, B., 2002, Lambda Tuning as a Promising Controller Tuning Method for the Refinery, Proc. AIChE AGM, March, Session 42 (available on Emerson website 2012 http://www2.emersonprocess.com/siteadmincenter/PM%20PSS%20Services%20Documents/Consulting%20Services/LambdaTuningasaPromisingControllerTuningMethodfortheRefinery.pdf).
Pax, R.A., 2001, Non-Contact Acoustic Measurement of In-Mill Variables of SAG Mills, Proc. SAG 2001, Vancouver, Vol. II, pp. 386–393.
Pinto, J., 2008, Pinto's Pointers, available: http://www.jimpinto.com/writings/pointers2008.html.
Pontt, J., Valderamma, W., Magne, L., and Pozo, R., 1997, MONSAG: In Sisteme PARA el Monitereo On-Line de la Carga en Molinos SAG, Proc. Workshop SAG '97, Chile.
Rajamani, R.K., and Herbst, J.A., 1990, Editors Control '90—Mineral and Metallurgical Processing, SME.
Rosenblum, F., Labuc, V., and Spira, P., 1884, A New On-Stream Particle Size Analyzer for Slurries, in Control '84, Ed. J Herbst, SME, pp. 15–20.
Ruel, M., 2011, Control System Performance Assessment—Best Practices, Proc. 43rd Can. Min. Proc., CIM, p. 347.
Samskog, P.O., Soderman, P., Storeng, U., Bjorkman, J., Guyot, O., and Broussaud, A., 1996, Model-Based Control of Autogenous and Pebble Mills at LKAB Kiruna KA2 Concentrator (Sweden), Proc, SAG 1996, Vancouver, pp. 599–616.
Sbarbaro, D., del Villar, D., 2010, Advanced Control and Supervision of Mineral Processing Plants, Springer, 1st Edition.

Sbarbaro, D., Ascencio, P., Espinoza, P., Mujica, F., and Cortes, G., 2008, Adaptive Soft Sensors for On-Line Particle Size Estimation in Wet Grinding Circuits, Control Engng. Practice, Vol. 16, pp. 171–178.
Shah, S.L., 2012, Searching for Nuggets of Gold in Your Process Data, Proc. XXVI IMPC, Delhi, Sept. 24–28, Paper no. 1112.
Van Drunick, W., and Moys, M., 2001, The Use of an Energy Balance to Measure and Control the Rheology of Mill Discharge Slurry, Proc. SAG 2001, pp. 304–316.
Van Nierop, M.A., and Moys, M.H., 1995, Measurement of Load Behaviour in an Industrial Grinding Mill, IFAC Automation in Mining, Mineral and Metal Processing, Sun City, South Africa.
Verhappen, J., 2011, Network Evolution: 25 Years of Connectivity, Manufacturing Automation, Sept.
Vien A., Fragomeni D., Larsen C., Fisher D., 1991, MOCCA: A Grinding Circuit Control Application, Proc. 23rd AGM CMP, pp. 1–18.
Vien, A., Palomino, J., Gonzalez, P. and Perry, R., 2000, Multiple Feeder Control, Proc. 32nd CMP, pp. 295–312.
Vignos, J.H., and Bruno, S.J., 1979, On-Line Instrument for Coarse Particle Size Distribution Measurements, SME Preprint 79–377.

ABOUT THE AUTHORS

Brian Flintoff is currently the senior vice president for technology development at Metso–Mining and Construction. He has spent about 39 years in the mineral processing industry in operations (Brenda Mines Ltd.), academia (University of Alberta, faculty of engineering), consulting (Brenda Process Technology and successors, AMEC Engineering), and technology management and development (Metso). His technical interests lie in the areas of process analysis, simulation, control, and training.

Olivier Guyot is currently global manager of the Advanced Process Control product line at Metso Mining and Construction Technology Development. Olivier holds an engineering degree from the Alès School of Mines in France. He has 24 years of experience in advanced control, simulation, and modeling of mineral and metallurgical processes. Guyot has successively been a software developer, project engineer, project manager and project coordinator for Metso Cisa where he created the OCS©, VisioFroth™, and VisioRock™ software.

Jeff McKay is the current manager of Expert Systems Global for Metso Cisa. McKay has a BS degree in metallurgical engineering from the University of Utah. He has 30 years of experience in minerals processing, including 15 years of column flotation experience and 20 years of Adaptive Expert control experience. McKay has designed and installed Expert control systems at numerous sites around the world in crushing, grinding, flotation, thickening and filtration, and kiln and induration control using a variety of advanced control systems, including the Metso Cisa OCS and software package. Jeff has also installed vision- and audio-based control systems at many plants around the world. He is currently responsible for training the Metso Cisa engineering staff in regard to Expert control and for advancing the control concepts used by Metro Cisa in their APC projects.

Andre Vien is currently the process engineering manager in the Technology Development at Metso–Mining and Construction. He has been working with the Metso Technology Development (and its predecessors) since 1988. During this time, he has worked on many data analysis, modeling, and advanced control projects. As part of his work, Vien developed several engineering tools for mass balancing, modeling, and data analysis. He has co-authored several papers and participated in many short courses. He is the founder of the computer-based mill operator training (CBT) program. His recent work includes process audits and metallurgical forecasting (life-of-mine modeling), as well as research and development inside Metso–Mining and Construction.

Advances in Discrete Element Method Application to Grinding Mills

Raj K. Rajamani, Samira Rashidi, and Nikhil Dhawan
Department of Metallurgical Engineering, University of Utah, Salt Lake City, Utah, United States

ABSTRACT: Discrete element method (DEM) has made significant impact in the design and operation of grinding mills. The internal charge motion of grinding balls and ore can be readily examined. As a result, this method enables one to design and examine mill internals via simulation. This manuscript details essential advances made with DEM. These include mill power draft calculations, shell lifter design, lifter wear prediction, pulp lifter simulation and ore particle breakage in grinding mills. The status of advancement in each of these topics is discussed.

INTRODUCTION

Grinding mill modeling has attracted the attention of a number of researchers ever since Bond published his method of energy calculation with an index called Bond work index (Bond 1961). In the 1970s, the population balance or selection and breakage function model was at the forefront of research (Fuerstenau et al. 1973). A number of advancements were made in the application of this model to ball mill grinding. However, the same models could not be extended to semi autogenous grinding. Simply carrying out pilot scale SAG grinding is a very tedious task. At this time, the discrete element method appeared on the scene. DEM helps in simulating a number of features of SAG mills and AG mills, not to mention balls mill as well. The principal advances made with DEM are discussed in this manuscript.

MILL POWER DRAFT

The discrete element method in its entirety was first published tin the civil engineering literature (Cundall and Strack 1979) from where it spread rapidly into many engineering disciplines including the mineral processing industry. In the 1970s the prevalent theme in the comminution literature was the advancement of Bond's formula (Bond 1961) for mill power prediction. The torque arm formula was the predominant theory (Hogg and Fuerstenau 1972; Guerrero and Arbiter 1960; Hlungwani et al. 2003). The torque needed to sustain the offset of the center of gravity of the charge from the mill center was being refined in many ways. The culmination of such efforts led to the more detailed empirical formula (Morell 1992) which included the toe and shoulder angle. The success of this model is due to its ability to define the charge boundaries. Yet all of these models were lacking in their ability to respond to change in lifter geometry or ball size distribution. This is where DEM made its mark, since it accounts for cascading charge, interpenetrating balls between layers in the cascade and also cataracting charge. Here the energy consumed in thousands of collisions of balls with other balls, mill walls and lifters is summed to arrive at mill power. Since the

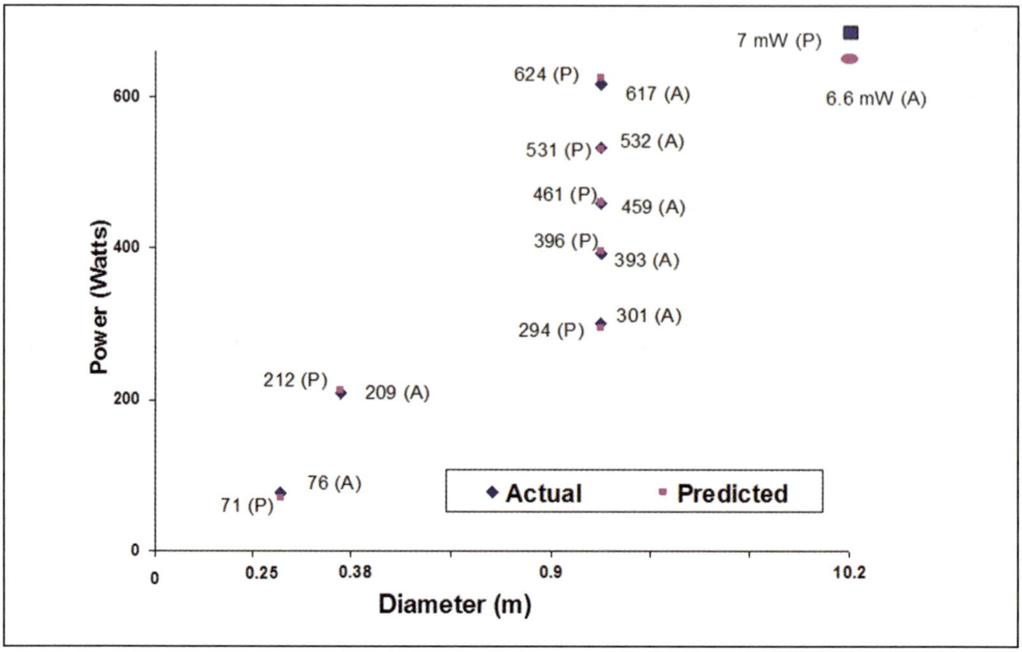

Figure 1. Comparison of actual power and predicted power draw using DEM for different mill diameters

geometry of mill internals and ball size distribution is taken into account explicitly, DEM computed power responded to variations in lifter design and ball size distribution.

The precursor to DEM is the computation of the trajectory (Powell 1991) of a single ball while it is in contact with the leading face of the lifter. The balance of forces due to gravity, centrifugal, sliding and rolling gave rise to accurate trajectories of a single ball. Even today this methodology is used in SAG mill lifter design. The single ball trajectory is a valuable tool for visualization of the parabolic trajectory of a ball and hence the strike zone on the toe of the mill (Sepulveda 2012). In fact, the best choice of lifter release angle can be determined using single ball trajectories. The best angle corresponds to the trajectory that can be maintained even as the lifter wears out (Royston 2007).

In DEM (Datta et al. 1999; Rajamani et al. 2003) the impact energy in each and every collision is summed for one or two full revolutions and is then converted to the mill power draft over the full mill length. While in 3D simulations the power can thus be computed directly, in 2D the power draft in a slice of the mill is computed and is then converted to power draft of the full mill length. The author has successfully computed SAG and ball mill power in hundreds of case studies besides verifying power prediction in laboratory size mills as shown in **Figure 1** (Rajamani et al. 2000). The capability of 2D DEM for power predictions (Nierop et al. 2001) at as high 120% critical speed was done in a 55 cm mill. Given the uncertainties in power measurements 2D DEM is a viable tool. However, one could always claim the superiority of 3D DEM over 2D DEM for power predictions (Cleary 1998) which cannot be refuted by simple arguments. Further with 3D DEM a slice of SAG mill (Cleary 2001a) can be simulated to predict full mill power.

A further advance in DEM model would be to change the shape of particles from disk or sphere to non-circular and non-spherical shape. When advancing the DEM model from 2D disk to 2D non-spherical to 3D slice to 3D full (Cleary et al. 2003) the prediction of shoulder angle, toe angle, power and vortex center increasingly improved. Obviously, these findings point in the direction of 3D DEM for full scale charge motion plus breakage studies. A typical screenshot of 2D DEM charge motion is shown in Figure 2.

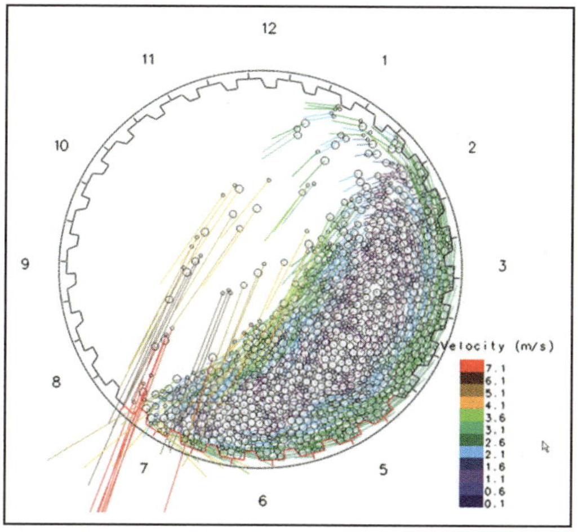

Figure 2. Typical screenshot of 2D DEM charge motion

The fundamental structure of DEM simulation rests primarily on the spring-dashpot contact model (Rajamani et al. 2000). There are spring constants, dashpot constants and friction parameters that must be set correctly. These parameters are difficult to determine experimentally and most researchers depend on the overall prediction of mill power as a goalpost for verification. Yet these parameters can affect normal and shear directional collisional energy on individual particles. In a careful photographic study (Chandramohan and Powell 2005) of binary collisions and particle bouncing on a plate, the shear mechanisms and sliding friction model imposed on the ubiquitous spring-dashpot model was found to be woefully inadequate. On the other hand a study of bouncing ball inside a mill shell (Dong and Moys 2002) found that spring-dashpot model is accurate enough as long as coefficient of friction and coefficient of restitution is correctly chosen. Regardless, the spring-dashpot model continues to be the working model in many DEM research groups.

LIFTER DESIGN

In the early 1960s, Art MacPherson investigated the effect of lifter spacing to height ratios and concluded that a ratio of 4:1 maximized grinding. Such designs were carried into Highland valley Copper, Copperton and other concentrators. While such designs lived up to the power draft expectations, at least one of them, Freeport's Grassberg operation reported severe damage to the SAG mill shell in the first few months after commission. DEM with its ability to animate the charge motion on a PC screen, as shown in Figure 2, comes to the rescue in such scenarios. It was found that the 14-inch tall lifters were causing direct ball strikes across the mill shell. This study would form the basis for future shell lifter designs. For the first time, the lift of charge, cascading, cataracting, packing and other aspects was viewed directly. In fact Minera Alumbrera mine (Sherman and Rajamani 1999) was the first mine, based on a DEM study incorporated 30 ° relief angle lifters, with considerable success. This study showed that the number of lifters needed for a mill of diameter D (in feet) was just D lifters instead of the "twice the diameter" rule that had been in existence. The key to lifter design was the fact that DEM was showing charge motion accurately. Hence, a number of researchers including the author (Cleary 2001a,b; Monama and Moys 2002; Hlungwani et al. 2003; Rajamani et al. 2000; Venugopal and Rajamani 2001) employed a variety of laboratory scale mills to verify DEM charge motion. At this time a spectacular advancement in filming the entire charge motion with X-ray and Gamma-ray was also developing (Kallon et al. 2011). After so much of verification, DEM has become a trusted tool for shell lifter design (Herbst and Nordell 2001; Rajamani et al. 1999) among liner manufacturers and plant process engineers.

LINER WEAR

Naturally, if DEM can track the influence of liner design it can also track liner wear. Liner wear life is an important to all concentrators. One could not have imagined modeling liner wear prior to DEM. The key information to wear modeling is in the heart of discrete element computations. They are the shear force, velocity and contact location on a lifter bar, shear energy at contact and normal energy at contact. Some use the well-known Archard's law (Powell et al. 2011; Herbst and Qiu 2007) and some use unique prescriptions of wear based on collision energies (Kalala et al. 2005 a, b). The sound concepts behind these wear models are reflected in the accurate wear profiles precited for a wave liner in a coal mill or block type lifters in copper ore grinding. In a carefully conducted study it was shown that the wear life of a 40 ft. SAG mill could be accurately modeled, and hence this study led to a unique design of a curve shaped lifter (Hart et al. 2006) with a lead angle of 28 degrees. It should be understood however the wear rate constant is unique to the ore slurry and metallurgy of the lifter bar.

Since steel loss is a significant contributor to cost of metal production, it would be obvious to account this loss on a quarterly basis. While DEM may be a predictive tool a practical measurement tool would be more preferable. To this end, today a system known as Mill mapper (Toor et al. 2011a) has the capability to map the entire interior of large mills. With this tool mill operations can now quantify the steel losses and take actions to reduce losses. It would be a matter of time before DEM is coupled with this measurement to take cost saving measures in plants.

PULP LIFTER FLUID FLOW

The next potential area of application of DEM is pulp lifter design. Plants are well aware of pooling issues that is detrimental to mill throughput. Furthermore, pulp lifters are inherently prone to carry over flow and flow back. Then there is the issue of radial pulp lifter versus curved pulp lifters; which one is suited for a particular SAG mill operation? Finally, the quantity of flow through grates into the pulp lifter is still lacking. Plants often increase open area of grates with the hope of increasing mill throughput. However, the back flow from the pulp lifter increases simultaneously thus nullifying the anticipated increase in throughput. DEM could be used to simulate slurry discharge profiles in the pulp lifter by modeling the slurry by a number of discrete slurry particles (Rajamani 2007). While valuable inferences in carry over flow can be drawn from such simulations, it lacks the viscous momentum transport model that is so essential to fluid flow. The obvious solution is to couple computational fluid dynamics (CFD) with DEM as has been done in the chemical engineering literature. The price to be paid for such rigor is computational complexity and so excessive computer time. The solution is smoothed particle hydrodynamics (SPH) algorithms. Just like DEM, SPH handles fluid flow via particles and hence is referred to as mesh less algorithm. These particles are given properties of viscosity and interaction with solid walls etc. In summary this is an ideal scheme to couple with DEM. Then the slurry flow through the porous ball charge (Cleary et al. 2006; Herbst and Nordell 2011) can be modeled and then the slurry can be followed during its transit through the grate and then into the pulp lifter until it exits via the discharger. Thus the momentum transfer from the ball charge to the slurry enables the computation of slurry flow in the mill discharge.

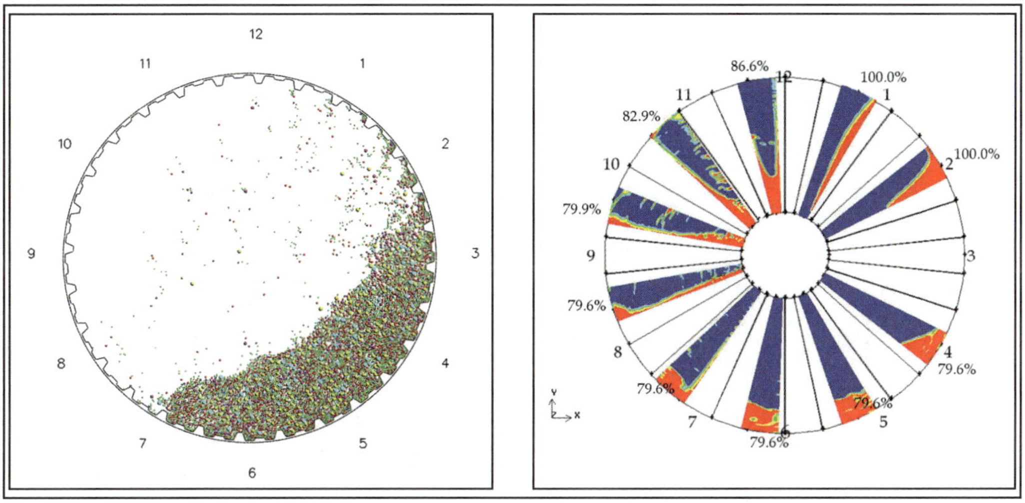

Figure 3. DEM charge profile and CFD prediction of free surface profiles as the mill discharge, Los Bronces SAG mill (diameter 8.26 m and length 4.19 m) (Rajamani et al. 2011)

Another alternative to DEM with SPH coupling is direct CFD simulation of a single blade in the pulp lifter. The charge profile in the mill interior is simulated with DEM. The profile of charge in the neighborhood of 7, 6, and 5 pm regions indicates the approximate level of slurry in the blade of a pulp lifter. It may be argued that slurry level above mill charge may vary considerably. Nevertheless, this approximation is quite reasonable. Then the full CFD simulation of the blade (Rajamani et al. 2011) is carried out; summing over all the blades gives the overall discharge through the mill. This effort has been very successful when compared with published information on six SAG mills; one example is shown in Figure 3. In summary, advances in slurry flow via DEM is continuing.

PARTICLE BREAKAGE IN MILLS

DEM can be readily employed to model breakage since the force of impact and energy of impact are the back bone of DEM algorithm. However, the intractable nature of ore particle breakage itself is a severe challenge. A fundamental theory of ore particle breakage under different impact loads is nonexistent today. Even if one existed it would not be sufficient since particles in a mill break under multiple point stresses. Nevertheless, describing the evolution of ore particles size distribution is the frontier of DEM applications to grinding mills.

The approach to describing particle fracture varies in its complexity. The prevailing charge can be assumed as rigid particles and the spectrum of impact energy in all of the collisions can be taken as the driving force for breakage of particles. It is well known in the modeling of breakage, breakage rate or selection function and breakage distribution function categorize the breakage into a set of size classes. Therefore, the formalism is to convert the impact energy spectrum into selection and breakage functions. Such formalism leads to a model that is akin to the selection and breakage model known popularly as population balance model of grinding. In such an approach the time evolution of size distribution was successfully predicted (Datta and Rajamani 2002). The same

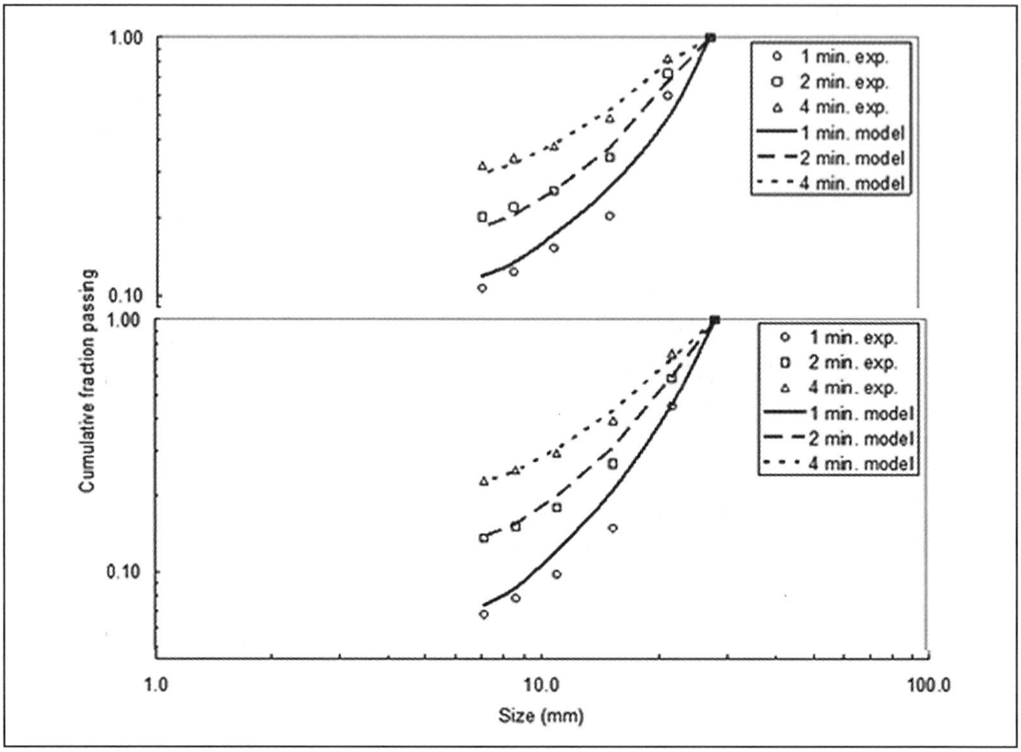

Figure 4. DEM-based prediction of ore size distribution in a 90 × 15 cm ball mill operated at 26% and 21% loading

methodology was extended to a pilot size mill and the predictions are shown in Figure 4 (Tuzcu and Rajamani 2011).

Yet another method to arrive at broken particle size distribution is to characterize breakage of particles in a drop weight apparatus. The results can be linked in a suitable functional form between broken particle size and applied energy. Then in DEM (Morrison and Cleary 2004; Cleary and Morrison 2011) the computed normal and shear energy of collision on individual particles can be used in conjunction with the derived functional form to determine progeny size distribution. A key expression for converting impact energy into probability of a particles survival (Vogel and Peukert 2004) is shown in Equation 1. A more convenient form of the above equation for DEM is shown in Equation 2.

$$s = 1 - \exp\left\{-f_{mat} * x * k\left(w_{m,kin} - w_{m,\min}\right)\right\} \quad \textbf{(1)}$$

$$s = 1 - \exp\left\{-b\sum_{i}\left(E_i - E_0\right)\right\} \quad \textbf{(2)}$$

where, s is the breakage portion, $w_{m,kin}$ is the single impact energy, $w_{m,min}$ is the no damage energy, x is the particle diameter, k is the number of impacts and f_{mat} is the material parameter, b is the material parameter, E_i is the ith collision energy and E_0 is the no damage energy.

These expressions together with a model of progeny distribution upon breakage is a convenient frame work in DEM since successive impacts can be recorded and so the damage on a single particle can be accumulated and be broken at the appropriate time (Morrison et al. 2011). Thus, it is possible to model both attrition and impact breakage in SAG/AG mills. Nevertheless, modeling the breakage due to accumulating damage and multi point contact breakage is a complex task. More detailed models quadruple the time investment in DEM computing.

A much rigorous approach is to consider the continuous spectrum of collision energies from which a fraction of the impact energy is absorbed by the particle. Once again, a continuous breakage function is defined for particles broken in this manner. Next a particle surface model irrespective of the collision energy is defined. Then the continuous impact energy distribution (computed with DEM) can be allocated to body and surface breakage and the respective progeny distributions are modeled to include in the appearance term of the mass balance. In essence such a micro scale model owing the scale employed is extremely complex in its mathematical formulation, yet it carries lot more detail into the computations (Tavares and Carvalho 2010; Carvalho and Tavares 2012). These models continue to improve as computational power advances every year.

Finally, if one wants to go into as much detail as following each and every particle undergoing breakage in the mill then the discrete grain breakage (Herbst and Potopov 2004; Herbst and Nordell 2001) model belongs in this category. DEM is used in the complete description of the grinding media. At any instant of time in DEM the particles are allowed to break into progeny particles according to the discrete grain model. Next, the progeny particles are included in the subsequent time steps in DEM. Although promising, this numerical technique requires enormous computational resources such as a CRAY-super computer. In addition, because each ore particle is followed in the algorithm a discrete grain liberation (Herbst and Qiu 2007) model too can be imbedded. Finally high computational resources enable one to include SPH for the simulations formulation of slurry flow. In summary there are a few different formulations of particle breakage in DEM. All of them are advancing year by year. As of today, there is no single formulation that can encompass all of the complexity of ore breakage in grinding mills.

MEASUREMENTS TO VALIDATE DEM

While DEM is progressing, there are two experimental techniques that closely mimic DEM. The first one is positron emission particle tracking (PEPT) (Parker et al. 1997) which is a technique for tracking granular particles inside the mill shell. The basis of PEPT is the labeling of a particle with a radio nuclide which decays via the emission of a positron, the antiparticle of an electron. While the positron decays two simultaneous gamma rays are emitted defining the line of emission. The PET camera surrounding the shell detects many thousands of lines of emission which is then processed to locate the coordinates of the particle precisely within the shell. PEPT is now (Kallon et al. 2011) capable of quantifying circulation rate of charge particles, the profile of the charge and the velocity of individual particles. This technology is evolving and perhaps someday it will be a tool for diagnostics on a plant scale.

Another technology that can take the benefit of DEM simulation is the mill-mapper: a 3D laser scanner mounted inside the mill collects the 3D volume map of the mill interior space. The point cloud is then analyzed to provide mill liner thickness, liner shape, liner weight, and mill volume. Since it provides liner height and cross sectional geometry of shell lifters, DEM can be combined with this data to assess the effective breakage regime during the life time of a lifter set (Toor et al. 2011b).

DEM COMPUTATIONS

Thus the DEM has grown to accommodate a number of aspects of tumbling mill operation. Further, the scale of simulation is at the level of plant scale mills in three dimensions. Obviously, such computations require enormous computing resources and computational time. Notwithstanding weeks and weeks of compute time, one employs multi-node processors, CRAY computer and parallel computing algorithms and so on. A recent advancement in this respect is GPU computing (Rajamani et al. 2011). Here the three-dimensional computing is done on the video card of the desktop PC. Commodity graphic cards deploy hundreds of processors in the computation. The algorithms are written in OpenCL or CUDA language, which deploy the calculations on hundreds of processors automatically. In fact a speed up factor of fifty was established for the Nvidia Geforce 580 graphics card over single CPU. The simulation of a SAG mill with 250,000 spheres was completed in 8 hours and the simulation of ball mill with one million spheres was completed in 27 hours. A typical snap shot of DEM-GPU combination for a plant ball mill is shown in Figure 5. GPU is a strong candidate for the future of DEM simulations since the card hardware seems to advance every six months.

CONCLUSION

DEM has advanced in the grinding mill field since its introduction in 1991. The prediction of mill power draft, charge motion and lifter design has matured to an advanced level. Particle breakage in mills is advancing rapidly despite the difficulty inherent in this phenomenon. DEM computations stretch the limit of available resources, yet progress has been made in this front also. In the near future DEM will become the standard tool for design of mills, if not already.

ACKNOWLEDGMENTS

Raj K. Rajamani acknowledges the many contributions of Barada Mishra during his years at the University of Utah and also acknowledges his graduate students referenced in this paper for their experimental and theoretical contributions.

REFERENCES

Bond, F.C. 1961. Crushing and grinding calculations. Allis-Chalmers, Publication No. 07R9235C.

Carvalho, R.M., Tavares, L.M. 2012. Predicting the effect of operating and design variables on breakage rates using the mechanistic ball mill model. Accepted *Minerals Engineering*.

Chandramohan, R. and Powell, M.S. 2005. Measurement of particle interaction properties for incorporation in the discrete element method simulation. *Minerals Engineering*. 18: 1142–1151.

Cleary, P.W. 1998. Predicting charge motion, power draw, segregation and wear in ball mills using discrete element method. *Minerals Engineering*. 11: 1061–1080.

Cleary, P.W. 2001a. Charge behavior and power consumption in ball mills: sensitivity to mill operating conditions, liner geometry and charge composition. *Int. J. Miner. Process*. 63:79–114.

Cleary, P.W. 2001b. Recent advances in DEM modeling of tumbling mills. *Minerals Engineering*. 14: 1295–1319.

Cleary, P.W., and Morrison, R. 2011. Understanding fine ore breakage in a laboratory scale ball mill using DEM, *Minerals Engineering*. 24: 352–366.

Cleary, P.W., Morrison, R., and Morrell, S. 2003. Comparison of DEM and experiment for a scale model SAG mill. *Int. J. Miner, Process*. 68: 129–165.

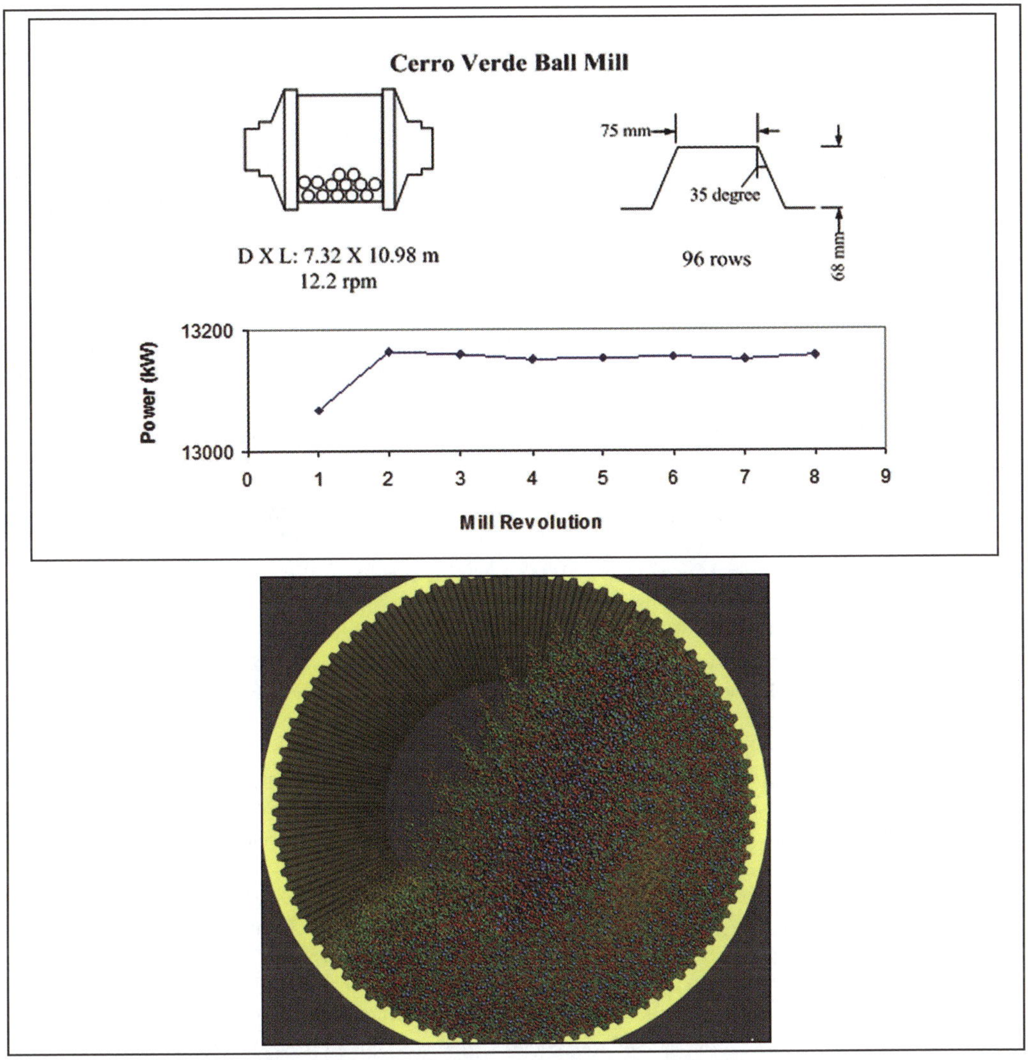

Figure 5. Simulated power draw of the Cerro Verde ball mill with 3D GPU snapshot

Cleary, P.W., Morrison, R.D. 2012. Prediction of 3D slurry flow within the grinding chamber and discharge from a pilot scale SAG mill. *Minerals Engineering*. 39, 184–195.

Cleary, P.W., Sinnott, M., and Morrison, R. 2006. Prediction of slurry transport in SAG mills using SPH fluid flow in a dynamic DEM based porous media. *Minerals Engineering*. 19: 1517–1527.

Cundall, P.A. and Strack, O.D.L. 1979. A discrete numerical model for granular assemblies. *Geotechnique*, 29: 47–65.

Datta, A., Mishra, B.K., and Rajamani, R.K., 1999. Analysis of Power Draw in Ball Mills by the Discrete Element Method. *Can. Met. Quart.* 38: 133–140.

Datta, A., and Rajamani, R.K., 2002. A direct approach of modeling batch grinding in ball mills using population balance principles and impact energy distribution. *Int. J. Min. Process.* 64: 181–200.

Dong, H., and Moys, M.H. 2002. Assessment of discrete element method for one ball bouncing in a grinding mill. *Int. J. Miner. Proc.* 65: 213–226.

Fuersteanu, D.W., Grandy, G.A., and Herbst, J.A. 1973. Population Balance Models for the Design of Continuous Grinding Mills. In *X International Mineral Process Congress*, Proceedings, Edited by M.J. Jones, pp. 375–385, London.

Guerrero, P.K., and Arbiter, N. 1960. Tumbling mill power at cataracting speeds. *AIME Transactions*, 217:73.

Hart, S., Nordell, L., Faulkner, C. 2006. Development of SAG mill shell liner design at CADIA using DEM modeling. In *International Autogenous and Semiautogenous Grinding Technology*, Edited by D.J. Barratt, M.J. Allan, and A.L. Mular. Vancouver, B.C. Canada.

Herbst, J.A., and Nordell, L. 2001. Optimization of the Design of SAG Mill Internals Using High Fidelity Simulation, In *International Autogenous and Semiautogenous Grinding Technology*, Edited by D.J. Barratt, M.J. Allan, and A.L. Mular. Vancouver, B.C. Canada.

Herbst, J.A., and Potapov, A.V. 2004. Making a discrete grain breakage model practical for comminution equipment performance simulation, *Powder Technology* 143: 144–150.

Herbst, J.A., Qiu, X. 2007. Validation of High Fidelity Simulation Wear Predictions in Various Types of Mineral Processing Equipment. In: *Proceedings of the 39th Annual Canadian Mineral Processors Conference*, Ottawa, Ontario, Canada, 459–471.

Hlungwani, O., Rikhotso, J., Dong, H., and Moys, M.H. 2003. Further validation of DEM modeling of milling: effects of liner profile and mill speed. *Minerals Engineering*. 15: 993–998.

Hogg, R., and Fuerstenau, D.W. 1972. Power relationships for tumbling mills. *Trans. SME-AIME*. 252: 418–423.

Kallon, D.V., Govender, I., and Mainza, A.N. 2011. Circulation rate modeling of mill charge using positron emission particle tracking. *Minerals Engineering*. 24: 282–289.

Katala, J.T., Bwalya, and M., Moys, M.H. 2005a. Discrete element method (DEM) modeling of evolving mill liner profiles due to wear: Part I. DEM validation. *Minerals Engineering*. 18: 1386–1391.

Katala, J.T., Bwalya, and M., Moys, M.H. 2005b. Discrete element method (DEM) modeling of evolving mill liner profiles due to wear: Part II. Industrial case study. *Minerals Engineering*. 18: 1392–1397.

Makokha, A.B., Moys, M.H., Bwalya, M.M., Kimera, K. 2007. A new approach to optimize the life and performance of worn liners in ball mills: Experimental study and DEM simulation, *Int. J. Miner. Processing*, 84, pp. 221–227.

Monama, G.M., and Moys, M.H. 2002. DEM modeling of the dynamics of mill startup. *Minerals Engineering*. 15: 487–492.

Morrell, S. 1992. Prediction of grinding mill power. *IMM Transactions Section C*, 101: 25–32.

Morrison, R., and Cleary, P.W. 2004. Using DEM to model ore breakage within a pilot scale SAG mill. *Minerals Engineering*. 17: 1117–1124.

Morrison, R., Cleary, P., Delaney, G., Cummins, S., Loveday, B., and Powell, M. 2011. Predicting the evolution of rock size and shape distributions using DEM based on different modes of breakage in AG and SAG mills, In *Proceedings of the International Autogenous Grinding, Semiautogenous Grinding and High Pressure Roll Technology*, Edited by K. Major, B. Flintoff, B. Klein and K. McLeod. Vancouver, B.C. Canada.

Nierop, M.A., Glover, G., Hinde, A.L., and Moys, M.H. 2001. A discrete element method investigation of the charge motion and power draw of an experimental two-dimensional mill, *Int. J. Miner. Process*. 61: 77–92.

Nordell, L.K. 2001. SAG and ball mill comminution rate and power analysis using 3-D discrete element method (DEM) to accounting for particle shape, size range and fluid rheology, In *International Autogenous and Semiautogenous Grinding Technology*, Edited by D.J. Barratt, M.J. Allan, and A.L. Mular. Vancouver, B.C. Canada.

Parker, D.J., Dijkstra, A.E., Martin, T.W., and Seville, J.P.K. 1997. Positron emission particle tracking studies of spherical particle motion in rotating drums. *Chemical Engineering Science*, 52: 2011–2022.

Powell, M.S. 1991. The effect of liner design on the motion of the outer grinding media in a rotary mill. *Int. J. Min. Process*. 31: 163–193.

Powell, M.S., Weerasekara, N.S., Cole, S., LaRoche, R.D. and Favier, J. 2011. DEM modeling of liner evolution and its influence on grinding rate in ball mills. *Minerals Engineering*. 24: 341–351.

Qiu, X., Potapov, A., Song, M., and Nordell, L. 2001. Prediction of Wear of Mill Lifters Using Discrete Element Methods, In *International Autogenous and Semiautogenous Grinding Technology*, Edited by D.J. Barratt, M.J. Allan, and A.L. Mular. Vancouver, B.C. Canada.

Rajamani R.K. 2007. Simulation of Charge Transport in the Pulp Lifter of a Semi Autogenous Grinding Mill. In *Proceedings of the Discrete Element Methods 07*. Brisbane, Australia.

Rajamani, R.K., Mishra, B.K., Venugopal, R., and Datta, A. 2000. Discrete element analysis of tumbling mills. *Powder Technology*. 109: 105–112.

Rajamani, R.K., Mishra, B.K., Joshi, A. and Park, J. 2003. Two and Three-Dimensional Simulation of Ball and Rock Charge Motion in Large Tumbling Mills. In *Discrete Element Methods*, Edited by B.K. Cook and R.P. Jensen. Virginia: American Society of Civil Engineers.

Rajamani R.K., Alkac D., Delgadillo J.A., Kumar P., Page D., Fillion M., and Pelletier S. 2011. Pulp-Lifter Flow Modeling Study in Pilot Scale Mills and Application to Plant Scale Mills. In *Proceedings of the International Autogenous Grinding, Semiautogenous Grinding and High Pressure Roll Technology*, Edited by K. Major, B. Flintoff, B. Klein and K. McLeod. Vancouver, B.C. Canada.

Rajamani R.K., Callahan S., and Schreiner J. 2011. DEM Simulation of Mill Charge in 3D via GPU Computing. In *Proceedings of the International Autogenous Grinding, Semiautogenous Grinding and High Pressure Roll Technology*, Edited by K. Major, B. Flintoff, B. Klein and K. McLeod. Vancouver, B.C. Canada.

Royston, D. 2007. Semi-autogenous grinding (SAG) mill liner design and development. *Min. and Mett. Processing*. 24: 121–132.

Sepulveda, J.E. 2012. Moly-Cop Tools public domain software.

Sherman, M., and Rajamani, R.K. 1999. The Effect of Lifter Design on Alumbrera's SAG Mill Performance: Design Expections and Optimization. In *Proceedings of the Canadian Mineral Processors Conference*, Ottawa, Canada, 255–266.

Tavares, L.M., Carvalho, R.M. 2010. A mechanistic model of batch grinding in ball mills, In *XXV International Mineral Processing Congress*, Brisbane, AUSIMM, 1: 1287–1297.

Toor, P., Perkins, T., Powell, M.S., Franke, J. 2011a. The Influence of Liner Wear on Milling Efficiency, Metallurgical Plant Design and Operating Strategies. In *MetPlant 2011*, Perth, Australia, pp. 193–212.

Toor, P., Franke, J., Perkins, T., Bird, M., Robertson, J. 2011b. Quantifying the influence of liner shape and mill filling for performance optimization. In *Proceedings of the International Autogenous Grinding, Semiautogenous Grinding and High Pressure Roll Technology*, Edited by K. Major, B. Flintoff, B. Klein and K. McLeod. Vancouver, B.C. Canada.

Tuzcu, E.T., and Rajamani, R.K. 2011. Modeling Breakage Rates in Mills with Impact Energy Spectra and Ultra-fast Load Cell Data. *Minerals Engineering*. 24: 252–260.

Venugopal, R., and Rajamani, R.K. 2001. 3D simulation of charge motion in tumbling mills by the discrete element method. *Powder Technology*. 115: 157–166.

Vogel, L., and Peukert, W. 2004. Determination of material properties relevant to grinding by practicable labscale milling tests. *Int. J. Min. Process.* 74: S329–S338.

ABOUT THE AUTHORS

Raj K. Rajamani received his bachelor's and master's degrees in chemical engineering from Annamalai University and Indian Institute of Technology, Kanpur, India. He earned his PhD in metallurgical engineering at the University of Utah. He has held a rese arch and teaching faculty position at the University of Utah since 1979 and is currently a professor of metallurgical engineering. In 2009, he received the Antoine M. Gaudin Award, presented by SME for his "seminal work in the application of discrete element methods in the modeling of charge motion in semiautogenous and ball mill grinding, and contribution to the basic science of comminution and classification." Rajamani's research also includes computational fluid dynamics of hydrocyclones, eddy current separation, and ultra fine grinding in planetary mills.

Samira Rashidi graduated from Tehran Polytechnic with a BS degree in mining engineering and an MSc degree in mineral processing. She has 5 years of experience working with Iran Mineral Sector Strategy, Iran Mine Standards and Criteria Project, and Iran Mineral Research and Application Company focusing on mineral processing, simulation, and modeling. She is currently a PhD student at the University of Utah. Rashidi is conducting her thesis in the field of comminution under the supervision of Raj Rajamani.

Nikhil Dhawan currently works as a graduate research assistant at the University of Utah. He is pursuing his PhD in metallurgical engineering with Raj Rajamani. He holds a BS degree in metallurgical engineering from Punjab Engineering college, India.

The Future of Comminution

Jens Lichter
Lyntek Incorporated, Lakewood, Colorado, United States

ABSTRACT: The challenges facing future comminution circuits are primarily centered on the reduction of the operating cost and the carbon footprint associated with the comminution process. This challenge is being addressed on numerous fronts. This includes more intelligent circuit designs that replace traditional comminution machines with more efficient options, fine tuning of existing comminution processes, reductions in the use of steel grinding media and wear liners, new emerging comminution technologies for the pretreatment of ore, improving circuit stability and control, and the reduction of total tonnages treated by sorting or pre-concentration of ores. A holistic approach that considers improvements in all of these aforementioned areas will ultimately drive the development of the comminution circuit of the future.

INTRODUCTION

The challenges facing future comminution circuits are primarily centered on the reduction of the operating cost, and the carbon footprint associated with the comminution process. Additional environmental challenges will include the reduction of water usage; however, the urgency to eliminate water from the comminution process will not be fully realized until a dry beneficiation process becomes feasible for metalliferous ores. Solutions are being developed on numerous fronts and research continues on new comminution processes, but the path from concept to implementation is long.

Therefore, the primary drivers in the development of future comminution circuits can be listed as follows:

1. Minimizing energy requirements—Comminution remains one of the primary contributors to the global consumption of energy, as well as being a significant cost contributor to the beneficiation process. The reduction of energy consumed is therefore one of the strongest drivers in the development of future comminution circuits.
2. Reduction in the usage of steel grinding media and wear liners—Steel media and metallic liners contribute significantly to the total carbon footprint associated with the comminution process, as well as the operating costs. Reducing wear and media consumption is therefore a key opportunity.
3. Circuit stability and control—Tonnage variations contribute significantly to inefficient processes. Equipment selections, as well as advanced control strategies, can help to minimize tonnage fluctuations, thereby significantly improving overall circuit efficiencies.
4. Optimization of existing comminution machines—The advancement of multiphysics simulation tools is opening up opportunities to optimize comminution processes at a

fundamental level. Generic solutions will make way to custom configurations, designed to optimize equipment for a specific duty.

5. Emerging comminution processes—These may be applied to pre-treat ores to reduce energy requirements and improve coarse ore liberation.
6. Ore sorting and pre-concentration—These processes will potentially reduce the total tonnages being treated, thereby reducing energy requirements per ton of metal produced.
7. Scale—As grades become poorer, the tonnages being treated increase. Machine sizes required to treat the larger tonnages will therefore continue to increase in size and capacity. This pressure will be felt not only on SAG and Ball mills, but also on HPGRs and stirred mills. Limits on machine sizes, as well as drive requirements, will ultimately be determined by the manufacturing limitations of foundries and fabrication facilities. These limitations may be solved by making changes in the design and the manufacturing methods of comminution machines.

These opportunities are discussed in more detail below, but it should be recognized that improvements in the overall efficiency of comminution processes can be found in many areas. The best approach will therefore be a holistic approach that considers the potential for improvement in all of these areas, and is used to design a circuit that attempts to optimize every aspect of the comminution circuit of the future. Improvements in the overall comminution circuit efficiency of up to 40% to 50% appear to be within reach; this should be the ultimate objective of the comminution circuit of the future.

ENERGY REDUCTION

Undoubtedly, the primary driver in the development of comminution circuits is the need to reduce the energy requirement of the comminution process. As important as the reduction of operating costs is, the desire for sustainable development and the reduction of the carbon footprint. The reduction of the carbon footprint (or the total energy requirement) of the comminution process takes into consideration not only the electrical power required for the comminution equipment, but also the energy required to support the process. Most notably, this is the energy required to produce the grinding media and the wear liners, but secondary equipment, including conveyors and classification equipment, are also significant contributors.

Approximately 4% to 7% of global energy use is attributed to the mining and minerals industries. One of the primary contributors is the comminution process consuming an estimated 0.5% to 1% of the total global energy consumption (Rábago et al. 2001; US DOE 2001; Daniel et al. 2010, 2011).

The per capita consumption of minerals worldwide is increasing and new deposits are trending towards harder ores that require finer grinding to achieve liberation. The fraction of global energy use attributed to comminution will therefore increase unless step changes are made to comminution circuits.

In recognition of the significant role played by comminution in the world's energy consumption, numerous efforts are underway to improve the overall efficiency of comminution circuits. A brief overview of some of these activities is provided below.

Comminution Circuits

One of the most active areas of investigation into the reduction of overall comminution energy is the redesign of the comminution circuit, with an emphasis on the use of more efficient comminution machines in place of traditional Semi Autogenous Grinding (SAG) and ball mills.

Alternate comminution machines are constantly under review but the gestation period from concept to a fully functional large tonnage comminution device is long. A timeline of 10 to 20 years or more for product development, and the eventual adoption of new technologies into comminution circuits, is typical. Fortunately, well-established technologies exist that have demonstrated considerable opportunity for substantial reductions in the total energy required for comminution. Primarily, these are High Pressure Grinding Rolls (HPGRs) and stirred mills.

High Pressure Grinding Rolls

The successful development of the HPGR for use with hard abrasive ores has provided a viable alternative to SAG mills for many circuits. The estimated reduction in the specific energy requirement of HPGR/ball mill circuits vs. SAG/ball mill circuits has been estimated at 10 to 25% (von Michaelis 2009, Marsden 2008, Rule et al. 2008). In addition to the direct energy saving, the elimination of SAG mill grinding media would contribute to an additional reduction of 10% to 15% of the total energy consumption, for a combined reduction in the carbon footprint of 26% to 39% (Daniel et al. 2010). HPGR availabilities and operating costs are now considered to be at least comparable with those of SAG mills for many applications.

Stirred Milling Technologies

Stirred milling technologies fall into two primary categories: screw type vertical mills, such as Metso's Vertimill™ and Nippon Eirich's Tower Mill™, and stirred media mills, which fluidize the grinding media, such as Xstrata's IsaMill™, Metso's SMD™, FLSmidth's VXPmill™, and Outotec's HIGMill™. These technologies are more commonly applied to regrind applications, but are being evaluated as possible ball mill replacements. These mills are generally recognized as being 25% to 50% more efficient than a ball mill in secondary or regrind applications, but are relatively untested as a replacement for a primary ball mill. The screw type vertical mills have an inherent advantage over stirred media mills in this role; they are generally more capable of handling coarse feeds and are more forgiving of changes in the feed conditions.

HPGR, Stirred Mill Circuits

Circuits consisting of HPGRs followed by screw type vertical mills have been suggested, and modeled in excess of 10 years with the promise of overall improvements in the energy consumption of the comminution circuit of up to 45% (Valery and Jankovic 2002). Figure 1 shows a typical circuit concept.

The challenges faced in replacing ball mills with screw type vertical mills, or stirred media mills, are primarily scale and the ability to accept coarse feed.

The largest ball mills currently built have an installed power of 24MW, compared with a maximum installed size of 3.4MW for the Vertimill™ and 3.0MW for the IsaMill™. Taking into consideration the improved efficiency of stirred mills, it would require approximately 5 to 7 mills

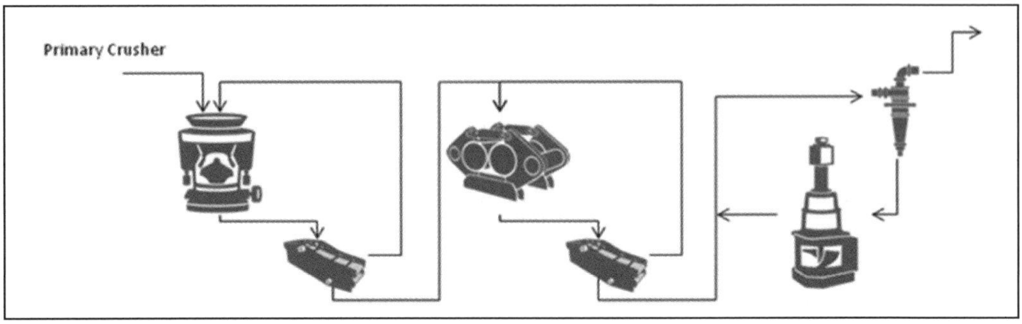

Figure 1. Typical high-efficiency circuit concept. Primary crushing followed by HPGRs and stirred mill technologies.

to replace a single large ball mill for high tonnage circuits. Both of these milling technologies are therefore being actively developed to increase the size and power draw of individual machines.

Feed size is another primary consideration. Stirred mills have primarily been used as secondary or regrind mills, with a few notable exceptions such as Red Dog and McArthur River, which utilize SAG mill/stirred mill circuits and incorporate careful control of the top size being fed to the mills. Screw type vertical mills are better able to handle relatively coarse feeds (~ minus 8mm) than stirred media mills, which prefer minus 1mm feed or less. The coarser feeds can be prepared in a single stage closed circuit HPGR, but finer feeds will require two stages of HPGRs. There are a number of efforts underway to evaluate HPGR/stirred mill circuits, but data is not readily available, and it is not clear what efficiency advantage stirred milling technologies will have over ball mills when utilized as a replacement for primary ball mills. When operating as a replacement for primary ball mills, a reduced efficiency advantage of ~20% is expected, compared to standard ball mill configurations.

The option of incorporating two stages of HPGRs to prepare feed for screw type vertical mills, or stirred media mills, would eliminate much of the concern over the effect of coarse feed on the efficiency and availability of the mills. Pilot tests have been run on platinum ore to evaluate two stages of HPGRs followed by an IsaMill (Ayers et al. 2008). In another study by Drozdiak (2011), pilot scale tests on a copper nickel sulfide ore compared specific energies for cone crusher/ball mill, HPGR/ball mill and HPGR/stirred mill circuits. The T80 to the ball mill after cone crushing was 2.12mm, the T80 to the ball mill after single stage HPGR was 1.6mm, and the T80 to the Netzsch mill after two stages of HPGRs was 0.34mm. The HPGR/ball mill circuit was shown to be ~8% more efficient than the crusher/ball mill circuit, and the HPGR/stirred mill circuit was ~9% more efficient than the HPGR/ball mill circuit. While it is probable that these circuits were not optimized, the work does show the feasibility of this type of circuit, as well as the potential for significant reductions in the total energy usage.

Good data from operating plant utilizing HPGR/stirred mill circuits is still scarce and generally not in the public domain. Much emphasis is therefore placed on simulated circuits in the evaluation of circuit options. Caution is required in the use of data from simulated circuits; the empirical and phenomenological models used for the stirred mills are not able to adequately calculate the effect of coarse feed on mill performance. As data becomes more available, the combination of HPGRs with stirred mills will become more common.

Additional Opportunities to Optimize Comminution Circuits

Kawatra and Eisele (2005) suggested that two of the main causes of energy loss in grinding circuits are underutilization of equipment due to process bottlenecks and overgrinding due to material being retained in the circuit after it has reached target size. Improving classification efficiency, as well as circuit stability and control, directly address these areas. Reducing media and liner consumption provides another opportunity to further reduce the carbon footprint of comminution circuits.

Classification Efficiency

Improving classification efficiency primarily reduces overgrinding, which is the wasted application of energy to provide a product that is finer than it needs to be for efficient separation. The most commonly designed cyclone classification circuits are based on single stage classification, with the cyclone O/F density being prescribed by downstream processes. Capital cost is the primary consideration after achieving the correct product P80. Alternative circuit options that remove some of the traditional constraints are, however, available in the design of classification circuits. McIvor (2012) describes six case studies indicating potential improvements in the comminution efficiency of ball mills circuits of 3% to 30%. The classification circuit changes described utilize a number of strategies; this includes the use of high efficiency cyclones, improving water balances, optimizing circulating loads, and using two-stage classification.

Another area worthy of consideration for low tonnage applications is the use of screens. Screen technologies are currently being developed in parallel to milling technologies, and polyurethane screen panels with apertures as fine as 75 microns are available. Screens have two significant advantages over hydrocyclones. Screens control the top size of the product, thereby minimizing recovery losses due to oversized particles in the product, while maintaining coarser P80s in the final product. Secondly, hydrocyclones have an inherent characteristic in that a significant fraction of final product (up to 50%) will report to the cyclone underflow (and therefore the mill feed) resulting in significant over-grinding. Screens can reduce this misplaced material with the potential of almost entirely eliminating it. Improvements of up to 50% in overall circuit performance have been reported, but these are unusual cases. Improvements of 20% are, however, feasible as a direct result of improved classification efficiency and the coarsening of the final product.

Circuit Stability

SAG/ball mill circuits, in particular, are prone to substantial variations in throughput as the ore characteristics change. As ore changes from softer to harder, the circuit will progress from being ball mill limited to SAG mill limited. This results in challenges not only in the balancing of the comminution circuit, but also causes disturbances in the remainder of the concentrator leading to recovery losses and therefore a higher energy demand per ton of metal produced.

The use of HPGRs in place of SAG mills provides an additional advantage with regard to circuit stability. HPGRs are essentially constant tonnage machines. Changes in the ore properties primarily affect the power draw of the HPGR, but tonnages remain far more consistent than would be the case with SAG mills. This inherent stability in the circuit throughput will significantly simplify process control and will result in overall improvements in average comminution efficiency as well as consistency in the recovery of minerals.

The use of variable speed drives for ball mills provides another opportunity to improve overall comminution circuit efficiencies. The cost of variable speed drives is reducing and has become an affordable option for smaller ball mills. Variable speed control is also inherent with ring motors (gearless drives), which are almost standard for large mills. Fixed speed ball mills are largely constant power devices and changes in the power draw can only be affected over time by reducing or increasing the ball charge. The ability to control the speed of ball mills to control grind for normal short term fluctuations will significantly reduce over-grinding; this provides another opportunity to substantially reduce the overall power requirement of the comminution circuit.

The use of advanced instrumentation and controls, including expert controllers with predictive capability, has been shown to improve overall circuit efficiency by as much as 5% to 15%. As mill sizes increase, and the number of parallel processing streams decreases, and the opportunity to incorporate expert control systems to comminution circuits increases significantly.

Grinding Media (Secondary Energy Consumption)

Steel media and wear liners are estimated to add over 50% to the total carbon footprint of the comminution circuit and can have an even greater influence on the operating costs (depending on the cost of power and delivered steel media). The use of HPGRs eliminates the grinding media consumed by the SAG mill, however, the ball mills or vertical stirred mills still require steel media.

AG/pebble mills offer an alternative to SAG/ball mill circuits, eliminate the need for steel media, and have significantly lower rates of liner wear. Recent installations, such as the Boliden Aitik expansion (Markstrom 2011), have shown that large format AG/pebble mills are a viable alternative to conventional SAG and ball mills. The Boliden circuit utilizes two 38' × 45' AG mills and two 30' × 38' pebble mills, in closed circuit with screw classifiers. These mills process up to a combine tonnage of 6,000 tph of copper ore. An AG/pebble mill circuit was selected in preference of a SAG/ball mill circuit after trade off studies indicated that the operating cost for an AG/pebble mill circuit would be $45 million less per annum, for an increase in the capital cost of $15 million. Pebble media generation for the Boliden AG and pebble mills was estimated using multiphysics tools. Predicted operating cost estimates, as well as media generation rates, were confirmed in the field.

The primary concerns with AG/pebble circuits have always centered on the susceptibility of AG mill tonnage to changes in the ore properties. Industries such as the North American iron ore industry, as well as the Scandinavian iron ore and copper industries, have however demonstrated convincingly that if there is a desire to do so, these circuits can be operated efficiently and produce substantial cost benefits to the user.

An interesting alternative to current HPGR/ball mill circuit concepts would be the evaluation of HPGR/pebble mill circuits. Special considerations would need to be given to the generation of pebble media for the pebble mills, by scalping the feed ahead of the HPGR. This type of circuit would combine the efficiency and fully autogenous operation of an HPGR with an autogenous pebble mill, and eliminate the need for steel grinding media. The incorporation of HPGRs would also eliminate most of that concern of the AG mills susceptibility to tonnage variations due to changes in the hardness of the feed ore. However, a brave soul would be required to be the first to install such a circuit!

Feed Optimization

The optimization of crusher or mill feed by improving fragmentation at the mine is well documented and has demonstrated substantial benefits to SAG/ball mill circuit capacities. These technologies continue to be developed. Products such as Orica's Vistis™ have more than twice the relative bulk strength of ANFO and therefore allow significantly higher energy intensities. These high energy explosives, combined with new blasting techniques, allow the controlled application of powder factors of $4kg/m^3$ and above. At these powder factors, reductions in the total energy requirement of SAG mill applications of 35% to 50% are indicated (Thompson 2012, Ziemski 2011).

Finer blasts also provide opportunities to HPGR circuits. Increased fines generation, and a finer top size in the feed, creates opportunities to reduce the product size from single stage HPGR circuits, which will benefit the subsequent grinding circuit.

DEVELOPMENT OF NEW COMMINUTION TECHNOLOGIES

There are few, if any, new technologies on the horizon, so it is best to refer to emerging technologies. The "new" machines currently being brought to market for use in metalliferous comminution circuits, either as production or pilot scale test units, build on existing concepts that have long histories. The two primary categories would be (1) machines with long operational histories in industrial minerals that are being migrated to metalliferous applications and (2) novel (currently unused) methods of comminution using predominantly non-mechanical methods of transferring energy to the ore.

Migrated Technologies

Because of a growing understanding of the need to be able to provide higher efficiency comminution options (other than ball mills), especially for regrind applications, a number of new machines have come to market. Well-established mills, such as the Metso's Vertimill™ and SMD™ and Xstrata's IsaMill™, are now being augmented by Outotec's HIGMill and FLSmidth's VXPmill, among others (Wang and Forssberg 2003). Collectively, these mills will undoubtedly be developed to be able to handle more tons, coarser feed, and to do so with acceptable media costs and maintenance requirements. Their use in future comminution circuits has been discussed above; application can be expected to increase.

Emerging Technologies

The most interesting of the emerging technologies fall into the category of augmenting existing comminution processes, rather than necessarily replacing them. While not new concepts, these processes are seeing renewed interest as the drive for more energy efficient comminution continues. The two most prominent augmentation techniques are High Voltage Electric Pulse, and Microwave radiation. These techniques are capable of independent size reduction, but they are currently not efficient in this role. Rather than replacing existing comminution machines, these techniques are predominantly being tailored to ore pre-treatment, with two specific objectives. Firstly to weaken rock by creating fractures that will result in a "softer" ore for subsequent comminution steps. And secondly, to specifically enhance liberation characteristics so that liberation occurs at coarser product sizes. The latter is achieved as a result of preferentially breaking along grain boundaries rather than across grains, as might be the case with mechanical fracture mechanisms. The use of Ultrasonic energy is also being investigated as a means to enhance mechanical fracture rates.

High Voltage Electric Pulse

High Voltage Electric Pulse devices such as the selFrag (selFrag AG, Switzerland) batch unit have been used to pre-treat ores with high voltage electric pulses. The feed size to the electric pulse units is relatively coarse, 10mm to 45mm, which makes the treatment well suited to HPGR feeds. Data currently available is for batch treatment of ores, but a pilot scale continuous unit is being developed. In work published by Wang et al. (2011), four parallel tests were conducted on ores subjected to electric pulse energy, and conventional crushing. Electrical energy applied for the electric pulse pre-treatment was of the order of 1 kWh/ton to 3 kWh/ton. Both JKRBT (JK Rock Breakage Tester) and BWI_R (Bond rod mill work index) tests were run on the selFrag and non-selFrag treated samples. Two parameters were compared to estimate the level of ore "softening" that resulted from the selFrag treatment

The first was the A*b parameter used to characterize the ore hardness in the JKSimMet comminution model. The second was the Bond Rod Mill Work Index. The testing showed a 9% to 52% increase in the A*b value (a higher value is an indicator of a softer ore) and a 0% to 24% reduction in the Bond rod mill work index.

Initial work has also demonstrated that improvements in recovery can also be achieved as a result of electric pulse per treatment, but data is specific to relatively coarse size fractions and it is unclear at what level these benefits will be realized in normal beneficiation processes.

When considering the "softening" effect, it is important to understand that this softening of the ore is not a fundamental change in the material properties of the rock, but an indicator of micro cracking along primary grain boundaries. As a result, tests such as the JKRBC, which specifically investigate energy required to first fracture, show a significantly larger effect than tests that require multiple fracture events such as a Bond test. It is reasonable to assume that the finer the product required, the less the softening effect will be. The concept of ore softening therefore has to be explored carefully, and be evaluated specifically to the final product size required for the process. The use of a softening factor in a simulator should be avoided unless the correct reduction ratio is used.

Microwave Radiation

Another area of research is the use of microwave radiation to weaken ores. Most early research concentrated on the feasibility of the use of continuous microwave emitters to assist comminution (Kingman et al. 2000, Vorster 2001, Wang and Forrsberg 2000), and were often combined with quenching to enhance the effect. These studies showed that substantial changes in the properties of ores could be achieved with microwave treatment. Reductions in the Bond Rod Mill Work Index of treated of up to 70% were observed. However, these results were obtained using continuous exposure to microwaves, and applied relatively massive amounts of energy, well in excess of normal energy consumed in comminution processes.

Recent research has concentrated on the use of pulsed microwave systems to affect substantial changes in the apparent hardness of minerals at modest energy requirements (Kingman et al. 2004). At the core of this work is the understanding that the power density is more important than the total energy applied, and that short bursts of very high power exposure can significantly weaken rock. Initial indications are that microwave treatment can significantly influence both the strength and liberation behavior of ores at economic energy inputs of less than 1 kWh/t. Multiple hardness indices were quoted, including Point Load testing, JKMRC Drop Weight testing, and batch ball mill grindability tests. The trends are similar to those discussed for the high voltage electric pulse testing in that substantial "weakening" of the ore is seen with the use of microwave pretreatment. The benefit is, however, most significant for coarse size fractions, with the benefits reducing as the product becomes finer. A drop in the uniaxial compressive strength of up to 50% was noted (a 100% "softening" of the ore at initial fracture), the A*b value derived from a drop weight test increased by 41%, and the breakage rate in a batch ball mill increased between 67% and 10%. Of note was the result from the batch ball mill tests; there was an indication of a diminished weakening of the ore as the product size reduced from a mean size of 18mm (67%) to 4mm (10%).

Degree of liberation was also measured using QEM*SEM analysis. This analysis showed significant improvements in liberation for the coarse size fraction (+500 microns) from 31.8% liberation for the untreated sample to 69.2% for the treated sample, but insignificant changes in the degree of liberation for the balance of the size fractions.

Ultrasonic Enhancement of the Comminution Process

Ultrasound assisted comminution relies predominantly on ultrasound augmented mechanical comminution devices. One example is the work by Gaete-Garreton et al. (2000, 2003) which applied ultrasound energy to an HPGR concept. In this design, the rolls apply mechanical stress to the rock, but one of the rolls is ultrasonically activated. The tests have demonstrated energy savings of up to 6% in HPGR type machines with the use of ultrasonic enhancement, but more interestingly, have also reported a very significant reduction in the wear rate of the rolls with roll wear rates being reduced by ~50% with the application of ultrasonic energy. This work, as well as the work by Yerkovic et al. (1993) has at least one significant departure from conventional HPGR applications. The gap in these laboratory devices is typically set at a fraction of the top size of the feed, and breakage is therefore more single-particle than packed bed. It was not clear from the publications whether any work on the application of ultrasonic enhancement of roll crushers was ever attempted for a packed bed scenario, where the damping of the ultrasonic waves by the bed is likely. HPGRs operate with a packed bed with the large majority of applied forces being antiparticle. It is therefore not clear to what extent these results would transfer to operating plant.

Optimizing Comminution Machines (aka Improving the Mouse Trap)

Multiphysics Modeling

Refinement of current comminution technologies is taking place as a result of both traditional incremental machine improvements and by the application of a rapidly developing suite of multiphysics tools. The latter, in particular, deserves additional discussion. Multiphysics tools are machine independent and represent the first true microscale model of a comminution machine.

The primary multiphysics tools include Discrete Element Method (DEM), Computational Fluid Dynamics (CFD), and fracture mechanics. These tools can be augmented with phenomenological models such as the Population Balance Model (PBM). Models such as these provide the ability to model almost any comminution device at the level of the particle–particle and particle–machine interactions. Improved code, rapidly increasing availability of multi-core computers, and clusters has significantly advanced the size and complexity of problems that can be analyzed. These tools have been used in dozens of studies to improve crusher and mill performance over the last decade.

DEM modeling provides a complete history of all impact events in a comminution device, typically limited to some minimum particle size. Fracture mechanics can then be applied to develop a probability of breakage (or a breakage rate) which is a unique function of the machine, how it is operated, the ore breakage characteristics, and the feed size of the ore. These models adequately consider both the volumetric capacity, as well as the power applied, by a crusher or mill. They therefore permit the detailed optimization of a comminution machine for a specific ore and duty.

There are limitations; these relate to the size of the problem to be modeled and the complexity of liquid/solid interactions. Where the size of the problem (computational effort required) is too great to look specifically at single particle breakage events, the DEM models can be combined with PBM models to provide the solutions required (Potapov et al. 2007).

Since the initial successful application of DEM in SAG mill liner design for the Alumbrera SAG mills (Sherman et al. 1999), where a 15% improvement in mill capacity was reported for liners designed with the aid of DEM modeling, the use of DEM for the design of mill liner profiles has become almost standard. The initial efforts concentrated on designing a liner profile that would provide a good charge trajectory in a mill. These early uses have now been augmented to include wear modeling. A particularly interesting development is the ability to evaluate the effect that liner profiles have on breakage rates, specifically of the coarse and critical size material in the SAG mills. Initial indications are that improvements of 5% to 25% are feasible for SAG and ball mill operations (Herbst and Lichter 2006). Mill discharge systems have also been extensively modeled to identify and eliminate volumetric restrictions, and to improve wear life (Lichter et al. 2011).

Multiphysics modeling has also been applied to crushers and HPGRs. A fully functional crusher model is available that can look in detail at the effect of crusher cavity design, crusher speed, and closed side setting on crusher performance (Lichter et al. 2008).

Figure 2 shows a sectional view of an HP100 crusher model using both DEM and fracture mechanics. Of particular interest with these models is the ability to evaluate system parameters such as crusher feed arrangement, crusher speed, and crusher throw, which are not variables that can be easily manipulated in the field.

A Multiphysics model of the HPGR is described in the work by Herbst et al. (2011). Like the crusher model, the effects of all primary design and operating parameters can be evaluated in a simulated environment. These models include task-specific ore characterizations tests developed from existing testing techniques.

The use of these tools goes beyond the optimization of existing installations. A significant advantage of a simulated environment is that the normal plant variability can be eliminated, thereby allowing an independent look at each of the operating parameters. In addition, simulations permit the optimization of machines during the design stage, and the detailed evaluation of multiple machine design parameters that are not available with traditional design tools.

Improving Crusher Performance

A possible alternative to HPGRs in comminution circuits (primarily for lower tonnage applications) is the use of multistage crushing to generate a fine feed suitable for subsequent ball or stirred milling. The current generation of tertiary cone crushers all rely on inter-particle crushing to significantly increase fines generation. The development of the new Sandvik Vibrocone™ crusher, which

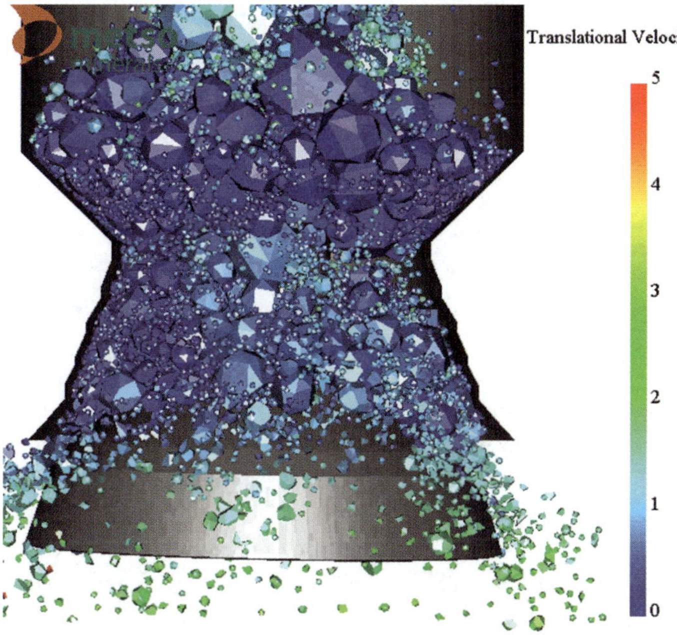

Figure 2. Sectional view of an HP100 DEM, breakage simulation

uses unbalanced weights to provide the crushing action, has the potential to significantly increase fines generation when compared to other crushing technologies.

Improving Ball Mill Efficiency

In parallel with the research and testing on the use of stirred mills as a replacement for ball mills, is the re-evaluation of ball mill efficiency. It has been understood for many decades that ball mill efficiency is a function of mill speed, charge level, and media size. The work published by Plavina and Clark (1986), together with data collected at Bougainville, showed convincingly that higher speed and higher charge levels do not necessarily translate into more tones and that a deeper understanding is required of the effect that operating variables have on mill performance. The Bougainville data indicated that for a change from a 41% charge level and 81% of Critical Speed (Cs), to a 32% charge level and 74% of Cs, the specific energy (kWh/ton) required to grind to an equivalent product was reduced by ~12%. Similarly it has been shown that changes in media size can significantly improve ball mill efficiencies and improvements of 5% to 15% in the mill efficiency are feasible.

Current efforts to improve ball mill operating efficiency include operating the mills at slower speeds, reduced charge levels, and with more care taken in the selection of media size and liner profiles. If the same care is taken when preparing ball mill feed, as is required when preparing feed for a stirred mill (both vertical screw and stirred media), then significant changes are possible to liner design and media size. Combined with slowing down the mill, and operating at lower charge levels, these factors will likely reduce the energy consumption of ball mills by ~20% or more.

Ball mills sized and operated to maximize efficiency will operate at lower tonnages than mills operated to maximize throughput. Ball mills optimized for efficiency will therefore be larger and have a higher capital cost than mills optimized for throughput. Adoption of ball mills optimized for efficiency rather than capital cost will therefore require either adequate field data or the broader application of models that that allow the evaluation of ore breakage parameters as a function of feed size, media size, mill speed and mill charge level. The question remains: Once installed, will the mills be operated to optimize efficiency or will the traditional pressures to maximize tons come into play?

ORE SORTING AND PRE-CONCENTRATION

Decreasing grades of ore deposits, combined with the need to process increased tones, have resulted in increased efforts to develop ore sorting techniques that can be applied to the coarse ore feed to the mill. If a significant fraction of the feed to the concentrator can be eliminated after primary or secondary crushing, then the overall energy requirements per ton processed from the mine can be reduced significantly. There are some "hidden" costs associated with the use of ore sorting, however, and these include lost revenue due to mined mineral being discarded and the additional cost of hauling and dumping the reject rock. Coarse ore haulage carries a significantly higher price tag than the pumping of slurries and can run as high as $2 per ton.

An alternative to pre-sorting is the early concentration of ores. Coarse separation devices, such as pressure jigs and coarse ore flotation, can be used to concentrate ores after an initial fine crush or coarse grind. Gangue material can then be rejected ahead of the fine grinding step, which consumes most of the power and grinding media. The gangue material can be pumped to the tailings facility, significantly reducing the cost of eliminating rejects.

This type of circuit would require an ore that allows separation and gangue rejection, at a coarse particle size, but can be considered as a viable circuit option if the mineralogy allows it.

WATER ELIMINATION—DRY COMMINUTION CIRCUITS

Water conservation is as critical an environmental issue as the conservation of energy and the reduction of the carbon footprint, and is particularly relevant to concentrators in arid areas. The use of HPGRs opens up new opportunities for dry grinding circuits as well, and can be used in conjunction with dry ball mills and high efficiency classifiers.

Another possibility is the development of the roller mill as a replacement for ball mills. Roller mills are recognized as being significantly more energy efficient than ball mills, but have not been suitable for abrasive ores. Lessons learned in ruggedizing HPGRs could provide the key to developing roller mills suitable for abrasive ores.

Dry comminution circuits have the possibility of being more efficient than wet comminution circuits primarily because of the efficiency of dry classification and the ability to remove finished product as it is created. Dry mills also have significantly reduced media and liner consumption rates, further adding to their benefit as regards the total energy equation. Moisture in the feed is, however, the controlling variable as the need to dry feed will rapidly eliminate any energy efficiency advantages.

In order to gain the most tangible benefit from the elimination of water from the comminution process, dry beneficiation processes must be developed for metalliferous ores. Research continues, but this remains the predominant challenge in the elimination of water from the beneficiation process.

SUMMARY

A broad array of opportunities exists, to significantly alter the comminution circuits of the future. To mention a few, these opportunities include reconfiguring circuits to incorporate higher efficiency comminution options, optimizing existing comminution machines, optimizing feed size preparation, pre-treatment of ores using emerging technologies, optimizing circuit stability and control, and the reduction of tons treated by pre-concentration of the feed ore.

Considerable work is still required before all of these circuit concepts are fully realized. Improved efficiency comminution circuits will typically require an increase in the capital cost, and the design risk associated with any new processes. In order to accelerate the widespread implementation of these opportunities, new modeling tools and ore characterization tests will need to be developed, and adopted, so as to better define the benefits at the design stage. The ability to justify the operating cost benefits of new comminution circuit options will accelerate the adoption of new processes. Failing that, the normal process of one or two trail installations followed by years of evaluation will ensue.

The potential prize is large and the absolute need to reduce the carbon footprint of the comminution process cannot be ignored. Overall reductions in the energy requirement of the comminution process of up to 30% to 50% appear to be attainable, with similar reductions in the operating costs. Pilot, or production scale, tests are underway on most of the options discussed; the future of comminution will be interesting indeed!

ACKNOWLEDGMENTS

The author expresses his gratitude for the assistance and input received from Robert Dunne and Brian Flintoff.

REFERENCES

Ayers, C., Knopjes, L., and Rule, C.M. 2008. *Coarser feed applications of MIG IsaMilling.* MEI Comminution 08 Conference, Falmouth, Cornwall, UK.

Daniel, M.J. and Lewis-Gray, E., 2011. *Comminution efficiency attracts attention.* Bulletin of the Coalition for Eco-Efficient Comminution (CEEC), presented at AusIMM, Nov 2011.

Daniel, M., Lane, G., and McLean, E. 2010. *Efficiency, economics, energy and emissions—Emerging criteria for comminution circuit decision making.* XXV International Mineral Processing Conference (IMPC), 2010 Proceedings. Brisbane, Australia, September 2010.

Drozdiak, J.A. 2011. *A pilot scale examination of a novel high pressure grinding roll/Stirred mill comminution circuit for hard rock mining applications.* Masters Thesis, University of British Columbia.

Gaete-Garretón, L.F., Vargas-Hernandez, Y.P., and Velasquez-Lambert, C. 2000. Application of ultrasound in comminution. Ultrasonics 38, pp. 345–352.

Gaete-Garretón, L.F., Vargas-Hernandez, Y.P., Chamayou, A., Dodds, J.A., Valderama-Reyes, W., and Montoya-Vitini, F. 2003. *Development of an ultrasonic high-pressure roller press.*, Chem. Eng. Science 58 (2003), pp. 4317–4322.

Herbst, J.A., and Lichter, J.K. 2006. *Use of multi-physics models for the optimization of comminution operations.* Advances in Comminution, Editor: S. Komar Kawatra. Published by Society for Mining, Metallurgy & Exploration.

Herbst, J.A., Mular, M.A., Pate, W.T., and Qiu, X. 2011. *Detailed modeling of an HPGR/HRC for prediction of plant scale unit performance.* Proceedings 2011 SAG Conference, Vancouver, BC, Canada, Paper 46.

Kawatra, S.K., and Eisele, T.C. 2005. *Optimization of comminution circuit throughput and product size distribution by simulation and control.* Final technical report, Michigan Technological University.

Kingman, S.W., Vorster, W., and Rowson, N.A. 2000. *The effect of microwave radiation on the processing of Palabora Copper Ore.* Min. Eng. 13 (3), pp. 313–327.

Kingman, S.W., Jackson, K., Cumbane, A., Bradshaw, S.M., Rowson, N.a A., and Greenwood, R. 2004. *Recent developments in microwave-assisted comminution.* Int J. Miner. Process 74(2004), pp. 71–83.

Lichter, J., Lim, K., Potapov, A., and Kaja, D. 2008. *New developments in cone crusher performance optimization.* Conference proceedings, Procemin 2008. pp. 71–80.

Lichter, J.K., Suazo, M., Noriega, R., and Murariu, V. 2011. *The application of multiphysics models for the design of mill discharge systems.* Proceedings 2011 SAG Conference, Vancouver, BC, Canada, paper 55.

Markstrom, S. 2011. *Commissioning and operation of the AG Mills at the Aitik Expansion Project.* Proceedings 2011 SAG Conference, Vancouver, BC, Canada, paper 62.

Marsden, J.O. 2008. *Energy efficiency and copper hydrometallurgy*, Hydromet. Phoenix, AZ.

McIvor, R.E. 2012. *Plant performance improvements using grinding circuit "classification system efficiency."* Proceedings, 2012 SME Annual Meeting, Seattle, WA, Pre-print 12-014.

Plavina, P., and Clark, G. 1986. *The selection, comissioning, and evaluation of a large diameter, variable speed ball mill at Bougainville Copper Limited.* Proceedings of the 13th CMMI Congress, Volume 4—Metallurgy.

Potapov, A., Herbst, J., Song, M., and Pate, W. 2007. *A DEM-PBM fast breakage model for simulation of comminution processes*, Proceedings of Discrete Element Methods, Brisbane, Australia.

Rábago, K.R., Lovins, A.B., and Feiler T.E. 2001. *Energy and sustainable development in the mining and minerals industries.* Special report commissioned by the Mining Minerals and Sustainable Development(MMSD) project of the International Institute for Environment and Development (IIED).

Rule, C.M., Minaar, D.M., and Sauermann, G.M. 2008. *HPGR—Revolution in Platinum?* Paper presented to SAIMM Platinum Conference—Platinum in Transition, Sun City, Botswana, October 6–9, 2008.

Sherman, M., and Rajamani, K. 1999. *The effect of lifter design on Alumbrera's SAG Mill Performance: Design expectations and optimization.* Paper #17, Mineral Processors Conference, Ottawa, Canada.

Thomson, S. 2012. Controlling fragmentation. *International Mining.* pp. 80–84.

U.S. Department of Energy. 2001. *Mining—Industry of the future*, Office of industrial technologies, office of Energy Efficiencyand renewable Energy, doe/GO-102001–1157.

Valery, W., and Jankovic, A. 2002. *The future of comminution.* 34th IOC on Mining and Metallurgy, Yugoslavia.

Von Michaelis, H. 2009. *How energy efficient is HPGR?* World Gold Conference 2009. The Southern African Institute of Mining and Metallurgy.

Vorster, W. 2001. *The effect of microwave radiation on mineral processing.* Ph.D. Thesis, University of Birmingham.

Wang, E., Fengnian, S., and Manlapig, E. 2011. *Pre-weakening of mineral ores by high voltage pulses.* Miner. Eng. doi:10.1016/j.mineng2010.12.011.

Wang, E., Fengnian, S., and Manlapig, E. 2012. *Factors affecting electrical comminution performance.* Miner. Eng. 34, pp. 48–54.

Wang, E., Fengnian, S., and Manlapig, E. 2012. *Mineral liberation by high voltage pulses and conventional comminution with same specific energy levels.* Miner. Eng. doi:10.1016/j.mineng2011.12.005.

Wang, Y., and Forssberg, E. 2003. *International overview and outlook on comminution technology.* Working paper, Research conducted at the Luleå Tekniska Universitet and published by the Swedish Mineral Processing Research Association.

Wang, Y.M., and Forssberg, E. 2000. *Microwave assisted comminution and liberation of solids.* Published in Mineral Processing on the Verge of the 21st Century Published by A.A Balkema, Netherlands, pp. 3–9. ISBN 905809 1724.

Yerkovic, C., Menacho, J., and Gaete, L. 1993. *Exploring the ultrasonic comminution of copper ores*, Min. Eng. Vol. 6, No. 6, pp. 607–617.

ABOUT THE AUTHOR

Jens Lichter graduated from the University of Natal, South Africa, in 1984, with a master's degree in chemical engineering. After starting his career managing pilot plant facilities for the Council for Mineral Technology, he has held various positions including managing the comminution business unit for Metso in South Africa; managing the process engineering group for Metso in York, Pennsylvania; and managing Metso's Technology Development group in Colorado. He is now a vice president at Lyntek Incorporated in Denver, Colorado. His activities have covered a broad spectrum of disciplines, including process plant design, comminution circuit design, and the development of advanced modeling tools for the mineral processing industry. Lichter has been active in projects on six continents, in both metallic and industrial minerals applications.

Separations

Innovations in Flotation Processing Technology—The Next Hundred Years

Michael G. Nelson
University of Utah, Salt Lake City, Utah, United States

ABSTRACT: Mineral flotation technology developed rapidly in the first half of the last century, with continuing innovation in the development of machines, design of circuits, and design and production of reagent continuing. Since about 1960, there has been continuing improvement, but less innovation. This paper considers the nature of and reasons for innovation in three technologies—aircraft engines, portable data storage, and flotation machines—and examines the opportunities for future innovation in mineral processing technology.

INNOVATION

Innovation is conveniently defined as the creation of better products, processes, technologies, or ideas that affect markets, governments, and society. Innovation differs from invention in that an invention may become an innovation if it is widely used, but an innovation may not be new or novel enough to be considered an invention. Similarly, innovation differs from renovation in that innovation signifies a substantial change as opposed to incremental improvements. This paper focuses on innovation as produced by technological change.

Christensen (1997) distinguishes between *sustaining* change and *disruptive* change:

> This study of technological change…revealed two types of technology change, each with very different effects on the industry's leaders. Technologies of the first sort sustained the industry's rate of improvement in product performance…and ranged in difficulty from incremental to radical. The industry's dominant firms always led in developing and adopting these technologies. By contrast, innovations of the second sort disrupted or redefined performance trajectories—and consistently resulted in the failure of the industry's leading firms.

In the author's experience, discussions among manufacturers and user of flotation equipment often return to the question of innovation, with the posing of such questions as "How big can these machines get," or "When will someone come up with a really *new* idea?"

Before considering the development of mechanical flotation technology, it is worthwhile to consider development of other technologies. The author has previously discussed the development of aircraft engines as an example of rapid innovation (Nelson 2011). **Figure 1** illustrates the presence of sustaining and disruptive change in this development. From 1920 to 1940, the development of piston-type, internal combustion engines proceeded quickly, with marked increases in engine size and engine power, but little significant change in basic technology. After jet engines were

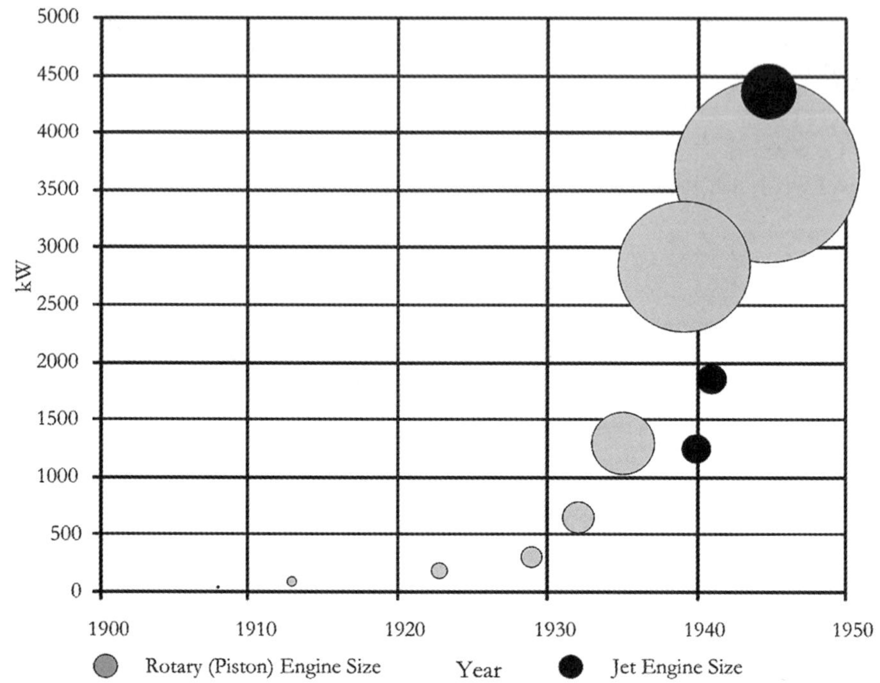

Source: Nelson 2011.

Figure 1. Sizes of piston-type and jet aircraft engines, with circle size proportional to engine weight

introduced in 1939, they rapidly surpassed internal combustion engines, producing much higher power with lower weight.

Christensen considered the development of computer disk drives in his analysis of sustaining and disruptive technology. His analysis began with the first, large disks developed by IBM between 1952 and 1956. These drives, which were about the size of a refrigerator, could store 5 Mb of information. As computers became smaller and non-integrated manufacturers entered the market, more than 100 companies began producing disk drives. Almost all of these companies failed because, according to Christensen, "Coping with the relentless onslaught of technology change was akin to trying to climb a mudslide raging down a hill. You have to scramble with everything you've got to stay on top of it, and if you ever once stop to catch your breath, you get buried."

The 'mudslide' of portable computer memory didn't stop in 1997. Figure 2 shows the development of such devices (not including external hard-disk drives) through 2004. The most recent development in USB drives, a 1-terabyte device offered by Victorinox as part of a Swiss Army knife, is not shown, because even with a logarithmic scale, its size renders the rest of the data too small to be seen on the graph.

Why did these technologies—jet engines and computer memory—develop so rapidly? There are of course many reasons, but the most obvious include the following:

- In both cases, there was a widespread, popular interest in the technology. In the first half of the 20th century, aviation was arguably the most exciting and glamorous technology; the same can be argued for computing in the last half that century.

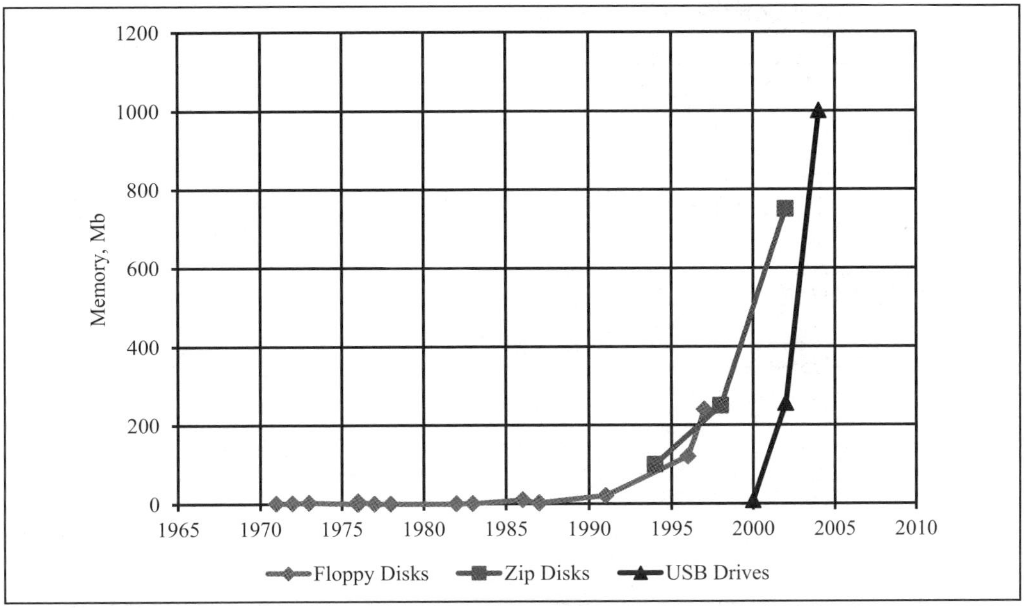

Figure 2. Capacity of computer memory devices, 1970–2005

- Interest from government entities—first from the military and later from space exploration—provided financial support and ready markets for new technology.
- Concurrent to the development of interest in these technologies, there was a rapid accumulation of technical understanding of the fundamental science pertinent to each. In case of aviation, knowledge of physics, fluid mechanics, thermodynamics, and of course aeronautical engineering were key; in computer technology, knowledge of solid-state physics, electronics, and other disciplines were similarly important.
- At least in part because of the reasons just listed, the right people were available at the right time.

FLOTATION TECHNOLOGY INNOVATION

In a previous analysis of the development of flotation technology, the author identified significant innovations in the field (Nelson 2011), as shown in Table 1.

Note that six of the 12 improvements in Table 1 occurred within 11 years (1903–1914) of the beginning of industrial-scale flotation in Australia. Again, the reasons for such rapid innovation are considered, and include the following:

- There was a general, widespread interest in mining. There was a popular perception that a career in mining was the route to prosperity. Many capable individuals entered the industry, and many prestigious universities offered degrees in mining and metallurgical engineering.
- As the industrial revolution continued in much of the world, there was an increasing demand for almost all mineral commodities.
- Large, unrecovered resources, such as the tailings at Broken Hill, awaited the development of suitable technology to allow profitable exploitation.

Table 1. Significant improvements in flotation technology

	Reference	Date	Affiliation	Country
Sulfide flotation by adhesion	Everson (1916)	1885	(Self)	USA
	Delprat (1903, 1904, and 1907) and Potter (1904)	1899–1904	BHP	Australia
Concentration in a froth	Sulman and Picard (1905), and Hebbard (1913)	1905	Minerals Separation	England/ Australia
Aeration and froth formation by agitation	Hoover (1914)	1909	Minerals Separation	Australia
Product stream retreatment	Hyde (1912)	1911	(Self)	USA
Air addition	Fagergren (1916)	1913	(Self)	USA
	Owen (1913)	1913	Junction North	Australia
Level control	Callow (1915)	1914	General Engineering	USA
Special impellers and stators	Booth (1936), Daman (1925 and 1934), Daman and Logue (1934), Fagergren (1929 and 1934), and Fahrenwald (1922, 1928, and 1934),	beginning ~1920 & continuing	Denver, Cyanamid, Galigher	USA
Standpipes and draft tubes	Drake (1939), Fagergren (1936), Daman Jr. (1966), and Reck (1970)	beginning ~1940 & continuing	General Engineering, Cyanamid, WEMCO	USA
Hydrodynamic design	Arbiter, Harris, and Yap (1969)	1969	Columbia University	USA
Large (>2.8 m^3) rectangular machines	Lynch et al. (2010)	beginning ~1962 & continuing	Denver, Galigher, WEMCO	USA
Large (>50 m^3) cylindrical machines	Maxwell (1971)	1972	Opemiska Copper	Canada
	Lynch et al. (2010)	1982	Outokumpu	Finland
Improved feed and discharge systems	Nelson et al. (2009)	beginning ~1995 & continuing	Dorr-Oliver, METSO, Outokumpu, WEMCO	Worldwide

Source: Modified from Nelson 2011.

- It was relatively easy to test new products and processes, because the high demand for commodities led to the availability of funding and sites for large-scale testing.
- Once again, the right people were available at the right time.

CURRENT INNOVATION—SUSTAINING OR DISRUPTIVE?

Of the innovations listed in Table 2 that took place after 1950, one of the most significant might be considered the development of bigger cells. It is worthwhile to consider how this occurred. In 1944, the largest Denver cell was the No. 30 Sub-A, which had a volume of 2.8 m^3. Denver's first 5.6-m^3 cell was simply two, 2.8-m^3 cells bolted together, with two mechanisms installed. **Figure 3** shows the introduction of larger rectangular cells in the 1960s. There is little evidence of disruptive innovation.

Table 2. Factors related to innovation in mining and mineral processing

1912	2012
A general, widespread interest in mining	There is widespread interest in the products and profits derived from mining, but (with notable exceptions) relatively little societal interest in development of new technologies.
An increasing demand for almost all mineral commodities	The increased demand from economic development in China, India, and elsewhere is well recognized, but has mainly been met by development of new deposits in remote areas, and by increasing the scale (thus decreasing the unit costs) of many operations.
The existence of large, unrecovered resources	Large, high-grade deposits are becoming increasingly difficult to find, and when they are found are often in remote locations, where logistics are difficult and risks are high.
The relative ease of testing new products and processes	The large scale of successful operations, in combination with the strong emphasis on control of operating costs, makes it difficult to conduct plant-scale tests of new technologies.
The availability of the right people at the right time	The 'graying' of the mining industry workforce has been extensively discussed, most recently by Brandon (2012). In addition, science and engineering graduates from many prestigious universities are drawn to the high salaries and bonuses of the financial 'industry.'

The introduction of cylindrical cells may be considered a disruption, however slight. The Maxwell cells, with volumes of 17 and 57 m^3, were introduced in 1971 and 1972 respectively. However, they were used only in a few Canadian mills. Outokumpu introduced its 60-m^3 TankCell in 1982, and larger sizes from Metso, Outokumpu, and WEMCO followed in the 1990s, as shown in Figure 4.

In preparation for the recent inaugural series of the G.D. Delprat Distinguished Lecture on Flotation in 2011, two manufacturers of large flotation machines were asked to list the most significant improvements or innovations in their respective machines in the last 10 years. The responses are summarized below.

FLSmidth

- Improved froth recovery, using new designs for froth crowders and launders.
- Flash flotation for rapid recovery of quick-floating minerals.
- Inert gas flotation for minerals that are easily oxidized.
- Mixed rows and circuits using induced and forced air machines to achieve improved grades and recovery throughout an entire plant.
- Decreased maintenance requirements through the use of external dart valves and standardized machine designs.
- Improved scaleup procedures, using computational fluid dynamics and hydrodynamic modeling.
- New research methods, including improved bubble sizing methods and isokinetic, high-speed cameras, for optimization of machine design.

Outotec

- Large cells that can accommodate high tonnage, high pulp densities, and coarse grinds, while also showing improved metallurgical performance.

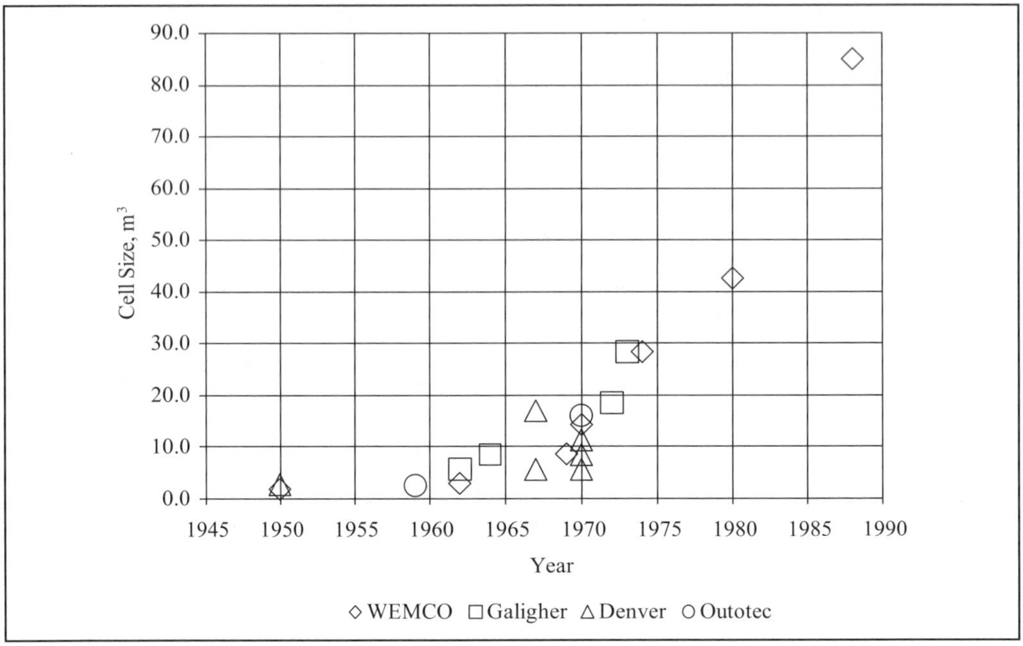

Source: Nelson 2011.

Figure 3. Sizes of rectangular flotation machines

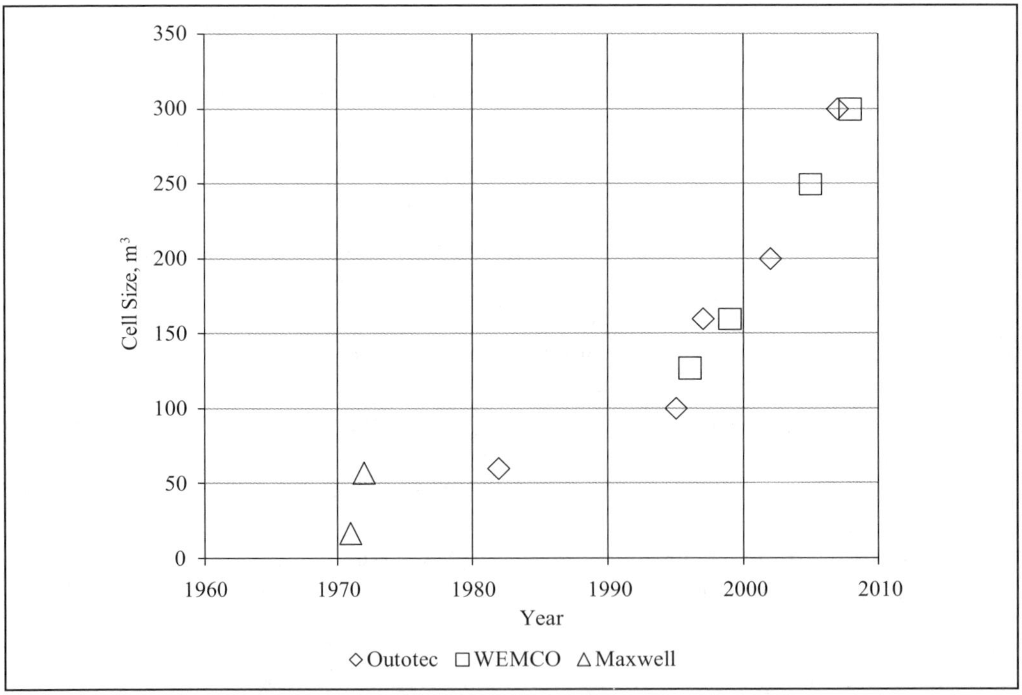

Source: Nelson 2011.

Figure 4. Sizes of cylindrical flotation machines

- Reliable scaleup methods, which are being used in the design 500-m^3 cell to be introduced 'soon.'
- The new FloatForce® mechanism, which provides increased recovery, improved pumping, lower power draw, and lower maintenance costs.
- An optional, auxiliary agitator to enhance flow in the upper part of cell.
- Successful application of flash flotation in large cells.

These are notable improvements, and have been well received in the industry. However, the author respectfully suggests that most are, in fact, sustaining, and not disruptive, representing a process in incremental improvement.

Other Recent Flotation Innovations

Outside the conventional, mechanical cells, some other flotation technologies should be considered, namely flotation columns, Jameson cells, and coarse-particle flotation machines.

Flotation columns are a variety of the pneumatic flotation technology, first used by Lyster and Bradford at Broken Hill (Lynch et al. 2010), and introduced on a large scale by Callow (1915) in a slightly different form. The first column was patented by Norris (1907), and variations patented by Towne and Flinn (1919) and Cole (1921) were tested and the Inspiration and Morenci mills in the 1920s. They produced high-grade concentrates, but were subject to the familiar problems of sanding, channeling, and plugging of air diffusers, and were abandoned before their designers could solve these problems.

More recent development of column cells started in the phosphate industry in the U.S. (Hollingsworth and McDonald 1960), and the copper industry in China at about the same time (Lynch et al. 2010). Development proceeded under the sponsorship of the Deister Concentrator Company, and later the Column Flotation Company of Canada. Columns are widely used today, primarily as cleaners in circuits that also employ mechanical cells. Difficulties are still encountered with the plugging of air spargers, regardless of design, and with maintaining correct feed slurry densities in complex circuits. Furthermore, in copper concentrators where columns are used for molybdenum recovery, losses of coarse molybdenite particles can be excessive.

The Jameson cell is a variation of the high intensity cells, in which intensive mixing of flotation pulp with air creates very small bubbles and enhances bubble-particle collisions. Such machines, described in detail by Lynch et al. (2010), achieve high intensity with either forced or induced air, and are capable of floating fine particles at high rates. The Jameson design began as a new sparger for tall columns, but the design soon migrated to short tanks, where it functioned more effectively. The bubbles in the Jameson cell have a mean diameter of 300 microns, compared with about 1,000 microns in conventional cells. Jameson cells have found application in the flotation of coal, industrial minerals, and metal ores, and also in the removal of organics in solvent extraction/electrowinning circuits.

In suitable applications, primarily flotation of fine particles, cleaning of rougher concentrates, or quick recovery of easily floated particles, flotation columns or Jameson cells can effectively replace mechanical cells, often saving floor space and reducing energy consumption.

Coarse particle flotation has received attention since the first investigations of Garrett (1933) at Broken Hill. The Denver Unit Cell, introduced in the early 1940s, was often installed at the ball mill discharge, sometimes in series with a jig (Lynch et al. 2010). The Unit Cell embodied the philosophy stated succinctly by Arthur Daman, president of the Denver Equipment Co., "Recover your mineral as coarse as possible and as soon as it is free."

This concept was resuscitated by Outokumpu (now Outotec) as the SkimAir® machine, and the recovery of coarse particles early in the flowsheet was called flash flotation (Tracy and Gillis 1986). Flash flotation is now widelty used, especially in treatment of ores bearing coarse particles of the valuable mineral constituents, most commonly gold, nickel, copper, and lead flotation operations. Residence times are typically short, in the order of a couple of minutes (Nelson et al. 2008).

Perhaps the most recent development in coarse particle flotation is the Hydrofloat cell (Kohmuench et al. 1999), developed at the Virginia Polytechnic Institute and State University, in cooperation with Eriez Magnetics, Inc. This machine combines the hindered-bed separators long used in mineral processing with particle flotation (Kohmuench et al. 1999). In traditional hindered separators low-density, coarse particles may be displaced to the high-density underflow, because of the accumulation of coarse low density particles at the top of the teeter bed. In the HydroFloat cell, the teeter bed is continuously aerated by injecting compressed air and frothing agent into the fluidization water. The air is dispersed into small bubbles by circulating the water through a high-shear mixer, and the bubbles become attached to the hydrophobic particles in the teeter bed, reducing their effective density. The action of the hindered bed forces the low-density agglomerates to overflow into the product launder, and particles that do not attach to the air bubbles report to the dewatering cone.

Twelve HydroFloat cells were installed at Florida phosphate plant, four for treating coarse particles and eight for ultra-coarse particles (Kohmuench et al. 2007). The coarse-particle units process 56% more material than the pneumatic cells cells they replaces, with consistent recovery and grade, and decreased reagent consumption. The eight ultra-coarse retrofits process 86% more material, with a 6.5% recovery increase and a 37% reduction in collector consumption. Feed grades were not given. Favorable results have been reported with the Hydrofloat cell in recovery of fine coal and carbon from waste streams (Parekh and Chen 2002), but thus far these are limited to laboratory tests.

ACHIEVING INNOVATION

Why has innovation to flotation machines proceeded so slowly of late? Table 2 shows a summary of the factors that led to innovation in flotation technology in the early 1900s, in the first column, with the author's appraisal of the current status of each factor in the second column.

The fourth item in Table 2, which relates to the ease of testing new products, bears further discussion. Product testing was key to the development of mineral processing technology in the first half of the last century. Perhaps the prime, but not the only, example, is the testing conducted under the direction of Louis Ricketts at Inspiration Copper in Arizona (Lynch et al. 2010). In 1913, Ricketts first constructed a 50-stpd (45-tpd), gravity-flotation test plant, where 10-st samples were tested. This was expanded in 1914 to a larger, 600-stpd (545-tpd) test plant, which cost $1 million—about $23 million in current dollars. The second plant quickly became "the Mecca of metallurgists and mill men." The most important factor in all testing was the determination of metallurgical efficiency; the cost of testing was secondary. The variables investigated included different types of mills and classifiers in grinding circuits, gravity and flotation machines in different circuits for concentration, reagents and points of addition in flotation circuits, and recovery of non-sulphide copper minerals.

The results of tests in the 600-stpd (545-tpd) plant were use to design a 15,000-stpd (13,636-tpd) production plant in 1915, which included the following equipment:

- Marcy ball mills and Dorr classifiers for grinding
- Rougher and cleaner circuits for flotation
- Double-deck Deister tables for scavenging coarse Cu from flotation tailings
- Eighteen flotation lines, each 900 stpd (818 tpd), with four lines of Callow cells, 13 lines of Inspiration cells (modified Callow cells), and one line of Minerals Separation cells

The production plant soon achieved 80% total copper recovery, with 90% recovery of sulphide copper, and produced a 30% copper concentrate from 1% ore. Plant capacity was soon increased to 21,000 stpd (19,091 tpd).

It is difficult to imagine that a $23-million 'test plant' would be constructed in the today's mining industry. Furthermore, testing new equipment in an existing plant is difficult and risky. Consider the Copperton Concentrator operated by Rio Tinto's Kennecott Utah Copper Company in Utah, which processes about 150,000 st/day (136,364 tpd), containing about 0.6% copper. The concentrator has five grinding lines. If arbitrary but reasonable assumptions are made—a 90% copper recovery, a margin of $4,400/t, and availability of 90%—approximately 13,000 t of copper, representing at least $57 million in net revenues, would be processed in a 3-month, in-plant test. If the testing resulted in a 2% loss in recovery, a financial loss of over $1 million would be incurred from unrecovered copper. Recovery of gold, silver, and molybdenum might also decrease, resulting in additional losses. This helps understand why plant operators are reluctant to allow in-plant tests.

Another factor that probably affects the development of flotation technology is also seen in the development of aircraft engines. The previous discussion noted how internal combustion engines followed a path of *sustaining* change, while the introduction of the jet engine produced a *disruptive* change. Since the first jet engine, change in aircraft engine technology has been almost entirely sustaining. Engines are bigger and more efficient, with many features that make them easier to operate and maintain, but the underlying technology remains the same. The same is true of flotation machines, and many other mineral processing technologies.

OBSERVATIONS AND CONCLUSIONS

The past one hundred have seen significant changes in mineral processing technology. Beginning with tumbling mills and flotation in the early 1900s (Lynch and Nelson 2010), and continuing to improved classification with hydrocyclones and the development of AG and SAG mills later on. However, improvements in the last 25 to 30 years have been less dramatic, primarily focusing on increasing equipment sizes and improving efficiencies. It may well be that, as with aircraft engines, the *disruptive* innovations have all been made.

What does this mean for the future of mineral processing technology? Many of the recent improvements in the area have been driven by the development of large, low-grade deposits, particularly of iron and copper ores. As deposits become more and more difficult to find, it will be necessary to look at alternatives. The mining of smaller orebodies, marine minerals, and even deposits in space are all being considered. In almost all cases, new processing technologies will be required.

One area where there appears to be an immediate opportunity is in pre-sorting of feed streams, with processes being more matched to the nature of the feed. Current, large-volume mills are designed to accommodate wide variations in feed. Smaller, more mineralogically complex deposits will likely require more 'precise' processes to achieve the required operating economies. Ultimately, it might be possible to use nanobots, similar to those being developed for use in medicine, to selectively extract minerals or metals in-situ, precluding many of the current mining and milling processes.

Mineral processors should therefore stay up to date with developments in miniature robotics, nanotechnology, and specialized materials, and constantly consider how such technologies might be used in the mills of the future.

REFERENCES

Arbiter, N., C.C. Harris, and R.F. Yap. 1969. Hydrodynamics of Flotation Cells. *Trans. SME* 244:134–148.

Brandon III, C.N. 2012. Emerging Workforce Trends in the U.S. Mining Industry. Littleton, Colorado: SME.

Booth, L.E. 1936. Aerating Machine. U.S. Patent 2,055,065. September 22.

Callow, J.M. 1915. Ore Concentrating Apparatus. U.S. Patent 1,104,755. January 12.

CFGear. 2012. What is the Biggest USB Flash Drive? 2011. http://www.cfgear.com/what-is-the-biggest-usb-flash-drive/. Accessed September 2012.

Christensen, C.M. 1997. *The Innovator's Dilemma—When New Technologies Cause Great Firms to Fail.* Cambridge, Massachusetts: Harvard Business School Press.

Cole, D. 1921. Apparatus for Separating or Concentrating Ores. U.S. Patent 1,375,211. April 19.

Daman, A.C. 1925. Flotation Machine. U.S. Patent 1,556,083. October 6.

Daman, A.C. 1934. Flotation Machine. U.S. Patent 1,942,803. January 9.

Daman, A.C., and L.H. Logue. 1934. Flotation Apparatus. U.S. Patent 1,985,153. December 18.

Daman, Jr., A.C. 1966. Development and Theory of the D-R* Denver Flotation Machine (Denver Bulletin No. F10-B121). *Deco Trefoil,* Nov-Dec:9–16.

Delprat, G.D. 1903. Extraction of Zinc, Lead and Silver Sulfides from their Ores. U.S. Patent 735,071. August 4.

Delprat, G.D. 1904. Apparatus for Use in Certain Processes of Extracting Sulfides from Ores. U.S. Patent 763,663. June 28.

Delprat, G.D. 1907. Ore Treatment at the Broken Hill Proprietary Mine. *Proc. AusIMM,* Paper 158, pp. 1–29.

Drake, L.D. 1939. Method and Apparatus for Diffusion Control. U.S. Patent 2,128,445. February 28.

Everson, J.L. 1916. Carrie Jane Everson, Inventor. *Mining American,* January 15.

Fagergren, W. 1929. Apparatus for Circulating and Distributing Flotation Pulp. U.S. Patent 1,736,073, November 19.

Fagergren, W. 1934. Aerating Machine. U.S. Patent 1,963,122. June 19.

Fagergren, W. 1936. Aerating Machine. U.S. Patent 2,054,249. September 15.

Fagergren, W., and W.D. Green. 1916. Ore Concentration. U.S. Patent 1,195,453. August 22.

Fahrenwald, A.W. 1922. Flotation Apparatus. U.S. Patent 1,417,895. May 30.

Fahrenwald, A.W. 1928. Flotation Practice in the Coeur d'Alene District, Idaho. *Trans. AIME* 1928:107–132.

Fahrenwald, A.W. 1934. Machine for Flotation of Ores. U.S. Patent 1,984,366. December 18.

Garrett, J.E. 1933. Flotation of Unclassified Ball Mill Discharge for the Recovery of Lead and Zinc Concentrates. *Proc. AusIMM* 92:475–499.

Hebbard, J. 1913. Evolution of Minerals Separation Process on Central Mine. *Proc. AusIMM* No. 10, pp. 75–128.

Hollingsworth, C.A. and C.J. McDonald. 1960. Concentration of Comminuted Materials. U.S. Patent 2,938,629. May 31.

Hoover, T.J. 1914. Concentrating Ores by Flotation. 2nd edition. London: The Mining Magazine.

Hyde, J.M. 1912. Art of Concentration of Mineral Substances. U.S. Patent 1,022,085, April 2.

Kohmuench, J.N., G.H. Luttrell, and M.J. Mankosa. 1999. Testing of the Hydrofloat Cell for Recovery of Coarse Phosphate, chapter in *Beneficiation of Phosphates—Advances in Research and Practice,* Patrick Zhang, Hassan El-Shall, and Ronald Wiegel, editors. Littleton, Colorado: SME.

Kohmuench, J.N., M.J. Mankosa, D.G. Kennedy, J.L. Yasalonis, and G.B. Taylor. 2007. Implementation of the HydroFloat Technology at the South Fort Meade Mine. *Min. Metall. Process.* 24(4):264–270.

Lynch, A.J., G.J. Harbort, and M.G. Nelson. 2010. *History of Flotation.* Carlton, Victoria: AusIMM, pp. 72–147.

Lynch, A.J., and M.G. Nelson. 2012. 1880–1930: The Golden Years of Mineral Processing Technology, in *Proceedings of XXV IMPC,* paper 121–00, XXV International Mineral Processing Congress, Brisbane, Australia, September 6–10, 2010 (published on CD-ROM).

Maxwell, J.R. 1971. The Development, Commercial Installation, and Operating Performance of Large Capacity Maxwell Flotation Cells. SME Preprint 71-B-16. Littleton, Colorado: SME.

Nelson, M.G. 2011. The AusIMM Metallurgical Society G.D. Delprat Distinguished Lecture 2011: Innovation in Flotation Technology. *MetSoc News (The newsletter of the Metallurgical Society of the AusIMM)* 3:1–4.

Nelson, M.G., F.P. Traczyk, and D. Lelinski. 2002. Design of Mechanical Flotation Cells, chapter in *Mineral Processing Plant Design—Operating Practice and Control,* D.R. Halbe, editor. Littleton, Colorado: SME.

Nelson, M.G., D. Lelinski, and S. Gronstrand. 2009. Design and Operation of Mechanical Flotation Cells, chapter in *Recent Advances in Mineral Processing Plant Design,* D.Malhotra, P.R. Taylor, E. Spiller, and M. LeVier, editors. Littleton, Colorado: SME.

Norris, D.H. 1907. Apparatus for Separating the Metallic Particles of Ores from the Rocky Constituents Thereof. U.S. Patent 873,586. December 10.

Owen, T.M. 1913. Commonwealth Patent 9,381 (reported in minutes of the Broken Hill Branch AusIMM, 1930).

Parekh, B.K., and Chen, Z. 2002. Recovery of Fine Coal from Waste Stream by Selective Flocculation Process. SME Preprint 02–006. Littleton, Colorado: SME.

Potter, C.V. 1904. Process of Separating Metals from Sulfide Ores. U.S. Patent 776,145. November 29.

Reck, W.H. 1970. Flotation Apparatus and Process. U.S. Patent 3,491,880, January 27.

Sulman, H.L. 1919. Flotation. *Trans. Inst. Min. Met.* 29:44–204.

Sulman, H.L., and H.F.K. Picard. 1905. Ore concentration U.S. Patent 793,808, July 4.

Towne, R.S., and F.B. Flinn. 1919. Process for Separating Ore Materials from Each Other. U.S. Patent 1,295,817. February 25.

Tracy, S.V., and J.M. Gillis. 1986. Flotation in Grinding Circuits. SME Preprint 86–316. Littleton, Colorado: SME.

Wikimapia. 2011. Kennecott's Copperton Concentrator. http://www.wikimapia.org/5395471/Kennecott-s-Copperton-Concentrator. Accessed September 2012.

ABOUT THE AUTHOR

Michael G. Nelson is associate professor and chair of the Mining Engineering Department at the University of Utah. He was previously professor of mining engineering at the University of Alaska, Fairbanks. He has worked for Kennecott Copper, Western Zirconium, CONSOL Energy, and EIMCO Process Equipment. Nelson is a member of SME; the Canadian Institute of Mining, Metallurgy and Petroleum; and several other societies and associations. He holds eight U.S. patents and one Canadian patent. He has also co-authored and contributed to several books, including three chapters in the third edition of the *SME Mining Engineering Handbook* and *History of Flotation*, published by AusIMM.

Major Innovations in the Evolution of Flotation Reagents

D.R. Nagaraj and Raymond S. Farinato
Cytec Industries Inc., Stamford, Connecticut, United States

ABSTRACT: The history of mineral flotation is a record of over a century of innovations, especially in the development of chemical reagents, the principles of their action, their manufacture and the role they play in sustaining mineral processing operations even as the quality of ores continues in overall decline. An analysis is given that organizes the evolution of flotation reagents into four historical and one emerging period. We emphasize the major innovations that have directed progress on this subject, especially those innovations that have positively impacted industrial scale applications.

INTRODUCTION

Since the advent of flotation in mineral processing, the growth in diversity and tonnage of chemicals used has mirrored those same features of the ores treated. Chemical flotation reagents, such as collectors, modifiers and frothers have been critical to mining separation technology for both sulfide and non-sulfide systems for over 100 years. Innovation in the development and use of chemicals has gone hand-in-hand with innovation in the development of flotation technology. These innovations were also greatly influenced by the general zeitgeist and paradigms existing in the broader field of chemistry at any one time. The introduction of chemicals made the flotation process practical and widely applicable; conversely, developments in flotation technology made possible the use of a wider variety of chemicals. The resulting positive feedback of this relationship led to the development of improved flotation operations.

Nagaraj and Ravishankar (2007) provided a comprehensive and critical review of the flotation reagents of industrial (practical) relevance developed and used over the past 100 years. In that review they gave a brief overview of the chronology of reagent development. That overview is the starting point in this paper for much of the discussion of flotation reagent innovations that either benefited existing processes or resulted in new processes/inventions. The historical analysis in the present paper has been expanded with new information and with an *emphasis on innovations and their impact on the mining industry*. Our emphasis will be on innovations that have been successfully applied in plant practice, and those that are poised for implementation.

Booth and Freyberger (1962) noted two distinct periods of development of flotation chemicals up to that time: (a) oil flotation (1860–1920) and (b) chemical flotation (1921–1960→). Nagaraj and Ravishankar (2007) highlighted three distinct periods and cast them in a slightly different light from their perspective to reflect the drivers and goals leading to innovations. Thus they noted: (a) oil flotation (1860–1920), (b) discovery and expansion of chemical additives (1921–1950), and (c) rational, targeted chemical design (1951–2007→). There were several periods of especially intense activity within the second and third epochs; specifically in the 1920s, 1950s and again in

Table 1. Periods in the history of flotation reagent innovations and developments of practical importance

Period	Activity	Example Reagents
1	1860–1920: Use of conventional and conveniently available basic chemicals	Oils, soaps, $CuSO_4$, lime, SO_2, Na_2CO_3, Na_2S, $KMnO_4$, cresylic acids, pine oil, eucalyptus oil
2	1921–1950: Discovery and expansion of the chemical palette; adoption of commercially available chemicals from other industry sectors	Collectors: thiocarbanilide, dyes, chelating agents, dithiocarbamates, amines, xanthates, xanthate esters, dithiophosphates, mercaptans, thionocarbamate, mercaptobenzothiazole, xanthogen formates, trithiocarbonates, petroleum sulfonates, alkyl sulfates, dialkyl sulfosuccinates Frothers: Short chain alcohols, ketones, aldehydes and esters Modifiers: dextrin, dyes, lignin, sodium cyanide, sodium silicate, tannin
3	1951–2000: Rational, targeted design of compounds for use specifically in flotation applications	Collectors: alkoxycarbonyl thionocarbamates and thioureas, dialkyl thionocarbamates, allyl alkyl thionocarbamates, dithiophosphinates, phosphonic acids, monothiophosphates, monothiophosphinates, sulfosuccinamates, allyl xanthate esters, dialkyl sulfides, alkyl hydroxamates Frothers: polypropylene glycols and ethers, alkoxy-paraffins, targeted formulations Modifiers: synthetic polymers with specific functional groups, modified polysaccharides, polyacrylates, zinc cyanide, etc.
4	1990-present: New methodologies and approaches; integration of process mineralogy; reagent optimization in plant practice	Holistic approach, advanced statistical methods, complex interactions, new chemical classes and technological approaches
5	Emerging: Coordinated efforts between academia and industry; top-down systems approach; emphasis on sustainability and targeted greener chemicals and processes	In development

the 1980s. A more refined analysis of the historical development presented here reveals arguably five periods of innovations and developments. These are summarized in Table 1.

It is likely that a fifth period is emerging at this time, where the emphasis is placed on increasing the sustainability of mining separation operations. These development activities incorporate the latest advances in computational modeling and automation technology combined with a more sophisticated understanding of chemical interactions in order to identify high-performance flotation reagents that effectively target specific minerals while reducing water and energy consumption and waste generation. We discuss here how much of this may be achieved through the use of greener chemicals and processes, and by adopting a more holistic approach to solving the problem.

This approach to flotation reagent design, selection and optimization is generally applicable to other mineral processing unit operations as well.

It is impossible to condense the 100 years of innovation in flotation reagents into a small number of pages without sacrificing on the details or restricting the discussion to a few innovations. We have chosen breadth over depth in this paper because we believe that this will provide the reader a more complete story and mitigate personal bias. As is often the case with any technology, innovations build on one another. A postmortem of major innovations invariably reveals that they occur through accretion of prior minor innovations. It is seldom that a single innovation is accorded a revolutionary status, though it is highly tempting to do so.We have also restricted the innovations to those that have made a *substantial impact on practice*. A discussion organized according to the chronology of the development of chemicals over the past 100 years highlights the remarkable innovations that have occurred in the discovery and/or application of chemicals in flotation. After about the early-1960s there is a dearth of critical reviews of flotation reagents (in the scientific literature) of the quality and completeness of those that appeared in *AIME Transactions* during the previous period.

PERIOD 1: THE EARLY PERIOD (1860–1920)

The most important innovation in this period was the introduction of air or in-situ generation of CO_2 as the third phase. This gave rise to the birth of froth flotation. This innovation removed the need for large quantities of oil as a buoyant separation medium, decreasing oil requirements from >10% to less than 1% of the weight of the ore. Then as the technology spread, the focus was to devise ways to reduce the oil consumption further, and improve efficiency and selectivity of separation. Oil consumption was reduced to 0.5 to 1 kg/t. With this reduction in oil consumption, inherent differences in the frothing and collecting powers of various oils were noted, which led to other innovations in terms of selective or differential flotation. For example, several important modifiers (chemicals intended to enhance or enable selectivity in separation) such as lime, sulfur dioxide, sodium carbonate, sodium sulfide, and copper sulfate were used to enhance separations. These inorganic modifier reagents were culled from much trial and error testing of virtually any cheap electrolyte that was available in those days, hardly an exercise in innovation; however, this activity did lead to major innovations subsequently in differential flotation schemes. For example it was discovered that alkaline circuits could provide much more performance leverage in terms of lower overall reagent consumption and facilitated the use of certain modifiers that were impossible to use in the previously popular sulfuric acid circuits where doses ranging from 5–100 lb/ton were typical. Indeed, by the end of the 1920s most of mineral flotation was conducted in alkaline circuits.

Another critically important innovation occurred during the period from 1913 to 1921 as the result of pioneering work of Perkins and his co-workers (Perkins 1917, 1919, 1921) who established that slightly-soluble, small organic compounds were effective collectors. They revealed numerous small organic compounds such as thioureas, thioalcohols, dyes, diazo and amine compounds, which were vastly different from the fatty oils then in use. There was a gradual realization among operators and experimenters that organic flotation reagents provided two different functions, both of which were important: (a) collection (i.e., differential separation of minerals into the froth phase), and (b) froth formation and stabilization. This realization led to a working classification of organic flotation agents into collectors and frothers. They also identified the essential molecular features of organic compounds that made them good collectors for sulfide minerals (i.e., organic, slightly water-soluble, chemically reduced, easily oxidized, and non-frothing) and the manner in which they cause mineral collection Later, Taggart et al. (1930) noted that "Soluble organic collecting agents function by reacting chemically with the surface substance of the mineral particle to be collected, thereby forming an organic reaction product that is insoluble in water in the concentration present, and adheres at the reacting surface and thereby changes that surface, for all practical purposes, to an organic surface with water-repelling properties." Perkins' innovations laid the foundation for a large number of innovations that occurred in the 1920s (next section), which led to the development of new and highly effective non-frothing collecting agents that are almost universally still used today. He determined that neutral or slightly alkaline conditions improved the efficiency of these collectors and that practice continues to this day.

PERIOD 2: DISCOVERY AND EXPANSION (1921–1950)

During this period in flotation history the chemical toolbox that people reached into to advance their technology was largely stocked from the *materia chemica* of other strong industries of that era, such as tanning, agriculture, petroleum, rubber, textiles (dyes), food, soaps, detergents and explosives. The availability of chemicals for flotation was subject to the influences of these other industries, as well as the prevailing state of war or peace. Nevertheless the phenomenal innovations and developments during this period firmly established the so-called era of *chemical flotation*.

The 1920s was quite a revolutionary period and phenomenal inventions and innovations in flotation occurred then. The most noteworthy of these was building on the pioneering work of Perkins and others (*loc cit*). This period was one of expanding on the concept of using *small* amounts of *short-chain* (C_2–C_6) *organic molecules* (many of which were *water-soluble*) as *collectors* (Table 2) for froth flotation of sulfide ores to replace large amounts of long chain ($> C_{12}$), relatively insoluble oils as collectors for froth flotation. The reagents named in many patents all contained either sulfur or nitrogen or both, the significance of which was noted much later after the work of Taggart et al. (1930). These elements are most preferred in reagents for sulfide systems.The work of Perkins, Corliss, Sayre, Vivian, Keller, Lewis, and Whitworth during the period 1917–1926 demonstrated the efficacy of these small molecules and revolutionized flotation. This was the beginning of so-called *chemical flotation*. Most notable of these collectors were: short chain amines (e.g., napthylamine, diaminobenzene), napthol, dyes, thiocarbanilide (diphenyl thiourea), and other chelating agents (all developed during 1917–1921); xanthates (1923–1925); potassium salt of dithiocarbamic acid (1924); and dithiophosphates in 1926 (especially the cresylic acid derivative; the sodium salts of aliphatic dithiophosphates were developed in 1928 to cater to the alkaline or lime circuits typically employed). Both xanthates and dithiophsophates found immediate acceptance in the industry and within a few years became the workhorses in sulfide flotation; they still are today. Dithiocarbamates, on the other hand, were not widely adopted (their usage is still relatively small). Thiocarbanilide became popular and was used widely for niche applications (e.g., improve Ag recovery) until about two decades ago. Although xanthates are most often accorded a unique status, this is not quite accurate. The foundation had been laid by the work of Perkins and others during the period 1917–1921. Dithiophosphates and dithiocarbamates, discovered about the same time as xanthates, are structurally very similar to xanthates. The real invention/innovation here was the recognition that certain sulfur compounds were most effective in sulfide mineral flotation.

The discovery and recognition of the efficacy of xanthates and dithiophosphates marked also the beginning of the search for other sulfur compounds. A natural source of such compounds was

Table 2. Water soluble collectors

Xanthates	Trithiocarbonates	Dithiocarbamates	Mercaptobenzothiazole
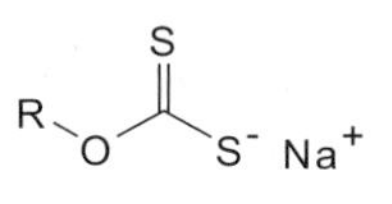	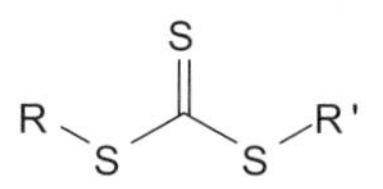	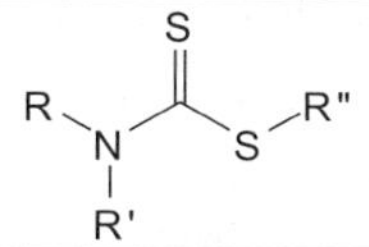	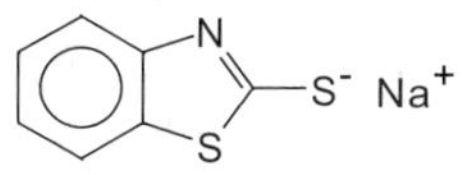

Dialkyl xanthogen formates	Dialkyl dithiophosphates	Dialkyl thionocarbamates	Thiocarbanilide
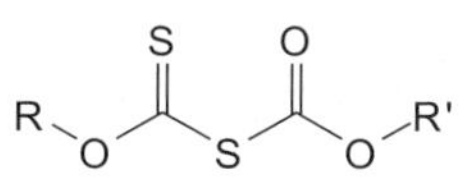	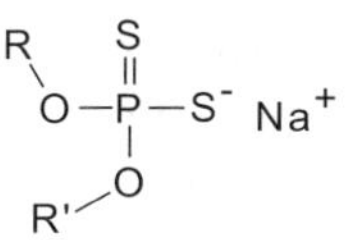	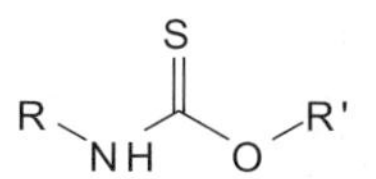	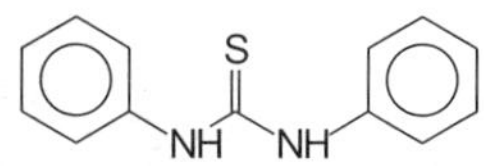

the then burgeoning rubber and agricultural industries. Douglass exploited these in a number of patents on, for example, alkyl xanthogen formates (1927), dialkyl thionocarbamates (1928a) and dialkyl trithiocarbonate (1928b). Bolton (1928) patented mercaptobenzothiazole, which gained wide application, and is used even today. Trithiocarbonates (whether dialkyl or monoalkyl) never quite enjoyed wide application perhaps owing to their instability, stench, and lack of any unique features over xanthate or dithiophosphate. Alkyl mercaptans were developed during this time, but they found only sporadic and limited use. Although thionocarbamates were patented in 1928 (Douglass (1928a) called it alkyl ester xanthogen alkyl amide; a somewhat different terminology than what is common today), they remained obscure until a manufacturing process patent from Dow in 1954. They became popular soon thereafter. Xanthogen formate enjoyed only limited use, perhaps reflecting its limited capability. Synergistic mixtures of collectors, e.g., those of thiocarbanilide, mercaptobenzothiazole and dithiophosphates, were developed and commercialized during the 1930s (Christmann and Jayne 1936; Christmann and Falconer 1936) and these and many other mixtures continue to be important today.

While companies like Du Pont did not commercialize the use of these compounds as flotation collectors, companies like American Cyanamid did. Much of the flotation reagent work that came out of American Cyanamid from the mid-1920s and beyond was likely accelerated by the diversity of chemistries and technologies that were brought under the same roof during periods of business diversification. While some flotation reagents patented early in the game derived from adjacent areas of research, these successes spawned more deliberate and specific for-purpose efforts.

The seminal work of Taggart and his co-workers in the 1920s, published in 1930, laid the foundation for identifying and establishing chemical functional group features of collectors, depressors, and frothing agents. This work coupled with the fundamental studies of Gaudin, Wark, Sutherland, and others was an important conceptual advance that supported future reagent development for both sulfide and non-sulfide mineral systems.

The focus on reagents in the 1920s and 1930s was on sulfide ores, and no impressive record of achievement exists in the treatment of non-sulfides. It was thought that non-sulfide separation was difficult. How could one float non-sulfides if all the collectors were floating sulfides away from them? However, this changed largely due to the work at the U.S. Bureau of Mines and other institutions (Gaudin and Glover 1928; Coghill and Clemmer 1935; Ralston 1938) beginning in the late 1920s in the flotation separation of oxide, carbonate and phosphate minerals from quartz and silicate gangues by means of anionic collectors such as fatty acids. The thrust of the innovation was identifying and developing suitable modifiers, conditions and schemes to achieve selective separations while employing rather non-selective collectors such as soaps and amines. Taggart (1951) noted: "after trying anionic and cationic collectors use relatively large quantities of silicate ion, sulfuric acid, or organic colloid for cleaning, purely on a cut-and-try basis. When everything else has failed, try hydrofluoric acid, calcium hypochlorite, alum or aluminum chloride, or hexametaphosphate. Chance is as good as anything else to determine the order of such tests" The 1930s and 1940s saw many developments on reagents and schemes for oxides; a general description can be found in Taggart (1951); Falconer and Crawford (1946); Rogers et al. (1946); Falconer (1960); Thom (1962). A brief summary of the chronology of development is given by Nagaraj and Ravishankar (2007; Table 1 therein). Research during this period lead to the identification of several new chemical classes with benefits as collectors for non-sulfide ores, including hydroxamates (Pöpperle 1941) and chelate forming reagents (Gutzeit 1946). American Cyanamid was the most prolific in this period (over 180 patents from 1921–1951) with the development of products for practical application including fatty acid polyamine condensates, alkyl guanidines, imidazolines, sulfosuccinates, and petroleum sulfonates, and schemes for the selective separations in a wide variety of non-sulfide systems including energy minerals. These reagents and schemes are still used today.

Once this era of chemical flotation was in full swing it became clear that there was not only benefit in designing more powerful and selective collectors, but the flotation process could also be ratcheted up through the judicious use of *modifiers* that were synergistic with those collectors and facilitated differential flotation. In 1922 Sheridan and Griswold (1922) produced landmark publications on the systematic study of inorganic modifiers, especially NaCN, which demonstrated that a small amount of cyanide caused a marked decrease in the floatability of zinc and iron sulfides compared to those of copper and lead, and revolutionized the processing of complex Pb-Cu-Zn-Fe ores. This second era, especially in the 1920–1940s saw other modifiers placed in the flotation reagent arsenal in addition to Na cyanide, such as synthetic dyes and a new class of modifiers drawn from natural products such as dextrin, tannin, starch, dyes and quebracho (Moyer 1958; Gutzeit 1946).

The structural features required for reagents to impart selectivity in separation were well articulated by Gutzeit in his influential paper (1946). He noted the importance of applying knowledge of specific chelate-forming organic analytical reagents, particularly in the selection of reagents for flotation, which included collectors and depressants. He noted that many of these specific reagents are uneconomical. Some of them could be used at low cost, if their efficiency were proved to be outstanding. Indeed this occurred much later (see next section).

PERIOD 3: RATIONAL, TARGETED DESIGN (1951–2000)

During this period innovations in flotation reagents were of quite a different character. There was a gradual change from "screening" and "expansion (exploratory)" type of research to efforts at better mechanistic understanding and designing *de novo* for specific separation challenges based on advanced understanding of organic, polymer and coordination chemistry. The aim of much of this led to rational targeted design, which exploited specific chemical interactions between reagents and minerals. This kind of inquiry applied equally to research on macromolecules—synthetic polymers were a burgeoning field during this period. Polymers offered unique features and advantages. There was growing recognition of the power and efficiency of polymers, and the great flexibility they offered for tailored design.

The conceptual formalism that became the most productive foundation for understanding, and producing guidelines for designing many of these specific interactions was the donor-acceptor model applied to the mineral systems (details found in Nagaraj 1988). The basis for this model evolved out of the acid-base concepts arising from the work of G.N. Lewis that first surfaced in the early 1920s (Lewis 1923; Jensen 1980). In that paradigm, reactions between acids (electron acceptors) and bases (electron donors) are viewed in terms of a redistribution of valence electrons. Lewis' ideas spurred developments in quantum mechanical formalisms and in empirical correlations to organize, explain and predict donor-acceptor interactions. These developments, specifically in the chemistry discipline, were applied to interactions in solution, such as in the works of Pearson (1963), and more recently in van Oss et al. (1987). These concepts were applied to mineral systems (minerals and metals are the acceptors; flotation reagents comprising the donors) during the period 1970–1985 (fully described in Nagaraj 1988). The donor-acceptor paradigm has become the conventional framework in which to understand and develop reagents that bind selectively to either metal ions in solution or mineral surfaces in suspension, and has enabled a greater degree of predictive power in practical systems. This scheme is typically applied to binary interactions between reagent and target, which is of course a simplification of the actual situation in ore flotation. There is no inherent flaw in extending application of the donor-acceptor method to more complex mixtures where competitive adsorption is the central feature.

Armed with the deeper understanding of reagent design based on the donor-acceptor model, a new suite of reagents was developed, especially in the late 1970s and early 1980s for both sulfide and non-sulfide systems. Like the 1920s, the 1980s was a period of intense activity in flotation reagent development (a review of literature shows that many companies were active during this period, including Cytec Industries, Inc. (formerly American Cyanamid Company), Dow Chemical (MRI), Phillips Chemicals, Clariant (formerly Hoechst), Henkel, Montan Chemicals, Elf Aquitaine, and others) and there was a significant investment in R&D in many companies and institutions. These activities ran parallel with research and development in surfactants and polymers in a variety of fields. Significant efforts were focused on developing reagent-mineral characterization data and extensive structure-property-performance relationships which led to numerous innovations. Selected examples of major innovations in this period will be highlighted below. These examples are but a small sample of a large body of available empirical structure-property-performance correlations that form the basis for design of new reagents for any given mineral separation challenge.

Some important examples of new classes of collectors developed in this period are: alkoxycarbonyl thionocarbamates, alkoxycarbonyl thioureas, dithiophosphinates, monothiophosphinates and monothiophosphates, dialkyl thionocarbamates, *allyl* alkyl thionocarbamates, *allyl* xanthate esters, alkyl hydroxamates, phosphonic acids, and sulfosuccinamates (Table 3). Examples of modifiers include synthetic polymers with mineral-specific functional groups containing the donors N, O and S for both sulfide and non-sulfide systems. Most of these are still in commercial use today. These innovations rested squarely on the conceptual innovation that emphasized the role of the *functional group* rather than attributes like charge (ionic, neutral), solubility, hydrolysable, oxidizable, etc. The power of this framework comprising the donor-acceptor model and functional group contribution to performance can be readily appreciated with a few examples of new sulfide collectors, non-sulfide collectors and synthetic polymers. These are largely taken from the following

Table 3. New classes of collectors

alkoxycarbonyl alkyl thionocarbamates	alkoxycarbonyl thioureas	dithiophosphinates	allyl alkyl thionocarbamates
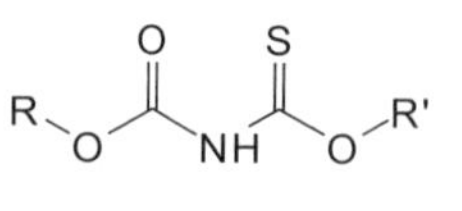	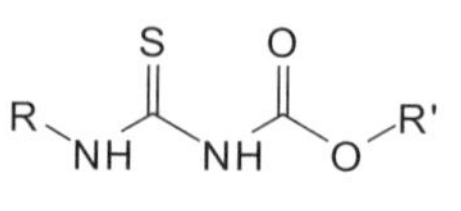	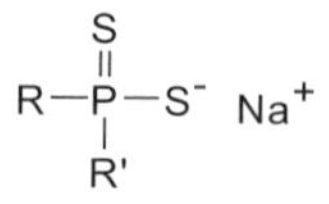	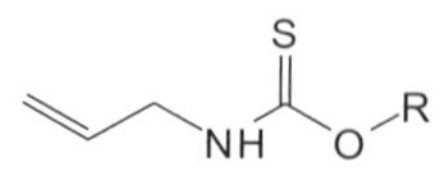
dialkyl thionocarbamates	allyl xanthate esters	alkyl phosphonates	monothiophosphates
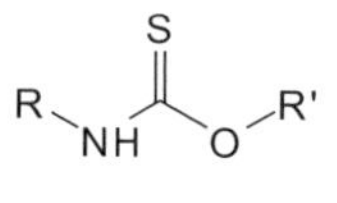	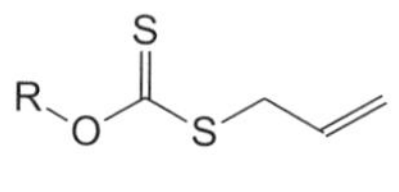	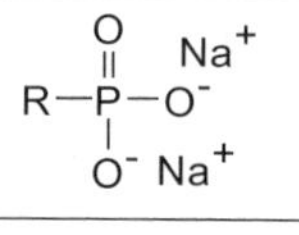	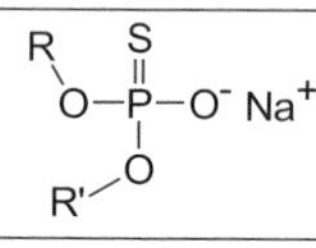

references: Nagaraj and Avotins 1988; Nagaraj et al. 1988, 1989; Nagaraj 1997, 2000; Farinato et al. 1993).

The innovative concept was that in systems where the driving force for adsorption of flotation reagents on minerals is *chemical*, the properties (strength and selectivity) could be designed by exploiting and manipulating type, position, and number of donor atoms, particularly by introducing electron withdrawing/donating groups, which additionally enabled chelate-forming functional groups, and by manipulating the substituents attached to the functional groups. A database of comprehensive structure-property-performance relationships becomes very important, and offers a wide scope for reagent design and control in these systems. Since chemical interactions have the highest adsorption energies, changes in structure of the reagent molecule can potentially result in large changes in the strength of adsorption, selectivity, interfacial properties, and flotation response. This has been clearly demonstrated for a large number of reagent families in flotation research and practice. Structure-property relationships take on added dimensions in polymers in terms of complexity and possibilities. In addition to changes in functional group, factors such as molecular weight, degree of substitution, location and distribution of functional groups become critical.

In the case of the alkoxycarbonyl thionocarbamates, the innovation was to reduce the pK_a of the traditional dialkyl thionocarbamate (e.g., isopropyl ethyl thionocarbamate, IPETC) molecule sufficiently to facilitate application in a lower pH circuit (<10.5 vs. 11–12 used in Cu flotation circuits) to realize significant savings in lime cost, while ensuring selectivity (against pyrite) and efficiency. This was achieved by introducing in the primary functional group (-O-C(=S)-NH) a strongly electron-withdrawing alkoxycarbonyl group on the N atom. This substitution not only decreases the pK_a of the molecule by several pH units, but also introduces an additional active donor, O, in the form of C=O, thereby extending the functional group to the more complex -O-C(=S)-NH-C(=O)-O, which forms a highly favored, six-membered chelate with Cu atoms (note: this is not possible with IPETC). A further innovation was the replacement of the O with a N donor in the basic thionocarbamate group, thus giving a thiourea functional group (-NH-C(=S)-NH), and extending this with the introduction of an alkoxycarbonyl group. The alkoxycarbonyl thioureas are structurally similar to the alkoxycarbonyl thionocarbamates, but their collector properties are often significantly different. The major advantages of the alkoxycarbonyl collectors relate to their efficient recovery of coarse sulfides and middlings, high selectivity against pyrite and allowing operation at low pH values that translate into significant savings in lime cost.

Enhanced performance can be often realized by merely changing the hydrocarbon (non-polar) part of the reagent molecule while keeping the functional group intact. For example, a slightly branched hydrocarbon group in a collector molecule can provide greater selectivity in flotation than a linear hydrocarbon group (e.g., diisobutyl dithiophosphate vs. di-sec-butyl dithiophosphate). Similarly, significant differences are observed between aryl and alkyl groups. Aryl dithiophosphate floats galena far better than an alkyl dithiophosphate. Three examples serve to illustrate the innovations in targeted design by modification of the substituent group attached to a functional group. The first pertains to dithiophosphinate vs. dithiophosphate. In the dithiophosphinate, the substituents are directly attached to the central P atom, whereas in dithiophosphate they are attached through an O, resulting in dramatic changes in collector properties. The dithiophosphinate is far more selective and specific for Cu, Ag, Au, PGM, Ni, Cd, Hg etc. The stability constant for the Cu complex for example is almost 15 orders of magnitude greater. The second example pertains to the thionocarbamates: when the isopropyl substituent in dialkyl thionocarbamate is replaced by the *allyl* group (with a double bond), which facilitates π-bonding, again this results in substantial changes in the collector property. The third innovative concept is one of transforming non-selective thiol collectors (such as xanthate and alkyl mercaptan) into very selective collectors for targeted minerals. This is an extreme case of manipulating substituents, viz. introducing a new substituent altogether. A good example is attaching an *allyl* group to one of the S atoms in xanthate resulting in S-allyl xanthate ester which is very selective for metallic Au and Au-alloys, Ag minerals, PGM alloys, and MoS_2, and extremely selective against Fe sulfide gangue, and penalty element (As, Bi, Sb) gangue. Similarly, attaching a substituent to an alkyl mercaptan results in a dialkyl sulfide, which is very selective for a few targeted mineral species (Klimpel et al. 1994).

Another fascinating innovative example is the seemingly simple replacement of one of the (two) S donors by an O donor in the dithiophosphate (or dithiophosphinate) functional group (>P(=S)S) which results in quite a dramatic change in collector properties. Thus the dialkyl *mono*thiophosphate is a strong collector for sulfides in an acid circuit (pH 2–7) in contrast to the corresponding dithiophosphate which is effective in an alkaline circuit (pH >8). Another unique attribute of the monothio compound is that it is a selective collector for metallic Au and Au-Ag alloys in the alkaline pH range. Similar differences exist for the corresponding mono- and di-thiophosphinates. The differences in the collector properties between the mono and dithiophosphates are attributed to tautomerism that exists in monothiophosphates.

Many innovations occurred in the development of efficient and economical *manufacturing processes* for certain flotation reagents, and these had a profound impact in the industry by catalyzing the much needed widespread applications of the specific reagents in many innovative ways. Three examples can be cited. First is the development of a manufacturing process for dialkyl thionocarbamate (Douglass 1928a) in 1954 by Dow (Harris and Fischback 1954). The second example is for the manufacture of dithiophosphinate by American Cyanamid using a novel process based on the unique phosphine chemistry which resulted in very high yields and high purity product. The third example pertains to hydroxamates which were first suggested for flotation in 1941 (Pöpperle 1941). Subsequently their utility was expanded in the 1950s through 1970s (Gorlovskii et al. 1961; Peterson et al. 1965; Fuerstenau and Pradip 1984), but plant usage was sporadic and minor, mostly in the old Soviet bloc. Two important innovations occurred in the 1980s. First was the use of alkyl hydroxamate for removing colored impurities from kaolin clays (Yoon and Hilderbrand 1986). The second was the development of a practical, safe and economical process for the manufacture of alkyl hydroxamates (Wang and Nagaraj 1989) which resulted in a clean liquid product that was stable and easy to handle.

There were several other innovations in non-sulfide flotation: the application of sulfosuccinamates for a variety of non-sulfide systems (for e.g., cassiterite, Arbiter and Hinn 1968); alkyl and alkylamino phosphonic acids for several difficult-to-float minerals (e.g., cassiterite; Wottgen and Lippman 1963); and alkyl morpholine for soluble salt separations (Keen and Opie 1962).

The major innovation in frothers, particularly for sulfide flotation, was the almost concurrent development of polyglycol ethers (Tveter 1952) and glycols (Booth and Dobson 1954). These have become indispensable in in the industry along with the alcohol frothers.

In terms of modifiers, the majority of developments had occurred in the second period (prior to 1951). Additional development occurred to address specific industry challenges. For example, (i) the development of Zn cyanide for silver-rich ores instead of sodium cyanide to reduce Ag losses in polymetallics ore systems (Allen et al. 1953); (ii) use of guar gum as effective depressant for talc (Drake 1961), and (iii) carboxymethyl cellulose as a depressant and dispersant (Gorlovskii 1961); (iv) sodium ferrocyanide for depression of Cu-rich minerals in Cu-Mo separation (Barker and Young 1953); and (v) several thiol-containing depressants for Cu-Mo separation (Gannon 1980; Parlman and Bresson 1982; Bresson et al. 1983; Nagaraj et al. 1986).

A major innovation, however, was the successful application of the donor-acceptor concepts (mentioned earlier) in the development of water-soluble, synthetic polymeric modifiers with targeted molecular architecture and mineral-specific functional groups capable of targeting specific sites on minerals (via complexation or chelation), though the first example of such modifiers dates back to 1956 (Aimone and Booth 1956). Although such a task appears rather straightforward in theory, a truly significant challenge was in developing robust, practical and economical manufacturing routes to produce polymers with consistent structural features. These modifiers offer many unique advantages over the traditional modifiers: non-hazardous, better dosage-performance and lower treatment cost (due to the efficiency of a large molecule), ease of handling, lower toxicity, low odor, ease of structural modification to suit different applications and ore variability, and consistency from batch to batch. Examples of successful products introduced to the market place (Nagaraj 1997, 2000) addressed the following: (i) phosphate beneficiation; (ii) rejection of iron sulfides and penalty elements in sulfide circuits; (iii) depression of Cu and Fe sulfides in Cu-Mo separation; (iii) rejection of Mg silicates and other non-sulfide gangue in sulfide circuits. Although these kinds of polymeric modifiers are currently used in the industry, our application knowledgebase is still limited and inadequate at this time. This is one great area for research.

PERIOD 4: NEW METHODOLOGIES AND APPROACHES FOR REAGENT SELECTION AND OPTIMIZATION (1990–PRESENT)

The innovations and successes in the rational targeted design of flotation reagents, with the enthusiasm and intense R&D activity in the 1980s, were beginning to hit a bumpy road by 1990 in terms of large scale implementation in the industry. There were many important factors (some impacting the whole industry and others impacting chemical usage) driving this situation, which resulted in a recognizable trend that exists even today. These factors were well articulated in several publications in an SME symposium in 1994 (Cappuccitti 1994; Klimpel 1994; Malhotra 1994; Nagaraj 1994), and discussed in greater detail more recently (Nagaraj and Ravishankar 2007, see Table 12 therein; Nagaraj 2010). There was a growing recognition that the traditional approach to flotation reagent selection and optimization had become somewhat arbitrary, reductionistic, and fraught with misunderstanding and largely based on personal experience, with no accepted standard of practice in place (Nagaraj and Bruey 2003; Nagaraj 2005). Such an approach is very costly because the failure rate is high in implementation particularly with new chemistries. Even when the best reagent is developed for an unmet need in the industry, it is deemed a failure if we fail to apply it properly on the large scale. This issue becomes acute when one considers the high overall cost ($5–10 million from concept to commercialization) and the long time required (5–10 years) in developing and implementing new reagents and technology. Compounding this problem is the exponential growth in environmental regulations and the high upfront cost of regulatory registration for new chemicals (as much as $2 million in 8 countries for a single chemical). Additionally, the plant metallurgist is presented with a large variety of chemistries, differentiated not only by major functionalities but also by many other features. This complexity naturally reflects the complexity of mineral flotation, but creates a challenge for reagent selection and optimization.

The critical innovation that emerged from this situation was the development, beginning in the mid-1990s, of a holistic approach, called Flotation Matrix 100™ to reagent selection and optimization (Nagaraj 2005, 2008, 2010; Franzidis 2005). Such an approach confronts and deals with the complexity of ore flotation systems, rather than circumvent it. The premise of the approach is that flotation outcome from ores is variable and is determined by complex interactions between chemical, physical and operational variables (or factors). This applies to flotation at any scale. Flotation is a statistical process and the outcome has significant non-linearity. Floatability and flotation outcome are not an inherent property of a given mineral or ore. It is a response to a large array of variables (i.e., flotation process is multivariate). A rule-based expert system reagent selection software, with twenty-four criteria matched against mineralogy and plant needs, was developed. Additionally, advanced statistical tools and techniques for dealing with variability and complexity, both at lab scale (design softwares) and plant scale (REFDIST, Bruey 1999) were developed. The major advances in quantitative mineralogy (e.g., Mineral Liberation Analysis) and advanced spectroscopic techniques for plant problem solving (e.g., ToF-SIMS) have been critical in the application of the holistic approach. On the whole, the holistic approach highlighted the ideas of best practices, plant needs and expansion of the knowledgebase, with the ultimate goal of developing a robust and optimized solution that is suitable for the dynamic and variable plant system and that has the highest probability of successful implementation. Although significant progress has been made in implementing the holistic approach in the past 10 years as plant managers continue to understand and embrace the holistic approach, there is currently no compendium of successful case studies in published literature, and it may be another decade before the approach becomes an industry standard.

Since the 1990s, development efforts have also incorporated advances in engineering and process mineralogy to provide solutions better optimized for in-plant applications. Attempts to model the kind of complex systems found in flotation operations have been reasonably successful at the engineering level (e.g., King 2001); *however, these models do not yet incorporate information on mineral surface chemistry and the interplay between multiple soluble components in a flotation pulp and in the froth phase.* Integrating such physical-chemical details into process models would constitute a significant contribution and a major innovation.

The importance of analyzing and modeling the froth phase was increasingly recognized during this period; however, a detailed picture of the physical micro-processes involved in the transport of asymmetric particles of different wettabilities through and by a network of liquid lamellae is still under construction. Froth phase global properties such as gas holdup, froth height, frothability, etc. or image analysis parameters (color, mobility, stability) have been used in an engineering level analysis; but clearly there is a need for a deeper understanding of this very important portion of the flotation process. Similarly, understanding the dynamics and complexity of water chemistry and modeling these features remains a great challenge (and certainly ripe for critical innovations). A host of sophisticated techniques to characterize ores and study the mineral-solution interface and interaction of reagents with mineral surfaces are now available and can provide much needed insight to unravel the complex systems of practical importance to the mineral processing industry. It is beyond the scope of this paper to dwell on all these important aspects.

In addition, bio-mineral processing (i.e., the use of microorganisms in mineral beneficiation) and biologically sourced reagents have also made significant advances in this modern period (Chandraprabha and Natarajan 2010); however, they are still in the laboratory scale.

PERIOD 5: EMERGING TRENDS—SUSTAINABILITY, GREENER CHEMICALS, INTEGRATED APPROACH, STRATEGIC COLLABORATION

Areas where new technology is required are greener chemicals that can address the industry's need for more robust, environmentally benign, energy- and water-efficient sustainable separation solutions. Most major mining companies include sustainability targets in their annual objectives today. Innovative improvements in flotation unit operations would support this goal. The importance of employing a holistic, or whole systems approach, for ore characterization and to develop truly novel, step-change mining chemical technology is central to this undertaking.

Compared to physical-mechanical methods, chemical solutions to mineral separation problems are much easier to implement and control. As a result, chemical solutions are now an integral part of sustainable mineral processing. Chemicals today play a critical role, not just in flotation, but in almost all areas of mineral separation as well as in waste treatment, environmental remediation, mine closures and improving water and energy efficiency. Thus, chemicals can provide tremendous benefits and have a significant impact on sustainable mineral development with regard to both environmental footprint reduction and efficient conversion of mineral resources into reserves. In essence, the opportunities for improvements in energy efficiency in mineral processing are quite large. The use of increasing amounts of high quality energy to solve our technical problems is not viable into the future. For example, trying to solve the problem of increasingly poor ore grade by simply increasing the mechanical energy density brought to bear on the problem is unsustainable. The use of efficacious chemicals to enhance separation effectiveness can greatly reduce the total energy required to turn a resource into a final product.

Several issues must be addressed, however, if development efforts in the coming years will once again reach the high level of productivity achieved in the two most active development periods (1920–1940 and 1980–2000). Firstly, research efforts must be both more coordinated and integrated so that experiments are designed appropriately to fill knowledge gaps. For example, information is needed on modifier and frother functions and the dynamics of real, complex separation systems in the plant. Secondly, cooperation among diverse groups of researchers in academia and industry is necessary to ensure that the right questions are asked, which will ultimately provide the knowledge and understanding that will help overcome longstanding problems such as the effect of water chemistry, the adverse effect of altered silicates and slimes, recovery of coarse and fine particles and difficult mineral separations.

Thirdly, a holistic approach to investigating the chemistry of flotation is necessary. Research in the past has followed a reductionist path, focusing on simplified and often idealized representations of parts of the mineral separation system. In most cases these studies do not shed enough light on how to improve practical, commercial scale plant operations. All three types of flotation reagents (collectors, frothers and modifiers) interact with one another, as well as with chemical, physical-mechanical and operational factors. Therefore none of these factors should be studied independently if a true understanding of these complex systems is to be gained. When the increasing complexity of ores, growing demand for mineral products, rising regulatory oversight and need for improved sustainability of mineral separation processes are also considered, it is clear that a paradigm shift in the research framework for developing chemicals for mining applications is critical for successful advancement of mineral processing. Serious recognition of this fact and positive action taken to face these challenges are beginning to lead us into the next, emerging phase of flotation reagent innovations.

CONCLUDING REMARKS

A major component of sustainable development is conversion of mineral resources into reserves—and chemicals play a critical role in this. Chemicals can have great benefits and an impact in all five major categories of sustainable mineral processing development: energy efficiency, water efficiency, waste management, SHE (safety, health and environment) and converting mineral resources into reserves. As high-grade feed stocks continue to decline, innovations in all three categories of flotation reagents (collectors, frothers and modifiers) and their optimized use will continue to play a major role in sustaining the mineral processing industry.

Use of chemical energy (in this case, the amount of energy used in forming an efficacious chemical reagent) can greatly reduce the mechanical energy required in some stages of separation processes. That is, in order to produce an incremental improvement in value mineral recovery, the amount of energy embodied in the chemical reagents employed is much less than the amount of mechanical energy required to produce an equivalent incremental improvement in recovery. Chemical solutions to mineral separation problems are much easier to implement (within existing infrastructure) and control than are physical-mechanical solutions. If an ore composition or even type changes you are not as likely to change the equipment as you are to change the functional chemicals. This inherent flexibility allows for greater responsiveness within the confines of the existing equipment. You can have lots of different chemicals in your toolkit, but usually only a modest number of mechanical and equipment options at hand to optimize your separation process for what may be a highly variable ore.

Historically, chemical innovations of practical importance have rested on four different pillars. Obviously a new chemical, class of chemicals or synergistic blend of chemicals constitutes the most visible form of an innovation in this context. A less appreciated type of innovation is an improvement in manufacturing a chemical. Unless there is sufficient availability of a sufficiently economical material in sufficient quantity, then a chemical it is not likely to be employed in any meaningful way in a commercial process. Innovations that open the door to practical application on the large scale typical of mineral processing operations form a second important pillar. The third pillar is comprised of those intellectual advances that open our eyes to a better way of conceptualizing the physical, chemical and engineering aspects of the separation process. Innovations in this category often rely quite heavily on innovations that form the fourth pillar—analytical tools and methods that allow us to develop a deeper, more sophisticated and quantitative picture of the flotation process.

The state of our ability to rationalize empirical observations and predict behavior of a process is a good indicator of our conceptual understanding. On this score there is still some open territory for research in flotation. Intellectual innovations are enormously important in catalyzing improvements in not only reagent development, but also in their optimal use. The history of flotation reagents contains many good examples of this. Flotation reagents developed for sulfide and non-sulfide mineral systems, and even the conceptual frameworks underpinning their development, are characterized by some significant differences. For example, concepts of donor-acceptor interactions and chelation were central to understanding the mechanism by which sulfide mineral collectors function; whereas much of the functioning of non-sulfide mineral collectors can be understood on the basis of electrostatic and electrodynamic interactions. The complexity of practical ore systems, compounded by the presence of intrinsic (e.g., water, dissolved salts) and purposely added substances (e.g., collectors, frothers, and modifiers), can present a formidable challenge to model even semi-quantitatively. However, it has become clear that a holistic understanding of systems of practical importance is an edifice worth scaling and necessary to make sufficient progress to keep up with sustainability goals.

REFERENCES

Note: There are some inconsistencies in the way references are catalogued in the electronic media; e.g., the Gutzeit reference appears as both 1946 and 1947.

Aimone, F.M. and Booth, R.B. 1956. Flotation of ores using addition polymers as depressants. US2740522.

Allen, C.F., Fitzsimmons, W.M. and Booth, R.B. 1953. Depression of sulfide minerals with zinc cyanide ammoniate. US2660306.

Arbiter, N. and Hinn, H. 1968. Beneficiation of cassiterite ore by froth flotation. GB1110643.

Barker, L.M. and Young, O.E. 1953. Flotation recovery of molybdenite. US2664199.

Bolton, E.K. 1928. CA281647.

Booth, R.B. and Dobson, J.M. 1954. Frothing agents for the flotation of ores and coal. US2695101.

Booth, R.B. and Freyberger, W.L. 1962. Froths and frothing agents. In *Froth Flotation, 50th Anniversary Volume*. Edited by D.W. Fuerstenau. New York: American Institute of Mining, Metallurgical, and Petroleum Engineers: 258.

Bresson, C.R., Parlman, R.M., and Kimble, J.B., 1983. BE895254.

Bruey, F. 1999. The REFDIST method for design and analysis of plant trials. SME Preprint 99–190.

Cappuccitti, F.R. 1994. Current trends in the marketing of sulfide mineral collectors. In *Reagents for Better Metallurgy*. Edited by P.S. Mulukutla. Littleton, CO: SME.

Chandraprabha, M.N. and Natarajan, K.A. 2010. Microbially-induced mineral beneficiation. *Min. Proc. Ext. Met. Rev.*, 31:1–29.

Christmann, L.J. and Falconer, S.A. 1936. Oxide flotation. US2029156.

Christmann, L.J. and Jayne, D.W. 1936. Flotation reagent. US2043192.

Coghill, W.H. and Clemmer, J.B. 1935. Soap flotation of the nonsulfides. In *Milling Methods*, Trans. AIME, 112, 449–465.

Douglass, W.A. 1927. US1652099.

Douglass, W.A. 1928a. US1674166.

Douglass, W.A. 1928b. US1659396.

Drake, R.T.N. 1961. CA616687.

Falconer, S.A. 1960. Minerals beneficiation–non-sulfide flotation with fatty acid and petroleum sulfonate promoters. *Trans. AIME* (Tech. Publ. 59B33): 207–215.

Falconer, S.A. and Crawford, B.D. 1946. Froth flotation of some nonsulphide minerals of strategic importance. *Trans. AIME*: 527–542.

Farinato, R.S., Nagaraj, D.R., Larkin, P., Lucas, J. and Brinen, J.S. 1993. Spectroscopic, flotation and wettability studies of alkyl and allyl thionocarbamates. SME Annual Meeting. Littleton, CO: SME. Preprint Number 93–168.

Franzidis, J.P. 2005. Integration of science and practice in mineral flotation. *Min. Mag.* May: 12–17.

Fuerstenau, D.W. and Pradip. 1984. Mineral flotation with hydroxamate collectors. In *Reagents in the Mineral Industry*. Edited by M.J. Jones and R. Oblatt. London: Institute of Mining and Metallurgy: 161–168.

Gannon, D.J. 1980. Hydrophilic thio compounds as selective depressants in the flotation separation of copper and molybdenum. US4196073.

Gaudin, A.M. and Glover, H. 1928. Flotation of some oxide and silicate minerals. Tech. Papers 1: 78–101.

Gorlovskii, S.I. 1961. The application of high-molecular-weight compounds to flotation of sulfide minerals. *Obogashch. Rud.* 6(5): 18–23.

Gorlovskii, S.I., Vlodavskii, I.K., Koltunova, T.E. and Shchukina, N.E. 1961. Flotation of minerals of the wolframite group. SU 135431.

Gutzeit, G. 1946. Chelate-forming organic compounds as flotation reagents. AIME Technical Publication No. 2077.

Harris, G.H. and Fischback, B.C. 1954. Process for the manufacture of dialkyl thionocarbamates. US2691635.

Jensen, W.B. 1980. *The Lewis Acid-Base Concepts: An Overview*, Wiley–Interscience.

Keen, J.L. and Opie, J.W. 1962. Flotation concentration of halite. US3032198.

King, R.P. 2001. *Modeling and Simulation of Mineral Processing Systems*, Butterworth-Heinemann.

Klimpel, R.R. 1994. A discussion of traditional and new reagent chemistries for the flotation of sulfide minerals. In *Reagents for Better Metallurgy*, Edited by P.S. Mulukutla. Littleton, CO: SME: 59–66.

Lewis, G.N. 1923. *Valence and the Structure of Atoms and Molecules*, The Chemical Catalog Co.

Malhotra, D. 1994. Reagents in the mining industry: Commodities or specialty chemicals? In *Reagents for Better Metallurgy*, Edited by P.S. Mulukutla. Littleton, CO: SME.

Moyer, S.P. 1958. Flotation: The art of utilizing modifying agents. *Mining Magazine* 98: 9–16.

Nagaraj, D.R. 1988. The chemistry and applications of chelating or complexing agents in mineral separations. In *Reagents in Mineral Technology*. Edited by P. Somasundaran and B.M. Moudgil.

Nagaraj, D.R. 1994. A critical assessment of flotation agents. In *Reagents for Better Metallurgy*, Edited by P.S. Mulukutla. Littleton, CO: SME.

Nagaraj, D.R. 1997. Development of new flotation chemicals. *Trans. Indian Inst. Met.* 50(5): 355–363.

Nagaraj, D.R. 2000. New synthetic polymeric depressants for sulfide and non-sulfide minerals. Proceedings of the XXI International Mineral Processing Congress, Rome, Italy: Elsevier. Developments in Mineral Processing Series 13: C8b-1–C8b-8.

Nagaraj, D.R. 2005. Reagent selection and optimization—the case for a holistic approach. *Minerals Engineering* 18(2): 151–158.

Nagaraj, D.R. 2008. New approach to reagent development and applications in the processing of base and precious metals ores. XXIV International Mineral Processing Congress, Beijing.

Nagaraj, D.R. 2010. Chemical aspects of sustainable mineral processing and a framework for research and technology transfer. XXV International Mineral Processing Congress, Brisbane.

Nagaraj, D.R. and Avotins, P.V. 1988. Development of new sulfide and precious metals collectors. Proceedings of the II International Minerals Processing Symposium, Eylul University, Izmir, Turkey, Izmir, Turkey: 399.

Nagaraj, D.R., Basilio, C.I. and Yoon, R.H. 1989. The chemistry and structure-activity relationships for new sulfide collectors. SME: Preprint Number 89–306.

Nagaraj, D.R. and Bruey, F.S. 2003. Reagent optimization in base metal sulfide flotation-pitfalls of standard practice. *Flotation and Flocculation—From Fundamental to Applications*; Proceedings of Strategic Conference and Workshop, Kona, Hawaii: Ian Wark Research Institute, University of South Australia, Adelaide, Australia: 257.

Nagaraj, D.R., Wang, S.S., Avotins, P.V. and Dowling, E. 1986. Structure-activity relationships for copper depressants. *Trans. Inst. Min. Metall.* 95: C17–26.

Nagaraj, D.R., Lewellyn, M.E., Wang, S.S., Mingione, P.A. and Scanlon, M.J. 1988. New sulfide and precious metals collectors: For acid, neutral and mildly alkaline circuits. XVI International Mineral Processing Congress: Elsevier.

Nagaraj, D.R. and Ravishankar, S.A. 2007. Flotation reagents—a critical overview from an industry perspective. In *Froth Flotation: A Century of Innovation* Edited by M.C. Fuerstenau, G.J. Jameson and R.-H. Yoon. Littleton, CO: SME. 375–424.

Parlman, R.M. and Bresson, C.R. 1982. Recovery of molybdenum by flotation with depressants. BE890660.

Pearson, R.G. 1963. Hard and soft acids and bases. Journal of the American Chemical Society 85(22): 3533–3539.

Perkins, C.L. 1917. US1240597, US1240598, CA180529, CA180528.

Perkins, C.L. 1919. US1302966, US1322816, US1394639, US1364304, US1364305, US1364306, US1364858.

Perkins, C.L. and Sayre, R.E. 1921. US1364307, US1364308, US1364859.

Peterson, H.D., Fuerstenau, M.C., Rickard, R.S. and Miller, J.D. 1965. Chrysocolla flotation by the formation of insoluble surface chelates. *Trans. AIME* 232(4): 388–392.

Pöpperle, J. 1941. Verfahren zur schwimmaufbereitung von erzen. DE700735.

Ralston, O.C. 1938. Flotation and agglomerate concentration of nonmetallic minerals. Bureau of Mines Report of Investigations–1938 RI 3397.

Rogers, J., Sutherland, K.L., Wark, E.E. and Wark, I.W. 1946. Principles of flotation—paraffin chain salts as flotation reagents. *Trans. AIME* (Mining Technology).

Sheridan, G.E. and Griswold, G.G.J. 1922. US1427235, US1421585.

Taggart, A.F. 1951. *Elements of Ore Dressing*. New York, Wiley.

Taggart, A.F., Taylor, T.C. and Ince, C.R. 1930. Experiments with flotation reagents. *Trans. AIME* 87: 285–368.

Thom, C. 1962. Standard flotation separations. In *Froth Flotation 50th Anniversary Volume* Edited by D.W. Fuerstenau. New York: AIME.

Tveter, E.C. 1952. Frothing agents for flotation of ores. US2611485.

van Oss, C.J., Chaudhury, M.K. and Good, R.J. 1987. Monopolar surfaces. *Advances in Colloid and Interface Science* 28(1): 35–64.

Wang, S.S.-N. and Nagaraj, D.R. 1989. Novel collectors and processes for making and using same. EP 311759.

Wottgen, E. and Lippman, C. 1963. Phosphonic acids as collectors for cassiterite. *Bergakademie* 15(12): 868–870.

Yoon, R.-H. and Hilderbrand, T.M. 1986. Purification of kaolin clay by froth flotation using hydroxamate collectors. GB 2167687.

ABOUT THE AUTHORS

D.R. Nagaraj is a senior research fellow at Cytec Industries R&D Center in Stamford, Connecticut; adjunct professor at Columbia University and the University of Queensland; and co-editor of *International Journal of Mineral Processing*. His technical background includes minerals processing, extractive metallurgy, mineral aquatic chemistry, organic/analytical/coordination chemistry, water-soluble polymers, surface and colloid chemistry, and electrochemistry. Nagaraj is well known in the mineral industry for his expertise in reagents and chemical aspects of minerals/metals separation. He has a B.Sc. (Hons) in chemistry, B.Eng. and M.Sc. in metallurgy from Indian Institute of Science, and Eng.Sci.D. in mineral engineering and surface chemistry from Columbia University. He has more than 90 publications and 28 patents.

Raymond S. Farinato is currently a senior research fellow in the In Process Separation Department of Cytec Industries Inc. (Stamford, Connecticut), and an adjunct professor in the Earth and Environmental Engineering Department at Columbia University. He began his industrial career in 1981 at American Cyanamid, which became Cytec Industries in 1993. Farinato received a BS degree in chemistry from Rensselaer Polytechnic Institute and a PhD degree in physical chemistry from the University of Massachusetts–Amherst. He was a post-doc in biophysics at the University of California at Berkeley and consulted at the Lawrence Berkeley Labs on radiation damage to DNA. He was also an assistant professor at the University of Massachusetts; a senior engineer at the Nearshore Environment Research Center in Tokyo, Japan; and, more recently, an adjunct professor at Columbia University where he continues to teach a graduate level course on colloid and interface science. At Cytec he has worked predominantly in the areas of water-soluble polymers, colloid chemistry, and adhesion science, with an emphasis on applications in the waste management, petroleum, and mineral processing industries. He is a founding member of the International Symposia on Polyelectrolytes and an editor of *Encyclopedia of Surface and Colloid Science.*

Innovations in Froth Flotation Modeling

Peter Amelunxen
Aminpro Chile, Santiago, Chile

Kym Runge
Metso Process Technology and Innovation, Brisbane, Australia

ABSTRACT: This paper reviews the key events in the history of flotation modeling from 1935 through 2013. The focus is on first order kinetics models with an emphasis on the commonly used methods in the design and engineering sector of extractive metallurgy. Some ideas for future direction are also provided.

INTRODUCTION

The practice of mathematical modeling far predates the development of the flotation process, and therefore, the history of flotation modeling, in many respects, is a history of flotation itself. Tracing that history through the past 100 or so years, one can identify several key developments in the science. Some ideas were spawned by developments from within the field of extractive metallurgy; for example, the adoption of compartmental models was probably driven, in large part, by the acceptance of column flotation in the 1980s. Other milestones were achieved not because of innovation from within the industry but by advances in technology or fields of science outside of mineral processing. For example, the single most important event in the history of flotation modeling could arguably be the widespread adoption of personal computers, an event that gave plant designers and operators access to more advanced flotation modeling algorithms. This not only gave production metallurgists powerful tools to reduce costs or optimize processes, but also gave academics feedback on model effectiveness, thereby accelerating the development of further improvements.

This paper reviews the key events in the history of flotation modeling from 1935 through 2013. Some ideas for future direction are also provided. The manuscript has been divided into the four main periods of flotation modeling:

1. The Early Years (1935–1962), models based on personal experience or empirical data
2. The Developmental Years (1962–1990), the beginning of phenomenological modeling
3. The Acceptance of Flotation Modeling (1990–2013), and the adoption of computers
4. The Future

THE EARLY YEARS (1935–1962)

The earliest patents relating to froth flotation date back to the last half of the 19th century, but it wasn't until 1904 that flotation was applied commercially at Broken Hill in Australia, and then in 1912 in Montana (Arbiter 2000). The first three decades of development were focused mainly on refining the equipment design and application, and by the early 1930s the equipment characteristics began to coalesce around the mechanically agitated cell concept in use today (Jameson 1992). During these early years, scale-up and testing was dominated by trial and error through pilot, industrial or semi-industrial continuous tests, such as those described extensively by Gahl (1916). One of the earliest—often cited as the first—published suggestions that the cumulative recovery of economic minerals in a batch test approximated an exponential equation was given by Garcia-Zuñiga (1935). The equation is often referred to as the "Garcia-Zuñiga" equation and is shown as follows:

$$R = 1 - e^{-kt} \quad \textbf{(1)}$$

where R is the metal recovery, k is a constant, and t is the flotation time.

Throughout the ensuing decades, many studies were conducted to refine the concept. It is impossible to summarize all the studies here, but many of them sought to relate the rate constant, k, to various process phenomena such as

- Attachment and detachment rates (Schuhmann 1942),
- Ore characteristics such as particle size (Gaudin et al. 1942, Morris 1952),
- Mineralogy and particle oxidation (Imaizumi and Inoue 1963),
- Operating conditions such as reagent dosage or reagent surface coverage (Schuhmann 1942), and
- Machine-dependent characteristics such as bubble saturation or impeller energy input (Sutherland 1948).

While this early work played an important role in the lead-up to the development of modern phenomenological modeling methods, they were of limited use for the purpose of plant design and optimization. This was mainly due to the fact that the input variables were difficult or expensive to quantify, and Equation 1 was only valid for modeling the behavior of a batch test. For lack of better alternatives, engineers often adopted empirical methods to convert laboratory data into predictions of industrial cell performance. These empirical methods ranged from simple heuristics based on personal experience to more "scientific" methods such as the time multiplier method. In the time multiplier method, the industrial flotation bank retention time is estimated by multiplying the time taken to obtain a given recovery in a laboratory batch test by factor between 2 and 3. Discussion of these empirical methods can be found in Barbery et al. (1986); an example is given by Lindgren and Broman (1976).

Some parallel advances in the field of chemical reactor engineering deserve mention. One of the earliest calculations of flow and kinetics in continuously fed stirred tank reactors (CSTRs) is provided by Ham and Coe (1918). In 1935—the same year that Garcia-Zuñiga proposed the exponential equation for modeling batch test recovery—the first discussion of short-circuiting in CSTR cells was published (MacMullin and Weber 1935). It wasn't until the 1950s, however, that the modern discipline of chemical reactor design began to take shape. P.V. Danckwert's comprehensive analysis of residence time distributions was perhaps the most influential technical contribution (Dankwerts 1953). The field was still in its infancy, however, and the First European Symposium on Chemical Reaction Engineering, held in Amsterdam in 1957, provided the cross-pollination of ideas necessary for the discipline to crystalize. This conference was ultimately credited by Octave Levenspiel as the inspiration for his "Chemical Reaction Engineering," which was published in

1962 and was to become one of the most cited works in chemical engineering history (Current Contents 1979).

THE DEVELOPMENTAL YEARS (1962–1990)

1962 was an eventful year, not just because of the Levenspiel text. It was the year of the Cuban missile crisis, the death of Marilyn Monroe, the birth of Leonard Hill, and the arrest of Nelson Mandela. It was also the 50th anniversary of froth flotation in the United States, and it coincided with the introduction of the earliest first order CSTR kinetics model in the field of extractive metallurgy:

$$R = \frac{k\tau}{1 + k\tau} \tag{2}$$

where k is the overall first order kinetic rate constant with respect to time, and τ is the mean residence time of the cell. Equation 2 is generally credited to Arbiter & Harris (1962) but is also described by Dorenfeld (1962). It marks a clear departure from empirical or heuristic scale-up methods used earlier, and although its adoption was neither immediate nor widespread, it clearly formed the basis of all phenomenological scale-up approaches that were to come. It was a milestone in froth flotation modeling, as most of the subsequent developments were evolutionary incarnations of this same approach.

Rate Constant Distributions

It has been recognized since at least 1963 that the heterogeneous nature of the ore feeding a flotation cell gives rise to differences in floatability between the particles (Imaizumi and Inoue 1963). Researchers proposed alternatives to the first order rate equations for plug-flow reactors (PFRs) and CSTRs, depending on the kind of model being proposed. Because the exact shape of the rate constant distribution could not be accurately measured, methods were proposed to approximate the shape by fitting the curve parameters to experimental data. Runge (2007) classifies the approach according to the method used to approximate the shape of the distribution. These are:

- Empirically derived distributions, in which the shape of the distribution is defined by a small number of parameters that can be derived from experimental data. Examples include the rectangular distribution (e.g., Klimpel 1980), the triangular distribution (e.g., Harris & Chakravarti 1970), and the normal distribution (e.g., Chander & Polat 1994).
- Empirically derived flotation rate components, in which the feed particles are grouped into discrete classes and the mass and flotation rate of each class are fitted to experimental data. The best example is the Kelsall model (1961).
- Property-based rate components, in which the feed is discretized into classes based on physical characteristics such as size class and mineral type. The flotation rate for each size class is then fitted independently to the experimental data. Examples of this approach have been provided by Thorne et al. (1976) and Kawatra et al. (1982).

The Importance of Froth Recovery

The 1962 Arbiter paper was significant not just for providing the first 1st order kinetics model of a CSTR. It also presents a rather sobering assessment of the roll of froth recovery in flotation kinetics, namely:

- "The presence of a froth phase in the semi-batch case…complicates rigorous handling of the equations unless there is no drop-out from the froth. This can only be assured by removing the mineralized froth as fast as it appears." (Arbiter and Harris 1962)
- "With respect to continuous flotation [it is shown] that if there is drop-out of floatable material from the froth, which is probable in most cases, the value calculated for the rate constant will depend on the rate of concentrate removal and on the froth volume." (Arbiter and Harris 1962)
- "No clear decision as to the validity of *any* rate equation is possible from available data. Detailed consideration of…these equations suggest that the problem has been oversimplified by disregarding the froth as a separate phase." (Arbiter and Harris 1962)
- "The complications introduced by fall-out from the froth…are serious obstacles to the use of simple first order equations." (Arbiter and Harris 1962)

Arbiter's exceptions were noted and some early simulators, such as the N.I.M simulator developed at the University of Witwatersrand (King 1976), began to include a term to account for froth recovery, denoted R_f, but the parameter often remained a user-specified input and no clear analytical solution to the problem would become available for many years. As a result, the flotation rate constant came to be called the "overall" flotation rate constant in recognition of the fact that it was the combined rate of particle recovery in the pulp zone and the recovery of those particles across the froth zone.

Modeling of Entrainment

Since at least 1972 it has been recognized that the hydraulic entrainment of particles into the froth is a linear or near-linear function of the water recovery into the froth (Johnson et al. 1972), and that finer particles have a higher probability of entrainment than coarser particles. The so-called classification function of entrainment was proposed; Figure 1 shows one of the earliest examples to appear in the literature. Later, Thorne et al. (1976) add that particle specific gravity also affects the probability of entrainment, with particles of lower specific gravity having a higher probability of entrainment for a given size class than those of higher specific gravity. For the purpose of scale-up, no model was yet available for estimating water recovery in industrial cells.

The Batch Test

The batch test has evolved little since the introduction of the Denver sub-A bench scale laboratory flotation cell, which by 1954 was already shipping with cell volumes rated for samples of 50-gram, 500-gram, 1,000-gram, and 2,000-gram masses. The early cells shipped with a sample of flotation reagents and a complementary vanning plaque (Denver Equipment Company 1954). Most of the original characteristics are still preserved in the modern Denver laboratory cell, and together with the Bond ball mill, few other examples can be found of laboratory mineral processing equipment with such longevity.

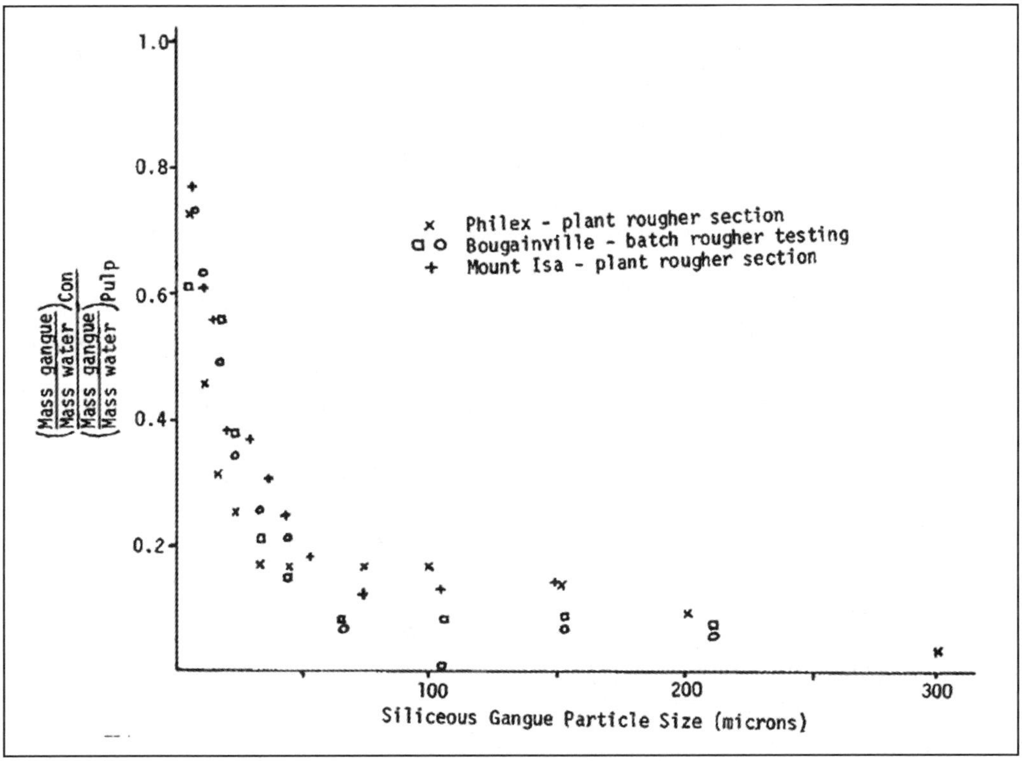

Figure 1. Observed entrainment classification functions (Johnson et al. 1972)

If the laboratory testing apparatus remained fairly constant, the modeling of the batch test was evolving. Engineers began to adopt alternative kinetics equations such as the Klimpel model or the Kelsall two-component equation discussed above. An additional fitted term in the first order (Equation 1) PFR equation was introduced to improve the fit; this was the so-called Agar parameter (Agar et al. 1980) and was meant to account for the time elapsed between when the air is turned on and the first froth is removed into the pan, and is defined by Φ:

$$R = \left(1 - e^{-k(t+\Phi)}\right) R_{\infty} \quad (3)$$

where R_{∞} is the flotation recovery at infinite time.

Early Process Models

Although, by the 1980s most plant operators and design engineers were still using empirical scale-up factors for the purpose of process design, many academic and research institutes were actively developing phenomenological process models for predicting flotation performance. The N.I.M. simulator at the University of Witwatersrand was only one such example—significant modeling work was also being done at the University of Queensland (Schwarz 2012) while McGill University was investigating reaction kinetics in the emergent subfield of column flotation (Dobby and Finch 1985).

The Impact of Column Cells on Flotation Modeling

The idea of column flotation has been around since the early 1900s and the first patent of a modern-looking countercurrent column cell was that of M. Town and S. Flynn, dated 1919 (Rubinstein 1995). While the early concepts were notable, limitations in air sparging and process control technology inhibited their adoption, and column flotation largely disappeared in favor of mechanically agitated cells. In the 1960s, Canadian researchers P. Boutain and R. Tremblay began tests with the first modern counter-current columns (Rubinstein 1995) and by the early 1980s the first units were being installed in production environments (Coffin 1982, Amelunxen 1985).

The column flotation cell was a key milestone in the maturation of froth flotation, not just because of its impact on process design and overall metallurgical efficiency. The use of much deeper froth beds than those traditionally used for mechanical cells, combined with the frequent use of wash water, gave impetus to the separation of froth recovery and entrainment effects from the traditional kinetics equation model. It is no accident that one of the earliest graphical depictions of the compartmental model (discussed below) can be found in a 1990s text on column flotation (Finch and Dobby 1990). Indeed, much of the early flotation modeling research in progress at the University of Queensland in the late 1980s was directed at column flotation cells (Schwarz 2012). These models would eventually evolve into one of the first commercially available flotation simulators.

THE ACCEPTANCE OF FLOTATION MODELING (1990–2013)

In preparing this manuscript the authors debated the title of this section. It could have been called "The Computer Era" of froth flotation, because the events after 1990 were driven largely by the availability and practicality of computers, and this, together with good research, is what made flotation models as advanced, accessible and accepted as they are today.

For the purpose of this narrative, this era of froth flotation modeling has been dated from the first shipment of Microsoft Windows 3.0 in May 1990 (Microsoft Corporation 2012). Even though Apple computers already existed, the open standard of IBM's hardware configuration and Microsoft's MS-DOS® platform, had previously generated an unstoppable economy of scale, and by 1990 IBM-compatible PCs had 80% of the market share. The release of a graphical user interface (GUI) for IBM-compatible PCs that professional or home users could understand and operate, combined with the integration of the Excel spreadsheet and other applications, provided the conditions required for widespread adoption of the flotation models that hitherto had been confined to academic and research institutions. Shortly thereafter, commercially available flotation models began to appear in the form of desktop applications. USIM-PAC®, developed by the BRGM, was one of the first general circuit simulators (Guillaneau et al. 1993). The Julius Krutschnitt Mineral Research Center's (JKMRC) JKSimFloat® model, shipped as an MS-DOS® version in 1993, was the first publicly available model focused solely on flotation modeling and optimization (Schwarz 2012). As flotation modeling began to gain mainstream acceptance, researchers continued to refine the underlying phenomenological models, and the use of computers also played a key role in this refinement. The refinements in model development that were yet to come were made possible, in part, because computers were now available to control the sensors, process the data, and evaluate the results. Some of the most significant updates are discussed below.

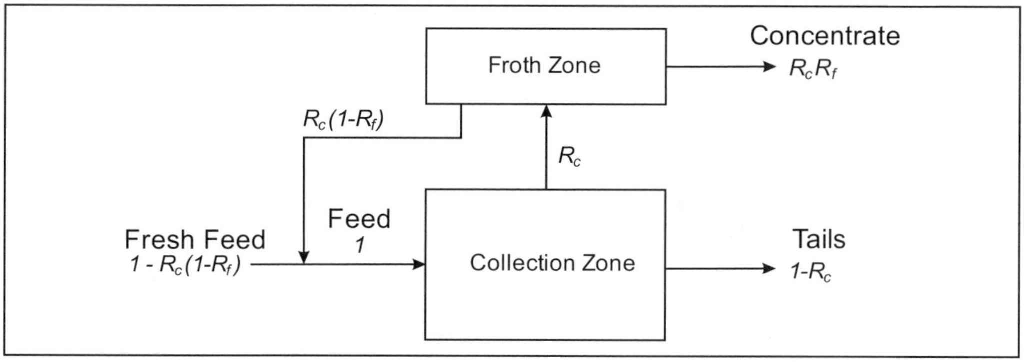

Figure 2. Schematic representation of the compartmental model

The Compartmental Model

The compartmental model (Finch and Dobby 1990) can be described schematically by depicting the froth flotation process as being comprised of two sub-processes—a pulp zone in which particles float due to collection, and a froth zone in which some particles drop back into the pulp and the rest are recovered to the concentrate (Figure 2).

If the froth zone recovery is defined as the fraction of collected particles that are recovered, then the overall flotation recovery, R_{flot}, can be defined by

$$R_{flot} = \frac{R_c R_f R_\infty}{R_c R_f + 1 - R_c} \tag{4}$$

where R_f is the froth recovery and R_c is the collection zone recovery, defined by the first order rate equation for either plug flow conditions or continuous fed reactors (Finch and Dobby 1990).

Entrainment

Degree of entrainment (*ENT*), sometimes referred to as the entrainment factor, is defined as the ratio of solids recovery by entrainment (R_s) to water recovery (R_w):

$$ENT = \frac{R_s}{R_w} \tag{5}$$

Previous work on entrainment, discussed above, identified the relationships between particle size, particle specific gravity, and degree of entrainment. Researchers began to fit empirical equations to the curves of particle size versus degree of entrainment and one notable model form is that of Savassi et al. (1998):

$$ENT_i = \frac{2}{e^{2.292\left(\frac{d_i}{\xi}\right)^{adj}} + e^{-2.292\left(\frac{d_i}{\xi}\right)^{adj}}} \tag{6}$$

where

$$adj = 1 - \frac{\ln\left(\frac{1}{\delta}\right)}{e^{\left(\frac{d_i}{\xi}\right)}} \tag{7}$$

and d_i is the particle size, ξ is the entrainment parameter (fitted), and δ is the drainage parameter (fitted). Alternatively, Yianatos and Contreras propose (2010) the following:

$$ENT_i = e^{-0.693\left(\frac{d_i}{\delta}\right)^{\Phi}} \tag{8}$$

where δ and Φ are fitted parameters. Note that both models are empirically-fitted curves and are of limited use for predicting plant-scale entrainment effects from batch test work. Zheng et al. (2006) show that the entrainment classification curve is closely related to the froth height and air rate in an industrial cell, with lower froth heights and higher air rates leading to higher degrees of entrainment for a given particle size class and specific gravity. Amelunxen and Amelunxen (2009) show how the Zheng et al. (2006) data can be used to derive a function relating the entrainment curve to the froth height and air rate, but as Yianatos et al. (2010) correctly point out, entrainment characteristics can change significantly from one system to another, and more work is required before an accepted plant-scale entrainment model is available.

The recovery of particles by entrainment was first incorporated directly into the first order CSTR model recovery model by Savassi (1998). The equation is now used by most plant simulators, and it is written as follows:

$$R_{tot} = \frac{k\tau(1 - R_w) + ENTR_w}{(1 + k\tau)(1 - R_w) + ENTR_w} \tag{9}$$

For batch tests, Equation 9 is rearranged and the term *kt/(1+kt)* is replaced by the plug-flow reactor model.

Machine Operating Characteristics

The 1990s saw the emergence of a range of tools designed to measure the gas dispersion properties in flotation machines. They are reviewed in detail by Gomez and Finch (2002, 2007) and include the following:

- Gas velocity sensors, which provide a direct measurement of superficial gas velocity, J_g.
- Bubble sizing tools, initially consisting of a bubble sampling device and laser or light-based volumetric determinations such as that developed by the University of Cape Town (Randall et al. 1989, O'Connor et al. 1990, Tucker et al. 1994). Due to improvements in digital photographic technology, the industry now favors the camera and viewing chamber methods.
- Gas holdup measurement devices and sensors, such as the conductivity probe developed at McGill University (Uribe-Salas et al. 1994).

Using these tools and others like them, researchers began to investigate the effects of cell operating conditions on flotation kinetics. Gorain et al. (1998) performed a series of tests with different impeller types and air rates, finding a linear relationship between the overall rate constant and the bubble surface area flux (S_b).

These findings are debated (Heiskanen 2000) but by-and-large validated by other researchers (see, for example, Hernandez-Aguilar et al. 2004).

Floatability Components and Kinetics Tracking

As kinetics-based computer models began to take shape, modelers needed methods for estimating the kinetics parameters of a blend of streams, or the residual kinetics of concentrates and tails after a flotation step. One of the earliest methods for calculating the rate constant of a blend of ores consisted of converting the first order rate constant into an additive parameter through the use of a transform function, calculating the weighted average transformed rate constant of the blend, and then performing the back-transform to determine the rate constant of the blend (Amelunxen and McCord 1995). In this approach, R_∞ is usually considered to be naturally additive. The concept is also applied in reverse to calculate the kinetics of the product streams of a flotation stage.

Gorain's et al. work, described above, together with the recent availability of computers, gave impetus to the alternative method of floatability components. After Gorain et al. (1998), the overall flotation rate constant is expressed as follows:

$$k = R_f P S_b \tag{10}$$

where P represents the natural, inherent floatability of a particle, given the chemical characteristics of the stream. Because P is a property of the ore (for a given set of chemical or pulp conditions), it can be tracked through a flotation circuit and therefore the metallurgical performance of subsequent stages can be modeled without having to perform detailed metallurgical testing (Harris 1997, Runge 2001).

Froth Recovery

Froth recovery is the black sheep of flotation modeling. At the industrial scale, it is often estimated by measuring the collection zone rate constant in the lab and then "tuning" the plant froth recovery, so that the modeled plant recovery matches the observed plant recovery. Because the froth recovery at the lab scale is not commonly known (most assume 100%) the resulting estimates of the collection rate constants can be questionable. For example, if the laboratory R_f is actually lower than the assumed 100%, then the true collection zone rate constants will be higher than the estimated values. This bias yields a "tuned" plant froth recovery value that is lower than that which would result by using the true collection zone rate constant. Clearly, direct measurement of the froth recovery would be extremely useful for resolving the controversy, and over the past 10 or so years, tools and methods to address these problems have begun to emerge.

Vera et al. (1999) demonstrate a method to infer the froth recovery in both industrial and laboratory cells. The method, illustrated graphically in Figure 3, consists of performing various batch tests at different froth levels to measure the overall flotation rate constant. The collection zone rate constant is then estimated by extrapolating the curve to a froth height of zero. Using this approach, the authors derived estimates of the froth recovery in the bottom-driven, high S_b, cell developed by the JKMRC (Vera et al. 1999). They report values of R_f ranging from 50% to 95%.

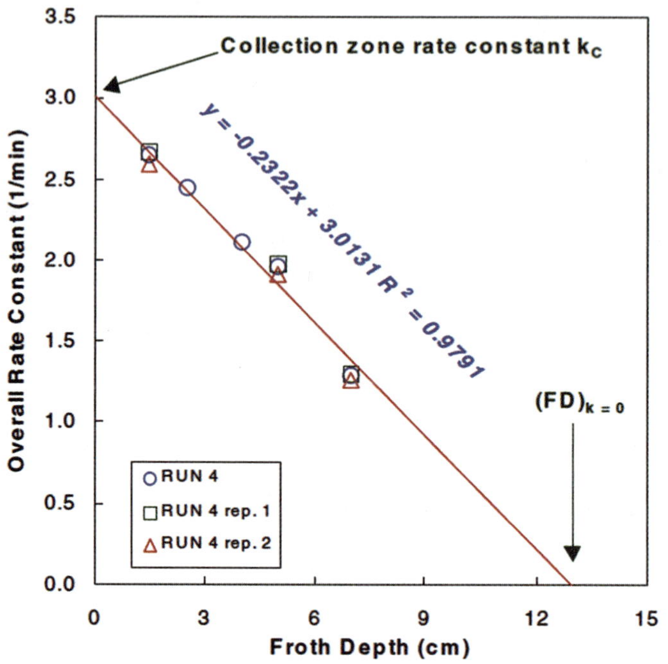

Figure 3. Estimation of collection zone rate constant, from Vera et al. (2001)

More recently, researchers (Savassi et al. 1997, Vera et al. 1999 Alexander et al. 2003, Seaman et al. 2004, Yianatos et al. 2008) have developed and utilized an alternative method for calculating the froth recovery. The approach consists of collecting bubble samples from the pulp phase and measuring or calculating the solids loading. Combined with air rate measurements, the method provides an estimate of the solids flow rate into the froth, allowing for the calculation of R_f from a mass balance around the froth zone. The method has led to some interesting findings. For example,

- Hadler and Cilliers (2009) used this technique, among others, to demonstrate that the air flow rate to a bank is closely related to the froth recovery and/or froth stability (as characterized by the air recovery parameter).
- Seaman et al. (2004) apply this method to estimate the froth recovery in a bottom-driven laboratory cell, yielding froth recovery values of around 30%—much lower than those derived using the variable froth depth method. They suggest that the discrepancy may be due to the fact that much of the bubble detachment process may occur at the pulp–froth interface (PFI), a region that can't be studied adequately by simply varying the froth height. This suggestion is supported by the results from chromite flotation trials performed by van Deventer et al. (2001).
- Tabosa et al. (2011) performed experiments with semi-industrial cells of different aspect ratios (length to height), concluding (among other things) that higher energy dissipation in the pulp zone may lead to froth or interface zone instabilities, causing reductions in froth recovery.

The above findings suggest that S_b, or one or more of the components of S_b, also plays a role in the particle recovery across the froth zone, pulp–froth interface, or both.

The authors expect that this area will see further developments in the near future as researchers continue to improve the technology for direct measurement of particle attachment and collection zone recovery, the froth recovery, and the parameters that describe the pulp–froth interface. This, in turn, may lead to a better understanding of some of the technical and process design issues that are currently much-debated—selective froth recovery, froth crowding effects, carrying capacity or lip loading constraints—concepts that, for lack of better alternatives, are currently modeled by empirical formulae or heuristics.

Water Recovery

Predicting the recovery of water to the concentrate of a flotation cell is not just necessary for entrainment modeling, but for estimated residence times of downstream stages as well. As such, water recovery is a key parameter in flotation circuit design, process control, and circuit optimization. Current water recovery models in practical or design applications are largely empirical in nature (Amelunxen and Amelunxen 2009a) and summarized descriptions can be found in Zheng et al. (2006). The empiricism persists (at least in non-academic circles), in part because the physical or system properties required by the fundamental or physics-based models—bubble diameter in the froth, froth zone pulp viscosity and the like—are too difficult or too expensive to justify in laboratories operating under commercial constraints, even if representative froth zone samples could be easily produced. Nevertheless, some currently employed scale-up methods do offer an option to use a physics-based water recovery sub model, such as that of Neethling et al. (2003), even if many of the input variables are given fixed or assumed values for greenfield applications.

Process Modeling and Scale-up

The differences in current scale-up models reflect the extent to which the various practitioners agree or disagree on the above concepts. There are far more similarities than differences. For example, all contemporary, validated flotation models assume 1st order kinetics, and all of them use the compartmental model, or a close analogue, for CSTRs. They all consider entrainment effects and plant scale froth recoveries, and most also recognize the importance of maximizing froth recovery at the laboratory scale. Most models consider rate constant distributions in some form or another, either through floatability component models (JKTech 2004), abstract or empirical distributions (Dobby and Savassi 2005), or property-based models (Amelunxen and Amelunxen 2009b).

The main differences between the current models lie in how each concept is incorporated in the model. For example, some groups have adopted bottom-driven cells to facilitate froth removal and to maximize S_b and froth recovery; others use standard Denver bench scale machines with forced air, a froth crowder, and much faster froth scraping rates and flotation times. Some models relate froth recovery and degree of entrainment to the froth depth and air rate; others leave the froth recovery and degree of entrainment up to the modeler. The most significant difference lies in the representation of the rate constant distribution—some of the alternatives currently in use include

- A single mean rate constant,
- A two-component (Kelsall) model,
- A rectangular distribution (Klimpel or equivalent),

- An empirical distribution, or
- A relationship with size class and mineralogy.

Each alternative has its own set of strengths and weaknesses, and its suitability depends on the application. A discussion and comparison of these strengths and weaknesses is presented elsewhere by Runge (2007).

Dynamic Models and Physics-Based Models

Up to this point, discussion on flotation models has focused on static, phenomenological models. In recent years the industry adoption rate of dynamic and physics-based flotation modeling software and methods has begun to increase. In practical applications, such as for the design and engineering of new concentrators, the use of these has been limited to the determination of required surge capacities, launder and pipe sizing, and process control requirements. Nevertheless, technology continues to evolve, both in computer processing power and laboratory methods (in the authors' opinion, the fidelity of laboratory data is the slower of the two). In the not-to-distant future it is possible that this class of model will become the more favored approach for process optimization and plant design. An excellent review of the approach is given by Herbst et al. (2005).

THE FUTURE

The authors speculate that the future of flotation modeling will be driven largely by technology, just as it has been previously. To illustrate, consider a thought experiment: a modern-day metallurgist goes back in time to 1962 with a modern-day laptop, complete with Matlab® and spreadsheet software. She goes to New York, where she meets one professor Nat Arbiter and gives him the laptop (she may also have to give him some computer lessons). What would the future have held?

The authors hope our future holds some of the following:

- Resolution of current limitations in process modeling, including more reliable models for industrial water recovery, froth recovery, and degree of entrainment based on parameters that can be measured in a laboratory.
- Improved understanding of physical flotation parameters, including turbulence, pulp viscosity, and froth viscosity, and how these effects can be used to customize the flotation cell design to a particular flow sheet application or set of feed stream characteristics.
- Cost reduction of current test procedures and scale-up methods through new technology, such as high resolution x-ray micro tomography (HRXMT), quantitative microscopy, quantitative image analysis, laser-based volumetric size analysis, and other methods
- Improved geometallurgical modeling through the use of high capacity computing and/or cluster computers, such that flotation optimization routines can be run for each block of ore, thereby enabling real-time mine planning optimization.
- Cheaper, smaller-scale ore characterization methods that will facilitate improved ore body characterization methods while still providing the high-fidelity flotation parameterization required for phenomenological and physics-based models to be run at the mine block level.
- Better incorporation of capital and operating cost models, so that flotation performance can be evaluated for cost optimization rather than metallurgical optimization
- Incorporation of upstream and downstream process models so that froth flotation optimization can be done at the enterprise level. This would require interfacing with the relevant

mining, milling, dewatering, water & tailings management, and environmental management models.

- Dynamic and physics-based models that are validated, easily-accessible, and robust enough for greenfield design applications.
- Happy, positive metallurgists who feel both challenged and pressured by all of the unknowns, yet can still find enough time to spend with their families.

REFERENCES

Agar, G.E., R. Stratton-Crawly, and T.J. Bruce. "Optimizing the design of flotation circuits." CIM Bulletin 73 (1980): 173–181.

Alexander, D.J., J.P. Franzidis, and E.V. Manlapig. "Froth recovery measurement in plant scale flotation cells." Minerals Engineering (Elsevier) 16 (2003): 1197–1203.

Amelunxen. "The mechanics of operation of column flotation machines." Proceedings of the Canadian Mineral Processors Conference. Ottawa: CMP, 1985.

Amelunxen, R., and P. Amelunxen. "Methodology for executing, interpreting, and applying kinetic flotation tests in scale-up, Part 1." Procemin 2009. Santiago, Chile, 2009a.

Amelunxen, P., and R. Amelunxen. "Methodology for using laboratory kinetic flotation parameters for plant design and optimization, Part 2." Procemin 2009. Santiago, Chile, 2009b.

Amelunxen, R., and McCord, T. " Expanded modal analysis to predict flotation performance at P.T. Freeport Indonesia," Proceedings of the SME Annual Meeting, Denver, Colorado, 1995.

Arbiter, N. "Development and Scale-up of Large Flotation Cells." Mining Engineering, 2000.

Arbiter, N., and C.C. Harris. "Flotation Kinetics." In Froth Flotation, 215–262. New York: AIME, 1962.

Barbery, G., M. Mourassa, and A. Maachar. "Chapter 28–Laboratory Testing for Flotation Design." In Design and Installation of Concentration and Dewatering Circuits. SME, 1986.

Coffin, V.L. "Column flotation at Mines Gaspe." CIM XIV International Mineral Processing Congress. Toronto: CIM, 1982.

Current Contents. "This Week's Citation Classic." Current Contents 49 (1979).

Dankwerts, P V. "Continuous flow systems. Distribution of residence times." Chem. Eng. Sci. 2 (1953): 1–13.

Denver Equipment Company. Denver Equipment Company Handbook. Handbook, Denver: Denver Equipment Company, 1954.

Dobby, G.S., and O.N. Savassi. "An advanced modeling technique for scale-up of batch flotation results to plant metallurgical performance." Centenary of Flotation Symposium. Brisbane, 2005.

Dobby, G.S., and J.A. Finch. "Particle collection in columns: gas rate and bubble size effects." Chemical Engineering Science 40, no. 7 (1985): 1061–1068.

Dorenfeld, Adrian C. "Chapter 15–Flotation Circuit Design." In Froth Flotation 50th Anniversary Edition, 365–381. AIME, 1962.

Finch, J.A., and G.S. Dobby. Column Flotation. Oxford: Pergamon Press, 1990.

Finch, J.A., J. Xiao, C. Hardie, and C.O. Gomez. "Gas dispersion properties: bubble surface area flux and gas holdup." Minerals Engineering, 2000: 365–272.

Gahl, Rudolf. "History of the flotation process at Inspiration." Arizona Meeting of the AIME. Tucson: AIME, 1916. 576–645.

Garcia-Zuñiga, H. "The efficiency obtained by flotation is an exponential function of time." Boletín Minero 47, no. Soc. Nac. Minera (1935): 83–86.

Gaudin, A.M., R. Schuhmann Jr., and A.W. Schlechten. "Flotation kinetics. II: Eth effect of size on the behavior of galena particles." Journal of Physical Chemistry, 1942: 902–910.

Gomez, C.O., and J.A. Finch. "Gas dispersion measurements in flotation machines." CIM Bulletin 95 (2002): 1066.

Gomez, C.O., and J.A. Finch. "Gas dispersion measurements in flotation cells." Int. J. Min. Process., 2007: 51–58.

Gorain, B.K., T.J. Napier-Munn, J.-P. Franzidis, and E.V. Manlapig. "Studies on impeller type, impeller speed, and air flow rate in an industrial scale flotation cell. Part 5: Validation of the k-Sb relationship." Mineral Engineering 16, no. 6 (1998): 739–744.

Guillaneau, J.-C., J. Villeneuve, S. Brochot, M.-V. Durance, and G. Fourniguet. "The supervisor of simulation: A step further to meet the process engineer needs." Proceedings of the XIX International Mineral Processing Congress. 1995.

Guillaneau, J.-C., M.-V. Durance, J. Libaud, P. Ollivier, and J. Billeneuve. "Computer-aided optimization of mineral processing plants, a case stud: Increasing the capacity of the shila gold mine, Peru." Proceedings of the SME Annual Meeting. Reno: SME, 1993.

Hadler, K., and J.J. Cilliers. "The relationship between the peak in air recovery and flotation bank performance." Minerals Engineering 22 (2009): 451–455.

Ham, A., and H.S. Coe. "Calculation of extraction in continuous agitation." Chem. & Met. Eng. 19 (9) (1918): 663.

Harris, C.C., and A. Chakravarti. "Semi-batch flotation kinetics: Species distribution analysis." Transactions AIME, 1970: 162–172.

Harris, M.C. A practical framework for flotation circuit modelling and simulation. SAIChE '97, 8th National meeting, SAIChE, Cape Town, 16–18 April 1997.

Heiskanen, K. "On the relationship between flotation rate and bubble surface area flux." Minerals Engineering 13, no. 2 (2000): 141–149.

Herbst, J.A., A.V. Potapov, W.T. Pate, and J.K. Lichter. "Advanced modeling for flotation process simulation." Centenary of Flotation Symposium. Brisbane, QLD, 2005. 111–119.

Hernandez, H., C.O. Gomez, and J. Finch. "Gas dispersion and de-inking in a flotation column." Minerals Engineering 16, no. 8 (2003): 739–744.

Hernandez-Aguilar, J.R., S.R. Rao, and J.A. Finch. "Testing the k-Sb relationship at the micro scale." Minerals Engineering, 2004: 591–598.

Honaker, R.Q., A.V. Ozsever, and B.K. Parekh. "Selective detachment process in column flotation froth." Centenary of Flotation Symposium. Brisbane, 2005.

Huber-Panu, I., E. Ene-Danalache, and D.G. Cojocariu. "Mathematical models of batch and continuous flotation." In Flotation (A.M. Gaudin Memorial Volume), 675–724. 1976.

Imaizumi, T., and T. Inoue. "Kinetic considerations of froth flotation." Proceedings of the 6th International Mineral Process Congress. Cannes, 1963. 581–593.

Jameson, G. "Flotation Cell Development." AusIMM Annual Conference. Broken Hill: AusIMM, 1992. 25–32.

JKTech. JKSimFloat User Manual Version 6.1. User's Manual, Brisbane: JKTech Pty Ltd, 2004.

Johnson, N.W., D.J. McKee, and A.J. Lynch. "Flotation rates of non-sulphide minerals in chalcopyrite flotation processes." Annual Meeting of the SME-AIME. San Francisco: SME, 1972. Preprint 72-B-24.

Kawatra, S.K., P.J. Suardini, and W.J. Whiten. "The computer simulation of an iron ore flotation circuit." Proceedings of the 14th International Mineral Processing Congress. Toronto, 1982. 10.1–10.19.

Kelsall, D.F. "Application of probability in the assessment of flotation systems." Transactions of the Institute of Mining and Metallurgy, 1961: 191–204.

King, R.P. "The use of simulation in the design and modification of flotation plants." In Flotation A.M. Gaudin Memorial Volume, by M.C. Fuerstenau, 937–962. New York: AIME, 1976.

Klimpel, R.R. "Selection of chemical reagents for flotation." In Mineral Processing Plant Design, 2nd Ed., 907–934. SME, 1980.

Levenspiel, O. Chemical Reaction Engineering. New York: John Wiley and Sons, 1962.

Lindgren, E., and P. Broman. "Aspects of flotation circuit design." Concentrates (Sala International Publication) 1 (1976): 6–10.

MacMullin, R.B., and M. Weber. "The theory of short-circuiting in continuous-flow mixing vessels in series and the kinetics of chemical reaction engineering." Transactions of the American Institute of Chemical Engineers 31, no. 2 (1935): 409.

Microsoft Corporation. www.microsoft.com. 2012. http://www.windows.microsoft.com.

Morris, T.M. "Measurement and evaluation of the rate of flotation as a function of particle size." T. AIME 193 (1952): 794.

Neethling, S.J., H.T. Lee, and J.J. Cilliers. "Simple relationships for predicting the recovery of liquid from flowing foams and froths." Minerals Engineering 267–287 (2003): 1123–1130.

O'Connor, C.T., Randall, E.W., and Goodall, C.M., "Measurement of the effects of physical and chemical variables on bubble size," International Journal of Mineral Processing, 1990, 28, 139–149.

Patwardhan, A., and R.Q. Honaker. "Development of a carrying-capacity model for column froth flotation." International Journal of Mineral Processing 77, no. 2 (2005): 65–79.

Polat, M., and S. Chander. "First order flotation kinetics models and methods for estimation of the true distribution of flotation rate constants ." International Journal of Mineral Processing 58 (2000): 145–166.

Randall, E.W., Goodall, C.M., Fairlamb, P.M., Dold, P.L., and O'Connor, C.T., "A method for measuring the sizes of bubbles in two- and three-phase systems, 1989, Journal of Physics, Section E, Scientific Instrumentation, 22, 827–833.

Rubinstein. Column Flotation Processes, Designs, and Practices. Amsterdam: Gordon and Breach Science Publishers S.A., 1995.

Runge, K. Modeling of Ore Floatability in Industrial Flotation Circuits. PhD Thesis Report, Brisbane: The University of Queensland, 2007.

Runge, K.C., M.E. Dunglison, E.V. Manlapig, and J.P. Franzidis. "Floatability component modeling–a powerful tool for flotation circuit diagnosis." Proceedings 4th International Symposium on Fundamentals of Minerals Processing. CIM, 2001. 93–107.

Savassi, O.N. Direct estimation of the degree of entrainment and froth recovery of attached particles in industrial flotation cells. Brisbane: University of Queensland, 1998.

Savassi, O.N., D.J. Alexander, N.W. Johnson, E.V. Manlapig, and J.P. Franzidis. "Measurement of froth recovery of attached particles in industrial cells." AusIMM Sixth Mill Operators Conference. Madang, Papua New Guinea: AusIMM, 1997. 149–155.

Schuhmann, R. "Flotation Kinetics. I: Methods for stead-state study of flotation problems." Journal of Physical Chemistry, 1942: 891–902.

Schwarz, S. "Personal Correspondence." 2012.

Seaman, D.R., J.-P. Franzidis, and E.V. Manlapig. "Bubble load measurement in the pulp zone of industrial flotation machines–a new device for determining the froth recovery of attached particles." In. J. Miner. Process., 2004: 1–13.

Sutherland, K.L. "Physical chemistry of flotation XI: Kinetics of the flotation process." Journal of Physical Chemistry, 1948: 394–425.

Tabosa, E., Runge, K., Crosbie, R., McMaster, J., and Hotham, P., "A study of the role of cell aspect ratio on flotation performance." Proceedings of Flotation 2011, Minerals Engineering International, Cape Town, 2011.

Tavera, F.J., C.O. Gomez, and J.A. Finch. "Estimation of gas holdup in froths by electrical conductivity: Application of the standard addition method." Minerals Engineering (Pergamon) 11, no. 10 (1998): 941–947.

Thorne, E.V., E.J.S. Manlapig, and A.J. Lynch. Modeling of industrial sulphide flotation circuits. Vol. 2, in Flotation, A.M. Gaudin Memorial Volume, edited by M.C. Fuerstenau, 725–752. 1976.

Tucker, J.P., Deglon, D.A., Franzidis, J.P., Harris, M.C., and O'Connor, C.T., An evaluation of a direct method of bubble size distribution measurement in a laboratory batch flotation cell, Minerals Engineering, 1994, 7(5–6), 667–680.

Uribe-Salas, A., C.O. Gomez, and J.A. Finch. "A conductivity technique for gas and solids holdup determination in three-phase reactors." Chemical Engineering Science 49, no. 1 (1994): 1–10.

van Deventer, J.S.J., D. Feng, and A.J. Burger. "The use of bubble loads to interpret transport phenomena at the pulp-froth interface in a flotation column." Chemical Engineering Science 56 (2001): 6313–6319.

Vera, M.A., J.P. Franzidis, and E.V. Manlapig. "Simultaneous determination of collection zone rate constant and froth zone recovery in a mechanical flotation environment." Minerals Engineering, 1999: 1163–1176.

Vera, M.A., J-P. Franzidis, and E.V. Manlapig. "The JKMRC high bubble surface area flux flotation cell." Minerals Engineering 12, no. 5 (1999): 477–484.

Vera, M.A., Z.T. Mathe, J.P. Franzidis, and M.C. Harris. "The modeling of froth zone recovery in batch and continuously operated laboratory flotation cells." Proceedings of the SME Annual Meeting. Denver: SME, 2001. Preprint 01–127.

Yianatos, J., and F. Contreras. "Particle entrainment model for industrial flotation cells." Powder Technology, 2010: 260–267.

Yianatos, J., F. Contreras, P. Morales, F. Coddou, H. Elgueta, and J. Ortiz. "A novel scale-up approach for mechanical cells." Minerals Engineering, 2010: 877–884.

Yianatos, J.B., M.H. Moys, F. Contreras, and A. Villanueva. "Froth recovery in industrial cells." Minerals Engineering 21 (2008): 817–825.

Zheng, X., J.-P. Franzidis, and N.W. Johnson. "An evaluation of different models of water recovery in flotation." Minerals Engineering 19 (2006): 871–882.

ABOUT THE AUTHORS

Peter Amelunxen completed his bachelor's degree in mining engineering at the University of Arizona and his master's degree in mineral processing at McGill University. He worked for 5 years for MinnovEX Technologies (now part of SGS) in the areas geometallurgy, process control, and process modeling and optimization. He then spent 5 years in with Freeport-McMoRan in roles of increasing responsibility at the Bagdad concentrator, then Climax Molybdenum Company, and finally the Cerro Verde concentrator in Peru. In 2008 Peter moved to Santiago to start a consulting business focused on geometallurgical studies for the optimization and design of crushing, grinding, and flotation circuits. He is a founding partner and serves as director of metallurgy for Aminpro Chile, the metallurgical testing arm of the business. He is also a partner and founding member of Project Nexus, an engineering company.

Kym Runge is a mineral processing engineer who specialises in flotation circuit analysis using modelling and simulation techniques. Her expertise in flotation was developed during a 10-year period at the JKMRC where she completed a PhD and later worked as a researcher and consultant. She also had a significant role in the development of JKSimFloat, a flotation simulation program. She then joined Metso's Process Technology & Innovation group where she has worked for more than 7 years as a flotation specialist performing consulting and research work. She continues to undertake this role at Metso but has recently joined the JKMRC, in a part-time capacity, to assist with the P9 research program.

Innovations in Flotation Plant Practice

Stephen Grano
Institute for Mineral and Energy Resources, University of Adelaide, Adelaide, Australia

Massimiliano (Max) Zanin
Ian Wark research Institute, University of South Australia, Mawson Lakes, Australia

Joe Pease
Xstrata Technology, Queensland, Australia

ABSTRACT: This paper discusses the drivers and imperatives for innovative changes in flotation plant practice over the last 30 years. Improvements in characterisation technology have led to improved understanding of contributions to inefficient mineral separation in flotation plant cells and circuits. In some cases this has also led to new process technology and approaches to targeted plant designs which may be applied globally. Technology improvements may drive further developments in characterisation tools and flotation technology.

INTRODUCTION

Innovations in flotation plant practice seem to be few and far between. However, many of the customs and practices we now take for granted have come from innovations in both characterisation and process technologies over the years. Rather than a detailed account of all the innovations, this paper is a broad overview of the major influences on fine and coarse particle recovery, flotation selectivity, and general approaches to address key processing problems. Acknowledgments are made to the many flotation experts who have contributed enormously over the years. The focus is principally sulphide mineral systems, though comments on fine and coarse particle recovery are also applicable to non-sulphide minerals.

In pre-1970s flotation plants treating 'complex' ores, for example fine grained lead zinc ores, typically achieved around 70% to 80% recovery into the flotation products. There was no way of systematically 'knowing' what was causing the low and often inconsistent recovery on a daily, let alone on an hourly basis. The advent of correct sampling protocols, in-stream metal analysis, size by size analysis and automated quantitative mineralogy have provided new tools and performance measures of flotation plants. This has allowed plant managers to systematically and scientifically pursue improvements in flotation recovery and concentrate grades.

Just as the flotation process was developed to recover mineral particles too fine for gravity separation, most of the recent developments in flotation have been driven by the inexorably growing need to recover particles at finer sizes. Flotation of fine particles is challenging, but is essential as we need to exploit increasingly finer grained ore bodies and make higher grade concentrates with lower impurities (e.g., arsenic). Particles that were once considered "slimes" are now standard particle sizes feed to many flotation circuits—and indeed often need further regrinding to enhance liberation.

Though ultrafine particles (below 10 microns) often perform poorly in existing circuits, a growing number of operations have shown that, with the right conditions and right plant design, they can perform extremely well. In turn, the lessons learnt from achieving good fines flotation may also have application and benefits for coarse particle flotation.

BASIC PRINCIPLES OF FLOTATION

This paper documents some innovations in plant practice informed by knowledge of how particles of different size and liberation class are recovered. Many of the conventional views of flotation are based on data similar to that shown in Figure 1. These basic principles may be summarised as follows:

1. *Fine value particles* less than 10 microns invariably have a lower flotation rate constant, and therefore lower recovery at a given residence time, as intermediate particles (20–40 microns) which is the ideal range for bubble-particle collision (Ralston and Crawford 1988). If these particles are sufficiently hydrophobic (contact angle >75°), with similar static contact angle value as the intermediate particles, then they will be recovered given adequate flotation residence time (Muganda et al. 2011), and froth recovery. This type of fine particle behaviour is shown in Figure 1a for a particular plant (Grano et al. 2007). The slower flotation recovery is due to the lower collision efficiency for fine particles (e.g., Dai et al. 2000). Given adequate pulp residence time fine, liberated, but adequately hydrophobic (contact angle >75°) fine particles will eventually be recovered into the concentrate to high value in roughing and cleaning circuits, under most hydrodynamic conditions.
2. *Fine value particles* less than 10 microns also have higher critical contact angle than intermediate sized particles (Chipfunhu et al. 2011), meaning that higher static contact angles are needed for fine particles to attach to bubbles in flotation. While the exact mechanism responsible for this behaviour is still unclear, the practical outcome is that there is sometimes observed a non-floating or slow-floating component of fine value (and gangue) particles that are only recovered by the entrainment mechanism even at very long flotation times under 'normal' hydrodynamic conditions in plant and laboratory cells. In these cases, the fine particle flotation rate may demonstrate multi-components. For fine particles below a critical contact angle attachment to bubbles is ineffective (Miettinen et al. 2010). This behaviour has been observed for fine galena, pentlandite and sphalerite. This fine particle behaviour is shown in Figure 1b for a particular plant. Increasing the pulp residence time in the plant by adding more cells (as an example) will not increase their recovery appreciably, except by the non-selective entrainment mechanism. For this type of fine liberated particle, approaches to increase the hydrophobicity (contact angle) of fine particles, which are below the critical contact angle, are required.
3. In many cases, it may be difficult to furnish sufficient hydrophobicity (high enough contact angle) on fine particles *uniformly*, particle to particle, above a critical value to ensure high recovery values (Muganda et al. 2011; Chipfunhu et al. 2011). Fine particles are prone to be more greatly affected by surface oxidation and surface contamination by hydrophilic species from other sources such as grinding media than intermediate sized particles (Grano 2009). In many cases, simply adding more collector may not redress the low contact angle on the deficient fine particles and, instead, may induce the flotation of, particularly, intermediate

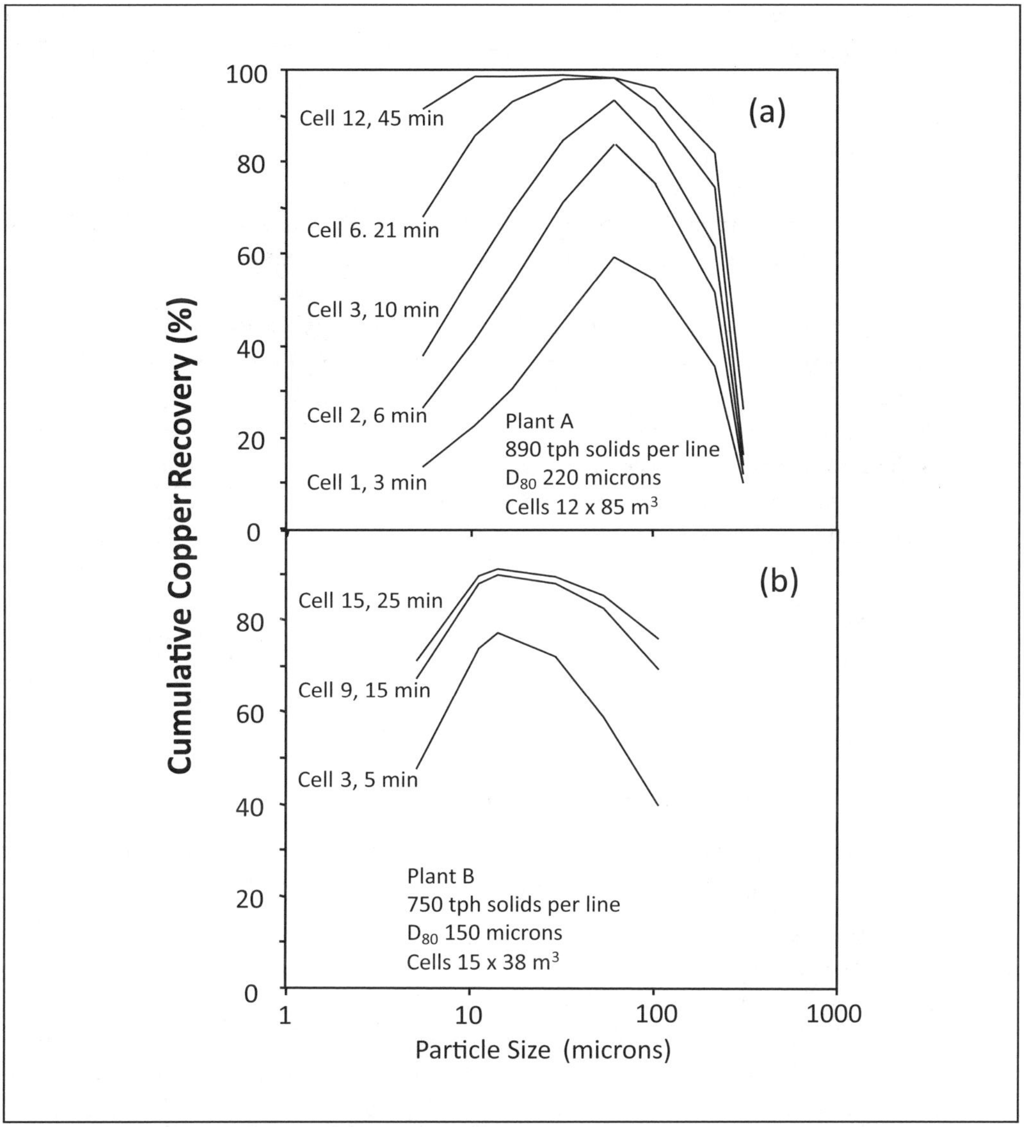

Figure 1. Copper recovery as a function of particle size for two different rougher circuits and for increasing cumulative flotation residence time (minutes) down the bank of cells. Cells 12 and 15 are the final cells in the rougher train in Plant A and B, respectively. Note the different final fine copper recovery relative to the intermediate size fractions for these plants.

size fractions of liberated iron sulphide minerals (pyrite, pyrrhotite) and low grade composite particles containing the value target mineral (Grano 2009). Rather than attempting to enhance the hydrophobicity of fine particles by adding more collector, approaches to remove or minimise the formation of hydrophilic surface coatings, which do not interact with collector, may need to be pursued as a first step towards increasing their recovery.

4. Coarse value particles are affected by the stability of the bubble-particle union which is, in turn, controlled by the hydrophobicity of the particle and its inertia (Schulze 1977). In

some cases, it is possible to recover coarse (+300 microns), liberated particles given adequate (high) hydrophobicity under 'normal' hydrodynamic conditions in cells (Muganda et al. 2011). For composite particles, the decreased size of the hydrophobic domain may reduce the wetting perimeter of the bubble on the particle and reduce the stability of the bubble-particle union (Xu et al. 2010). This phenomenon also holds for intermediate sized particles with a contact angle below a critical value. In some cases, it may be possible to increase the recovery of coarse composite particles by additional collector but the increase will be limited for the case of very low grade composites having low surface exposure of the hydrophobic domain and, instead, may induce the flotation of liberated gangue sulphide particles particularly in intermediate size fractions. Another option to increase the recovery of coarse particles is to reduce turbulence and energy dissipation in the flotation cell (Kohmuench et al. 2010).

5. We now have powerful tools to understand and diagnose the performance of different sizes and liberation classes of particles. However, the crucial point is that particle behaviour is not fixed; it can be changed by appropriate liberation and surface preparation. Measuring the flotation rate for liberated 10-micron particles in roughing tells us how the particle performs under those conditions—conditions which inevitably are a compromise to suit the *average* rougher feed particle size, not the individual size fractions. This tells us nothing about how much better those particles could perform if the conditions were tailored for them specifically. So the new analytical tools must be used to develop the circuit design, not to simply measure it. A model that assumes fixed flotation rates to develop circuit design may condemn the plant to mediocre performance.

DRIVERS AND IMPERATIVES OF PLANT PRACTICE

Given the basic principles of flotation behaviour of particles as a function of size and liberation class outlined above, the following principles of flotation plant practice can be established:

- *Minimise primary grinding*, to reduce energy consumption in the main process stream and to minimise the generation of fine particles in roughing flotation (Pease 2010). The primary grind should be as coarse as possible without affecting recovery in roughing. Coarsening the primary grind may have additional benefits to the floatability of the fine particles for different reasons. An important factor for the recovery of coarse, composite particles is detachment in the collection and froth zones. The locking characteristic of the value (hydrophobic) and gangue (less hydrophobic or hydrophilic) minerals is critical as this controls the tenacity of particle attachment to bubbles. How the hydrophobic domain is distributed across the composite particle surface, within a liberation class, may also be important.
- The coarse, composite particles collected in the roughing stage need to be reground for mineral liberation before cleaning; not after cleaning by regrinding the cleaner tailing (Pease 2010). If composite particles have been intentionally recovered in roughing, then the valuable minerals must be liberated from the gangue minerals before they have the opportunity to report to the final concentrate where they would lower the concentrate grade. To make very clean concentrates, it is usually necessary to regrind the coarse size fraction of the whole rougher concentrate, since composite particles are recovered even in the early stages of rougher flotation and not just towards scavenger cells.

- Regrinding should be carried out in an environment which introduces the minimum amount of metal hydroxides from media and other sources such as process water (Pease 2010). In regrinding, there is a large increase in mineral surface area and often an increase in pulp temperature, factors which together may contribute to surface precipitation during regrinding, especially when the process water is saturated in a particular species (e.g., calcium sulphate). Fine particles of value (and gangue) are more adversely affected by surface precipitation from process water and grinding media inputs. Grinding with fully inert grinding media is the best way to minimise surface contamination of fine value mineral particles in regrinding. Pulp temperature and process water composition are important in regrinding, particularly for very fine regrind particle sizes.
- By adequately regrinding the rougher concentrate prior to cleaning, there will be less circulating load which usually contributes enormously to volumetric flow rates of slurry (Pease 2010). For circuits with inadequate liberation in the cleaner feed, excessive dilution water and depressant may be required in cleaning to maintain the target final concentrate grade. This will cause loss of fine value minerals from the cleaning circuit from reduced residence time and reduced hydrophobicity. By reducing the volumetric flow and grade of the cleaner tailings it becomes possible to target recovery from this stream directly without recycling to the primary rougher stage.
- By considering the mill to smelter energy balance and energy use, much higher grade concentrates may be targeted, devoid of composite particles, which may attract a premium price due to lower energy consumption, lower emissions and costs in smelting (Pease 2010).

HISTORICAL PERSPECTIVE

Flotation was developed when existing techniques—gravity separation—could not recover value from the finer grained ores that needed to be developed. As the need to exploit ever finer grained ores increases, flotation needs to be applied to ever finer particle sizes. Particles that were once considered "slimes" and almost untreatable are now standard—the entire concentrate production of some concentrators cannot start until all particles are below 10 microns. The average particle size recovered into the McArthur River concentrate is 2.5 microns (Pease et al. 2006). Therefore developments in flotation circuits have been dominated to now by the need to understand the liberation, surface preparation, and hydrodynamics of ultrafine particle flotation.

Though Taggart may have regarded sub 37 micron particles as "slimes," Taggart nevertheless understood the need to tailor conditions for different mineral classes; for example, in "sand-slimes" circuits (Taggart 1920). This concept appears to have been forgotten in the trends to "simplify" and standardise circuits, and to use ever-larger flotation cells to treat a mix of mineral classes resulting from combining fine or reground cleaner tailings with primary mill product in rougher flotation. The modern innovations in circuit design—staged grinding and flotation tailored to mineral size and liberation classes—are, of course, not so much an innovation but a revival of the basic principles outlined above.

As finer ores make flotation more dependent on the correct liberation and surface chemistry, so we have developed much better tools to assist us: quantitative mineralogy, analytical surface chemistry, and numerical modelling techniques. These are essential and valuable tools, but they must be used and combined correctly.

The impact of grinding environment on flotation chemistry has been increasingly recognised, since it is more important for finer particle sizes. It is clear that the critical surface chemistry of fines recovery must be considered as part of a complete system including grinding and flotation. These developments also have lesser, but still significant, benefits at coarser particle sizes. The impact of different grinding environments in AG, SAG and ball mills has been studied (Grano et al. 1994). High Intensity Conditioning was developed to improve the surface condition of fine particle after grinding with steel media (Bulatovic and Salter 1989). Subsequently, high intensity inert grinding was developed to provide both the mineral liberation and surface preparation in one step (Johnson 2006).

EVOLUTION TO CURRENT PRACTICE

This section attempts to generalise how current flotation practice has evolved over the last three decades, and maps out some potential new developments in the coming decade. Each of tables 1–3 shows research outcomes in red, implications and adaptations to plant practice in yellow, and potential new developments in blue. Emphasis is placed on how the basic principles embedded in research outcomes have implications in plant practice and how it has been applied.

Table 1 shows how characterisation tools, such as the development of quantitative mineralogy and in-stream bulk chemical analysis prior to 1990, have led to improved process understanding, particularly in relation to primary and regrind particle size and new process control strategies. The use of quantitative mineralogy to characterise and subsequently aid optimisation of the flotation process is now standard in most flotation plants (Sutherland and Gottlieb 1991).

Characterisation tools also extend to in-situ and ex-situ surface characterisation of minerals which have led to enhanced understanding of the chemical domain (pH-Eh) needed for optimum selectivity (Woods and Richardson 1986), intentional or accidental activation by metal species (Finkelstein 1997), and the presence of interfering hydrophilic layers. Hydrophilic surface layers may be derived from process water inputs as an example. New collectors have been developed based on completely new chemistries which have shown better selectivity (Sheridan et al. 2002) and precious metal recovery (Basilio et al. 1992). New surface analysis techniques have shed light on surface heterogeneity between particles, for the first time explaining why some particles are recovered and others are not, even within the same size fraction and mineral liberation class (Brito e Abreu and Skinner 2011).

In more recent years, there has been a strong focus of the pivotal role of the froth phase in controlling mineral recovery (froth recovery) and water recovery by the entrainment mechanism. These were preceded by developments on froth washing and column froths in the 1980s (Espinosa-Gomez et al. 1988). Tools to characterise the froth behaviour in terms of its stability (Barbian et al. 2003), froth bubble size, and froth velocity (Broussaud et al. 2011) have led to new strategies to distribute frothers (Hadler et al. 2005), as well as to optimise the froth phase recovery with respect to air flow rates (Hadler et al. 2009) and froth depth (Hadler et al. 2012). Over the next decade, new technology will allow surface characterisation of individual mineral particles within the pulp, in-stream liberation analysis, and in-situ characterisation of the particle laden gas-liquid interface (e.g., film elasticity) as a means to control frother additions.

Table 2 shows the development of industry problem motivated research which has focussed on fine and coarse value mineral recovery and selectivity. The importance of the collision mechanism to explain the lower flotation rate of fine particles, known over 30 years ago, led to the research and

Table 1. Influence of characterisation tools on industry practice

	Decade →				
Focus	**Pre-1970–1980**	**1980–1990**	**1990–2000**	**2000–2010**	**2010–2020**
Bulk Chemical Analysis	Development of In-Stream XRF Probes	Process Control Loops to control reagent addition based on in-stream analysis			In-Situ surface hydrophobicity measurements
Quantitative Mineralogy	Correct sampling for Process Control measures	Quantitative Electron Microscopy Qem-Scan Mineral Liberation Analyser	Primary grind size to achieve adequate liberation for rougher recovery	Regrind size to achieve adequate liberation for final concentrate grade	In-situ and on-line liberation measurements
Mineral Surface Chemistry	Collector adsorption mechanisms	Eh/pH domain for collector adsorption, New collector chemistry developments for selectivity and gold recovery	Phosphinate, Thionocarbamate, and thiourea collectors with enhanced selectivity against gangue sulphide minerals	Phosphinate, Thionocarbamate, thiourea collectors with enhanced selectivity against gangue sulphide minerals	Thin film drainage rate and bubble-particle contact for individual mineral phases, static & dynamic contact angle in-situ measures
Surface Analysis & Characterisation	Infrared spectroscopy and electro-kinetics	Collector adsorption mechanisms, Intentional and Incidental heavy metal ion activation	X-Ray Photoelectron Spectroscopy	Reagent and process water schemes to control surface layers, oxidation	Wider use of Reverse Osmosis water in flotation
				ToF-SIMS, LIMS surface analysis techniques.	Mineral and particle size specific surface information
Froth Phase	Water recovery and entrainment mechanism for fine particles	Column Cells Froth Washing	Froth Image Analysis Froth process control loops	Froth phase models Froth recovery Water recovery and Entrainment models	Frother selection based on function and action in froth and collection zones
		New frothers developed—less volatile and molecular structure tailored to give desired frothing properties	Froth and pulp phase characterisation tools to select frothers based on desired action	Double launders in flotation cells to reduce froth transport distance and retention time Froth stability characterisation	New methods to recover froth phase from plant cells

Red = research, Yellow = plant implications, blue = future.

Table 2. Industry problem led research and influence on flotation practice

Focus	Decade →				
	Pre-1970–1980	1980–1990	1990–2000	2000–2010	2010–2020
Fine (<10 μm) Particle Recovery	Hydrodynamics—Collision Efficiency and Frequency dominated	High Turbulence Bubble –Particle Contact Reduced Bubble Size Interpretation of recovery by particle size curves	Jameson Cells New spargers to deliver small bubbles in conventional column cells	Selective Fine particle Coagulation for Paramagnetic Minerals	Selective fine particle coagulation for other minerals
	Hydrophobicity—Attachment Efficiency dominated	Hydrophilic Surface Coatings on Fine Particles Interpretation of recovery by particle size curves	High Intensity Conditioning, high intensity inert grinding (e.g., IsaMill) for attrition surface cleaning Polymeric Dispersants	Grinding Environment Control, Water and Media inputs	Wider use of Reverse Osmosis water in flotation especially in cleaning and regrinding. Use of saline water in roughing
Coarse Particle (+150 microns) Recovery	Hydrodynamics—Stability Efficiency dominated	Particle-Bubble Detachment Model Interpretation of recovery by particle size curves		Low Energy Dissipation Bubble-Particle Contact Co-current separation to reduce turbulence	Scavenger cells exclusively treating Coarse particles under tailored physical and chemical conditions
	Hydrophobicity Detachment dominated	Particle-Bubble Detachment Model Interpretation of recovery by particle size curves		Enhancing the hydrophobicity of coarse composite particle	Control of rheology and particle separation prior to flotation of coarse composites
Selectivity against Gangue Sulphide Gangue			Surface analysis confirms copper and lead activation of gangue sulphide surfaces	Environmentally benign complexants to prevent activation Chelating Polymers	Flotation and conditioning gas to control oxidation state

Red = research, Yellow = plant implication, blue = future.

Table 3. Integration of flotation with other unit operations and plant optimisation

	Decade →				
Focus	**Pre-1970–1980**	**1980–1990**	**1990–2000**	**2000–2010**	**2010–2020**
Integration of Flotation with Comminution Technology		Flash Flotation in Grinding Circuit	Benefits of Fully Semi-Autogenous Milling on Flotation Chemistry	Rolls Crushing to Flotation	Selective Liberation within Flotation Circuit by staged use of inert attrition mills Open circuit flotation tailings Separate flotation of fine fraction of Run of Mine ore to avoid activation
Integration of Flotation with other Process Technology in Metal Production			Gravity recovery gold potential identified in grinding circuits	Gravity–Flotation recover gravity gold in flotation regrinding circuits Hydrometallurgy–Flotation integration	Energy balance across grinding, flotation, roasting, smelting or leach, solvent extraction, electro-win, or other new approaches
Plant Optimisation Tools			Flotation Models–Gas Dispersion–flotation rate relationship, Bubble-particle flotation models	Flotation Simulators Geometallurgy Flotation gas profiling in flotation cells	Floatability components determined from in-situ and on-line measurements

Red = research, Yellow = plant implication, blue = future.

development of gas sparger technology in conventional column cells and high turbulence flotation cells such as the Jameson Cell in the 1980s (Jameson 1988). Understanding of the fundamental mechanisms of particle capture by bubbles also led to improved interpretation and eventual wide spread use of recovery by particle size curves (e.g., Trahar 1981) (e.g., **Figure 1**). The high critical contact angle to initiate fine particle (below 10 microns) attachment to bubbles led to approaches to mechanically or chemically clean fine particles to enhance their hydrophobicity. The Isa Mill or other high intensity inert-media stirred mills are able to clean the surfaces of sulphide minerals by the attrition mechanism, whilst also avoiding contamination of the surfaces by hydrophilic iron hydroxide.

In more recent years, coarse particle recovery has become a focus, for which detachment from bubbles is an important mechanism (Rodrigues et al. 2001), a particular issue for composite particles on which there may only be low exposure of the hydrophobic domains. For these particles, the small perimeter of the three phase line of contact controls the stability of the union of the composite particle to the bubble in the pulp phase (Xu et al. 2010). Over the next decade, the very significant research and plant results on selective coagulation of paramagnetic minerals in fine particle size fractions (Engelhardt et al. 2005; Rivett et al. 2007) could be extended to other mineral systems.

Finally, Table 3 shows developments and progress on the integration of flotation technology with other unit operations, both upstream and downstream of flotation itself. Flash flotation was implemented in the 1980s and the benefits of autogenous milling on fine particle recovery were noted subsequently (Grano et al. 1994). The more recent uptake of High Pressure Grinding Rolls may have benefits beyond lower energy consumption in grinding for reasons related to the particle fracture mechanism. The potential for the recovery of gravity recoverable gold in the regrind circuit of flotation plants was noted in the 1980s and 1990s. The integration of flotation with other processing options, such as hydrometallurgy, is expected to be a strong focus over the next decade, driven by the high operating and capital cost of grinding for liberation in flotation.

Flotation models and simulators have now become standard practice for flotation operators wishing to optimise their circuits and to predict future performance using geometallurgical approaches. The flotation models are based on separating the collection and froth zones, measures of froth zone recovery (Runge et al. 2010), conservation of the floatability components (Alexander et al. 2000), and understanding of the relationship between gas dispersion in cells and the flotation rate constant (Gorain et al. 1998; Deglon et al. 1999). A challenge for the models is to independently quantify the floatability components under conditions of varying hydrophobicity, a challenge which may be overcome in the next decade. Importantly, practitioners must realise that models are only as good as their data assumptions. If a particle is modelled as slow floating, perhaps because it is a low grade composite or because it has a hydrophilic surface coating, then the model will regard it as slow floating in any circuit configuration. Instead, the task of the metallurgist is to understand why the particle is slow floating, and to try to change its flotation rate by changing liberation class or surface hydrophobicity. In short, a computer model may just accept particle behaviour. A metallurgist sets about to understand and change particle behaviour.

In summary, trends suggest that size by size analysis, quantitative mineralogy, advanced characterisation tools and process models will be used together to design integrated grinding and flotation circuits of the future. The objective will be to achieve the correct liberation in the correct stage of the flotation circuit to minimise the overall energy consumption and installed grinding equipment cost. Grinding mills with inert grinding media will be embedded within flotation circuits at different points to provide both the right liberation and the right surface preparation in the right part of the circuit. New approaches to recover composite particles in roughing flotation (froth and collection zone) will allow the primary grind size to be as coarse as possible within the limits of the conventional flotation process. Rapid flotation methods such as the Jameson cell will be used to recover fast floating particles at the highest possible grade prior to cleaning in conventional cells (Seaman et al. 2011). Increased residence time, magnetic coagulation, and approaches to enhance hydrophobicity will be used to increase the flotation rate of slower floating fine particles in cleaning circuits. There will be continuing emphasis on using recycled, low quality water, and/or saline water in flotation (Corin et al. 2010). Flotation equipment will be combined to exploit the best characteristics of each cell type. The trend towards open circuiting the cleaner tailings will continue with either finer regrinding or additional cleaner scavenger flotation capacity.

Flotation circuits will be made simpler and designed to address the fundamental requirements of the mineral particles. The physical and chemical conditions will be tailored for the different mineral classes. Understanding the needs of individual mineral classes will allow them to be dealt with directly and with minimal recirculation. The flotation circuits will behave more predictably without troublesome feedback loops, allowing much easier control. Simple process control strategies will then be more effective—stabilising control, levels, flow rates, reagents, grinding and particle size control. Variations in the feed will be minimised by understanding the mineralogy of different ore sources. Mine production schedules and blending strategies can minimise short term variations in the feed. Dynamic process models can be applied to the varying feed to predict performance and optimise the flotation circuit well in advance of the feed changes (Nees and Gamarano 2011). With short term variation in the feed under tighter control, optimising process control can then be tackleled armed with the dynamic process models.

IMPEDIMENTS TO TECHNOLOGY INTRODUCTION

Simplistic and Inadequate KPIs and Organisational Silos

Most organisations attest to the "mine to metal" concept, but the concept is rarely applied (Pease 2010). KPIs are set for mines, concentrators and smelters that encourage operators to stay within their silo—cost per tonne being a typical example. Each unit process is usually encouraged to optimise their own process in isolation of the impact on upstream or downstream processing. Of course low cost per tonne of ore is a valid objective, but only if it is kept in the context of other KPIs that recognise mineral behaviour and impact on the overall process economics. The KPIs that affect mineral behaviour and optimise overall profit are usually ignored because they are difficult to measure and may not be well understood—feed variability, grind size, liberation, and surface chemistry.

Simplistic Management Techniques

Management techniques like statistical control (e.g., Six Sigma) can be useful tools, but they cannot replace understanding the needs of individual mineral classes. They must be informed by the right mineral science and circuit design, not replace them.

Improper Use of Modelling and Simulation

Just as organisations can lock-in mediocre performance by inadequate KPIs and management techniques, metallurgists may make the same mistake by inappropriate use of models and simulation

(Pease 2010). If a particle has a low flotation rate, the task is to understand why. What can be done to enhance liberation, surface preparation, chemistry and circuit design to change this flotation rate need to be the focus. To merely accept the low flotation rate then model it in multiple circuits may miss opportunities to increase recovery.

Reliance on Larger Equipment Rather Than Tailored Design

Larger equipment undoubtedly improves productivity (Coleman and Dixon 2010). Another source of productivity is to deal with minerals in the correct manner at the right point in the circuit. Combining many streams into a mix of different mineral classes with circulating loads with compromised pulp chemistry is an inferior solution, no matter how large the flotation cells are. Productivity improvements may mean smaller grinding mills and smaller flotation cells distributed around the circuit to deliver the right degree of liberation under the right flotation conditions as the minerals require. Flexible plant layouts and strategies will be needed.

Outsourcing Design

The above problem is compounded by outsourcing plant design before the mineralogical needs are completely understood and communicated by the owner. Without adequate mineralogical and metallurgical guidance to tailor the design, the safe path for the engineering company is to apply standard designs and equipment selections. This is unlikely to lead to innovation, or optimising conditions for different orebodies. The (flexible) plant design should be informed by new mineralogical information as it becomes available.

THE FUTURE

Or back to the future? Research and industry have developed very powerful analytical and processing tools. These tools have allowed significant improvements in the difficult field of fine particle flotation. Applying these developments in relatively simple circuits allows high grades and recoveries to be made from extremely fine grained and complex ores—excellent flotation results are routinely achieved in the sub-10-micron fraction. Applying some of these same techniques to coarse particle flotation may also improve performance. Tailoring the chemical and physical conditions for different particle size fractions should be a future target.

To achieve the full potential, and to optimise the entire processing chain, requires a true mine to metal approach, including management KPIs. The high cost of fine grinding, and the needs of fine particle flotation, has revived the concept of staged grinding and flotation circuits. The new analytical tools in mineralogy and surface analysis now allow us to custom design circuits to suit the needs of the minerals—to use the minimal amount of equipment to grind and separate the correct minerals in the correct place. These same concepts could be applied for flotation circuits globally, to develop more efficient concentrators, with smaller equipment items, less equipment, lower circulating loads, combining technologies, and producing higher grade concentrates. In turn, this will improve the efficiency of downstream processing, in hydrometallurgy or pyrometallurgy, truly optimising the mine to metal chain. Thus while further research and developments are desirable, the most immediate improvements may come from more widely combining and applying the developments of the last 30 years.

ACKNOWLEDGMENTS

The authors express their appreciation and acknowledgment to the many flotation practitioners, researchers, technology and engineering companies who have contributed enormously to flotation plant practice over the years. It simply is not possible to name all of the valuable contributions in the space available for this paper. Rather than a detailed account of all the innovations, a broad overview was attempted addressing the major influences on fine and coarse particle recovery as well as selectivity, and broad approaches to address key processing problems.

REFERENCES

Alexander, D.J., Runge, K.C., Franzidis, J.P. and Manlapi, E.V., 2000. The application of multi-component floatability models to full-scale flotation circuits. Australasian Institute of Mining and Metallurgy Publication Series (6), pp. 167–177.

Barbian, N., Hadler, K. and Cilliers, J., 2006. The froth stability column: Measuring froth stability at an industrial scale. Minerals Engineering, Selected papers from the Centenary of Flotation Symposium, 5–9 June 2005, Brisbane, Australia 19(6–8): 713–718.

Basilio, C.I., Kim D.S., Yoon, R.H. and Nagarj, D.R., 1992. Studies on the use of monothiophosphates for precious metals flotation. Minerals Engineering, 5, 397–409.

Brito E Abreu, S. and Skinner, W., 2011. ToF-SIMS-derived hydrophobicity in DTP flotation of chalcopyrite: Contact angle distributions in flotation streams. International Journal of Mineral Processing, 98: 35–41.

Broussaud, A., Guyot, O., Monredon, T. and Valdbenito, M., 2011. Leading edge froth vision technology, In: 8th International Mineral Processing Seminar (Gecamin Ltda. Publ.), Santiago, Chile, pp. 273–284.

Bulatovic, S.M., and Salter, S.H., 1989. High intensity conditioning: a new approach to improve flotation of mineral slimes. In: Conference of Metallurgists. Halifax: Canada Publisher, pp. 182–197.

Chipfunhu, D., Zanin, M. and Grano, S., 2011. The dependency of the critical contact angle for flotation on particle size–Modelling the limits of fine particle flotation. Minerals Engineering 24(1): 50–57.

Coleman, R. and Dixon, A., 2010. Tried, tested, and proven—300 m^3 flotation cells in operation. In: Proceedings of the XXV International Mineral Processing Congress (IMPC) 2010: Smarter Processing for the Future, AUSIMM, 3429–3440, Brisbane, Australia, XXV International Mineral Processing Congress (IMPC) 2010 (Ralph Holmes Eds).

Corin, K., Reddy, A., Miyen L., Wiese J. and Harris, P., 2010. The effect of ionic strength of plant water on valuable mineral and gangue recovery in a platinum bearing ore from the Merensky reef, In: Proceedings of the XXV International Mineral Processing Congress (IMPC) 2010: Smarter Processing for the Future, AUSIMM, 1807–1814, Brisbane, Australia, XXV International Mineral Processing Congress (IMPC) 2010 (Ralph Holmes Eds).

Dai, Z., Fornasiero, D. and Ralston. J., 2000. Particle-bubble collision models–A review. Advances in Colloid and Interface Science 85: 231–256.

Deglon, D.A., Sawyerr, F. and O'Connor, C.T., 1999. A model to relate the flotation rate constant and the bubble surface area flux in mechanical flotation cells. Minerals Engineering 12(6): 599–608.

Engelhardt, D., Ellis, K. and Lumsden, B., 2005. Improving fine sulfide mineral recovery–plant evaluation of a new technology, Australasian Institute of Mining and Metallurgy Publication Series, Centenary of Flotation Symposium, pp. 829–834.

Espinosa-Gomez, R., Finch, J.A. and Johnson, N.W., 1988. Column flotation of very fine particles. Minerals Engineering 1: 3–18.

Finkelstein, N.P., 1997. The activation of sulphide minerals for flotation: a review. International Journal of Mineral Processing 52(2–3): 81–120.

Gorain, B.K., Napier-Munn, T.J., Franzidis, J. -P., Manlapig, E.V. 1998). Studies on impeller type, impeller speed and air flow rate in an industrial scale flotation cell. Part 5: validation of k-Sb relationship and effect of froth depth. Minerals Engineering 11(7): 615–626.

Grano, S., 2009. The critical importance of the grinding environment on fine particle recovery in flotation. Minerals Engineering 22(4): 386–394.

Grano S.R., Wong, P., Skinner, W., Johnson, N.W. and Ralston, J., 1994. The effect of autogenous and ball mill grinding on the chemical environment and flotation of the copper ore of Mount Isa Mines Limited, In: III Latin-American Congress on Froth Flotation (University of Concepcion Publ.), Concepcion, Chile, pp. 351–388.

Grano, S.R., Akroyd, T. and Mular, M.A., 2007. A model study of copper rougher recovery optimization at PT Freeport Indonesia. Aus. Inst. Min. Metall. Ninth Mill Operators' Conference 19–21 March 2007, Fremantle, WA, pp. 25–37.

Hadler, K., Aktas, Z. and Cilliers, J.J., 2005. The effects of frother and collector distribution on flotation performance. Minerals Engineering 18(2): 171–177.

Hadler, K. and Cilliers J.J., 2009. The relationship between the peak in air recovery and flotation bank performance. Minerals Engineering 22(5): 451–455.

Hadler, K., Greyling, M., Plint, N. and Cilliers, J.J., 2012. The effect of froth depth on air recovery and flotation performance. Minerals Engineering 36–38(0): 248–253.

Jameson, G.J., 1988. New concept in flotation column design. Minerals and Metallurgical Processing 5(1): 44–47.

Johnson, N.W., 2006. Liberated 0–10 μm particles from sulphide ores, their production and separation—Recent developments and future needs. Minerals Engineering 19(6–8): 666–674.

Kohmuench, J., Mankosa, M., Yan, E., Wyslouzil, H., Christodoulou, L. and Luttrell, G., 2010. Advances in coarse particle recovery –Fluidised bed flotation, In: Proceedings of the XXV International Mineral Processing Congress (IMPC) 2010: Smarter Processing for the Future, AUSIMM, 2065–2076, Brisbane, Australia, XXV International Mineral Processing Congress (IMPC) 2010 (Ralph Holmes Eds).

Miettinen, T., Ralston, J. and Fornasiero D., 2010. The limits of fine particle flotation. Minerals Engineering 23(5): 420–437.

Muganda, S., Zanin, M. and Grano S., 2011. Influence of particle size and contact angle on the flotation of chalcopyrite in a laboratory batch flotation cell. International Journal of Mineral Processing 98(3–4): 150–162.

Nees, M.R. and Gamarano, T.V., 2011. The use of advanced simulation software for the development of the Virtual Process Plant, In: 8th International Mineral Processing Seminar (Gecamin Ltda. Publ.), Santiago, Chile, pp. 343–351.

Pease, J.D., 2010. The elephant in the mill, In: Proceedings of the XXV International Mineral Processing Congress (IMPC) 2010: Smarter Processing for the Future, AUSIMM, 123–132, Brisbane, Australia, XXV International Mineral Processing Congress (IMPC) 2010 (Ralph Holmes Eds).

Pease, J.D., Curry, D.C. and Young, M.F., 2006. Designing flotation circuits for high fines recovery. Minerals Engineering 19(6–8): 831–840.

Ralston, J. and Crawford R., 1988. The influence of particle size and contact angle in mineral flotation. Int. J. Miner. Process. 23: 1–24.

Rivett, T., Wood, G. and Lumsden, B., 2007. Improving fine copper and gold flotation recovery–a plant evaluation. Australasian Institute of Mining and Metallurgy Publication Series, 9th Mill Operators' Conference, pp. 223–228.

Rodrigues, W.J., Leal Filho, L.S. and Masini, E.A., 2001. Hydrodynamic dimensionless parameters and their influence on flotation performance of coarse particles. Minerals Engineering 14(9): 1047–1054.

Runge, K., Crosbie, R., Rivett, T. and McMaster, J., 2010. An evaluation of froth recovery measurement techniques, In: Proceedings of the XXV International Mineral Processing Congress (IMPC) 2010: Smarter Processing for the Future, AUSIMM, 2313–2324, Brisbane, Australia, XXV International Mineral Processing Congress (IMPC) 2010 (Ralph Holmes Eds).

Schulze, H.J., 1977. New theoretical and experimental investigations on stability of bubble/particle aggregates in flotation: A theory on the upper particle size of floatability. International Journal of Mineral Processing 4(3): 241–259.

Seaman, D., Manton P. and Griffin P., 2011. Separation efficiency improvement of a low grade copper-gold flotation circuit, In: 8th International Mineral Processing Seminar (Gecamin Ltda. Publ.), Santiago, Chile, pp. 273–284.

Sheridan, M.S., Nagaraj, D.R., Fornasiero, D. and Ralston, J. 2002. The use of a factorial experimental design to study collector properties of N-allyl-O-alkyl thionocarbamate collector in the flotation of a copper ore. Minerals Engineering, 15, 333–340.

Sutherland, D.N. and Gottlieb, P., 1991. Application of automated quantitative mineralogy in mineral processing. Minerals Engineering 4(7–11): 753–762.

Taggart, A.F., 1920. Flotation. Journal of the Franklin Institute 189(4): 485–498.

Trahar, W.J., 1981. A rational interpretation of the role of particle size in flotation. International Journal of Mineral Processing 8(4): 289–327.

Woods, R. and Richardson P.E., 1986. The flotation of sulfide minerals–electrochemical aspects. Advances in Mineral Processing–A Half-Century of Progress in Application of Theory to Practice: 154–170.

Xu, D., Wang, W., Ametov, I., Fornasiero, D. and Grano, S.R., 2010. The detachment of coarse composite particles from bubbles, In: Proceedings of the XXV International Mineral Processing Congress (IMPC) 2010: Smarter Processing for the Future, AUSIMM, 2601–2610, Brisbane, Australia, XXV International Mineral Processing Congress (IMPC) 2010 (Ralph Holmes Eds).

ABOUT THE AUTHORS

Stephen Grano is the executive director and professor of the Institute for Mineral and Energy Resources at the University of Adelaide. The Institute covers the spectrum from geoscience to energy technology. Grano is an experienced and internationally recognized metallurgical engineer with nearly 30 years of postgraduate experience. Experience in mineral flotation has been gained at Mount Isa and during his time at the Ian Wark Research Institute with whom he now collaborates.

Massimiliano (Max) Zanin is senior research fellow at the Ian Wark Research Institute, University of South Australia. He is a mineral processing engineer with more than 10 years of research experience in minerals processing, waste treatment, and sustainable use of resources. Over the last several years, he has been senior researcher on several minerals industry and ARC Linkage funded projects (including the AMIRA International projects P260 and P541 on minerals flotation and flotation froths). Zanin is heavily involved in technology transfer to industry, conducting several on-field studies at the major flotation operations worldwide.

Joe Pease is chief executive of Xstrata Technology, a developer and provider of specialist minerals technology and services. Pease has 30 years of experience in operating, managing, and supporting concentrators and smelting plants. He is a director of AMIRA International, the Julius Kruttschnitt Mineral Research Centre, the Wark Institute, and the Centre for Eco Efficient Comminution, and he is a member of the Minerals Tertiary Education Council in Australia.

Ore Sorting

Bo R. Arvidson
Bo Arvidson Consulting LLC, Lakewood, Colorado, United States

Hermann Wotruba
RWTH Aachen University, Aachen, Germany

ABSTRACT: While manual ore sorting has been reported since the beginning of mining history, significant advances have been made recently in various sensing technologies in recent years. As the equipment/systems relative costs are reduced, machine sorters are showing an increased impact on the mining industry, specifically in areas such as uranium and magnesite processing. This paper reviews the initial adaptation of food sorters to optical ore sorters, specific radiometric sensing sorters for uranium ores to the modern photometric, and several sensing technologies based sorters.

INTRODUCTION

Ore sorting is typically used for three purposes:

1. Preconcentration, primarily of low-grade ores
2. Producing final quality material
3. Sorting a feed into a high-grade and a low-grade product

By rejecting waste grade material at low overall cost, an ore may be upgraded to a degree that makes it profitable (Arvidson 1987). In some cases, the waste may even contain a detrimental component, which by its elimination, makes subsequent processing possible (Bibby 1982).

The final quality material may be for direct sale or for use in a downstream process, such as grinding for filler/powder, or calcination/fusion/smelting.

The third type of application is intended for materials where there may be no waste or relatively little waste to be rejected resulting in two valuable products which may then proceed to different downstream processing lines. Advancing technologies are likely to increase the use of this application.

A large range of sensing technologies and fast complex computations capabilities continue to be developed and will be mentioned in the emerging technologies section. New ore sorting possibilities are being continuously developed.

HISTORY

In the 1940s, food sorters came on the market. These were based on optical recognition technologies. It was soon realized that machine sorters could also be used for ores. By making some models more robust, modified peanut and grape sorters were applied to rock salt and talc sorting. One of the first large installations was for magnesite ore sorting in Greece. Diamond sorters were also developed, first using optical methods. However, since the global food sorting market grew very rapidly, the main focus of development was in that area rather than for ores.

Special sorters for diamonds using their x-ray fluorescence property (and in some situations luminescence often induced by ultraviolet light) soon overtook manual sorting and became an industry standard method. After preconcentration by heavy media density separation, sorters processing wet material upgraded products prior to drying and final automated dry sorting. Color sorting of the diamonds was done manually.

The development of uranium ore sorters began with the first machines installed at the Mary Kathleen mine in 1960 and later upgraded with high-capacity sorters in 1979. Several such machines were used to upgrade uranium ores, which otherwise would have been uneconomical. Several radiometric sorters were also used for gold ores in South Africa. Uranium associated with gold was used as a "handle" to recover the gold values.

Photometric sorters using a fast scanning laser beam were first used for gold ores in South Africa in 1972 (Kidd 1983). Such high-capacity machines changed the magnesite ore industry in Greece and were later used also in the Australian magnesite ore processing plants. The sorters were useful in several different but fairly unique applications such as some gold ores, limestone, phosphate, wolframite and other valuable minerals in quartz, talc and spodumene ores. One large installation was for lignite coal in Hungary.

Accelerating computer development in the 1990s resulted in possibilities for fast processing of huge, complex data sets at rapidly decreased cost making more expensive, dedicated hardwired processing technologies too costly. New developments resulted in less expensive machines that are more economical to use in an increasing number of ore sorting situations. Various sensors became more reliable and new waste recycling technologies were being adapted to ore sorting applications.

FIRST APPLICATIONS OF AUTOMATED ORE SORTING

The first optical sorters were intended to replace human labor to sort discrete particles that had distinct visible differences (Arvidson 2002). Material was fed in channels where each particle was seen by a scanning device to detect optical differences (light vs. dark), the signal over or under a threshold value fed to the mechanism that would kick the fast-moving particle out of the natural trajectory. Both mechanical plungers or flaps and air jets were used, the latter type dominating.

High-capacity optical sorters using a fast-scanning laser beam were purpose designed and constructed for heavy duty ore sorting (Schapper 1977). No longer was it necessary to feed the size fractions through channels. Each particle was tracked on a fast moving (4 m/s) belt and those determined to be satisfying criteria for ejection were accurately sized and located for a precisely tuned air blast. The fast processors were hard-wired, allowing moderate flexibility of the photometric parameter settings.

Although the first generations of advanced optical ore sorters could not come close to the capability of the human eye and brain to recognize subtle nuances in color, shape, and contrast within a particle, they were successful in many applications, most notably in sorting magnesite ore at the Fimisco operation in Greece. As the magnesite is brightly white and the unwanted material, mainly serpentine in large veins, is brown to black, sorting based on reflectivity was relatively easy.

When laser sorters became available, they were quickly accepted by the South African gold industry and further developed within the Greek magnesite industry. A pioneering large installation for the magnesite ores of Grecian Magnesite was implemented in 1977 and has been in full production to this day. This installation utilizes 11 laser sorters and has recently installed 5 line-scan camera sorters for fine sized material. There are other laser sorter installations, such as the 4 sorters at the Oroszlany mine in Hungary, but most consist of fewer units.

Radiometric sorter pioneering company Mary Kathleen Uranium Ltd operated over short time periods, first 1958 to 1963 and then 1976 to 1982. In both periods, radiometric sorters played an important role separating the highly acid consuming calcite waste from the uranium product.

One of the earliest radiometric sorter machines was used for many decades at the Schwartzwalder uranium mine until its closure in 1995. Energy Fuels used a late model for a mine operation in Arizona with great success. A coarse size fraction was upgraded substantially while rejecting low-grade waste thereby eliminating shipping costs to a remote process plant in Utah. Nearly 30 radiometric sorters were in operation at one time, recovering uranium minerals and some also gold as mentioned in the prior section.

INDUSTRY IMPACT IN THE 20TH CENTURY

Ore sorting was generally an uncommon technology in the mining industry before the 21st century. Several reasons for this are suggested.

- Ore sorting is generally best used in preconcentration applications, the most logical position is in the earliest part of processing, i.e., after crushing of run-of-mine ore or sometimes after secondary crushing. The issues are
  a. The technical responsibility area before the ore was fed into a process plant was typically related to mining. Ore sorting technology was generally unknown to mining engineers.
  b. If a process plant did not have secondary crushing, only primary followed by SAG (Semi-Autogenous Grinding), there was often poor economy to retrofit such an operation with additional crushing, washing, screening, conveying.
  c. Mining schools did not teach ore sorting to a significant level if at all within their mineral processing curricula. Hence, little information about sorting was available also to mineral processing professionals.
- Ore sorting coarse rock may cause a recovery reduction of the valuable mineral(s), even slightly. Hence, it was frequently thought that the overall recovery would be reduced. That was a false perception in many cases, but remained an obstacle for more in-depth investigations. It was shown in some cases, that the more uniform and higher feed grade for downstream processing could improve processing recovery compared to not sorted feed, ultimately improving overall recovery or often maintaining that same level at reduced overall cost.
- Most advanced ore sorting technologies were owned by a single large mining company for a long period of time. A former company manager stated that it was never their intent to make the ore sorting subsidiary a profitable business, but rather to secure that the technologies remained available within the group of companies.
- For some intended technology developments, sensor reliability was a great challenge. In initial commercial attempts, it was a requirement to have highly skilled operators and very stable power supplies which was not possible in some countries.

Nevertheless, ore sorting was very important in the cases where it was applied and highly profitable for the companies using them.

TECHNOLOGY DEVELOPMENTS

In the mid-80s, recycling of household waste was moving ahead in Europe, especially in Germany, encouraged by a new legal framework. It was no longer permitted to dump waste, and it was difficult to burn waste with a content of glass, metals, paper, cardboard or plastics. Hundreds of new recycling plants had to be built, and most of them worked by handpicking only. Productivity was low, labor costs were high, and the working conditions were not seen as acceptable for developed countries.

Contrary to the minerals industry, the recycling industry at that time was dominated by small and medium sized enterprises, owned by entrepreneurs comfortable making fast and sometimes risky decisions. Spin-offs from universities and other research institutions in the field of computer technology became available. Image processing was popular, which soon was applied to the emerging recycling industry. There was no "traditional" technology established and the recycling industry was free to choose new standards and eager to adapt technology. Computing power increased dramatically and new concepts of computer based data processing and sorting algorithms were developed. This led to the introduction of advanced sensor based sorting, first mainly optical sorting (with line scan cameras) and inductive sensors (metal detectors) in industrial recycling in a short period of time.

After technologies were improved and refined, they became useful for the ore processing industry. The prior arguments against employing them; not the right sensors available, low throughput, too complicated, too delicate, no skilled personal, etc. were all reduced or eliminated by the thousands of applications in the recycling industry. Hence, the exploitation of improved technologies was a natural consequence and is now an expanding technology area within the minerals industry.

Basics and Systems

Sensor-based sorting is a combination of a detection by a sensor and an ejection of single particles. Like hand-picking, the ejection or acceptance of a particle is based on a yes/no decision, which is generated by the evaluation of acquired data through a sorting algorithm. In other separation and concentration methods, the force, which moves, holds or deflects the particle, is directly coupled to the physical characteristics, such as magnetic properties or density. In sensor based sorting, physical property and separation action are not directly linked. The measurement of one or more properties takes place, after which it is decided by data processing, if the property characterizes the particle as to "eject" or to "accept." The whole system of sensor based sorting comprises of four sub-processes:

1. Feed preparation
2. Material presentation
3. Detection and evaluation
4. Ejection

Feed Preparation

Feed preparation is needed to ensure a condition of the feed material suitable for the chosen sensor and machine type. This includes screening to produce a narrow particle size distribution, and a treatment of the particle surface by de-dusting, scrubbing, washing, de-watering and/or drying. Few sensors need additional feed preparation processes. Sortable particle sizes are influenced by the spatial resolution of the particulate sensor and by necessary degree of liberation. Currently, material

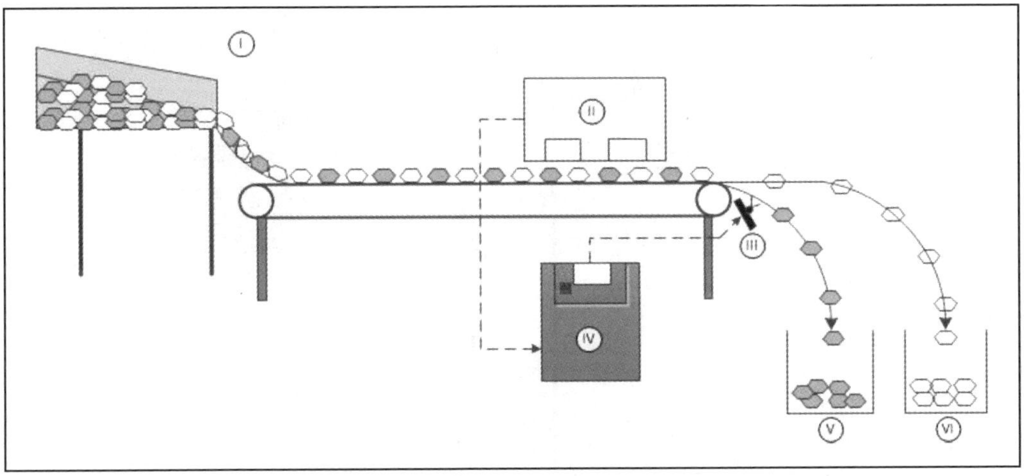

Figure 1. Belt-type sorter: (I) material presentation by vibrating feeder and belt, (II) sensors, (III) ejection nozzle array, (IV) CPU, (V & VI) accepted and ejected products, respectively

from 1 mm to about 300 mm (12") can be processed on a sensor based sorter. As the technology works with a single-particle layer feed, and the number of single particles which can be detected, evaluated, and ejected per time unit is limited, particle size has a major impact on throughput. At maximum particle sizes, a throughput of up to 300 ton per meter sorter width per hour is possible, but at particle sizes of a few millimeters, only a few tons per hour on the same machine width can be treated.

Material Presentation

The particles must be presented to the sensor(s) as single particles, accomplished by spreading and accelerating them along two or more conveying steps. For some sensors and detection modes, it is necessary to isolate each particle from its neighbors (single particle mode), which reduces the covering of the available space and decreases throughput. The particles have to follow defined paths between detection point and ejection point to avoid misplacement. For material presentation, the two basic current sorter machine types use either a vibrating feeder followed by a fast conveyor belt (belt-type sorter) or a vibrating chute followed by a sliding chute (chute-type sorter), see Figures 1 and 2.

Detection and Evaluation

The detection of material properties of minerals and ores can be made along the whole electromagnetic spectrum, see Figure 3. Not all possible wave lengths are employed in industrial scale yet.

Sensors can be classified into a few main groups:

1. Sensors for the detection of surface properties
2. Sensors for the detection of properties representing the whole particle
3. Sensors to detect secondary material properties like color, reflectivity, brightness, conductivity, density, magnetic susceptibility, gamma ray emission
4. Sensors for the detection of primary material properties such as elemental or mineralogical composition

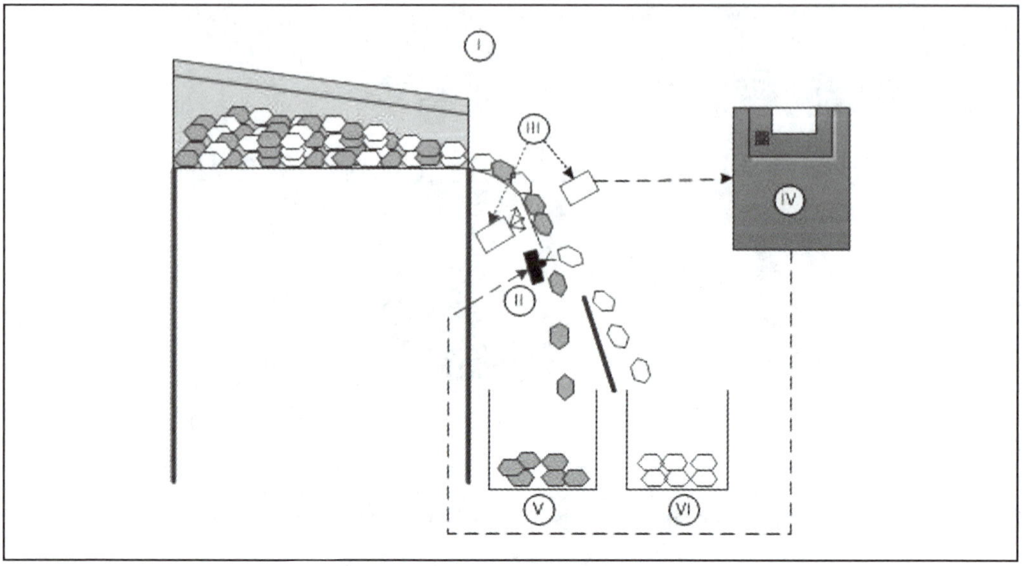

Figure 2. Chute-type sorter: (I) material presentation by vibrating feeder and sliding chute, (II) sensors, (III) ejection nozzle array, (IV) CPU, (V & VI) accepted and ejected products, respectively

Spectrum	[m]
Gamma-radiation	10^{-12}
	10^{-11}
X-ray	10^{-10}
	10^{-9}
	10^{-8}
Ultraviolett (UV)	10^{-7}
Visible light (VIS)	10^{-6}
Near Infrared (NIR)	10^{-5}
	10^{-4}
Infrared (IR)	10^{-3}
Microwaves	10^{-2}
	10^{-1}
	10^{1}
Radio Waves	10^{2}
	10^{3}
Alternating Current (AC)	10^{4}

Sensor/Technology	Material Property	Sorter application
RM (Radiometric)	Natural Gamma Radiation	Radioactive Ores
XRT (X-ray Transmission)	Atomic Density	Base and Heavy Metal Ores Precious Metal Ores Industrial Minerals, Coal, Diamonds, Scrap Metals
XRF	Visible Fluorescence under X-rays	Diamonds
COLOR (CCD Color Camera)	Reflection, Absorption, Transmission	Base Metal Ores Precious Metal Ores Industrial Minerals Diamonds, Glass
PM (Photometric)	Monochromatic Reflection/Absorption	Industrial Minerals Precious Stones Diamonds
NIR (Near Infrared Spectrometry)	Reflection, Absorption	Base Metal Ores Industrial Minerals Plastic Paper, Cardboard
IR (Infrared Camera)	Heat conductivity, heat dissipation	Base Metal Sulphide Ores Precious Metal Ores Industrial Minerals Graphite, Coal
EM (Electro-Magnetic Sensor)	Conductivity, Permeability	Base Metal Sulphide Ores Scrap metals

Figure 3. Sensors along the electromagnetic spectrum (Wotruba 2011)

Until recently, all properties which could be detected were "secondary"; properties like color, transparency, brightness, conductivity etc. These are not specific for a certain mineral or grade as there are minerals which can be found in a variety of colors, but a certain degree of conductivity can be related to various sulfides. For some applications, the detection of secondary properties is not sufficient.

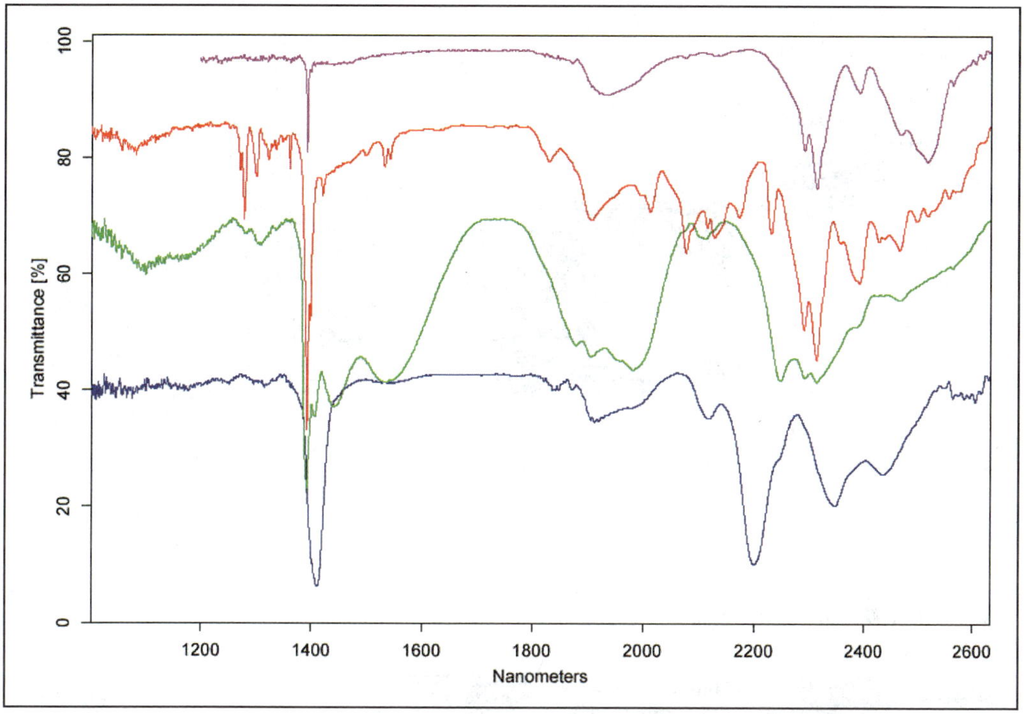

Figure 4. NIR-Spectra of milky quartz (blue), calcite (pink), talc (red), and chlorite (green)

There are sensors, which are able to detect "primary" properties like absorption of distinct wavelengths in the near-infrared (NIR)-spectrum or conduct direct elemental analyses by a XRF (x-ray fluorescence) sensor. With a NIR-sensor, NIR-active minerals can be identified selectively. This technology was used for decades in geological exploration. A NIR-Sorter of the most recent generation can scan and evaluate several hundred thousand NIR-spectra per second. With this sorter type, it is for example possible to distinguish between white quartz, calcite and talc, see Figure 4 (Robben 2011).

An electromagnetic machine sorting principle is shown in Figure 5. An example of elemental composition sorting is shown in Figure 6. Note that reproduction in black and white does not permit full justice to these illustrations.

The following sensors are currently applied in industrial scale (numbers within brackets refer to the sensor types listed above):

- Optical (line-scan camera; monochromatic laser scanner) (I, II)
- NIR (near-infrared spectroscopy) (I, IV)
- XRT (x-ray transmission (II, III)
- Inductive (II, III)
- XRF (x-ray fluorescence with photo-multiplier (for diamonds) (I, III)
- XRF (x-ray fluorescence for elemental analysis (I, IV)
- Radiometric sensor (gamma-ray spectrometer, scintillometer) (II, III, IV)

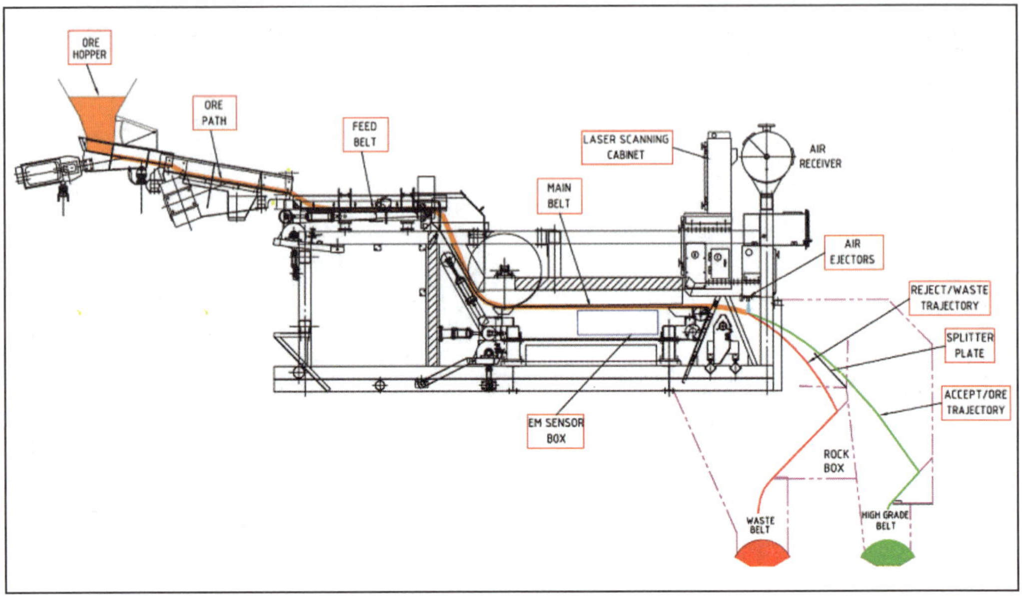

Figure 5. Principle of a sorter machine based on EM sensing

Figure 6. Illustration of XRT sorting—Sudbury nickel ore

Currently known applications in the minerals industry:

1. Optical Sensors (line scan cameras)
    a. Industrial minerals (calcite/limestone, dolomite, quartz, feldspar, talc, rock salt, etc.)
    b. Precious stones (diamonds, emeralds, tanzanite)
    c. Metal ores (chrome, gold, nickel, platinum, copper)
2. NIR
    a. Talc
    b. Colemanite (boron ore)
    c. Limestone
3. XRT
    a. Scheelite, wolframite
    b. Coal
    c. Diamonds
    d. Iron ore
4. Inductive
    a. Sulfide ores (nickel, copper, etc.)
5. XRF with photomultiplier
    a. Diamonds
6. XRF (elemental analysis)
    a. Sulfide ores (nickel, copper, zinc)
    b. Precious metals (gold, platinum)
    c. Chromite
    d. Manganese
    e. Iron ore
7. Radiometric
    a. Uranium ores
8. Monochromatic Laser Reflection
    a. Magnesite
    b. Limestone
    c. Talc

Sensors in laboratory, prototype and pilot scale (numbers within brackets refer to the sensor types listed above):

- Optical spectral resolution sensors (I, III)
- Sensors for fluorescent minerals (optical, with spectral resolution, laser induced) (I, III)
- Microwave excitation in combination with thermal infrared sensor (II, III)
- Rahman-spectroscopy-sensor (I, IV)
- Terahertz-transmission (II, III)
- LIBS (Laser Induced Breakdown Spectrometry). There is already in commercial uses for fast on-belt analysis of several elements, including such light ones as Ca and Mg. Hence, this technology may be adapted for ore particle sorting.

Ejection Methods

Ejection is either done by air jets from arrays of high-speed compressed air valves or electrically, pneumatically, or hydraulically driven flaps or plungers. Water valve system were abandoned very early. It is also possible to conduct bulk-sorting of a whole material stream by conveyor-flaps/dividers or reversible conveyors.

Summary of Possible Ore Sorter Applications (Examples)

Contrary to classical separation technologies for coarse sizes such as density separation or magnetic separation (e.g., for iron ores), sensor based sorters can separate particles with very small differences in their properties. This enables a range of applications:

- Waste elimination or pre-concentration before further milling and concentration
- Preparation of a final product or production of different qualities (diamonds, precious stones, industrial minerals)
- "Near-to-face processing" (use of sensor based sorting underground and in satellite mining)
- Re-processing of old waste dumps and coarse processing of other products
- Elimination of noxious minerals from plant feed to improve flotation or leaching or from coarse waste for improved deposition quality (e.g., pyrite-containing rocks)
- Diversion of plant feed to separate processing lines according to grade, mineralogical composition or contaminants

Sensor-based sorting is basically a dry process (some versions need a certain amount of water or steam for feed preparation/surface cleaning). This enables its use in desert, arid or permafrost areas.

Ore sorting-plants can be built very compact, for example in containers for mobile or semi-mobile applications. They can be fit into a tight space in underground installations. Waste can be dumped directly underground and used for backfilling, and shaft capacity for the ROM can be increased.

EMERGING APPLICATIONS AND ONGOING DEVELOPMENT

Increasing throughput is a constant requirement from all operations of massive resources like copper or iron ore. At current throughputs, the required capacity can only be provided by a number of parallel sorters. Increase of throughput can be done by optimizing material presentation, i.e., faster presentation speeds, hardware and software.

In the field of sensors, there are numerous simultaneous developments. For many applications, the line-scan camera is sufficient. When optical properties differ relatively little, laser based sorters can be an option. As a sensor for primary surface properties, NIR sensors are increasingly used. Additionally, "hyper-spectral imaging" technology, which is used in airborne geological measurements, is evaluated for sorting and quality control applications. In both types of multi-spectral detection, the critical point is data filtering and evaluation to reach a sorting decision in near real time.

A good potential for sorting applications is related to XRF sensors. Although it is a particle surface detection method, it comes close to the minerals processor's dream: To sort according to chemical composition and grade. Compared to its use for laboratory analysis and process control, the required measuring time is extremely short despite low spatial resolution of currently several centimeters, which leads to low signal-to-noise ratios. Nevertheless, it is possible to measure heavy elements, not based on the absolute, but relative counting rate. The lowest detection limit is reported to be about 0.1%. A disadvantage is the relatively low throughput of the actual industrial machines, where the particles are presented in a row of channels to the detectors, see Figure 7. At least 200 XRF sorters have been put in operation in the last two decades; an example is shown in Figure 8.

For metal ores and coal, XRT has shown very good results in industrial plants, like in the Mittersill scheelite mine (Austria) (Mosser 2010). It can identify elements according to their atomic density. Especially heavy elements, like tungsten, absorb x-rays nearly completely. Tungsten minerals like scheelite or wolframite can be recognized as black or grey dots within a rock and can be identified by their contrast to the matrix. The separation of light coal and heavy stone works well (and in relatively dry condition). First trials with diamonds have shown good results. The potential for sulfide and metal oxide ores (cassiterite, tantalite, etc.), and also for industrial minerals like barite or fluorite has been successfully evaluated in laboratory tests, but is still not used industrially.

Figure 7. XRF sorter with channel feed and mechanical flaps

Figure 8. Six XRF sorters working in parallel

Because of the very short measuring times compared to laboratory use, it is difficult to detect low grade values, such as for precious metal ores or other low-grade metal ores. To distinguish between ore and waste or between low and high grade ore it is still possible using "indirect" detection. Indirect detection means that not the target mineral or metal is detected, but some other "indicator mineral," or generally, an "indicator property" tells the sorter if this is a piece of ore or waste. One example is the detection of ore by alteration minerals, which can be detected by NIR sensors without being able to measure e.g., the copper content.

Combinations of sensors are possible. The combination of optical and NIR/optical is already the state of the art, as well as optical/inductive or XRT/inductive sensors. More than two sensors are currently not used in order to limit the effort and time for data processing. Whenever possible and acceptable, the aim is to use only one sensor to keep the set-up as simple and fast as possible.

As mentioned above, LIBS (Laser Induced Breakdown Spectrometry) could be the next sorter development. A future project is pattern recognition. To distinguish, for example, between fine grain, coarse grain or layered rocks is an easy task for the human eye, but a challenging one for image processing.

System Integration

Sensor-based sorting should play a major role in "mine to mill integration" (Bamber 2008, Kleine 2010):

- It may be integrated into the mining and backfill methods
- Adaption of downstream processes
- Consideration in the first planning steps of the processing plant and the mine
- Consideration of ore sorting in pre-feasibility and feasibility studies

SUMMARY

Several factors inhibited the wide-spread use of ore sorting technologies during several decades until the turn of the century. Entrepreneurs originated the first industrial purpose-built sorters, such as the uranium ore sorter. Food sorting machines were also adapted to heavier duty uses as required to stand up to conditions in ore processing plants. A major advance was achieved with the fast scanning laser sorter, although the number of applications was limited. Just over 30 sorter machines were in use for magnesite, limestone, phosphate, tungsten ore, lignite coal, talc, a gold ore in an underground mine, a lead-zinc ore in a similar situation and spodume-containing pegmatite ore until about 1990.

When microprocessors, advanced line-scan cameras and improved light sources became available, machine construction cost was substantially reduced. This made it possible to extend sorting to lower capacity sorting, such as for smaller particles than what prior art machines could do economically. Several small, entrepreneurial type companies entered the ore sorting market. In addition, sensor technologies were further developed, driven by the need to replace labor in government mandated waste recycling programs. With a broad industrial basis for such advances, adaptation to ore sorter applications became a natural consequence, just as in the beginning of the ore sorter history when food sorters formed the industrial technological basis. Therefore, it is anticipated that ore sorting will no longer be a niche unit operation in the mining industry.

ACKNOWLEDGMENTS

Illustrations were from personal files, Internet publications, and some from unnamed but gratefully acknowledged ore sorter vendors.

SELECTED REFERENCES

Arvidson, B.R. 1987. Economics and Technical Features of Preconcentration using Sorting, *Preprint No. 87–125, SME Annual Meeting,* Denver, Colorado, February 24–27, 1987.

Arvidson, B.R. 2002. Photometric Ore Sorting, *Mineral Processing Plant Design, Practice, and Control, Proceedings,* p 1033–1048.

Bamber, A.S. 2008. Integration Mining, Pre-Concentration and Waste Disposal for the Increased Sustainability of Hard Rock Metal Mining. PhD thesis: University of British Columbia, Vancouver, Canada.

Bibby P.A. 1982. Preconcentration by Radiometric Ore Sorting, *The Aus. I.M.M. North West Queensland branch Operators' Conference, September 1982.*

Kidd, D. 1983. Photometric Ore Sorting, Preprint No. 83-92, SME Annual Meeting, Atlanta, Georgia, March 6–10, 1983.

Kleine, C. 2010. Added value to the mining industry by the integration of sensor based sorting, *Aachen international mining symposia, Mineral Resources and Mine Development, Institute of Mining Engineering I, May 26th and 27th 2010,* 411–434.

Mosser, A. 2010. Operational Experience with XRT-Sorting for Pre-concentration of Tungsten Ore. *Proceedings- Sensor Based Sorting 2010.* Aachen, Germany.

Robben, M. 2011. Near-Infrared Sorting for Minerals. *In: Pretz, Thomas, Wotruba, Hermann; Nienhaus, Karl, Applications of Sensor-based Sorting in the Raw Material Industry;* Shaker Verlag ISBN-10: 3844005854.

Schapper, M.A. 1977. Beneficiation at Large Particle Size, using Photometric Sorting Techniques. *Presented at 2nd IFAC Symposium on automation in Mineral and Metal Processing, Johannesburg, September, 1976, and Australian Mining,* April, 1977.

Wotruba, H. 2011. Harbeck, H., Sensor-Based Sorting, *Ullmann's Encyclopedia of Industrial Chemistry, 7th Edition, and DOI: 10.1002/14356007.b02_18.pub2.*

ABOUT THE AUTHORS

Bo R. Arvidson has degrees from the Royal Institute of Technology (KTH), Stockholm, Sweden, and has more than 45 years of experience in the minerals industry. In 1974, he joined Sala International and was assigned to Sala Magnetics Inc. in the USA, and later became the technical director of the group. He has also worked for Amax Extractive R&D in Colorado, USA, and the Bateman Company. At Bateman he developed new high-intensity magnetic separation technologies and associated applications. He later returned to Colorado and built up Bateman's business specializing in the new magnetic separation area, ore sorting, and additional technologies such as linear screens and inter-stage carbon screens. In 1990 he formed his own company, International Process Systems Inc. (INPROSYS), which successfully developed, manufactured, and marketed advanced high-intensity magnetic separators. He was engaged in eddy-current separator technology, high-efficiency air classifier development, and photometric sorting, and developed more than 60 new industrial applications. The company was sold to Outokumpu Technology (now Outotec) in 2001 and he managed the merged Physical Separation group for 4 years. He then began his own consulting business.

Hermann Wotruba has a degree in mining engineering from the RWTH Aachen, Germany, where he also obtained his PhD in mineral processing. After 9 years working in industry, he went back as a full professor and head of the mineral processing group in 1998, which is his current position. He has extensive experience in the area of mineral processing and environment with project experience in many regions like South America, Southern Africa, and Asia. His work includes mineral processing base studies to flow-sheet development, equipment and plant design, plant commissioning and optimization, and environmental management with special emphasis on dry mineral processing and sensor-based sorting. He is one of the world's leading experts in small-scale and artisanal cleaner mining and processing technology and member of the International Mineral Processing Council.

Magnetic Separation

Bo R. Arvidson
Bo Arvidson Consulting LLC, Lakewood, Colorado, United States

Daniel Norrgran
Eriez Magnetics, Erie, Pennsylvania, United States

ABSTRACT: Large-scale magnetic separation processes became viable with the advance of commercial permanent magnets in the 1940s. Magnetite ores could then be efficiently processed at low cost. Weakly magnetic, relatively high-value minerals could be processed by the turn of the 19th century with low-capacity electromagnetic separators and from the 1930s with induced roll electromagnetic separators. Thirty years later, high-intensity magnetic separators for high-capacity processing of non-magnetite iron ores became available, complementing other beneficiation methods. Shortly thereafter, high-gradient versions of high-intensity magnetic separators were developed, impacting the kaolin clay processing industry and later some other minerals applications. More recently, a further version of the high-gradient magnetic separator (HGMS) type has benefitted the non-magnetite iron ore industry.

In 1981, the first high-intensity magnetic separator based on powerful rare-earth magnets was developed and with further advances changed the industrial minerals industry. Superconducting magnets for ore processing were introduced in commercial scale in the early 1980s and later generations changed the kaolin clay processing industry again. The paper will give an overview of the historical developments of magnetic separation for minerals, the impact on the ore processing industry as well as major application examples. An overview of designs is included, but detailed explanations of the operational features and comparisons will be given only as examples because some technologies tend to replace prior art and the interested reader may easily find plenty of literature on such subjects. Other important magnetic separation uses, such as removal of tramp iron, metals recycling and filtration of waste water are not included.

INTRODUCTION

Two basic types of magnetic separation technologies are used in the ore processing industry:

1. Low-Intensity Magnetic Separators (LIMS)
2. High-Intensity Magnetic Separators (HIMS), with Wet High-Intensity Magnetic Separators (WHIMS) as the most common type

There are different definitions with regard to the magnetic intensity level used for classification of the separators. Some LIMS have been developed into enhanced intensity levels and are then called Medium-Intensity Magnetic Separators (MIMS). HIMS are sometimes used for applications requiring moderate magnetic force strengths as well. To add to the complexity, some of the HIMS, usually using what is called high-gradient matrices, were called High-Gradient Magnetic Separators and the acronym HGMS was proposed and used.

It is suggested that the magnetic intensity level for LIMS would be max. 2,000 gauss (0.2 tesla) measured as close as possible to the magnetic poles in the separation space (being about the maximum achieved with ceramic magnets alone), MIMS 2,000 to 6,000 gauss, while HIMS, WHIMS HGMS would be rated at minimum 6,000 gauss in the working space without any collecting matrix in the gap between poles. Commercial superconducting HGMS were measured up to 50,000 gauss (5 tesla), while at least one conventional (copper coil) HGMS measured up to 24,000 gauss (2.4 tesla).

For magnetite ore applications, the magnetic intensity level can vary from around 800 gauss to 1,800. Removing very weakly magnetic components from kaolin clays typically requires a minimum of 18,000–20,000 gauss (1.8–2 tesla), common for the production of paper filler or coating products. Each application may be optimized by varying the magnetic field intensity, the collecting matrix, magnet configuration and many other processing parameters.

HISTORY

It is well known that one of the greatest inventors of all time, Thomas A. Edison (1890), attempted to develop a commercial iron ore process in the 1880s using some of his inventions including electromagnetic separators. The venture eventually failed, but other people developed electromagnetic separators that were successfully used in small scale applications, such as the Wetherill separator at the turn of the 19th century (invented by John Price Wetherill) and the induced roll separator developed in 1926. Large capacity separators became viable for magnetite ores when the AlNiCo magnet was invented. It was developed in 1931, and the first known use for magnetic drum separators was in the 1940s. When ceramic ferrite magnets came on the market, these replaced AlNiCo magnets due to their much improved coercivity, i.e., resistance to demagnetization. That enabled new efficient magnetic circuit designs. The size of the drum separators grew to 1.22 m (48") diameter and over 3 m (10 ft) width in the early 1970s and has stayed at that maximum size until recently.

High-intensity magnetic separators for continuous operation, primarily for dry processing of industrial minerals, were developed in many countries from the 1940s. These were of the electromagnetic induced roll type, generically called IMR (or sometimes IRM). Most had limited unit capacities and could process only fine size particles, typically less than 1 mm (<16 mesh Tyler).

The first Wet High-Intensity magnetic separators (WHIMS) were batch operating units and had also relatively limited capacity. One of the pioneers was S.G. Frantz, who developed batch operating magnetic filters in the 1930s, which could remove moderately magnetic particles from slurries. Continuously processing WHIMS machines came on the market in the 1960s and by early 1970s, the Jones type WHIMS was used in very large industrial plants. By that time, the first batch operating High-Gradient Magnetic Separators (HGMS) had entered the kaolin clay processing area. The latter was also developed into continuously operating units, with improvements and modified concepts that continued into the 1990s.

Advances in superconducting and rare-earth alloy magnet technologies resulted in industrial separator developments in the 1980s. In the next section, the most important of these and their applications until the 1980s will be reviewed and then the following section will discuss later events.

MAGNETIC SEPARATOR DEVELOPMENTS AND MAJOR APPLICATIONS UNTIL 1980

Small-scale operations such as separating wolframite from cassiterite at 10 tons per day can hardly be considered having a major impact on the ore processing industry, although it was very important for the individual operations when such processing became available in the 19th century. Therefore, only major developments will be discussed here. The first such remarkable event was the American development of the induced magnetic roll (IMR) separator by Fred R. Johnson of the Exolon Co. A typical IMR with a central electromagnetic coil, double start feeding and two high-intensity separation stages is shown in Figure 1.

Figure 1. Typical IMR

Additional variations of this concept in the 1940s resulted in numerous applications. A driving force was the need for Titanium for jet engines and nuclear components. Beach sands were easily processed using the induced roll separators and are still used today to recover ilmenite and leucoxene, as well as related uses in the beach sand industry. In addition, weakly magnetic iron ores were processed by large capacity purpose-built IMR plants.

Removing iron-bearing minerals from glass sand and ceramic raw materials to produce higher quality products was also a major application area. An important limitation for processing minerals was the top feed particle size, which was restricted to around 1 mm due to the small air gap between the magnetic circuit poles and the rotating roll required to achieve a strong magnetic field. There were a few applications involving up to 2 mm size particles. A Russian version is known for handling such cases. It had an extremely powerful magnetic system to overcome the air gap issue and to enable stronger magnetic forces for processing small particles.

The next major magnetic separation development that changed the minerals industry was the drum separator for magnetite ores. As massive high-grade ores were being depleted, the demand for concentrating fine-grained and low-grade magnetite ores increased. Also, reducing slag forming constituents, especially in emerging blast furnace and steel making methods, became more important. High-grade concentrates for sinter feed, pellets for blast furnace feed, and eventually also for DRI (direct reduction iron) processes became available. Figure 2 shows one large LIMS installation for wet processing.

Dry LIMS were also developed in the 1950s for iron ores. These are less common than the several different types of wet LIMS, but important when dry processing is more useful than wet. Also, when it is possible to reject low-grade waste in coarse particle size range, the dry LIMS, or a version that uses a conveying belt enveloping the drum can be beneficially employed with a great economic advantage.

Figure 2. LIMS installation producing magnetite

Figure 3. Dual horizontal ring WHIMS

Although huge resources of hematite ores with sufficiently high Fe grade to be used as lump ore feed for blast furnaces exist, by the late 1960s, the incentive for large-capacity WHIMS to process fine fractions to increase the Fe grade drove developments. One of the first big scale installations was at Itabira, Brazil, where 24 Jones-type separators with 80–100 tph capacity each processed itabirite, a form of specular hematite that is relatively highly paramagnetic. This German made separator type became practically an industry standard for hematite and limonite/goethite iron ores around the world, with copies made in Slovakia, China and eventually other countries. Figure 3 shows a high-capacity WHIMS with a dual ring configuration containing a matrix of grooved plates, one of the most common types.

In the industrial minerals area, batch operating HGMS of increasingly larger size changed the kaolin industry in the United States, United Kingdom, and what was then Czechoslovakia. The concept of using high-gradient generating matrix of filamentary material originated from MIT (Massachusetts Institute of Technology) in collaboration with the J.M. Huber Company in Georgia, USA. A special electromagnetic design, also an MIT concept used in "magnetic bottles" to contain high-temperature plasma, could generate a 20,000 gauss (2 tesla) field in the empty space for the matrix, which was about 3 times greater than other conventional magnet designs at

Figure 4. World's strongest continuously operating HGMS under assembly

the time. Very strong magnetic forces were generated when using steel wool matrices, that could capture extremely weakly magnetic components, identified as iron stained anatase, which discolored the kaolin product (typically 90% < 2 µm particle size) unless removed. Magnetic Engineering Associates manufactured a High-Gradient Magnetic Separator (HGMS) prototype, which was installed in 1968. It led to larger machine designs, which were capable of processing 20 to about 80 tph. Another industry standard was born that continued with further development of the magnetic circuit, see as discussed in the next section.

A continuously working HGMS development was initiated to make it possible to process ores with a high content of weakly magnetic materials, i.e., more than 10% by weight of magnetics in the feed. The world's largest HGMS with over 7-meter diameter ring could process more than 1,000 tph of fine weakly magnetic particles when fully equipped. One version with a smaller diameter, which can generate 15,000 gauss (1.5 tesla) in the working volume, is shown in Figure 4.

TECHNOLOGY DEVELOPMENTS AFTER 1980

Two technology developments impacted the magnetic separator technologies in the early 1980s. Permanent magnet alloys involving rare-earth (R-E) elements increased the magnetic flux level dramatically; see Figure 5. Initial designs used Sm-Co alloys with a strength level of 26 MgOe energy product. In the mid-1980s Nd-Fe-B magnets became available, which were stronger. Today, magnets with over 45 MgOe energy product and very high coercivity (allowing high processing temperatures) are utilized in the most powerful designs. Rare-Earth Roll (RER) magnetic separators and soon also the Rare-Earth Drum (RED) type came quickly on market. An example of the pioneering RER separator is shown in Figure 6.

Superconducting magnet developments driven mainly by medical instrumentation demand, resulted in a large industrial base for the special alloys manufacturing and cooling systems required to maintain zero resistivity in the coils. As soon as high-availability systems were feasible, the thermal (copper coil) magnetic circuit was replaced by the superconducting system, which required only a small fraction of the power. An example of the first such HGMS for kaolin clay is shown in Figure 7.

A further development was a semi-continuous superconducting semi-continuous HGMS operating at 5 tesla (50,000 gauss) initially invented in the UK to be used for Cornwall clays. The further enhanced version is shown in Figure 8. It became the new industry standard for beneficiation of kaolin clays and talc.

In addition, a mechanical design development that combined the best continuously moving matrix concept (a vertical ring), a specially-structured filamentary matrix and a jigging motion has

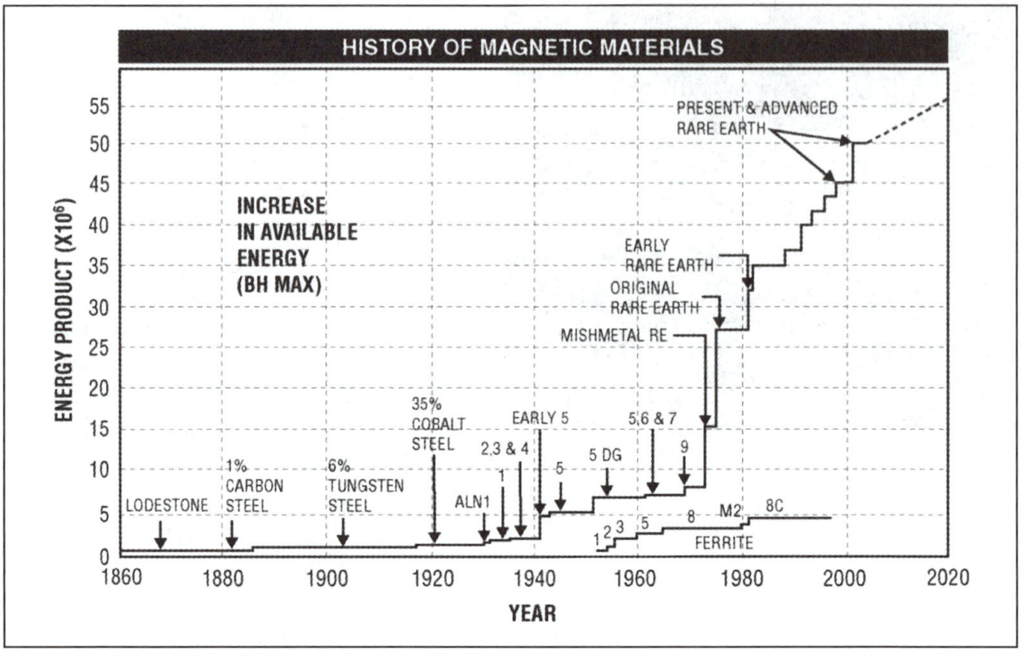

Figure 5. Development of magnetic materials over the years

Figure 6. A permanent magnet roll separator

Figure 7. Superconducting HGMS for kaolin clay processing

advanced the WHIMS type separator to a higher level (Dahe 2009). It is improving the processing of all non-magnetite iron ores by enhanced separation selectivity and even achieving industrial minerals cleaning not previously economically feasible. Yet operationally, it is more available than has been experienced with prior art WHIMS. Industrial machines are shown in Figure 9.

CURRENT MAIN TECHNOLOGIES AND APPLICATIONS

The Wetherill separator type is still in use today in various forms for separating high-value minerals, typically after calcination of the ore feed. If cross belt separators for metals and tramp iron removal from conveyor belts are ignored, the known industrial applications are mainly for tin ores, removing wolframite and other moderately magnetic materials from cassiterite.

Rare-Earth Drums (RED) are becoming prevalent for the concentration of some paramagnetic minerals. These medium-intensity separators are used to magnetically collect hematite or ilmenite at relatively high capacity (Norrgran et al. 2002, Gover et al. 2011).

Although IMR separators are still in use, the R-E type separators are usually the high-intensity separator of choice for new operations (Arvidson 2001). In many applications, the greater separation efficiency and the lower operating cost have motivated the replacement of IMR with both RER and RED. One typical example is replacing 6 IMR with 3 RER for the same plant feed rate while boosting the high-quality feldspar product recovery by 12%. In addition, numerous new applications have been made viable with RER and RED. Without going into much detail, the illustration in Figure 10 explains one of the major reasons for the possibility to achieve separations of coarser particles (currently even up to the 50–75 mm (2"–3") range than previously possible and the greater separation efficiency: The absence of a separation air gap. Yet, the magnetic force is very high. By

Figure 8. Superconducting magnetic separators for clays

Figure 9. Vertical ring HGMS processing iron ore

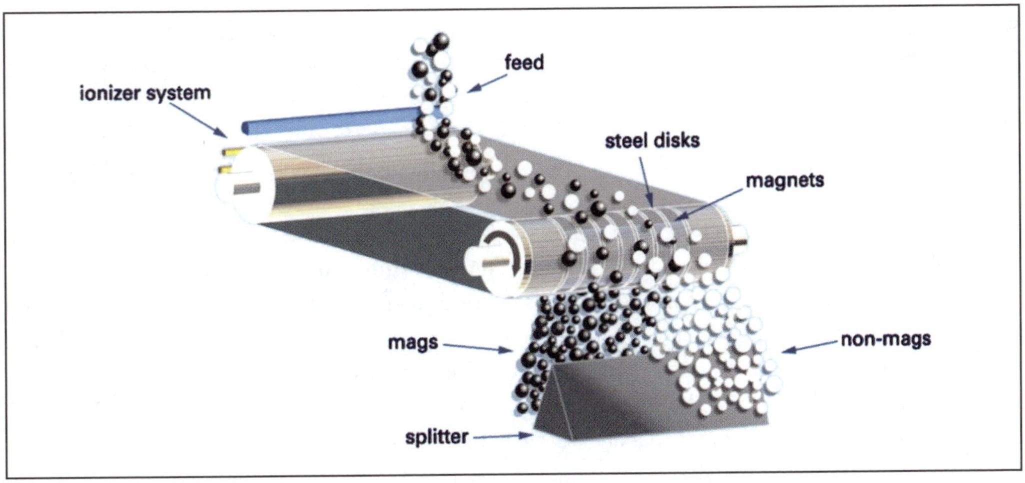

Figure 10. Basic principle of RER operation

Table 1. Qualitative ranking of magnetic separator applications

Ranking	Ore Type	Examples	Separator Type
1	Iron ore (Paramagnetic)	Hematite	WHIMS, HGMS, dry RE separators
2	Magnetite	Taconite	LIMS wet or dry
3	Beach sands	Heavy mineral concentrates	LIMS, WHIMS, IMR, dry RE separators
4	Clay and talc	Kaolin clay, talc	HGMS
5	Filler materials	Calcium carbonate	Dry RE separators
6	Glass and ceramic feedstock	Silica, feldspar, nepheline syenite	LIMS, WHIMS, IMR, HGMS, dry RE separators
7	Refractory feedstock	Alumina, chromite, magnesite	LIMS, IMR, dry RE separators
8	Non-ferrous metal ore	Sulfide minerals, wolframite, cassiterite	WHIMS, HGMS, dry RE, Wetherill

using high competing centrifugal forces balanced against the exceedingly strong magnetic forces, sharper separations can be achieved than in magnetic separators with a limited centrifugal force due to particles bouncing against poles, even at moderate roll speeds and to eddy current problems.

Another feature that enabled an important break-through in one application is the possibility for avoiding all contact with any metal in the separator. The application is production of ultra-high purity quartz (UHPQ) from natural ores (e.g., approx. 2–5 ppm and even less metal element impurities). More than 90% of UHPQ on the market has been processed in specialized RER.

For an effort to rank the magnetic separation applications for ores in some order of importance, see Table 1. The acronyms used for the various technologies are shown also. There may be exceptions to both the order and the indicated technologies.

EMERGING DEVELOPMENTS AND APPLICATIONS

The largest magnetic separation capacity application by commodity is iron ores. Very few magnetite iron ores can be utilized without magnetic separation. There is a global trend towards leaner ores (Arvidson 2012). Since the largest portion of traded iron ores consist of hematite, the declining feed grades of major new iron ore resources inevitably will require more processing to maintain required product quality for the feed to iron and steel making processes. In addition, a large portion of new resources expected to be developed have challenging characteristics, such as very fine liberation sizes and detrimental elements (Arvidson 2012). Hence, the demand for increasingly sophisticated iron ore processing technologies will be prevalent, with and magnetic separation is a very as an important component.

Low-intensity drum type magnetic separators in iron ore concentration are becoming increasingly larger to contend with high capacity applications. Drum diameters are increasing to 1.5 m (60") and drum widths exceed 4 m (160"). Larger drum diameters themselves have some effect on the unit capacity, but the operating gap of the separator is more important and needs to be enlarged to enhance unit capacities. New designs in magnetic elements using rare earth permanent magnets have been developed to extend the magnetic field through increased operating gaps.

Wet High-Intensity Magnetic Separators have followed the path of magnetic drum separators. They have become increasingly larger to match the increasing scale of production.

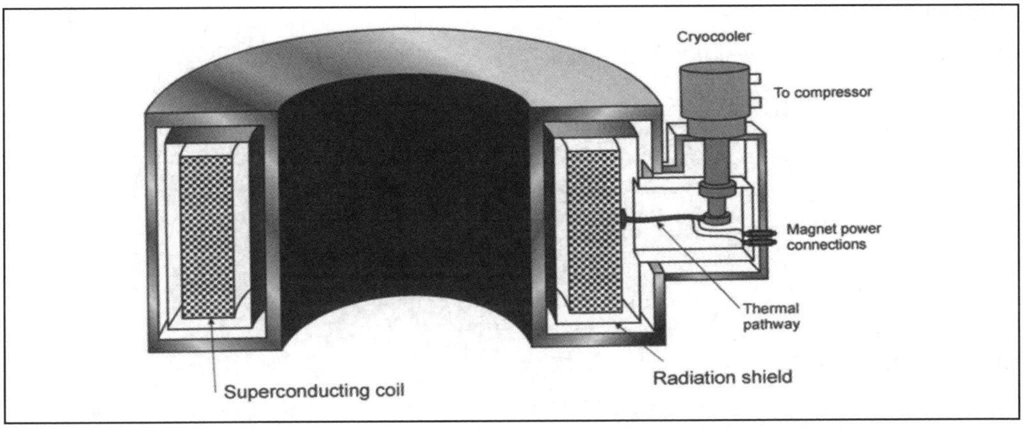

Figure 11. Schematic representation of a superconducting electromagnet operating at 7 degrees K through conduction

For other than iron ore commodities, the global supply scene is not expected to change as much. Magnetic separation will remain an important part of the technologies to be used. Demand for "exotic" elements, such as rare-earth group metals, tantalum, ultra-high purity quartz is expected to increase more than most other commodities, and magnetic separation is typically a critical part of the often rather complex processing schemes.

Besides developing larger models of the current machines, one technical development deserves to be mentioned, the cryogen-free superconducting coil system (Norrgran et al. 2009). This is the most recent technology for superconducting magnetic separators and it focuses on the simplicity of the design. Whereas the original superconducting separators required that the electromagnetic coil be cooled directly with liquid helium, the most recent technology operates cryogen-free. The separator utilizes a niobium-titanium conductor to generate the magnetic field. A schematic representation of this technology is illustrated in **Figure 11**. The coil is housed in a cryostat which uses a vacuum insulation to thermally shield the coil. Cooling is accomplished by conduction with the use of a cryocooler. The cryocooler leads are cooled to superconducting temperature with helium in a closed loop system with a helium refrigeration unit. The cryocooler has a copper thermal pathway that subsequently cools the coil by conduction to superconducting temperatures. The cryocooler reaches a temperature of 5 degrees K and the coil reaches a temperature of 7 degrees K through conduction.

INDUSTRY IMPACT

Tens of thousands of low-intensity magnetic separators and thousands of high-intensity magnetic separators are in use today. Ever-improving designs and better quality components have driven newer equipment to high standards with regard to efficiency and reliability. In a processing plant, the availability of magnetic separators is often on a higher level than other unit operations. The possibility to economically process ever leaner ores and some with challenging characteristics such very fine size particles and hitherto detrimental components, enhances resources that can be used for the benefit of improving conditions for humankind (Arvidson 2012).

SUMMARY

Magnetic separation is one of the fundamentally important technologies for ore processing. It may have not attracted as much academic attention as processing technologies of chemical nature in the past and in current education, but it is hoped that this paper will be helpful to better understand its importance and encourage more important developments that are still possible in this area.

Several examples in the history of magnetic separators show that important innovations have occurred when technologies related to magnet systems have been developed in other research or industrial areas than ore processing, and then these were adapted for use in magnetic separators. Entrepreneurs from Edison, Wetherill and to those living today have been the main drivers.

ACKNOWLEDGMENTS

The authors have used some information available on the Internet, in their personal files, in referenced publications, as well as some pictures from vendors not to be named but gratefully acknowledged.

REFERENCES

Arvidson, B.R. 2001. Economic Benefits of Using Advanced Magnetic Separation Technology, *Industrial Minerals' Processing for Profit 2001,* March 25–27, 2001, Atlanta, GA, USA.

Arvidson, B.R. 2012. Processing High-Grade Concentrates from Challenging Low-Grade Iron Ore Deposits, *paper to be presented at 'Advances in Iron Ore Beneficiation' at the International Symposium of NMD-ATM 2012,* Jamshedpur, India, Nov. 16, 2012.

Dahe X. 2009. Application of Slon Magnetic Separators in Modernising the An Shan Oxidised Iron Ore Processing Industry, *AUSIMM Iron Ore Conference,* Perth, WA, 2009.

Edison T.A. 1890. Magnetic Separator, US patent 430,280, June 17, 1890.

Gover D., Berton A., Norrgran D. 2011. Magnetic Concentration of Hematite at Wabush Mines–Plant Scale Application of Rare Earth Permanent Magnetic Separators. *SME Annual Meeting. R-H Yoon Symposium: Magnetic and Electrostatic Separation February 27–March 2, 2011,* Denver, Colorado.

Norrgran, D., Mankosa M.J. 2002. Selection and Sizing of Magnetic Concentrating Equipment Plant Design/Layout, Mineral Processing Plant Design Update. Vancouver, British Columbia. October, 2002.

Norrgran D., Ashton D., Fears P. 2009. Installation and Operation of the Eriez Cryogen-Free Superconducting Magnet at Sibelco. *Mineral Processing Plant Design—An Update Conference.* September 30–October 3, 2009,Tucson, Arizona.

ABOUT THE AUTHORS

Bo R. Arvidson has degrees from the Royal Institute of Technology (KTH), Stockholm, Sweden, and has more than 45 years of experience in the minerals industry. In 1974, he joined Sala International and was assigned to Sala Magnetics Inc. in the USA, and later became the technical director of the group. He has also worked for Amax Extractive R&D in Colorado, USA, and the Bateman Company. At Bateman he developed new high-intensity magnetic separation technologies and associated applications. He later returned to Colorado and built up Bateman's business specializing in the new magnetic separation area, ore sorting, and additional technologies such as linear screens and inter-stage carbon screens. In 1990 he formed his own company, International Process Systems Inc. (INPROSYS), which successfully developed, manufactured, and marketed advanced high-intensity magnetic separators. He was engaged in eddy-current separator technology, high-efficiency air classifier development, and photometric sorting, and developed more than 60 new industrial applications. The company was sold to Outokumpu Technology (now Outotec) in 2001 and he managed the merged Physical Separation group for 4 years. He then began his own consulting business.

Daniel Norrgran, a native Minnesotan, earned his BS degree in physics from Hamline University in St. Paul, and an MS in mineral engineering at the University of Minnesota in Minneapolis, where the emphasis on his work was the electrostatic and high-intensity magnetic separation of paramagnetic minerals. He then earned his MBA from Penn State University. He was a prolific writer, authoring dozens of technical papers on magnetic separation advancements that were published in professional trade journals around the world. Dan was employed at Eriez for 27 years, most recently serving as a manager of the Minerals & Materials Processing Division, where he traveled extensively worldwide in growing the company's global equipment sales to the mining, coal processing, industrial minerals, and other related industries into Eriez's largest served markets. He was a respected member of several professional associations including the Canadian Institute of Metallurgy, the International Minerals Association, and the Society for Mining, Metallurgy & Exploration, where he served as National Committee Chairman. In 2004, he received Eriez's highest honor, Person of the Year, in recognition for "managing the Mineral Processing Division to its present key position within the Worldwide Operations."

Innovations in Electrostatic Separation

Steve Hearn
Huntsman Corporation, Woodlands, Texas, United States

ABSTRACT: The paper discusses the different ESS techniques that evolved and their relevance in the mineral processing industry as well as current developments.

INTRODUCTION

Electrostatics, in a layperson's experience, can be illustrated, for example, with charging a pen by rubbing it on cloth then using it to attract small pieces of paper. This party trick for kids found its way into a practical means of separation, not only of minerals, but also metals from plastics and even for separating various food stuffs.

Early patents were in fact held by the Quaker Co. (yes, those same folks that produce your breakfast oats) and also the Gilroy Co. Gilroy? Those folks who travel to that city in California may have come across their garlic festival. Those early E.S. separators made use of conductivity differences between the oil laden garlic flesh and its dry skins. Over the years ESS has also been applied to separating pecan nuts from shells and sorting tea leaves, but early work with minerals seems to date from a Mr. Huff who built a mill in Wisconsin for base metal separation in 1908 (Wentworth 1913).

Any dry mixture which contains particles of different conductivity (affected by moisture or perhaps oil content) is potentially amenable to this method of separation.

Although the title of this paper is Electrostatic Separation, there are several techniques employed industrially. Broadly, the methods employed are conductor/non-conductor separations by corona charging or triboelectric separation using a precharging method. Both techniques will be examined separately.

ELECTROSTATIC SEPARATION

In an electrostatic (or electrodynamic) separator, particles are fed over an electrically grounded steel drum and then subjected to corona charging by electron bombardment, Figure 1. This is usually affected by means of a high voltage (~50kV) DC charge typically to a tungsten wire stretched full width of the drum.) This charge causes the particles to attach to the drum. Conductive particles gradually lose their charge to the drum (because of the grounding) and are flung off, whereas the nonconductive material adheres longer, as it does not readily lose that charge. Splitters which are installed underneath the drum are adjusted to collect the conductors falling off the drum separately from the non-conductors. Removal of these non-conductors is often helped by a shaped static or attracting electrode, Figure 2.

After receiving a reversing charge (from an AC charged electrode) the remaining non-conductors also fall off the drum, aided by gravity and usually roll brushing. Brushing is also

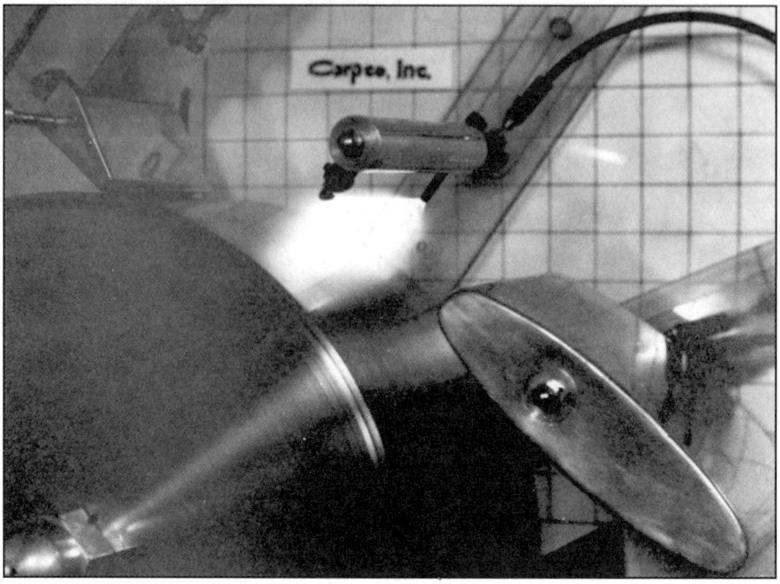

Figure 1. Corona discharge from electrode wire (Hoyles 1998)

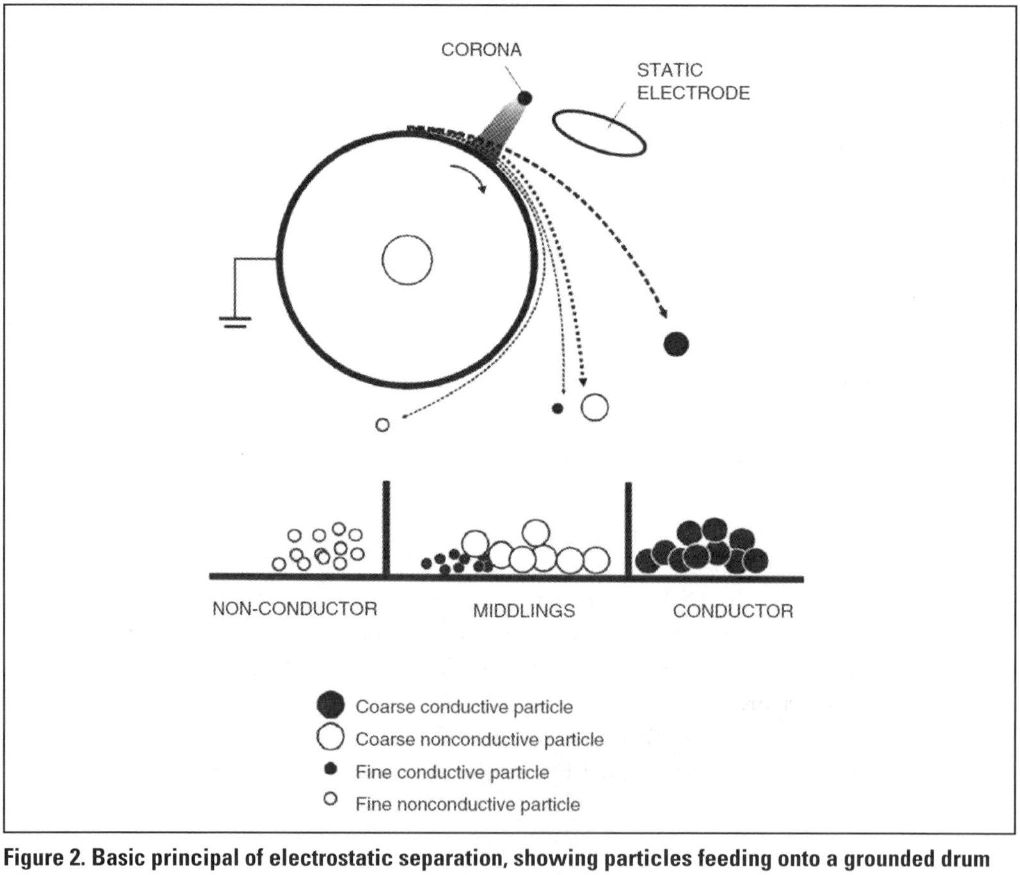

Figure 2. Basic principal of electrostatic separation, showing particles feeding onto a grounded drum

important for its overall effect in cleaning the drum for its return journey, so to speak as a dirty drum will be electrically insulated such that other material will not stick.

Applications

Early patents refer to the use of ESS in phosphate separation and iron ore is also important but the main application for ESS remains minerals sands. This separation process in its purest form relies on the relative electrical conductive/non-conductive difference of particles within the mineral suite. Specifically the conductive properties of ilmenite (the major TiO_2 bearing mineral) as well as rutile (under certain temperature conditions) and monazite contrast with non-conductive silica and zircon also typically present in the sands.

Much of the world's supplies of mineral sands come from Australia, so it's perhaps no coincidence that much recent innovation has come from that country. In fact early separators were developed (from those used in the food industry) and were employed back in the 1950s in Florida and New Jersey. (The former Carpco Co., of Jacksonville Florida, was a post-war pioneer in this field.)

Traditionally this method of separation works best with washed and classified feed, which exists naturally with mineral sands, occurring as they do as beach deposits (no natural fines remain). These heavy sands are usually then further wet gravity separated, removing much of the silica, dried, magnetically then electrically upgraded. Ignoring these basic principals, particularly that of cleanliness, usually leads to disaster as we can see in a "how not to" case study below.

Millions of tons of hematite ore concentrate have also been updated by this method at the Cliffs Wabush Mine in Labrador since the 1960s. More recently ESS has been incorporated into the chromite recovery circuit in Coos Bay, Washington, again after the critical washing step imparted by gravity concentration.

Aside from these actual mineral separations, ESS was successfully employed in the 1980s at several precious metal smelters, recovering metal prills from slag, typically after the Merrill Crowe process. This method was employed at a number of U.S. gold mines including FMC's Paradise Peak mine in Nevada and Brohm Resources in South Dakota, for example, with gold and silver recoveries from slag in excess of 90%.

The process involved dry grinding to liberate the metallic prills, de-dusting then ESS rather than the more usual wet milling and gravity tabling. These separated dry prills could immediately be resmelted (Maki and Taylor 1987).

HOW NOT TO APPLY ELECTROSTATIC SEPARATION

Iron ore recovery has taken place in Sweden north of the Arctic Circle for many years, mainly under the auspices of the Swedish Government. In wishing to exploit a hematite deposit some years ago, the company decided to utilize electrostatic separation like they'd seen successfully employed at the Wabush Mine in Canada (Lawver 1969). Remember those basic principles of feed cleanliness? Well, the company obviously didn't and fed a dry ground ROM ore to the separators. Quickly, very quickly, the fines and clays present with the hematite ore stuck to the steels drums. Without an effective cleaning method (imagine clay building up on conveyor belt idlers?), the separation very quickly stopped as the drums were in fact by then electrically insulated.

It is not known how long the plant operated (days only perhaps?) but it remained for many years a silent testimonial to the folly of not looking at all the properties of a mineral suite being taken into a processing plant, **Figure 3**.

Figure 3. ESS for iron ore in N. Scandinavia

ELECTROSTATIC SEPARATION IN THE 21ST CENTURY

Ore Kinetics' Corona Stat appears to be dominant in the electrostatic mineral separation market across the globe, particularly in South Africa and Australia. Many of the improvements that Corona Stat features seem to address the fundamentals of good operating practice; Figure 4. These would include moisture control, good feeding systems, roll and electrode cleaning (Gates 2004).

Ore Kinetics is a proponent of controlling the mineral grain environment by adjusting mineral temperature thereby ensuring the cumulative exposure to moisture/temp versus time is maintained as a constant, meaning more uniform particle conductivity. Previously, plant operations accepted that variable conductivity is an inevitable fact and process variables such as roll speed and/or voltage was used to mask this conductivity change. This concept of environment control allows for optimized machine variables to be used which consistently mean more uniform separation with reduced size effects taking place.

The improved wire and roll cleaning technologies perform similar functions with respect to roll availability. The wire is maintained as a "smooth cylinder" by removing contaminates that induce non-uniform ionization points along the wire. Typically these would cause excess current draw and "arcing" when separation momentarily ceases and in a worst-case situation, roll contamination effectively insulates the roll and prevents separation at all.

TRIBOELECTRIC SEPARATION

Triboelectric separation relies on electrical charge differences induced in non-conductive materials by surface contacting. Oftentimes this is no more sophisticated than heating and tumbling of the particles, whereby some of the particles gain electrons and hence a negative charge. Others, with a

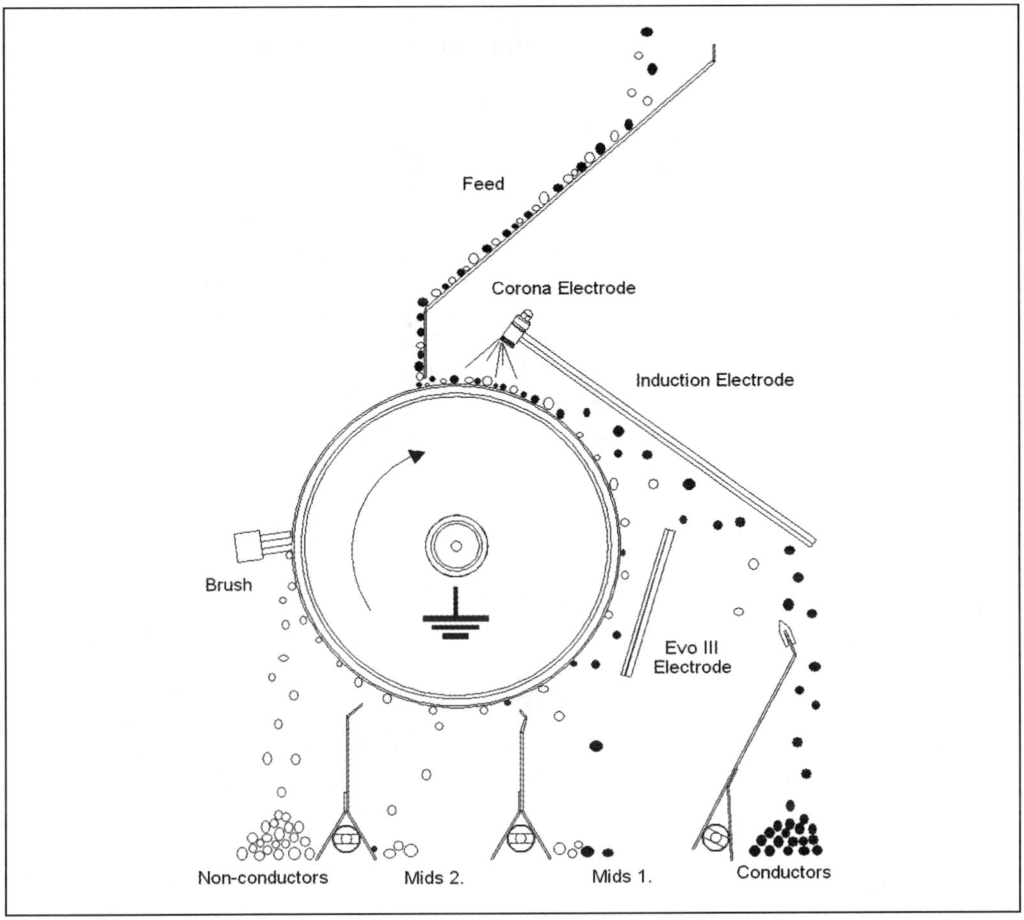

Figure 4. Corona Stat roll configuration (Gates 2004)

lesser tendency to acquire these electrons, will tend to become positively charged. This differential charge can often be enough to effect a separation.

A simple triboelectric separator consists of vertical plates or tubes (2–3 m or 6–10 ft long) between which the precharged particles are allowed to fall. Because either side is oppositely charged, particles are attracted to or deflected toward these plates or rolls. Rolls are often used as they can be rotated, facilitating cleaning. Static plates can accumulate particles sticking to them and eventually will be become electrically insulated whereby separation will cease, Figure 5.

Generally "free fall separation" such as is described above is limited to separating particles sized >75 microns, due to effects of wind. The attractive forces being relatively weak are not effective on particles >500 microns, as gravity would exert a stronger pull than would any electrostatic attraction.

Particles sticking to the electrodes presented challenges to equipment manufacturers, who strived for many years to find innovative cleaning solutions. Ideas ranged from vibration, to wiping the electrodes. Perhaps the most effective (at least when tube type electrodes were employed) was to use rotating brushes against the electrodes.

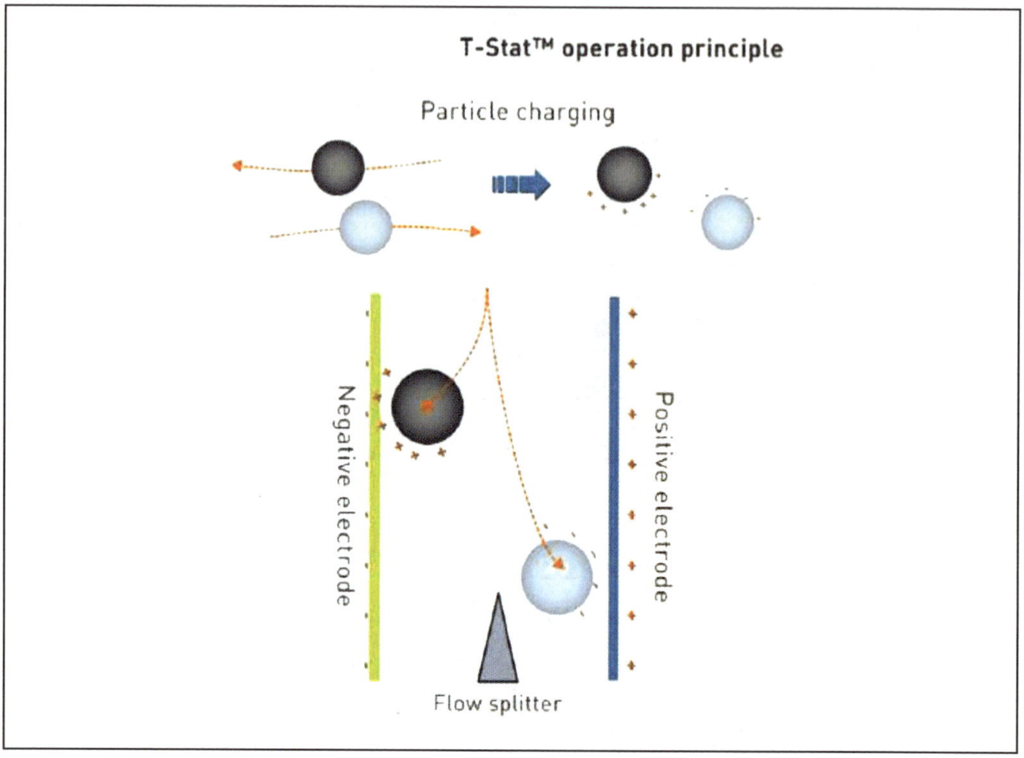

Figure 5. Triboelectric separation (Outotec brochure 2008)

Application

The use of triboelectric separation for separating hard salts began in the late 1940s. It was found that salts, potassium chloride and sodium chloride would become differentially charged by contact electrification. This effect was enhanced by high temperature treatment, elevating the mixture to 300–700°C, followed by cooling the mixture to 100–200°C. International Minerals and Chemical Corporation (now Mosaic) was the first to use this procedure in the early 1950s in a production plant located in Carlsbad, New Mexico, but have since discontinued the technique (Northcoat and LeBaron 1958). It remained to Kalle and Saltz in Austria to solve some of the inherent problems of salt separation and this was achieved later by better roll cleaning, for example. The author understands that a Canadian potash producer may now be employing this methodology currently although details are not available.

INNOVATION IN TRIBOELECTRIC SEPARATION

The ST separator is a more recent horizontal version of the type of separator described above and was originally developed in 1989. The horizontal belt system in essence, performs the same function as the vertical electrodes and also provides the transportation system, so to speak, after particle separation has occurred. The open mesh belts are moving in opposite directions so that material collected by each discharges at opposite ends of the separator (Gasiorowski et al. 2012).

Most of the installations described by the manufacturers relate to fly ash separation from carbon in the form of unburnt coal. The carbon receives a positive charge and the ash (and other

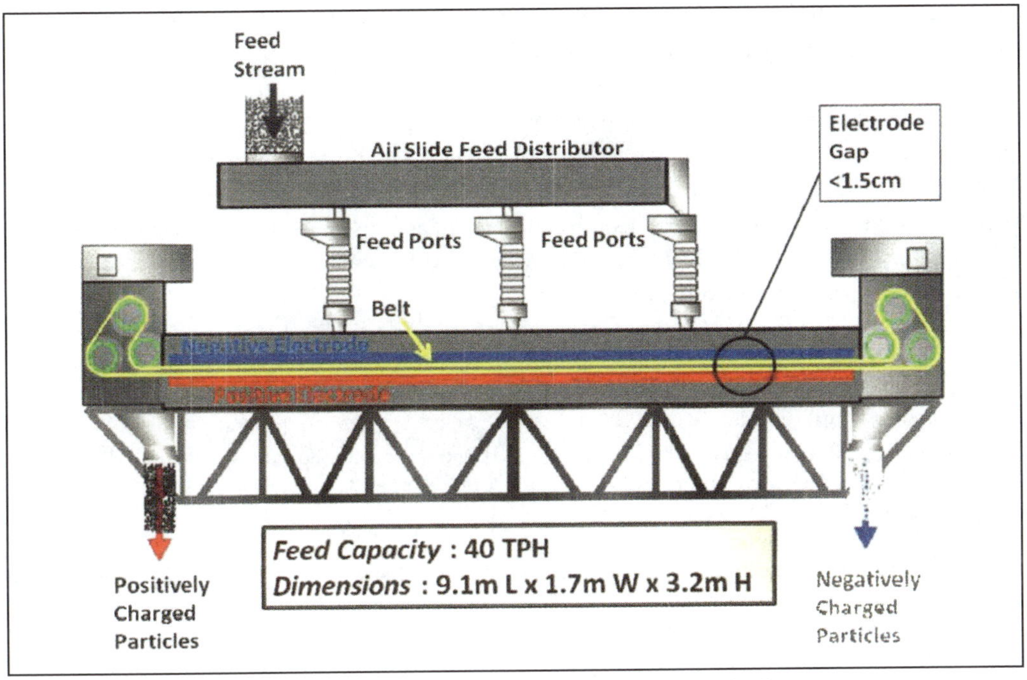

Figure 6. STI Separator configured for ash separation (Gasiorowski et al. 2012)

mineral present) becomes negatively charged. The economic driver is to reduce the carbon content of the ash to the extent that is had sufficient strength to be utilized as a cement additive for example. Unburnt carbon can of course be recombusted. STI claim that up to 40 tph feed (typically power plant non-combustible waste) can be handled by a machine of the dimensions shown in Figure 6.

SO WHAT'S REALLY INNOVATIVE IN ELECTROSTATIC SEPARATION?

In keeping with the theme innovations, this paper seeks to provide comment on various types of separation technique with particular attention being paid to innovative technology and its actual application.

There are many intriguing patents out there, but few of these novel ideas seem to have been commercially applied.

Coating techniques would appear to augment natural conductive or non-conductive properties (Cytec is active in this field although it seems a committed industrial user is still a way off [Maki and Taylor 1987]).

Broadly, it is the opinion of this author that there is little that is truly new in ESS over the last few decades. What is marketed as innovative almost invariably involves bigger, more power, better feed systems or state of the art process control for example. It is remarkable how much technology waxes and wanes in and out of popularity so to speak over the years and it seems to be the same with electrostatic separation.

Corona Stat now has a dominant position with respect to minerals sands recovery and separation, but the technique still employs similar grounded drums, charging and static electrodes to those used, say, in the Carpco separators of the 1950s. It is the innovations of application and cleaning of these electrodes that we see which has led to the company's success and dominant market position.

Triboelectric separation has seen broader application perhaps, being employed recently to great effect in recovering useful fly ash, recycling and in a few "pure" mineral separations such as potash. Here, the STI technology seems radically different from anything seen previously. Undoubtedly an economic driver such as an industry requiring large tonnages to be treated is missing. Maybe coal cleaning, for example, might in the future provide the driver which would lead to a more widespread employment of triboelectric separation?

REFERENCES

Bada, S.A., Falcon, R.M.S., and Falcon, L.M., 2010. Potential for ESS in the Upgrading of RSA Fine Coal SAIMM, pp. 691–700.

Gasiorowski, S.A., Bittner, J.D. and Hrach F.J., 2012. Triboelectric Belt Separator for Beneficiation of Fine Minerals. Canadian Minerals Processing Conference proceedings. Jan. pp. 399–408.

Gates, P., 2004, US Patent Office, 6723938 pp. 1–6.

Hoyles, P. 1988. Private communication.

Knoll F.S. and Taylor J.B., 1986. Selection and Sizing of Electrostatic Separation Equipment, SME pp. 208–205.

Lawver, J.E., 1969. State of the Art of Electrostatic Separation of Minerals, Journal of the Electrochemical Society, pp. 57C–60C.

Maki, T.D. and Taylor J.B., 1987. Precious Metal Slag Treatment Using an Electrostatic Separator Small Mines Development in Precious Metals, pp. 245–252.

Northcoat, E., and LeBaron, I.M., 1958. Application of Electrostatics to Feldspar Separation, Mining Engineering, pp. 1087–1093.

Outotec Bochure, 2008, "More out of Ore 2008."

Ravishankar, S.A. and Kolla, H., 2009. Chemically Enhanced ESS. 7th Int. Heavy Mineral Conf. SAIMM pp. 203–206.

Wentworth, H., E.S., 1913. Concentration or Separation of Ores AIME, Jan., pp. 411–426.

ABOUT THE AUTHOR

Steve Hearn graduated with a BSc in metallurgy from Nottingham University, United Kingdom. He moved to South Africa, working for several years with E.L. Bateman (a major engineering company engaged in the process engineering field) on many mining projects during his career there. In 1986 he joined Carpco, a company in N. Florida later acquired by Outotec, which specialized in physical separation, including electrostatic separation, for industrial mineral processing. Hearn is currently located in Denver where he is based as a metallurgist for Huntsman, supporting the company's specialty chemical group for mining applications.

Density-Based Separation Innovations in Coal and Minerals Processing Applications

Rick Honaker
University of Kentucky, Lexington, Kentucky, United States

Robert C. Dunne
Newmont Mining, Denver, Colorado, United States

Kevin Galvin
University of Newcastle, Callaghan, New South Wales, Australia

ABSTRACT: Density-based separators have been used as a preferred method of concentrating minerals and upgrading coal for several centuries. However, significant technological advances continue to be developed and introduced into commercial practice. The main focus of development has been the improvement in separation efficiency over a larger particle size range and the ability to achieve effective density-based separations for particles as small as 10 microns. The success of enhanced gravity separators in the recovery of ultrafine particles has resulted in their implementation into grinding circuits in an effort to prevent overgrinding and reduce circulating loads while also reducing the consumption of the chemicals used in conventional recovery processes and circuits. A review of recent innovations in density-based separations is provided in this publication along with the impact on the strategies and circuits used to process a wide range of minerals and coal.

INTRODUCTION

Gravity separation is second only to hand sorting in being the oldest of methods used to concentrate minerals and upgrade coal. The use of jigging screens to achieve density-based separations for agricultural purposes was reportedly practiced as far back as the ancient Egypt era. Agricola recorded the use of jig screens and sluices for the separation of heavy metals in the sixteenth century (Hoover and Hoover 1950). The industrial revolution which occurred from 1760 to 1840 resulted in a demand for large-scale production of minerals and coal thereby fueling the ingenuity and energy of innovators to develop density-based separators that were used to recover minerals from relatively rich ore deposits. The alpha units of modern day separators such as continuously operating jigs and shaking tables were developed, patented and commercialized in the late nineteenth century. Inventions that provided enhanced gravity separations in a mechanically-applied centrifugal field were patented in the 1890s and served as the basis of recently commercialized technologies.

The early twentieth century brought continued development of high capacity, density-based separators for coarse particle concentration such as the Baum jig and the Chance Cone. Several dry separation technologies were developed and commercialized in the period of 1920 to 1940 mainly for the treatment of coal. As the ore bodies became lower grade and underground mechanized coal mining became more prevalent, development activities shifted somewhat to technologies capable of treating fine (1 × 0.15 mm) particles. Flowing film separators such as the spiral concentrator and Reichert cone concentrator was introduced and realized wide spread commercial use in the middle part of the twentieth century. In addition, hydraulic fluidized-bed classifiers used for particle size separations were proven to provide effective density-based separations for a wide variety of fine materials when operated under certain conditions.

The late twentieth century was the era of enhanced gravity separator (EGS) development due to a desire to provide a low-cost, highly efficient recovery of ultrafine heavy metals such as gold and tin. Mechanisms commonly employed in conventional technologies such as jigging, flowing film sluices, fluidized-beds and shaking tables were applied in devices that achieve a separation in a centrifugal field created by mechanical action. The EGS units have been commercially applied to the recovery of a wide variety of different metallic minerals present in particle sizes as fine as 10 microns.

Burt (1984, 1986, 1999) provided a very detailed overview and description of the various commercially available density-based separators. The purpose of this publication is to provide a review of the innovative developments that have occurred over the past century along with their commercial applications.

WATER-BASED DENSITY SEPARATORS

Pulsed Devices (Jigs)

Recorded history indicates that jigging processes have been applied to ore dressing as far back as the Grecian time period. Descriptions and figures in *De Re Metallica* (Hoover and Hoover 1950) denote the use of jigging to upgrade sulfide ores in the sixteenth century which typically involved hand held jigging screens. Mechanization of the jigging process with production levels typical of current industrial units were developed and commercialized in the nineteenth century and their use maximized during the early twentieth century. When high grade deposits of minerals were depleted in the middle twentieth century, jigs units were commonly replaced with heavy media separators or froth flotation to handle the more finely disseminated ores. However, jigs remain a popular unit for the treatment of coarse iron ore, fine minerals and coal.

All jigs utilize a screen, either fixed or moving, and a means to provide pulsations through a particle bed which results in particle separation on the basis of density by hindered settling and consolidated tricking. Heavy mineral particles pass through the screen while light particles overflow a weir. In 1848, the first fully continuous jig was introduced by Berard which utilized a mechanical piston to generate the pulsation and suction strokes. The mechanical plunger-type jig was highly popular throughout the nineteenth century for the treatment of both minerals and coal. For mineral separations, the Harz Jig was the most popular in the early twentieth century (Taggart 1945). As shown in Figure 1a, the jig utilized a plunger mounted eccentrically on a wheel to provide a relative uniform pulse through a static screen and particle bed. Several compartments were used with separate plunger units to allow independent control. The primary application of the Harz jig was gold but was also applied to lead, lead-zinc, manganese, tungsten and sulfide-gold ores (Cope 2000). The Pan-America jig (Taggart 1945) developed in the 1930s utilized a flexible diaphragm in the cone section of the jig directly below the fixed screen which was manipulated by a mechanical arm mounted eccentrically to provide an up-and-down movement of the cone and thus a pulsation

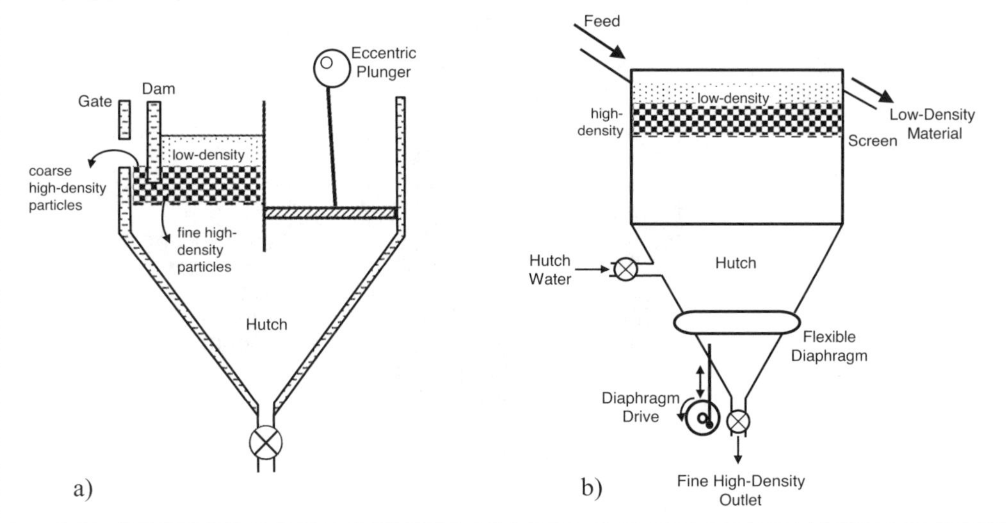

Figure 1. Mineral jigs: (a) Harz jig and (b) Pan-America jig (after Taggart 1945)

source to the particle bed (Figure 1b). Hutch water was added to reduce the speed of the suction stroke and allow addition time for particle stratification based on density.

The Baum jig was a significant development introduced in 1892. The plunger used in the Harz jig and other units was replaced with compressed air which was added and exhaust through the use of valves. The improved control of the pulse cycle in larger compartments allowed more efficient separation to be achieved over a wider range of particle size at higher throughput capacities (Sanders et al. 2002). The Baum jig became the most popular unit throughout the middle part of the twentieth century with bed widths of around 2.5 meters. The jig cycle typically follows a sinusoidal form and has been the subject of significant investigation in an effort to optimize performance (Tanaka et al. 1990; Ilijima et al. 1998).

As a result of the demand to increase the throughput of the Baum jig, the difficulty of creating a uniform bed pulse across a large particle bed was realized which caused issues with product grade. This led to the development of the Batac jig in 1947 (Takakuwa and Matsumura 1954) which uses compressed air injection in a chamber directly below the fixed jig screen instead of a separate chamber. As a result, efficient separations have been realized with industrial units up to seven meters wide. The benefits of the Batac jig include the ability to achieve separation density values as high as 7.85 have been summarized by Sanders et al. (2002). The Apic jig is a relatively recent advance in under-pulsed jigs like the Batac jig which provides enhanced control of the air flow into and out of the chambers and improved gate control for removing the low-density particles (Loveday and Jonkers 2002).

Near face removal of coarse rock from coal was the subject of an extensive study conducted by German researchers in the 1980s. The objective was to reduce the cost and increase production of underground operations that were 610 m deep (Sanders et al. 2000). The result was the development of the Humboldt ROMJIG. As shown in Figure 2, the ROMJIG utilizes a moving jig screen which provides the jigging through mechanical motion provided by a hydraulic arm. In this way, the feed end of the screen plate moves up and down while the discharge end is pivoted. The jig

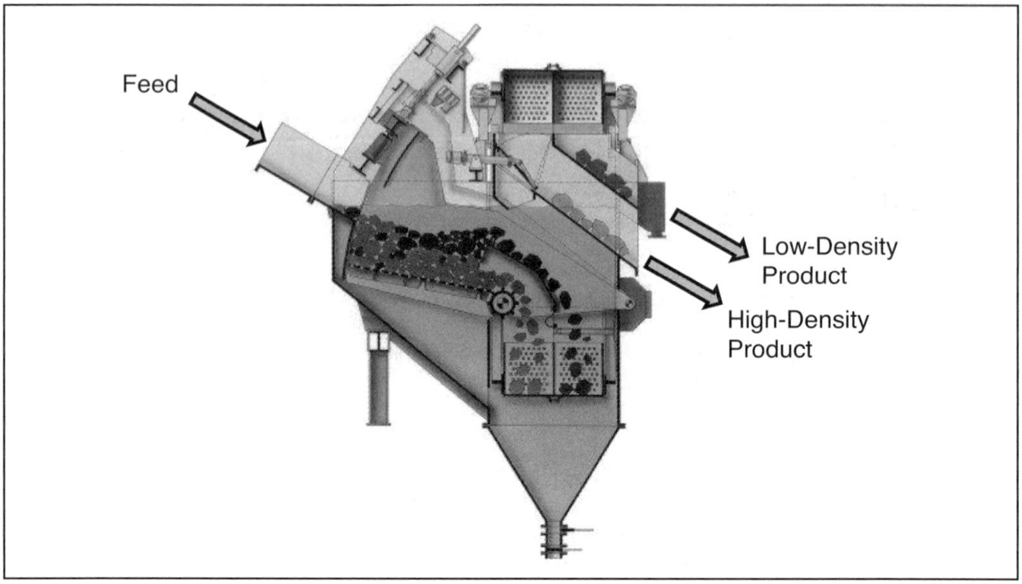

Figure 2. Schematic of the ROMJIG (Sanders et al. 2003)

plate is a screen panel with 15-mm slotted openings. The operational characteristics include mass feed rates up to 350 tph, the maximum and minimum feed particle sizes are 400 mm and 30 mm, respectively, low water requirements of around 185 lpm and no air requirements.

Another recent jig development that utilizes a moving screen is the In-Line Pressure jig (IPJ). The feed slurry enters through the top center of the unit into a chamber above the jig screen that is totally enclosed and filled with particles and water. The jig screen is attached to a hydraulic ram that moves the screen in an up and down motion. As such, the particle bed is lifted to a controlled height during each cycle and subsequently allowed to settle at a rate subject to the particle density. The high density particles pass through ragging material and the screen which is constructed of wedgewire. The unit is capable of treating particle sizes up to 30 mm. Industrial applications include diamonds, gold, silver, lead/zinc, tantalum, tin, copper, manganese, cobalt and garnet (Gray 1997).

Flowing-Film Separators

When a thin film of water flows across a flat surface, a fluid shear occurs at the interface between the surface and the water which creates a fluid velocity gradient. The velocity increases from the surface and through the flowing film. High-density particles within the thin film settle through the high velocity streamlines and into the low velocity region at the solid surface whereas low-density particles are unable to settle out of the high velocity streams and thus report with the majority of the fluid flow.

One of the earliest forms of flowing film separators was the buddle which was primarily used in the early nineteenth century for the treatment of metallic ores such as tin, lead and zinc. Buddles were typically circular measuring 4.3 to 6.7 m in diameter and built into the ground at a depth of about 0.5 m. The inside surface was constructed of brick or stone and sloped downward from the center to the outer circumference. The mineral slurry was feed in the center in a manner that distributed the slurry as a thin film flow. As the flow moved down slope, the high-density particles

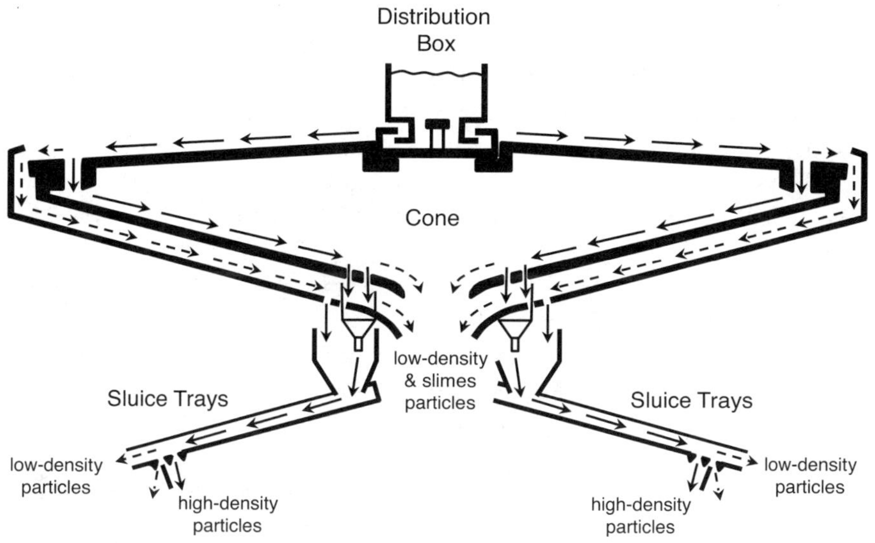

Figure 3. Reichert cone concentrator (Reichert 1965)

settled out to the bottom of the buddle and the low-density particles along with the slimes were carried outward toward a small sluice gate. Brushes were used to stir the settled solids and release any entrapped low-density materials into the high velocity streams. After a period of time, the flow was stopped and the settled solids removed and subsequently re-cleaned in another buddle. The most noted disadvantage of the buddle was the number of re-cleaning stages needed to achieve the desired final product grade (Davies 1902).

Reichert Cone Concentrator

The modern day buddle is the Reichert Cone Concentrator which was invent by Ernst Reichert and Mineral Deposited Limited in 1965 (Reichert 1965). The unit was developed in respond to a need for high capacity gravity concentration for heavy mineral sand applications (Ferree 1973). The Reichert Concentrator is a series of inverted cones having a diameter of 2 or 3.5 m. As shown in Figure 3, feed is distributed at the center of the cone and the flowing film moves downward across the gently sloping surface toward the outer circumference of the cone. The high density particles settle and form a particle bed along the cone surface which is removed by a 0.635-cm wide slot located in the cone surface along the circumference. The low-density material along with the majority of the fluid overflows the edge of the cone. Both the high-density and low-density streams are retreated in a second portion of the cone which slopes downward toward the center axis of the unit. The high-density material collected from the two-stage cleaning is removed and re-treated in pinched sluice trays. Multiple cones could be used along the same axis to achieve the desired upgrading and recovery with as many as a twelve used in a single unit. The Reichert cone treated 45 to 75 t/h of –1.5 mm material and achieved effective concentration to a particle size of 45 microns (Ferree and Mashburn 1981). The optimum slurry solids concentration was 55%–65% by weight. Operational issues noted as a disadvantage were the limits on upgrading after several cleaning and scavenging stages and the inability to observe the separation.

Vibrating Separators

The addition of a vibration motion to improve the efficiency of flowing film separators was the subject of a significant number of developments in the late nineteenth century. Frue (1874) added a lateral shaking motion to a vanner which provided a separation by feeding slurry containing ground ore down a gently sloped, moving conveyor belt. High-density particles settled onto the belt against the force of 12 to 24 lpm of water being added across a 1.2-meter-wide belt to wash the light particles down slope in the opposite direction of belt movement. The top size treated was around 300 microns.

Shaking tables have been the most popular of the vibrating separators and remain a common unit today for the recovery of gold, tin and other heavy metal minerals in the particle size range of 1.65 × 0.074 mm and coal in the 6.7 × 0.15 mm size range. The most common shaking table is the Wilfley table (Wilfley 1910). The oblong table has riffles that extend from the table deck and run across most of the deck length. Feed and wash water enters nearly perpendicular to the riffles. High-density particles settle through the flowing film and concentrate behind the riffles while the low-density particles are washed over the riffles. The vibration motion moves in the same direction as the riffles, which creates a driving force for the settled solids to report to the shorter edge of the table.

Spiral Concentrators

The spiral concentrator has been widely used in industry to achieve efficient fine particle density-based separations since 1943 when the Humphreys spiral was introduced. The separation is a result of a circulating flow shown in Figure 4 and designated as stream C which occurs when slurry is introduced into a downward spiraling trough. Light particles are lifted by the circulating flow from the inner part of the trough and carried to the outer area of the trough whereas high density particles settle onto the trough surface and move toward the inner part of the trough. Internal channels along the vertical axis and splitters located at the end of the trough are used to collect the light and heavy particle streams.

The concept of spiral concentration for coal cleaning was the subject of an invention introduced by Pardee (1937). The invention is described as one in which particles are separated by

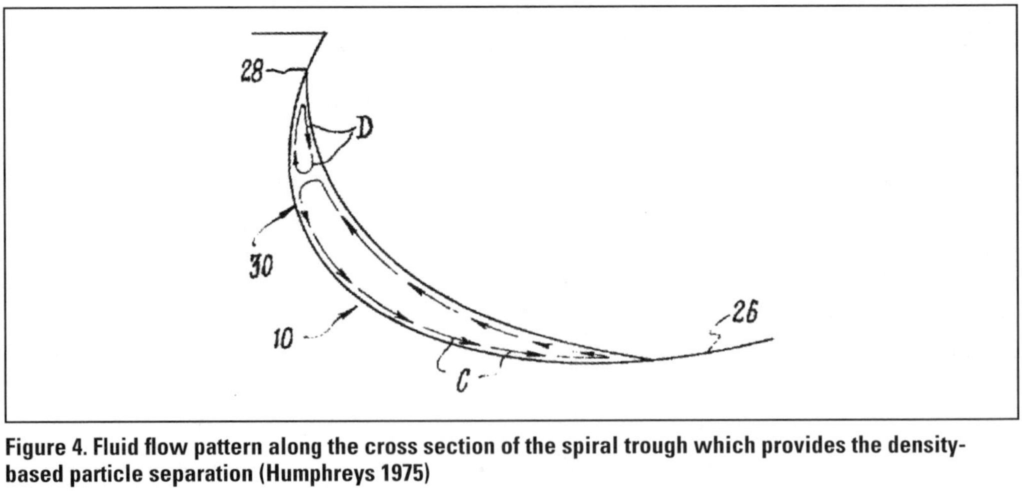

Figure 4. Fluid flow pattern along the cross section of the spiral trough which provides the density-based particle separation (Humphreys 1975)

differences in density as a slurry flows downward along a helical path. Humphreys (1943) followed with a spiral concentrator which incorporated the use of wash water to enhance the separation and the location of draw-off points located at various points along the bottom of the trough to remove the high density particles. The initial use of the spiral was the recovery of chromite from gold-placer sand tailings (Hubbard et al. 1953) but soon grew to several other mineral applications as described by Thompson (1958).

Subsequent developments involved efforts to improve process efficiency and increase the number of potential applications. Oberg (1955) introduced the concept of adding reagents and pre-aerating the feed to a spiral in an effort to decrease the effective density of a targeted heavy mineral component through the creation of bubble-particle aggregates. As such, the spiral could be used to separate a high density mineral component from other mineral components of similar density. Phosphate rock can be separated from silica using this method (Thompson 1958).

The use of wash water was viewed as a problem associated with the original spiral due to scarcity in some locations and costs associated with dewatering and treatment. Close (1963) introduced a spiral that utilized channels constructed within the helical trough to improve separation of silica from heavy minerals without the need for wash water addition. Humphreys (1975) followed with a modification to his original design which removed the need for wash water and in large part represents the current spiral design.

Commercial applications for several decades after their introduction were generally reserved to relatively easy mineral separations such as enriched chromium placer ores, iron ore and heavy mineral sands. The application limitations were mainly due to their construction from cast iron or cement which limited the ability to vary the profile and pitch of the trough. Around 1980, the advancement of construction materials led to spirals being constructed of fiberglass and spray coated with polyurethane. The new material construction allowed alterations to the trough geometry to address the needs of more difficult separation applications. The lightweight material construction permitted the entwining of two and three spiral starts around one central vertical stick or axis. For fine coal applications, the trough was designed to allow higher particle retention times by making the trough wider, changing the trough profile and decreases the vertical pitch (Richards et al. 1985). Higher throughput capacities per start were realized for mineral applications from the re-design of the trough. A summary of the spiral trough changes has been previously described by Richards and Palmer (1997).

Studies conducted by Luttrell et al. (1998) revealed that significant improvements in separation efficiency can be realized when re-treating light particle stream of a spiral concentrator in a second spiral unit in a rougher-cleaner arrangement. The gain in efficiency was associated with the outer secondary circulating stream labeled D in Figure 3. The study found that fine high-density particles that enter in the outer circulating stream do not report to the lower part of the trough and thus are collected in the low-density particle stream. In response to the finding, spiral manufacturers developed a reflux box that is placed within the spiral trough about 3 to 4 turns down from the spiral feed point. As shown in Figure 4, the reflux box capture the material in the middle to outer part of the trough, re-mixed and re-distributes the material back onto the trough. At the same location, a high-density particle stream is captured and placed into an inner channel where it is transported as a final product. The modified unit, known as a compound spiral, improves quality in coal applications (Bethell and Arnold 2003; Luttrell et al. 2007) and increases product recovery for heavy mineral applications (Figure 5).

Figure 5. Repulping box of a SX7 compound spiral concentrator

Fluidized-Bed Separators

Fluidized particle bed separators (FBS), also known as teetered-bed or hindered-bed separators, have been used for over a century in coal and mineral processing plants, primarily for particle size separations. FBS units can be operated in a manner that provides a very efficient density-based separation by utilizing the finest high density particles in the feed stream to create an autogenous dense medium.

One of the oldest and perhaps most popular teetered-bed separator is the Stokes unit. Fine particles, typically having a particle size in the range of 1 × 0.15 mm, are fed in a slurry form into a center well located at the top of the unit. Upon entry, particles begin to settle against an upward flow of fluidization water which is added through a distribution plate at the bottom of the unit. The upward water velocity is set at a rate that equals the settling velocity of the finest (e.g., 0.21 × 0.15 mm), high density particles in the feed. As a result, the particles in this particle class become suspended and create a particle bed that cumulates upward until it reaches a controlled height, which is typically a short distance from the bottom of the feed pipe. A pressure tape extends vertically through the Stokes unit to measure the pressure created by the particle bed density and height. When the bed level reaches the control height as indicated by the bed pressure, a controller opens the discharge spigot to allow high-density reject particles to pass through and report to the underflow discharge pipe. The control is set to allow continuous adjustment to the underflow rate based on the amount of high-density particles reporting in the underflow stream. The low-density particles are unable to penetrate the high density particle bed and are carried by the fluidization water into the overflow stream of the separator.

The positive features of FBS units include the ability to achieve efficient density-based separations at relatively high mass flow rates (10–20 $t/h/m^2$), the capability adjusting on-line to changes in feed characteristics as well as variations in feed rate, and their overall simplicity of operation. Their inherent disadvantages are the need for tight top size control in the feed to prevent the recovery of low-density oversize in the underflow stream, the need of a mechanical underflow valve and control system that requires maintenance, the use of clean water in the injection system to prevent plugging and the existence of components that are susceptible to wear in the bottom of the units.

There are a number of different commercial units that operate in much the same manner as the Stokes classifier including the Floatex (Elder et al. 2001; Mankosa et al. 1995), Lewis hydrosizer (Lewis 1990), Linatex hydrosizer (Deveau and Young 2005), Allflux separator (Short et al. 2001) and the Hydrosort (Doerner 1997). The Allflux separator is a unique unit in which two stages of density-based separations using fluidized particle beds occur in a single unit (Short et al. 2001). The feed is injected tangentially into the center of a cylindrical tank where the coarse, high-density particles settle to an underflow stream against a rising flow of fluidization water in a centrifugal field. The low density particles and the fine high density particles overflow a partition and enter an outer separation chamber where the fine, high density particles are fluidized and flow into a second underflow stream. The low density particles overflow into an outer weir. Reported applications include sand and gravel operations to remove low density contaminates and coal cleaning.

One of the problems associated with conventional FBS units is the feed injection into the center of the unit at a depth equivalent to about one-third of the total height of the unit. This design feature causes turbulence to occur in a zone in which the high density particles, large and small, are settling toward the fluidized particle bed. In addition, the upward velocity of water is increased above the feed injection point due to the additional hydraulic loading which could result in the elutriation of high density particles into the overflow stream with the low-density particles. To counter this problem, Mankosa and Luttrell (1999) equipped an FBS system with a unique feed system that gently introduces the feed slurry across the top of the separator using a transition box, which is reflected in the commercial name for the technology, i.e., the Crossflow Separator (Figure 6). The feed transition box increases the area of the feed introduction to the full width of the separator which reduces velocity and thus turbulence. The feed flows smoothly over the top of the separator and into the overflow launder. As a result of the feed not being submerged into the teetered-bed unit, changes in feed slurry characteristics do not impact the separation performance. The upward velocity within the separator is constant throughout the vertical plane of the cell. A baffle plate is also located at the discharge end of the feed well to prevent short-circuiting of solids to the overflow launder.

An additional improvement involves the inclusion of the slotted plate above a series of bars through which the fluidized water is injected into the cell. The fluidization water passes through large diameter (>12.5 mm) holes at intervals greater than 150 mm. The water injection holes are simply used to introduce the water while the slotted plate provides dispersion of the water. This modification essentially eliminated problems associated with the plugging of distributors or pipes.

An in-plant test program conducted in a heavy mineral sands application revealed a significant increase in mass feed flux rates while maintaining the heavy mineral content in the concentrate at 76% (Dunn et al. 2000; Kohmuench et al. 2001a). A 2.75 × 2.75 m^2 unit was installed in a coal preparation plant to treat 200 t/h of 1 × 0.15 mm coal. After optimization, the unit consistently generated acceptable product grade while recovering 97.5% of the combustibles (Kohmuench et al. 2006).

The Reflux Classifier is established in coal preparation (Bethell 2012) and currently being evaluated for its application in mineral processing to recover minerals including the concentration of chromium and iron. The Reflux Classifier consists of a lower fluidization zone, and an upper system of parallel inclined channels as shown in Figure 7a. A side view of a commercial unit, Figure 7b, shows that there is a cylindrical fluidization zone measuring 2,000 mm in diameter and 2,100 mm

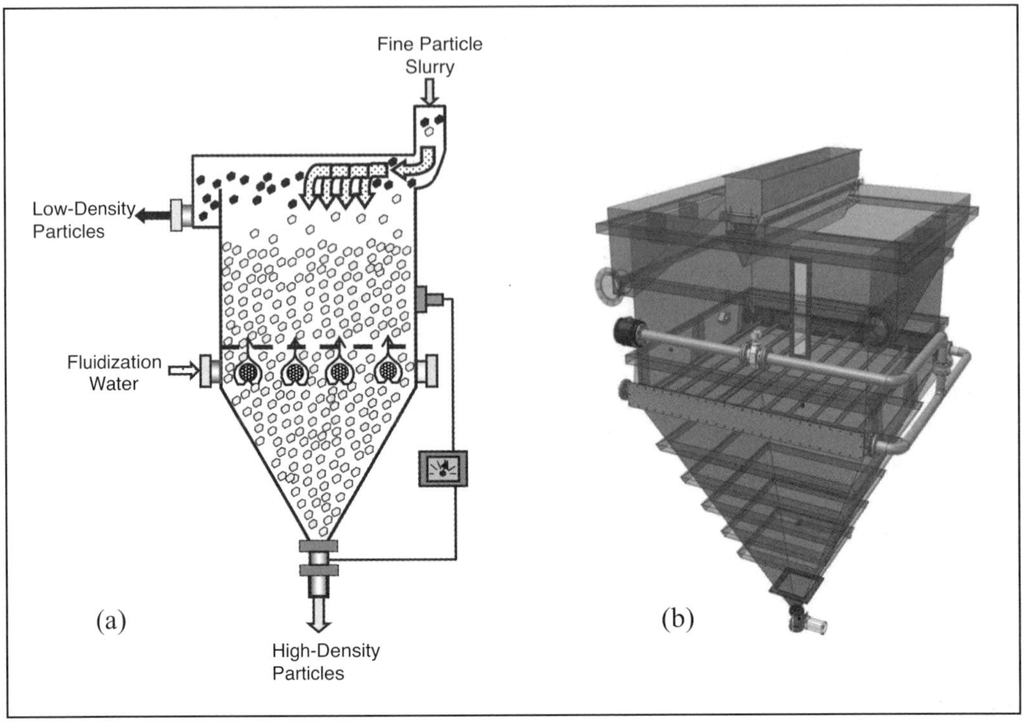

Figure 6. Schematics of the CrossFlow fluidized bed separator: (a) two-dimensional view showing the basic operating principles and (b) three-dimensional view of a commercial unit (courtesy of Multotec Process Equipment (Pty) Ltd)

in height. The nearly square upper section has a cross-sectional area of 2,306 × 2,297 mm^2 and a height of 1,336 mm.

Boycott (1920) was the first to observe the enhanced segregation achieved by inclining a suspension. Ponder (1925) and Nakamura and Kuroda (1937) developed a simple kinematic relationship to describe this effect. More recently, Galvin et al. (2009, 2010) discovered the benefits of using closely spaced inclined channels, which promote laminar flow and a high shear rate at the plate surface. A new separation mechanism, referred to as the "laminar-shear mechanism," develops in the inclined channels. Relatively fine dense particles segregate from the flow, sliding downwards and returning to the lower fluidized bed zone. However, lower density particles over a broad range of particle size continue to convey upwards through the inclined channels. The relatively large, low-density, particles experience inertial lift, and hence fail to segregate onto the inclined channels. These relatively coarse particles become exposed to the higher fluid velocities, conveying easily to the system overflow.

Galvin (2012) reported on the application of the Reflux classifier for upgrading chromium which led to a commercial installation. In this application, the focus was on particles larger than 106 microns. Unlike spirals concentrators, which usually involve three to five stages followed by a desliming device, the Reflux Classifier requires one stage of cleaning and achieves simultaneous desliming. The operation is flexible, and can readily be applied to much finer particles, or much coarser particles. The inclined channels also provide a significant capacity advantage over conventional

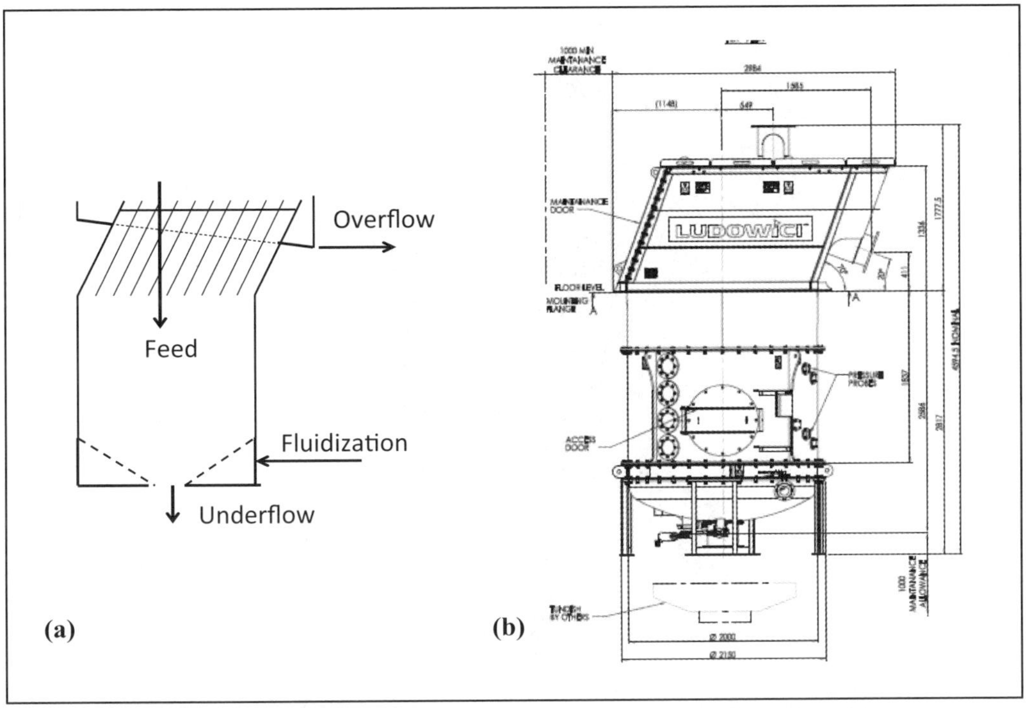

Figure 7. Schematics of (a) a representation of a Reflux Classifier showing the fluidization zone and system of inclined channels and (b) the side view of a commercial unit (courtesy of Eriez Manufacturing)

fluidized beds, especially as the particle size decreases. Moreover, the system is insensitive to low feed pulp density and can be deployed directly after a classifying screen without the need for thickening cyclones.

A commercial installation at a U.S. coal preparation plant utilizes two Reflux classifiers to treat nominally 1 × 0.15 mm bituminous coal at a mass feed flow rate of 110 t/h and a volumetric flow rate of 230 m^3/hr. Excellent ash reductions are reportedly being achieved when treating two different coal sources with significantly different cleanability characteristics (Ghosh et al. 2012).

The effective particle size ratio for achieving a density-based separation using a conventional fluidized bed separator is in the range 3–4:1. FBS are typically optimized to ensure against the loss of low density material to the underflow stream based on a given top size. In an effort to increase the effective particle size ratio, Mankosa and Luttrell (2002) developed a process referred to as the HydroFloat separator in which air bubbles are injected into the fluidized particle bed. The air bubbles collide and attach to the coarse, low density particles, which are hydrophobic naturally or by pretreatment with a collector. As a result, the effective density of the particle-bubble aggregate is sufficiently low to allow the aggregate to move upward through the particle bed and into the overflow stream with the low-density particles. Using this concept, the effective particle size range that can be effectively treated in a fluidized bed separator is around 6:1 (Kohmuench et al. 2001b; Luttrell et al. 2006).

The HydroFloat process has been commercially successful in the treatment of phosphate as a substitute for coarse particle flotation. The benefits realized from the installation of the process at a

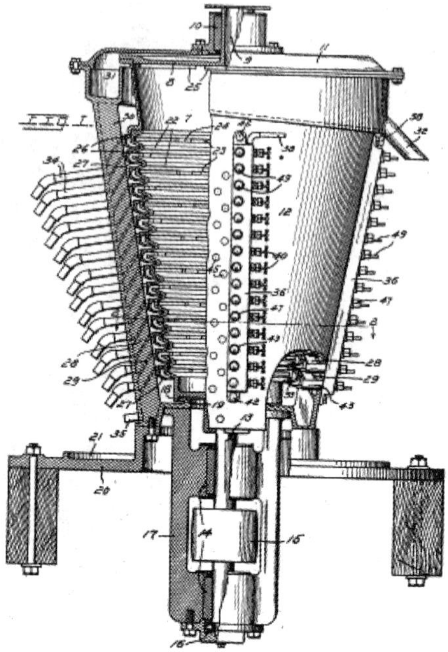

Figure 8. Early development of a continuous enhanced gravity separator (Eccleston 1923)

phosphate plant were an increase in the overall throughput capacity and a significant reduction in reagent consumption (Kohmuench et al. 2007). Other applications include coal cleaning and the recovery of diamonds.

Enhanced Gravity Separators

Enhanced gravity separation involves the application of centrifugal forces in order to target relatively fine particles. Particles that would normally settle according to Stokes' law experience a centrifugal force with an effective acceleration of G times the normal acceleration, g, due to gravity. This G force produces a significant increase in the terminal velocity of each particle. Moreover, the settling regime shifts towards the intermediate regime (Vance and Moulton 1965) in which the terminal velocity scales directly with the particle diameter. This means the dependence of the particle velocity on the particle size decreases, which in turn, leads to an enhanced level of separation performance.

The need to utilize centrifugal force to recover pulverized or granular minerals from ore based on density difference was the focus of a significant amount of development as early as the late 1800s even though large scale commercial applications did not occur until nearly a century later. Seymour (1893) patented a process to separate metals from ore using a rotating bowl in which mercury was retained by centrifugal action and used to capture the heavy metals. Feed slurry entered the process in the bottom center of the spinning bowl and distributed outwardly along the circumference of the bowl. The high-density metal particles settled into the mercury while the light particles remained in the water film and overflowed the top rim of the bowl. After a given time period, the feed slurry was stopped and the heavy particles removed from the bowl. A series of other semi-continuous centrifugal units were developed which utilized the principles of flowing film and riffling in a spinning bowl (Ponten 1910; Bradbury 1912; MacNicol 1935).

Eccleston (1923) followed with the development of the first fully continuous enhanced gravity separator which used a series of riffles along the circumference of a bowl shaped as a truncated cone as shown in **Figure 8**. High-density particles settled against the riffles edges where an inward stream of fluidization water was injected. A small opening behind each riffle allowed the high-density particles to be continuously discharged while the low-density particles overflowed each riffle and reported to the overflow stream exiting the top rim of the bowl.

A number of similar types of enhanced gravity separators (EGS) were developed and commercially utilized in Russia and China throughout the late 1900s. As summarized by Burt (1984), EGS units having capacities from 0.5 to 14 t/h where utilized for the recovery of gold, cassiterite, tungsten, iron, cinnabar and other heavy minerals.

The developments that led to the EGS units commonly used in the industry today occurred over last two decades of the 20th century and detailed reviews of the technologies have been provided by Luttrell et al. (1995) and Cole et al. (2012). The technologies vary by their separation mechanism and the magnitude of the applied centrifugal field. Each process is reportedly effective over a particle size range of about 1 mm to 10 microns. For material having a heavy mineral content less than about 1% by weight, semi-continuous units are recommended. Fully-continuous units are available for feed streams containing more than 1% heavy minerals. Mass throughput capacities are as high as 1000 t/h for semi-continuous units and 400 t/h for continuous units.

A review of the historical development of the Falcon concentrator has been provided by McAlister and Armstrong (1998). The essential feature of a continuous Falcon Concentrator is a vertically-aligned, open-topped bowl shaped as a truncated cone which is mounted on a rotating shaft. A centrifugal force up to 300 *g*'s can be produced to cause deposition and stratification of the fine particles against the inside of a smooth centrifugal bowl. Mass transport chutes and valve-nozzle assemblies placed at equal distances along the top circumference of the bowl allow continuous discharge of the heavy particles into an underflow launder. The continuous unit has the capability of yielding 40% of the feed mass flow to the underflow stream.

The first commercial installation of a continuous Falcon concentrator involved the scavenging of the final tailings stream at Tantalum Mining Corporation of Canada Limited. The unit produces a concentrate containing 70% metal with a mass yield of 30%. A significant amount of the recovery improvement was due to the capture of a majority of the tantalum finer than 20 microns (McAlister and Armstrong 1998).

A semi-continuous unit of the Falcon Concentrator was originally designed with the same smooth bowl construction used for the continuous unit. However, in an attempt to improve separation performance, a fluidization zone was added to the upper section of the bowl. The unit operates with a continuous flow of feed slurry for a period of 5 minutes to several hours depending on the amount of high-density material in the feed. The applied centrifugal force can be varied from 50 to 200 g's. A range of commercially available units provides throughput capacities from 1 to 400 t/h. The first application of a semi-continuous Falcon concentrator was at the Blackdome processing facility in 1986 where the unit was used to recover gold from a shaking table tailings stream (McAlister and Armstrong 1998). Concentrates with up to 30% gold metal were obtained from a single pass through the unit and gold recovery increased by 6%. Cole et al. (2012) estimated that 723 semi-continuous units have been placed into service worldwide of which 45 are units with capacities greater than 200 t/h.

A recent development in the Falcon concentrator line is a semi-continuous unit that is focused on the treatment of particles having a size between 37 and 3 microns. The target application is scavenging of the slime reject streams. The unit has the ability to apply centrifugal fields up to 600 g's and is equipped with a system that can vary the overflow lip width which allows a large range of high density material to be recovered from a single treatment cycle. The maximum throughput capacity available to 2 t/h (Anon. 2013).

The concept of the Knelson concentrator was patented in 1986 by B.V. Knelson and has been applied in over 1700 industrial operations worldwide on a variety of materials. Both the continuous variable discharge (CVD) and semi-continuous units employs fluidized particle bed separation concepts in a mechanically applied centrifugal field. The Knelson concentrator operates by introducing water through a series of fluidization holes located in rings that circle the circumference of a bowl. The bowl, which has a truncated cone shape, is rotated at speeds that provide a centrifugal field up to 200 times gravity. Feed slurry is introduced through a tube that directs the material toward the bottom center. Upon reaching the bottom, the slurry is accelerated outward and up the cone wall toward the rings. The fluidization water entering in the rings provides an inward velocity that allows the creation of a fluidized particle bed comprised of heavy particles. The high-density particles that pass through the fluidized particle bed are extracted in a CVD unit through a series of controlled pinch valves located along the circumference located in the center of a concentration ring. The discharge ports are designed to handle mass flows equivalent to 1%–50% of the feed.

For semi-continuous units, the heavy particles settle into the bottom of each slot and retained in this position until the feed slurry flow is turned off. Once the feed and bowl is stopped, the bowl is rinsed to remove the high-density particles. These units are recommended when the percentage of high density material in the feed is less than 0.5% by weight.

The semi-continuous units are more popular in industry with around 1700 placed in service whereas approximately 65 CVD units exist in mineral separation operations (Cole et al. 2012). There are numerous publications describing the variation commercial applications. A range of unit sizes exists which offer mass throughput capacities up to 1000 t/h for semi-continuous units and 100 t/h for continuous units.

A centrifugal jig known as the Kelsey jig was commercialized in 1992 (Beniuk et al. 1994). A review of developments in the technology was provided by Richards and Jones (2004). The Kelsey jig consists of a series of hutches which are rotated about a central feed pipe. The unit is capable of generating centrifugal fields up to 100 g's. A cylindrical wedge wire screen is mounted across the top of each hutch to retain ragging material. Feed slurry enters the unit through the central feed pipe and flows outward across the bed of ragging. Hutch water is added under pressure and the water pulsed by means of push arms acting on a flexible diaphragm. The pulsations create oscillations in the bed that differentially accelerate particles based on differences in density. Low-density particles flow across the ragging material and overflow the top of the unit, while high-density particles pass downward through the ragging/screen and are discharged through actuated valves. Key operating parameters include rotational speed, pulse rate, stroke length, specific gravity and particle size of the ragging material, and the hutch water addition rate. In most cases, the unit forms its own ragging material from coarser and heavier feed particles. The need to constantly replenish the bed of ragging appears to be the major shortcoming of this particular design.

The Kelsey jig has been successfully used commercially for the concentration of gold, mineral sands, nickel, tantalum, tin, tungsten, and platinum. Specific applications include fine gold and metal sulfides recovery from leach tailings, magnesium mineral removal from nickel sulfide concentrates, and high grade zircon recovery from plant tailings (Richards and Jones 2004). Approximately 34 units have been installed in operating plants worldwide with 10 plants associated with the processing of tin and heavy mineral sands (Cole et al. 2012).

Perhaps the most efficient enhanced gravity separator is the Multi-Gravity Separator (MGS) which utilizes table riffling principles (Mozely 1990; Tucker et al. 1992). The unit consists of three main components, i.e., cylindrical rotating drum, internal scraper network and variable-speed differential drive. Selective separations of fine particles are achieved along the internal surface of the rotating drum using the same basic principles employed by a conventional shaking table. However, by replacing the table surface with a rotating drum, the particles are subjected to many times the normal gravitational pull. This feature allows the MGS to separate much finer particles than would otherwise be possible using conventional flowing-film separators.

The Graviton is a centrifugal device that incorporates typically eight Reflux Classifier units with a system of parallel inclined channels aligned at 20° to the horizontal. This arrangement, which involves closely spaced channels, minimizes the effects of the Coriolis Force (Schaflinger 1990). Galvin and Dickinson (2012) examined the classification of fine silica particles. The key discovery was that the laminar-shear mechanism evident under static conditions ($G=1$) occurred at significant values of G with the selective effects of inertial lift evident. Moreover, it was found that the benefits of the inclined channels and the G forces actually multiply. For example, with $G=73$, a remarkably high capacity advantage in excess of 1000 fold was realized. The preliminary work was conducted using a small-scale Reflux Classifier within a 2-m-diameter centrifuge. The work is yet to be taken to pilot or full scale. However, the findings indicate the potential for a significant advance in gravity separation over the next few years.

PNEUMATIC DENSITY-BASED SEPARATIONS

The use of air as a medium to achieve density-based separations was the focus of significant development in the early part of the twentieth century. The primary application was in coal cleaning with industrial installations mainly in the U.S. and Europe. The estimated amount of coal cleaning by pneumatic processes in 1939 was 30 to 40 million tons annually (Gaudin 1939). According to Arnold et al. (1991), the amount of coal processed in the U.S. through dry cleaning plants reached a peak in 1965 at 25.4 million tonnes. The largest dry-based cleaning plant processed 1,270 t/hr of minus 19 mm coal using a total of 14 cleaning units.

Pneumatic density-based separators have been widely applied in the food processing industry, the recycling industry for copper recovery from used electric wire and the separation of minerals. Jarman (1942) reported a number of potential applications for mineral separations including the recovery of mica from silica sand, gold from sand, fluorspar from quartz and calcite and garnet from kyanite. In one application, an 'air flotation' table was used to concentrate WO_3 in an 1 × 0.5 mm particle size fraction from 4% to 64% with only a trace of WO_3 in the tailings which accounted for 85.8% of the feed.

The pneumatic technologies incorporate the same basic mechanisms utilized in wet separators including (i) dense medium separations, (ii) pulsated air jigging, (iii) riffled table concentration and (iv) air fluidized launders.

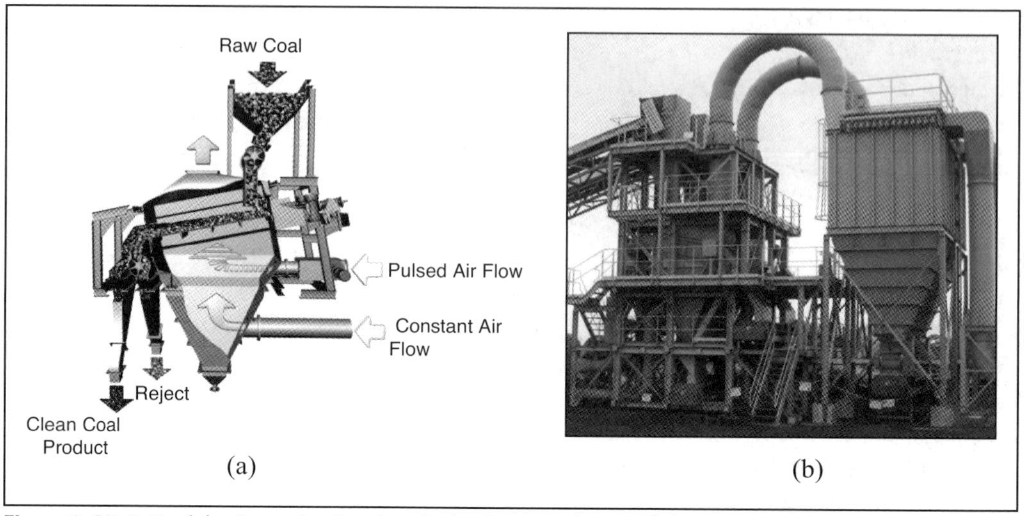

Figure 9. Allair jig: (a) schematic showing the basic operation and (b) photograph of full-scale unit at a coal processing facility (Snoby et al. 2009)

Air Jigs

The early development of pneumatic devices using jigging principles resulted in a number of technologies including the Plumb jig, the Hooper jig and the Revelart-Berry jig (Gaudin 1939). These technologies either used pulsating air through a sieve or a moving sieve in a steady stream of air to generate the jigging action. The most commonly used air jig in the U.S. was the Stump, Super Air Flow jig. It included a vibrating bed, a single fan with a mechanism for producing pulsations and a multiple sink product discharge mechanism. Recent modifications to the Stump jig have enhanced the operational characteristics and performance of the unit. The changes include (Kelly and Snoby 2002)

- A star gate to meter homogeneous feed evenly across the entire width of the jigging bed;
- A constant velocity fan (working air) to loosen the bed of material;
- Installation of external vibrators to enhance particle transport in the jig;
- Addition of a pulse air fan to optimize stratification by offering independent control of stroke frequency, amplitude, and acceleration; and
- Application of an automatic bed level control system to achieve a constant separation performance and product quality over large variations in the feed ash content.

The air jig unit shown in Figure 9 has an automatic bed level sensor that utilizes nuclear density instrumentation. The information from the nuclear density instrument is used to activate a refuse discharge star gate valve to remove refuse at a rate proportional to the amount of impurities entering in the feed. By using only one refuse removal mechanism, the air jig constantly maintains a reserve layer of high density material over the outlet. This concept keeps the lower density coal particles at a fixed distance from the screen bed, thereby minimizing misplacement.

There are several industrial installations using air jigging technology worldwide including operations located in Spain, India, Brazil and the U.S. Although applicable for any light weight material, most installations involve coal cleaning for the purposes of ash and/or sulfur reduction.

Commercial air jig units are capable of treating up to 60 t/h of coarse (75 × 12 mm) coal and up to 40 t/h of fine (12 × 1 mm) coal (Snoby et al. 2009).

Air Tables

Riffled tables using air to create a fluidized bed of high-density fine particles were the most commonly used pneumatic separators in the early 1900s. Common technologies were the Birtley and Sutton-Steel tables. These technologies were found to provide effective density separations for coal coarser than 6 mm and minerals having a particle size as small as 0.3 mm (Jarman 1942). Throughput capacity was around 5 t/h per table. Recent modifications to the air tables include vertical suspension of the table, modern dust collection technology and automation instrumentation which increased throughput capacity to 120 t/h per table when treating coal in the particle size range of 75 × 6 mm (Lu et al. 2003).

The separation data obtained from coal studies conducted in China indicate that the unit has the potential to provide an effective separation for particles as coarse as 80 mm to a lower size limit of around 3 mm. The operational data also indicate that the process is relatively insensitive to surface moisture up to a value of about 7–10% by weight. Performance evaluations conducted on full-scale units indicate the ability to provide a relatively high separation density (RD_{50}) of around 2.0 RD while achieving probable error (Ep) values that range from 0.15 to 0.25 (Lu et al. 2003). A detailed evaluation conducted in the U.S. using a 5 t/h air table unit found that the table has the ability to provide significant upgrading for all ranks of coal (Honaker et al. 2008).

Air Dense Medium Fluidized Bed

A technology that applied a dense medium comprised of air and magnetite to clean coal was initially developed by Fraser and Yancey (1925). However, there were no commercial installations of air dense medium fluidized bed (ADMFB) processes until 1994 when a 50 t/h unit was built in China to process 320,000 tons/year of coal having a particle size range of 50 × 6 mm (Lubin 1998). A review of various ADMFB processes developed at lab or pilot scale was provided by Sahu et al. (2009). Luo and Chen (2001) developed an ADMFB process that provides efficient separations at two different relative densities in the same unit. A negative aspect of ADMFB processes is that surface moisture must remain less than 4% to minimize magnetite loss.

SUMMARY AND FUTURE DEVELOPMENTS

A significant number of new developments in gravity separation have occurred over the last 100 years which have provided a very positive impact on the profitability of the mining industry worldwide. During this period, many novel concepts were overlooked and under-utilized as a result of declining high grade ore deposits and the resulting emphasis on technologies that exploit other physical property differences such as froth flotation and magnetic separations. However, concerns over the capital and operating costs of these systems in addition to the environmental impact of chemical additions have resulted in a rejuvenation in the interest of applying gravity-separators for the treatment of ultrafine material.

The development of high-capacity density separators that are effective for treating a broad range of ores and coal over an expanding particle size range has been the primary focus over the past century. Commercial success has been realized where emphasis was the maximization of process efficiency and ensuring low maintenance requirements. Concerns with the disadvantages over competitive technologies like froth flotation have led to a significant number of developments in fine particle separation technologies involving flowing film and fluidized bed principles. Technologies have also been successfully commercialized that simultaneously exploit differences in particle surface chemistry and density to achieve improved process efficiencies. Improving process efficiencies at high mass flow rates over a broader range of particle sizes will continue to be the drivers for future technology development.

Many of the enhanced gravity separators have been commercialized over the past three decades despite the fact that the basic concept was the subject of several inventions in the late nineteenth and early twentieth centuries. The reason for the recent developments is the interest in utilizing gravity separators to economically concentrate minerals such as gold at particle sizes approaching one micron. New developments are focusing on providing improved separation efficiency at higher throughput capacity and even finer particle sizes with lower grade materials. Effort is also being focused on using modified conventional technologies under natural gravity conditions to achieve effective density-based separations in an effort to avoid the expense and maintenance required of a relatively large, high speed centrifugal separators.

Developments in new and improved dry, density-based separation technologies are expected to continue as many of the new deposits and mining operations are being developed in arid regions at a time when water supply is a general concern worldwide. Recent developments have mainly focused on updating and improving technologies that were originally introduced in the early twentieth century. These efforts will continue as researchers attempt to develop methods to enhance the knowledge concerning the impacts that geometric configuration and operating parameter values have on separation performance using advanced tools such as computational fluid dynamics and dynamic discrete element particle models. Enhancing performance by exploiting multiple material property differences will also likely be another area of focus. For example, combining dry density-based separation principles with the optical sorting technology currently being commercially evaluated could potentially lead to a new innovation. The inability to provide an effective density separation for particles finer than 1 mm will also likely be the subject of future developments.

REFERENCES

Anon., 2013. Sepro Minerals Systems. www.seprosystems.com/mining-equipment/gravity-concentrators/gravity-concentrators-falcon-uf.html.

Arnold, B.J., Hervol, J.D., and Leonard, J.W., 1991. Dry particle concentration, *Coal Preparation,* 5th (ed. J.W. Leonar), SME.

Benuik, V.G., Vadeikis, C.A., and Enraght-Moony, J.N., 1994. Centrifugal jigging of gravity concentrates and tailings at Renison Limited. *Miner Eng,* 7(5/6), p. 577.

Bethell, P.J., 2012. Dealing with the challenges facing global coal preparation. *Coal Prep. Soc. Amer. J.,* 11(1), p. 26.

Bethell, P.J., and Arnold, B.J., 2003. Comparing a two-stage spiral to two-stages of spirals for fine coal preparation, *Advances in Gravity Concentration* (ed. R.Q. Honaker and R. Forrest), SME.

Boycott, A.E., 1920. Sedimentation of blood corpuscles. *Nature* 104, p. 532.

Bradbury, W., 1912. Ore-separator. U.S. Patent 1,042,194.

Burt, R.O., 1984. *Gravity Concentration Technology.* New York: Elsevier.

Burt, R.O., 1986. Selection of gravity concentration equipment. In *Design and Installation of Concentration and Dewatering Circuits* (ed. A.L. Mular and M.A. Anderson), SME, p. 194.

Burt, R.O., 1999. The role of gravity concentration in modern processing plants. *Miner. Eng.*, 12(11), p. 1291.

Close, F.A., 1963. Spiral Concentrators. U.S. Patent 3,099,621.

Cole, J., Dunne, R., and Giblett, A., 2012. Review of current enhanced gravity separation technologies and applications. In *Separation Technologies for Minerals, Coal, and Earth Resources* (ed. C.A. Young and G.H. Luttrell), SME.

Cope, L.W., 2000. Jigs: the forgotten machine. *Engineering & Mining Journal.* August. 30.

Davies, E.D., 1902. *Machinery for Metalliferous Mines: A Practical Treatise for Mining Engineers,* New York: Van Nostrand.

Deveau, C. and Young, S.R., 2005. Pushing the limits of gravity separation. Annual meeting SME, Preprint 05-84.

Doerner, K., 1997. Producing high-quality sands with sophisticated technology in the new sand beneficiation plant of Pinnow gravel works. *Aufbereit Tech Miner Process* 38(1), p. 27.

Dunn, P.L., Stewart, S.O., Kohmuench, J.N., and Cadena, C.A., 2000. A hydraulic classifier evaluation: upgrading of heavy mineral concentrates. SME Preprint Number 00-155, SME.

Elder, F., Kow, W., Domenico, J. and Wyatt, D., 2001. Gravity concentration—a better way (or how to produce heavy mineral concentrate and not recirculating loads. *Proceedings of the International Heavy Minerals Conference.* AusIMM, pp. 115–118.

Eccleston, C.W., 1923. Centrifugal separator. U.S. Patent 1,473,421.

Ferree, T.J., 1973. An expanded role in minerals processing seen for Reichert Cone. *Min. Eng.,* 29.

Ferree, T.J., and Mashburn, L.F., 1981. Fine gold recovery with a Reichert cone—A case history, *SME-AIME Fall Meeting,* Preprint No. 81-417.

Fraser, T., and Yancey, H.F., 1925. The air-sand process of cleaning coal. U.S. Patent 1534846.

Frue, W.B., 1874. Improvement in ore-washers. U.S. Patent 158,057.

Galvin, K.P., 2012. Application of the Reflux Classifier for measuring gravity recoverable product. In *Separation Technologies for Minerals, Coal and Earth Resources* (ed. C.A. Young and G.H. Luttrell), SME.

Galvin, K.P., and Dickinson, J., 2012. Particle transport and separation in inclined channels subject to g forces. *Chem. Engr. Science,* 87, p. 294.

Galvin, K.P., Walton, K., and Zhou, J., 2009. How to elutriate particles according to their density. *Chem. Engr. Science.* 64, p. 2003.

Galvin, K.P., Walton, K., and Zhou, J., 2010. Application of closely spaced inclined channels in gravity separation of fine particles. *Miner Eng* 23, p. 26.

Gaudin, A.M., 1939. *Principles of Mineral Dressing,* McGraw-Hill, New York.

Ghosh, T., Patil, D. Honaker, R.Q., Damous, M., Boaten, F., Davis, V.L. and Stanley, F., 2012. Performance evaluation and optimization of a full-scale reflux classifier. *Coal Prep. Soc. of America J.* 11(2) p. 24.

Gray, A.H., 1997. InLine Pressure Jig—An exciting, low cost technology with significant operational benefits in gravity separation of minerals, *Proceedings of the AusIMM Annual Conference,* p. 259.

Honaker, R.Q., Saracoglu, M., Thompson, E., Bratton, R., Luttrell, G.H., and Richardson, V., 2008. Upgrading coal using a pneumatic density-based separator. *Inter. J. Coal Prep. Util.,* 28(1), p. 51.

Hoover, H.C. and Hoover, L.H., 1950. *De Re Metallica.* New York: Dover Publications.

Hubbard, J.S., Humphreys, I.B. and Brown, E.W., 1953. How Humphreys spiral concentrator is used in modern dressing practice. *Mining World,* 15(6), p. 40.

Humphreys, I.B., 1943. Helical chute concentrator and the method of concentration practiced thereby. U.S. Patent 2,431,559.

Humphreys, I.B., 1975. Helical chute concentrator and method of concentrating. U.S. Patent 3,891,546.

Iijima, Y., Shimoda, Y., Kubo, Y., Jinnouchi, Y., Sakata, S. and Konisha, M., 1998. Improvement of Baum jig into vari-wave type in Ikeshima colliery, *XII International Coal Preparation Congress,* AusIMM, p. 257.

Jarman, G.W., 1942. Special methods for concentrating and purifying industrial minerals. *AIME Transactions,* p. 304.

Kelley, M., and Snoby, R., 2002. Performance and cost of air jigging in the 21st century. In *Proceedings of the 19th International Coal Preparation Conference,* Lexington, Kentucky, p. 177.

Kohmuench, J.N., Mankosa, M.J., Luttrell, G.H., and Adel, G.T., 2001a. A process engineering evaluation of the CrossFlow separator. SME Preprint Number 01-80, SME.

Kohmuench, J.N., Luttrell, G.H., and Mankosa, M.J., 2001b. Coarse particle concentration using the hydrofloat separator. *Miner. Metall. Proc.,* 18(2), p. 61.

Kohmuench, J.N., Mankosa, M.J., Honaker, R.Q. and Bratton, R.C., 2006. Applications of the CrossFlow teetered-bed separator in the U.S. coal industry. *Miner Metall. Proc. J.* 23(4), p. 187.
Kohmuench, J.N., Mankosa, M.J., Kennedy, D.G., Yasalonis, J.L., Taylor, G.B., and Luttrell, G.H., 2007. Implementation of the HydroFloat separator at the South Fort Meade mine. *Miner Metalll Proc,* 24(4), p. 264.
Lewis, R.M., 1990. The Lewis hydrosizer. NC State Minerals Research Laboratory Publication No. 81-12-P.
Loveday, G. and Jonkers, A., 2002. The Apic jig and the Jigscan controller take the guesswork out of jigging. *Proceedings of XIV International Coal Preparation Congress*, SAIMM, p. 247.
Lu, M., Yang, Y. and Li, G., 2003. The application of compound dry separation technology in China. *Proceedings of the 20th Annual International Coal Preparation Exhibition and Conference*, Lexington, Kentucky.
Lubin, W., 1998. Mechanism and application of coal dry beneficiation with air-dense medium fluidized bed. *J. Cent. South Univ. Techn.* 5(2), p. 100.
Luo, Z., and Chen, Q., 2001. Dry beneficiation technology of coal with an air dense-medium fluidized bed. *Inter. J. Min. Proc,* 63:167.
Luttrell, G.H., Honaker, R.Q., and Phillips, D.I., 1995. Enhanced gravity separators: new alternatives for fine coal cleaning. *Proceedings of the 12th International Coal Preparation Conference*, Lexington, KY.
Luttrell, G.H., Kohmuench, J.N., Stanley, F.L. and Trump, G.D., 1998. Improving spiral performance using circuit analysis. *Miner Metall Pro,* 15(4), p. 16.
Luttrell, G.H., Westerfield, T.C., Kohmuench, J.N., Mankosa, M.J., Mikkola, K.A., and Oswald, G., 2006. Development of high-efficiency hydraulic separators. *Miner Metall. Proc.,* 23(1), p. 33.
Luttrell, G.H., Honaker, R.Q., Bethell, P. and Stanley, F., 2007. Design of high-efficiency spiral circuits for preparation plants, *Designing the Coal Preparation Plants of the Future* (ed. B. Arnold, M. Klima and P. Bethell), SME, pp. 73–88.
MacNicol, A.N., 1935. Centrifugal separator. Australian Patent 22,055/35.
Mankosa, M.J., and Luttrell, G.H., 1999. Hindered-bed separator device and method. U.S. Patent 6,264,040.
Mankosa, M.J., and Luttrell, G.H., 2002. Air-assisted density separator device and method. U.S. Patent 6,425,485.
Mankosa, M., Stanley, F. and Honaker, R.Q., 1995. Combining hydraulic classification and spiral separation for improved efficiency in fine coal cleaning. *High Efficiency Coal Preparation* (ed. S.K. Kawatra) SME, pp. 99–108.
McAlister, S.A., and Armstrong, K.C., 1998. Development of the Falcon concentrator. SME Preprint Number 98-172, SME.
Mozely, R., 1990. Minerals separator, U.S. patent application 4964845.
Nakamura H., and Kuroda K., 1937. La Cause de l'acceleration de la vitesse de sedimentation des suspensions dans les recipients inclines. *Keijo Journal of Medicine* 8, p. 256.
Oberg, F.N., 1960. Spiral Concentrator Apparatus. U.S. Patent 2,952,360.
Pardee, F., 1937. Concave spiral separator. U.S. Patent 1937.
Ponder P., 1925. On sedimentation and Rouleaux formation. *Qrtrly J. of Exper. Physiology* 15, p. 235.
Ponten, A., 1910. Centrifugal sluicing-machine. U.S. Patent 980.001.
Reichart, E., 1965. Method and apparatus for the wet gravity concentration of ores. Australian Patent 628811965.
Richards, R.G., and Jones, T.A., 2004. The Kelsey centrifugal jig—An update on technology and application, SME Preprint 04-021. SME.
Richards, R.G. and Palmer, M.K., 1997. High capacity gravity separators—A review of current status. *Miner Eng,* 10(9), p. 973.
Richards, R.G., Hunter, J.L. and Holland-Batt, A.B., 1985. Spiral concentrators for fine coal treatment. *Coal Preparation,* 1, p. 207.
Sahu, A.K., Biswal, S.K., and Parida, A., 2009. Development of air dense medium fluidized bed technology for dry beneficiation of coal—A review. *Inter. J. Coal Prep. Util.,* 29(4), p. 216.
Sanders, G.J. and Ziaja, D., 2003. The role and performance of Romjig in modern coal preparation practice, *Proceedings of the 20th International Coal Preparation Conference,* Lexington, KY, p. 207.

Sanders, G.J., Gnanaiah, E.U., and Ziaja, D., 2000. Application of the Humboldt de-stoning process in Australia. *Proceedings of the Eighth Australian Coal Preparation Conference*, Australian Coal Preparation Society.

Sanders, G.J., Ziaja, D. and Kottman, J., 2002. Cost efficient beneficiation of coal by Romjigs and Batac jigs, *Proceedings of XIV International Coal Preparation Congress*, SAIMM, p. 395.

Schaflinger, U., 1990. Review Article: Centrifugal separation. *Fluid Dynamics Research,* 6, p. 213.

Seymour, C.E., 1893. Means for and process of separating metals from ores. U.S. Patent 489,101.

Short, M.A., Snoby, R.J. and Jungmann, A., 2001. Beneficiation of 3mm × 0.15mm fine coal with two-stage hindered settling equipment. SME Preprint Number 01-1. SME.

Snoby, R., Thompson, K., Mishra, S., and Snoby, B., 2009. Dry jigging coal: Case history performance. SME Preprint 09-052, SME.

Taggard, A.F., 1945. *Handbook of Mineral Dressing*, New York: John Wiley and Sons.

Takakuwa, T. and Matsumura, M., 1954. A contribution toward the improvement of the air pulsated jig, Proc. 2nd Inter. Coal Preparation Congress, Essen: Steinohlenbergbauverein Paper A118.

Tanaka, M., Jinnouchi, Y., Sawata, Y., and Kawashima, S. Effect of the wave pattern of pulsation of an air-pulsated jig. *Proceedings of 11th International Coal Preparation Congress*. Tokyo, p. 139.

Thompson, J.V., 1958. The Humphreys spiral concentrator: its place in ore dressing. *Min Eng* p. 84.

Tucker, P. Chan, S.K., Mozely, R.H., and Childs, G.J., 1992. Modeling of the Multi-Gravity separator. *Les Techniques*, p. 45.

Vance, W.H., and Moulton, R.W., 1965. A study of slip ratios for the flow of steam-water mixtures at high void fraction. *A.I.Ch.E.J.,* 11(6), p. 1114.

Wilfley, A.R., 1910. Concentrating table. U.S. Patent 157,287.

ABOUT THE AUTHORS

Rick Honaker is the chair and professor of the Mining Engineering Department at the University of Kentucky (UK). He served as an assistant and associate professor of mining engineering at Southern Illinois University–Carbondale from 1991 to 1999. Honaker joined the faculty at UK as an associate professor in 2000 and promoted to the full-professor rank in 2006. During his tenure as a faculty member at two institutions, Honaker has developed an extensive research program in coal and mineral processing with a significant amount of work focused on fine particle processing and applied surface chemistry. His research findings have been reported in more than 140 technical publications and reports.

Robert C. Dunne is the group executive–metallurgical fellow at Newmont Mining Corporation, Denver, Colorado. He is a graduate of the University of the Witwatersrand, South Africa, and is an adjunct professor at the Julius Kruttschnitt Mineral Research Centre, Brisbane, Australia. Over the last 30 years, Dunne has worked for a number of mining companies including Anglo American, Anglovaal, and Newcrest Mining, and also spent time at the Western Australian School of Mines lecturing and at Mintek, South Africa, where he was involved in introducing new technologies into the mining industry. He has authored and co-authored more than 50 papers in a variety of fields, including mine-to-mill, comminution, gravity separation, flotation, gold hydrometallurgy, water treatment, and metallurgical management.

Kevin Galvin is professor of chemical engineering and director of the Center for Advanced Particle Processing and Transport at the University of Newcastle, Australia. Prior to joining the University in 1993, he spent 10 years with BHP Research and 3 years at Imperial College where he earned his PhD. Galvin is also the inventor of the Reflux Classifier and related technologies.

Innovations in Dense Medium Separation Technology

Tim Napier-Munn
Julius Kruttschnitt Mineral Research Centre, University of Queensland, Australia

Jeremy Bosman
Pesco, South Africa

Peter Holtham
Julius Kruttschnitt Mineral Research Centre, University of Queensland, Australia

ABSTRACT: This paper presents a brief account of the 100-year history of practical dense medium separation in mineral and coal processing, and identifies the key developments over that period in equipment, equipment scale, dense media, process design, and modeling and control. Early developments related to mechanical design of static bath separators, and the use of natural materials as media. Ferrosilicon for high density separations first appeared in the 1930s, and a major advance occurred in the invention of the dense medium cyclone by Dutch State Mines in the 1940s, which has dominated DMS processes ever since. The most important innovation in the last 50 years has probably been the major increase in the scale of equipment, in particular that of cyclones. Other improvements have been largely incremental.

INTRODUCTION

Dense medium separation (DMS) is a process by which minerals of different densities are separated by placing them in a bath of a liquid or suspension of intermediate density, allowing them to sink or float according to density, and then removing the low and high density products for sale, disposal or further processing. This can be done either under gravitational force in a so-called 'static' bath or under centrifugal force in a 'dynamic' vessel such as a cyclone. The low and high density products are sometimes referred to as 'float' and 'sink' respectively. In coal, with separations in a low density range (typically less than RD 1.5), the float is the concentrate (clean coal), whereas in most metalliferous and diamond operations, with separations in a higher density range (typically greater than RD 2.6), the sink is the concentrate or high grade product.

The term 'DMS' is distinguished from the more general term 'gravity concentration' by the presence of the separating liquid or suspension of density intermediate between that of the low and high density minerals. In gravity concentration, by contrast, the separating fluid is usually water or air which are both less dense than all the mineral particles present, and separation is achieved by some form of differential motion through this fluid in devices such as jigs or spirals.

A DMS enthusiast might be forgiven for claiming that it is the most important unit operation in mineral processing, based on tonnage treated. Although it is difficult to prove, it is likely that the widespread use of DMS in upgrading bulk commodities such as coal and iron ore, plus its role in the pre-concentration of minerals such as diamonds and base metals, means that the total tonnages treated by DMS are larger than those treated by the more glamorous flotation process. However flotation attracts at least an order of magnitude more investigative interest among researchers than does DMS, and it may be time to redress this balance somewhat, particularly as anecdotal evidence strongly suggests that the understanding of the DMS process by operators and designers is now at an all-time low.

In this paper we give a brief description of the development of the process from its beginnings as an industrial process and attempt to show the innovation processes at work in its evolution.

A BRIEF HISTORY OF THE DMS PROCESS

In one sense DMS is the simplest of all density separation processes and the one that can be most closely duplicated in the laboratory (Burt 1984). Placing a mixture of sand and wood chips in a beaker of water, stirring, and allowing it to settle will demonstrate the principle. The first reported attempts to apply DMS did use pure liquids as the dense medium, but because of the high densities of mineral particles the liquids were generally exotic, expensive and mostly toxic in various ways, and in due course they were replaced by suspensions of dense solids in water.

The other main developments related to the separating vessel, including the distinction referred to earlier between static bath separators used for coarser material (>5mm) and dynamic separators used for finer material, and the regeneration of the medium.

The following selected chronology is adapted from Burt (1984), Leonard (1991), and Holtham (2006) together with the authors' personal knowledge.

1858	Bessemer's patent covered the use of chlorides of iron, manganese, barium and calcium as medium, but there was no serious industrial application until the 20th century.
1911	Du Pont developed the use of chlorinated hydrocarbons to achieve higher densities.
1911	The Chance process patent covered the use of sand/water medium in a cone separator (Chance 1924). We would probably call this a teeter bed separator today.
1921	First Chance industrial process for cleaning anthracite.
1922	First experiments with magnetite medium for coal cleaning (Conklin process).
1931	First general use of DMS plants.
1930s	DMS plants for minerals using galena, in the UK and USA.
1937	First use of FeSi in an ore concentrating plant in the USA, with medium recovery using cross-belt magnetic separators.
1940	First US ferrosilicon patent (Wade 1940).
1940	American Cyanamid Co. introduces DMS process for coal cleaning including magnetic recovery of magnetite medium.
1938	Tromp process first to use magnetite medium commercially in Germany.
1942	Dense medium cyclone patented by Dutch State Mines, Holland.
1946	First use of dense medium cones in diamond processing, in South Africa (Chaston and Napier-Munn 1974).
1947	Establishment of Dutch State Mines' subsidiary Stamicarbon.
1955	First use of dense medium cyclones in diamond processing, in Tanzania, using magnetite medium and wide-angle cyclones (Chaston and Napier-Munn 1974).
1955	First use of atomizing process by Knapsack (Hoechst) to produce spherical FeSi for high density separations (Rodis 1955).

1960s Use of tetrabromoethane (TBE) advocated for treating tin ores (Brien and Pommier 1964) and potash ores (Tippin and Browning 1966). Never widely used.
1960s Development of Dyna Whirlpool and Vorsyl dynamic separators.
1970s Development of the Tri Flo dynamic separator.
1970s First physical tracers used to routinely determine process efficiency in diamonds.
1980s First physical tracers used to routinely determine process efficiency in coal.
~1992 First use of cyclone diameters >1,000mm in coal preparation (current maximum 1,500mm).
~2005 First use of cyclone diameters >610mm in mineral separations (current maximum 800mm).

SOME INNOVATIONS THAT HAVE WORKED

The Dense Medium Cyclone Story

Until the late 1930s all dense medium processes used static bath separators. Folklore then has it that in 1939 the Dutch State Mines company (DSM) was running a shallow bath dense medium plant in Holland for the cleaning of coal using a loess (clay) medium which was thickened in a cyclone. When the cyclone blocked it was noticed that the overflow pipe was full of clean coal, suggesting that somehow the thickening cyclone was concentrating the coal. This observation led to further investigation and ultimately the development of the dense medium cyclone (DMC) by DSM. In fact Holtham (2006) points out that the truth may be more prosaic. Two apparent references by DSM staff to the origins of the process suggest that it was the result of systematic investigation. Driessen (1945), in one of the first technical descriptions of the process, stated (in reply to a questioner) that 'the idea that a separation of coal and slate in a normal cyclone thickener, as used at the loess washery, might be possible came first, and afterwards the truth of it was verified by experiment.' And Dreissen (1973) stated that '30 years ago DSM...started on a very extensive research program aimed at developing a coal separation process which was to have a better separations efficiency for particle sizes down to 0.5 mm than had the equipment used so far...' by which he meant the cyclone washer.

Whether the DMC arose from serendipity or a systematic investigation, or both, DSM clearly saw the potential of the invention and patented it in 1942, and the rest as they say is history. An extensive R&D program was conducted at the DSM Central Test Station at the Emma coal mine during the German occupation of Holland, to develop the technology. This included the invention and patenting of the sieve bend as a method of draining the cyclone products, the development of appropriate DMS flowsheets, rubber liners to accommodate wear, and the optimization of the cyclone geometry and dimensions including the introduction of a vortex finder. Several different shapes were investigated, some not unlike those of devices subsequently developed and patented by others, and one of the present authors (TJNM) remembers visiting the DSM scrap yard in 1972 where many of the discarded prototypes were lying. The result was the geometry which is in almost universal use today in the processing both of coal and mineral ores around the world. There are good reasons why the present standard 'DSM design' works so well and has stood the test of time.

The patents were vested in a wholly-owned subsidiary company established in 1947, Stamicarbon,* and licensed to a number of engineering companies around the world. They

* Stamicarbon still exists, having undergone several business transformations from coal and coke to chemicals, and is now principally a supplier of urea plants.

represent one of the few examples of mineral processing patents which have earned substantial revenues for the inventors. Stamicarbon accumulated a database of performance data from pilot plants and from the full scale plants which they and their licensees built, initially mostly in coal applications but then increasingly in minerals. The database was used to develop regression models of the process which allowed them to give process guarantees to their customers, increasing their competitive advantage. Plant design and operation was codified in the famous 'DSM Manual,'* issued to licensees, probably one of the most illegally copied manuals in the history of mineral processing. The last edition was issued in the 1990s.

DMCs offer several advantages over static bath separators, including the ability to treat fine material (generally down to 0.5mm, and to zero in a few coal preparation cases though this requires special medium cleaning provisions), a relatively small foot-print, and mechanical simplicity. The maximum particle size able to be treated is limited to about ⅓ the diameter of the inlet and is therefore directly related to the cyclone diameter itself.

An interesting discrepancy exists in the pressure head recommended by the DMS manual for coal (low density) and mineral (high density) applications. 9D is the recommended head for coal (where D = cyclone diameter), but for minerals heads in the range 15D–40D are recommended, and engineering companies have largely followed these recommendations to this day. There is no known reason why the heads proposed for low and high density separations should be different and we speculate that the higher head was proposed for minerals as a safety precaution due to the lack of experience with high density separations when mineral applications were first introduced in the early 1950s. At least two experimental studies at full scale have suggested that 9D is also a satisfactory head for mineral applications (Napier-Munn et al. 2009). Adoption of this value in minerals would lead to significant capex and opex savings.

The main evolution in cyclone since the DSM patents lapsed has been in improved inlet geometries, the use of more effective wear materials such as Nihard, polyurethane and ceramic liners, and a large increase in scale (discussed further below). Evolute inlet designs have been found to offer reduced wear rates, especially on the vortex finder, and higher capacities. New cyclone geometries have been tried (e.g., Rong and Napier-Munn 2000). However current cyclone geometry and DMS process design differ little from the standards developed by DSM more than half a century ago, a remarkable testament to the effectiveness of their innovation strategies, engineering skills and business model.

Other Dynamic Vessels

Several other dynamic separators have been developed over the years. They are reviewed in Holtham (2006) and Wills and Napier-Munn (2006) and include the following.

- *Dyna Whirlpool:* An inclined cylindrical vessel developed in the USA in the early 1960s for coal and later used for minerals.
- *The Tri-Flo:* A multi-stage separator similar to two Dyna Whirlpools bolted together in series, developed by Ferrara at the University of Trieste in Italy in the late 1970s.
- *Vorsyl:* A cylindrical device developed by the National Coal Board in the UK in the late 1960s. A 720mm device has a capacity of 120 t/h in coal cleaning.

* The full title is *Guide to the Calculation of Cyclone Washeries and Heavy Medium Float-and-Sink Washeries.*

- *Larcodems:* Also similar to the Dyna Whirlpool, developed by British Coal as an efficient replacement for the Baum jig. Kumba's iron ore concentrator at Sishen in South Africa has employed a 1,200mm Larcodems separator treating 600–800t/h of –90+6mm material (Myburgh 2006).
- *High cone angle cyclones:* These have been used sporadically over the years to achieve higher cut-points for a given medium density than the standard DSM design (e.g., Chaston and Napier-Munn 1974).

These devices claimed some advantage over the industry standard DSM cyclone and although some have a reasonable installed base, none have proved as popular or enduring as the DSM cyclone.

Static Bath Separators

The early dense medium processes were of the static bath type, in which the separation takes place under gravitational forces: the high density particles sink in the bath and the low density particles float, the separation taking place at approximately the density of the medium. Most of the development over the years was in the mechanical arrangement by which feed was introduced to the bath and the products removed. This led to a number of proprietary devices manufactured by different companies including the Wemco cone, Wemco drum, Drewboy, Teska, Barvoy and Norwalt baths (Burt 1984; Wills and Napier-Munn 2006), trough-type separators, and many others. The design innovations were developed principally in the USA and Europe (mostly Germany and the UK). Most of these devices are used in coal preparation, though the DM drum is used for coarse concentration in iron ore, magnetite and other minerals.

Static baths have a relatively large footprint and cannot effectively concentrate particles below about 5 mm in size because of the very slow settling rates of fine particles. Dynamic separators are required in this duty, but static baths can accommodate very large particles where liberation characteristics permit this.

Increased Equipment Scale

As in other branches of mineral processing, the large increase in scale of equipment is probably the biggest single innovation in recent decades. This is driven mainly by capital and operating cost advantages, and there is always a challenge in preserving process efficiency whilst enjoying the benefits of economies of scale. Much work has gone into evaluating the efficiency of large devices as they come on stream to provide the comfort that process efficiency has not been lost (e.g., Lee et al. 1995).

In DMCs minerals have lagged coal in the race for scale; 800mm is probably the biggest DMC currently used in non-coal applications. However, in coal preparation 1,000mm DMCs were introduced in the early 1990s and the largest unit installed to date is 1,500mm. Osborne (2010) has documented the reduced flowsheet complexity and drop in unit capex in the coal industry as the larger equipment has been introduced over the last 30 years (Table 1). Similar advantages were reported for the trial of an 800mm unit replacing 4 × 400mm cyclones in zinc-lead pre-concentration in Xstrata's plant at Mt. Isa (Napier-Munn et al. 2009). In particular distributors have been shown to be biased, causing inefficiencies in DMC performance. Eliminating distributors from the flowsheet therefore has cost, operability and efficiency benefits.

Table 1. Economies of scale in the coal industry for a 1,000 t/h plant (Osborne 2010)

Equipment Year	1977	1987	1997	2007
No. modules	6	4	2	1
No. deslime screens	6	4	2	1
No. and size DM cyclones	12 × 500mm	8 × 660mm	2 × 1,200mm	1 × 1,500mm
No. drain and rinse screens	24	16	4	2
Total no. items	164	110	90	82
Capex	\$26m	\$23m	\$20m	\$18m

The Dense Medium

The first dense medium suspensions were those easily available on individual mine sites (e.g., clay, barytes, galena, pyrite) but later magnetic media such as natural magnetite and manufactured milled ferrosilicon (FeSi) were introduced for coal and mineral separations respectively. Ferrosilicon is believed to have first been used in the US in 1937 and provided the capability to process minerals at higher separating densities than those used for coal because of its higher density (6.7–7) than that of magnetite (~5). The first US ferrosilicon patent in 1940 was taken out by a Minnesota University researcher for a material grading 82% Fe and 13–14% Si, the latter being the key to its corrosion resistance (Wade 1940). Magnetic media were easy to clean and recover using simple magnetic separation, unlike galena and pyrite which had to be regenerated using flotation.

At high solids concentration suspensions exhibit high apparent viscosity which impedes the separation process. This is why FeSi is used in preference to magnetite for mineral separations. However milled FeSi itself cannot be used beyond a medium density of about 3.1, and this led to the development of atomized FeSi by Knapsack (Hoechst) in Germany (Rodis 1955). This material has a rounded shape which results in reduced viscosity and the ability to reach medium densities up to about 3.8, a density which has been used in iron ore. Knapsack also developed a ferrochrome medium for even higher densities (over 4) but this was rarely used.

Work is currently going on to develop a spherodized magnetite capable of operating at medium densities of about 2.8 RD (compared with a maximum of 2.3 RD with conventional magnetite) (Blair 2010). If successful this product might be about half the cost of FeSi for the same duty.

FeSi corrosion is an important source of medium loss. Stewart and Guerney (1998) reported a method of monitoring corrosion on-line and inhibiting its progress through aeration and the addition of sodium nitrite.

Process Design

Figure 1 shows the flowsheet for a typical modern dense medium process. Most of the equipment is devoted to the cleaning, regeneration and recycling of the medium by screening, magnetic separation and densification.

The two largest components of operating cost in DMS are power for pumping and the consumption of medium, so the flowsheet has evolved over the years to minimize these costs especially through reduced medium consumption. The original medium densifiers were spiral classifiers but these have largely been replaced by dynamic densifiers, usually simple pipe-like devices but sometimes cyclones, which have a much smaller footprint and can react much faster to process changes (Bosman 2007). Cyclone concentration has been introduced between primary and secondary

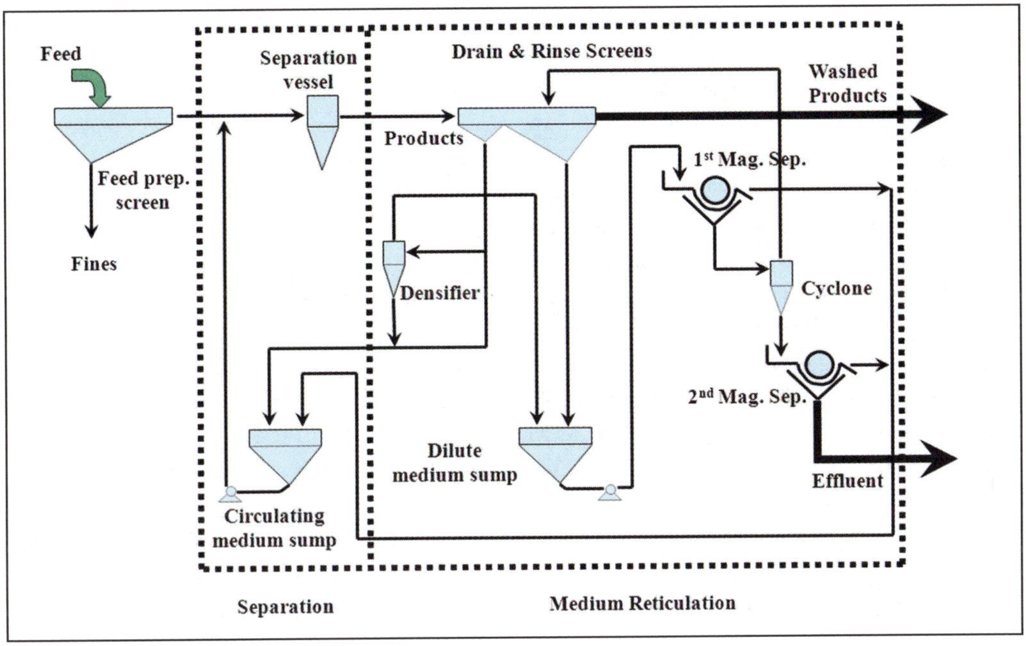

Figure 1. Dense medium process flowsheet

magnetic separation to improve medium recovery in the second stage. A mixture of sprays, flood boxes and rising screen panels has been introduced to improve medium washing on the vibrating screens.

A choice is available between pump feeding and gravity feeding a DMC, in the latter case via a specially designed head tank. Pump feeding requires less building head room but is subject to higher pump wear and thus variations in cyclone performance. As history has turned out most coal preparation DMCs are pump fed and most mineral DMCs are gravity fed. There does not seem to be any rational reason why this dichotomy should prevail.

There have been intermittent attempts over the years to use cyclones for the treatment of very fine feeds (down to 75μm or even to zero), particularly in coal (e.g., Woodruff and Corbridge 1989), but they have not enjoyed wide acceptance. Here the challenge is mostly in medium recovery for which magnetic separators have been used in preference to screens.

Other incremental improvements over the years, some of which have profited from developments in other branches of mineral processing, include improved wet magnetic separators for medium recovery, the use of variable-speed drives on feed pumps, and the introduction of banana screens for medium drainage and washing.

Modeling, Process Control, and Evaluating Process Efficiency

Regression modeling based on data collected in pilot and full scale plants is a useful way of predicting process performance in process design and optimization. As noted earlier Stamicarbon developed its own process models which it used to give process guarantees. Public domain DMC models include those of Wood et al. (1987) for coal preparation and Dunglison and Napier-Munn (2000) for both coal and mineral separations. A DM drum model was developed by Baguley and Napier-Munn (1996).

Computational fluid dynamics (CFD) modeling of DMCs has now reached the stage where it is producing plausible and useful insights into the process (e.g., Narasimha et al. 2007). In particular it has been used to confirm that involute inlets have superior characteristics over tangential inlets.

The main control variable in DMS is the medium density, which controls the density of separation, and from the earliest days this was managed by over-densifying the return medium and then adding water to achieve the required density. When on-line density gauges became available, particularly nucleonic gauges, these took over from manual density measurement and simple PID control was (and still is) used to achieve a density set-point.

Model-based stabilizing and optimizing control, using steady state or dynamic process models to infer process performance from simple on-line measurements (e.g., density of product media), has been attempted many times but is not yet widely used (Holtham 2003; Firth et al. 2010; Meyer and Craig 2010). There is however considerable potential for this approach, and in general process control of dense medium plants is, in most applications, still relatively primitive.

The only true way to fully evaluate a dense medium process is to take samples of the product, fractionate them in heavy liquids, and use the resulting density distributions to construct the separation or Tromp curve. This is time-consuming, expensive, and uses toxic heavy liquids. Physical density tracers were developed as an alternative, first in diamonds in the 1970s (Napier-Munn 1985) and then in coal (Davis et al. 1985; Lee et al. 1995). They have been used routinely in some cases though the method is limited by the difficulty of collecting the product tracers. Painted ore particles have also been used (Napier-Munn et al. 2009). The automatic detection and counting of radiotracers has been proposed but not yet implemented, though experiments have been conducted with magnetic tracers.

SOME INNOVATIONS THAT HAVE NOT WORKED (YET)

Heavy liquids: As noted earlier heavy liquids have been proposed for use in DMS (Brien and Pommier 1964; Tippin and Browning 1966) and certainly have some process advantages such as lower viscosity and no segregation of the medium. However their cost and toxicity have ruled them out except in laboratory applications.

Magnetic fields: Investigations have been conducted to see if the application of magnetic fields could improve the efficiency of DMS (Vatta et al. 2003), and attempts have been made to enhance the separating effect of a DMC with magnetic media using a magnetic field applied around the body of the cyclone (e.g., Svoboda et al. 1998). However none of these have found industrial application.

Rheology modifiers: Because the rheology of medium suspensions is important in determining process efficiency, attempts have been made to modify the rheology using additives, mostly to reduce the viscosity of high density suspensions or those contaminated by clays (e.g., Aplan and Spedden 1964) or to maintain the stability of the suspensions (e.g., Valentyik 1972). However the cost of the reagents has precluded their industrial use.

Dry processes: These are superficially attractive, particularly for arid regions, for the rather obvious reason that they do not require water. However there are disadvantages. Air being much less dense than water will generally require more power to achieve a given separation, and dust is often a problem, as is damp feed. Dry analogues of some dense medium processes such as fluid bed and pinched sluice separators have been tried using ferrosilicon medium, but have not yet been successful in production applications (Napier-Munn and Morrison 2003) though several development programs are currently underway.

SUMMARY AND CONCLUSION

The history of industrial dense medium separation in mineral and coal processing is about 100 years old. The innovation in the early days was related to the mechanical design of static bath separators, mostly in coal, some of which was driven by the need for vendors to market proprietary designs. Early media were natural materials found on the mine sites (e.g., sand, galena, and pyrite) but manufactured ferrosilicon appeared in the 1930s and allowed both easy regeneration by magnetic separation and the attainment of the high medium densities needed for mineral concentration. Interestingly milled ferrosilicon seems to have been invented by a US academic researcher, though the atomized material was developed by a German commercial company.

A major DMS innovation was the invention and development of the DM cyclone, possibly serendipitously. Certainly the inventors, Dutch State Mines, made best use of it with a thoroughly researched industrial solution and an effective business model including the close management of patents and other intellectual property.

There have been no major innovations in the last 50 years other than the large increase in the scale of equipment, particularly cyclones, that has also been apparent in other branches of mineral processing. Cyclone diameters in coal preparation have increased 5-fold in that time, suggesting a 25-fold increase in capacity, with attendant advantages in capex, opex, operability and flowsheet simplicity. The cyclone diameter increase in mineral applications has been only half that in coal, for reasons which are not understood. Incremental improvements have included the improved management of medium cleaning and regeneration, and the implementation of automatic process control. Many of these have benefited from advances in other branches of mineral processing or other enabling technologies.

ACKNOWLEDGMENT

The authors acknowledge the assistance of Will Blair of FerrOz in clarifying some aspects of the history of ferrosilicon as a dense medium.

REFERENCES

Aplan, F.F., Speden, H.R. 1964. Viscosity control in heavy media suspension. *VII IMPC*. New York, Part II, September, pp. 103–113.

Baguley, P.J., Napier-Munn, T.J. 1996. Mathematical model of the dense medium drum. *Trans. Inst. Min. Met.* 105: C1–C8.

Blair, W. 2010. Private Communication (FerrOz).

Bosman, J. 2007. The selection and application of densifiers in dense medium circuits. *Minerals Engineering* 20 (4).

Brien, F.B., Pommier, L.W. 1964. Results of an investigation of the use of heavy liquids in a cyclone for concentrating values from tin ores. *SME Transactions* 229(December): 335–341.

Burt, R.O. 1984. Gravity concentration technology. *Development in Mineral Processing* 5: 605. Elsevier.

Chance, T.M. 1924. Application of sand flotation process to the preparation of bituminous coal. *Transactions of AIME* 70, 740–749.

Chaston, I.R.M., Napier-Munn, T.J. 1974. Design and operation of dense-medium cyclone plants fpr the recovery of diamonds in Africa. *Journal of the Southern African Institute of Minerals and Metallurgy* 75(5): 120–133.

Davis, J.J., Wood, C.J., Lyman, G.J. 1985. The used of density tracers for the determination of dense medium cyclone partition characteristics. *International Journal of Coal Preparation and Utilization* 2: 107–125.

Dreissen, H.H. 1973. Stamicarbon's heavy-medium cyclone process for the separation of minerals and coal. Stamicarbon memo 13 Stc/73, October.

Driessen, M.G. 1945. The use if centrifugal force for cleaning fine coal in heavy liquids and suspensions with special reference to the cyclone washer. *Journal of the Institute of Fuel* XIX (105): 33–45.

Dunglison, M.E., Napier-Munn, T.J. 2000. The application of a new model of the dense-medium cyclone. In *Dense Media 2000: 7th Samancor Symposium on Dense Media Separation*. January 30, Benmore, kwa Zulu natal, R.S.A: Samancor Limited, pp. 119–135.

Firth, B., Holtham, P.N., O'Brien, M., Hu, S., Dixon, R., Burger, A., Sheridan, G. 2010. Joint evaluation of monitoring instruments for dense-medium cyclones. ACARP Report C17037. November.

Holtham, P.N. 2003. On-line dynamic modelling of dense medium cyclones. ACARP Report C12001. November.

Holtham, P.N. 2006. Dense medium cyclones for coal washing. *A Review of Transactions of AIME* 59(5): 521–533.

Holtham, P.N., Wood, C.J., Hammond, R. 1995. Operating and performance evaluation of the 1150 mm primary dense-medium cyclone at Wakworth Mining. *Proceedings of the 7th Australian Coal Preparation Conference*. Edited by Smitham, J., Mudgee, NSW, Paper B1: Australian Coal Preparation Society, September.

Leonard, E.J., ed. *Coal Preparation*, 5th ed. Littleton, CO: SME.

Meyer, J.W., Craig, I.K. 2010. The development of dynamic models for a dense-medium separation circuit in coal beneficiations. *Minerals Engineering* 23(10): 791–805.

Myburgh, H.A. 2006. The influence of the quality of ferrosilicon on the rheology of dense-medium and the ability to reach higher densities. *Journal of the Southern African Institute of Minerals and Metallurgy* 106((November): 726–732.

Napier-Munn, T.J. 1985. Use of density tracers for determination of the Tromp curve for gravity separation processes. *Transactions of IMM* 94: C47–C53.

Napier-Munn, T.J., Gibson, G., Bessen, B. 2009. Advances in dense-medium cyclone plant design. *Proceedings of the 10th Mill Operations Conference*. Adelaide, Australia: AusIMM, October.

Napier-Munn, T.J., Morrison, R.D. 2003. The potential for the dry processing of ores. *Proceedings on Water in Mining*. Brisbane, Australia: AusIMM, October. pp. 247–250.

Narasimha, M., Brenan, M.S., Holtham, P.N., Napier-Munn, T.J. 2007. A comprehensive CFD model of dense-medium cyclone performance. *Minerals Engineering*. 20 (4).

Osborne, D. 2010. Value of R&D in coal preparation development. *Proceedings of the XVI International Coal Preparation Congress*. Lexington, KY, April. pp. 845–858.

Rodis, F. 1955. Production and properties of ferrosilicon for heavy media separation. *Proceedings of the III International Mineral Dressing Congress*. Goslar, Germany, May.

Rong, R., Napier-Munn, T.J. 2000. 8th Australian Coal Preparation Conference. Port Stephens, Australia: Australian Coal Preparation Society, pp. 112–126.

Syoboda, J., Coetzee, C., Campbell, Q.P. 1998. Experimental investigation into the application of a magnetic cyclone for dense-medium separation. *Minerals Engineering* 11(6): 501–509.

Tewart, K.J., Guerney, P.J. 1998. Detection and prevention of ferrosilicon corrosion in dense-medium plants. *6th Mill Operations Conference*. Madang, Australia: AusIMM, October. pp. 177–183.

Tippin, R.B., Browning J.S. 1966. Heavy liquid cyclone concentration of New Mexico potash ores. *Transactions of SME* (December): pp. 360–366.

Valentyik, L. 1972. The rheological properties of heavy media stabilized by polymers. *Transactions of AIME*. 252(March): 99–105.

Vatta, L.L., Kekan, R., Bongani, R., Myburgh, I., Svoboda, J. 2003. *Physical Separation in Science and Engineering* 12(3): 167–168.

Wade, H.H. 1940. Gravity separation of ores (Patent for FeSi, 82% Fe, 13–14% Si). U.S. Patent 2,206,980. Wills, B.A., Napier-Munn, T.J. 2006. *Wills' Mineral Processing Technology*. 7th Edition, Butterworth-Heinemann.

Wood, B.A., Davis, J.J., Lyman, G.J. 1987. Towards a medium behaviour based performance model for coal-washing DM cyclones. *Dense Medium Operations Conference*. Brisbane, Australia: AusIMM, July. pp. 247–256.

Woodruff, D.B., Corbridge, C. 1989. Treatment of fine coal and minerals using dense-medium cyclones. *Mines and Quarry* 18(1–2): 90–94.

ABOUT THE AUTHORS

Tim Napier-Munn is a metallurgist with degrees from Imperial College, London, and Witwatersrand University, Johannesburg, including two research degrees in studies of the dense medium cyclone. He has spent his 40-year career in applied R&D for the mineral industry, working for De Beers in South Africa and for universities in England and Australia. He joined the Julius Kruttschnitt Mineral Research Centre (JKMRC) in 1985, becoming its director in 1997 and inaugural managing director of JKTech Pty Ltd in 2001. He retired from these two roles in 2004 and now works part time for the JKMRC and consults through JKTech. He has given his professional development course on statistics for metallurgists more than 130 times around the world and is writing a book on the subject. His research interests have included dense medium separation, comminution and classification, modelling of mineral processes, and the application of statistics to mineral processing.

Jeremy Bosman holds degrees in chemical engineering and operations research. He initially worked in the South African iron and steel industry and then spent 2 years with Blue Circle Cement. In 1987, he joined Multotec in their cyclone division and worked there for 16 years and was responsile for all aspects of classification and dense medium cyclones. In 2003, he formed his own consulting company, PESCO. He is also a senior lecturer at the University of Pretoria where he does postgraduate lecturing in minerals processing and sampling.

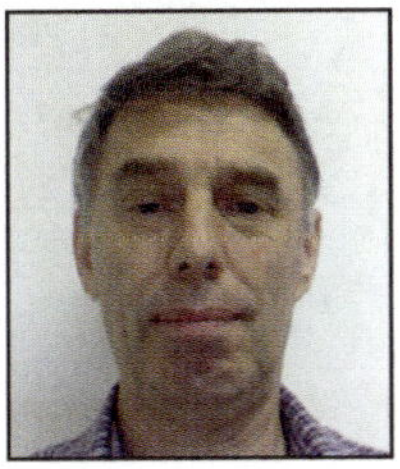

Peter Holtham graduated from the University of Leeds in 1973 with a BSc degree in applied mineral sciences. From 1973 to 1977 he worked as a metallurgist for Roan Consolidated Mines in Luanshya, Zambia. He completed an MSc degree in control engineering at the University of Manchester in 1979 and then joined De Beers in Namibia as a process control metallurgist. In 1984, he was appointed as lecturer in mineral processing in the School of Mines at the University of New South Wales. Since 1992, he has been at the University of Queensland's Julius Kruttschnitt Mineral Research Centre, where his research interests lie in dense medium and gravity concentration processes, and coal preparation.

Innovations in Coal Processing

Gerald H. Luttrell
Dept. of Mining & Minerals Engineering, Virginia Tech, Blacksburg, Virginia, United States

ABSTRACT: Coal processing, which is also commonly referred to as washing, cleaning or preparation, has historically been viewed within the mineral processing community as a low-tech industry that uses only mature technology. This perception is inaccurate, however, since market demands for improved product qualities coupled with increasingly challenging reserves have forced the industry to develop and deploy a new generation of high-efficiency technologies. Modern coal processing facilities now incorporate a sophisticated array of advanced solid-solid and solid-liquid separation processes to remove unwanted impurities such as ash, sulfur, and moisture from run-of-mine coals. The processing circuitry incorporated by these state-of-the-art facilities now rivals most mineral and chemical plants in terms of design challenges and operational complexity. This article provides a brief overview of the current state of the coal processing industry and highlights noteworthy examples of recent technological advancements that have recently been applied in industrial operations.

INTRODUCTION

Coal processing is the method by which mined coal is upgraded in order to satisfy size and purity specifications dictated by a given market. These facilities make use of low-cost, solid-solid and solid-liquid separation processes that remove impurities such as waste rock and water from run-of-mine coals. For coal-fired power stations, these impurities reduce coal heating value, leave behind an undesirable ash residue, contribute to gaseous and particulate emissions of environmental concern, and increase the cost of coal transportation (Doherty 2006). The presence of mineral impurities can also influence the suitability of coal for high-end uses such as the manufacture of metallurgical coke or generation of petrochemicals and synthetic fuels. Unwanted surface moisture, which directly impacts utilization by reducing heat value and increasing transportation costs, can create serious handling and freezing issues for industrial consumers. Consequently, all coal supply agreements impose strict limitations on the purity levels of purchased coal.

The methods and processes used to upgrade coal have changed dramatically during the past century (Carris 2007). In the early 1900s, manual sorting methods such as hand picking were used by the industry for quality improvement. This labor-intensive approach was soon made obsolete by mechanized processes that reduced misplacement and increased productivity. Many of these early facilities cleaned only the coarser particles and either recombined untreated fines back into the coal product or discarded the fines as waste. These historic periods were followed by many decades of technology development that allowed operators to partially or completely upgrade of the entire size range of mined coals (Osborne 2012). Today's modern coal processing facilities, some of which handle thousands of tons per hour of feed coal, are as functionally complex as their counterparts in either the mineral beneficiation or chemical processing industries. This level of

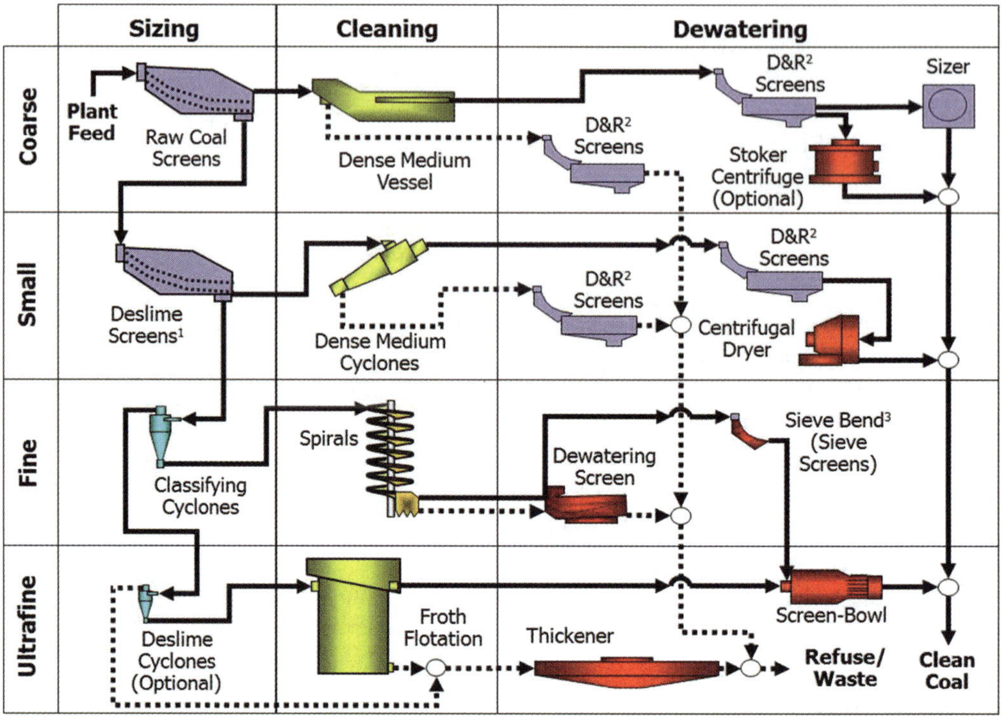

Notes: 1 - Raw coal and deslime screen decks may be on the same machine, 2 – D&R screens primarily used to recover dense medium, 3 - Sieve bends may also be combined with clean coal classifying cyclones.

Figure 1. Simplified flowsheet for a modern coal processing facility

design sophistication is dictated by increasingly competitive markets that require plants to operate at very high levels of processing efficiency which, in turn, has pushed coal producers to abandon old technology and to develop and adopt new processes and practices for cleaning run-of-mine coals. In fact, an efficiency improvement that yields a seemingly insignificant gain of just one ton per hour of metallurgical coal valued at \$150/ton corresponds to a financial benefit of over \$1 million annually for a coal processing facility (i.e., 1 ton/hr × 8,760 hr/yr × 80% runtime × \$150/ton = \$1,051,200).

The push for enhanced efficiency has forced plant designers to develop flowsheets that incorporate several parallel processing circuits, often three or more, which are specifically configured to optimize the upgrading of each size fraction. The multiple circuits are needed to maintain efficiency since the processes employed in coal processing each have a limited range of applicability in terms of particle size. For example, Figure 1 provides a simplified flowsheet for a modern "generic" plant that includes four independent circuits for treating coarse (>10 mm), medium (10–1 mm), small (1–0.15 mm) and fine (<0.15 mm) coal. In the U.S., the coarse and medium size fractions are typically treated using dense medium vessels and dense medium cyclones (DMCs), respectively. The small coal fraction, which cannot be handled by dense medium processes due to practical constraints associated with recovery of the circulating medium, can be treated by a variety of water-based gravity separation processes including water-only cyclones, spirals, teeter-bed separators or a combination of these unit operations. The finest size fraction, which is too small for gravity separation processes, is usually upgraded by froth flotation. The entire fine coal slurry may be treated by flotation, deslimed to remove a substantial portion of ultrafine (<35–40 microns) solids prior to flotation, or discarded as waste in some older plants. Dewatering of the coal products is typically performed using centrifugal basket-type dryers for the coarser factions and screenbowl centrifuges for the finer fractions. Vacuum filters are utilized in a few plants, but are difficult to justify unless a thermal dryer is available to ensure that moisture specifications can be met. Unfortunately, site specific variations in the liberation characteristics and mineralogical properties of different run-of-mine coals preclude the adoption of a standard flowsheet that is appropriate for the entire industry. Differences in circuit layouts are typically justified by flowsheet designers based on both technical and financial considerations, which attempt to balance the need to accommodate a specific raw coal size distribution or washability against any undesirable increase in capital, operating or maintenance costs. Operator preferences and vendor biases also appear to contribute to the large variations that are observed in how coal is cleaned.

The United States currently operates 292 coal preparation plants in 13 states according to the latest annual census of coal processing plants conducted by Coal Age (Fiscor 2012). The facilities, which are primarily located in the eastern coalfields, represent a total installed feed capacity of nearly 200,000 tons per hour. Assuming around-the-clock operation, this production level equates to a maximum fleet capacity of about 1.7 billion tons of feed annually. As shown in Figure 2, more than half of the plants are located in the two eastern states of Kentucky (24.7%) and West Virginia (26.7%). The plants located in these two states also represent 58% of the total installed feed capacity. The newest and most modern coal processing facilities make use of a variety of innovative technologies that have resulted in unprecedented levels of processing efficiency. Several of the most noteworthy innovations are described in the following sections, although the discussion has been limited only to commercial systems which have recently been applied within the industry on a production scale. Additional descriptions and details related to most of these innovations have been described in two outstanding books entitled *Designing the Coal Preparation Plant of the Future*

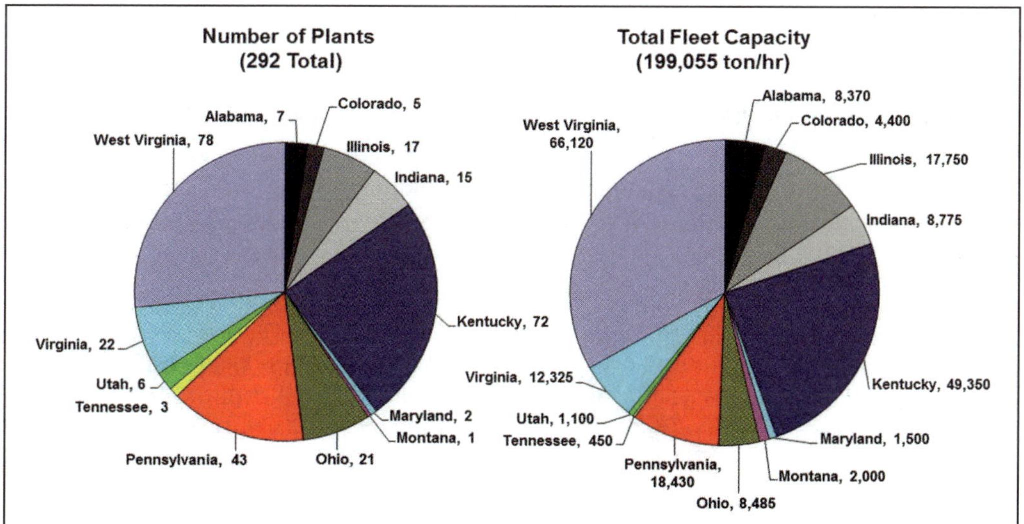

Figure 2. U.S. coal preparation plant census (after Fiscor 2012)

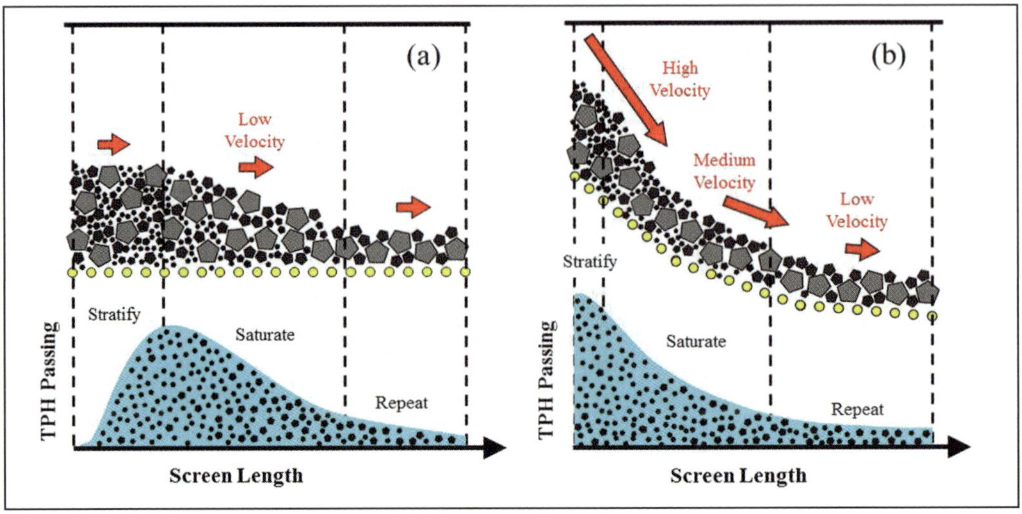

Figure 3. Material movement on (a) conventional flat-panel and (b) multi-slope banana screens

and *Challenges in Fine Coal Processing, Dewatering and Disposal*, which were published in 2007 and 2012, respectively, by the Society for Mining, Metallurgy & Exploration (SME).

INNOVATIONS IN PARTICLE SIZING

The first step in coal processing involves crushing and sizing of the run-of-mine coals into acceptable size classes. The selected topsize should balance improvements in liberation with reductions in performance created by the need to treat a higher percentage of the feed tonnage in less efficient fine coal circuits. The vast majority of equipment used in coal sizing operations is mature technology and includes well known unit operations such as vibrating screens and classifying cyclones. However, a few notable shifts in equipment design have recently occurred within the industry, including the introduction of several types of innovative screening and classification systems.

Screening Technology

The most notable innovation in screening technology involved a move away from conventional flat-panel screens to multi-slope "banana" screens. Banana screens have become the equipment of choice for new coal sizing and desliming applications, and limited drain/rinse applications. The increased popularity is largely due to the high unit capacity offered by these machines. As shown in Figure 3a, feed solids enter a conventional flat panel screen at a relatively low velocity, which creates a thick bed of solids. Although sufficient open area may exist for particle passage, many of the fines cannot pass since they become trapped in the upper portions of the thick bed of coarser solids. In contrast, banana screens incorporate decks with increasingly flatter slopes along the screen length, as illustrated in Figure 3b. Depending on the application, the slope of the first section will typically be 30–35 degrees, while the slope of the last section will typically be 0–15 degrees. Feed solids introduced onto the steep first section move very fast and form a thin bed. The thin bed allows fines to be immediately exposed to the screen surface so they can quickly stratify and pass to undersize. In coal applications, typically 55–65% of the fines pass through the first third of the screen length for banana screens. The velocity of the bed slows down as the solids reach the shallower middle decks

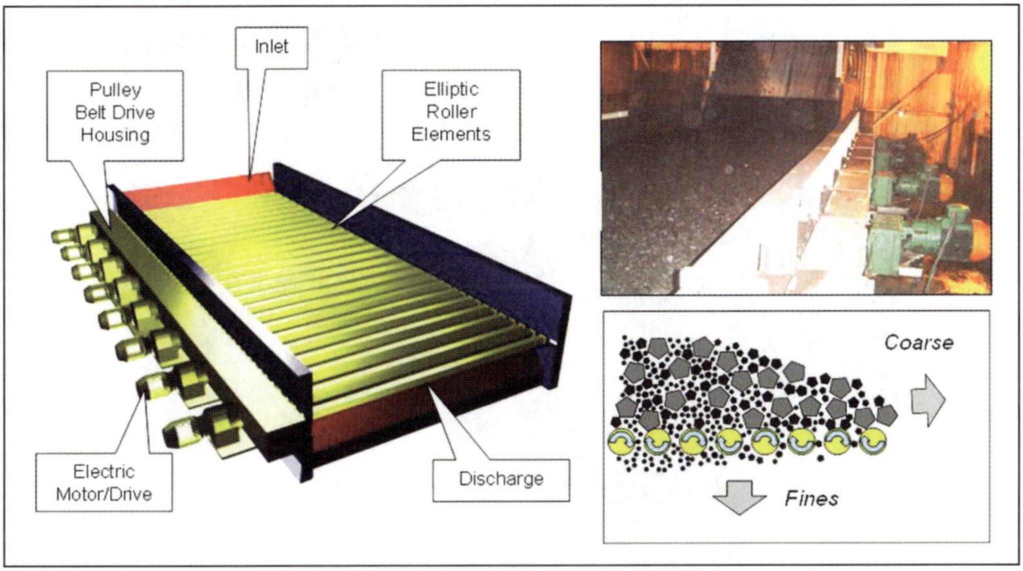

Figure 4. Roxon™ roller screen for dry coal sizing (Courtesy of Sandvik)

so that selective sizing of near-size particles can be initiated. The last section, which has the lowest slope, provides repetitive attempts for the passage of near-size particles due to the lower velocity and longer retention time. These phenomena allow banana screens to outperform conventional screens for coal feeds containing a high proportion of fines. When properly installed and sized, they can also be used successfully for drain-and-rinse screening applications.

Another notable innovation in coal sizing involves the emergence of dry screening systems. Dry screening is typically utilized only for partial washing plants where coarser particles are subjected to density separations and all or some portion of the fines bypass wet processing. This approach avoids the need to wet the fine fractions that contribute the vast majority of the moisture in the final marketed product. Since standard vibrating screens tend to blind when attempting to dry size damp run-of-mine feeds, operators have been forced to utilize either roller screen technology, such as the Roxon™ system (Figure 4), or flip-flop motion machines, such as the bivi-TEC™ system (Figure 5). The Roxon screen employs a series of elliptical rollers to retain and transport oversize particles, while fines pass through gaps between the rollers. This technology has been successfully demonstrated at two eastern plants operated by Consol Energy (Yanchak and Metzler 2006) and one western plant operated by Arch Coal (Bethell and Barbee 2007). The bivi-TEC™ screen uses a floating frame that is isolated from the screen body by rubber springs. The screen body is excited by freely rotating eccentric weights that induce tensioning and relaxing of flexible urethane panels. The unique configuration creates about 50 G's of tossing motion to the material bed, while only about 2 G's of force is applied to the screen body. This technology has been successfully used in a western coal processing operation by Arch Coal (Bethell 2012).

Desliming Technology

The desliming of coal products is one of the most important sizing activities performed in coal processing. Poor removal of slimes often allows some percentage of ultrafine high-ash particles to

Figure 5. Bivi-TEC® screen for dry coal sizing (Courtesy of Aggregates Equipment, Inc.)

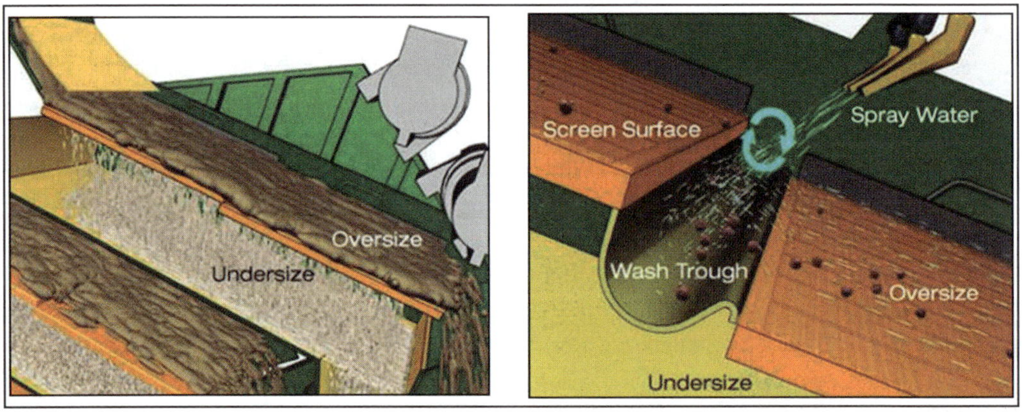

Figure 6. Working features of the StackSizer™ screen (Courtesy of Derrick Screens)

report to the clean coal product, thereby reducing quality by increasing both ash and moisture levels (Bethell and Luttrell 2005). The primary method used within the industry to combat this problem is multistage sizing equipment. These include two-stage fine-wire sieve bends, which are now common in the industry, and two-stage classifying cyclones (Firth and O'Brian 2003). The later approach has been shown to achieve particle size separations of 35–40 microns while limiting ultrafine bypass to <10% in the oversize (underflow) product (Honaker et al. 2007). When compared to sieves, the primary disadvantage of two-stage classification using cyclones is the particle density effect that tends to concentrate ash-bearing minerals in the underflow product.

Another new technology that has recently been adopted by the coal industry for slimes control is the Derrick StackSizer™ (Brodzik 2007). This innovative technology (**Figure 6**) consists of multiple (up to five) screen decks operating in parallel having fine urethane screen surfaces with openings in the range of 75 to 100 microns. The entrapped fines in the coarse overflow are released by repulping the initial overflow material with wash water in a trough between the upper and lower section of the screen deck. This repulping process allows the fines to find their way through the screen openings, thereby minimizing ultrafine bypass to the overflow product. Industrial demonstrations of this technology have confirmed the superior slimes removal capability of this technology for size separations as small as 75 microns (Hollis 2006; Mohanty et al. 2002, 2009).

INNOVATIONS IN SOLID-SOLID SEPARATION

The most important function of a coal processing facility is the separation of carbonaceous material from waste rock. This separation is typically accomplished using low-cost processes that exploit differences in physical properties, such as density and wettability, which vary with mineral content. Several innovations, largely led by equipment manufacturers, have recently occurred within this particular segment of the coal processing industry.

Dense Medium Separations

The most popular and efficient method for treating run-of-mine coals coarser than about ¼ inch (6.3 mm) is the dense medium vessel. This density-based separator consists of a large open tank through which a dense suspension of finely pulverized magnetite in water is circulated. Magnetite can be readily recovered and reused using magnetic separators. Low-density coal particles introduced into the suspension float to the surface of the vessel where they are transported by the overflow onto a collection screen, while high-density waste rock sinks to the bottom of the vessel where it is collected and discarded by a series of mechanical scrapers or lifters. The washed coal and rock products pass over drain-and-rinse (D&R) screens to wash the magnetite medium from the surfaces of the products and dewater the particles. Dense medium cyclones (DMCs) are commonly used to treat particles of coal and rock that are too small to be efficiently separated in a static vessel. These high-capacity devices also use an artificial magnetite-water medium to separate low-density coal from high-density rock, but the rate of separation is increased by the centrifugal effect created by passing medium and coal through one or more cyclones.

Most of the recent innovations related to dense medium separation have involved the utilization of improved wear-resistant materials of construction and the development of larger high-capacity units. For example, the industry now commonly utilizes DMCs as large as 55 in. (1.4 m) in diameter (Ziaja and Yannoulis 2007). In addition to offering a substantially higher capacity, these large diameter units are also capable of treating a larger particle topsize, e.g., 3–3.5 inches (75–90 mm), which can in some cases simplify plant circuitry by obviating the need for a static vessel (de Korte and Engelbrecht 2007). The use of fewer large DMCs can also offer improved levels of efficiency by minimizing differential cutpoints and avoiding surging issues that commonly occur when utilizing banks of multiple smaller DMCs (Wood 1997). Another recent advancement in dense medium separation is the development of improved magnetic separators. Technologies such as the Eriez self-leveling drum offers substantial improvements in operational simplicity and performance (Bethell and Barbee 2007). This new drum-style separator incorporates design features such as stronger magnets (950 Gauss interpole) and a wider magnetic arc (up to 140 degrees), which has increased the capacity of a single unit to >120 m^3/hr per meter of width, while reducing losses of magnetic solids down to below 0.15 kg/m^3 of treated slurry.

Figure 7. Two-stage compound spiral circuitry

Water-Based Gravity Separations

Coal particles that are smaller than about 0.5–1.0 mm are generally too small to be effectively handled by the drain-and-rinse screens required by dense medium processes. Therefore, particles in the size range between about 0.2 and 1 mm are commonly processed using water-based gravity separators. The most popular technologies used for this purpose are water-only cyclones and spirals, although advances in teeter-bed technology have created a renewed interest in this process. Water-only cyclones (WOCs), which are similar to classifying cyclones, separate coal and rock within a self-generated medium of solids and water. Newer units utilize a truncated wide-angled conical bottom to emphasize density partitioning and suppress size effects. This well-known conventional process has been largely displaced in the U.S. coal industry by spiral separators. Spiral technology consists of a corkscrew-shaped trough supported along a central pole. As feed slurry passes down the trough, the resultant flowing film segregates particles according to density, although variations in particle size and shape also influence the separation.

Perhaps the most notable innovation in this important segment of fine coal processing has been the industrial recognition of the need for multi-stage separators and circuitry. WOCs and spirals are often employed in two stages or in combination with each other to improve performance. For example, both fundamental studies and experimental field assessments have shown that conventional single-stage spirals suffer from poor separation efficiencies since the products typically contain misplaced particles of relatively pure rock and coal. When using a single-stage of spirals, operators are forced to either discard the middlings and sacrifice coal yield or retain the middlings and accept a lower coal quality. To overcome this limit, two-stage spirals have been developed in which the clean coal and middlings products are retreated by the secondary stage of spirals (Figure 7). This arrangement improves quality by reducing the density cutpoint and improves efficiency by rejecting a greater proportion of the misplaced entrapped rock (Bethell and Arnold 2003). Studies have shown that the middlings stream from the secondary spirals should be recycled back to the circuit feed to further improve the separation efficiency (Luttrell et al. 1998). In a recent industrial installation, this configuration provided a sharpness of separation more than 20% greater than that achieved using a traditional two-stage circuit without recycle (Luttrell et al. 2007). The major spiral manufacturers now offer compound spirals that mimic the rougher-cleaner circuit by repulping the primary clean and midds products part way down the same spiral run (Figure 7). The efficiencies afforded by this new technology rival those once believed to be attainable only through the use of dense medium separations (Zhang et al. 2008). The coal industry has also recently adapted spiral technology for treating very fine particles (Honaker et al. 2007), perhaps as small as 0.05 mm, for cases in which froth flotation is deemed impractical (e.g., oxidized coal). While the outcomes from these niche applications have been mixed, several very successful installations of fine spirals have been reported in cases involving the pretreatment of flotation feeds to remove pyritic sulfur (Chafin et al. 2012). These applications utilize plant classification circuits, such as cyclones and sieves, to create streams enriched in fine high-density pyrite that is then rejected by the fine "desulfurization" spirals. This pretreatment minimizes the recovery of pyritic sulfur, and particularly pyrite containing inclusions of carbon, that often floats in flotation processes that rely on differences in particle wettability for selectivity.

Another important innovation that is currently taking place in coal processing is the industrial deployment of advanced teeter-bed separators. Conventional teeter-bed designs utilize an upward flow of elutriation water to create a fluidized bed of hindered-settling particles. The resultant teeter-bed carries low-density particles upward into an overflow launder, while denser particles settle through the fluidized bed and are discharged out the bottom. Unfortunately, the processing efficiency in coal applications can be quite low due to wide variations in the solids content and size distribution of the feed. Several new technologies, such as the Eriez CrossFlow™ Separator and the Ludowici Reflux Classifier™, have been specifically developed to combat these problems. Figure 8 provides a schematic comparison between the CrossFlow™ separator and a traditional teeter-bed unit (Kohmuench et al. 2002). The tangential feed system used by the CrossFlow™ ensures that the entire separation chamber is left virtually undisturbed, allowing for a constant flow regime. The improved feed injection system provides an increase in both the separation efficiency and throughput capacity, which results in significantly lower operating demands in terms of power and water for the CrossFlow™ design. Another innovative technology, the Reflux Classifier™, represents a radical deviation from other types of teeter-bed separators (Nguyentranlam and Galvin 2001; Galvin et al. 2002). This unique version of lamella technology (Figure 9) uses a series of internal inclined plates to accelerate segregation of high-density particles back into the teeter-bed and to enhance conveying of low-density particles to the clean product overflow. Recent data (Galvin 2012; Ghosh et al. 2012) from full-scale installations in the coal industry indicate that this process offers excellent separation efficiencies (i.e., Ep values less than 0.1) over a wide range of particle sizes (i.e., 2 × 0.125 mm). The unit has also been shown to be capable of operating with high hydraulic loadings due to the much higher sedimentation area (Galvin et al. 2005, 2010).

Froth Flotation

Froth flotation is by far the most widely practiced method for upgrading very fine coal particles (Laskowski et al. 2007). Particles as large as 0.5 mm can be treated in coal flotation plants, although the industry typically uses this process only for particles with topsizes smaller than about 0.1–0.2 mm (100 to 65 mesh). Froth flotation exploits inherent differences in the surface wettability

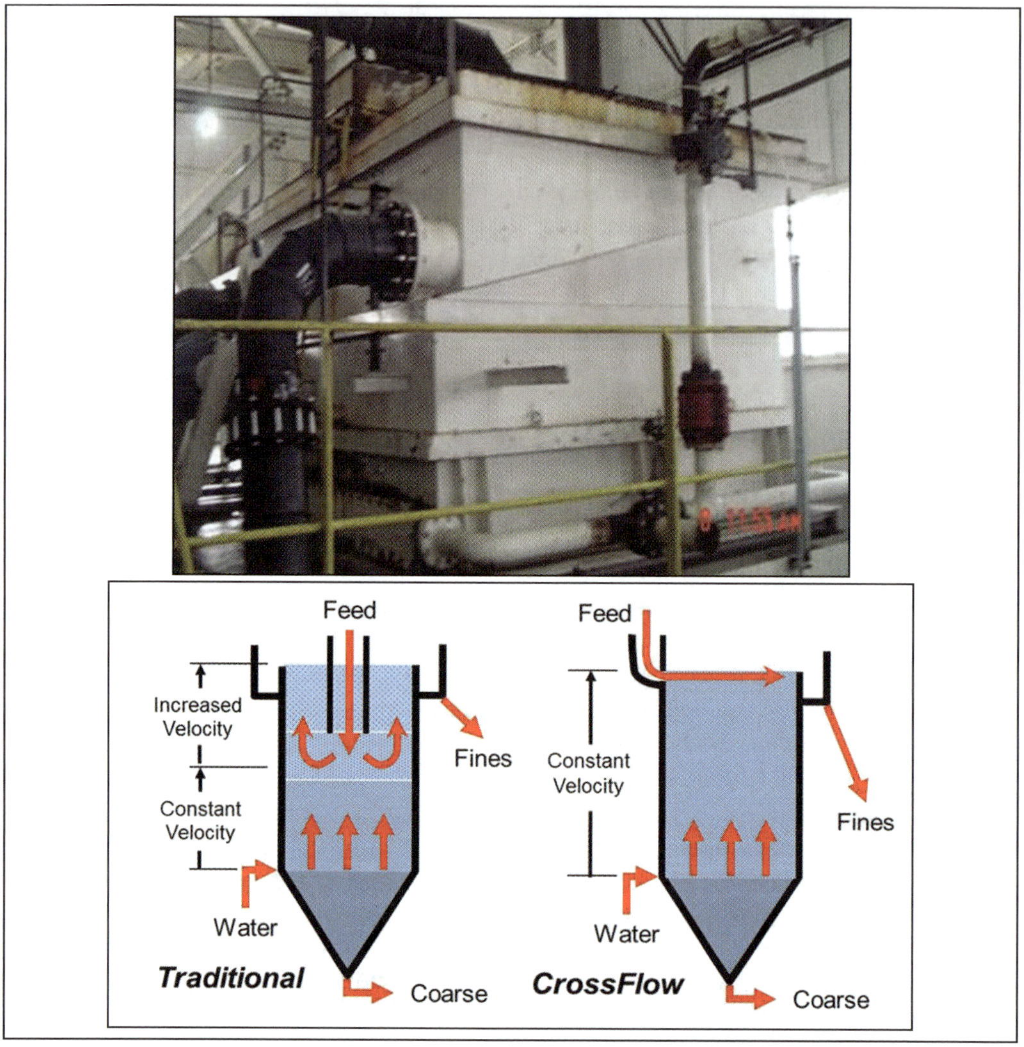

Figure 8. Comparison of a standard top-fed teeter-bed unit and tangential-fed CrossFlow™ separator

of organic matter, which is naturally hydrophobic; and mineral matter, which is hydrophilic. Most coals treated by coal processing facilities can be floated using only a frothing agent, although hydrocarbon collectors, such as diesel fuel or fuel oil, are often added to improve flotation kinetics (Laskowski 2001). In the United States, older plants made use of conventional mechanical stirred-tank flotation machines. However, since the 1990s, column flotation machines have become more widely accepted for upgrading of fine coal streams (Davis et al. 1995). This popularity can be largely attributed to the ability of columns to remove high-ash clays from the froth product via the addition of wash water to a relatively deep froth. The elimination of the high-ash component from the froth product allows a commensurate increase in the density cutpoint in dense medium circuits. This improvement permits the plants to move closer to the optimum condition of all cleaning circuits operating at the same incremental ash value (Abbott 1982). In addition, the coarsening of the plant product has the added benefit of reducing total plant moisture via recovery of coarse product

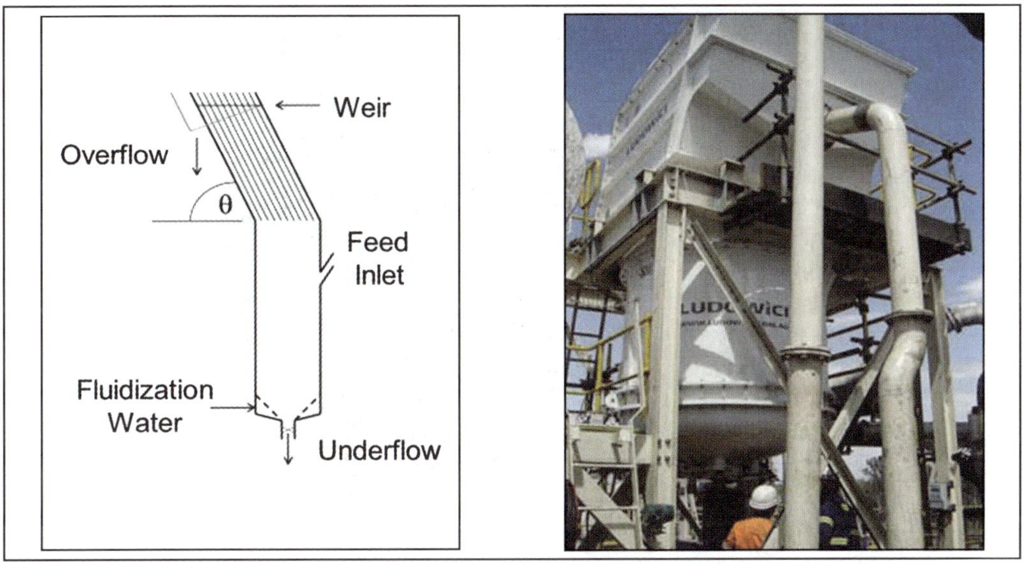

Figure 9. Lamella concept for a Reflux Classifier™ (Courtesy of FLSmidth Ludowici)

from the dense medium circuits and simultaneous elimination of ultrafine clay particles from the froth circuit.

The majority of column flotation installations incorporate either the Microcel™ or CoalPro SlamJet™ gas sparging systems. The SlamJet™ technology, which uses a series of removable air-lances equipped with a single discharge orifice located at the end of the sparger, injects air at high velocity into the column cell to create and disperse fine bubbles. This technology was further refined when CPT updated the SparJet™ spargers by incorporating a self-closing mechanism that eliminates any backflow of slurry into the aeration system. The Microcel™ technology uses a centrifugal pump to circulate slurry from the lower section of the column through parallel in-line static mixers into which compressed air is injected. The intense shear rate within the static mixers produces very small bubbles and high levels of turbulent energy dissipation needed for fast flotation kinetics. Another aeration system that has recently begun to make inroads into the coal industry is the CavTube™ sparger (Kohmuench et al. 2012). As shown in Figure 10, this technology forces a mixture of air and slurry through a small tube at a velocity sufficient to generate hydrodynamic cavitation. The cavitation phenomenon has been shown to improve flotation kinetics and reduce reagent demand, which is believed to be due to in-situ bubble formation on hydrophobic particles (Zhou et al. 1997). Full-scale tests conducted in the coal industry showed that pretreatment of feed coal slurry using this aeration system increased coal recovery by nearly 20 absolute percentage points (Honaker et al. 2007).

While there are hundreds of successful installations of column cells in the coal industry, this technology poses challenging design issues arising from the massive size of the tanks and the associated structural foundation loads. To address this problem, the coal industry has recently moved towards the use of another new flotation system known as the StackCell™ technology (Figure 11). This machine makes use of a pre-aerated high-shear feed canister that provides efficient bubble-particle contacting, thereby substantially shortening the residence time required for coal collection

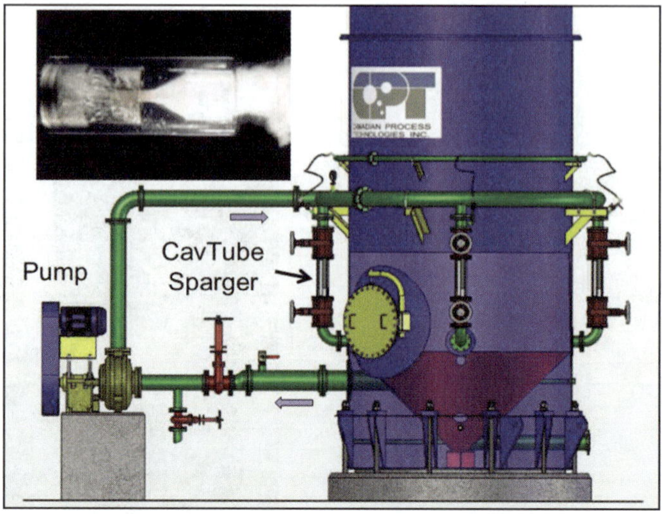

Figure 10. CavTube™ sparger system for fine coal flotation (Courtesy Eriez Flotation Division)

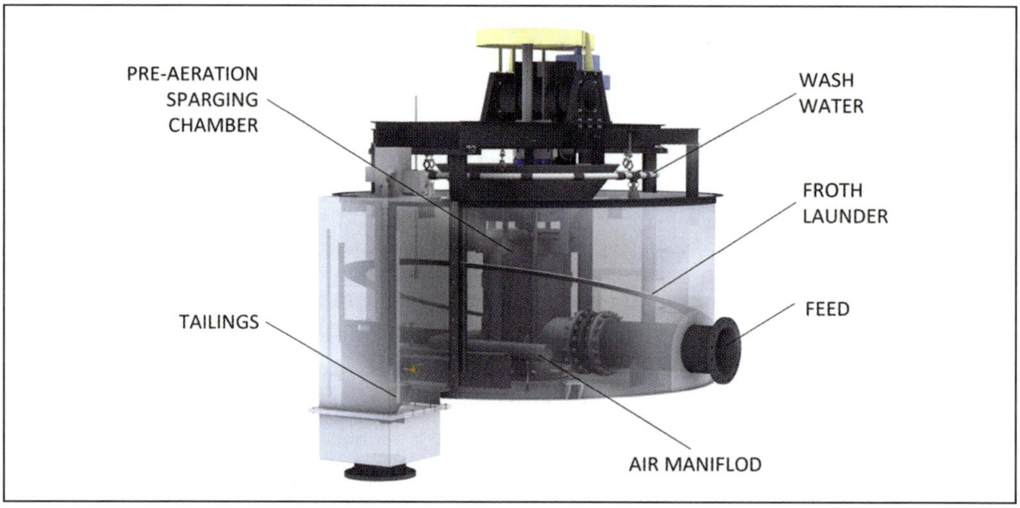

Figure 11. Transparent view of a StackCell™ flotation machine (Courtesy of Eriez Flotation Division)

and virtually eliminating most of the column height (Kiser et al. 2012). Much like column flotation, the StackCell™ technology makes use of a froth washing system to avoid the hydraulic carryover of ultrafine high-ash slimes into the froth product. Consequently, industrial installations of the technology have been shown to be capable of attaining column-like cleaning performance within a very compact volumetric footprint. Other potential advantages of the process include low air pressure requirements, low capital and installation costs, and increased flexibility in plant retrofit applications (Kohmuench et al. 2010).

One last noteworthy innovation in coal flotation is the recent development of environmentally friendly flotation collectors. While most operations still use traditional collectors such as fuel oil, diesel fuel or kerosene to improve kinetics, some states have imposed regulations that

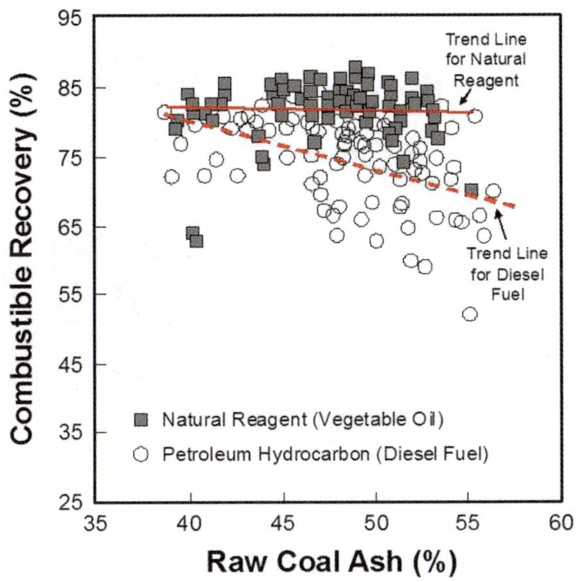

Figure 12. Comparison of natural and petroleum-based flotation collectors (after Skiles 2003)

prohibit the addition of hydrocarbon collectors when fine refuse is pumped underground for disposal. In response, several chemical companies have developed a new generation of reagents for the coal mining industry that is capable of meeting stringent environmental limits. For the case of coal flotation collectors, these reagents include several patented blends of chemicals formulated from proprietary additives and/or natural products such as blends of canola oil, vegetable oil, and soybean oil. Industrial trials (Skiles 2003) indicate that many of these new products can match, and in some cases exceed, the performance of existing hydrocarbon collectors (Figure 12). In one noteworthy example, the switch from diesel to a new collector called Nalflote® increased flotation recovery by 9.12%, adding 7,760 tons of additional clean coal per month to the plant product (www.nalco.com).

Dry Cleaning Operations

Although coal processing has traditionally been performed using water, there has been renewed interest in recent years in developing dry processes for coal upgrading (Arnold et al. 2003). Dry processes are well suited for processing coals in western coalfields where water is scarce. Although much of these reserves do not require processing other than size control, increased levels of rock dilution have recently been observed in these regions due to increasingly challenging reserves and the use of less selective large-scale mining equipment (Bethell 2007). Dry deshaling also has the potential to be adapted for removal of high-ash rock at production faces to minimize transportation costs (Honaker et al. 2006). For these reasons, several dry coal cleaning technologies, such as pneumatic jigs and air tables, have begun to be appear at several mine and utility sites in the United States (Mohanty et al. 2010). For example, the FGX air table technology (Figure 13) has been successfully applied to upgrade run-of-mine coals of different ranks in the U.S. and abroad (Lu et al. 2003, Li and Yang 2006). The separator consists of a vibrating perforated deck through which fluidization

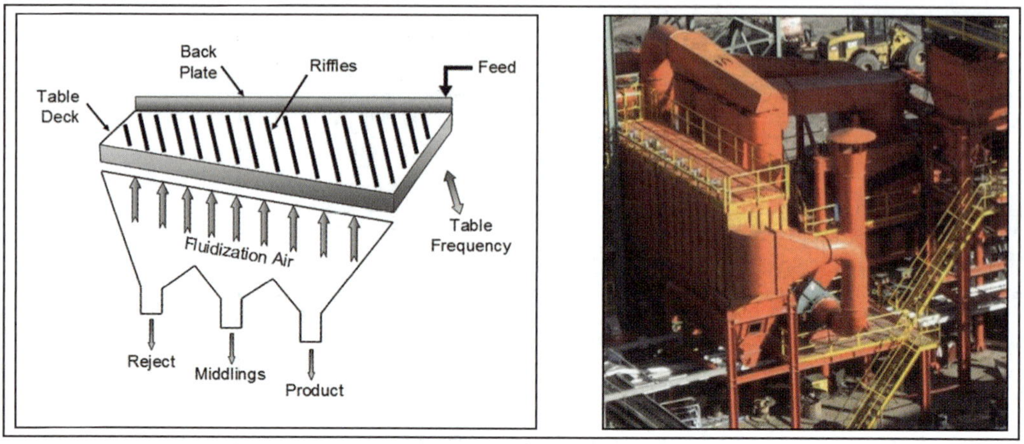

Figure 13. Working features of an FGX dry coal air table

air is passed. Data obtained when treating 50 × 6 mm bituminous coal indicate that up to 90% of high-density rock can be rejected with negligible losses of valuable coal (Honaker et al. 2008). Similar good results have also been reported for the Allair™ Jig manufactured by AllMinerals Ltd. (Kelley and Snoby 2002). The air jig (**Figure 14**) has been successfully applied in several applications within the U.S. for coal cleaning (Weinstein and Snoby 2007). The development of automated sorters, which use optical, electromagnetic, or x-ray detection to identify and extract rock from coal, also show considerable promise for dry coal cleaning (Jong et al. 2003), although none are currently in use in the U.S. coalfields.

INNOVATIONS IN SOLID-LIQUID SEPARATION

The final step in coal processing is the removal of surface moisture from the cleaned coal products. Several different types of mechanical dewatering methods are required to attain optimal performance for each size class (Arnold 1999). The removal of water from coarser (>5 mm) coal is carried out using simple screens. For particles down to about 1 mm, centrifugal dryers that use either a rotating scroll or vibratory action to transport solids are commonly used to gain lower product moistures. For finer particles (<1 mm), screenbowl centrifuges are commonly employed to provide low product moistures (Mohanty et al. 2008), although some ultrafine coal can be lost as waste effluent using this technology. Filtration processes, primarily disc filters, are used to dewater fine coal in cases where high coal recovery is desirable, although at the expense of higher product moistures (typically 25–30%). Thermal dryers can also be used to reduce coal moisture to very low levels if dictated by market demands, although the use is declining due to high costs and environmental emission issues. Solid-liquid separation is also an important issue for water conservation and waste disposal, and typically involves clarification technologies such as thickeners, belt filter presses and plate-and-frame pressure filters (Orr 2002).

Existing technologies for fine coal dewatering tend to produce unacceptably high moistures (25–30%) or intentionally sacrifice ultrafines (<0.04–0.05 mm) as waste in an attempt to lower moisture. Therefore, several innovations have recently been introduced in the coal processing industry to improve the performance of conventional dewatering equipment (Meenan 2005). One of the most notable of these is the commercial development of chemical dewatering aids. These

Figure 14. Working features of the Allair™ jig (Courtesy of AllMineral)

specialty reagents are designed to reduce water surface tension and/or increase the surface hydrophobicity of coal so as to enhance water shedding (Yoon and Basilio 1997; Basim and Yoon 1998). Industrial applications indicate that dewatering aids can reduce cake moistures in vacuum filtration by 30–50%. Field data also indicate that the reagents can significantly increase throughput by reducing cake formation times in vacuum filtration. A reduction in vacuum filter power demand has also been reported in some industrial applications involving dewatering aids. Another application of innovative reagents in fine coal dewatering involves the addition of shear-resistant polymers to improve the recovery of ultrafine coal in screenbowl centrifuges. While screenbowl centrifuges typically provide lower product moistures, this technology also typically achieves a lower overall recovery of coal fines (80–90%) compared to filters (95–99%). A standard rule-of-thumb used by industry is that screenbowl centrifuges recover only about half of particles finer than about 40–50 microns. Early attempts to improve the recovery of fines by adding conventional flocculants to the screenbowl feed have been largely unsuccessful (Miller and Wen 1984). However, recent demonstrations (Burchett et al. 2006) showed that solids recovery could be dramatically improved using a redesigned feed tube to introduce specially formulated shear-resistant polymer directly into the low-solids clarified pool within the rotating bowl (Figure 15). The structurally-modified synthetic polymer, which was developed by Nalco Chemical, works when added directly to the rotating bowl since it avoids premature contact between the polymer and the vast majority of the feed solids, and provides an ideal low-solids environment for full chemical activation. Plant demonstrations indicate that this modification can greatly reduce, and in some cases entirely eliminate, the loss of ultrafine solids in screenbowl centrifuges. The primary downside of polymer injection is that any increase in ultrafine coal recovery also produces a corresponding increase in product moisture.

Since most of the issues related to moisture removal involve ultrafine particles (Le Roux et al. 2005), several innovative technologies for treating this ultrafine material have recently been introduced into the coal industry. One such technology developed by Decanter Machine, Inc. is

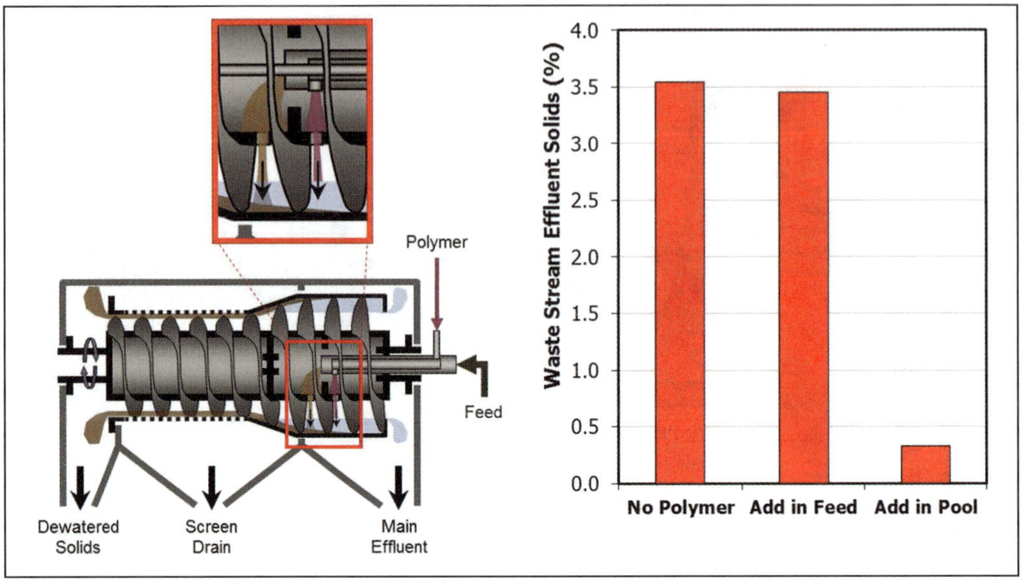

Figure 15. Polymer injection technology for screenbowl centrifuges

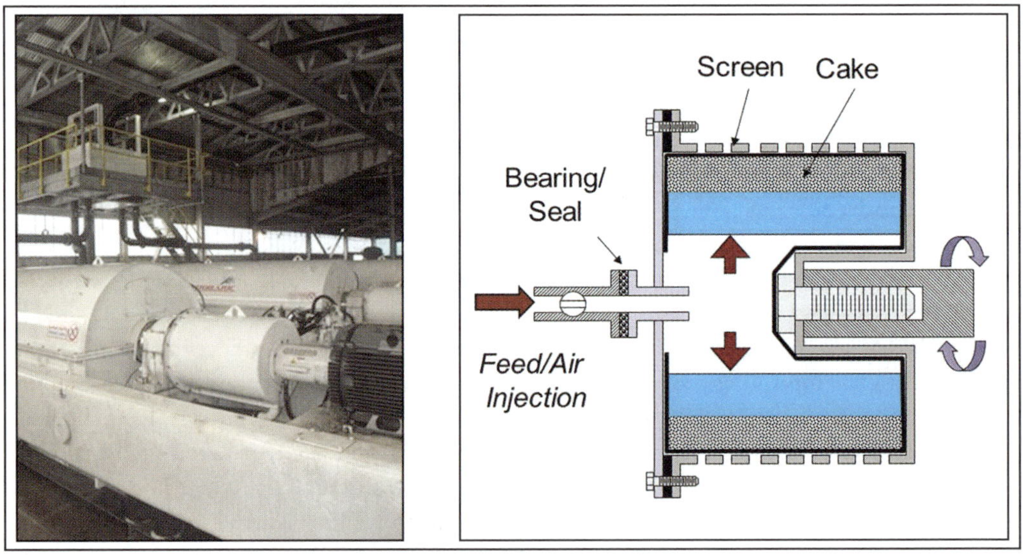

Figure 16. Operating principle of the Centribaric™ dewatering technology

the Centribaric™ centrifuge (Figure 16). This novel dewatering device combines centrifugation and pressure filtration within a single process to substantially reduce moistures over what can be achieved using conventional dewatering systems (Yoon and Asmatulu 2002; Keles 2010). Data obtained from plant demonstrations (Franklin et al. 2012) showed that this advanced technology achieved 30–50% lower moisture values when compared to conventional dewatering processes. Moreover, this technology was shown to be capable of maintaining very high recoveries (>95%) of ultrafine coal. The process also offers several significant benefits over existing dewatering processes including no chemical consumption, no heat input and low power consumption. Competition for this technology includes plate-and-frame filter presses, which have been installed at several industrial sites and have been shown to provide similar results in terms of coal recovery and moisture removal (Patwardhan et al. 2006; Prat 2012). Unfortunately, cost comparisons for these competitive processes have not been reported.

SUMMARY

The historical view of coal processing as a low-tech industry is no longer valid as the industry now utilizes a sophisticated array of advanced solid-solid and solid-liquid separation processes. These modern state-of-the-art facilities now rival most mineral and chemical plants in terms of design and operational complexity. Noteworthy innovations that have already taken place in the industry include new methods and equipment for coal sizing, cleaning and dewatering. While these technologies will continue to have a significant impact on the cost, recovery and quality of coal produced in the United States, the continued development of new separation technologies and practices will be needed to ensure that the industry remains competitive in increasingly global markets.

REFERENCES

Abbott, J., 1982. The optimisation of process parameters to maximise the profitability from a three-component blend, In 1st Australian Coal Preparation Conf., April 6–10, Newcastle, Australia, pp. 87–105.

Arnold, B.J., 1999. "Simulation of dewatering devices for predicting the moisture content of coals," Coal Preparation, Vol. 20, No. 1–2, pp. 35–54.

Arnold, B.J., Hervol, J.D. and Leonard, J.W., 2003, "Dry Particle Concentration," Coal Preparation 5th Edition, Society for Mining, Metallurgy & Exploration Inc., Littleton, CO, Chapter 7, Part 3, pp. 486–496.

Basim B.G. and Yoon, R.H., 1998. "Dewatering fine coal using novel methods," Preprint No. 98-161, SME Annual Conference and Exhibit, Orlando, Florida, March 9–11 1998.

Bethell, P.J., 2007. "Coal Preparation—Current Status and the Way Ahead," Invited Presentation, National Commission on Energy Policy (NCEP), Denver, Colorado, September 17, 2007, 31 pp.

Bethell, P.J., 2012. "Dealing with the challenges facing global fine coal processing," Challenges in Fine Coal Processing, Dewatering and Disposal (M. Klima, B. Arnold and P. Bethell, Eds.), Society for Mining, Metallurgy & Exploration Inc. (SME), Inc., Englewood, Colorado, pp. 33–45.

Bethell, P.J., and Arnold, B.J., 2003. "Comparing a two-stage spiral to two stages of spirals for fine coal preparation," Advances in Gravity Concentration (R.W. Honaker and W.R. Forrest, eds.), Society for Mining, Metallurgy & Exploration Inc., Littleton, Colorado, 196 pp.

Bethell, P.J., and Barbee, C.J., 2007. "Today's coal preparation plant—A global perspective." Designing the Coal Preparation Plant of the Future (B.J., Arnold, M.S. Klima, P.J. Bethell, Ed.), Society for Mining, Metallurgy & Exploration Inc., Littleton, Colorado, pp. 9–20.

Bethell, P.J., and Luttrell, G.H., 2005. "Effects of ultrafine desliming on coal flotation circuits," Proceedings, Centenary of Flotation Symposium, Brisbane, Australia, Paper No. 38, June 2005, pp. 719–728.

Brodzik, P., 2007. "Application of Derrick Corporation's StackSizer technology in clean coal spiral product circuits," Designing the Coal Preparation Plant of the Future (B.J., Arnold, M.S. Klima, P.J. Bethell, Ed.), Society for Mining, Metallurgy & Exploration Inc., Littleton, Colorado, pp. 89–96.

Burchett, R.T., McGough, K.M., Luttrell, G.H., 2006. "Improved screen-bowl centrifuge recovery using polymer injection technology," Coal Age, Vol. 111, No. 8, pp. 22–26.

Carris, D.M., 2007. "A historic perspective," Designing the Coal Preparation Plant of the Future (B.J., Arnold, M.S. Klima, P.J. Bethell, Ed.), Society for Mining, Metallurgy & Exploration Inc., Littleton, Colorado, pp. 3–8.

Chafin, C., Bethell, P.J. DeHart, G. and Corder, R., 2012. "Maximizing fine pyrite rejection at the Arch Coal Leer Plant," Challenges in Fine Coal Processing, Dewatering and Disposal (M.S. Klima, B.J. Arnold, P.J. Bethell, Ed.), Society for Mining, Metallurgy & Exploration Inc., Littleton, Colorado, pp. 67–78.

Davis, V.L., Jr., Bethell, P.J., Stanley, F.L. and Luttrell, G.H., 1995. Plant practices in fine coal column flotation, High-Efficiency Coal Preparation, Society for Mining, Metallurgy & Exploration Inc., Littleton, Colorado, pp. 237–246.

De Korte, G.J. and Engelbrecht, J., 2007. "Dense medium cyclones," Designing the Coal Preparation Plant of the Future (B.J., Arnold, M.S. Klima, P.J. Bethell, Ed.), Society for Mining, Metallurgy & Exploration Inc., Littleton, Colorado, pp. 61–71.

Doherty, M., 2006. "Coal quality impact on unit availability and emissions" Technical Presentation, American Electric Power Site Visit, Asia Pacific Partnership on Clean Development and Climate, Columbus, Ohio, October 30–November 4, 2006, 10 pp.

Firth, B. and O'Brian, M., 2003. Hydrocyclone circuits, Coal Preparation—A Multinational Journal, Vol. 23, pp. 167–183.

Fiscor, S., 2012. "U.S. preparation plant census," Coal Age, October 2012, pp. 22–34.

Franklin, J., Schultz, W. and Estes, T., 2012. "Centribaric operations update," Challenges in Fine Coal Processing, Dewatering and Disposal (M.S. Klima, B.J. Arnold, P.J. Bethell, Ed.), Society for Mining, Metallurgy & Exploration Inc., Littleton, Colorado, pp. 303–308.

Galvin, K.P., 2012. "Development of the Reflux Classifier," Challenges in Fine Coal Processing, Dewatering and Disposal (M.S. Klima, B.J. Arnold, P.J. Bethell, Ed.), Society for Mining, Metallurgy & Exploration Inc., Littleton, Colorado, pp. 159–185.

Galvin, K.P., Callen, A.M. and Spear, S., 2010. "Gravity separation of coarse particles using the Reflux Classifier," Minerals Engineering, Vol. 23, No. 4, pp. 339–349.

Galvin, K.P., Callen, A.M., Zhou, J., and Doroodchi, E., 2005. "Performance of the Reflux Classifier for gravity separation at full scale," Minerals Engineering, Vol. 18, No. 1, pp. 19–24.

Galvin, K.P., Doroodchi, E., Callen, A.M., Lambert, N. and Pratten, S.J., 2002. "Pilot plant trial of the reflux classifier," Minerals Engineering, Vol. 15, Issues 1–2, pp. 19–25.

Ghosh, T., Patil, D., Honaker, R.Q., Damous, M. Boaten, F., Davis, V.L. and Stanley, F.S., 2012. "Performance evaluation and optimization of a full-scale Reflux Classifier," Proceedings, Coal Prep 2012, Lexington, Kentucky, April 30–May 3, 2012, pp. 67–82.

Hollis, R., 2006. "Blue Diamond Coal Company's Leatherwood preparation plant undergoes major circuit modification," 24th International Coal Preparation Conference, Lexington, Kentucky, pp. 95–112.

Honaker, R.Q., Luttrell, G.H. and Lineberry, G.T., 2006. "Improved coal mining economics using near-face deshaling," Minerals and Metallurgical Processing Journal, Vol. 23. No. 2, pp. 73–79.

Honaker, R.Q., Saracoglu, M. Thompson, E. Bratton, R Luttrell G.H. and Richardson, V., 2008. "Upgrading coal using a pneumatic density-based separator," International Journal of Coal Preparation and Utilization, Vol. 28, No. 1, pp. 51–67.

Honaker, R.Q., Boaten, F. and Luttrell, G.H., 2007. "Ultrafine coal classification using 150 mm gMax cyclone circuits," Minerals Engineering, Vol. 20, No. 13, pp. 1218–1226.

Honaker, R.Q., Jain, M., and Saracoglu, M., 2007. "Ultrafine coal cleaning using spiral concentrators," Minerals Engineering, Vol. 20, No. 14, pp. 1315–1319.

Honaker, R.Q., Luttrell, G.H. and Bethell, P.J., 2007. "Status of current coal preparation research," in Designing the Coal Preparation Plant of the Future (B.J. Arnold, M.S. Klima and P.J. Bethell, Eds.), Society for Mining, Metallurgy & Exploration Inc., Littleton, Colorado, pp. 181–198.

Jong, T.P.R., Mesina, M.B. and Kuilman, W., 2003. "Electromagnetic Deshaling of Coal," Physical Separation in Science and Engineering, Vol. 12, No. 4, pp. 223–236.

Keles, S., 2010. "Fine coal dewatering using hyperbaric filter centrifugation," PhD Dissertation, Virginia Tech, Blacksburg.

Kelley, M. and Snoby, R., 2002. "Performance and cost of air jigging in the 21st century," Proceedings, 19th Annual International Coal Preparation Exhibition and Conference, Lexington, Kentucky, pp. 175–186.

Kiser, M., Bratton, R.C., Luttrell, G.H., Kohmuench, J., Yan, E., Christodoulou, L., Davis, V.L., and Stanley, F.L., 2012. "StackCell flotation—A new technology for fine coal recovery," Challenges in Fine Coal

Processing, Dewatering and Disposal (M.S. Klima, B.J. Arnold, P.J. Bethell, Ed.), Society for Mining, Metallurgy & Exploration Inc., Littleton, Colorado, pp. 81–94.

Kohmuench, J.N., Mankosa, M.J., Luttrell, G.H., and Adel, G.T., 2002, "Process engineering evaluation of the CrossFlow separator," Minerals and Metallurgical Processing, Vol. 19, No. 1, pp. 43–49.

Kohmuench, J.N., Mankosa, M.J. and Yan, E.S., 2010. "Evaluation of the StackCell Technology for Coal Applications," International Journal of Coal Preparation and Utilization, Vol. 30, No. 2–5, pp. 189–203.

Kohmuench, J.N., Yan, E.S. and Christodoulou, L., 2012. "Column and nonconventional flotation for coal recovery—Circuitry methods and considerations," Challenges in Fine Coal Processing, Dewatering and Disposal (B.J., Arnold, M.S. Klima, P.J. Bethell, Ed.), Society for Mining, Metallurgy & Exploration Inc., Littleton, Colorado, pp. 187–209.

Laskowski, J.S., 2001. Coal flotation and fine coal utilization, Developments in Mineral Processing, Elsevier Science Publishing, Amsterdam, The Netherlands.

Laskowski, J.S., Luttrell, G.H., and Arnold, B.J., 2007. "Coal flotation," Froth Flotation: A Century of Innovation, 100th Anniversary Commemorative Volume of Flotation (M.C. Furstenau, G.J. Jameson and R.-H. Yoon, Eds.), Society of Mining Engineers, Littleton, Colorado, 2007, 891 pp.

Le Roux, M., Campbell, Q.P., Watermeyer, M.S. and de Oliveira, S., 2005. "The optimization of an improved method of fine coal dewatering," Minerals Engineering, Vol. 18, No. 9, pp. 931–934.

Li, G. and Yang, Y., 2006, "Development and application of FGX series compound dry coal cleaning system," China Coal, Technology Monograph of the Tangshan Shenzhou Machinery Co., Ltd., pp. 17–28.

Lu, M., Yang, Y. and Li, G., 2003. "The application of compound dry separation technology in China," Proceedings, 20th Annual International Coal Preparation Exhibition and Conference, Lexington, Kentucky, pp. 79–95.

Luttrell, G.H., Honaker, R.Q., Bethell, P.J., and Stanley, F.L., 2007. "Design of high-efficiency spiral circuits for coal preparation plants," in Designing the Coal Preparation Plant of the Future (B.J. Arnold, M.S. Klima and P.J. Bethell, Eds.), Society for Mining, Metallurgy & Exploration Inc., Littleton, Colorado, pp. 73–87.

Luttrell, G.H., Kohmuench, J.N., Stanley, F.L. and Trump, G.D., 1998. Improving spiral performance using circuit analysis, SME Annual Meeting and Exhibit, Orlando, Florida, March 9–11, Preprint No. 98–208, p. 7.

Meenen, G., 2005. "Implications of new dewatering technologies for the coal industry," Invited Lecture, Center for Advanced Separation Technologies (CAST) Workshop, July 26–28, 2005, Virginia Tech, Blacksburg, Virginia, 14 pp.

Miller, K.J. and Wen, W.-W., 1984. "Effect of operating parameters and reagent addition of fine coal dewatering in a screen bowl centrifuge," Pittsburgh Energy Technology Center Report, DOE/PETC/TR-85/1 (DE85002885), p. 19.

Mohanty, M.K., Palit, A. and Dube, B., 2002. "A comparative evaluation of new fine particle size separation technologies," Minerals Engineering, Vol. 15, pp. 727–736.

Mohanty, M.K. Zhang, B. and Akbari, H. 2010. "Evaluation of the FGX dry separator for cleaning Illinois basin coal," Draft Final Technical Report, Department of Commerce and Economic Opportunities/Illinois Clean Coal Institute, Project No. 08-1/4.1A-4, January.

Mohanty, M.K., Zhang, B. and Geilhausen, R., 2009. "On-site evaluation of the Stack Sizer in conjunction with the Falcon Concentrator," Final Technical Report, Department of Commerce and Economic Opportunities/Illinois Clean Coal Institute, Project No. 07-1/9.1B-2, April.

Mohanty, M.K., Zhang, B., Khanna, N., Palit A., and Dube B., 2008. "Screen-bowl centrifuge: A high efficiency particle size separator," Minerals and Metallurgical Processing, Vol. 25, No. 2, pp. 61–67.

Nguyentranlam, G. and Galvin, K.P., 2001. "Particle classification in the reflux classifier," Minerals Engineering, Vol. 14, No. 9, pp. 1081–1091.

Orr, F.M. (Committee Chair), 2002. Coal Waste Impoundments: Risks, Responses and Alternatives, Committee on Coal Waste Impoundments, National Research Council, National Academy Press, Washington, D.C., 230 pp.

Osborne, D., 2012. "Milestones in fine coal cleaning development," Challenges in Fine Coal Processing, Dewatering and Disposal (M.S. Klima, B.J. Arnold, P.J. Bethell, Ed.), Society for Mining, Metallurgy & Exploration Inc., Littleton, Colorado, pp. 3–32.

Patwardhan, A., Chugh, Y.P., Arnold, B.J., Terblanche, A.N. 2006. "Dewatering ultrafine clean coal in a T.H. filter press," Coal Preparation, Vol. 26, No. 1, pp. 33–54.

Prat, G., 2012. "Dewatering fine coal and tailings with a filter press," Challenges in Fine Coal Processing, Dewatering and Disposal (M.S. Klima, B.J. Arnold, P.J. Bethell, Ed.), Society for Mining, Metallurgy & Exploration Inc., Littleton, Colorado, pp. 279–292.

Skiles, K.D., 2003. "Search for the next generation of coal flotation collectors," Proceedings, Coal Prep 2003, 20th Annual International Coal Preparation Exhibition and Conference, April 29-May 1, Lexington, Kentuck, pp. 187–204.

Weinstein, R. and Snoby, R., 2007. "Advances in dry jigging improves coal quality," Mining Engineering, Vol. 59, No. 1, pp. 29–34.

Wood, C.J., 1997. "Coal preparation expertise in Australia: In-plant issues and the potential impact of broader applications," Proceedings, Coal Prep '97, Lexington, Kentucky, pp. 179–198.

Yanchak, D. and Metzler, S.P., 2006. "Dry screening at small sizes using elliptical roller screens," Proceedings, Coal Prep 2006 Conference and Exhibition, Lexington, Kentucky, pp. 37–40.

Yoon, R.-H. and Asmatulu, R., 2002. "Methods for improving centrifugal filtration," U.S. Patent No. 6,440,316, Acc. 27, August 2002.

Yoon, R.-H. and Basilio, C.I., 1997. "Chemical-mechanical dewatering process," U.S. Patent No. 5,670,056, September 1997.

Zhang, B., Yang F., Sahoo, P.K., Mohanty M.K. and Zhang, X. 2008. "Optimization of a compound spiral for cleaning high sulfur fine and ultrafine coal," Proceedings XXIV International Mineral Processing Congress, Beijing, September 24–28, Vol. 2 pp. 653–662.

Zhou, Z.A., Xu, Z., Finch, J.A., Hu, H. and Rao, S.R., 1997. "Role of hydrodynamic cavitation in fine particle flotation," International Journal of Mineral Processing, Vol. 51, No. 1–4, pp. 139–149.

Ziaja, D. and Yannoulis, G.F., 2007. "Is there anything new in coarse of intermediate coal cleaning?" Designing the Coal Preparation Plant of the Future (B.J., Arnold, M.S. Klima, P.J. Bethell, Ed.), Society for Mining, Metallurgy & Exploration Inc., Littleton, Colorado, pp. 43–59.

ABOUT THE AUTHOR

Gerald H. Luttrell is the E. Morgan Massey professor of mining and minerals engineering at Virginia Tech. Since joining the faculty in 1986, he completed more than $14 million in externally sponsored R&D projects, obtained 15 patents, authored more than 200 publications in journals and proceedings, and submitted more than 200 technical reports. His professional honors include the PCMIA Stephen McCann Educational Excellence Award (1995), Henry Krumb Lecturer (2001, 2009), Percy Nicholls Award (2005), Frank F. Aplan Award (2007), and the Robert H. Richards Award (2012). On February 6, 2013, he was elected as one of the newest members of the National Academy of Engineering. Luttrell's research has contributed to the development of a variety of innovative technologies for coal preparation and mineral processing. He actively promotes technology transfer and has presented more than 100 short courses and workshops for the mining industry.

A Review of Industrial Innovations in Phosphate Processing

Patrick Zhang
Florida Industrial and Phosphate Research Institute, Bartow, Florida, United States

Regis Stana
R Squared S, Inc., Lakeland, Florida, United States

Hassan El-Shall and Brij Moudgil
University of Florida, Gainesville, Florida, United States

ABSTRACT: This paper reviews significant technological developments in phosphate processing, which have been applied on an industrial scale. It covers phosphate mineral processing, phosphoric acid manufacturing, analysis and process control, and waste treatment and utilization. Nothing improved the industry efficiency more dramatically than the development and commercialization of the Crago "Double Float" flotation process for processing low-grade siliceous phosphate ores. The development and adoption of the direct-reverse flotation process for removing dolomite from sedimentary phosphate has the potential to at least triple the world phosphate reserves. Installations of on-line analyzers based on the NMR (Nuclear Magnetic resonance) and LIBS (Laser Induced Breakdown Spectroscopy) technologies have resulted in significant efficiency improvements. Huge energy saving and environmental benefits have been achieved by the transformation of phosphoric acid manufacturing from the thermal method to the "wet acid" process, which was made possible by industrialization of the dihydrate and hemihydrate processes. Recovery of uranium from phosphoric acid played an important role in increasing the world energy resource. Large scale utilization of phosphogypsum has taken off in many parts of the world. The concept of "green mine" (comprehensive recovery of mineral resources and complete waste utilization) has become a reality at some phosphate complexes.

INTRODUCTION

Phosphate is a non-renewable resource essential for plant growth and crop production, and therefore vital to feeding the fast growing population of the world. Phosphate also plays an important role in many industrial applications, such as green energy batteries, animal feed supplements, detergents, human foods (including dairy, meat, bakery products and soft drinks), metal surface treatment, corrosion inhibition, flame retardant, water treatment, pharmaceuticals and personal care products, asphalt modification, latex paint, oil drilling, paper processing, cotton processing, cement, and ceramic production.

Like many essential materials, phosphate must be mined and processed. In all the above listed applications, phosphorus in the upgraded phosphate ore must be first extracted either in elemental form or as phosphoric acid, and then processed into various products. Elemental phosphorus is produced by reducing phosphate, with a simplified form of $Ca_3(PO_4)_2$, using carbon at high

Table 1. World phosphate rock use distribution

Use	Percent
Phosphoric acid manufacturing	71
Single super phosphate (SSP) production	13.5
Others, including animal feed, fused magnesium phosphate (FMP), triple super phosphate (TSP), mono-potassium phosphate (MKP), nitrogen-potassium-phosphorus (NPK), and nitrogen-phosphorus (NP)	10
Elemental phosphorus production	5
Direct application	0.5

Table 2. World phosphate rock production by country in thousand tons

Country	2006	2007	2008	2009	2010
Finland	858	831	780	658	817
Russia	10,813	10,937	9,810	9,538	10,844
USA	30,464	30,231	30,850	26,609	25,244
Brazil	5,801	6,095	6,344	5,949	5,693
Peru					791
Egypt	2,177	2,504	3,179	3,708	3,435
Morocco	27,098	27,638	24,198	18,163	25,655
South Africa	2,664	2,553	2,287	2,237	2,499
Tunisia	7,812	8,005	7,623	7,298	8,132
Israel	2,907	3,050	3,034	2,530	3,078
Jordan	5,881	5,546	6,148	5,271	6,529
Saudi Arabia					121
Syria	3,580	3,678	3,221	2,147	3,765
India	1,533	1,507	1,485	1,588	2,000
China	56,879	62,666	61,800	64,500	69,100
Australia	1,843	2,174	2,492	2,113	2,095
Others	9,577	10,312	11,163	12,052	14,812
Total	168,354	176,220	175,010	162,733	182,111

temperature, while sulfuric acid is manufactured by reacting phosphate with a strong acid with sulfuric acid being the most widely used (Beker 1989).

Of the phosphate ore mined worldwide, approximately 90% is used to produce phosphate fertilizers, with 5% for making animal feed supplements and 5% for industrial applications. A more detailed phosphate rock use distribution is shown in Table 1 (Prud'homme 2005).

There are two major types of phosphate deposits, sedimentary and igneous, with the sedimentary type accounting for over 90% of the total world phosphate resources (Notholt et al. 2005). Major igneous phosphate deposits are found in Brazil, South Africa, Russia, Mexico and Canada. Morocco, China, U.S., Tunisia, Jordan and Saudi Arabia host most of the world's sedimentary phosphate deposits. Currently, there are about two dozens of countries that produce phosphate rock. Table 2 lists some major phosphate producers with corresponding production data from years 2006–2010 (IFA 2012).

With the depletion of the limited high grade phosphate deposits, nearly all the phosphates currently mined require concentration prior to use for phosphoric acid manufacturing or elemental P production. Major concentration methods include flotation, magnetic separation and calcination.

Phosphate in the naturally occurring phosphate ores exists in several variations of the very stable apatite compound. In phosphoric acid manufacturing or elemental P production, the strong bonding in the apatite is broken by either chemical leaching with a strong acid or thermal reduction by carbon. Treatment with sulfuric acid to produce phosphoric acid is the most widely used method of accomplishing this breakdown. The first phosphoric acid plant was built in Germany in about 1870. In the U.S., phosphoric acid was first manufactured in 1890.

ADVANCES IN FLOTATION

With a few exceptions where high grade phosphate ore can be utilized directly with minimal processing such as washing, crushing and sizing, all phosphate plants in the world use the flotation technology for upgrading phosphate minerals.

The "Double Float" Process for Siliceous Phosphate

The true revolution in phosphate beneficiation was accomplished by Arthur Crago (1942) using his "double float" flotation process. A simplified flowchart of the Crago process is shown in Figure 1. The major advantages of this process include the following:

- Capable of producing high quality products from low grade feeds
- Easy to operate
- Somewhat tolerant to slime
- Tolerant to feed grade fluctuation

Since its broad adoption in the 1940s, the Crago process has been perfected by engineers and researchers within the phosphate industry. Many of the improvements are best summarized by Lawver's four laws (Oswald 1993; Gruber 1999). These "laws" include the following:

1. Deslime well.
2. Size well.
3. Conduct high-solids conditioning prior to rougher flotation.
4. Maintain a steady feed rate.

A more recent development is the successful use of sodium silicate as a sand depressant in the fatty acid flotation step. If less sand is reported to the rougher concentrate by adding a depressant, amine consumption in the cleaning flotation step can be reduced significantly thus reducing operating costs, as amine is the most expensive reagent used in the Crago process. Another significant benefit is a higher grade product. Although sodium silicate had been known for its capacity to depress silica, its application in the Crago phosphate flotation process was not successful until Robert Snow (1979, 1987) discovered all the "tricks" including the perfect addition time of less than 30 seconds before the finish of conditioning, the optimal dosage, and the appropriate concentration of sodium silicate.

Today, the Crago process is yielding phosphate rock products that meet the stringent requirements for DAP (diammonium phosphate) production from flotation feeds with as low as 4% P_2O_5, giving an overall all recovery of up to 90%.

The fundamentals underlying the Crago process are discussed most comprehensively by Guan (2009). Increased understanding of the theoretical aspects will lead to further improvements of this significant process in the future.

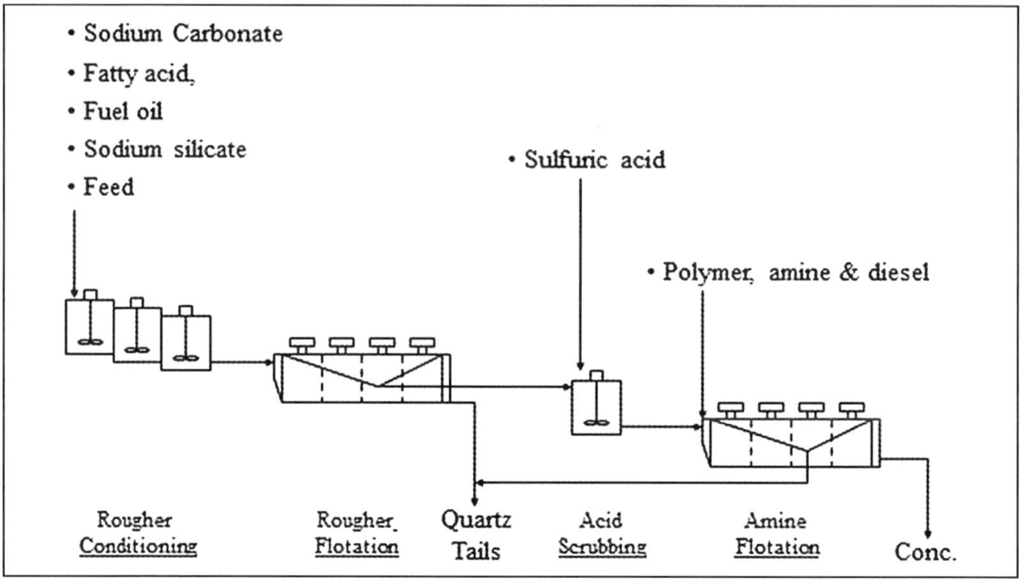

Figure 1. The "Double Float" process for siliceous phosphates (Guan 2009)

Direct-Reverse Flotation Process for Carbonaceous Ores

In the acidulation process for manufacturing phosphoric acid, MgO-containing minerals are fully dissolved and usually coexist in phosphoric acid by acid hydrolysis reaction to form $Mg(H_2PO_4)_2$. Therefore, it is not easy to separate the dissolved Mg from phosphoric acid. This is the main reason why magnesium causes various problems in the acid plant. $Mg(H_2PO_4)_2$ can greatly increase the viscosity of phosphoric acid, causing difficulty in ion diffusion and local concentration differences in the acidulation process, thus affecting the uniform growth of calcium sulfate crystals. High viscosity also reduces the filtration rate in separation of phosphoric acid from phosphogypsum.

The magnesium present in the reactor neutralizes some of the first hydrogen ions thus reducing the concentration of hydrogen ions in solution and seriously affecting the kinetics of phosphate rock acidulation. If one wants to maintain a certain H^+ concentration by increasing the amount of sulfuric acid, the SO_4^{2-} concentration would be too high leading to small and needle shaped calcium sulfate thus reducing filtration rate. Another obvious detrimental effect of MgO is the increased use of sulfuric acid and the need for higher amount of defoamer due to the carbonate content in dolomite.

Dolomite in phosphate is a worldwide problem. All the major phosphate producing countries are faced with this problem. It was estimated that phosphorus reserves tied with high carbonate ores are far more than those easy to float (Saint-Guilhem 1975; Table 3).

Separations of salt-type minerals such as apatite, calcite, dolomite, barite, fluorite and scheelite from oxides and silicates have been successfully carried out by flotation methods. However, separation of these minerals from each other is often difficult because these minerals contain the same or similar alkaline earth cations, which cause them to exhibit similar surface (electrical) behavior. Systematic research in carbonate phosphate ore beneficiation started in the early fifties (Borisov 1956). Since then, research efforts in this area have been extensive (Gu et al. 1999; Gruber et al. 2001; Gao et al. 2003; Zhang et al. 2002). Some of the techniques investigated include the

Table 3. Classification of phosphate reserves (millions tons of elemental P)

Region	Floatable Ores	High-Carbonate Ores	Upgradable Ores by Sizing
USA	1100	1900	
Africa	400	7500	300
Russia	2000	2000	

following: (1) direct flotation of phosphate with the addition of a carbonate depressant, (2) reverse flotation of the carbonate gangue while depressing phosphate, (3) physical methods such as gravity separation, scrubbing-desliming, and heavy media separation, (4) calcination, and (5) acid leaching.

Limited successes were achieved in separating dolomite from phosphate in the past. For example, in Florida USA, a physical separation method was developed and commercialized (Lawver et al. 1982; Wiegel and Hwang 1984). Heavy media cyclones were used for processing the high dolomite portion of the ore body. Using this technique, a concentrate with 1.2% MgO was obtained from feeds averaging 3% MgO. This operation was, however, terminated about 10 years ago due to high operating costs and loss of phosphate.

It is only during the past two decades that great progress has been achieved in commercializing dolomite separation technologies based on flotation. These developments mostly occurred in China, where a majority of the phosphate deposits are dolomitic (Wu and Gao 1995; Zheng and Gao 1996; Li 2012). The success of these processes relies mostly on the use of reagents–depressants such as S series, and collectors named as PA series. Even though the compositions of these reagents are proprietary, the published information suggests that the major components of S series depressants are reactant of formaldehyde and sulphonate naphthalene and phenol (Han 2010). The dolomite collector "PA series" was invented in China, but similar reagents were prepared by US companies in collaboration with Chinese researchers (Gao 2003). These collectors are possibly mixtures of various proportions of vegetable fatty acids, surfactants containing groups such as sulphonates, succinates, etc.

Direct Flotation

In this process, separation of dolomite from phosphate is achieved by direct flotation of phosphate with an effective dolomite depressant. Figure 2 shows a flowsheet of this process for a plant with a capacity of 1.5 million tons per year (Gao and Gu 1999). Successful commercialization of this process depended on development of a selective dolomite collector PA42 and a power depressant S711 for both dolomite and silica (Han et al. 2010; Li 1987; Zhong 1985).

The process is simple and has been used by large phosphate plants in China. The reported data indicated that a concentrate of 30–34% P_2O_5 and 1.5–2.0% MgO at a recovery of 80% could be obtained from a feed containing 15–18% P_2O_5 and 4–6% MgO (Han et al. 2010). It should be mentioned that the concentrate grade is relatively low with low selectivity.

Reverse or Double Reverse Flotation

Under this scheme, dolomite is first floated under mild acidic conditions using fatty acid soap selective reagent (PA 31) as carbonate collector. If the silica content in the flotation feed is high, a silica flotation step is followed, which is the case in double reverse flotation. Figure 3 shows a flowsheet of this scheme for a plant with a capacity of 2.5 million tons per year (Gao and Gu 1999). The

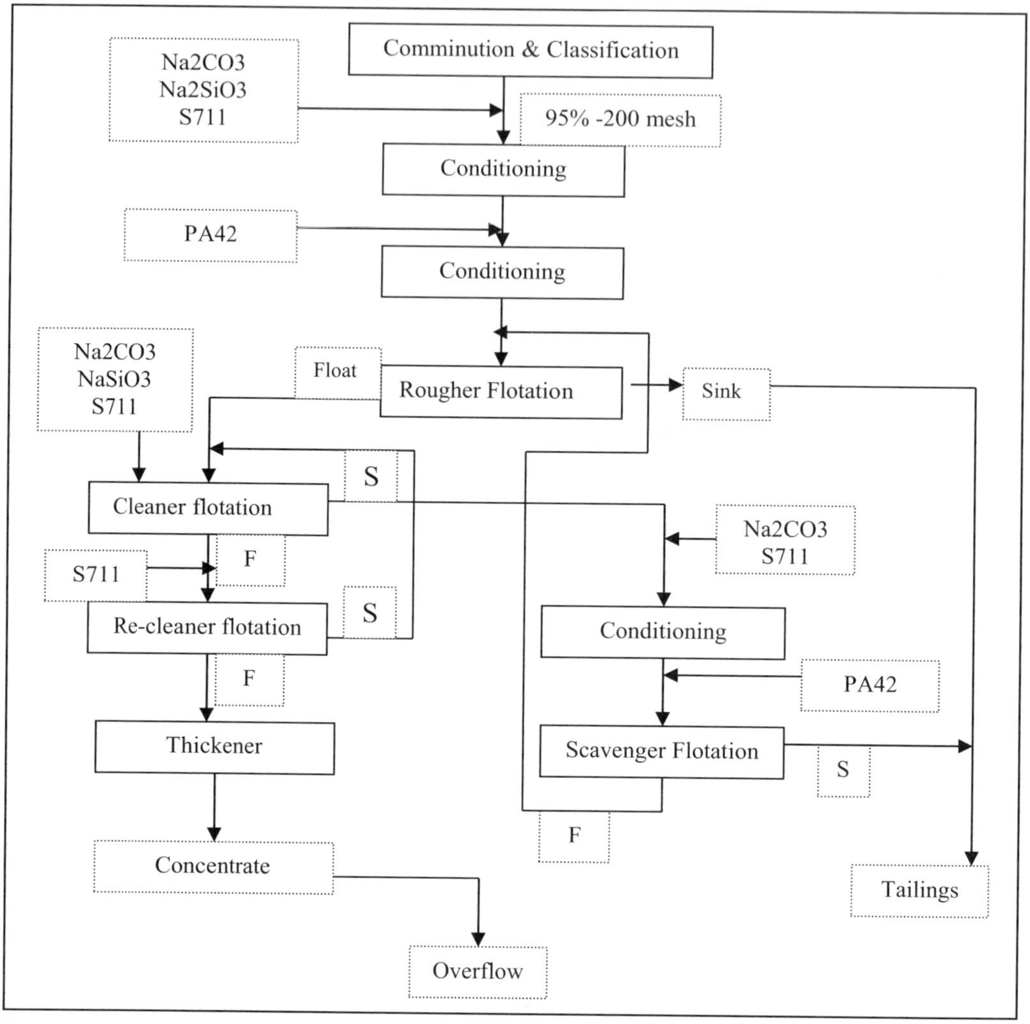

Figure 2. Direct flotation flowsheet for dolomite removal—Dayukou plant, Hubei, China (Gao and Gu 1999)

carbonate removal is over 85% and the MgO content in the phosphate concentrate can be reduced to below 1%.

Direct-Reverse Flotation

This process is gaining more market share rapidly and is favored by both researchers and engineers in the field (Wi et al. 2010 2012; Luo and Liu 2012; Qian et al. 2010; Han et al. 2010; Zheng et al. 2006). **Figure 4** shows a flowsheet of this scheme for a plant with a capacity of 2.5 million tons per year (Luo and Liu 2012).

The YP series or Custofloat collectors are prepared specifically for this application by ArrMaz Custom Chemicals, A USA Company. It is claimed that the raw materials are available and can be disperse easily in water as it is based on saponified fatty acid chemistry (Wang et al. 2010).

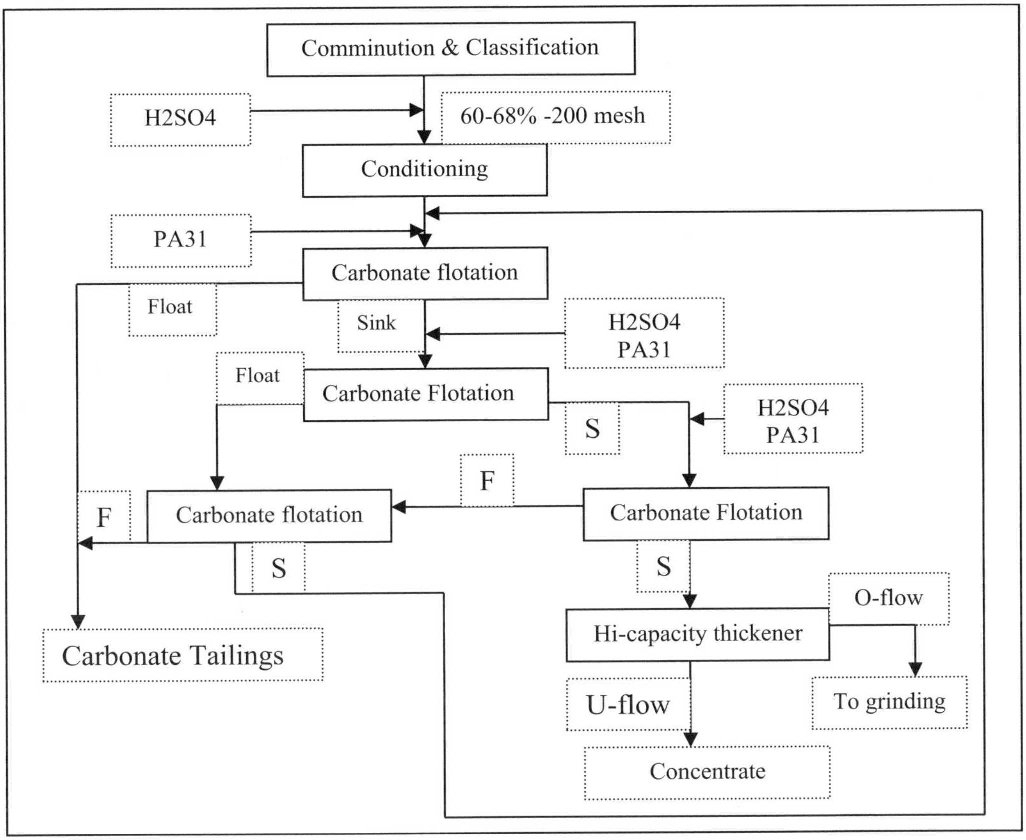

Figure 3. Reverse flotation flowsheet for dolomite removal—Wengfu plant, Guizhou, China (Gao and Gu 1999)

Like the Crago process for siliceous phosphate, the direct-reverse flotation process will likely become the dominant industrial practice for carbonaceous ores, because it has the following distinct advantages:

- It is suitable for flotation feed of varying grades.
- It usually does not require depressants that may have environmental consequences.
- It is easy to operate.

As an example of the good performance of this process, Zheng (2006) reported that a feed containing 21.3% P_2O_5 and 4% MgO could be beneficiated to a concentrate of 31.7% P_2O_5 and 0.8% MgO at a recovery of over 82%.

PROCESSING OF IGNEOUS PHOSPHATE

Although a majority of the world phosphate deposits are of sedimentary type, igneous phosphate ore is predominant in such important phosphate producing countries as Russia, Brazil, South Africa and Canada.

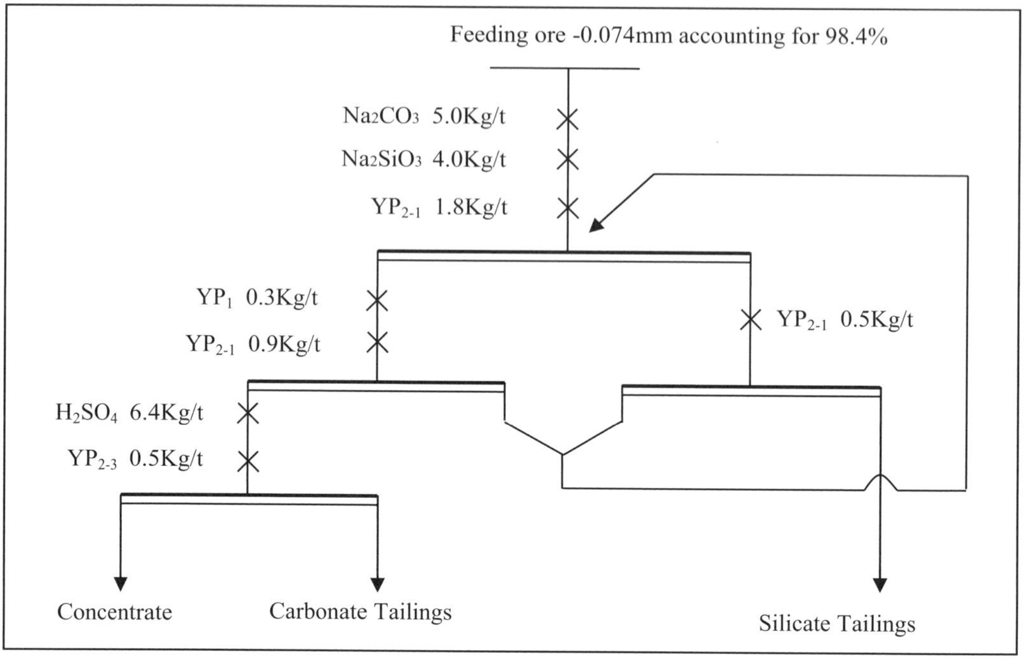

Figure 4. Direct-Reverse flotation flowsheet for dolomite removal—Jinning plant, Yunnan, China (Luo and Liu 2012)

Magnetic Separation

Although magnetic separation was not originally invented for processing phosphate, its application in processing igneous phosphate has played a critical role in sustaining the phosphate fertilizer industry in several countries. At the Kapuskasing Phosphate Operations, Canada, magnetic separation is carried out on the final flotation product. The re-cleaner flotation concentrate is pumped to a high gradient magnetic separator (HGMS). The HGMS reduces the iron content from approximately 5% to less than 2% (Rifaat and Grimm 2006). Nearly all phosphate beneficiation plants in Brazil have some magnetic separation circuits (Guimaraes et al. 2006).

ON-LINE ANALYSIS AND PROCESS CONTROL

On-line analysis and process automation are powerful tools for achieving optimal process efficiency. The rewards of an effective on-stream analysis technique coupled with a modern process control scheme include improved recovery of P_2O_5, reduced chemicals consumption and improved concentrate grade. As the quality of phosphate reserves declines, improvements in plant efficiency become more critical. Progress in automatic control of phosphate beneficiation plants has been slow. This is due to the following difficulties in analyzing the plant slurry: (1) multi-element analysis (P_2O_5, CaO, SiO_2, and MgO) is required for meaningful process control, (2) the slurry is a heterogeneous system with about 20–30% solids, and (3) the major elements to be analyzed, phosphorus and calcium, are low in atomic number thus preventing the use of X-ray fluorescence technique, which is widely adopted in other mineral industries.

Plant Installations of Nuclear Magnetic Resonance (NMR) Analyzer

Significant breakthroughs in on-line analysis started when the Florida Institute of Phosphate (FIPR) (currently the Florida Industrial and Phosphate Research Institute) funded a collaboration effort between PCS Phosphate and Harrison Cooper Systems, Inc. to evaluate an NMR based analyzer for on-line analysis of phosphate slurry (King and Cooper 1990; Cooper 1991; Cooper and Vedova 1991). This effort eventually accomplished the first phosphate flotation control based on on-line flotation feed analysis With a significant pay-off as supported by plant operation data (Shoniker et al. 1998; Shoniker 2004). Benefits included up to 5% increase in recovery, higher concentrate grade, and lower reagent costs. Since then, a new generation of NMR analyzer for phosphate has been commercialized (Davis 2006) and many new units have been installed in Florida and other parts of the world for better control of phosphate processing.

Development and Commercialization of Laser-Induced Breakdown Spectroscopy (LIBS)

In the mining of phosphates, all too often non-mineralized material is mined and mineralized material is not mined, because the dragline operator is unable to distinguish between the mineralized and non-mineralized material at the mine face. It is clear that mining costs would be significantly reduced if the miners could "see" better where the phosphate really existed. LIBS (Laser Induced Breakdown Spectroscopy) technology is being developed by FIPR that has the potential to change this situation in the near future.

Laser-induced breakdown spectroscopy is a type of atomic emission spectroscopy which uses a highly energetic laser pulse as the excitation source. The laser is focused to form a plasma, which atomizes and excites samples. In principle, LIBS can analyze any matter regardless of its physical state, be it solid, liquid or gas. Because all elements emit light of characteristic frequencies when excited to sufficiently high temperatures, LIBS can (in principle) detect all elements, limited only by the power of the laser as well as the sensitivity and wavelength range of the spectrograph and detector. In practice, detection limits are a function of (a) the plasma excitation temperature, (b) the light collection window, and (c) the line strength of the viewed transition. LIBS makes use of optical emission spectrometry and is to this extent very similar to arc/spark emission spectroscopy.

A LIBS unit was developed by FIPR and is already being used successfully at a Florida mine to analyze pebble size rock on a moving belt to make discard decisions (Gaft et al. 2002–2010; Groisman et al. 2012). The LIBS unit has been in operation for over four years and has had a significant effect on rock quality. Most importantly, the phosphate industry has found the analyzer to be extremely useful. According to one estimate, one installation at a mine with high-dolomite pebble could ahave positive economic impact of approximately two million dollars per year (Gaft et al. 2010).

The same technology is being developed to provide analysis of "in situ" materials from a distance to the dragline operator. While still under development, it is expected that it will provide information from a distance of 50 meters or more. The actual application could take several forms depending on the effective distance, accuracy and safety issues. It potentially could look at a mine face/cut and provide an accurate picture of not only the mineralized zone but the expected rock quality. The only place where the unit would not provide accurate information is when the mining cut is under water (Gaft and Stana 2012).

Table 4. Advantages of the three major processes for phosphoric acid manufacturing

Process	Advantages
Dihydrate	• Long track record of experience • Predictable performance • High capacity relative to equipment size • Moderate recovery and sulfuric acid requirement • Proven potential for recovery of uranium by-product • Best for recovery of fluosilicic acid by-product
Hemihydrate	• Minimum capital cost • Energy benefit from needing little or no steam to concentrate acid • Eliminate 27–42% evaporators • Usually eliminate rock grinding • Low cooling water requirement • Moderate phosphate recovery • Added recovery benefit where gypsum water is recirculated • Low sulfuric acid requirement • Easy to run and maintain; tolerant of process upset • Higher analysis fertilizer due to purer acid
Hemi-Di	• 98–99% P_2O_5 recovery • C Very low sulfuric acid requirement • C Energy benefit from needing little or no steam to concentrate acid • C Eliminate 27–42% evaporators • C Usually eliminate rock grinding • C Low cooling water requirement • C Gypsum purity is suitable for making a variety of by-products • C Potential for enhanced uranium recovery (to be confirmed) • C Higher analysis fertilizer

If this technology becomes fully operational, the "new pictures" of how the phosphate formations really exist could significantly alter the way the phosphate is mined, thereby leading to lower cost of phosphate mining and processing.

MANUFACTURING OF PHOSPHORIC ACID AND FERTILIZERS

There are two primary routes for recovering the phosphate values from phosphate rock for making fertilizers or producing phosphorus chemicals. In the thermal process, phosphate is reduced by carbon at high temperature to elemental phosphorus gas and condensed in water. However, the s predominant process is the "wet acid" process in which phosphate rock is acidulated with sulfuric acid.

The majority of today's phosphoric acid plants are based on either the dihydrate or hemihydrate processes mainly developed or modernized by Dorr, Prayon, St. GobaidRhone Progil, Fisons, Jacobs Engineering Group, Kellog-Lopker, Nissan, Mitsubishi, and BreyedHeurty (LookChem 2012). Another popular process in recent years has been the Hemi-Di process (Gobbittt 2011). The advantages of each process and the general selection criteria for them are well presented by John Wing (2006, 2008), as summarized in Table 4.

As of 2010, there are 13 phosphoric acid production facilities that have at least one reactor that utilizes one of the hemi production processes and about 8% of the wet process phosphoric acid is produced by one of these methods (IFDC 2011).

UTILIZATION OF PHOSPHOGYPSUM

Depending on the process used, the wet phosphoric acid manufacturing process produces about 4.3 to 4.9 tons of phosphogypsum (PG) for each ton of P_2O_5 produced. Since the current world production of phosphate rock is about 170 million tons and at least 70% of the rock is used for wet acid production, the minimum annual PG production is 160 million tons, assuming that the average rock grade is 30% P_2O_5 and that the average PG generation is 4.5 tons per ton of P_2O_5. There are also billions of tons of PG in the existing stacks worldwide. Utilization of this huge amount of by-product would not only benefit the world economically, but also would reduce or eliminate the environmental impacts of stacking or disposing of PG into the ocean.

Extensive research on PG use has been conducted by FIPR I (FIPR 2012). However, the most active industrial applications for PG use are taking place in Brazil (Jacomino 2010) and China (Wu 2010). A comprehensive review on PG use research and applications has been published recently (Zhang and Stana 2012).

Phosphogypsum Use in Agriculture

PG provides three of the major elements required by plants including calcium, sulfur and phosphorus (Kost at al. 2010). The vital roles that calcium plays in plant growth are delineated thoroughly by Sumner (1995). Numerous studies have demonstrated that use of PG enhances root growth thus helping plants absorb other nutrients, especially N (Jacomino 2010). Field demonstration data showed that application of PG more than doubled peanuts yields in Georgia, USA (Sumner 1995); increased alfalfa hay yield on a highly weathered acid soil by more than 33% (Sumner 1990); and produced 100% more apple in Brazil (Pavan et al. 1987).

Phosphogypsum Use in Construction

Many uses have been found for phosphogypsum as construction materials, such as cement retarder, wallboard, plasterboard, building blocks, stucco, plaster, road base materials, and building structure. With the exception of road base materials, many of these uses require high temperature treatment, usually by calcination, to convert dehydrate PG into at least hemihydrate.

Cement

There are three major approaches for utilizing PG to make cement. The first approach is direct use as PG based cement mortars. Hemihydrate PG was found to be more suitable than dehydrate PG for this purpose (Chang and Lin 1986). The second approach is high temperature treatment to convert dehydrate PG into hemihydrate, the treated PG is then used as cement retarder. Due to its energy consumption, this method has an economic disadvantage against natural gypsum. Another approach involves recovering sulfur from PG and using the clinker as the major component for making cement. A major phosphate producer has invested in two production lines for producing cement retarder using PG, each line with a capacity of 400,000 tons/year (He 2010). This plant, with trwo lines in operation, will consume 700,000 tons of PG per year.

PG Building Blocks and Wall Boards

In this application, dehydrate PG is first calcined and converted into hemihydrate as the major raw material for making building blocks. Other ingredients include admixture such as retarder and WRA (Water Reducing Admixture); additive, such as fly ash and cement; light weight aggregates;

and water. These type of blocks are usually used for non-load-bearing partition walls. Touted as "green" building material (Shen 2010), this PG use is developing rapidly in China. Wengfu Group's current plant produces 500,000 square meters of PG blocks, and has plans to expand in the near future (He 2010). In March 2012, Wengfu started a plant for producing paper covered gypsum board with a capacity of 30 million square meters consuming 300,000 tons of PG annually. Another company invented and commercialized a process for making PG blocks without converting dihydrate PG into hemihydrate (Wang 2010).

PG Use as a Road Base Material

The FIPR Institute pioneered the research and demonstration on PG use as a road base material. The FIPR studies spanned nearly 30 years, covering testing of various mixtures of PG with cement, leaching studies, risk assessments, and environmental monitoring studies (Lloyd 1985; Chang 1989). Conclusions from those studies may be summarized as follows:

1. PG use as a road base material poses no environmental and human health problems.
2. PG is a superior road base materials.
3. Using PG is less costly than other road base materials.

Phosphogypsum Use as a Chemical Raw Material

Although various chemicals can be produced from PG, only sulfur recovery and ammonium sulfate manufacturing have been seriously pursued and commercialized.

Sulfur Recovery

In terms of chemistry, sulfur recovery from PG can be classified as two basic processes (Wheelock et al. 1987). The following reactions take place in the so-called Process 1:

$$CaSO_4 + 2C \rightarrow CaS + 2CO_2 \quad \textbf{(1)}$$

$$CaS + H_2O + CO_2 \rightarrow CaCO_3 + H_2S \quad \textbf{(2)}$$

$$2H_2S + 3O_2 \rightarrow 2SO_2 + 2H_2O \quad \textbf{(3)}$$

$$2H_2S + SO_2 \rightarrow 3S + H_2O \quad \textbf{(4)}$$

It can be seen that Process 1 generates calcium carbonate as the major by-product.

Process 2 generates lime (CaO) as the major by-product, with the following major reaction:

$$2CaSO_4 + C \rightarrow CaO + 2CO_2 + SO_2 \quad \textbf{(5)}$$

Recovery of sulfur from PG has been investigated extensively and practiced in several countries. As a result, many options are available to suit economic conditions and market demands for by-products in different regions and countries (Clur 1988; Kendron and Lloyd 1988; Sliger 1988; Ragin and Brooks 1988; May et al. 1988; Kouloheris 1982, 2008; Kuehle and Knocsel 1988). Although the price for sulfur is the most significant factor in deciding whether to invest in a sulfur recovery plant from PG, environmental benefits and consideration for resource conservation are gaining more importance in making decisions. The latter is the case in China, where large plants are in operation or being built to recover sulfur from PG (He 2010; Wu 2010).

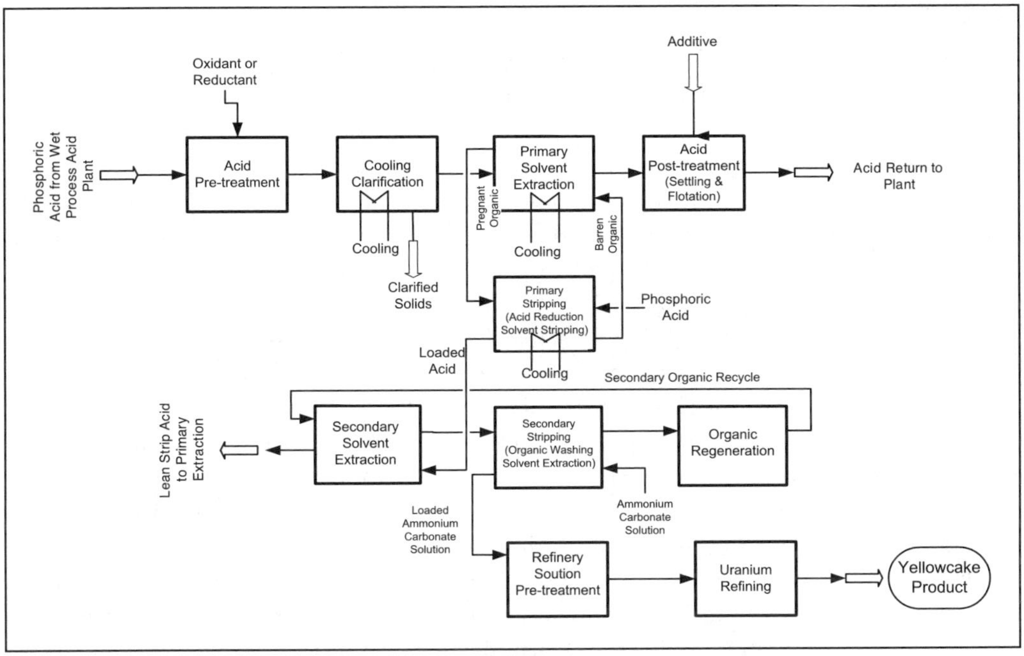

Figure 5. Overall process for uranium recovery from phosphoric acid (Astley and Stana 2012)

Ammonium Sulfate Production from PG (Merseburg Process)

This process for ammonium sulfate production can be expressed by the following two chemical reactions (Slack 1968):

$$2NH_3 + CO_2 + H_2O \rightarrow (NH_4)_2CO_3 \quad \textbf{(6)}$$

$$CaSO_4 + (NH_4)2CO_3 \rightarrow (NH_4)_2SO_4 + CaCO_3 \quad \textbf{(7)}$$

The Wengfu Group has started production of ammonium sulfate and light CaO using PG as the raw material (He 2010). The plant is designed with two production lines each with a capacity of 250,000 tons of ammonium sulfate per year.

RECOVERY OF URANIUM FROM PHOSPHORIC ACID

Although commercial recovery of uranium from phosphoric acid had its ups and downs in the past (Astley and Stana 2012), and there is no plant in operation today, the technology and industrial practice are well developed and will likely be revived in the future. As a matter of fact, the International Atomic Energy Agency (IAEA) is leading a worldwide program to rebuild uranium recovery from phosphoric acid plants with improved efficiency (Moussaid 2012). This program is inspiring renewed research efforts in this field and related extraction of rare earths from phosphoric acid (Zhang 2012; Gado 2012; Chtara 2012; Zhang 2012). The primary technology for uranium extraction from phosphoric acid is based on solvent extraction with DEPA/TOPO as the extraction reagent system (Hurst 1976,1977). DEPA stands for di(2-ethylhexyl) phosphoric acid while TOPO represents trioctyl phosphine oxide. A typical overall flowsheet for producing uranium from phosphoric acid is shown in **Figure 5** (Astley and Stana 2012).

CLOSING REMARKS

The Crago "double float" process for siliceous phosphate and the direct-reverse flotation process for carbonaceous ore will continue to dominate phosphate beneficiation schemes for decades to come. Process improvements may be accomplished by implementing better process control, development of more selective and less expensive reagents, and optimization of air bubbles for flotation.

Innovative combinations of the dihydrate and hemihydrate processes can still achieve dramatic results in phosphoric acid manufacturing (Theys 2011), not only from the standpoint of the wet acid process itself, but also in consideration of phosphogypsum disposal and utilization, uranium recovery and rare earths extraction.

Sustainable development is being embraced by both the industry and R&D community. Recent industrial practices have shown the possibility of processing phosphate with nearly zero waste accumulation. Impressive results have been achieved by some of the "green" mines, where all solid, liquid and gas wastes are either used or recycled (Wang 2011). This trend will continue.

The thermal process may come back to play a significant role in processing the low grade phosphate resources. The much anticipated large scale demonstration of the Improved Hard Process speaks to this fact (Hard 1983; Walters 2011; JDCPhosphate 2012).

In China's Guizhoug Province that currently produces about 20 million tons of phosphate rock per year, nearly 40% of the PG produced is being consumed by various uses, and their goal is to reach 100% PG utilization by 2025 (Chemall.com 2012). It may be possible for PG to totally replace mined gypsum for various uses in the future in some regions, if not worldwide.

Other significant innovations in phosphate processing have been discussed in the literature. These include applications of flotation column and the HydroFloat separator for improved coarse and fine particle recovery (Mankosa et al. 2002; Kohmuench et al. 2012; Zhang and El-Shall 2012); transition from heated flotation to flotation at ambient temperature in China (Qian et al. 2010); recovery of phosphate from slime for nearly 30% of the total rock product at some Brazilian plants (Guimaraes et al. 2006). Advances in filtration and heat recovery are other developments that are critical for achieving high efficiency in today's wet acid chemical plants.

ACKNOWLEDGMENTS

The authors acknowledge the partial financial support of the Center for Particulate and Surfactant Systems (CPaSS—A NSF I/UCRC) and the Particle Engineering Research Center (PERC) at the University of Florida. The views expressed are those of the authors and do not necessarily represent the views of the CPaSS industrial members.

REFERENCES

Astley, V. and Stana, R. 2012. Recovery of uranium from phosphoric acid: history and present status. In *Beneficiation of Phosphates: New Thought, New Technology, New Development*. Edited by P. Zhang, J. Miller and H. El-Shall. Society for Mining, Metallurgy & Exploration Inc., Englewood, Colorado.

Becker, P. 1989. *Phosphates and Phosphoric Acid*. New York: Marcel Dekker, Inc.

Borisov, V.M. 1956. Conditions for the beneficiation of difficult-to-enrich phosphate ores of the Kara-Tau deposits. [In Russian]. *Khim. Prom.* pp. 13–19.

Chang, W. 1989. *Phosphogypsum for Secondary Road Construction*. Final Report to Florida Institute of Phosphate Research. FIPR Publication No. 01-033-077. Bartow, Florida.

Chang, Wen F. and Lin, K.T. 1986. A comparative study on strength properties of cement mortars using phosphogypsum: hemihydrate vs. dehydrate. In *Proceedings of the Third Workshop on By-Products of Phosphate Industries*. FIPR Publication No.01-031-046. Bartow Florida.

Chemall.com, 2012. Acceleration in Comprehensive Utilization of Phosphogypsum. http://newscenter.chemall.com.cn/NewsArticleg.asp?ArticleID=271874. Accessed October 2012.

Chtara, M.C. 2012. Rare earth elements in Tunisian phosphate and phosphoric acid. Presented at the IAEA Regional Training Workshop: Uranium Resources Assessment and Recovery from Phosphate and Rare Earth Elements Ores, Cairo, Egypt, June 17–21.

Clur, D.A. 1988. Fedmis sulfuric acid/cement from phosphogypsum process. In *Proceedings of the Second International Symposium on Phosphogypsum*. FIPR Publication No. 01-037-055. Bartow, Florida.

Cooper, H. 1991. An investigation of potential for improved efficiencies in phosphate rougher flotation through on-line BPL measurements. Final Report to Florida Institute of Phosphate Research. FIPR Publication No. 04-032-091. Bartow, Florida.

Cooper, H.R. and Vedova, R. 1991. Phosphate beneficiation control with on-stream slurry analysis. In *Mintech '91*. New York: Sterling Publications International.

Crago, A. 1942. Process of concentrating phosphate minerals. U.S. Patent 2,293,640.

Davis, V.E. 2006. Process nuclear magnetic resonance (NMR) analyzers—benefits for phosphate beneficiation plants. In *Beneficiation of Phosphates: Technology and Sustainability*. Edited by P. Zhang, J. Miller, H. El-Shall, and R. Stana. Society for Mining, Metallurgy & Exploration Inc., Littleton, Colorado.

FIPR, 2012. Chemical Processing & Phosphogypsum Publications http://www1.fipr.state.fl.us/FIPR/FIPR1.nsf/a1380a2dc3df745f85256b4b006398eb!OpenView. Accessed October 2012.

Gado, H.S. 2012.The Egyptian experiment for the uranium extraction from phosphoric acid. Presented at the IAEA Regional Training Workshop: Uranium Resources Assessment and Recovery from Phosphate and Rare Earth Elements Ores, Cairo, Egypt, June 17–21.

Gaft, M. and Nagli, L. 2002. *Laser-Induced Time-Resolved Luminescence and Laser-Induced Breakdown Spectroscopy for Evaluation of Phosphates with High Dolomite Content*. Final report to Florida Institute of Phosphate Research. FIPR Publication No. 04-065-192. Bartow, Florida.

Gaft, M. and Stana, R. 2012. *Remote Real-Time Analyses of Phosphate Rock*. Final report to Florida Institute of Phosphate Research. FIPR Publication No. 04-032-091. Bartow, Florida.

Gaft, M., Brestel, M., Sapir-Sofer, I., and Modiano, H. 2005. *Development of LIBS Module for Dolomite Concentration Evaluation in Phosphates*. Final report to Florida Institute of Phosphate Research. FIPR Publication No. 04-067-210. Bartow, Florida.

Gaft, M., Stana, R., Sapir-Sofer, I. and Modiano, H. 2007. *LIBS Module for Dolomite Content Evaluation on a Conveyor*. Final report to Florida Institute of Phosphate Research. FIPR Publication No. 04-067-224. Bartow, Florida.

Gaft, M., Stana, R., Dvir, E., Modiano, H., Schone, Y., and Zuarez, H. 2010. Laser-induced breakdown spectroscopy (LIBS) for online analysis of phosphates: Long-term stability test. In *Beneficiation of Phosphates: Technology Advance and Adoption*. Edited by P. Zhang, K. Swager, L. Leal Filho, and H. El-Shall, SME, Littleton, Colorado.

Gao, Z. and Gu, Z. 1999. Plant practices of phosphate beneficiation in China. *In Beneficiation of Phosphates: Advances in Research and Practice*. Edited by P. Zhang, H. El-Shall and R. Wiegel. Society for Mining, Metallurgy & Exploration Inc., Littleton, Colorado.

Gao, Z.Z., Zheng, S.B., Guan, C., and Hwang, C. 2003. *Optimizing the Formulation for Dolomite Collector "PA-31" Using Raw Materials from the United States*. Final Report to Florida Institute of Phosphate Research. FIPR Publication No. 02-150-197. Bartow, Florida.

Gobbittt, J. 2011. YARA Hemihydrate (HH) and Hemidihydrate (HDH) processes for phosphoric acid production. Presented at the 1st International Symposium on Innovation and Technology in the Phosphate Industry, Marrakech, Morocco, May 9–13.

Groisman, Y., Nagli, L., and Gaft, M. 2012. On-line laser induced breakdown spectroscopy (LIBS) analysis in the phosphate industry. In *Beneficiation of Phosphates: New Thought, New Technology, New Development*. Edited by P. Zhang, J. Miller and H. El-Shall. Society for Mining, Metallurgy & Exploration Inc., Englewood, Colorado.

Gruber, G.A. 1999. Anionic conditioning for phosphate flotation. *In Beneficiation of Phosphates: Advances in Research and Practice*. Edited by P. Zhang, H. El-Shall and R. Wiegel. Society for Mining, Metallurgy & Exploration Inc., Littleton, Colorado.

Gruber, G., Zheng, S.B., and Hwang, C. 2001. *A Pilot Scale Demonstration of the IMC/CLDRI/FIPR Flotation Process for Florida High-MgO Pebble*. Final Report to Florida Institute of Phosphate Research. FIPR Publication No. 02-133-178. Bartow, Florida.

Gruncharov, IV., Pelovski, Y., Dombalov, IV., Kirilov, Pl., and Videnov, N. 1988. Processing of phosphogypsum to sulfuric acid. In *Proceedings of the Second International Symposium on Phosphogypsum*. FIPR Publication No. 01-037-055. Bartow, Florida.

Gu, Z.X., Gao, Z.Z., and Hwang, C. 1999. *Development of New Technology for Beneficiation of Florida Dolomitic Phosphate Resources*. Final Report to Florida Institute of Phosphate Research. FIPR Publication No. 02-129-167. Bartow, Florida.

Guan, C. 2009. Theoretical background of the Crago phosphate flotation process. *Minerals and Metallurgical Processing Journal*. 26(2): 55–64.

Guimaraes, R.C., Peres, A.E.C., and Araujo, A.C.D. 2006. Column flotation: A key to the production of apatite concentrates from slimes. In *Beneficiation of Phosphates: Technology and Sustainability*. Edited by P. Zhang, J. Miller, H. El-Shall, and R. Stana. Society for Mining, Metallurgy & Exploration Inc., Littleton, Colorado.

Han, Y., Lu, Zhenfu, Li, Y., and Zhu, Y. 2010. Phosphorous resources beneficiation and research advances in China. In *Beneficiation of Phosphates: Technology Advance and Adoption*. Edited by P. Zhang, J. Miller, K. Swager, L. Leal Filho, and H. El-Shall. Society for Mining, Metallurgy & Exploration Inc., Littleton, Colorado.

Hard, R.A. and Mu, J.J. 1983. Method of producing phosphorus pentoxide in a kiln with reduced carbon burnout. U.S. Patent 4397826.

He, H.M. 2010. Strategy for safe PG stacking and utilization. Presented at 2010 China International Symposium on Phosphogypsum (PG) Comprehensive Utilization Technology Development and Promotion. Beijing, China, June 8–9.

Hurst F.J. 1976. Recovery of uranium from wet-process phosphoric acid by solvent extraction. SME Preprint No. 76-B-66. Littleton, CO: SME.

Hurst F.J. 1977. Recovery of uranium from wet-process phosphoric acid by solvent extraction. *Trans. AIME, Mineral Processing*. 262: 240–248.

IFA, 2012. International Fertilizer Association Production and Trade Statistics: Phosphate Rock. http://www.fertilizer.org/ifa/HomePage/STATISTICS. Accessed October 2012.

IFDC, 2011.*Worldwide Phosphoric Acid Capacity Listing by Plant*, September. International Fertilizer Development Center (IFDC). Muscle Shoals, Alabama.

Jacomino, V. 2010. Brazilian experience in PG use. Presented at the 2010 China International Symposium on Phosphogypsum Comprehensive Utilization Technology Development and Promotion, June 9–10, Beijing, China.

JDCPhosphate Inc. 2012. Breakthrough Process–Technical Grade Phosphoric Acid Produced from Low Grade Phosphate Ores, Silica and Petroleum Coke Enhances Phosphate Sustainability www.jdcphosphate.com/pdfs/jdcflyer.pdf. Accessed October 2012.

Kendron, T.J. and Lloyd, M. 1988. Phosphogypsum to sulfuric acid with cogeneration—A competitive edge. In *Proceedings of the Second International Symposium on Phosphogypsum*. FIPR Publication No. 01-037-055. Bartow, Florida.

King, J.D. and Cooper H.R. 1990. Development of nuclear magnetic resonance slurry analyzer for phosphate mineral beneficiation process. In *Control '90–Mineral and Metallurgical Processing*. Society for Mining, Metallurgy & Exploration Inc., Littleton, Colorado.

Kohmuench, J.N., Yan, E.S., and Christodoulou, L. 2012. Column and non-conventional flotation for phosphate recovery. In Beneficiation of Phosphates: New Thought, New Technology, New Development. Edited by P. Zhang, J. Miller and H. El-Shall. Society for Mining, Metallurgy & Exploration Inc., Englewood, Colorado.

Kost, D., Chen, L., and Dick, W. 2010. What is Gypsum and What is Its Value for Agriculture? http://library.acaa-usa.org/asp/libraryhome.asp. Accessed October 2012.

Kouloheris, A.P. 1982. *Evaluation of Fluid Bed Decomposition of Phosphogypsum.* Final Report to Florida Institute of Phosphate Research. FIPR Publication No. 01-002-002. Bartow, Florida.

Kouloheris, A.P. 2008. *Evaluation of Potential Commercial Processes for the Production of Sulfuric Acid from Phosphogypsum.* Final Report to Florida Institute of Phosphate Research. FIPR Publication No. 01-002-001. Bartow, Florida.

Kuehle, K.H. and Knoesel, K.R. 1987. Energy saving process for thermal decomposition of phosphogypsum and other calcium sulfates for the production of H_2SO_4 and cement clinker by applying the circulating fluid bed. In *Proceedings of the Second International Symposium on Phosphogypsum.* FIPR Publication No. 01-037-055. Bartow, Florid.

Lawver, J.E., Wiegel, R.L., Snow, RE., and Huang, C.L. 1982. Beneficiation of dolomitic Florida phosphate reserves. Presented at the XIV International Mineral Processing Congress. Toronto, Canada, October.

Li, X.K. 1987. Research on flotation depressor S-711 and black liquor of cotton paper-pulp making. *Industrial Minerals & Processing.* 3: 31–34.

Li, Y. 2012. Technology development and practice for beneficiating Yunnan collophanite. In Beneficiation of Phosphates: New Thought, New Technology, New Development. Edited by P. Zhang, J. Miller and H. El-Shall. Society for Mining, Metallurgy & Exploration Inc., Englewood, Colorado.

Lloyd, M. 1985. *Phosphogypsum: A Review of the Florida Institute of Phosphate Research Programs to Develop Uses for Phosphogypsum.* FIPR Publication #01-000-035. Bartow, FL.

LookChem, 2012. Manufacture of phosphoric acid. www.lookchem.com/Chempedia/Chemical-Technology/Inorganic-Chemical-Technology/2915.html. Accessed October 2012.

Luo, L. and Liu, X. 2012. Relations between MgO occurrence in phosphorite and its removal by flotation process. In Beneficiation of Phosphates: New Thought, New Technology, New Development. Edited by P. Zhang, J. Miller and H. El-Shall. Society for Mining, Metallurgy & Exploration Inc., Englewood, Colorado.

Mankosa, M.J., Kohmuench, J.N., Luttrell, G., Gruber, G., and Shoniker, J. 2002. *In-Plant Testing of the HydroFloat Separator for Coarse Phosphate Recovery.* Final Report to Florida Institute of Phosphate Research. FIPR Publication No. 02-137-188. Bartow, Florida.

Martin, H.S. 1933. Milling methods and costs at No. 2 concentrator of the Phosphate Recovery Corporation. In *Transactions of the American Institute of Mining Engineers.* 2: 339–354.

May, A., Rice, D.A., Carter, D.C. Jr. 1988. Recovery of sulfur from phosphogypsum Part 2.—Conversion of sulfide to sulfur. In *Proceedings of the Second International Symposium on Phosphogypsum.* FIPR Publication No. 01-037-055. Bartow, Florida.

Moussaid, M. 2012. UxP–Uranium eXtraction from Phosphates and Phosphoric Acid. www.uxponline.com/Home.aspx. Accessed October 2012.

Notholt, A.J.G., Sheldon, R.P., and Davidson, D.F. 2005. *Phosphate Deposits of the World, Vol. 2: Phosphate Rock Resources.* Cambridge University Press.

Oswald, G.R. 1993. Fatty acid phosphate conditioning and flotation—plant practice. In *Beneficiation of Phosphates: Theory and Practice.* Edited by H. El-Shall, B. Moudgil and R. Wiegel. Society for Mining, Metallurgy & Exploration Inc., Littleton, Colorado.

Pavan, M.A., Binghan, F.T. and Peryea, J. 1987. Influence of calcium and magnesium salts on acid soil chemistry and calcium nutrition of apple. *Soil Science Society of America Journal.* 51 (6): 1526–1530.

Prud'homme, M. 2006. Phosphate production in the new economies. Presented at the FIPR and IAEA Collaboration Meeting, Bartow, Florida, October 3–4.

Qian, Y., Huang, Z., and Zheng, S. 2010. Laboratory and pilot-plant flotation tests on Anning phosphate ore at ambient temperature. In *Beneficiation of Phosphates: Technology Advance and Adoption.* Edited by P. Zhang, J. Miller, K. Swager, L. Leal Filho, and H. El-Shall. Society for Mining, Metallurgy & Exploration Inc., Littleton, Colorado.

Ragin, M.M. and Brooks, D.R. 1988. Recovery of sulfur from phosphogypsum Part 1—Conversion of sulfate to calcium sulfide. In *Proceedings of the Second International Symposium on Phosphogypsum.* FIPR Publication No. 01-037-055. Bartow, Florida.

Rifaat, R. and Grimm, D. 2006. Mineral processing of phosphate rock at Agrium Kapuskasing phosphate operations. In *Beneficiation of Phosphates: Technology and Sustainability*. Edited by P. Zhang, J. Miller, H. El-Shall, and R. Stana. Society for Mining, Metallurgy & Exploration Inc., Littleton, Colorado.

Saint-Guilhem, R. 1975. La Valorisation des Phosphates Mineraux Pauvres a Gangue Carnobatee. Reprients No. 11. Congres International de Valorisation des Minerais Seminaire sur.

Shen, C.X. 2010. Characteristics of gypsum block & prospect of gypsum block industry in China. Presented at the 2010 China International Symposium on Phosphogypsum Comprehensive Utilization Technology Development and Promotion. Beijing, China, June 9–10.

Shoniker, J. 2004. Use of NMR technology for on-line phosphorus analysis. Presented at the 19th Regional Phosphate Conference, Lakeland, Florida, October 15–16.

Shoniker, J., Vedova, R., and Vaughn, R.L. 2012. PCS Phosphate White Springs automatic control and on-stream analysis innovations have pay-off in big gains. www.hrcsystems.com/Analyzers/Phospholyzer/Phospholyzer_Presentation/phospholyzer_presentation.html. Accessed October 2012.

Slack, A.V. 1968. *Phosphoric Acid.* Volume 1. New York: Marcel Dekker, Inc.

Sliger, A.G. 1988. The M.W. Kellogg Company KEL-S process. In *Proceedings of the Second International Symposium on Phosphogypsum.* FIPR Publication No. 01-037-055. Bartow, Florida.

Snow, R. 1979. Beneficiation of phosphate ores. U.S. Patent 4,144,969.

Snow, R. 1987. Sodium silicate as a phosphate flotation modifier. U.S. Patent 4904375.

Sumner, M. 1990. *Gypsum as an Ameliorant for the Subsoil Acidity Syndrome.* Final Report to Florida Institute of Phosphate Research. FIPR Publication No. 01-024-090. Bartow, Florida.

Sumner, M. 1995. Literature *Review on Gypsum as a Calcium and Sulfur Source for Crops and Soils in the Southeastern United States.* Final Report to Florida Institute of Phosphate Research. FIPR Publication No. 01-118-118. Bartow, Florida.

Theys, T. 2011. New developments in Dihydrate-Hemihydrate processes. Presented at the 1st International Symposium on Innovation and Technology in the Phosphate Industry, Marrakech, Morocco, May 9–13.

Walters, M. 2011. A technical review of the improved Hard process. Presented at the 1st International Symposium on Innovation and Technology in the Phosphate Industry, Marrakech, Morocco, May 9–13.

Wang, G.W. 2010. Development and commercialization of comprehensive phosphosgypsum utilization at Guizhou Kailin Group. Presented at the 2010 China International Symposium on Phosphogypsum Comprehensive Utilization Technology Development and Promotion. Beijing, China, June 9–10.

Wang, G.W. 2012. Guizhou Kailin Group green mine construction and industrial developing practice. Presented at the 6th International Conference on Phosphate Beneficiation, Kunming, China, March 6–11.

Wang, G., Li, Y., Gu, Z., and Partin, D. 2010. Development of Custofloat® collectors for Processing Yunnan Sedimentary Dolomiteic phosphate ore, Proceeding of Beneficiation of Phosphates, Technology Advances and Adoption, SME, pp. 169–180.

Wei, Y., Sandenbergh, R., Li, D., and Li, X. 2010. Distribution of fatty acid collector and its effect on direct reverse flotation of a phosphate ore. In *Beneficiation of Phosphates: Technology Advance and Adoption.* Edited by Zhang, P., Swager, K., Filho, L.L., and H. El-Shall. Society for Mining, Metallurgy & Exploration Inc., Littleton, Colorado.

Wei, Y., Liu, X., Li, W., Li, F., and Sandenbergh, R.F. 2012. Distribution of fatty acid collector and its effect on direct reverse flotation of a phosphate ore. In *Beneficiation of Phosphates: New Thought, New Technology, New Development.* Edited by P. Zhang, J. Miller and H. El-Shall. Society for Mining, Metallurgy & Exploration Inc., Englewood, Colorado.

Wheelock, T.D., Fan, C.W. and Floy, K.R. 1987. Desulfurization of phosphogypsum. In *Proceedings of the Second International Symposium on Phosphogypsum.* FIPR Publication No. 01-037-055. Bartow, Florida.

Wiegel, R.L., and Hwang, C.L. 1984. A predictive model for heavy media cyclone separation of phosphate pebble from dolomite. SME Preprint No. 84-648. Littleton, CO: SME.

Wing, J. 2006. The Hemi era. Presented at the Clearwater Conference, American Institute of Chemical Engineers, Clearwater Beach, Florida, USA, June 10.

Wing, J. 2008. Selecting a phosphoric acid process. Presented at the Clearwater Conference, American Institute of Chemical Engineers, Clearwater Beach, Florida, USA, June 7.

Wu, L. and Gao, Z. 1995. *Commissioning Operation of Wengfu Phosphate Beneficiation Plant in the Capacity of 2.5 Million t/a of Feed Ore.* [In Chinese]. Final Report to CLDRI. Lianyungang, China.

Wu, X.Y. 2010. Current situation of the comprehensive utilization of PG in China and the prospects of the Chinese market. Presented at the 2010 China International Symposium on Phosphogypsum Comprehensive Utilization Technology Development and Promotion. Beijing, China, June 9–10.

Zhang, P. 2012. Sustainable, comprehensive utilization of phosphate resources. Presented at the IAEA Regional Training Workshop: Uranium Resources Assessment and Recovery from Phosphate and Rare Earth Elements Ores. Cairo, Egypt, June 17–21.

Zhang, P. 2012. Recovery of critical elements from Florida phosphate: Phase I. Characterization of rare earths. Presented at ECI International Conference: Rare Earth Minerals /Metals—Sustainable Technologies for the Future, San Diego, USA, August 12–17.

Zhang, P. and El-Shall, H 2012. Recent technological breakthroughs in phosphate processing. *Minerals and Metallurgical Processing Journal,* 29 (1), 36–46.

Zhang, P. and Stana, R. 2012. Phosphogypsum management and utilization: A review of research and industry practice. In *Beneficiation of Phosphates: New Thought, New Technology, New Development.* Edited by P. Zhang, J. Miller and H. El-Shall. Society for Mining, Metallurgy & Exploration Inc., Englewood, Colorado.

Zhang, P., Snow, R.E., Bogan, M. 2002. *A Screening Study on Phosphate Depressants for Beneficiating Florida Phosphate Minerals.* Final Report to Florida Institute of Phosphate Research. FIPR Publication No. 02-101-183. Bartow, Florida.

Zheng, S. and Gao, Z. 1996. *Report on Commissioning Operation of Dayukou Phosphate Beneficiation Plant in the Capacity of 1.5 million t/a of Feed Ore.* [In Chinese] Final Report to CLDRI. Lianyungang, China.

Zheng, S., Cao, X., Ge Z., and Song W. 2006. Beneficiation studies on sedimentary siliceous and calcareous phosphate ores in China. In *Beneficiation of Phosphates: Technology and Sustainability.* Edited by P. Zhang, J. Miller, H. El-Shall and R. Stana. Society for Mining, Metallurgy & Exploration Inc., Littleton, Colorado.

Zhong, J.G. 1985. On effect of some modifiers in phosphorite flotation. *Industrial Minerals & Processing.* 3: 41–45.

ABOUT THE AUTHORS

Patrick Zhang received his Ph.D. degree in metallurgical engineering from University of Nevada, Reno; M.S. degree from the Institute of Process Engineering, Chinese Academy of Science; and B.S. degree from Northeastern University. He has 20 years of experience in phosphate research and 6 years of research experience in gold mining and cyanide waste treatment. Since 1993, Zhang has served as the director of Research-Beneficiation & Mining with the Florida Industrial and Phosphate Research Institute. In this post, he is responsible for the mining and beneficiation research programs, including strategic planning, soliciting and evaluating research proposals, managing contract research projects, and directing in-house research programs. He has led the institute's beneficiation program to global prominence by conducting eight in-house projects, managing dozens of research contracts, chairing seven international conferences, authoring numerous technical papers and book chapters, and delivering keynote/plenary speeches at various conferences.

Regis Stana received his B.S., M.S., and Ph.D. degrees in chemical engineering from the University of Pittsburgh. As a former director of technology development for IMC-Agrico, Stana was responsible for forming a group to develop advanced control systems, automated mining equipment, new methods for clay disposal, and new fertilizer products. In his professional career, Stana was also responsible for process improvement, and cost reductions in existing phosphate fertilizer and uranium recovery plants. Stana is currently retired but serves part time as an industry professor at the University of Florida Particle Engineering Research Center and as a consultant to the phosphate industry. He holds 20 patents and has authored more than 50 publications. Stana has received numerous honors, including Chemical Engineer of the Year for Central Florida (1989), FES Outstanding Technical Achievement (1991), Outstanding Paper award at AIChE Spring Meetings (1992, 1995), Elected Fellow Engineer for AIChE (1992), and Hero of the Industry award (1998).

Hassan El-Shall is currently an associate professor of materials science and engineering and associate director for research at the Particle Engineering Research Center of the University of Florida. He has also served as associate director for beneficiation research at the Florida Institute of Phosphate Research from 1986–1992. He served as assistant professor of mineral processing at Montana Tech from 1980 to 1986. He has more than 40 years of experience in research (both basic and applied) and management. His major research interests include interfacial phenomena and its applications in mineral industry, and waste treatment in various industries. He expanded his research efforts to include mineral modifications for medical applications. El-Shall has been an active member of SME since 1977. He has published more than 165 publications, including 11 patents, 9 edited books, 1 authored manual, 19 chapter contributions to books, over 50 refereed articles in prestigious journals such as *Powder Technology*, *Mineral and Metallurgical Processing*, *International Journal of Mineral Processing*, and others, in addition to many industrial reports.

Brij Moudgil is a distinguished professor of materials science and engineering and director of the Particle Engineering Research Center (PERC) and the Center for Particulate and Surfactant Systems (CPaSS—NSF I/ UCRC) at the University of Florida in Gainesville. Moudgil also serves as the director of the UF Mineral Resources Research Center. He received a B.E. (metallurgy) degree from the Indian Institute of Science (IISc), Bangalore, and M.S. and Eng.Sc.D. degrees in mineral engineering from the Henry Krumb School of Mines, Columbia University in New York. Moudgil has published more than 300 technical papers and has presented over 450 papers at scientific meetings and seminars including 100 invited/plenary talks. He has been awarded 14 patents and has edited 10 books. His research and professional leadership accomplishments are recognized by several major awards including his election to the National Academy of Engineering. He has also been elected as foreign fellow of the Indian National Academy of Engineering. He is the recipient of the 2003 Distinguished Alumni Award of the Indian Institute of Science, Bangalore. He served as 2006 president of SME.His current research interests include engineered particulate systems for enhanced performance in advanced materials and minerals, nano- and biotechnologies, microelectronics, photocatalytic degradation of hazardous microbes, nanotoxicity, and greener reagents and particle technologies.

Mineral Sand Separations

D. Erik Spiller
Tetra Tech, Inc., and Colorado School of Mines, Golden, Colorado, United States

C. Alex Norgren
Tetra Tech, Inc., Golden, Colorado, United States

ABSTRACT: The separation and upgrading of heavy mineral sands is accomplished by a combination of wet and dry processing. The wet side comprises liberation by scrubbing, size classification, and various techniques of gravity separation to produce a bulk (total) heavy mineral concentrate. The dry side, treating size classified total heavy mineral concentrate, principally involves the use of magnetic and electrostatic separation to produce marketable concentrates of rutile, ilmenite, zircon, monazite, and other heavy minerals. These various separating unit operations were invented, developed and/or adopted as the result of issues experienced at the operational level but ultimately driven by market forces.

INTRODUCTION

The evolution of heavy mineral sands processing in a broad sense resembles that of human history. In addition to the great driver of change, economics, there have been occasional benevolent drivers such as the holistic benefit to society or simply a pure desire to increase knowledge. Other drivers on the broad horizon are more negative, but nonetheless drivers, such as war, accidents, ignorance, fear of change, and prejudice. These drivers or a combination thereof have theoretically played a role in the development of heavy minerals sands processing.

Heavy mineral sands comprise a diverse minerals group, with a range of physical characteristics and economic values, except they possess a specific gravity of 2.9 or greater. The most important economic minerals in the group are ilmenite, rutile, zircon, and monazite, each possessing its own set of characteristics which differentiates one mineral from the others.

Heavy mineral sand (HMS) deposits are credited to the deposition/concentration of the heavier particles due to currents and sedimentation on the world's ocean beaches. As a result, HMS is mined on, as the eco-groups often claim, pristine coastlines or in areas where ancient oceans once existed. In fact, titanium was first discovered by an amateur geologist by the name of William Gregor in 1791 who encountered the mineral ilmenite in sands near a creek bed via use of a magnet (Emsley 2001). It is likely that William was the first to perform any "modern" processing of HMSs and actually recover a discrete component in that group, something that would not be done in earnest until nearly 200 years later; once other aspects of human technology evolved to warrant harvesting ilmenite.

WAR AND CONSUMPTION OF HEAVY MINERALS

To elaborate on the point of war, as this is not one readily covered in other sections, the first major forays into HMS mining and processing can be attributed to the demands of World War II. The industry was born at a time when the market was eager to pay top dollar for the metals in mineral sands used to make radio components and welding rods needed for the manufacturing of all steel warships (Wright et al. 1986). In the 1940s, the impact on US mineral imports that began simply during World War I and grew substantially in the next big war resulted in the government's decision to encourage domestic HMS sources. This encouragement directly precipitated the very first adaptation of two major sand processing unit operations; the Humphreys spiral and Carpenter's electrostatic High Tension Roll (HTR). Therefore, the invention of both machines can be attributed to war's demand for minerals/metals. These two technology improvements alone allowed commercialization of the metals derived from HMS which played important roles in the course of history over the 20th century (and on) with resulting conflicts characterized by weapon's technologies derived of HMS products (Carpenter et al. 1953).

In the ensuing Cold War period, the titanium derived from the HMSs was used in instruments of war such as jet fighters and submarines, while zircon would find use in nuclear reactors constructed in pursuit of atomic weapons central to that war. Similarly, metals derived from HMSs were pivotal to the developments made in the space race, another proxy battleground for the Cold War (Shepherd 1986). Titanium, which had prior to the 20th century only seen very limited use as a pigment in the form of titanium dioxide (pigment through the ages) was bestowed with the distinction of being a "strategic resource" worthy of stockpiling in both the former USSR and USA. More recently, rare earth elements (REEs) have begun receiving similar "strategic" designation and one source of these elements is monazite which occurs to some extent in most all HMS suites. REEs are essential to modern electronics used by our military and society. Recent speculation that the world's supply of REEs could be restricted (China) for any number of reasons including a proxy or actual war, has renewed interest in HMS processing to produce monazite.

While the present and the future may follow a different trend, there is no denying that war strongly increased the demand for HMS primary and derived products, as well as forcing the advancement of associated technologies (Wright et al. 1986).

THE ROLE OF PRICE

While hardly unaffected by the effects of head grades, the price of mineral sands products also played a role in the development of the industry. During the early stages of HMS history, prices were reportedly quite favorable, and indeed spurred eager exploitation of deposits in Australia (Wright et al. 1986). Coupled with high head grades, lax product requirements, and seemingly insatiable demand this represented something of a golden era for the industry until the mid-1950s, after which the most noticeable innovations took place out of sheer necessity to survive when all but the demand reversed. The failure of many operations during this time can be attributed to failure to adapt new techniques. The price of ilmenite per ton on an inflation adjusted basis is shown below in **Figure 1** (U.S. Geological Survey 2012).

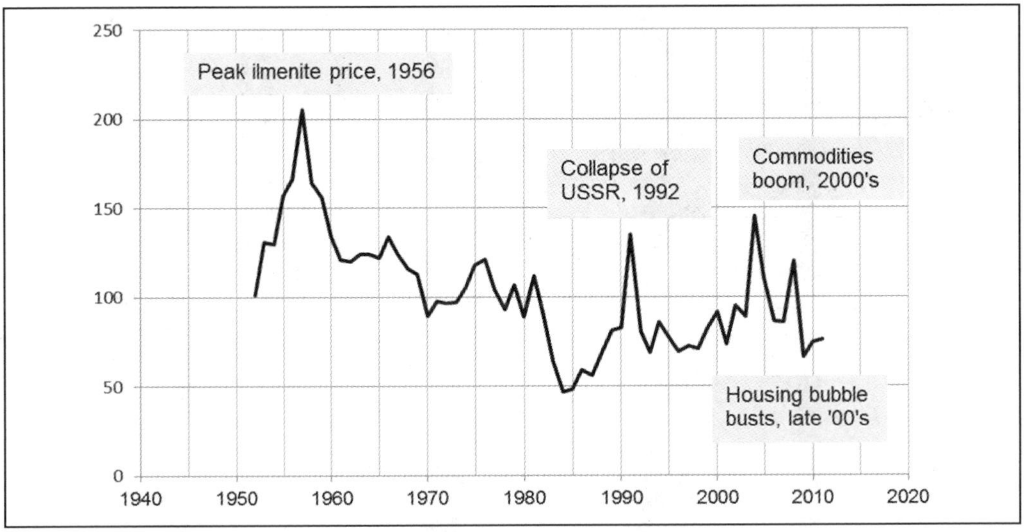

Figure 1. The price of ilmenite over time in constant 1998 US dollars (U.S. Geological Survey 2012)

OVERVIEW (THE TIMELINE)

It is important to realize that the HMS industry has not at all times had the luxury of adopting the most metallurgical efficient processing machines. This is because of the need to process higher throughputs at lower feed grade just to meet production goals. As a result, while more metallurgical efficient machines certainly existed, it could take many, many more of these low throughput machines to match the production rate of a slightly less efficient unit. While stubbornness to use what was familiar indeed played a role in equipment selection, most equipment was selected on basis of ease of operation, head grade, and throughput requirements. As the developments can be difficult to follow, an approximated timeline has been constructed and is shown in Figure 2 detailing the general period between the year 1935 and present, along with parallel timelines of mining and technology methods, targeted minerals and the geo-political climate.

GRAVITY CONCENTRATION

General Trends

The first mineral sands mining operations actually targeted placer gold, tin, or platinum that occasionally can be found in such deposits. During the period between 1870 and 1900, sands were generally mined by hand, with the occasional helpful horse, and processed through sluices, over shaking tables (Wright et al. 1986) and perhaps via amalgam plates. The other heavy minerals occurring in the sand were of little or no commercial value, let alone worthy of separating out specific minerals from the larger heavy mineral suite. In the case of Australia, it wasn't until 1935 that anyone began mining beach sands for specific heavy minerals. During this period, the sands were again mined by hand and horse, and processed via sluices, amalgam plates, and shaking tables, the specific process driven somewhat by the presence of placer gold. The resulting bulk HMS concentrate was

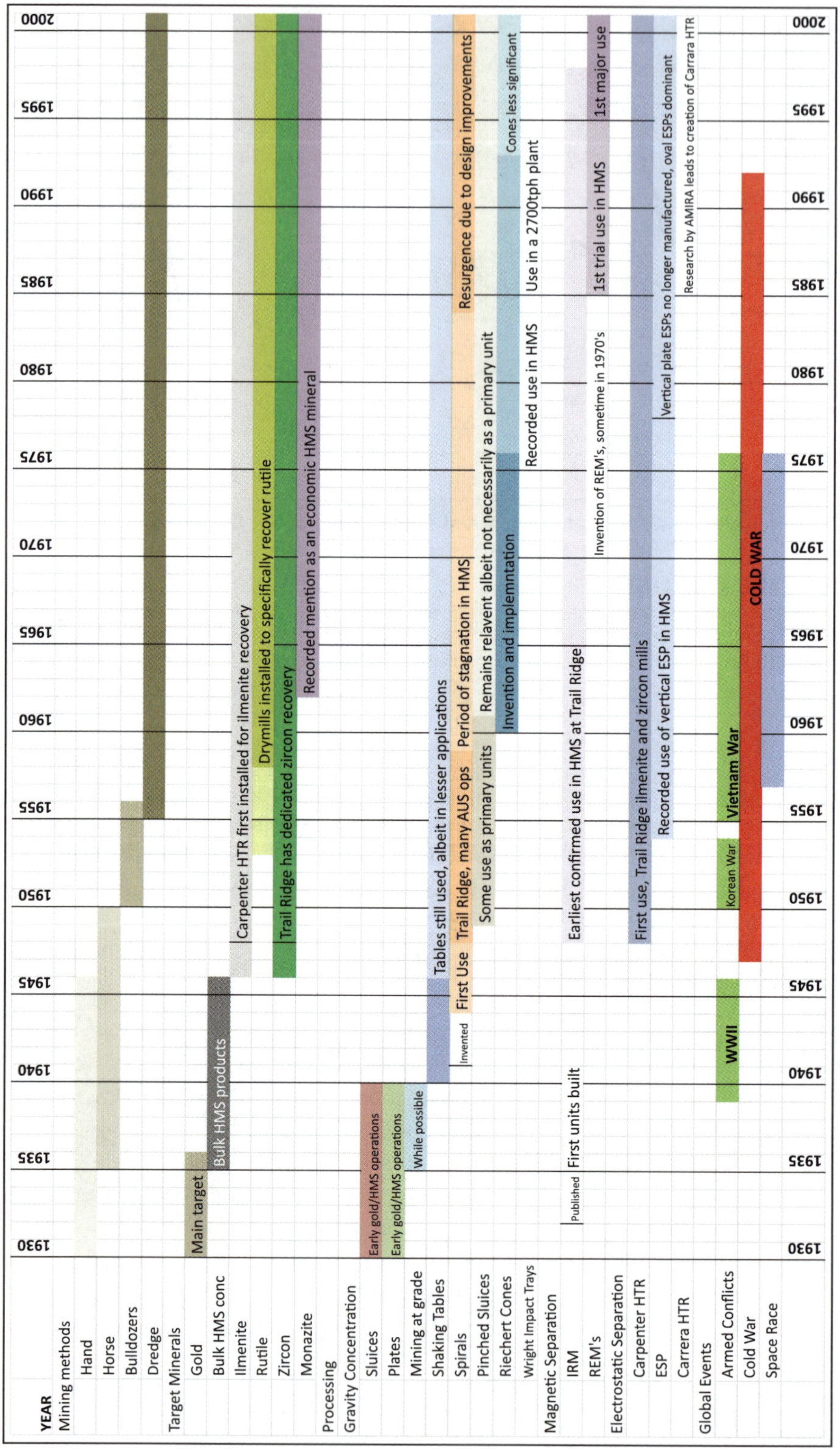

Figure 2. Timeline showing world events and separation equipment usage

Figure 3. Trail Ridge Florida, spiral array (Carpenter et al. 1953)

bagged and exported, requiring no further processing beyond drying. In some cases, the high head grades coupled with the relatively lax heavy mineral product grade requirements meant very little processing to separate the individual minerals was needed at all. However, in one notable case, a clean zircon product was produced from the HMS bulk concentrate by subjecting it to hot soap flotation. However, this "easy" approach to HMSs would not last, as the head grades began to fall considerably. More importantly by the mid-1940s the growth of a market eager to buy heavy minerals spurred the development and implementation of higher capacity and more selective processes.

In general, and in terms of primary HMS processing, the industry followed a pattern of depleting head grades driving higher "low cost" production rates. The first "new" high capacity method adopted by the industry was the Humphreys spiral at an operation in Florida in 1944 (Carpenter et al. 1953), pictured in Figure 3. This device held as a mainstay until the mid-1950s with the adaptation of pinched sluices in various forms (Wright et al. 1986). Note the complex feed distribution and single-start spiral design (Carpenter et al. 1953).

In the early 1960s, Ernst Reichert in Australia, made a pivotal observation that significantly changed HMS processing for years to come. "[Reichert] concluded that one major defect with pinched sluices was the side wall effect of the sluice" (Burt and Mills 1984). This lead to development of the Reichert Cone, which when combined with the need to achieve high throughputs, became dominant over spirals. For reference, unit throughputs on the order of 100 to as much as 350 tonnes per hour (t/h) were possible with a Reichert Cone compared to perhaps 0.5 t/h on a single start Humphreys spiral. The comparative ease of use coupled with the high throughputs lead to more effort being placed on development of Cones over spirals, which cemented their dominance until early to mid-1980s. By this time, spiral development had caught up with Cones, and in certain situations could metallurgically outperform the Reichert Cone.

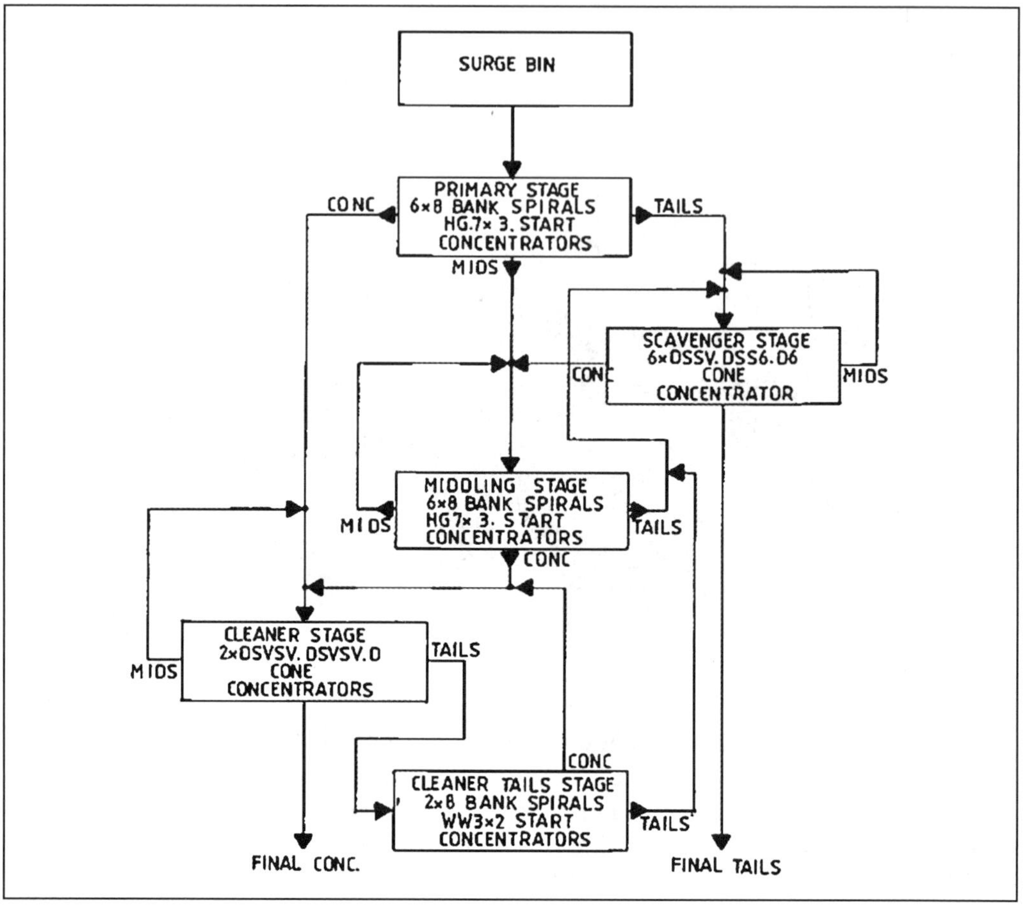

Figure 4. Flowsheet 1984 Western Sands Ltd. expansion of the gravity concentration plant immediately following the mining and screening facility. This flowsheet contains both spirals and Reichert cone concentrators operating in a complementary manner (Wright et al. 1986).

Eventually, the selection of spirals or Cone technology became dependent on the nature of occurrence (mineralogy) of the ore processed, and individual plant site-specific considerations with each method having their own accepted merits (Wright et al. 1986). An example of a flowsheet for a gravity concentration plant is shown in Figure 4. This being said, spiral concentrators are now dominant over Cones at present in virtually all new plant engineering designs. In addition, other particle classification and gravity concentration machines, such as the Floatex density separator and the Kelsey jig, have been implemented in concert with spirals and Cones (Walklate and Fourie 2006).

It is important to note that the machines selected for discussion within this paper are not inclusive of all the machines that existed or do exist, as some were simply made obsolete within such short time periods that they saw little adaptation by the HMS industry, or did not represent significant shifts in technology.

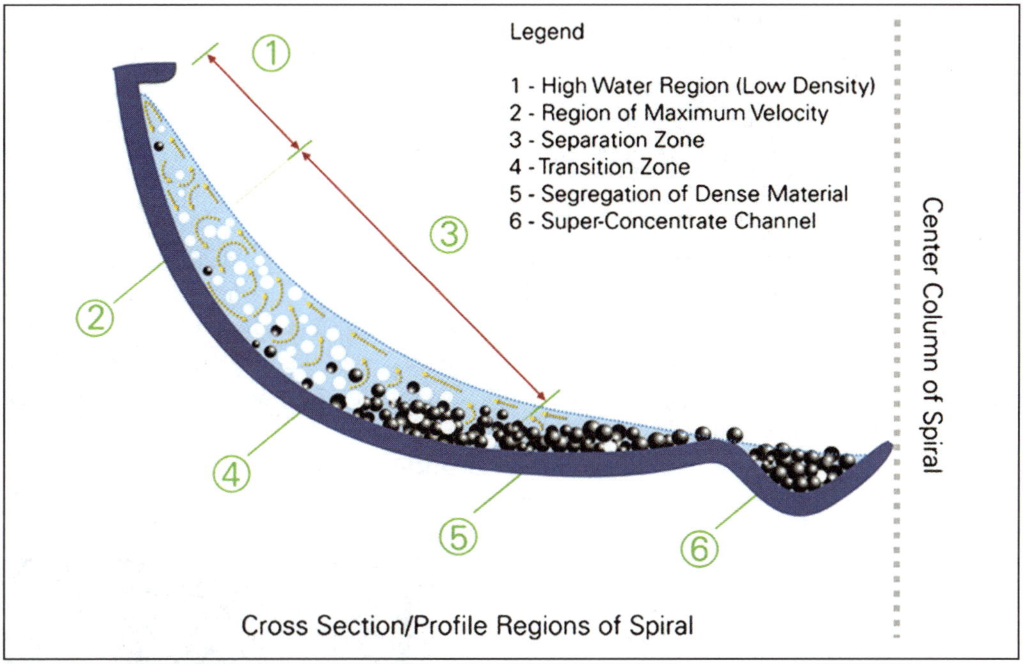

Figure 5. Cross section of spiral concentrator with separating regions (Courtesy of Outotec)

Spiral Concentrator

The spiral separator is a flowing film device configured as a multi-turn smooth bottom helical sluice of modified semi-circular cross section. The generally accepted operating principal of a spiral is shown in Figure 5.

Spirals were the first significant HMS processing device adopted for their high ratio of concentration conducted at significant production rates over the 1940s status-quo systems; albeit at levels that would be dwarfed by newer spiral versions (designs) forty years later. The drive to the "new" spiral concentrator can in part be attributed to depletion of HMS deposits with head grades so high that in essence no processing beyond harvesting was needed to produce a market-ready concentrate. The next-richest deposits required some processing, and the spiral proved to be an inexpensive, easy to operate, and easily installed solution. Additionally, a spiral could be readily improvised on a mining site with worn out truck tires.

Of note, the adoption of spirals was out of necessity to achieve production goals in the light of reducing head grades rather than an optimization consideration (Reaveley and Ritchie 1986; Wright et al. 1986).

While early models indeed worked, they were prone to issues. Very early machines evolved from truck tires to concrete versions and eventually to cast iron construction. Although the cast iron versions were friendlier than concrete spirals, they were still heavy and unwieldy with respect to evolving mobile and dredge mounted equipment engineering. These early spirals, primitive by today's standards and far from optimized, were sometimes surprisingly well performing, i.e., the Wyong spiral.

Another inherent problematic issue with spiral concentrators is that they cannot be miniaturized for design research, i.e., to this day the technology necessitates the construction of full scale prototypes for testing (Reaveley and Ritchie 1986). Research continues at present with modest success to create applicable mathematical models that, ideally, can be used to tailor spirals for individual applications (Jancar et all. 1995). Furthermore, months of effort to construct prototypes can be made worthless within the first five minutes of testing upon the discovery that it won't work at all (Reaveley and Ritchie 1986).

Despite these difficult developmental issues, design improvements did in fact occur which enabled spirals to be re-adopted by the HMS industry and to eventually supersede the apparent dominance of the Reichert Cone (discussed later). This reemergence can be attributed to multiple developments, such as the creation of wash-waterless spirals, which addressed the often problematic clogging issues experienced in wash waster delivery systems in the older models, as well as reducing overall plant water consumption.

Another important development was the advancement of spiral construction materials (the leap from cast iron to fiberglass and polyurethane) and manufacturing techniques such that spirals could have two, three or even four starts per column. This product engineering significantly improved unit capacity over the old, single start models by virtue of geometry, thus becoming more competitive with the inherently high-capacity Cones in terms of processed tonnes per area of floor space. Just as importantly, the metallurgical efficiency of spirals was improved enough that a plant would not have to sacrifice performance by switching away from Cones. This was partly achieved with an increase in spiral length, with five turn designs evolving in certain applications to seven turn designs leading to increased residence time and improved metallurgical performance. The combination of these factors allowed spiral concentrators to muscle their way back into the industry (Reaveley and Ritchie 1986). Considering spirals had to be capable of competing against Reichert Cones with unit capacities of up to 350 t/h, this was no small feat (Wright et al. 1986). Presently, spirals are the dominant gravity processing device for heavy mineral sands, essentially eliminating the use of Cone concentrators in the mid-1990s (Erik Spiller, personal observation). A photograph of modern seven-turn two and three start fiberglass-polyurethane spiral concentrators is presented in Figure 6.

Figure 6. Modern seven-turn spirals shown here in triple start (foreground) and double start (behind) configurations (Courtesy of Outotec)

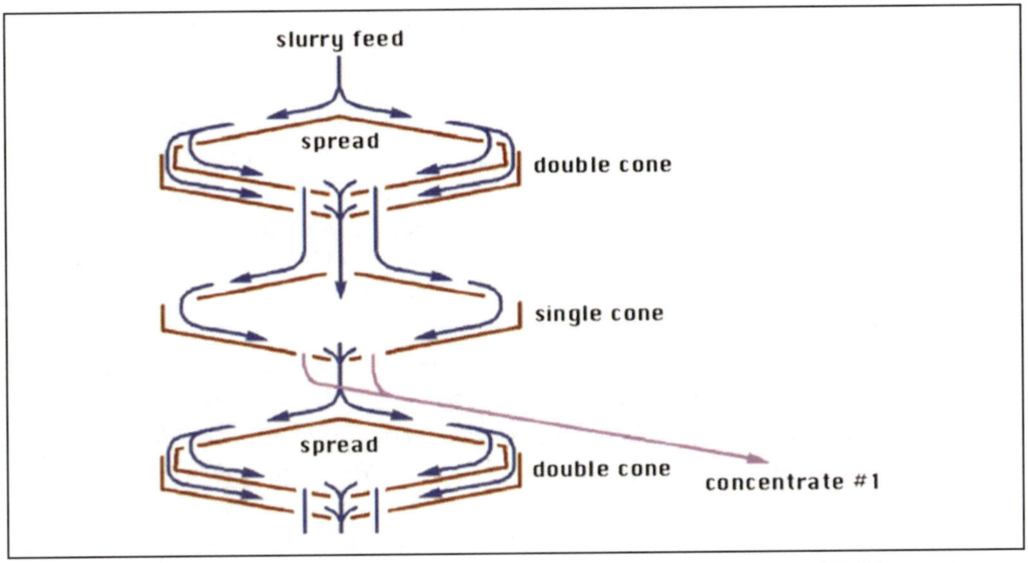

Figure 7. Internal flow and separation mechanism of a Reichert Cone Concentrator (Walklate and Fourie 2006)

Reichert Cones (an Evolution of the Pinched Sluice)

As previously noted, in the early 1960s, Ernst Reichert in Australia, made a pivotal observation that significantly changed HMS processing for years to come. "[Reichert] concluded that one major defect with pinched sluices was the side wall effect of the sluice" (Burt and Mills 1984). His substituted design was relatively efficient, extremely high capacity, and compact. The unit had no moving parts, a long life (little to no maintenance required), a small footprint, and was light enough for dredge mounting. Operating costs were very low; the only requirement was to pump the feed slurry to the top of the rougher Cones and perhaps some interstage pumping. It was reported that operating costs on the order of a few cents, perhaps \$0.12 to \$0.18 (US dollars) per tonne (Spiller, personal experience) through at least Rougher-Cleaner-Scavenger stage processing, could be expected. It was indeed an innovation that occurred as the result of economic pressure for higher processing rates and hence lower cost; both cultivated by genius.

The Reichert Cone is a flowing film devise wherein the feed pulp is evenly distributed over a 2-meter (later a 3.5-meter) diameter Cone to the periphery where it falls onto the upper surfaces of two parallel inverted Cones. The internal flow configuration of a typical Cone separating mechanism is shown in Figure 7. As the pulp flows towards the apex of the inverted Cones, the film/pulp becomes deeper and deeper as the Cone circumference becomes smaller and smaller; the heavies are preferentially flowing at the bottom of the pulp while the lighter particles report to the upper layer. The lower strata of the flowing film encounters an annular slot and since the heavies are nearer the bottom, they tend to pass through the slot while the lights pass over the slot. It is a relatively inefficient separation but the efficiency is improved by repeated processing through a stack of Cones as depicted in the Figure 7.

The domination of Cones over spirals during the period between 1960 and early 1980s can also be attributed the comparatively more rapid development of the technology (Wright et al. 1986; Walklate and Fourie 2006) compared with spirals. Reichert Cones were also growing in size, with diameters of up to 3.5 m as late as the mid-1980s, as opposed to earlier 2-m models. This enabled a potential growth in throughput up to 350 t/h per unit (Burt and Mills 1984). However, by this time, spiral technology was becoming competitive once again, thus the large diameter Cones saw less application then they otherwise may have. The last recorded sale of a Reichert Cone occurred in 1996; however this was only used as a gold scavenger, effectively marking the end of the Cone era.

Floatex Density Separator and Similar Units

The density separator, i.e., Floatex, is perhaps the best indication of the future of the HMS gravity separation philosophy. Driven by a desire to produce higher grade concentrates at lower cost, the density separating units are finding places within flowsheets treating difficult ores. While individually the separators provide adequate performance without significant operational problems, they have been shown by several sources to be best applied in concert with spirals as they exhibit complementary behavior (Elder et al. 2001). Additionally, when used in combination with spirals, they can produce significant Capital Expenditure (CAPEX) savings over spiral-only circuits. In some cases, it is possible to simultaneously improve both grade and recovery through such combinations (Elder et al. 1999).

Kelsey Jig (Gravity Separation with the Benefit of Centrifugal Forces)

The Kelsey Jig is accepted as a very efficient gravity concentration devise, i.e., demonstrating very precise metallurgical separation of minerals having relatively close specific gravities sometimes complicated by particle size. They have been applied at the cleaner stage on HMS wherein their centrifugal forces (enhanced) gravity separation design provides value. Their innovation driver was to make marketable mineral sands products from difficult ores. However, the complicated mechanical system and high maintenance costs coupled with a relatively low capacity has made them a niche machine. With a multitude of mechanical parts, the operational issues are such that they are often abandoned outright (Walklate and Fourie 2006).

MAGNETIC SEPARATION

Low Intensity Magnetic Separation

Low intensity magnetic separation (LIMS) was the vehicle by which the discovery of ilmenite was made; due to amateur geologist in the 1791 using a magnet on some sand in England. As such, this interestingly makes it the earliest recorded use of magnetic separation for HMS processing, although not on an industrial scale (Emsley 2001).

Wet High Intensity Magnetic Separators (WHIMS)

WHIMS has and still is used in a primary capacity in many HMS operations downstream of the gravity concentration plant as a means of recovering the ilmenite early on. This is not always the case, as alternatives including the use of electrostatic or other magnetic separation techniques may prove more suitable dependent on the HMS feed composition and throughput requirements.

Most developments of WHIMS occurred outside the field of HMS processing, with the resulting devices being adopted by the HMS industry when a new plant was created. However, in one noteworthy exception, collaboration between Reading and the developing Richard Bay HMS operation in the 1970s directly resulted in the creation of a 16-pole WHIMS device, which represented a significant increase in unit throughput from 5 t/h to 35 t/h per machine as well as a corresponding improvement in ilmenite extraction. The decision to implement WHIMS in the circuit was necessitated by both metallurgical and throughput requirements, which respectively ruled out the use of cross-belt separators and induced roll magnetic separators (Blendulf et al. 1986).

Induced Roll Magnetic (IRM) Separators

Induced roll magnetic separation (IRM) was first conceived in 1926 by Samuel Frantz, who began manufacturing the devices in 1935. These devices make use of electromagnets, and thus have significant energy consumption associated with their use (Invention of IRMS, 2012).

It is unclear exactly when IRMs were first adopted by the HMS industry, however they were used as early as 1948 in Florida's Trail Ridge HMS operation for the recovery of ilmenite from HMS concentrates, and are shown in Figure 7 alongside the original Carpenter high tension roll electrostatic separators installed in the Trail Ridge ilmenite dry mill (Carpenter et al. 1953).

Rare-Earth Magnetic (REM) Separators

Prior to developments in the field of material science, there did not appear to be any viable permanent magnet material that could also produce the comparatively high magnetic fields required for HMS processing. However, in the 1970s, such a magnet became a possibility with the creation of rare earth magnets. While it would still take longer to see applications in the HMS industry, developments were made during this period which would result in some prior optimization, such as the discovery in 1981, which demonstrated the ratio of magnet to steel thickness that would produce the highest possible steel magnetization (Arvidson and Rademeyer 1997).

A key distinction between IRM and REM roll separators, aside from the change in magnet type, was the former's magnetic air gap, and consequently a sensitivity to particle size. This air gap had to be sized for the intended particle size, however the laws of magnetics dictate that the strength of a magnetic field decreases rapidly with distance. As such, only smaller magnetic fields could be used for processing larger particles with an IRM separator. The REM roll separator has no such air gap, and thus is better suited for coarser particle sizes than the IRM separator (Arvidson and Rademeyer 1997).

As a result of this development, in 1985 the technology surrounding REM roll separators had advanced significantly enough that it saw attempted applications in HMS operations in South Africa and Australia. To say the application was successful would depend largely on one's definition of the word, as while the performance was lack-luster, the information gathered from the "successful" trial indicated that improvements were still needed. Such improvements included the interconnected factors of higher roll speeds, thicker belts, and their resulting impacts of higher feed rates and longer belt life. While REM roll separators and REM drum separators saw small scale applications in the HMS industry, generally in the form of retro-fitting, it would take ten years until, in 1995, the HMS industry would adopt an REM separator in any primary capacity. This supremacy of the REM separators was cemented in 1997 when an REM roll separator outright replaced an existing cross-belt in an Australian HMS operation (Arvidson and Rademeyer 1997).

Another noteworthy implication of REM roll separators is that tighter separation leads to significantly smaller, and in some cases the creation of virtually no middlings when compared to IRM separators for the same feed material. Similarly, it is also possible to deliberately create a discrete middling stream, should one have both the reason and suitable feed with which to do so. The greater implication of the former case is the simplification of flowsheets and possibly the outright elimination of processing equipment, which can have significant economic effects (Arvidson 2001). A photograph of a dry rare-earth roll magnet in operation is shown in Figure 8.

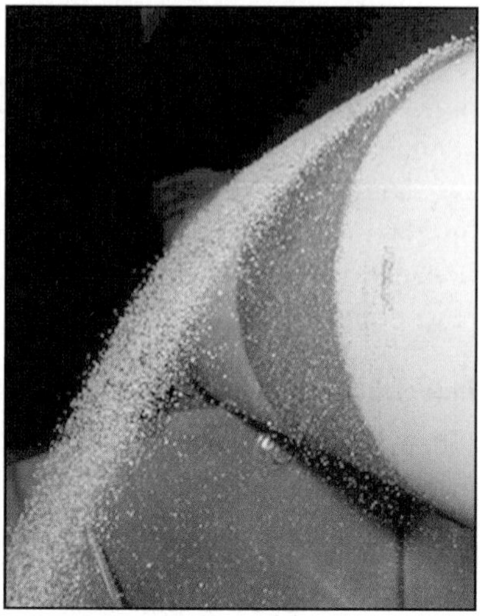

Figure 8. A rare-earth roll magnet in operation (Courtesy of Outotec)

In summary, REM separators were capable of offering noticeable and dramatic improvements, or at the least performing comparably, over older models such as IRM separators or cross-belt magnets in virtually all aspects such as CAPEX, OPEX, product grade, recovery, throughput, versatility, separation efficiency and floor space. These improvements can be further magnified should it lead to improvements in downstream operations such as high tension roll separators.

Issues

Rare earth magnetic devices need to be kept cool, which is especially problematic should the feed material be hot and dry, perhaps as a result of being previously processed in a dryer; as there are cases where dried HMSs have demonstrated better performance than slurries. Should they experience prolonged exposure to temperatures of 100°C or more without receiving supplemental cooling, permanent loss of magnetic field strength may result. On a less fundamental level, manufacturing techniques need to be capable of meeting tighter tolerances, failure to do so could result in notably non-uniform behavior and poorer performance. This problem has slowed the development of larger diameter dry rare-earth drum magnetic separators, which in turn restricts the unit capacity (Arvidson 2001).

A problem specific to HMS processing is that zircon can develop a static charge that results in it misreporting to the magnetic fraction as it can cling to the Kevlar belt. Solutions to this problem include the use of "corona charging equipment ('ionisers')" (Arvidson 2001) or switch the belt material to a thin PVC (Marcos and Gilman 2007).

Possible Future of Magnetic Separation

This field holds significant potential as developments are ongoing. In addition to REMs serving in a capacity as wet high intensity magnetic separator (WHIMS) (Arvidson 2001), there is also the idea of a continuous WHIMS utilizing a spiral magnetic field (Botsch and Schönert 1995). Additionally, small scale units of a new device, known as a low intensity rotating magnetic field (LIRMF) separator, are under development and may have ilmenite applications (Allen 2000). New advances in material science may also change the face of magnetic separation just as significantly as the development of REMs separators.

Figure 9. A photo of the Trail Ridge zircon mill showing the Carpenter HTRs, along with the downstream IRM separators (Carpenter et al. 1953)

ELECTROSTATIC SEPARATION

The HMS industry adopted electrostatic separation as early as 1948 with the implementation of the Carpenter HTRs in the Trail Ridge facility (Carpenter et al. 1953). Since then, the two dominant forms of electrostatic separators used have been a combination of high tension rolls (HTRs) and electrostatic plate separators (ESPSs.) While other forms of electrostatic separation exist or are under development, they have not seen significant use in the HMS industry, barring the case of the Carrara HTR as discussed later on (Germain et al. 2003).

One of the primary problems often encountered with electrostatic separation was the sizable amount of middlings that could be produced. Naturally these middlings would be recirculated to improve recovery, although this sometimes resulted in a much higher amount of recirculation than fresh feed, which had significant implications on the plant throughput (Germain et al. 2003).

High Tension Roll Separator (HTRS)

The first electrostatic roll-type separators were used in the 1920s (Knoll and Taylor 1986); however, it wasn't until 1948 that these separators were first applied to HMS separation by James Carpenter in an operation at Trail Ridge Florida as shown in Figure 9 (Carpenter et al. 1953).

The Australian HMS industry had widely adopted the use of electrostatic separation by 1954, becoming the first Australian mineral industry to do so, primarily "for the separation of rutile from zircon" (Hudson 1954). By this time, electrostatic plate separators were also in use in the HMS industry (Nairn 1964). HTRSs would undergo various improvements, such as optimizing electrode placement; however, the biggest development occurred in the very late 20th century as a result of significant research into the fundamental physics behind electrostatic separation (Knoll and Taylor 1986).

Electrostatic Plate Separators

The electrostatic plate separator (ESPS) saw significant evolution from its 1950s incarnation. This was manifested not only with the replacement of the vertical plate electrode with the more recent oval shape, but also in that by 1974 the flat vertical plate variant seemingly ceased to be manufactured.

The specific use of the ESPS would actually serve as a herald for shifts in the design philosophy later experienced in the gravity concentration area. This is due to the fact that many operations had come to recognize the complementary nature of the HTRS and the ESPS, such that ESPSs were frequently used to clean HTRS zircon product. The acceptance that one unit was better suited than the other for a particular stage in itself represents a growing sophistication in HMS plant design (Dyrenforth 1980).

Carrara HTR

As mentioned in the discussion of HTRs, its biggest development occurred rather late in the 20th century. With the recognition that electrostatic separators had an inherent sensitivity to particle size range, as well as the physics underpinning this shortcoming, a new unit was developed. As has been previously noted, HTRSs and ESPSs had been used in a complementary manner, so what followed with the application of a plate electrode into an HTR's body was not particularly farfetched as much as it was a bold development. Additionally, thanks to a better understanding of the physics, this electrode was covered in an insulating material, thus any conducting particle wouldn't automatically loose its charge by contacting it, as shown in **Figure 10**. The marriage of the two units netted a vastly more efficient process with significantly less sensitivity to the feed size distribution than either one could achieve before. Specifically, the new unit was able to produce a sharper separation, with noticeable reduction in the resulting middling streams that would otherwise require recirculation, resulting in better capacities (Germain et al. 2003).

SUMMARY

The development of heavy mineral sands processing has evolved significantly over the 20th century, and will likely continue to do so well into the 21st in ways we may not even have fully conceived of yet. In the face of fluctuating prices and decreasing head grades experienced throughout the century coupled with more stringent product grade demands and environmental regulations, the HMS processing industry has developed some of the most cost efficient operations out of necessity to survive. In the realm of gravity separation, it has become clear that one size does not fit all, and that various methods may work better in concert than in blanket application. The most readily adapted gravity separation methods appear to have been those that offer improvements in simplicity of operation (provided all other aspects are comparable.) In the realm of both electrostatic and magnetic separation, development has largely been guided by efforts to gain a better understanding of the fundamentals involved and apply those results to the creation of new devices.

REFERENCES

Allen, N.R., 2000. Low intensity rotating magnetic field separation. In *Proc. Int. Con. Min. Proc. Extr. Metal. MINPREX 2000,* p. 303.

Arvidson, B.R., 2001. The many uses of rare-earth magnetic separators for heavy mineral processing. *International Heavy Minerals Conference 2001*, p. 131–136.

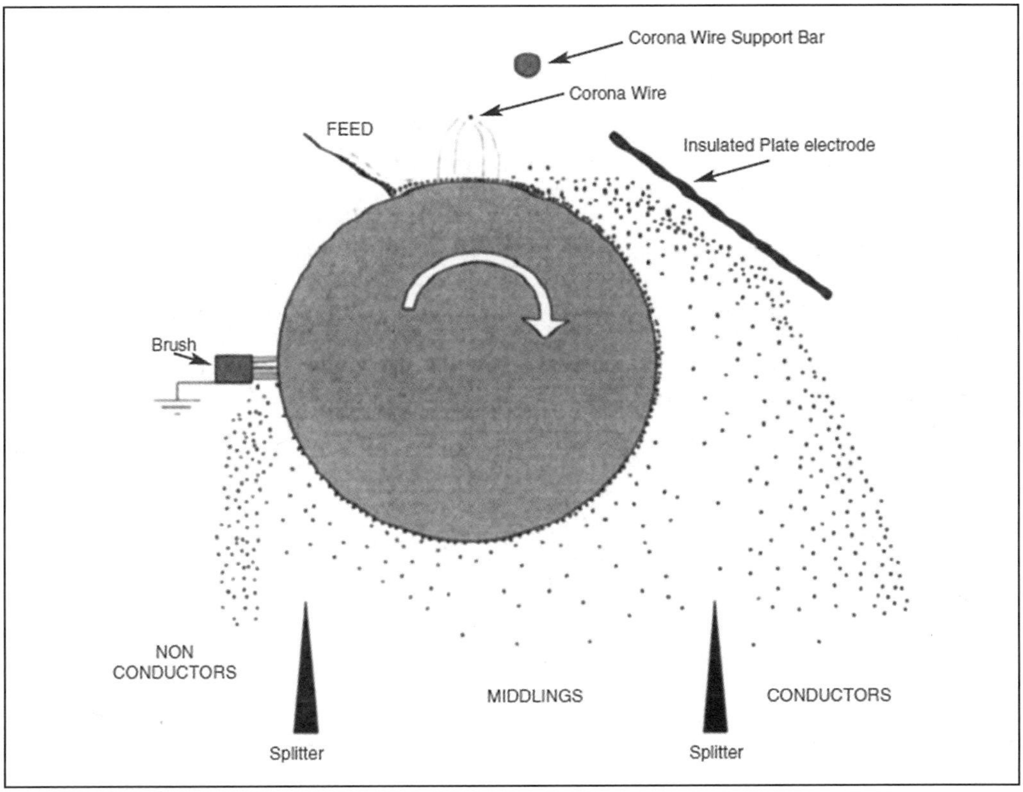

Figure 10. A simplified schematic of a Carrara HTR separator (Germain et al. 2003)

Arvidson, B.R., and Rademeyer L., 1997. Rare-earth magnetic separators for mineral sands applications *Heavy Minerals 1997,* South African Institute of Mining and Metallurgy. p. 129.

Blendulf, G., Dawson, N., and James, D., 1986. Utilisation of Readings wet high intensity magnetic separators at Richards Bay Minerals. *Proc.: Australia: A Source of Ilmenite, Rutile, Monazite and Zircon Conference,* AusIMM, p. 81.

Botsch, M. and Schönert, K., 1995. Continuous high intensity magnetic separation with a rotating spiral, *Proceedings of the XIX Int Min Proc. Congress—Physical & Chemical Processing, 2,* p. 185.

Burt, R.O. and Mills, C., 1984. Gravity concentration technology. *Developments in Mineral Processing,* Elsevier, New York, p. 221–316.

Carpenter, J.H., Detwiler, J.C., Gillson J.L., Weichel, Jr., E.C., and Wood, J.P., 1953. Mining and concentration of ilmenite and associated minerals at Trail Ridge, Fla. *Min. Eng.,* 5, 8, p. 789–795.

Dyrenforth, W.P., 1980. Electrostatic separation, *Minerals Processing Plant Design,* 2nd ed. (ed. A.L. Mular and R.B. Bhappu), SME-AIME. p. 479–489.

Elder, J., Venkatraman, P. and Hearn, S., 1999. Application of Floatex density separator for the heavy mineral sands industry. *Heavy Minerals 1999,* SAIMM, p. 113.

Elder, J., Kow, W., Domenico, J. and Wyatt, D., 2001. Gravity concentration—a better way (or how to produce heavy mineral concentrate and not recirculating loads). *International Heavy Minerals Conference 2001.* p. 115.

Emsley, J., 2001. Nature's building blocks: an A–Z guide to the elements, Oxford University Press, p. 451–452.

Germain, M., Lawson, T., Henderson, D.K. and MacHunter, D.M., 2003. The application of new design concepts in high tension electrostatic separation to the processing of mineral sands concentrates. *Heavy Minerals 2003,* SAIMM, p. 101.

Hudson, S.B., 1954. Recent investigations in Electrostatic Separation. *Proc. AusIMM.* p. 149–161.
Invention of IRMS, 2012, http://www.sgfrantz.com/sgfbio.htm.
Jancar, T., Fletcher, C.K., Holtham, P.N. and Reizes, J.A., 1995. Computational and experimental investigation of spiral separator hydrodynamics, *Proc XIX Int Min Proc Congress—Physical & Chemical Processing. 2.* p. 147.
Knoll, F.S. and Taylor, J.B., 1986. Selection and sizing of electrostatic concentrating equipment, *Design and Installation of Concentration and Dewatering Circuits* (ed. M.A. Anderson), SME, p 208–225.
Marcos, P.S., and Gilman, S.K., 2007. "Old tricks for new dogs," areas for focus in mineral sands processing, *6th International Heavy Minerals Conference 'Back to Basics,'* SAIMM, p. 55.
Meijer, R.J., 1998. Heavy minerals: from "Edelstein" to Einstein. *J. Geochemical Exploration*, 62, p. 81–103.
Nairn, P.R., 1964. Comparison of ore dressing methods used by ilmenite producers in Western Australia. Perth, *Proc. AusIMM,* Paper No. 4.
Reaveley, B.J. and Ritchie, I.C., 1986. The development of high efficiency spiral separators. *Proceedings Australia: A Source of Ilmenite, Rutile, Monazite and Zircon Conference*, AusIMM, p. 87.
Shepherd, M.S., 1986. Australian heavy mineral reserves and world trends, *Proceedings: Australia: A Source of Ilmenite, Rutile, Monazite and Zircon Conference*, AusIMM, p. 61.
Titanium Dioxide Whites—WebExhibits. 2012, Pigments through the ages http://www.webexhibits.org/pigments/indiv/history/tiwhite.html.
U.S. Geological Survey, 2012, Titanium Mineral Concentrate statistics.
Walklate, J.R. and Fourie P.J., 2006. A history of gravity separation at Richards Bay Minerals, *SAIMM,* 106, p. 741.
Wright, D., Richards, R.G. and Cross, M.S., 1986. The development of mineral sands separation technology—solving today's problems, anticipating tomorrow's. *Proceedings: Australia: A Source of Ilmenite, Rutile, Monazite and Zircon Conference.* AusIMM, p. 107.

ABOUT THE AUTHORS

D. Erik Spiller is a metallurgical engineer from the Colorado School of Mines and a Qualified Professional (QP) member of the Mining and Metallurgical Society of America (MMSA) and a registered (QP) member of SME. Spiller has more than 40 years of relevant experience working as a metallurgical engineer in the mineral resource industry. During his career, Spiller has held responsible positions in process research, process development, engineering, and senior management in research institutions, engineering houses, equipment companies, and operating companies. In addition, since 2008 he has served as an appointed research professor at the Colorado School of Mines; this was preceded by 15 years as an adjunct instructor at the same institution. Spiller lectures in mineral beneficiation and directs graduate students conducting metallurgical research in areas of mineral processing.

C. Alex Norgren is a recent metallurgical engineer graduate (2011) from the Colorado School of Mines. He worked for GeoBiotics before joining Tetra Tech, where he has worked on copper, magnetite, gold, molybdenum, potash, and uranium processing. His nearly 2 years at Tetra Tech has introduced him to flowsheet development and metallurgical balance modeling, especially as related to comminution, salt chemistry, and particulate separation.

Innovations in Liquid/Solid Separation for Metallurgical Processing

Mick L. McCaslin, Keith J. Mounteer, and Michael Phillips
FLSmidth, Midvale, Utah, United States

ABSTRACT: This paper summarizes innovations in major liquid/solid separation equipment used for metallurgical processing. Included are developments in thickeners and select pressure and vacuum filters, as well as key appurtenances for each. Process, operation, and maintenance details are minimal.

INTRODUCTION

The authors begin with an unverified assertion: humankind has engaged in liquid/solid separation for nearly as long as Homo erectus has been, well, upright. In fact, their predecessor may have separated liquids from solids by a simple expedient. It takes little imagination to picture a nominally cogent Pliocene era biped lounging about the watering hole, waiting for sediment to settle before he (or she) slakes their thirst; having found that they preferred their water without dirt and patience gave the solids time to drop out of suspension. Whether they knew it or not, they can lay some claim to having invented the thickener. What might they have done with flocculants?

It is conceivable that some benighted paleoanthropologist has produced a paper asserting that early man separated liquids from solids for specific purposes. Considering that our example has little to do with metallurgical processing, the authors did not attempt scholarly support for our hypothesis. However, it is known that Agricola referenced various liquid/solid separation techniques in processing ores and the testing thereof (Agricola 1556). So at the very least, it can be established that liquid/solid separation has been used in metallurgical processing for 450 years. If today's liquid/solid separation student spends minimal time flipping through *De Re Metallica*, they will clearly see that there have been thousands of innovations in the field since Georgius' day.

That observation alone was enough for the authors to reduce scope for the crux of this document to the time period after World War II. Considering that the term "liquid/solid separation" can refer to a bit of filter paper in a lab, screens in water lines, cartridge filters in chemical lines, and the largest thickeners in the world; we deemed it prudent to narrow our focus a bit more. Therefore, this paper addresses innovations in the thickeners, pressure filters, and vacuum filters used regularly in metallurgical processes to dewater millions of tons of concentrate and tailings in mills and concentrators around the globe.

We consider those technologies that have stood the test of time. New liquid/solid separation devices are often touted as panaceas and the failures of these inventions are many. Successes rarely extend beyond niche applications and those are seldom in the metallurgical processing field.

Therefore the topics selected are as follows:

- High Capacity Thickeners
- High Density & Deep Cone Thickeners
- Flocculant Development
- Thickener Feed Dilution
- Thickener Drives
- Horizontal Filter Presses
- Vertical Filter Presses
- Rotary Vacuum Disc Filters
- Horizontal Belt Vacuum Filters
- Filter Media

If we fail to address a process near and dear to the reader's heart or discuss a favorite liquid/solid separation technology, we beg forgiveness. We hope that our attempt to take a huge subject to a manageable level and target the largest possible audience proves sufficient excuse.

If there is an innovation theme that bridges all of the liquid/solid separation technologies listed, it is one common to many mining technologies: "Go big or go home."

HIGH CAPACITY THICKENERS

Thickening is widely applied in mineral processing to increase solids concentration, often prior to a filtration step. Thickeners typically have a feedwell in the center to reduce kinetic energy of incoming feed (which is also a common place to add flocculant) and rakes to ensure steady movement of thickened solids to the outlet.

John Van Nostrand Dorr patented the first thickener with rakes (Dorr 1907). Despite his chemical training, Dorr's patent makes no mention of flocculants or other means of chemically enhancing settling rate, but today it is unlikely that any mill would build a thickener without considering those advantages. Innovations in flocculants drove development of "high rate" or "high capacity" thickeners. Those we address separately, but to further this discussion it is sufficient to say that the magic of chemistry permitted a smaller thickener to process the same tonnage as a conventional thickener. Still, design innovations in the thickeners themselves leveraged the chemical advancements or reduced challenges brought about by them.

Some years back "conventional thickeners" tended to mean thickeners that did not use flocculants (Schoenbrunn and Laros 2002), whereas high capacity thickeners did. This changed after Fredrick Eis filed US patent #3,523,889 (Eis 1968). The key technology referenced in the patent was mixing of incoming feed and flocculant. Broadly written, as some would argue, the disputes over its validity led manufacturers to pursue alternate feedwell designs and other advances to capitalize on the effects of synthetic flocculants. It also contributed to a shift in terminology where today "conventional thickener" implies that the design includes no special means to optimize the performance of flocculants (Schoenbrunn and Laros 2002).

Increased tonnage in a smaller footprint, particularly in processes with elevated specific gravity, means higher loads on drive mechanisms and rakes. We address drives separately, but stronger rakes are required for High Capacity Thickeners than conventional thickeners. The relatively high specific gravity solids present in metallurgical processes also demand more robust rakes than less demanding applications, such as municipal sludge.

Aside from simply building stouter rakes, other key design features deal with specific problems. Innovations include minimizing surface area in high-scale applications such as alumina refining and including spikes on blades to re-suspend heavily thickened materials like magnetite. Streamlined designs reduce torque on both drive and rakes, and posts (termed pickets or Thixoposts) create space between the blades and rake arms for sticky and/or viscous applications (Schoenbrunn and Laros 2002).

Conventional thickeners may be outdated, but it is clear that High Capacity Thickeners are the norm in metallurgical processes where very high underflow densities are not required.

HIGH DENSITY AND DEEP CONE THICKENERS

The British National Coal Board (NCB) built the first Deep Cone Thickeners (DCTs) in the 1960s. A number of collieries installed them for dewatering coal refuse. While failing in consistency, suitable underflow for depositing onto conveyors, they did produce remarkably dense underflows. Denver Equipment Company licensed the technology but had limited commercial success (Schoenbrunn 2007).

Advancements in flocculants, feed dilution, and drives (addressed separately) made High Density and Deep Cone technologies practical. Of course, the basic tank design is also a key ingredient. Both thickener types use tank design to create deeper mud beds with higher mud residence time for dewatering.

Tall sidewalls in High Density Thickeners permit 2 to 3 m (6.6–9.8 ft) bed depths, as compared to 1 m (3.3 ft) bed depths in conventional and High Capacity Thickeners. Floor slopes are in the range of 1:6, which is similar to traditional thickeners. Conversely, Deep Cone Thickeners utilize high sidewalls and steep cone bottoms. The latter are in the range of 30–60°.

Robinsky pioneered the use of this technology for tailings disposal beginning in 1973 (Robinsky 1978). His experience built on the National Coal Board work in Deep Cone Thickeners.

In about 1978, Alcan began applying the high-density concept to alumina red mud in Jamaica by increasing sidewall height, albeit in a rake-less flat-bottomed tank. Underflow densities increased to the range of 25–27% by weight (wt%), which is not notable except when compared to the 15–17 wt% that the plant achieved previously. Subsequently adding rakes led to underflow densities of 30–32 wt% (Schoenbrunn 2007).

Alcan built on Robinsky's work by applying Deep Cone Thickeners to alumina tailings in 1987 (Schoenbrunn 2007).

Outokumpo and EIMCO Process Equipment Company developed competing High Density Thickener designs in the late 1980s. In 1996, EIMCO licensed the Alcan technology, leading to marketing of Deep Cone Thickeners outside of coal and alumina (Schoenbrunn 2011).

Higher underflow densities also drove rake design. Rakes are more robust and have lower profiles to reduced load. Another universal modification is addition of pickets that extend above the rake arms into the mud bed and create channels to aid in water removal (Schoenbrunn 2007).

FLOCCULANT DEVELOPMENT

The science (many plant operators and engineers will say that polymer treatment is an art at best) of conditioning water, wastewater, and slurry streams has been in use since the 1800s. It can improve settling and hence dewatering, as well as prevent scale buildup. Operators used lignins found in common potatoes in steam engine boiler water. The treatment slowed calcium carbonate scale buildup on boiler tubes. Scaling led to tube failures with spectacular destruction of the steam engine, and unfortunately any operators in the vicinity.

Thickener applications often involve handling particles finer than 0.1 microns. These particles remain in suspension primarily due to electrostatic charge, often negative, causing them to repel each other and hinder gravity settling. References cite early 20th century use of starch and guar as coagulants to neutralize these charges allowing the fine particles to collide and stick together (Gardner and Ray 1939; Dahlstrom 1989; Schoenbrunn 2011). The resulting heavier agglomerates, referred to as floc particles, will readily settle. If allowed to settle in a quiescent liquid the floc particles will remain together, but if subjected to excessive agitation, they break apart. Breaking the floc particles reduces overall mass, which negatively affects subsidence velocity, meaning longer settling time and larger basins.

Designers must answer numerous questions. "Should polymer be fed diluted and, if so, how dilute?" "Is the proper point of application upstream or downstream of the feed pump?"

Economical application of flocculants requires constant attention. Any change in type of feed solids, solids concentration, particle size distribution, etc. will require adjustments in flocculant rates, concentration, feed points, and possibly the type used. A reliable supplier of these materials is invaluable in controlling costs and ensuring performance. Buying on price and frequently putting the polymer contract out for bid invites short-term gains and long-term performance issues. Poor application often eliminates front-end savings very quickly.

Given the complexities outlined in the preceding paragraphs, the argument that polymer conditioning is little more than an art has merit. Chemical companies continue to advance the "science," occasionally develop new compounds in solving a specific challenge, and produce new flocculants with some consistency.

Organic flocculants were the norm until advances in the 1950s led to synthetic flocculants (Ballantine et al. 2011; Linke and Booth 1959). Due to their efficacy, synthetics have largely supplanted organics (Schoenbrunn 2011). Development of, and advances in, synthetic flocculants are the most significant innovations in the last 50 years.

THICKENER FEED DILUTION

A majority of minerals processing applications require some type of thickening. Historically the goal was to introduce feed streams to the thickener at the highest solids concentration possible. Operators believed that higher feed concentration yielded higher underflow concentration, but the practice delivered marginal outcomes. They then used organic flocculants to increase the underflow concentrations, but with mixed results. Plant operation is dependent on thickener performance and when thickeners did not produce the required underflow concentration; operators tried increasing flocculant dosage. They added more reagents to the same injection points and at additional locations. Sometimes the only positive result was to increase the income of the flocculant provider.

Consumption of flocculants could be one of a plant's largest operating costs. Operators and plant designers pressed thickener manufacturers to improve operation and reduce flocculant usage.

Equipment manufacturers who could guarantee thickener underflow at reduced treatment costs would have a significant competitive advantage. Thickener feed dilution was advanced in pursuit of that end, actually diluting the feed-stream instead of maximizing solids concentration. Supernatant from the thickener or plant water is suitable for the purpose.

Dilution has resulted in numerous advantages and increased application of flocculant. More dilute feed results in more efficient contact between the solids particles and the flocculant stream and better distribution of feed solids. Significant reductions in mixing intensity are possible.

Concentration of solids in the settling zone is lower, resulting in decreased interference between adjacent solids particles. Settling velocity is not decreased which eventually results in a greater underflow concentration without an increase in unit size.

Finally unit sizes are reduce due to improved thickening and higher underflow concentrations. A smaller thickener can now yield underflow concentrations previously requiring much larger units.

The first dilution design used windows in the feedwell walls in the 1950s. The density differential between the feed stream and the clarified liquor drew supernatant into the feedwell (Schoenbrunn 2011). Today the method is still used.

Different pumped arrangements have evolved in which pumps return liquor from the overflow launders or from the upper portion of the thickener itself. While easily controlled, the major drawback to the pumping approach is the cost, maintenance, and complexity of additional pumps in the system. This is particularly true when dilution water flow is high.

Another method of dilution, patented by EIMCO Process Equipment Company (Wood et al. 1994), is to use an eductor in the feed line. Feed flowing through the eductor draws in diluting supernatant from the upper reaches of the thickener. Flocculant addition and thorough mixing occurs before the feed reaches the feedwell. Proper eductor design considers the amount of dilution required, but operators adjust with feed velocity.

Thickener manufacturers seek new ways to control solids content of the incoming feed, balance that against flocculant addition, consider mixing schemes, look to avoid disrupting the floc structures and the already settled particles, and minimize the cost of the equipment. Meanwhile their attorneys guard against possible patent infringement from the scientists' latest ideas. The search continues.

THICKENER DRIVES

The single most significant advancement in thickener drives came in the early 1970s with the adoption of large diameter precision bearings for the main gear. Previously large thickener drives used strip liner bearings. Hardened steel rings (strips) mounted in the bearing housing by the drive manufacturer formed the bearing races, with bearing balls placed between the "strips." Demanding metallurgical processes seldom use strip liner bearings, but because of their low cost, that is not the case in all applications.

Precisions bearings are constructed to tighter tolerances, take much higher thrust and overturning loads, do not permit ball-to-ball contact, and can tolerate higher radial load. All of these features facilitated larger thickeners in heavy metallurgical slurries.

For many years, the largest main gears used in thickeners were 3 m (120 in.) in diameter. Early incarnations were segmented gears, whereas ring gears from a single forging came about in the 1980s. This greatly improved concentricity and permitted higher loads. Currently, forged ring gears as large as 3.56 m (140 in.) are available.

Used widely and for years, multiple pinions increase torque applied to the main gear and hence the rakes. To ensure one pinion did not lead the others and take a larger percentage of the total load, electric drives typically used high-slip motors. Another approach was to use hydraulic drives, which balance automatically.

Advances in variable frequency drives, and ability to link them, eliminated any concerns about imbalance, but more importantly brought about better control. They can reverse rakes easily and automatically speed up as torque increases.

We can expect future innovations in drive size, advanced control, torque monitoring, and reliability.

HORIZONTAL PRESSURE FILTERS

Traditional horizontal filter presses (plates arrayed along a horizontal axis) have been around since the middle part of the 1800s (Margueritte 1874) or earlier. Their greatest strength is that they can dewater any filterable slurry.

Because filter cloth is sandwiched between two plates with recessed chambers formed into them, the greatest weakness of the traditional filter presses is that production must stop while operators open the press to permit filter cake discharge. That is time-consuming and labor intensive, particularly since discharge of the cake means cleaning the sealing surfaces before closing the press to put it back in service.

Beginning in the 1950s and 60s, filters became more automated, material advancements permitted handling high acid or base solutions, and other innovations led to less operator involvement.

Introduction of plastic plates, typically polypropylene, also occurred in the 1960s. These were lighter and less expensive, but still handled the necessary pressures. They could filter aggressive streams and maintain a satisfactory service life. Development of membrane plates helped produce filter cakes in difficult applications.

Ironically, the horizontal configuration failed to capture much of the metallurgical processing market. The vertical presses (addressed separately) dominated largely due to mechanical configuration advantages, despite the fact that their history is less than half that of the more traditional filter presses.

Automation bridged the gap beginning in the 1990s. Programmable logic controllers (PLCs) led to methodical and consistent performance of a sequence of operations. These include washing sealing surfaces between cycles without operator involvement and controlling cloth-shaking systems to assist cake discharge, which also reduces recirculating load to the process. Control schemes extended to feed pumps to produce uniform cakes.

While the horizontal filters have not caught up in the number of installed units on concentrates and tailings, they have caught up in process performance. These machines successfully challenged the once-dominant vertical presses. Operators and designers consider them a better value for many applications.

Current trends are for larger filters that require less floor space per area of filtration. Automation developments will improve cloth life and reduce downtime. Cost per ton is the performance basis for comparing filters. Production is king and total life-cycle cost drives the development of flow sheets. Reliability and availability are key parameters to plants wishing to maximize their throughput.

VERTICAL PRESSURE FILTERS

A vertical configuration (plates arrayed along a vertical axis) was among the many innovations to minimize the time and labor involved in operating filter-presses. A 1959 US patent by Gene Hirs (Hirs 1955) first documented the configuration. Several variants followed (Davis 1971; Ostrop

1974), but the first widely successful manufacturer in metallurgical processes was Larox. In the mid- to late 1970s, Larox installed these units in copper, lead, nickel, platinum, palladium, and zinc.

Like horizontal presses, recessed plates close to form a chamber. Similarly, filter media between the plates captures solids during the filtration cycle. However, when the filtration cycle is complete, the press opens automatically and the filter cloth advances over rolls to discharge cake. The cloth advance also permits automatic washing both sides of the filter media, prior to the press closing and continuing operation.

In the typical vertical filter press design, slurry enters each chamber. Using compressed air, an elastomer diaphragm attached to the top of each chamber expands to press against the slurry, forcing liquid through the filter cloth, which then exits at ports in the upper and lower plates. Solids of course remain behind and form the filter cake. At the conclusion of the filtration cycle the chamber opens, the cloth advances and transports the filter cake out of each chamber to discharge. Cloth washing occurs simultaneously. Because the machines use one continuous filter cloth, the filtration areas effectively advance one chamber and the press closes to repeat the cycle.

A variant patented in 1985 in Japan (Yagishita et al. 1985) introduced slurry at the top of the chamber. Feed stopped before the chamber was completely filled and compressed gas forced liquid through the filter media and out ports in the bottom plate only.

Made commercially viable by Pneumapress in the 1990s; the current configuration uses separate filter cloths for each chamber, permitting isolation of one chamber in the event of cloth failure. It also has the advantage of discharging all filter cake on one side of the machine.

Vertical filter presses generally have the highest capital cost per unit area of filtration. However, when looking at total life-cycle cost, they are often second only to automated horizontal filter presses and can be a better option with filter cakes that are difficult to discharge.

ROTARY VACUUM DISC FILTERS

One of the most frequently used vacuum filters in metallurgical processes is the rotary vacuum disc filter, originally patented in 1918 by Orange James Salisbury (Salisbury 1918). The filtering component is a row of discs on a horizontal shaft comprised of individual segments or sectors. Sectors make up each disc with drainage grids to allow passage of the filtrate. Filter media covers the grids on each sector and captures solids. The center shaft is hollow and rotates discs into and out of a tank containing slurry.

The rotary vacuum disc filter's great advantages are continuous filtration, high solids throughput, and low cost and footprint per filtration area. Advances in the last 50–60 years have largely been in terms of size. Disc diameters as large as 6 m (19.7 ft) have been operated successfully, with total filtration areas approaching 200 m^2 (2,153 ft^2). Filters operate with up to 20 discs, albeit in diameters smaller than 6 m.

A recent innovation is simplified bayonet mounting of disc sectors. Rim retaining devices that permit quick and easy removal/installation are also a recent advancement.

Adequate suspension of heavy mineral slurries has been one of the major problems in operation of disc-type vacuum filters. A common method involved a horizontal rotating shaft with paddles located between each of the filter discs. The method was quite successful but required frequent shut down for maintenance.

A recent design feeds slurry through nozzles located between each filter disc. The nozzle injects new slurry into the mass already in the vat, dissipating energy of the incoming slurry and maintaining heavy particles in suspension.

Rotary vacuum disc filters were widely used for concentrate filtration through the 1960s and 1970s. They were largely supplanted by pressure filters in the 1980s and 1990s, but are now being used in tailings filtration where final cake moisture is less of a factor than for many concentrate applications.

HORIZONTAL BELT VACUUM FILTERS

In 1916, John M. Callow filed a patent for a "Filter-Belt" that qualifies as a horizontal belt vacuum filter in which vacuum is applied directly to a traveling belt made of canvas or other filter fabric (Callow 1916). That was well before the filters became common in the 1960s. Advancing the filter belt while vacuum is applied was the fatal flaw in the referenced patent. Any filter cloth would tear in two rather than advance. Even overcoming that problem, abrasion would likely have reduced any filter cloth to shreds in a matter of hours.

The concept of the "Filter-Belt" had several potential advantages, one being the ability to displace wash soluble solids or other liquids from the filter cake with a minimum of wash liquor. High wash efficiency is still the single greatest advantage of horizontal belt vacuum filters.

Two technological leaps made horizontal belt vacuum filters viable. The first is credited to John Gunnar Wallny, who in 1932 applied for a German patent wherein he added an "elastic" (e.g., rubber) carrier belt under the filter media (Wallny 1936). Holes in the rubber belt transferred vacuum to the cloth and permitted drainage of the filtrate. The design made the machine marginally viable by effectively removing the vacuum load from the filter cloth and reducing the overall vacuum load while applying it to the stronger carrier belt. In a 1943 patent (referencing an earlier Swedish submittal of 1937), Wallny mentions refinements that describe the basics of today's horizontal belt vacuum filter in which the holes in the carrier belt are in a straight line. That modification would reduce vacuum load on the belt to a minimum. Wallny also mentions "grooves and ribs" in the belt to facilitate drainage of the filtrate (Wallny 1943).

Further investigation ties Wallny to the phosphate industry in Landskrona, Sweden. The manufacturer "Landskrona" was an early pioneer in wet phosphoric acid filters. Filtres Philippe in France was also an early manufacturer and they claimed installations beginning in 1948. The early units reportedly had frequent rubber belt failures. While not confirmed, the topic is a reasonable introduction to the second technological leap.

Research and application just before, during, and after WWII substantially improved the life of elastomers and their viability in industrial uses (Holden 2000; Bhowmick and Stephens 2001). Horizontal belt vacuum filters were one such use. The leading supplier in North America during the late 20th century was EIMCO Process Equipment Company who entered the market in 1961.

The number of installations in the 1960s and 1970s exploded. Improvements came fast during the period and involved designs to minimize wear between the rubber carrier belts and vacuum boxes, means to prevent slurry running off the sides of the filters, methods to evenly distribute slurry and wash water, advancements in rubber belt widths, etc.

A major variant developed in the 1980s, generally known as indexing belt filters, shared similarities with Callow's original patent. The designs eliminated the rubber carrier belt and greatly reduced the cost of the machines by turning off vacuum when pan or cloth moved relative to each other. Poor suitability for aggressive duty and high belt speeds minimizes their application in metallurgical processing.

In 1986 Delkor, previously a licensee for Filtres Philippe, came on the scene as a major manufacturer by installing 26 belt filters at the Nchanga Tailings Leach Plant in Zambia of 80 square meters filtration area (Katoto et al. 1996; Chisakuta et al. 2005); the largest ever made at the time. These used a single rubber carrier belt as opposed to multiple belts to reach a 3-m (9.8-ft) width, which also meant they were the largest single-belt machines. This was a major milestone in the race to build ever-larger machines.

Innovations in support deck design followed with introduction of wear belt decks and roller decks in the 1990s. A single prevalent design for carrier belt support on either side of the vacuum box has not emerged.

Uses for belt filters are tailings dewatering, hydrometallurgical applications, and zinc processing.

FILTER MEDIA

For at least 100 years, cotton, wool, and paper were the materials for industrial filter media. All are still used to some degree, but significant advancements with artificial fibers started in the 1940s (Billings and Wilder 1970). Early fibers were multifilament acrylic and options expanded in the 1950s and 1960s to include nylon, polyester, polypropylene, and others. Materials like polytetrafluoroethylene (PTFE) and polyetheretherketone (PEEK), while having been around for some years, are too expensive for widespread use; although other fibers coated with PTFE reduces blinding and improves filter cake release.

The new materials offered better wear characteristics, improved strength, and greater stability. The latter was important for applications where the filter media moves independent of other support, such as vertical filter presses and horizontal belt vacuum filters. In many applications, the new fabrics were less prone to blinding than their natural counterparts. Whether or not all of the advantages were available in a given process, the new fabrics offered superior life in a majority of applications. Longer life translated to reduced downtime.

For filtration of finer particles, it took quite some time for the synthetics to compete with natural fibers. Abrasion resistance, strength, stability, and even reduced blinding are of little value if the media cannot capture a sufficient quantity of solids. However, beginning in the 1990s advances in weaving and finishing improved filtering characteristics to where synthetic filter cloth is almost universally applied. There are non-woven synthetics, although woven media is ubiquitous in metallurgical processes.

Plain weaves were the most common in the early synthetics. It offered most of the advantages mentioned, but had limited filtration capability. Twill weaves released filter cakes well, but satin weaves and double layer weaves were options that greatly affected filter performance. The former offered high airflow and a slick surface on the cake side for exceptional cake release. The latter provided better filtration and high strength (Hoijer and Grimm 2011).

Finishing options can fine-tune the filter media's performance. Calendaring (passing cloth between cold or heated rolls under very high pressures) flattens the fabric and is desirable for three reasons:

1. It smoothes the cloth surface to give better cake discharge,
2. It tightens the fabric to decrease porosity, and
3. It locks the filaments of loosely woven monofilament fabrics.

Heat setting after weaving can give dimensional stability to synthetic fibers. Water repellent finishes and anti-static treatments are also possible, although rarely of benefit to the metallurgical processes addressed in this text.

CONCLUSIONS

Prophets of future innovations in liquid/solid separation are many, but typically not very successful at prognostication. If the past is any indication, we can be sure that larger thickeners and filters are waiting in the wings. Other possible advances include the following:

- Incremental changes in High Capacity Thickeners, but a significant step change is unlikely. Today's young engineers will probably not refer to what are now High Capacity Thickeners as "conventional" or "low capacity" during their careers.
- Things like low-drag rakes and improved picket designs are possible in High Density and Deep Cone Thickeners, as are better methods for getting underflow out of the thickener. Some innovations will likely come from designers of paste transportation systems that will in turn permit the thickener suppliers to push the envelope further.
- Flocculant Development will certainly continue, very often for specific customer's needs. Is there a revolution in the offing that will dramatically improve settling? Doubtful.
- We can expect improvements in Thickener Feed Dilution to reduce capital and operating costs, while maintaining dilution and improving flexibility for the operator.
- Thickener Drives will benefit from improved torque monitoring, better control, and automation; leading to greater reliability. Costs may also come down with advances in materials.
- Horizontal Filter Presses are most likely to improve through advances in materials and automation. Better control, faster cycle time, and extended wear are all areas where manufacturers will focus development efforts.
- Vertical Filter Presses will also benefit from material and automation improvements. However, it will be interesting to see if the compressed gas design dominates or the traditional membrane; or if they both carve out commercially viable markets.
- There are new applications for Rotary Vacuum Disc Filters, meaning manufacturers will invest in development, which was minimal for several decades. Improvement efforts will focus on filter segment mounting, retention, and alignment; but also cake discharge and slurry suspension.
- Horizontal Belt Vacuum Filters will likely see advances in materials like drainage belt elastomers and vacuum box wear surfaces. Designs for belt support decks alongside the vacuum boxes still differ considerably. In the coming years, we can expect one configuration to emerge.
- The largest potential in Filter Media is probably in material innovation rather than new weaves or finishing methods. Those could be in the base materials themselves or in coatings.

The authors hope that readers find the preceding treatise valuable. We focused on innovation and limited detailed discussion of topics like function, design, operation, and maintenance. Other sources cover those details well. However, there is no substitute for a qualified expert's involvement in your project.

REFERENCES

Agricola, G., 1556. *De Re Metallica.* Translation Herbert Clark Hoover. 1912. Lou Henry Hoover. 1950 Edition. New York: Dover Publications, Inc.

Ballentine, F., Lewellyn, M.E. and Moffatt, S.A., 2011. Red Mud Flocculants used in the Bayer Process. In *Light Metals 2011.* Edited by Stephen J. Lindsay. TMS. pp. 107–108.

Bhowmick, A and Stephens H. 2001. *Handbook of Elastomers.* 2nd Edition. New York: Marcel Dekker, Inc. Basel: Marcel Dekker AG. p 818.

Billings, C.E. and Wilder, J., 1970. *Handbook of Fabric Filter Technology, Volume I.* Fabric Filter Systems Study. Bedford, Massachusetts: GCA Corporation. pp. 1–26.

Callow, J.M., 1916. *Filter-Belt.* US Patent No. 1,201,021.

Chisakuta, G.C., Banda, M. and Chabinga, M., 2005. *Process Developments at Tailings Leach Plant.* Paper at the Third Southern African Conference on Base Metals. The South African Institute of Mining and Metallurgy. p 380.

Dahlstrom, D.A., 1989. Liquid-Solid Separation—Challenges in the Fast Lane. In *Challenges in Mineral Processing.* 1st ed. Edited by K.V.S. Sastry and M.C. Fuerstenau. Society for Mining, Metallurgy & Exploration Inc. (SME). p 468.

Davis, S.S., 1971. *Pressure Filter.* US Patent No. 3,760,945.

Dorr, J.V.N., 1907. *Decanting Apparatus.* US Patent 867,958.

Eis, F.G., 1968. *Method and Apparatus for Separating Liquids from Solids.* US Patent No. 3523889.

Gardner, G.R. and Ray, K.B., 1939. *Flocculation and Clarification of Slimes with Organic Flocculants.* Technical Publication No. 1052. AIME. p 290.

Hirs, G., 1955. *Filter Apparatus.* US Patent No. 2,867,325.

Hoijer, J. and Grimm, R., 2011. *The Black Art of Filter Media.* Issue 28. Outotec Australia Quarterly E-Newsletter.

Holden, G., 2000. *Understanding Thermoplastic Elastomers.* Cincinnati: Hanser Gardner Publications, Inc. München: Carl Hanser Verlag. pp. 4–5.

Katoto, W.K., Rao A.S., Kasongamulilo, V.S. and Lumbwe, H.T., 1996. *Expanding TLP Copper Production by Heap Leaching.* Paper at 1996 Annual Meeting of Society for Mining, Metallurgy, and Exploration, (SME). p 1.

Linke, W.F. and Booth, R.B., 1959. *Physical Chemical Aspects of Flocculation by Polymers.* AIME Transaction Vol. 217. AIME. pp. 364–371.

Margueritte, L.J.F., 1874. *Improvement in Obtaining Sugar from Molasses.* US Patent No. 148375.

Ostrop, E., 1974. *Plate Filter with Horizontally arranged Bilaterally acting Filters.* US Patent No. 3,968,040.

Robinsky, E.I., 1978. *Tailings Disposal by the Thickened Discharge Method for Improved Economy and Environmental Control.* Proceedings of the Second International Tailing Symposium, Denver, U.S.A. pp. 75–92.

Salisbury, O.J., 1918. Rotary Suction-Filter. US Patent No. 1259139.

Schoenbrunn, F. 2007. *A Short History of Deep Cone Thickener Development.* Australian Centre for Geomechanics. pp. 51–56.

Schoenbrunn, F. 2011. *Dewatering to Higher Densities; An Industry Review.* Australian Centre for Geomechanics. pp. 19–24.

Schoenbrunn, F. and Laros, T., 2002. Design Features and Types of Sedimentation. In *Mineral Processing Plant Design, Practice, and Control Proceedings.* 1st ed. Edited by A.L. Mular, D.N. Halbe, D.J. Barratt. Littleton, CO: SME. pp. 1331–1341.

Wallny, J.G., 1936. *Apparatus for Filtering and Washing Mixtures of Solid and Liquid Substances.* US Patent No. 2,034,784.

Wallny, J.G., 1943. *Apparatus for Filtering and Washing Solid Matter.* US Patent No. 2,314,294.

Wood, L.J.A., Turner, S.D., Laros, T.J., and Emmett, R.C., 1994. *Self-Diluting Feedwell for Thickener Dilution.* US Patent No. 5,643,463.

Yagishita, A., Iriyama, Y. and Osuga, J., 1985. *Automatic Filtration Dewatering System.* US Patent No. 4,994,870.

ABOUT THE AUTHORS

Mick L. McCaslin is director of the Minerals Testing Center at FLSmidth in Midvale, Utah. He is a registered professional engineer and has more than 30 years experience in plant operations and maintenance, equipment supply, and systems design. Previously he worked in copper milling, smelting, and refining; vacuum and pressure filtration; and industrial centrifugal pumps. In his current role, he is privileged to lead a team of world-class mineral processing experts in merging the laboratories of three legacy businesses, growing two more primary laboratory groups from the ground up, and building a one-of-a-kind minerals and metallurgical testing facility.

Keith J. Mounteer recently retired from FLSmidth after a combined 35 years of service. He spent more than 10 years in research and development where he was awarded a U.S. patent for designing a gas-dissolving vessel for a dissolved air flotation system, as well as writing an article on biological composting of waste solids published by the Environmental Protection Agency. The majority of time at FLSmidth, Mounteer worked as a product manager and sales support in the Vacuum Filtration Group. During a 10-year break from FLSmidth, he was awarded a Betz Eagle for the highest increase in sales of water and waste water treatment products in his assigned sales area, which at that time covered the western portion of the United States.

Michael Phillips is a senior applications engineer at FLSmidth in Midvale, Utah. He is a registered professional engineer and has more than 20 years of experience in equipment supply, systems design, and plant operations. Previously he worked in vacuum and pressure filtration, as well as shipbuilding, oil field operations, and electronics manufacture. In his current position, he evaluates flowsheets for various applications, specifying valves, instruments, and pumps, and works on specific projects in the R&D arena.

Hydrometallurgy

Innovative Strategies for Copper Hydrometallurgy

J. Brent Hiskey
University of Arizona, Tucson, Arizona, United States

ABSTRACT: Hydrometallurgy currently accounts for approximately 21% of the world primary production of copper. Chile and the United States in 2010 produced 2.09 and 0.43 million mt, respectively, of high grade copper by leaching/solvent extraction/electrowinning methods. These numbers reflect a remarkable growth in copper hydrometallurgy over the past few decades. This growth can be directly attributed to a number of key innovations. Of these, copper solvent extraction certainly qualifies as a hallmark innovation. A review of these innovations shows that technology is quite fluid, that it diffuses and circulates between various mining and processing sectors. Furthermore, it normally requires a long incubation period for the complete acceptance and implementation of new technology.

INTRODUCTION

What do we mean by innovation? A discussion on innovation would take a very long time and we would probably never arrive at a consensus of what it is really all about. It has different meanings to different people and institutions. Peter Drucker (1998) provides a very insightful definition for innovation. He states that innovation "*…is the means by which the entrepreneur either creates new wealth-producing resources or endows existing resources with enhanced potential for creating wealth.*"

Over the past decade many businesses have recognized the need for technological innovation and have assimilated the discipline of innovation into their corporate culture. Successful management of innovation can offer a sustained competitive advantage in a global marketplace. There are many dimensions to innovation; however, there are primarily two key elements to the process. The first is creative discovery which is commonly related to pure R&D. The second element is generally known as deployment or implementation (tech transfer) of discovery. Examples of successful R&D proliferate the technical and patent literature; unfortunately, positive examples of tech transfer are more limited.

The global mining industry is faced with numerous challenges that seem to intensify with time. Batterham (2008) has presented a set of concerns that represent formable challenges to the industry including: supply and demand; ore grade depletion; mineral complexity; the handling, disposal, and minimization of waste; water availability; energy; and social, environmental, and policy concerns. If we examine ore grade depletion for copper we see a steady decline in average copper yield (*grade × recovery*) Hiskey (1994). In 1904 when Daniel C. Jackling advanced his great industrial innovation of open pit mining at Bingham Canyon, the average yield of copper ores mined in the United States was about 2%. Copper recoveries were relatively low at the time and the ore grade was correspondingly quite high. At present, the average yield of concentrated ore is 0.43% (Edelstein 2012). The situation with ore grade depletion is even more pronounced, since metallurgical recoveries have increased markedly over the years. Many involved with the mineral and metal extraction enterprise are faced with the difficult question: "*Is it ore, or is it waste?*" This is especially true for large copper porphyry deposits which lack a sharp distinction between waste and ore. Fortunately, hydrometallurgy is ideally suited for ores containing dilute mineral phases. This paper examines key technological innovations that have contributed to the remarkable growth in hydrometallurgical recovery of copper. Modern copper hydrometallurgy mainly encompasses heap and stockpile leaching, solvent extraction, and electrowinning (L/SX/EW). Significant innovative contributions to each operation will be discussed in order.

INNOVATIVE STRATEGIES IN COPPER HYDROMETALLURGY

Information provided by the U.S. Geological Survey on current world copper production indicates that hydrometallurgy accounts for 21% of primary production (Edelstein 2012). The total electrowon copper produced in the world in 2010 was estimated to be 3.31 million metric tons. As presented in Table 1, Chile is clearly the largest producer of copper by hydrometallurgical methods, accounting for nearly 63% of the world total EW production.

The growth in EW copper production in Chile has been nothing short of phenomenal during the past two decades as depicted in **Figure 1**. By comparison, the EW production in the United States has been relatively constant at approximately 500,000 mt during this period. The EW figure is an excellent measure of the total hydrometallurgical contribution to copper recovery, since the amount of primary copper produced by cementation is considered small. Modern copper hydrometallurgy in essence represents an integrated system of leaching (heap and stockpile leaching),

Table 1. World EW copper production in 2010

Country	EW Production (1000 mt)	% World Total EW
Chile	2,089	63.1
United States	430	13.0
Congo	262	7.9
Peru	153	4.6
Zambia	150	4.5
Mexico	80	2.4
Laos	64	1.9
Other	82	2.5

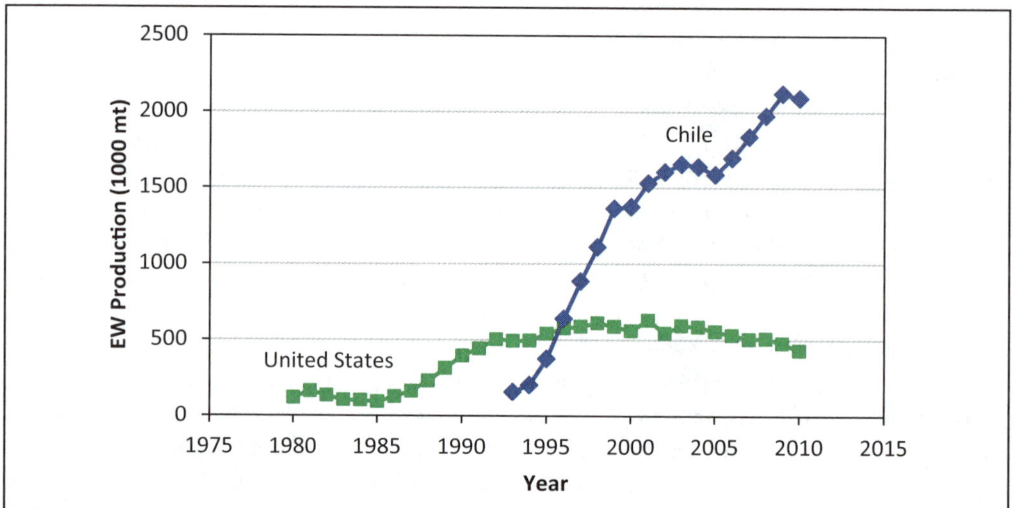

Figure 1. EW copper production from Chile and the United States

solution concentration and purification (SX), and metal recovery (electrowinning). The hydrometallurgical growth spurt observed for the United States in the mid to late 1980s resulted from the adoption of innovations in all three unit operations. These innovations took about a decade before they appear in the copper mines in Chile. This shows that there is a natural diffusion and circulation of innovations within a particular industry.

Important Innovative Advances in Heap and Stockpile Leaching

Basic hydrometallurgical techniques for the extraction and recovery of copper have been used for millennia. The ancient recovery of copper by solution mining in China was detailed by Pu (1982). It was reported that Liu-An (177–122 BC) authored a book titled "Huainanius" which documented the general observation of copper cementation on iron. It is believed that rudimentary elements of heap leaching may have been practiced anciently, but they were lost in the dim recesses of antiquity. In the middle ages, Agricola (1556) described the recovery of alum by heaping leaching of aluminous rock. Heap leaching was employed on a relatively large scale at Rio Tinto, Spain in mid 18th century (Taylor and Whelan 1942). At Rio Tinto, it is interesting to note that heaps containing massive sulfides were designed and operated with careful consideration of ore placement, solution application rate, air circulation, and temperature control.

The development of modern copper heap leaching practice reveals an exchange and movement (i.e., diffusion and circulation) of technology between various mining interests. By the 1960s copper dump (stockpile) leaching was well established in the Western United States (Sheffer and Evans 1968). There were certain practical aspects of dump leaching technology that were quite developed (e.g., basic knowledge of bacterial leaching, reaction chemistry, dump construction, and water management). In the late 1960s, there was a impetus in the exploration and development of processing methods for the treatment of low-grade gold and silver resources. Heap leaching was an obvious processing approach. The precious metal mining industry borrowed and adapted key elements from copper dump leaching to advance precious metal heap leaching practice. They subsequently made substantial refinements and improvements to heap leaching. For example, the advantages of using mechanical stacking systems, advanced liner and leak detection systems, ore agglomeration, and drip irrigation are noteworthy in this regard. Then in the early 1980s, the copper industry adopted parts of heap leaching technology from the gold industry to design highly efficient and productive heap leaching operations. The innovative advances in copper heap leaching have been more evolutionary and multifaceted than being a single/revolutionary development and for the most part have depended on technology diffusion. In this section, advances in ore stacking, agglomeration, and drip irrigation are discussed.

Ore Stacking

Scheffel (2002) has presented a very comprehensive discussion of copper heap leaching design and practice. Many early heap leaching operations constructed the ore pile using simple conveyor assemblies, truck haulage, and possibly front-end loaders. As discussed by Chamberlin (1981) each technique created problems with respect to heap porosity and permeability. The truck placement method was especially problematic. To avoid heavy equipment being in direct contact with the pad and liner system, first lifts were built by end dumping from the truck and advancing progressively across the entire width of the pad. The surface of the heap was exposed to multiple cycles of heavy truck traffic. This resulted in severe surface compaction of the ore. Attempts to break up the compaction layer by ripping were marginally effective and most operations experienced poor and inconsistent leach performance because of the reduced permeability. Clark and Hawkins (1988) categorized the issues as follows:

- Compaction by heavy equipment
- Size segregation
- Compression of bottom material
- Generation and migration of fines

These problems can cause channeling of leach solution with the attendant loss of solution contact with large volumes of ore. One approach that partially alleviated these problems was the construction of heaps with various configurations of conveyor/stacker systems. The mobile conveyor system installed at Pegasus' Florida Canyon gold mine was an early example (Clark and Hawkins 1988).

Continued refinements in mobile conveying and stacking have resulted in a truly innovative concept for copper heap leaching. The early evolution of portable conveyor systems involved what is now commonly referred to as "grasshoppers." These units were portable, but not self-propelled. They required substantial downtime to relocated and reposition. Modern automated Super Portable technology which features, all-track propulsion, on-board generators, and fully articulated head and tail systems reduce the downtime, increase the maneuverability and overall flexibility of material handling. These factors enhance the general profitability and efficiency of heap leaching. The Safford heap leaching project of Freeport-McMoran employed a Super Portable Stacker system in its design (Carter 2007; Graber 2012). Crushed and agglomerated ore is delivered to the leach area by a 7,000 ft overland conveyor. The material is then distributed on the pad by an elevating conveyor/stacker arrangement. The system includes: a series Super Portable elements comprising horizontal conveyors, a feed conveyor, and radial stacker. Each module has twin powered and steerable crawler tracks with mobility in both forward and reverse directions. The stacker is equipped with 20 ft extensions or stingers which retract as a row of new material is laid down. Once the last row is completed (stinger fully retracted), a section of horizontal conveyor is removed and the stacker retreats to a new position and the stringer extended to start a new lift. This process repeats itself until a leach cell is completed. Figure 2 shows the conveyor/stacker system at Safford.

Super Portable conveyor/stacker systems have markedly improved the overall efficiency of materials handling during heap construction. Furthermore, these innovations have removed many of physical problems associated with permeability. Undoubtedly, the presence of clay particles and excessive fines along with chemical factors such as rock decomposition and decrepitation can still aggravate permeability and porosity problems.

Acid Cure and Agglomeration

Bouffard (2005) authored an excellent review of the fundamental and practical considerations regarding agglomeration for both precious metal and copper ores. It was stated that "*Agglomeration was a technical and economical breakthrough technology for heap leaching of clayey ore and ore containing high fines content.*" Historically, the research by investigators at the U.S. Bureau of Mines in Reno, Nevada (Heinen et al. 1979), is significant with respect to this innovation. This group used agglomeration techniques to enhance percolation rates during the heap leaching of clayey gold ores. These early discoveries were important, but in terms of the overall innovation process, the sustained efforts by McClelland (1981, 1983, 1986,) as a technology transfer champion cannot be overlook.

Figure 2. Super portable system using retreat stacking

This innovation has greatly benefited heap leaching of precious metal and copper ores. Bouffard (2005) classified the benefits into three categories:

1. Heap physical structure
   – Segregation is minimized
   – Migration of fines is reduced
   – A more porous heap is created
2. Leach chemistry
   – Solution/mineral contact is increased
   – Bioleaching environments are improved
3. Environmental impact
   – Rinsing efficiencies are better
   – Dust emissions are decreased

With respect to copper, Bouffard (2005) summarized agglomeration data and information for 19 heap leaching operations. Copper ores are typically agglomerated without the aid of supplemental binders, using only water and additions of sulfuric acid. The data tabulated indicate H_2SO_4 additions ranging from 2 to 60 kg/t. and final moistures of approximately 7 to 10%. Cure times were typically 1 to 2 days, but could go as high as a month. As mentioned by Moats and Janwong (2008), the binder of choice for copper ore agglomeration is concentrated sulfuric acid. Indeed, the sulfuric acid ore particle reactions are critical in establishing the solid bridges between particles. The mechanism involves a chemical welding of particles as decomposition and dehydration occur. The general concept of agglomeration and acid curing of ores in metal extraction processes is not new. Lankenau and Lake (1973) described the use of acid cure procedures for the treatment of low grade uranium ores in one example and low grade copper ores in another. In the case of a copper ore, concentrated sulfuric acid was sprinkled on the ore and water added to maintain approximately 10% moisture content. The ore was mixed periodically and allowed to cure for 10 days. A thin layer (TL) leaching method was invented by Johnson (1977) that involved adding strong sulfuric acid (as high as 60% H_2SO_4) to the ore in a mixing drum and allowing the ore to cure for 2 to 3 days. It was found that as the ore cures dehydration and induration take place (i.e., the ore particles become dry and hard) and that the resulting agglomerates were strong enough to support the weight of a column of stacked ore. The TL acid cure approach was used at the Lo Aguirre mine where a mixed oxide/sulfide ore was treated with 25 kg/mt of H_2SO_4 and 40 kg/mt of water (Domic and Brimm 1980). Acidulated ore was allowed to cure for 24 hr before leaching counter-currently in two stages. Eventually, this process evolved into a biologically assisted thin layer leaching of acid cured agglomerates.

A relatively recent illustrative example of agglomeration in copper ore leaching is the Collahuyasi Copper mine in northern Chile. Run-of-mine oxide ore is crushed to –9 mm in three stages and then agglomerated with 28–35 kg/mt of H_2SO_4 and 80–100 kg/mt water. The ore pellets are then allowed to cure for 1 day. It is reported that copper recoveries in the range of 68–85% can be achieved in 60 days of leaching. Even with this level of success there is still little know about the agglomeration process.

Moats and Janwong (2008) presented an overview of crushed ore agglomeration. Various agglomeration mechanisms were discussed in terms of the role of binders and particle–particle interactions. Unquestionably, the group at Michigan Tech (i.e., Lewandowski, Kawatra, and co-workers; 2006, 2007) has contributed significantly to our understanding of the phenomenological details, mechanisms, and process parameters for agglomeration. In addition these investigators have helped establish testing procedures and protocols for the evaluation of binders. The research of Lu, Driesinger, and West-Sells is significant in terms of our understanding of the reactions involved in the acid curing/agglomeration process. Agglomeration of an ore sample from the Zaldivar mine with a P_{80} ~12 mm was investigated with and without the addition of water for various dosages of H_2SO_4. It was conclusive that superior leach results were achieved with the addition of water. Importantly, their XRD results on fine particles confirmed chemical reaction with specific rock forming mineral phases with the H_2SO_4/H_2O system. Clinochlore, $(Mg^{2+},Fe^{2+})_5Al_2Si_3O_{10}(OH)_8$, was found to exhibit significant reaction with the H_2SO_4/H_2O treatment. There was suppression and significant line broadening of x-ray peak for clinochlore. Moreover, The XRD pattern confirmed the presence of hydrated metal sulfate salts, which could possible serve as chemical welding agents for particle–particle bonding. Morphology studies using SEM analysis proved that small particles were bonded to large particles with the addition water during agglomeration. These are definitely important contributions to our understanding of acid/cure agglomeration; nonetheless, there are still opportunities for research and there is also a need to develop quantitative measures for quality control.

There is a recent innovation that deserves mention at this point. Kodali et al. (2011) have performed a laboratory evaluation of an acid resistant "stucco" binder for agglomeration during copper heap leaching. The type of stucco used was calcium sulfate hemihydrate ($CaSO_4{\cdot}\frac{1}{2}H_2O$). Ideal agglomerates were formed using 3–5% stucco binder and 7–10% acid solution. The agglomeration mechanism involves the formation of gypsum as the calcium sulfate hemihydrate undergoes hydration according to the following reaction:

$$2CaSO_4{\cdot}\tfrac{1}{2}H_2O + 3H_2O \rightarrow 2CaSO_4{\cdot}2H_2O$$

Gypsum crystals form a network of solid bridges between the ore particles. A quantitative evaluation of performance measures such as pellet strength were not report; however, there was evidence that fines would be effectively immobilized by this technique.

Drip Irrigation

Of special importance in every heap and stockpile leaching operation is solution application. In general, leaching solutions transport lixiviant to the mineralized values contained within an ore fragment and wash dissolved values from the heap or stockpile. Leaching solution has been applied to the surface of heaps and stockpiles by a variety of methods including the following;

- Ponding/flooding
- Trickle
- Multiple low-pressure sprinklers
- Single high-pressure sprinklers
- Subsurface injection
- Drip irrigation

The evolution of solution distribution technology for leaching has followed that used for agricultural irrigation both in terms of methods and materials.

Scheffel (2002) briefly described the primary solution application methods used in copper heap leaching. The selection of a method for a given operation depends on careful consideration and knowledge of climate, heap or dump height, surface area, scale of operation, permeability, rock particle size and mineralogy. The plastic Wobbler® is accepted as the industry standard for multiple low-pressure sprinkler systems. These systems are probably preferred as the principal method when certain conditions permit and for some applications such as side slope leaching. Brierley and Brierley (1999) suggest that sprinklers and wobblers can be used in locations where there is a net positive water balance with respect to evaporation losses; however, Chamberlin (1988) says that drip irrigation has an advantage when operating in regions where evaporation is high. Other advantages of drip irrigation are as follows:

- No problems with drifting spray mist
- Employees can work on the systems without protective rain suits
- Less ice build-up and fewer freeze-up during cold weather periods
- More control at very low application rates
- Diminished damage to surface agglomerates

Drip irrigation delivers leach solution through plastic tubes that lie on or under the surface. The tubes have small openings called emitters at evenly and closely spaced intervals along distributor lines. The type of emitter may be either an in-line type or one that resides flat against the tubing wall. Both designs incorporate a labyrinth structural pattern which promotes turbulent flow within the channels thereby reducing plugging. Nonetheless, drippers can clog with either suspended solids or by the build-up of precipitated salts.

Drip emitters deliver leach solution in a controlled manner at low application rates. For emitters on a 0.6 m grid, application rates would range from 0.28 cm hr^{-1} (0.001 gpm/ft^2) to 2.2 cm hr^{-1} (0.008 gpm/ft^2). There is some criticism that drip emitters have poor lateral coverage, leaving volumes of ore not contacted by the lixiviant. This isn't a problem when considering over-stacking and

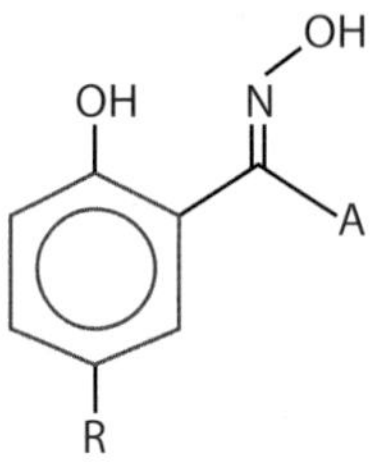

Reagent	R	A	Type
LIX 65	$C_{12}H_{25}$	C_6H_5	Ketoxime
LIX 65N	C_9H_{19}	C_6H_5	Ketoxime
SME 529, LIX 84-1	C_9H_{19}	CH_3	Ketoxime
LIX 860-1, LIX 622	$C_{12}H_{25}$	H	Aldoxime
P-1, LIX 860N-1	C_9H_{19}	H	Alcoxime

Figure 3. General structure of oxime molecules used for copper recovery (after Kordosky 2002)

the secondary leaching of lower lifts. In general, drip systems have helped optimize metal recovery and have allowed copper heap and stockpile leaching to operate in extremely dry climates.

Copper Solvent Extraction—A Hallmark Innovation

It is generally recognized that the most significant innovation in copper hydrometallurgy is that associated with copper solvent extraction. There are four areas where SX technology has advanced with regard to copper processing: (1) reagent development; (2) equipment design; (3) circuit configuration; and (4) accelerated coalescence and crud control. Unquestionably the hallmark innovation pertains to copper solvent extraction chemistry and reagent development. The development and application of copper selective extractants is a fascinating story. It is a story full of pure innovative discovery and perseverance. A historical account of this story has been documented by several authors (House 1981; Kordosky et al. 1999; and Kordosky 2002). There are two main innovators that must receive credit for what we now take for granted as modern copper SX. One individual was directly responsible for the pure R&D and the other was a true visionary that saw the commercial importance of these discoveries.

At General Mills Chemicals, Joe House put together the Minerals Development Group in 1960 to explore opportunities for solvent extraction other than uranium. Copper which at the time was being recovered by the ancient art of cementation was a logical area for the group to study. So the team of Joe House, Don Agers, and Ron Swanson embarked on a mission to synthesis an extractant molecule for copper. A survey of the copper industry at the time indicated that many viewed copper solvent extraction with skepticism and were often downright hostile towards this technology. Nonetheless, the research moved forward and by late 1962, the General Mills group had synthesized a a-hydroxy-oxime molecule that showed promise (Swanson 1965; House 1989; Kordosky 2002). A patent was filed by Swanson on February 25, 1963. This reagent was announced as LIX® 63. Following closely behind were the introduction of LIX 64 and LIX 64N. These reagents were evaluated in pilot plants operated by Bagdad Copper and Duval Corp. in 1965.

Maxie Anderson was the pioneering visionary that commercialized the application of copper solvent extraction at Ranchers Exploration and Development Corporation's Bluebird Mine near Miami, Arizona. Ken Power, one of the early champions of copper SX, was the General Manager of the Bluebird operation. He described in considerable detail the operation of the L/SX/EW plant which began operation in March 1968. The startup of the Bluebird SX plant used LIX 64, but was soon replaced by LIX 64N when it was introduced 1968. Approximately 4,100 mt of cathode copper were produced between July 1968 and June 1969 (Power 1970). Subsequently, copper SX plants were built by Bagdad Copper Corporation at Bagdad, Arizona (1970), by Capital Wire and

Cable Corporation for their scrap copper leaching plant near Casa Grande, Arizona (1970), and by N'Changa Consolidated Copper Mines (ZCCM) in Zambia (1974) (Scheffel 2002).

Kordosky (2002) states that LIX® 64N was a blend of LIX 65N (a 2-hydroxy-5-nonylbenzophenone oxime) as the active extractant molecule and a small catalytic quantity of LIX 63. The basic structure of the hydroxyl oxime molecule is depicted in **Figure 3**. This structure is the backbone for two separate classes of extractants: the aldoximes where A is a hydrogen atom or ketoximes where A is either a methyl or phenyl group.

It is important that we recognize other early developments of copper selective extractants. Kelex 100, a 8-hydroxy quinolone extractant, was introduced by Ashland Chemical Company in 1968. This was an especially strong extractant for copper and as such was promoted as being effective in extracting copper from highly acidic aqueous leach solutions. This reagent never gained commercial acceptance because it was extremely difficult to strip and it extracted large quantities of acid. To Ashland's credit, they introduced the concept of adding modifiers to improve the loading/stripping performance of these extractants. Hartlage and Cronberg (1973) patented the use of alkylated phenols as additives to increase stripping efficiency and to improve phase separation.

In 1973, Shell International Chemicals started manufacturing SME 529 as a commercial copper extractant. The active molecule in SME 529 was a 2-hydroxy-5-nonylacetophoneone oxime (van der Zeeuw 1976). Side products formed during the synthesis of this reagent were problematic which resulted in poor reagent performance if not scrupulously purified. Szymanowski (1993) reports that the only operation to use the SME 529 reagent was ZCCM in Chingola, Zambia.

Researchers at Imperial Chemical Industries (ICI) synthesized a number derivatives of salicylaldoximes as copper extractants during the period of February to August 1972. They settled on a 5-nonyl salicyaldoxoime designated as P-1 as the most suitable reagent. It took approximately 3 years to fully develop the manufacturing process. In 1974, Birch described the properties of this extractant which was produced under the AGORA® trademark. As early as 1973, Dalton conducted a series of experiments investigating the effect of nonyl phenol as a modifier (resulting U.S. Patent 4,142,952). It was observed that the net transfer capacity of P-1 for copper was increased with the addition of nonyl phenol. P-1 with additions of nonyl phenol was finally marketed by ICI under the Acorga P-5000 series of reagents. The P-5100, P-5200, and P-5300 extractants where based on a ratio of nonyl phenol to active oxime molecule of 1:1, 1:2, and 1:3, respectively. In 1976, the solvent extraction facility built at Kennecott's Ray Mine complex immediately started operation with these reagents.

Henkel acquired General Mill Chemicals Inc. in 1977 and continued a forward-thinking effort in the development of copper SX reagents. In 1979, Henkel introduced LIX 622 a tridecanol modified aldoxime with a 12 carbon alkyl group. This reagent was first use in 1981 at the Pinto Valley SX-EW plant of Cities Services Copper Corp.

This essentially closed the chapter on the development and application of first generation solvent extraction reagents. Since the early 1980s, the innovative strategy has been the use of non-modified mixtures of aldoximes and ketoximes. These mixtures are determined by performance and not structure. Today, the formulation of the current generation of solvent extraction reagents is highly advanced. Aldoxime/ketoxime blends are custom tailored to the specific plant and operational design requirements. Virnig and Kordosky (1994) have reviewed the current practice involved in the development of copper selective reagents. The development of LIX® 34 was used to illustrate the how reagent performance is optimized with regard to structure/properties. LIX® 34,

8-(decyl(methyl) benzenesulfonamido)quinoline, is a unique molecule that will extract copper at low pH with no iron loading.

Innovative Aspects of Copper Electrowinning

Copper electrowinning has been practiced on an industrial scale for nearly 100 years. Early copper electrowinning operations have been previously reviewed by Hiskey (1999). The basic features of copper electrowinning have not changed noticeably since its early development. Nonetheless, there are a number of recent innovative measures that have contributed to copper electrowinning technology. These innovations have improved the overall efficiency and productivity of the tankhouse operation, reduced energy, and improved cathode quality. Noteworthy innovations in copper EW include: permanent cathode technology, advanced anode materials, electrolytic additives, electrolyte circulation, mist control, cell construction, and plant automation. Permanent cathode technology as an innovation greatly enhanced both efficiency and productivity, and cathode quality. The modern copper electrowinning tankhouse produces cathodes that can complete around the world with the best electrorefined copper. This was not always the case. In fact, EW cathodes had a certain stigma when early SX reagents were being developed, and this negativity persisted for a long time.

Permanent Cathode Technology

The development and application of permanent cathode technology will be briefly reviewed in this section. Permanent cathode technology is another example of an innovation moving from one technology sector (electrorefining) to another (electrowinning). This innovation started with two different systems, the ISA PROCESS™ and the KIDD PROCESS®. The ISA PROCESS™ was invented by Mount Isa Mines and was first installed at their Copper Refineries Pty. Ltd. tankhouse in March 1979 (Hoey et al. 1987). The U.S. Patent for this invention was filed on February 9, 1979, and covered the use of stainless steel starter sheets. To a large extent, the motivation for this innovation was the costly and labor intensive aspects associated with the preparation and handling of copper starter sheets. This advancement also incorporated the basic design features of the cathode stripping machines developed by the Hikosima Zinc Refinery of Mitsui Mining and Smelting Co. (Hoey et al. 1987).

The KIDD PROCESS® was developed by Falconbridge Limited for their Kidd Creek copper refinery at Timmons, Ontario. Conversion of the tankhouse to the new process was completed in late February of 1987 (Kemp et al. 1991). The permanent cathode technology included a new type of stainless steel mother blank and a novel 12 position cathode stripping carousel. The major difference between the ISA PROCESS™ and the KIDD PROCESS® concerns the bottom of the cathode. The early ISA system used wax bottom masking to prevent copper deposition, thus producing 2 separate cathode sheets (one on each side of the stainless steel mother blank). By contrast, the KIDD system left the bottom exposed which created two connected sheets. As noted above, the stripping machines were of different design (ISA was a rectilinear configuration and KIDD was a carousel).

Several authors have summarized the benefits and advantages of permanent cathode technology (Hoey et al. 1987; Pariani 1987). The most significant are as follows:

- Reduced operating labor costs
- Improved cathode quality

- Higher capacity
- Better industrial management
- Greater safety
- Lower metal inventory
- Increased operational flexibility
- Better energy utilization

One of the earliest installations of the ISA PROCESS™ in a L/SX/EW facility was at Magma Copper Company's San Manuel complex in 1986 (Jenkins and Eamon 1990). The 2006 survey of world electrowinning operating data indicated that for the respondents: 21 used the ISA Process, 10 used the Kidd Process, 4 used the Outotec EW cathode technology, and 3 African plants used some generic form of permanent cathode system (Robinson 2007). The total installed capacity utilizing the ISA and Kidd systems was approximately 2.3 million mt per annum.

Aslin et al. (2010) has reviewed the current status and recent developments in cathode stripping machines (CSM). When Xstrata acquired Falconbridge in 2005, the intellectual know-how and design experience for the ISA and KIDD systems were merged. One of the issues with existing systems was the reliability of deposit separation using a point of force application. Xstrata Technology has invented a new separation process employing a high degree of automation and advanced robotics. This has resulted in a more accommodating and universal CSM.

CONCLUSIONS

There are a number of key innovations that have contributed to the remarkable growth in copper production by hydrometallurgical methods. All aspects of copper heap leaching practice have depended on critical inventions and pieces of technology. For the most part, the innovative process took decades to complete. Since the timeline for technological innovation is so long, mining companies need to maintain a sustained discipline of innovation. Energy, water, and waste disposal are common elements that challenge all mining operations and their global competitiveness. Water will surely be an issue in the future for everything we do hydrometallurgically. For example, in many locations, sea water or domestic wastewater (greywater) may be our only option for leaching. This would be a game-changer and the innovation gears would need to start moving.

REFERENCES

Agricola, G., 1556. De Re Metallica, Translated by Herbert C. Hoover and Lou Henry Hoover, 1912. 1950 Edition, Dover Publications, Inc., New York. p. 568.

Aslin, N.J. et al. 2010. Developments in Cathode Stripping Machines—An Integrated Approach for Improved Efficiency. In *Proceedings Copper 2010 Vol. 4—Electrowinning and refining*. Hamburg, Germany: GDMB.

Batterham, R.J. 2008. Current Practices Shaping the Science and Practice of Mineral Processing. *Proceeding of XXIV Mineral Processing Congress (IMPC) 2008*, Beijing, China, 24–28 September 2008, pp. 55–59.

Birch, C.P. 1974. The evaluation of the new copper extractant P-1. In *Proc. Int. Solv. Extr. Conf. ISEC '74*. Edited by G.V. Jeffrey. London: Society of Chemical Industry.

Bouffard, S.C. 2005. Review of Agglomeration Practice and Fundamentals in Heap Leaching. *Mineral Processing & Extractive Metallurgy Rev.*, 26: 233–294.

Brierley, C.L. and Brierley, J.A. 1999. Bioheap Processes—Operational Requirements and Techniques. In *Copper Leaching, Solvent Extraction, and Electrowinning Technology*. Edited by G.V. Jergensen II. Littleton, CO: SME.

Carter, R.A. 2007. Safford Moves Toward Startup. *WOMP The Mining E-Journal.* 2007 Vol. 4: http://www.womp-int.com/archive.htm.

Chamberlin, P.D. 1981. Heap Leaching and Pilot Testing of Gold and Silver Ores. Mining Congress Journal. April 1981: 47–51.

Chamberlin, P.D. 1988. Applying Leach Solutions to Heaps, *Randol Gold Forum 1988.* Golden, CO: Randol International Ltd.

Clark, M.L. and Hawkins, G.P. 1988. New Material Handling System for Heap Leach Stack Construction. SME Preprint No. 88-91. Paper presented at the SME Annual Meeting Phoenix, AZ Jan 25-528, 1988.

Dalton, R.F., 1979. Copper Extraction with Salicylaldoxime-P-Nonylphenol Mixtures. U.S. Patent 4,142,952. March 6, 1979.

Domic, E. and Brimm, E.O. 1980. TL Leaching: Experimental Studies for the Lo Aguirre Leach Project. In *Leaching and Recovering Copper from As-Mined Materials.* Edited by W.J. Schlitt. Littleton, CO: SME.

Drucker, P.F. 1998. The Discipline of Innovation, *Harvard Business Review.* November-December 1998:3–8.

Edelstein, D.L. 2012. Copper. 2010 Minerals Yearbook. United States Geological Survey, Washington, D.C.

Graber, G.M. 2012, Private Communications September 2012.

Hartlage, J.A. and Cronberg, A.D. 1973. Solvent Extraction Process for the Recovery of Copper Values. U.S. Patent 3,725,046. April 3, 1973.

Heinen, H.J., McClelland, G.E. and Lindstrom, R.E. 1979. Enhancing Percolation Rates in Heap Leaching of Gold-Silver Ores, *U.S.B.M. Report of Investigations 8388,* 20 p.

Hiskey, J.B. 1994. In-situ leaching recovery of copper: what's next? *Hydrometallurgy '94.* London: Chapman & Hall.

Hiskey, J.B. 1999. Principles and Practical Considerations of Copper Electrorefining and Electrowinning. In *Copper Leaching, Solvent Extraction, and Electrowinning Technology.* Edited by G.V. Jergensen II. Littleton, CO: SME.

Hoey, D.W. et al. 1987. Modern Tankhouse Design and Practices at Copper Refineries Pty. Ltd. In *The Electrorefining and Winning of Copper.* Edited by J.E. Hoffmann et al. Warrendale, PA: TMS.

House, J.E. 1981. The development of the LIX reagents. 1981 Gaudin Lecture, SME-AIME Annual Meeting, Chicago, IL. *Minerals & Metallurgical Processing* 6(1):1–6.

Johnson, P.H. 1977. Thin Layer Leaching Method. U.S. Patent 4,017,309. April 12, 1977.

Kemp, D.J., Matwijenko, O. and Scott, J.D. 1991. Development of the Kidd Process permanent stainless steel cathode technology at Falconbridge Limited, Kidd Creek Division. In *Proc. Copper 91 Int. Conf., Vol. III–Hydrometallurgy and Electrometallurgy of Copper,* Edited by W.C. Copper et al. New York: Pergamon Press.

Kodali, P. et al. 2012. Evaluation of stucco binder for agglomeration in the heap leaching of copper ore. *Minerals Engineering,* 24(2011): 886–893.

Kordosky, G.A. 2002. Copper recovery using leach/solvent extraction/electrowinning technology: Forty years of innovation, 2.2 million tonnes of copper annually. *The Journal of the South African Institute of Mining and Metallurgy.* Nov/Dec 2002, pp. 445–450.

Kordosky, G.A., Sudderth, R.B. and Virnig, M.J. 1999. Evolutionary Development of Solvent Extraction Reagents: Real-Life Experiences. In *Copper Leaching, Solvent Extraction, and Electrowinning Technology.* Edited by G.V. Jergensen II. Littleton, CO: SME.

Lankenau, A.W. and Lake, J.L. 1973, Process for Heap Leaching Ores. U.S. Patent 3,777,004. December 4, 1973.

Lewandowski, K.A. et al. 2006. Determination of Acid Resistance of Copper Ore Agglomerates in Heap Leaching. In *Sohn International Symposium—Advanced Processing of Metals and Materials Vol. 6—New, Improved and Existing Technologies.* Edited by F. Kongoli and R.G. Reddy. Warrendale, PA: TMS.

Lewandowski, K.A. and Kawatra, S.K. 2007. Agglomeration for Copper Heap Leaching. In *Cu 2007—Vol IV/Book 1 The John E. Dutrizac Symposium on copper Hydrometallury.* Edited by P.A. Riveros et al. Montreal, Quebec: MetSoc-CIM.

Lu, J., Dreisinger, D., and West-Sells, P. 2007. Acid Curing and Agglomeration. In *Cu 2007—Vol IV/Book 1 The John E. Dutrizac Symposium on copper Hydrometallury.* Edited by P.A. Riveros et al. Montreal, Quebec: MetSoc-CIM.

McClelland, G.E. 1986. Agglomerated and unagglomerated heap leaching behavior is compared in production heaps. *Mining Engineering*, July 1986. pp. 500–503.

McClelland, G.E. and Eisele, J. 1981. Improvements in Heap Leaching To Recover Silver and Gold From Low-Grade Resoruces. *U.S.B.M. Report of Investigations 8612*, 26 p.

McClelland, G.E. and Hill, S.D. 1981. Heap Leaching Gold-Silver Ores with Poor Percolation Characteristics. *Gold and Silver Leaching, Recover and Economics.* Edited by W.J. Schlitt, W.C. Larson, and J.B. Hiskey. Littleton, CO: SME.

McClelland, G.E., Pool, D.L. and Eisele, J.A. 1983. Agglomeration-Heap Leaching Operations in the Precious Metals Industry. *U.S.B.M. Information Circular 8945*, 16 p.

Moats, M.S. and Janwong, A. 2008. The Art and Science of Crushed Ore Agglomeration for Heap Leaching. *Hydrometallurgy 2008—Proceedings of the Sixth International Conference.* Edited by Courtney A. Young et al. Littleton, CO: SME.

Pariani, R.L. 1987. Mechanization of Full Deposit Stripping in Electrowinning. In *The Electrorefining and Winning of Copper*. Edited by J.E. Hoffmann et al. Warrendale, PA: TMS.

Power, K.L. 1970. Operation of the First Commercial Copper Liquid Ion Exchange and Electrowinning Plant. In *Copper Metallurgy*. Edited by R.P. Ehrlich. New York: AIME.

Pu, Y.D. 1982. The History and Present Status of Practice and Research Work on Solution Mining in China. In *Interfacing Technologies in Solution Mining*, Edited by W.J. Schlitt and J.B. Hiskey, Hoboken, New York: AIME.

Scheffel, R.E. 2002. Copper Heap Leach Design and Practice. In *Mineral Processing Plant Design, Practice, and Control*. Edited by A.L. Mular, D.N. Halbe, and D.J. Barratt. Littleton, CO: SME.

Sheffer, H.W. and Evans, L.G. 1968. Copper Leaching Practices in the Western United States. *U.S.B.M. Information Circular 8341*, 57 p.

Swanson, R.R. 1965. Liquid-Liquid Recovery of Copper Values using a-Hydroxy Oximes. U.S. Patent 3,224,873. December 21, 1965.

Szymanowski, J. 1993. *Hydroxyoximes and Copper Hydrometallurgy*. Boca Raton, Florida: CRC Press. Inc.

Taylor, J.H. and Whelan, P.E. 1942. The leaching of cuperous pyrites and the precipitation of copper at Rio Tinto, Spain. Inst. Min. Metall. Bulletin 457: 1–36.

Van der Zeeuw, A.J. 1976. Shell Metal Extractant 529—A New Selective Extractant for Copper and Nickel. In *Extractive Metallurgy of Copper—Volume II Hydrometallurgy and Electrowinning*. Edited by J.C. Yannopoulos and J.C. Agarwal. New York: TMS-AIME.

Virnig, M.J. and Kordosky, G.A. 1994. Reagent Development in the 90s—A Prespective. In *Reagents for Better Metallurgy*. Edited by P.S. Mulukutla, D. Malhotra, and B.A. Hancock. Littleton, CO: SME.

ABOUT THE AUTHOR

J. Brent Hiskey is professor of mining and geological engineering and Cimetta chair in the College of Engineering at the University of Arizona. From 1999–2009 he served as associate dean for research and administration in the College of Engineering. Prior to joining the University of Arizona in 1984, he was manager of metallurgical research at Kennecott's Process Technology Center. Hiskey received a BS in metallurgical engineering (1967), MS (1971), and PhD (1973) from the University of Utah. He has held research positions with US Steel and Alcoa and was an assistant professor in metallurgical engineering at the New Mexico Institute of Mining and Technology. His primary research interests are in the area of hydrometallurgy with special emphasis on leaching, cementation, ion exchange and solvent extraction, and the investigation of electrochemical processes involved in hydrometallurgical reactions. Current research activities include gold and silver leaching and recovery, the extractive metallurgy of complex copper ores, copper electrorefining and electrowinning, treatment of aqueous effluents, and novel approaches to metal recovery from aqueous solution. Hiskey has edited or coedited a number of books: *Gold*

and Silver Leaching, Recovery and Economics; Au and Ag–Heap and Dump Leaching Practice; and Interfacing Technologies in Solution Mining. In 1993, he edited the SME/TMS book *Hydrometallurgy: Fundamentals, Technology and Innovation.* Hiskey received the SME Arthur F. Taggart Award in 1974, and was the 1993 recipient of the AIME James Douglas Gold Medal which recognized his achievements in chemical metallurgy. In 1994, he was elected a Distinguished Member of SME. He received the SME Milton E. Wadsworth Extractive Metallurgy Award in 2003 and was recognized as the Henry Krumb Lecturer of the American Institute of Mining and Metallurgical Engineers for 2005–2006. He was awarded the The Minerals, Metals & Materials Society EPD Distinguished Lecture Award in 2010. Hiskey was elected to the National Academy of Engineering, Class of 1997.

Innovations in Gold and Silver Extraction and Recovery

John O. Marsden
President, Metallurgium, Phoenix, Arizona, United States

S. Andrew Sass
President, Sass Engineering, Inc., Fremont, California, United States

ABSTRACT: There are few, if any, branches of extractive metallurgy that can lay claim to a greater degree of innovation than in gold and silver extraction. Since the first use of oil-treated sheep fleeces for enhanced gold recovery around 1,300 BC, innovators around the world have strived to devise more effective ways to recover precious metals. This paper reviews the major innovations that have occurred in gold and silver extraction, with a focus on developments that have occurred in the modern era (1972–2012). Key factors that led to the successful implementation and widespread commercialization are discussed. Finally, likely areas of attention for future innovations are explored.

INTRODUCTION

Many of the methods currently used for gold and silver extraction are based on techniques that have been known, or established, for centuries. Gravity concentration, amalgamation, cyanide leaching, roasting, chlorination, zinc precipitation and carbon/charcoal adsorption are all processes that have been used at least 100 years ago, and combinations of these remain as the basis for most gold and silver recovery flowsheets today. The ingenuity of the early pioneers in developing technology with such longevity is admirable. However, the past forty years (1972–2012) have witnessed a period of exceptional innovation following the end of the official fixed gold price in the USA. These developments have helped to make gold and silver extraction safer, more efficient, and with reduced impacts and risks on human health and the environment.

This paper reviews the major technological innovations that have occurred in gold and silver extraction. For convenience, the history of technology development in gold extraction can be divided into four main eras:

- Early History Pre-Cyanidation (<1889)
- Cyanidation (1889–1971)
- Modern Era (1972–2013)
- The Future (>2013)

This review focuses on the Modern Era (1972–2013), with some consideration of Cyanidation. For information on Early History and Pre-Cyanidation, the reader is referred to the original historical texts (Agricola 1912, Eissler 1891, Rose 1896) and historical reviews (Marsden & House 2006, Shoemaker 1984, Habashi 1987, McNulty 1989). The final section of the paper discusses and speculates upon likely areas of innovation in the future.

Important context for innovation in gold and silver extraction is the metal price history, and how this has evolved over time. Over the past century, the gold price has experienced three major peaks: 1934–1945, 1979–1996, and 2006–Current. On a 2012 U.S. dollar basis, the gold price averaged around \$500/oz (\$18.93/oz in dollars of the day) around the turn of the 19th century, increasing to about \$550/oz (\$34.36/oz) during the 1934–1945 period, \$800/oz (\$394/oz) during the 1979–1996 period, and averaging about \$1,100/oz (\$990/oz) during the 2006–2012 period. All of these peak periods were interspersed with extended periods where the gold price averaged between \$200–400/oz. With the exception of the period around WWII, all of the peak periods were associated with significant development and innovation in gold (and silver) extraction technology.

CYANIDATION (1889–1971)

The solubility of gold in alkaline cyanide solutions was recognized as early as 1783 by Scheele (Sweden) and was studied in the 1840s and 1850s by Elkington and Bagration (Russia), Elsner (Germany) and Faraday (England). In fact, Elkington obtained a patent for the use of potassium cyanide solutions for electroplating of gold and silver, but its potential application to the extraction of these metals from ores and concentrates was not recognized at that time. In 1846, the chemistry of gold dissolution in aerated cyanide solutions was investigated and reported by Elsner to be as (Elsner 1846):

$$2Au + 4KCN + O + H_2O = 2AuK(CN)_2 + 2KOH \quad (1)$$

It can be seen from Equation 1 that the role and importance of oxygen in the reaction was recognized well before any commercial process was developed, even though the reaction mechanism is now better understood (Marsden & House 2006).

At about that time, it was well-known that the addition of small amounts of potassium cyanide to amalgamation drums assisted with the recovery of gold and silver into the amalgam. However, an important "adverse" side effect of this reagent addition was the loss of some metal values to the solution. Still, no connection was made as to how the dissolution of gold and silver in cyanide solution might be applied for their extraction and recovery. Amalgamation continued to be used as a primary method of recovery. Other methods of gold and silver extraction in use included; gravity concentration (sluices, rockers, strakes etc.), chlorination, and recovery as by-products from direct smelting and refining of high grade gold ores/concentrates, and high grade copper and lead ores/concentrates (Schnabel, 1921, Rose 1896).

In 1867, Rae obtained a patent in the USA for cyanide leaching of gold and silver ores, with recovery of gold and silver from solution by electrolysis (Julian & Smart 1904). This process was never applied and, interestingly, Rae appeared to acknowledge that the method for cyanide leaching of gold and silver ores had been anticipated prior to his patent. The early form of the cyanidation process, as it is now known, was developed and patented between 1887 and 1888 by MacArthur and the Forrest brothers, two Glaswegian physicians who converted a room in their surgery into a small metallurgical laboratory. The first patent considered a method for cyanide leaching of gold and silver ores (of perhaps somewhat dubious novelty) and the second, and most important, patent considered the precipitation of gold and silver from solution using "finely-divided zinc." There was much discussion of this invention among those involved in operating and developing gold districts around the world. The process was rapidly commercialized, first at the Crown Mine (New Zealand) in 1889. The technology spread like wildfire with applications at Robinson Deep (South Africa) in

1890, Mercur (Utah, USA) and Calumet (California USA) in 1891, El Oro (Mexico) in 1900, and La Belliere (France in 1904).

The development of cyanidation was particularly timely to the exploitation of the Witwatersrand pebble-quartz conglomerate ores in South Africa, following the famous gold discovery in the Transvaal in 1886. These ores had lower grades that those previously worked and much of the gold occurred as fine grains within hard siliceous host rock. In South Africa, the cyanidation process incorporating zinc cementation, replaced or supplemented gravity concentration techniques and generally increased gold recoveries from about 70% to 95%, rescuing a declining industry. This is indicated by the tenfold increase in South African gold production from less than 300,000 oz in 1888 to over 3 million oz in 1898 (Stanley 1987).

The point to be made here is the cyanidation process was not so much invented as it evolved over a period of some 40–45 years to the point at which the process could be commercially applied with reasonable certainty of a successful outcome. The key factor here was not the lack of any ability to effectively dissolve gold and silver in cyanide solution, and no significant developments dissolution chemistry occurred. However, the use of readily available, and relatively cheap, "finely-divided zinc" for precipitation was the key enabler. This provided a fast, simple, efficient and cheap method to recover gold and silver from solution, and to generate a product that could be readily filtered, dried and processed further to produce doré bullion. This was quite different to the more complex process that had been proposed by Rae in 1867. Rae's process probably would have worked, but at a much higher cost. In fact, the rather similar Siemens-Halske process based on electrolytic recovery of gold and silver directly from cyanide solutions was developed and implemented in several plants between 1894 and 1898. However, the process was never competitive (probably due to low current efficiency) and, to this day, the electrolytic recovery of gold and silver from dilute solutions remains elusive.

Successful commercial scale applications of cyanidation paved the way for many additional developments. Early Witwatersrand operations became a hotbed for technology and innovation, and these developments revolutionized the gold industry worldwide. The earlier generation of flowsheets employing crushing, screening, manual sorting, stamp milling, gravity concentration and amalgamation were replaced by cyanide leaching and zinc precipitation circuits. Many of the early cyanidation plants employed separate sands and slimes treatment, with the size separation made using dewatering cones, rakes or spiral classifiers. Sands and slimes were leached separately in vats on a batch basis, with double- and triple-leaching steps sometimes employed to achieve reasonable extraction efficiency. Milling in cyanide solution, which is commonly applied today, was tested on the Witwatersrand in 1892 and was first applied commercially in New Zealand in 1897 (Hamilton 1920).

The introduction of tube milling in 1904 represented a major step forward allowing significantly higher tonnages to be processed, compared to stamp milling. In addition, tube mills allowed the generation of products with much finer and more uniform size distributions, leading to improved liberation (Rickard 1907). Tube milling was, of course, utilized widely in mineral processing, and resulted in many subsequent developments such as rod, ball, pebble, autogenous and SAG milling.

Between about 1904 and 1908, cyanidation plant design and operation was revolutionized by several important equipment developments (Bain 2010). In the United States, particularly in the Black Hills of South Dakota, mechanical advances for separation of solids from the gold-bearing solutions were developed. Dorr's thickener and the first CCD circuit were implemented, along with Butters' vacuum filter and Oliver's segmented drum filter. The disk filter followed later (Dorr & Bosqui 1950).

Elsewhere, other developments occurred, including the Browns' air-agitated leaching tanks (New Zealand) and "Pachuca" tanks (Mexico). Air agitation in tall, narrow tanks is still used as the primary means of suspending and mixing solids particles in some plants, although large, cost-effective, mechanically-agitated systems are now widely preferred. While these developments had application to many areas of the mining industry (and to the recovery of metals and minerals other than gold and silver), they had an enormous impact on both the scale and efficiency of the cyanidation process for gold and silver. The combination of tube milling, air-agitated leach tanks, continuous thickening and filtering equipment resulted in the gradual shift to "all-slimes" treatment, rather than separate sands and slimes treatment plants. Between 1890 and 1918, most cyanidation circuits treated the sands and slimes separately. By 1946, there were 53 large cyanidation plants operating on the Witwatersrand of which 29 were all-slimes treatment plants.

The importance of aeration and the role of oxygen, or a suitable oxidant, were well-known to early cyanidation plant operators. Julian and Smart reported extensively on this issue, including; (i) consideration of pre-aeration to satisfy the oxygen demand of reactive sulfide minerals and organic matter prior to leaching, (ii) the use of strong oxidizers to accelerate leaching rates, and (iii) self-regulating air aspiration systems in the suction side of centrifugal slurry pumps (Julian & Smart 1904). These are all leaching modifications that have been "re-invented" several times in the Modern Era.

While the role of gravity concentration diminished after the introduction of cyanidation, many plants continued to use gravity techniques to recover coarse gold-silver particles. Amalgamation continued to be used to treat gravity concentrates, however by 1922 the practice of direct amalgamation of mill product was largely replaced by the use of corduroy strakes prior to amalgamation. This pre-concentrated the feed and significantly reduced the amount of mercury required, as well as reducing exposure to mercury. This practice, encouraged largely for health and security reasons, lead to many other advances in gravity concentration equipment, such as jigs, Johnson drums, and shaking tables for coarse gold recovery down to about 100–200 µm. Surprisingly, amalgamation mills were running into the 1970s and the process is still in use today by "artisanal" miners in parts of South America and Africa.

Throughout the development and implementation of cyanidation, the recovery of precious metal values from cyanide solutions received much attention. The efficiency of cementation for precipitation of gold and silver using zinc shavings was improved significantly by three developments:

1. Use of zinc dust rather than zinc shavings in zinc boxes, as devised by Merrill (Rickard 1907),
2. De-aeration of pregnant solutions (using the Crowe vacuum de-aeration tower), and
3. The addition of small quantities of a soluble lead salt to solutions prior to contact with zinc (Clennell 1915).

The use of fine zinc dust greatly increased precipitation kinetics and improved zinc utilization, resulting in less residual zinc in the precipitate and decreasing costs. De-aeration significantly improved the single-pass efficiency of the process, resulting in much lower barren solution gold and silver concentrations and reducing zinc consumption by side reactions. These improvements had a significant effect on the operating costs for the overall cyanidation process and they were rapidly incorporated into the design of subsequent cyanidation plants in South Africa, the USA, Australia and New Zealand. Even today, where zinc precipitation is applied, it is generally referred to as the "Merrill-Crowe" process after the pioneers of these improvements.

Between 1910 and 1930, flotation was introduced for the treatment of base metal sulfide ores. This quickly led to flotation being used for the concentration of gold-bearing sulfide minerals and free gold in ores that were suited to such treatment. Early examples of flotation for gold and silver recovery included the Empire Mine (California) and Mount Morgan (Queensland, Australia). Flotation produced marginally higher gold recoveries at significantly lower cost, enabling lower-grade sulfide ores to be treated economically.

During WWI, potassium cyanide became scarce and was replaced by calcium cyanide, also known as black cyanide. Although not a dark period, the innovations that occurred from the 1920s to 1960s were mainly associated with development of larger machinery and achieving higher throughput rates. Advances in comminution equipment aided gold and silver extraction, as well as production of most other metals. The two World Wars disrupted production, and fixed gold prices in times of rising production costs provided less incentive for innovation. It was, however, a period where the seeds of the innovations of the Modern Era were sewn. Several excellent texts on South African gold and silver metallurgical practice during this period are available in the literature (King 1949, Adamson 1972, and Stanley 1987).

THE MODERN ERA (1972–2013)

The major innovations of the modern era reviewed in this section are summarized in Table 1.

Activated Carbon and the Carbon-in-Pulp Process

The application of activated carbon for the recovery of gold and silver from cyanidation plant slurries gained worldwide commercial acceptance in the 1980s. Like the cyanidation process in the late 1800s, the history of use of activated carbon for gold and silver recovery began many decades earlier. The adsorption of precious metals from aqueous solutions onto activated carbon was first noted in the early 19th century. A process using wood charcoal to recover gold from chlorination leach solution was patented in the late 1800s (Davis 1880) and was applied at Mount Morgan (Australia). Charcoal was reportedly used at other operations in Australia for gold recovery, e.g., Yuanmi and Sons of Gwalia. Similarly, in the 1890s, activated carbon was considered as a possible alternative to electrowinning and zinc cementation for the Witwatersrand gold plants in South Africa. However, at that time, the only known way of recovering the gold was by combustion of the carbon and smelting of the resulting ash. The inability to re-use the carbon, coupled with the ease, simplicity and relatively low cost of the zinc cementation technology, made the process uncompetitive.

In the 1920s, the United States Bureau of Mines (USBM) studied the use of charcoal for gold and silver recovery from alkaline cyanide solutions and further work was done by Chapman and Potter at the University of Arizona in the 1930s. The first plant to use granular activated carbon, rather than charcoal, in a carbon-in-pulp-type configuration was probably the 250-tpd San Andreas de Copan Plant (Honduras) in 1949, where the loaded carbon was sold to a smelter for gold and silver recovery. In 1950, the Getchell Mine in northern Nevada operated a 500-tpd CIP plant, and a patent was granted to McQuiston and Chapman in 1951 for a CIP technique. Around this time period, much important work was conducted at the USBM in Reno and Salt Lake City, notably by Zadra, Salisbury and Ross, to devise a process for gold and silver recovery from carbon which would allow the carbon to be re-used.

Activated carbon was used commercially in 1952 at the Carlton Mine in Cripple Creek (Colorado), which operated until 1961 when the plant was closed due to declining ore grades

Table 1. Summary of major innovations in gold and silver extraction in the modern era (1972–2013)

	Innovation	Specific Technology	Pre-Cursors	Critical Enablers
1	Carbon Au-Ag Recovery Systems	CIP (Carbon-in-Pulp)	RIP-Former USSR Carbon use at San Andres de Copan with combustion of carbon and Au-Ag recovery from ash	Carbon Elution Interstage Screens Carbon reactivation Coconut Shell (hard) carbon
2	Heap Leaching	Cyanide leaching of crushed ore on lined heap leach pad systems	Copper stockpile leaching Copper heap leaching	Agglomeration technology CIC (more efficient than Zn ppt) Solution distribution and application systems
3	Sulfide Oxidation	Pressure oxidation	Zinc pressure oxidation	Low cost oxygen supply (VSA) Materials of construction Flash/splash technology
		Circulating Fluidized Bed (CFB) Roasting and Fluidized Bed Roasting (FB)	Edwards Roasting	Hot gas-particle handling systems Fluidized bed technology (BFB) Circulating Fluidized Bed technology (CFB)
		Biological oxidation	Copper Sulfide Biological Leaching	High oxygen mass transfer reactor development for atmospheric systems
4	Centrifugal Gravity Concentrators	Knelson Concentrator Falcon Concentrator	Spirals, Jigs, Tables, Sluices	Centrifugal bowl technology
5	Ultra fine grinding of flotation concentrates (as an alternative to full oxidation)	Fine Grinding Technology for sulfide concentrates	English China Clays mill Netsch mill	Stirred milling technology Isamill, Metso SMD Detritor, Metprotech
6	Cyanidation Improvements	Oxygen enhancement	Air injection in pachucas and mechanically agitated tanks	Reasonable cost oxygen supply (or hydrogen peroxide in a few selected cases)
		Intensive Cyanidation	Leaching with high NaCN Oxygen enhanced leaching EW of high grade Au-Ag solutions developed for CIP/CIC	Production of high grade gravity concentrates, especially from centrifugal concentrators (see item 4)
7	Cyanide Recovery (and Treatment of High Copper Feed Materials)	SART	AVR Cold stripping of carbon to remove Cu (El Indio)	Copper sulfide precipitation
8	Cyanide Destruction	Inco SO2-Air Hydrogen Peroxide	Environmental regulation Bird Mortality Cyanide code	Development work by CANMET
9	Integrated Flowsheets	CIP/CIL with Heap Leaching and CIC	Hybrid CIP-CIC circuits Hybrid heap leach-mill circuits	Carbon Adsorption Systems; Heap Leaching

(table continues)

coupled with the low fixed price of gold. An important feature of the Carlton flowsheet was the use of a reactivation kiln for carbon regeneration. The combination of efficient carbon regeneration and elution made carbon adsorption methods competitive with the Merrill-Crowe process, and this application was an important pre-cursor to the widespread development of CIP.

The Homestake Mine (South Dakota) was looking for a safer and environmentally-sound alternative to amalgamation. In 1973, the conventional slime treatment and zinc precipitation circuit was successfully replaced with carbon-in-pulp, becoming the largest scale application of carbon adsorption at the time and marking the beginning of the modern gold era. At this point, both the Council for Mineral Technology (later to become Mintek) and Anglo American Research Laboratories (AARL) in South Africa started to perform significant research and development work to understand the CIP process, and to adapt it to treat South African ores. As part of this work,

Table 1. Summary of major innovations in gold and silver extraction in the modern era (1972–2013) (continued)

	Innovation	Key Attributes	Downsides/Disadvantages	Spin Offs, Enhancements	Key Players
1	Carbon Au-Ag Recovery Systems	Eliminates the need for slurry filtration Eliminates/replaces Merrill-Crowe Zn precipitation	Gold losses onto fine carbon (minor)	Carbon-in-columns (CIC) Carbon-in-Leach (CIL) Kemix Pumpcell reactors	USBM AARL Mintek
2	Heap Leaching	Extraction of gold and silver from ores at ROM, coarse or crush sizes (no grinding required) Low capital and operating cost	Typically lower Au-Ag extractions achieved compared with milling and agitated leaching processes	ROM stockpile leaching Leaching in cold climates	USBM Newmont
3	Sulfide Oxidation	Efficient and complete oxidation of sulfides No sulfur dioxide emissions Arsenic and Mercury fixed	Requires autoclave technology High pressure system Materials of construction critical Titanium fire risk	Hot lime boil Pre-acidulation Carbonate addition	Homestake Getty/Barrick Sherritt Gordon
		Efficient and complete oxidation of sulfides and organic (preg rob) carbon	Locking/encapsulation of Au/Ag can occur Gas handling and sulfur dioxide capture required	Dry grinding	Lurgi, Newmont Barrick, Placer KCGM
		Lower cost (capital and operating) compared with roasting and pressure oxidation	Typically less efficient sulfide oxidation (slower, less complete) compared with roasting and pressure oxidation	Biological Heap Leach (Newmont); Geobiotics GeoCoat Process	Gencor, Bactech Mintek
4	Centrifugal Gravity Concentrators	Ability to recover Au particles down to about 25 microns into low mass concentrate	Limited capacity per unit	Intensive cyanidation of concentrates - Gekko In-line Reactor and Consep Acacia	Knelson Falcon Laplante
5	Ultra fine grinding of flotation concentrates (as an alternative to full oxidation)	Ability to grind sulfide concentrates down to 80% <7-15 microns at energy input of 30-80 kWh/ton (depending on mineralogy, feed size distribution)	Relatively high operating cost due to energy and grinding media requirements; however lower cost compared with sulfide oxidation methods	Ceramic grinding media	Xstrata Metso
6	Cyanidation Improvements	Ability to maintain high dissolved oxygen levels required for efficient cyanide leaching, even in the presence of reactive (oxygen-consuming sulfides, eg. Pyrrhotite)	Cost of oxygen or other oxidant	Enhanced oxygen sparging systems	Various
		Ability to maintain high cyanide and dissolved oxygen concentrations in solution at elevated temperature and achieve high Au-Ag extractions	Cost of reagents, energy for temperature (minor)	Extension to treat a wider range of feed stocks (e.g. scrap, electronic materials, other concentrates etc.	Consep Acacia Gekko
7	Cyanide Recovery (and Treatment of High Copper Feed Materials)	Removal and recovery of copper from cyanide solution; removal, recovery and recycling of NaCN	Cyanide losses, consumption Additional operating cost of SART system	Extension to treat a wider range of Cu-bearing feed materials	SGS Elbow Creek
8	Cyanide Destruction	Destruction of free and WAD cyanide <50 mg/L	Cost of sulfur dioxide/hydrogen peroxide		Inco
9	Integrated Flowsheets	Efficient handling of solutions, carbon through common system	None		Various

Davidson patented the AARL elution process in 1973, which rapidly became an established and popular carbon elution method worldwide. Largely as a result of these efforts, a small (90,000 tpy) CIP plant was installed to treat tailings at Modderfontein (South Africa) in 1978. In 1980, following the successful application at Modderfontein, three larger plants were installed in South Africa at President Brand (to treat calcine), Randfontein Estates, Western Areas, RM3 (Rand Mines), and Beisa (Boydell 1984, Laxen 1982). The floodgates opened and between 1981 and 1984, 11 major CIP and CIL plants were commissioned in South Africa alone, with each processing in excess of one million tons per year of material. The first CIP plant in Australia was commissioned at Kambalda in 1981, followed by plants at Noresman and Haveluck. The Jerritt Canyon Mine was commissioned in 1981 and several other CIP/CIL plants were commissioned in the USA and elsewhere in the early 1980s.

The key enablers for the successful development of CIP and CIL technology were

- Development of inter-stage screens (to retain carbon within each tank, allowing slurry to pass) that could handle the slurry flow rates at the desired slurry density and viscosity,
- Development of effective carbon elution techniques (e.g., Zadra, AARL), and
- The ability to regenerate the carbon using acid washing and thermal reactivation techniques, allowing the carbon to be re-used.

Many players in the industry worked on the development of inter-stage screens and numerous options and alternatives were tried and tested, modified, then tried and tested again. Initial CIP plants relied on external screens which required pumping of the entire slurry flow from tank to tank. The development of screens located within agitated slurry tanks provided many advantages. The key issue was how to prevent blinding of the screen over time, which could be as short as five minutes for the earliest models. Self-cleaning, air-swept, and mechanically-swept devices and both in-tank and external screens were employed with varying results. NKM (North Kalgoorlie Mines), modified-NKM, and Derrick screens are now largely preferred over the earlier Kambalda, EPAC and other configurations.

Quickly the ore leaching and carbon loading kinetics were understood, and the separate leaching and CIP steps were combined which reduced 'preg-robbing' of gold and silver by carbonaceous ore constituents. This combined circuit was referred as Carbon-in-Leach (CIL). However, the disadvantages of CIL were increased carbon inventory and gold-in-circuit inventory, lower gold loadings, and potentially higher carbon losses.

By the early 1980s, CIP/CIL was the flowsheet of choice. Despite lower silver loadings on carbon, it was being applied to treat primary silver deposits as well as gold. The demand for activated coconut shell carbon lead to shortages with vendors scrambling to add more production capacity to meet demand.

As the application of the CIP process gained momentum, a number of associated technical developments took place. For example, carbon manufacturers developed improved activated carbon products with improved activity and better abrasion resistance (specifically for gold and silver extraction). Methods for monitoring and controlling carbon quality were borrowed from the sugar industry. Improved approaches to pumping, handling and screening loaded, and treatment of stripped and virgin carbon lead to improved efficiencies. In early CIP/CIL circuits, carbon consumption was significant, around 200–300 g/t. An important development for reducing carbon consumption was the use of large radius sweeps and bends in piping, low friction piping, and recessed impeller pumps. The introduction of curved blade, large diameter, low-tip speed, agitators in CIP/ CIL significantly helped to decrease carbon degradation, as well as saving energy and cost. Pre-conditioning of carbon followed by screening and rejection of fines prior to addition into the CIP circuit helped to reduce gold on fine carbon losses significantly. Currently, many operations achieve 30–50 g/t carbon consumption with the application of best practices.

The development of the Kemix Pumpcell by Anglo American in South Africa in the late 1980s provided a lower capital cost option for CIP, effectively providing a high-intensity slurry-carbon contactor for improved adsorption kinetics (i.e., less volume and retention time required). The tanks could be arranged in a carousel-type configuration to minimize the movement of carbon through the system. This has some advantages for some applications and a significant number of Kemix Pumpcell systems have been applied commercially around the world (Whyte 1990).

Metallurgists in the former U.S.S.R. implemented Resin-in-Pulp (RIP) technology in the late 1960s and 1970s, whereby ion exchange resin beads were used to perform the same role as activated charcoal. A number of plants were reportedly built in Russia, Uzbekistan, Kazakhstan and Kyrgyzstan, for example at Muruntau (Uzbekistan), but most had problems with floating of the resin beads, resin degradation and resin losses. Also, complex elution and regeneration systems were required for the removal of base metals, the recovery of gold and silver, and for the removal of other contaminants. Subsequently significant work has been conducted on the development of effective resin systems in the West with considerable success (Fleming & Cromberge 1984). There are commercial installations of resin systems at Golden Jubilee (South Africa) and Penjom (Malaysia). However, these systems have yet to be more broadly applied commercially in the West.

Heap Leaching

Heap leaching of gold and silver ores was developed primarily by Heinen, Lindstrom, McClelland and others at the U.S. Bureau of Mines in the late 1960s and early 1970s as a low cost treatment option for low grade hydrothermal oxide ores in Nevada. Undoubtedly, the successful application of heap and stockpile leaching to extract copper from low grade dumps and stockpiles in Spain in the early part of the 20th century and following that in Arizona, New Mexico, and Utah, was an important influence on (and early motivator of) these efforts.

The first large-scale cyanide heap leach for gold and silver was commissioned at Newmont's Carlin Mine (Nevada) in 1971, processing material with grades below the established mill cut off grade. This was followed by applications at Cortez and Smoky Valley, now Round Mountain, in Nevada in the late 1970s (Heinen et al. 1980). Thereafter, the use of heap leaching boomed throughout the western United States with operations in Nevada, New Mexico, California, Utah, Montana, Idaho, and South Dakota.

Key enablers for heap leaching included the following.

- Effective agglomeration procedures for clay-bearing ores to provide solution permeability for a sustained period time. The main features of these agglomeration procedures included
  a. A suitable binder, such as lime, cement or combinations thereof;
  b. The correct moisture content for effective agglomeration, usually 6–10% water by weight; and
  c. Effective mixing of the ore, moisture and binder(s) to make stable, high quality, agglomerates using rotating drums or multiple conveyor drop points.
- Effective liner systems based on polyethylene material (e.g., HDPE/LDPE), compacted clay, and/or composite liners for containment of ore and solutions (both to maximize metal recovery and for environmental protection). PVC was also tried but proved to be unstable to ultraviolet light exposure.
- Stacking equipment that could place crushed ore onto a heap without creating excessive compaction and to maintain effective permeability within the heap.
- Methods to rip the heap to increase permeability prior to placing subsequent lifts.
- Solution distribution and irrigation systems to apply cyanide solution at the desired rates (5–15 $L/hr/m^2$ or 0.002–0.006 gpm/ft^2) to avoid breaking down the agglomerates and to avoid fine particle migration (to prevent plugging of the heap). Drip irrigation provided benefits for reduced evaporation and wind losses in the semi-arid regions where heap leaching

was most popular. This technology came from agricultural applications and was pioneered by Herzog, Wilder and Dixon at Coeur-Rochester in Nevada in 1987 (Arnold et al. 1990).

While Merrill-Crowe zinc precipitation technology was available for recovery of gold and silver from cyanide solution, heap leaching took advantage of the parallel development of carbon adsorption technology. Large volumes of solution could be processed effectively with up-flow Carbon-in-Columns (CIC) systems, while avoiding the need for costly solution clarification or filtration. CIC systems were quickly and widely adopted in the early 1980s. The same carbon elution and regeneration technologies as those developed for CIP were used to treat the loaded carbon from CIC. Where appropriate, common carbon elution and regeneration systems were used at operations which had both agitated leach-CIP and heap leach-CIC flowsheets. The importance of the synergies between CIC and CIP systems cannot be overstated. The Chimney Creek operation (Nevada), commissioned in 1987, took full advantage of these synergies and combined the best aspects of CIP, CIC and counter-current decantation in a "double-thickener" circuit utilizing CIC to treat thickener overflow solution and CIP to recover gold and silver from thickener underflow slurry (Mansanti 1988). This configuration improved overall recoveries, maximized metal loadings on carbon, and carbon stripping efficiency. Interestingly, the Chimney Creek plant successfully eluted gold and silver values from fine carbon using a Funda pressure leaf filter (down to <150 g Au/t) and it was the first plant to use IBM-XT personal computers as the process control system with "massive" 40 MB hard drives.

The sharing of facilities for CIP/CIL- and CIC-generated carbon, resulted in substantial capital and operating savings. For small operators, stripping and reactivation of loaded carbon was often available through third parties in nearby locations. Operators with multiple operations, such as Newmont and Barrick in Nevada, established centralized carbon handling facilities to further reduce costs and improve security.

As heap leaching technology matured, improved methods for testing of ore response to the heap leaching, testing of agglomerate strength and stability, testing of ore permeability under load, and other tools and procedures evolved.

The over-whelming success of heap leaching in the western United States was related to the suitability of the climate, the nature of the hydrothermal breccias-type ores commonly encountered (with gold and silver mineralization located as largely liberated particles along cracks and fissures, accessible to solution), the development of effective agglomeration procedures, and innovative operators. Gold recoveries ranging from 50 to 80% were achieved (depending largely on the ore size distribution and the ore liberation characteristics) at much lower capital and operating costs compared with milling-CIP plants. In addition, start-up times were relatively short, which led to the use of heap leaching to generate revenue early in the life of a project to help fund the construction of additional facilities and mining operations. This technology converted many millions of tons of waste material into ore (Marsden 1996).

In 1989, approximately 36% (112 tons) of U.S. gold production resulted from heap leaching, with operations ranging in size from 500 to about 10,000 tpd. Global gold production via heap leaching was approximately 136 tons, or just over 8% of the total. By 2005, worldwide heap leach gold production had increased to 236 tons, or 9.6% of production, with the U.S. share of this production declining to 70 tons. In 2005, the largest heap and stockpile gold-silver leaching operation in the world was Newmont's Yanacocha (Peru) which produced a staggering 90 tons of gold that year (Gold Fields Mineral Services 2006).

The fundamentals of gold and silver heap leaching have not changed significantly over the past 30 years or so, although the breadth and extent of applications have grown significantly. Several significant evolutionary developments that have occurred include the following.

- Heap leaching in extreme cold climates; for example:
  a. Zortman-Landusky operated a cold climate run-of mine and crushed ore heap leach in northern Montana successfully in the 1980s and 90s, pioneering the way for later cold climate applications (Marsden et al. 1993).
  b. The Fort Knox operation in Alaska has operated a successful run-of-mine heap leaching operation on low grade ore since late 2009. Winter temperatures reach down to –50°F. Leaching is conducted on a year-round basis with ore stacking during the warmest 8 months of the year (Kinross Fort Knox 2012).
  c. The Jinshan Gold Mine in China treats 20,000-tpd of run-of-mine ore via heap leaching with temperatures reaching –32°C in winter (Keane 2007).
  d. Common features of these cold climate heap leaching operations are the use of in-heap pregnant solution storage, rather than external storage ponds, and the direct delivery of barren onto the heap. This serves to maintain temperature in the circulating solutions and avoid freezing during the severe winter months.
- Heap leaching of high silver ores using high cyanide concentrations (≥1 g/L NaCN).
- Advanced modeling and simulation of heap leaching systems, leading to improved operation and control (PERI-University of Greenwich, Rio Tinto, University of British Columbia, AMIRA, others).
- Targeted solution injection to access and recover residual metal values in heaps and stockpiles, e.g., Hydro-Jex Technology (Seal 2007).

Sulfide Oxidation and "Refractory" Ore Pre-Treatment

Pressure Oxidation

From the mid-1970s through the 1990s, major efforts were directed at the treatment of so-called "refractory" ores—material not effectively treated by simple cyanidation, flotation or gravity concentration techniques. Historically (prior to 1970), severely refractory ores and concentrates, whether due to their sulfide, carbonaceous, telluride components, or combinations thereof, were roasted to completely oxidize the refractory components and render the contained gold leachable. Examples of such conventional roasting operations included; Fairview (South Africa), La Belliere (France), Getchell (Nevada), Mount Morgan (Australia), and Campbell Red Lake and Giant Yellowknife (Canada).

In 1971, the Carlin Mill (Nevada) successfully started to treat highly carbonaceous ("preg-robbing") ore with low sulfide content (<0.5% S) by chlorination, followed by cyanidation and zinc precipitation (later replaced by CIP). In 1977, the process was altered to a "double-oxidation" circuit by incorporating a pre-treatment step using heat and sodium carbonate to pre-oxidize sulfide minerals ahead of chlorination, thereby reducing chlorine consumption. This process was later used at Jerritt Canyon (Nevada) but, due to rising costs, was subsequently discontinued at both locations in the 1980s in favor of the original simpler chlorination process. Modern roasters are now used at both of these properties, as discussed later. Chlorination was applied at Emperor Mines (Fiji) from 1932 until the early 1980s to treat gold-bearing telluride flotation concentrates (McQuiston & Shoemaker 1975).

The use of pressure oxidation to treat gold and silver-bearing sulfide ores was investigated in the 1970s and 1980s. This work built upon the established use of pressure oxidation technology that had been applied successfully for many years in the zinc and nickel industries. Large scale pilot "demonstration" autoclaves were installed at Western Deeps in South Africa. In 1985, an acidic pressure oxidation plant using horizontal multi-compartment autoclaves was installed at the Homestake-McLaughlin Mine (California) to treat a pyritic ore. This was followed in 1986 by a pressure oxidation circuit to treat gold-bearing sulfide flotation concentrates at Sao Bento (Brazil). In 1988, an alkaline pressure oxidation facility was commissioned at Mercur (Utah). In the latter case, non-acidic conditions were required because of the high carbonate content (16% CO_3) of the ore. The acid generated by sulfur oxidation with an autoclave reacted with carbonate materials generating carbon dioxide gas, diluting the oxygen and requiring venting which removed heat and oxygen from the vessel. The McLaughlin flowsheet handled a lower amount of carbonate content by contacting the ore with acid produced from the oxidation step prior to the autoclave (termed "pre-acidification").

These plants established pressure oxidation as a viable method for treating a range of refractory ores and concentrates, with high gold recoveries. As a result, a number of large pressure oxidation plants and expansions were commissioned between 1988 and 2000, including Goldstrike, Getchell, Lone Tree, and Twin Creeks (all in Nevada), Campbell Red Lake and Con (Canada), Lihir and Porgera (Papua New Guinea), and Macraes (New Zealand).

The hot slurry discharge from pressure oxidation was typically cooled by "flashing" to atmospheric pressure. The flash steam was typically used to preheat the feed slurry. Starting with the McLaughlin autoclaves, the use of multiple flashing and heating stages, sometimes called "flash-splash," was an innovation that was designed using principles applied elsewhere in the chemical industry. Another development was to raise the sulfur content of the incoming ore to provide enough fuel to achieve fully autogenous operation, with no externally generated steam required to maintain operating temperature. Between 4.5–5.5% sulfide sulfur is generally the minimum requirement for autogenous operation, although this depends on the feed mineralogy.

The advantages of this technology over conventional roasting were

- Greatly reduced off-gas handling requirements (vent gas handling rather than roaster off-gas handling and treatment);
- Encapsulation of gold values in the calcine avoided;
- Arsenic was fixed as stable ferric arsenate (assuming sufficient iron present), rather than being volatilized in the roasting process;
- Higher gold recoveries; and
- The ability to accommodate slurry feed and avoid the need for dry grinding.

However, the disadvantages of pressure oxidation were the loss of most of the silver to jarosite formation (which is essentially insoluble in cyanide solution), the higher capital cost, and the higher oxygen requirement and cost.

In the late 1980s, a hot lime boil process was developed to address the issue of silver jarosite formation (Berezowsky and Weir 1989). This consisted of agitating the pressure oxidation discharge slurry for 1–2 hr at 80–90°C with lime to maintain a pH of about 11. Silver extractions were significantly increased, typically from 5–20% to over 80%. In some cases, this procedure also increased gold extraction due to the release of gold occluded in jarosite and iron precipitates. However, there was a cost associated with this process (energy for heat and agitation, lime consumption), although the heat could be obtained from the pressure oxidation step if the feed contained sufficient sulfur.

Following the successful installations to treat ores and concentrates, a number of important improvements and enhancements were developed:

- The use of an enlarged first compartment (to provide longer first compartment residence time and/or to reduce steam consumption by providing hot re-circulated slurry to mix with raise incoming slurry to reaction temperature)
- Improved materials of construction, including the use of ferralium and other alloys in place of titanium to reduce the risk of fires
- Improved autoclave lining systems to increase autoclave operating availability, allowing for longer time between inspections and maintenance
- Developments in high pressure slurry pumping at elevated temperatures to allow for more recovery of flash steam
- Development of a process to prevent significant jarosite formation by controlling the amount of free acid in the autoclave, using carbonate addition where necessary (Simmons 2003)
- Washing of ore feed to remove chloride and thereby reduce or eliminate the dissolution of gold as a chloride in the autoclave. This was an important development for some carbonaceous ores, e.g., Macraes (New Zealand)
- Injection of oxygen from a flange below the agitators to reduce the titanium piping requirements within the vessel

Roasting

In parallel with developments in pressure oxidation technology, significant improvements were made in roasting technology which brought low-cost roasting back into favor in the late 1980s. The implementation of fluidized bed (FB) and circulating fluidized bed (CFB) reactors with effective gas scrubbing and cleaning systems resulted in plants with significantly improved temperature control, better control of gas composition, and much lower particulate and sulfur dioxide emissions than conventional roasting technology described above. Between 1988 and 1990, whole ore roasters were commissioned at Big Springs, Cortez and Jerritt Canyon (all in Nevada), Golden Bear (Canada), and at Gidji/Kalgoorlie Consolidated Gold Mines (W. Australia). A large whole ore roaster utilizing dry grinding was installed at Newmont's Carlin operation (Nevada) in 1994 and other installations followed at Syama (Mali), Minahasa (Indonesia), and Goldstrike (Nevada).

The main advantages of roasting over pressure oxidation were

- Destruction of carbonaceous ("preg-robbing") components of the feed;
- Near-complete oxidation of sulfide minerals in very short retention times; and
- The ability of the CFB technology to avoid over-roasting of fine particles, due to the nature of the cyclone hot-gas separation technology employed.

However, the disadvantages of the new, improved roasting technology included

- The energy and cost required to eliminate the moisture from the ore or concentrate (i.e., by dry grinding, drying, or slurry feed to the roaster),
- The need to fix sulfur dioxide in the bed and/or gas scrubbing requirements,
- The need to either fix arsenic in the roaster bed or remove it from the gas phase, and
- Cost of gas handling and scrubbing systems to remove undesirable constituents.

Biological Oxidation

Throughout the 1970s and 1980s, researchers at Cardiff University (Wales), University of British Columbia (Canada), University of New Mexico (USA) and Gencor (South Africa), among many others, investigated bacterial oxidation to treat sulfide refractory gold and silver ores. In 1984, a 0.75-tpd pilot scale biological oxidation plant was commissioned at Fairview (South Africa) to treat gold-bearing sulfide concentrate using Gencor's BIOX technology. That same year, in an unrelated development, Equity Silver Mines (British Columbia, Canada) commissioned a 2-tpd agitated tank biological oxidation pilot plant, which indicated the global interest in this technology for gold and silver extraction. In 1986, a 10-tpd plant was commissioned at Fairview (South Africa) to process flotation concentrates and this operated successfully for many years. A biological oxidation circuit was installed at Sao Bento (Brazil) in 1991 to partially oxidize sulfide concentrate ahead of the pressure oxidation circuit to increase throughput. A larger, 720-tpd agitated biological oxidation system was installed to treat sulfide flotation concentrates at Ashanti-Obuasi (Ghana) in 1994. This application was successful and was subsequently expanded to 960-tpd, firmly establishing biological oxidation as a viable option for refractory gold and silver sulfide concentrates. Additional installations followed at Harbour Lights, Wiluna, Youanmi and Beaconsfield (all in Australia), Tamboraque (Peru) and Laizhou (China).

In 1990, a 1,500-tpd whole ore biological oxidation plant was started up at Tonkin Springs (Nevada), although the operation was subsequently shut down due to financial problems. As such, the whole ore biological oxidation in an agitated system has not yet been proven commercially. Newmont operated a biological heap leach at Carlin with some success in the 1990s. However, despite significant interest and efforts by many, biological heap leaching of sulfide gold and silver ores has not been widely commercialized.

An important attribute of materials that are most suitable for treatment by biological oxidation is the ability to liberate a substantial portion of the contained gold (and silver) without needing to achieve complete, or essentially complete, oxidation of the refractory sulfide minerals. Ores and concentrates containing gold and silver mineralization either hosted within more reactive sulfide minerals or located along the surfaces, within fractures and grain boundaries of sulfide minerals, are most amenable to biological oxidation. Unlike pressure oxidation and roasting, which achieve complete or near-complete oxidation within short retention times (minutes to hours), biological oxidation in agitated tank systems require several days (typically 3–5) to achieve acceptable levels of sulfide oxidation and gold-silver liberation. For biological heap leaching systems, retention times of at least 90 days are required.

The major advantages of biological oxidation over pressure oxidation and roasting are

- Operation at atmospheric pressure and moderate temperatures,
- Relatively simple flowsheet and less complex equipment,

- Lower capital cost for unit of throughput, and
- Potentially lower operating cost.

However, the disadvantages are the typically lower gold extraction achieved and the less robust operating control capability due to the longer residence time and slower process response to changes.

To put all of these oxidative pre-treatment developments into context, by 2004, approximately 127 tons of gold (5.2% of global production) was produced using pressure oxidation technology, compared with 88 tons (3.5% of global production) by roasting and 26 tons (1.1% of global production) by biological oxidation (Marsden 2006). The great majority of gold production is still produced by cyanide leaching, supplemented by gravity and flotation technologies.

Gravity Concentration

Gravity concentration has been applied for gold and silver recovery for millennia. Historically, gravity concentration has been used in conjunction with mercury amalgamation. In the late 1800s and early 1900s, gravity concentration equipment included sluices, strakes, trommels, "long toms," corduroy blanks, and gold pans. Amalgamation was incorporated into early flowsheets for treatment of Witwatersrand ores. The development of more sophisticated equipment for enhanced gravity separation such as jigs, spirals, shaking tables and Reichart cones improved the efficiency of this method of recovery over time. By the mid-1970s, many gold flowsheets still commonly employed one or more forms of gravity concentration. However, the use of amalgamation continued to fall out of favor due to human health and environmental risks. Either the concentrates had to be sufficiently high grade to allow direct smelting of the product to doré bullion (requiring substantial upgrading and a significant loss of values to the gravity tailings), or lower grade concentrates had to be treated separately by leaching or other means (e.g., amalgamation). Because of the high grade and generally coarse particle size distribution of the gold and silver values, often associated with reactive heavy sulfide minerals (e.g., pyrrhotite, marcasite, pyrite, arsenopyrite), leaching of such concentrates was typically fairly inefficient.

In 1978, Byron Knelson produced the first model of the Knelson centrifugal concentrator after visiting the Yukon alluvial gold fields in the early 1970s and deciding that there had to be a better way to recover gold (Knelson 1990). The unit was successfully tested on a number of gold and silver ores in Canada, USA and Australia with encouraging results. The centrifugal concentrator was designed to accept minus 6 mm feed and was capable of recovering gold particles down to 25 µm. In some applications, gold particles 10 µm and below were recovered (albeit with lower efficiency). It took some time for the gold industry to recognize the importance of this development. In 1984, one investigator in Australia observed that "current information regarding the results and applications show that the Knelson Concentrator represents a new era in free gold recovery" (Harris 1984). This was a very astute and prophetic observation at the time. A number of Knelson Concentrators were installed in the 1980s in Australia and Canada, but it wasn't until Andre Laplante developed his Gravity-Recoverable Gold (GRG) testing and evaluation procedure following his investigations into gravity gold recovery at Les Mines Camchib in the late 1980s/early 1990s that the commercial applications really took off (Laplante 1990, Laplante 2000). Laplante's innovative approach introduced a sound scientific rationale to the evaluation of gravity gold recovery which, up to that point, had been difficult to demonstrate based on the poor reliability of laboratory testwork procedures and interpretation of results. In 1996, Knelson introduced a 300-tph semi-continuous concentrator

(XD48) which was shortly followed by the 650-tph unit (XD70). Up to that point, centrifugal concentrators had fairly limited capacity, but these larger units provided options for larger gold ore milling operations. Successful installations included Rosebery, Sunrise Dam, and Kundana (all in Australia), Ashanti-Obuasi (Ghana), Morila (Mali), Freeport-Grasberg (Indonesia), Porgera (Papua New Guinea), Campbell (Canada), Serra Grande (Brasil) and Rio Narcea (Spain).

The Falcon Gravity Concentrator, a competitor to the Knelson units, was introduced in the early 1990s. Installations of 330-tph "semi-batch" units at Alumbrera (Argentina) and Telfer and Cadia (Australia) were completed in the late 1990s and early 2000s. The Knelson and Falcon centrifugal concentrators have provided the operator with excellent options for gravity gold (and silver) recovery.

Another important development that contributed to the success of the centrifugal concentrator technology was the parallel development of intensive cyanide leaching processes to replace the conventional, and out of favor, amalgamation process. Intensive cyanide leaching to treat gravity concentrates was investigated extensively in the late 1970s and early 1980s in South Africa by Anglo American Research Laboratories (AARL) and Mintek, among others. Intensive leaching operates on the principle of achieving fast leaching rates by (i) achieving elevated dissolved oxygen concentrations using a pressurized system, (ii) the addition of a suitable oxidant, (iii) elevated temperature, (iv) relatively high cyanide concentration, and (v) combinations of these conditions. The first commercial application of intensive cyanidation of gravity concentrates was at the President Brand plant in 1986 (South Africa). However, leaching of gravity concentrates had probably been tried at other operations prior to that time. During the 1990s and early 2000s, two other commercial processes emerged; the Consep Acacia reactor and the Gekko In-line Leach Reactor or "ILR" (Watson 2002, Longley 2003). As an example, the Acacia process uses cyanide concentrations of up to 15–25 g/L, 3–4 g/L caustic, and 2–10 g/L of a suitable oxidant. Leaching is carried out at 50–65°C.

These intensive leaching systems have a number of important advantages over (mercury) amalgamation:

- The elimination of mercury (health and environmental risks)
- The ability to achieve high gold (and silver) recoveries within relatively short leaching times, i.e., 18–24 hr
- Compatibility of the resulting gold and silver-bearing solution with electrowinning systems to recover precious metals from high grade carbon strip solutions
- Ability to supply intensive cyanide leaching systems as a package that can be pre-assembled and delivered to remote locations quickly and efficiently

The development of centrifugal concentrators and effective intensive cyanide leaching systems, coupled with Laplante's approach to the evaluation of ores for their amenability to gravity concentration techniques, have dramatically changed the approach to gravity gold and silver recovery over the past 25 years.

Flotation and Ultra-Fine Grinding of Concentrates

Flotation has been applied as an effective method of pre-concentration prior to cyanide leaching and/or oxidative pretreatment processes since the earliest days of flotation in the 1910s and 1920s. The large scale application of flotation at Paracatu (Brasil) is a good example of this, where approximately 56,000 tpd (20 M t/yr) of ore with a head grade of about 0.4 g/ton are processed using a combination of gravity concentration, flotation and cyanide leaching (Kinross 2012). Flotation concentrates are leached in a dedicated hydrometallurgical (CIL) facility while the flotation tailings are discarded without leaching.

With the exception of some evolutionary developments in reagents and flotation conditions and progressively larger flotation cells, the basic flotation process has remained largely unchanged from the early configurations. Xanthates are still used as the primary collector for gold and silver-bearing sulfides. A number of special collectors have been developed for gold extraction and these have been applied with success in a number of applications. An excellent review of flotation for gold and silver recovery is available in the literature (Kappes et al. 2011).

The application of inert gas-assisted flotation (using nitrogen) by Santa Fe Pacific Gold at the Twin Creeks and Lonetree operations (Nevada) provided a 5 to 15% boost in gold recovery to the concentrate and increased the rate of recovery (Kappes 2007).

The development of fine grinding equipment using horizontal and vertical-shaft stirred mills has had an impact on the application of flotation and the manner in which flotation concentrates are treated. Fine re-grinding of flotation concentrates is not a new concept; for example Anglovaal utilized fine grinding of a pyritic flotation concentrate at Loraine in South Africa in the 1970s (Bath et al. 1973). However, the adaptation and introduction of the Xstrata IsaMill (developed from the Netsch mill), the Metso SMD Detritor (developed from the English China Clays mill), and the Metprotech mill as options for fine grinding of flotation concentrates in the late 1980s paved the way for more efficient grinding down to finer sizes, around 80% passing 10–15 µm and below, hence the term "ultra-fine" grinding. Also, further development of the Vertimill following many successful tertiary grinding and regrinding installations has led to its consideration for ultra-fine grinding applications (down to about 80% <15 µm). Other fine grinding mills are in development.

The pursuit of ultra-fine grinding was important for both base and precious metals recovery and the developments have essentially paralleled each other. The ability to economically grind to such fine sizes presented the opportunity to liberate precious metals from refractory sulfide ores without the need for oxidative treatment. In 2001, Kalgoorlie Consolidated Gold Mines (W. Australia) installed an ultra-fine grinding circuit to treat refractory sulfide gold ore to supplement the roaster capacity. The ground product was cyanide leached, achieving a gold recovery of over 90% (Ellis 2003). This was a significant development as it was the first commercial application to avoid the need for oxidative pretreatment. Ultra-fine grinding has subsequently been installed at Kumtor (Kyrgyzstan) in 2005, Pogo (Alaska) in 2006, and Lake Cowal (Australia) in 2006. The product grind sizes are 80% <12 µm (95% <20 µm), 80% <10 µm (95% <17 µm), and 80% <15 µm, respectively, for these operations. Specific energy requirements to grind pyrite concentrate from approximately 80% <75 µm to 80% <7–15 µm vary between about 30 and 60 kWh/ton depending on the concentrate mineralogy and the specific product size. This power requirement increases to 50–80 kWh/ton as concentrate feed size increases to 80% <106 µm. Including the cost of grinding media (typically Colorado River Sand or equivalent, or suitable ceramic media), and assuming a power cost of $0.06/kWh, the total costs for fine grinding vary between $3.00–8.00/ton processed (for a 30–80 kWh/t range).

Cyanide Destruction

In parallel with the resurgence of the gold mining industry between 1972 and 1990, there was a period of increasing environmental awareness and a major movement toward control of factors affecting the environment, particularly in developed countries, but also worldwide. This has continued into the 21st century following the Kyoto Protocol (1997), with initiatives focused on the environmental impacts of mining including the Global Mining Initiative (1999–2002) and the International Council on Mining and Metals (established in 2002). A major effort with respect to the gold extraction industry was the publication of the International Cyanide Management Code (2002), to which most of the major gold and silver producers that use cyanide have committed to follow. This code was developed by the International Cyanide Management Institute, a non-profit organization set up under the United Nations Environment Program and the International Council on Metals and the Environment. All of this activity represented significantly increased emphasis on the control and treatment of gold extraction by-products and effluents, which should be considered as an integral part of gold extraction processes.

There are many alternatives for detoxification of cyanide-containing solutions. Where applicable, the preferred method is to allow the cyanide concentration to decay naturally through the CIP/CIL circuit to the point at which it reaches levels acceptable for discharge to the tailings containment facility. There are many operations that are able to meet strict discharge limits to tailings facilities without the need for any other form of cyanide destruction than natural degradation over time. However, these operations carefully manage cyanide concentrations down the leaching and CIP/CIL circuit, as well as wash ratios in thickeners, using re-circulated, reclaimed or fresh water in the circuit. The cyanide degrades further over time in the tailings facility and the understanding of such degradation processes has improved significantly over the past 25–30 years (Whitlock & Mudder 1986), including natural degradation of free, WAD and total cyanide species, thiocyanate and cyanate. The use of tailings thickeners and, where necessary, tailings filtration can assist with recovering and recycling cyanide-bearing solution. All of these practices help to reduce cyanide naturally within the overall extraction circuit.

Where the above methods are not sufficient to meet the Cyanide Code guidelines (e.g., ≤50 mg/L weak acid dissociable cyanide in tailings storage facilities) and/or regulatory environmental requirements, other methods of detoxification must be used. In the early 1980s, Inco (Sudbury, Ontario) developed and patented a method using sulfur dioxide and air, with copper as a catalyst, which was applied successfully at a number of Canadian operations. The primary reaction is as follows:

$$CN^- + SO_2 + O_2 + H_2O = CNO^- + H_2SO_4 \quad (2)$$

To achieve an acceptable reaction rate, addition of 30–90 g/t of copper (usually copper sulfate) is required. Either sulfur dioxide gas or sodium metabisulfite can be used as the sulfur dioxide source, with adequate air injection into the slurry to achieve the required SO_2/O_2 ratio. In practice, air flow rates of 1–2 L/min per liter of solution are required with 3–4 kg SO_2 required per kg CN^-, and sometimes higher than this. The oxidation is performed between pH 7.5–9.5, with pH 9 close to optimal. Lime must be added for pH control due to the formation of sulfuric acid.

After almost thirty years of application at operations throughout the world, the use of sulfur dioxide-air has become the preferred and most cost-effective method of cyanide destruction where natural degradation is not adequate.

Many other methods have been tested and used commercially with varying degrees of success. Hydrogen peroxide and Caro's acid (hydrogen peroxide and sulfuric acid) have both been used successfully at a variety of operations. The advantages are that free and WAD cyanides are converted to cyanate very rapidly due to the strongly oxidizing conditions. This can be useful in situations where rapid detoxification is required or where there is limited retention time available in existing equipment to achieve detoxification. The disadvantage is the relatively high cost of reagents, which is typically much more costly than the sulfur dioxide-air process to achieve similar residual free and WAD cyanide concentrations. However, one specific high-efficiency version of this general approach is the Maelgwyn Mineral Services process using the Aachen reactor, high-energy mass transfer and oxygenation, and activated carbon as a catalyst (Adams 2011).

An interesting application of Caro's acid is the "LIXKill" process developed by Barrick Gold whereby free and WAD cyanide, and thiocyanate, species are destroyed in the tailings reclaim solution before it is fed to the mill and/or autoclave circuit. This prevents gold dissolution in the grinding and pressure oxidation circuits, resulting in less adsorption on carbonaceous constituents of the ore (Pekrul 2007).

Biological oxidation for cyanide destruction in low cyanide concentration applications has been used very successfully at the Homestake Mine (South Dakota) since about 1984, and has since been applied at several other Homestake operations (Whitlock and Mudder 1986). This technology has proven capable of producing final effluent solution that can meet discharge limits, and the capital and operating costs are reportedly competitive with the sulfur dioxide-air process.

Alkaline chlorine and/or hypochlorite oxidation was used in the early days of cyanidation, at least partly due to the fact that cyanidation replaced chlorination leaching, and the chemistry and handling of chlorine was well understood at that time. Very rapid oxidation of free and WAD cyanide species is possible with this system. The use of chlorine/hypochlorite was tested and applied at a number of gold and silver operations in the late 1980s and early 1990s. Subsequently, this approach has generally gone out of favor (compared to the sulfur dioxide-air process, and the other processes described above) because of the preference not to introduce foreign chloride ions into the system, difficulties in controlling the process for optimum chemical utilization to meet the required effluent cyanide concentrations, and the high consumption and cost of chlorine/hypochlorite (i.e., 8–24 kg chlorine per kg CN).

Water conservation is, and will continue to be, an area for innovation. Dry stacked tailing, such as used at La Coipa (Chile), has additional benefits of cyanide recycling and reduced environmental concerns for groundwater contamination. Use of brackish and saline water is commonplace in Western Australia and is currently extending to applications in Chile and Peru.

Cyanide Recovery and Gold-Copper Ore Treatment

Some degree of cyanide recovery and recycling has been practiced for many years in the form of reclaiming solutions for re-use in the leaching process, for example from tailings facilities and from tailings thickeners. However, technology for the separation of free cyanide from solution to allow recycling in a concentrated form has been known for over a century. The method of acidifying cyanide-bearing solutions to generate hydrogen cyanide gas, the removal of such hydrogen cyanide from solution by volatilization with air, followed by adsorption and recovery of hydrogen cyanide into a caustic scrubber system has come to be known as the "AVR" (Acidification, Volatization, Recovery) process. An early form of the process was applied at both the Santa Gertrudis mill and the Guerrero mill of Compania de Real del Monte y Pachuca in 1924, accomplishing a considerable savings in cyanide consumption (Bryan et al. 1934). This general approach was subsequently applied commercially at Flin Flon (Canada) to treat cyanide-bearing process solution from about 1930 to 1975. AVR technology was applied at Golconda (Australia) in 1987. The process was subsequently further developed and significantly improved by incorporating the use of high efficiency packed-bed adsorption towers for improved stripping of hydrogen cyanide either from the solution or slurry phase directly, thereby avoiding solid-liquid separation (Goldstone & Mudder 1988). This approach was applied in the early 1990s at Golden Cross (New Zealand), and Delamar Silver Mines (Idaho), with more recent installations at Rio Paracatu (Brasil) and Cerro Vanguardia (Argentina). Despite the viability and cost-effectiveness of the technology, it has not been applied widely throughout the industry. This is largely as a result of the effectiveness of the simpler cyanide recycling procedures described above (e. g. tailings thickening, reclaim solution collection and recycling), and the effective management and control of cyanide concentrations within agitated leaching, CIP/CIL, and heap leaching circuits, attributable to the ingenuity, creativity and diligence of plant operating staff. In some cases, agitated cyanide leaching operations have been able to achieve final effluent discharge to tailings storage facilities <50 mg/L WAD cyanide without the use of any cyanide detoxification technology.

The need to be able to effectively treat ores containing significant amounts of cyanide-soluble copper has been recognized for many years, even back to the earliest days of cyanidation in the early 1900s. In the modern era (1972–2012), the Telfer (Australia), Ok Tedi (Papua New Guinea), and El Indio (Chile) operations were all learning how to deal with the presence of significant copper in the feed to their cyanide leaching circuits in the late 1970s and early 1980s. These plants all processed ores with variable cyanide-soluble copper, with up to 500–700 g/t cyanide soluble copper in feed. The ability to prevent copper loading onto activated carbon by maintaining a sufficiently high free cyanide concentration by keep the majority of the copper as the $Cu(CN)_4^{3-}$ complex was recognized in the 1980s. Where this was not sufficient to avoid significant copper loading on carbon, other techniques were required. One approach applied at El Indio was the use of a cold cyanide elution step to remove copper loaded onto carbon prior to hot elution, avoiding the co-elution of copper with gold and silver in the conventional (Zadra/AARL) elevated temperature elution step. This allowed the copper eluate solution to be treated and disposed of separately, and avoided the bulk of the copper from reporting to the doré bullion.

An important development for the treatment of copper-bearing ores was the development of the Sulfidization, Acidification, Recycle and Thickening ("SART") process. The process involves precipitation of copper sulfide (theoretically Cu_2S) by the addition of sulfuric acid and sodium hydrosulfide (NaSH) which releases the complexed cyanide as hydrogen cyanide (HCN). The solids are separated from the solution phase by thickening and filtration to produce a copper sulfide solid for sale or for further treatment. The thickener overflow solution is re-neutralized with lime and the contained hydrogen cyanide is converted back to calcium cyanide, suitable for recycling and re-use in the cyanide leaching circuit. The basic steps in the SART process date back to the 1940s, however the first recognition that this could be developed into an effective process to treat copper-bearing gold and silver ores was probably by George Potter (Potter et al. 1986). The process was subsequently further developed and improved, specifically for the Lobo-Marte project in Chile (MacPhail et al. 1998), but the first commercial application was at the Telfer operation (W. Australia) in 2005 to treat barren solution from leaching of a pyritic flotation concentrate. SART plants have subsequently been installed at Yanacocha (Peru), Maricunga (Chile), Lluvia de Oro (Mexico), and Gedabek (Azerbijan) to treat pregnant gold-bearing solutions produced by heap leaching.

Another approach to the application of SART that is currently being considered is the combination of SART with AVR. This approach would volatize the hydrogen cyanide with air, followed by re-adsorption of hydrogen cyanide into a smaller volume of scrubber solution to provide better water balance control. There will no doubt be other advances and improvements in SART and AVR technology that will provide lower cost and more efficient options for cyanide recovery and recycling. These developments will be critical to allow efficient processing of copper-bearing materials by cyanide leaching in future.

Mineralogy

During the past 15 years, significant innovations have occurred in process mineralogy. At the forefront of this work has been the development of automated scanning electron microscopy techniques (e.g., QemSCAN, developed by Intellection and the Mineral Liberation Analyzer, or MLA, developed by the JK Institute of Technology, Australia). These techniques are now well known to most in the industry, but the significance of being able to perform accurate, quantitative mineralogical analysis on representative samples of ore, intermediate processing products and residues from projects and operations cannot be over-stated. The individual mineral grain identification and quantification, size-by-size analysis, and mineral liberation/locking analyses that can be generated have revolutionized the approach to design and optimization of mineral and metal extraction worldwide.

Other advanced mineralogical techniques are also now available that provide important diagnostics for gold and silver recovery optimization, troubleshooting and process design (Chryssoulis 2005, Dimov 2011). The following summarizes some of the specific capabilities that are available:

- SEM/EDX provides quantification of gold/silver particle bulk composition.
- Secondary Ion Mass Spectrometry (SIMS) provides quantification of submicroscopic gold/silver as colloidal inclusions and/or solid solution within selected sulfide minerals.
- Time of Flight–Secondary Ion Mass Spectrometry (TOF-SIMS) provides identification and quantification of species present on the surface of gold and gold-silver alloy particles, e.g., coatings, residual reagents.
- Dynamic Secondary Mass Spectrometry (D-SIMS) is similar information to TOF-SIMS but with deeper surface penetration.

Other more specialized techniques (Mossbauer Spectroscopy, Auger Emission Spectrometry, and Laser Ion Mass Spectrometry) are available for investigating specific problems and issues in gold extraction.

THE FUTURE (>2013)

Some of the likely areas of focus for future innovation efforts in gold and silver extraction are reviewed in this section. This is not an all-inclusive list and there are many exciting opportunities that are being investigated and explored around the world at the time of preparing this paper.

Gravity Concentration

With the successful application of centrifugal gravity concentrators around the world, it is likely that attention will focus on the development of larger units (that can handle a larger proportion of ball mill discharge or cyclone underflow through a single unit) and units that can achieve continuous operation. These improvements will allow this technology to be applied to progressively lower grade ores.

Alternative Leaching Reagents and Systems

Since the inception of the cyanidation process described earlier in the late 1800s, cyanide has been used almost universally because of its relatively low cost, great effectiveness for gold and silver dissolution, selectivity for gold and silver over other metals, and effectiveness for loading onto (and stripping off) activated carbon. Also, despite some concerns over the toxicity of cyanide, it can be applied with little risk to human health and the environment. The oxidant most commonly used in cyanide leaching is oxygen, usually supplied from air, which contributes to the attractiveness of the process.

Since the mid-1970s alternative leaching reagent schemes to cyanide have been investigated for any or combinations of the following reasons:

- Environmental pressures, and in some cases restrictions or limitations, may make the application of cyanide difficult in certain locations,
- Some alternative reagent schemes provide faster gold (and/or silver) leaching kinetics,
- Several can be applied in acidic media, which may be more suitable for refractory ore treatment, and
- Some are more selective than cyanide for gold and silver over other one or more specific metals of concern, e.g., copper.

Some of the more important reagent systems that have been investigated (or re-investigated) include; chlorine-chloride, thiosulfate, thiocyanate, thiourea, ammonia, ammonia-cyanide, alkaline sulfide, bromine-bromide, iodine-iodide, and other halide combinations. All of the alternative reagent schemes have some significant disadvantages and, at this time, none appears to be widely applicable at least not without further significant advances in the technology. However, thiosulfate has emerged as the front runner of the alternative schemes, with the specific currently-recognized potential application to treat carbonaceous preg-robbing ore types. The reason for the particular applicability of the thiosulfate system is that the thiosulfate gold and silver complexes are not readily adsorbed onto carbonaceous constituents of the ore. This property in itself presents a problem for metal recovery from thiosulfate leach solution in that carbon adsorption is not suitable. However, other options have been investigated with considerable success including resin adsorption/desorption and direct precipitation from solution (e.g., using copper powder). The chemistry of thiosulfate sytems, which is more complex than cyanide leaching, has been discussed in detail in the literature (Breuer & Jeffrey 2003, Marsden 2006).

In the 1980s, an in-situ thiosulfate-based process (developed and promoted by Richard Ellwanger) was tested on the Witwatersrand in an underground mine for leaching of fines and washings within underground stoping areas (Marsden 2006). However, this technology was not successfully commercialized. In the mid-1990s Newmont operated a small scale heap leach (0.75 million tons) on biologically-oxidized ore using ammonium thiosulfate (Wan 2003). A relatively small scale operation was developed and commissioned at La Colorada in Sonora, Mexico, using copper cementation for gold and silver recovery (Yen et al. 2001). A significant amount of excellent research and development work has been conducted by Barrick Gold Corporation (and a predecessor company, Placer Dome), Newmont Gold, CSIRO (Australia), SGS Lakefield Research, and others, and this work is continuing. While there is a wealth of information available on the process research and laboratory scale development (most of this conducted since 1995), there is not yet public information available on large scale commercial demonstration or full scale applications of this technology. The key issue is whether this technology can be implemented successfully and cost effectively on a large commercial scale, with acceptable reagent consumptions and competitive metal extractions. There is evidence to suggest that this may be close for specific and highly selective applications.

Pressure ("Flash") Cyanidation

Emerging technology in the copper industry to apply pressure leaching to treat copper concentrates has resulted in the development of a group of processes that operate in the temperature range of 150–170°C. These include processes developed by CESL Technology, Sherritt-Dynatec, Freeport-McMoRan (formerly Phelps Dodge) and Anglo American. These processes all generate significant amounts of elemental sulfur in the pressure leaching residue. When treating materials that contain economically attractive amounts of gold and silver, processing using conventional cyanide leaching and CIP/CIL recovery generally results in high cyanide and lime consumptions. One option to process these materials is ammonium or calcium thiosulfate leaching (Marsden 2009), as discussed above. Typically, gold and silver extractions are at best equal to cyanide leaching extractions, but reagent consumptions and costs could be significantly lower, although the process technology has yet to be proven at a commercial scale.

Another option to recover gold and silver from elemental sulfur-bearing residues has been developed by CESL Technology using a short retention time ('flash") pressure cyanidation process (Robinson 2011). Super-atmospheric conditions are applied in conjunction with oxygen injection into the feed to achieve fast gold (and silver) leaching rates at elevated cyanide and oxygen concentrations. Retention time of around 60–90 minutes is required to achieved acceptable gold and silver extractions, compared to the 24 hrs typically required in conventional cyanide leaching circuits. This approach results in significantly reduced cyanide and lime consumptions compared to cyanide leaching under conventional conditions for the longer retention time.

This technology could also potentially be applied to treat the flotation concentrates that have been subjected to ultra-fine grinding to liberate gold and silver from refractory sulfide minerals. It is possible that for some ore types and materials, this could be a lower cost alternative to oxidative pretreatment. Cyanide and lime consumptions would be expected to be significantly lower using short retention time pressure cyanidation compared to conventional cyanide leaching.

Other Leaching Improvements

There are many challenges remaining in gold and silver extraction. The search for methods to leach gold minerals that are refractory to conventional cyanide leaching continues, for example gold-bismuth (Au-Bi) and gold-bismuthinite ($Au\text{-}Bi_2S_3$) intergrowths, and gold-telluride minerals such as; petzite ($AuAg_3Te_2$), calaverite ($AuTe_2$), sylvanite ($Au,AgTe_2$), and others.

Oxidative Pretreatment

Pressure oxidation and CFB roasting technologies continue to be employed successfully throughout the gold industry worldwide, as evidenced by the recent commissioning of the Pueblo Viejo pressure oxidation circuit in the Dominican Republic. However, the search for lower cost (capital and operating) options to oxidize sulfide minerals continues. While biological oxidation has been applied successfully commercially, there is still a desire to develop technology that can achieve effective sulfide oxidation under atmospheric conditions with shorter residence times, less energy input, and lower reagent consumption (e.g., Albion/NenaTech process, University of British Columbia Galvanox process).

Biological oxidation will likely continue to be applied to treat concentrates in situations where partial oxidization of gold-silver bearing sulfide minerals to liberate a substantial proportion of the metal values.

Other Potential Developments

- Direct electrowinning of gold and silver from dilute aqueous leach solutions using three-dimensional electrodes.
- Improved and lower cost cyanide destruction techniques, including, for example:
  - High energy super-oxygenation system plus activated carbon catalyst to get from 130 to <50 mg/L CN.
- As the distribution and diversity of gold and silver production around the world increases, it is expected that precious metal refineries will need to adapt to accept a wider range of feed materials with varying metal content, for example,
  - concentrates and materials containing base metals, platinum group metals, and other elements; and
  - increasing rates of production of electronic scrap, batteries, and high grade automobile scrap materials.

DISCUSSION

An important lesson from all of the major innovations in gold and silver extraction is that innovations are never "eureka" moments, but rather result from a sustained period of testing, investigating, modifying and improving a particular technological approach to a problem. In the case of cyanidation, carbon adsorption, heap leaching and refractory ore treatment processes, the technology had been known and versions of each process had been patented, tested and tried for several decades. Those who successfully commercialized these innovations learned from the prior versions of the technology, borrowed from other branches of the industry (and in some cases from other industries), and improved the application of the technology with often simple modifications. The first-user recognized the benefit of the emerging technology over the incumbent process, they were persistent and relentless in their pursuit of successful commercialization, and in all cases they relied on innovative and tenacious process operators (not the inventor or researcher) to implement the technology effectively. Success was not intuitively obvious in these efforts, and in most cases there were several failures or, at best, marginal and/or small applications of the technology that preceded widespread commercialization.

SUMMARY

A period of major innovation in gold and silver extraction occurred following the un-fixing of the gold price to the U.S. dollar in 1973. The rapid increase in the gold price that followed motivated major increases in gold production worldwide, with both Greenfield projects and Brownfield expansions. Exploration efforts increased geometrically with many of the finds being lower grade, finely-disseminated, and often preg-robbing or refractory in nature. This, in turn, provided significant impetus for innovations, resulting in the widespread commercialization of carbon recovery systems (CIP, CIL, CIC), heap leaching and oxidative pretreatment processes, centrifugal gravity concentrators, ultra-fine grinding and leaching of flotation concentrates, cyanide recovery, recycling and destruction technologies. Once again, the industry is currently in a period of strong precious metal pricing where a fresh round of innovation and development in the gold-silver industry should be possible. Many opportunities for innovation remain in gold and silver extraction. It is important that the industry (gold producers, equipment and reagent suppliers, and research academia) take advantage of this to prepare for the next period of depressed precious metal pricing.

ACKNOWLEDGMENTS

The authors thank Rob Dunne and Jim Arnold for helpful discussion on key issues. As always, the opinions expressed are those of the authors, and they do not reflect the opinion of any company or operation discussed within the text.

REFERENCES

Adams, M. & Glen, S. 2011. Pilot-scale and Continuous Bench Testing of the MMS CN-D™ Cyanide Destruction Process on an Australian Gold Plant Tailings Effluent. Proc. World Gold 2011. 50th Conference of Metallurgists, CIM, Montreal, Quebec, Canada. pp. 501–510.

Adamson, R.J. 1972. Gold Metallurgy in South Africa. Chamber of Mines of South Africa, Johannesburg.

Agricola. 1912. De Re Metallica. Translated H.C. Hoover & L.H. Hoover. The Mining Magazine, London (Original publication date 1556).

Arnold, J.R., Keane, J.M., Lofftus, V.G., & Ahlness, J.K. 1990. Gold Heap Leach Information Exchange. Proc. Gold 1990 Symposium. Edited D.M. Hausen et al. SME, Littleton, CO. pp. 283–311.

Bain, H.F. 1910. More Recent Cyanide Practice. Mining & Scientific Press. San Francisco.

Barter, J., Lane, G., Mitchell, D., Kelson, R., Dunne, R., Trang, C. & Dreisinger, D. 2001. Cyanide Management by SART. Proc. Cyanide: Social, Industrial and Economic Aspects, Ed. C.A. Young. The Minerals, Metals & Materials Society, Warrendale, PA. pp. 549–562.

Bath, M.D., Duncan, A.J., & Rudolph, E.R. 1973. Some Factors Influencing Gold Recovery by Gravity Concentration. J.S. Afr. Inst. Min. & Met., June. Johannesburg, South Africa. Discussion following paper by S.E. De Kok, pp. 378–381.

Berezowsky R.M.G.S. & Weir, D.R. 1989. Refractory Gold: The Role of Pressure Oxidation. Proc. World Gold 1989 Symposium. Ed. Bhappu, R.B. and Harden, R.J. American Institute of Mining, Metallurgical, and Petroleum Engineers, New York, NY. pp. 295–304.

Boydell, D.W. 1984. Developments in Plant Practice for the Recovery of Gold in South Africa. The AusIMM Perth and Kalgoorlie Branches, regional Conference on Gold Mining, Metallurgy & Geology. AusIMM, Melbourne, Australia.

Breuer, P.L. and Jeffrey, M.I. 2003. A Review of the Chemistry, Electrochemistry, and Kinetics of the Gold Thiosulfate Leaching Process. Hydrometallurgy, 2003, Vol. 1 pp. 139–154.

Bryan, R.R. and Kurylam, M.H. 1934. Milling and Cyanidation at Pachuca, Annual Meeting of AIME 1934.

Chryssoulis, S.L. and McMullen, J. 2005. Mineralogical Investigation of Gold Ores. Developments in Mineral Processing, Vol. 15. Ed. M.J. Adams. Elsevier B.V.

Clennell, J.E. 1915. The Cyanide Handbook. Second Edition. McGraw-Hill Book Company, Inc. New York.

Dimov, S.S. and Hart, B.R. 2011. Applications of Microbeam Analytical Techniques in Gold Deportment Studies. World Gold 2011–Proc. 50th Annual Conf. of Metallurgists, CIM, Montreal, QC, Canada.

Dorr, J.V.N. and Bosqui, F.L. 1950. Cyanidization and Concentration of Gold and Silver Ores, Second Edition McGraw-Hill, New York.

Eissler, M. 1891. The Metallurgy of Gold. Third Edition. Crosby, Lockwood and Son, London.

Ellis, S. 2003. Ultra Fine Grinding—A Practical Alternative to Oxidative Treatment of Refractory Gold Ores. AusIMM.

Elsner, L. 1846. Observations on the Behaviour of Pure Metals in an Aqueous Solution of Cyanide. Journal fur Praktische Chemie 37 (1): 441–446, Germany.

Fleming, C.A. and Cromberge, G. 1984. The Extraction of Gold from Cyanide Solutions by Strong and Weak Base Anion Exchange Resins. J. South African Inst. Min. & Met. 84 (5) pp. 125–137.

Gold Fields Mineral Services 2006). Gold Survey 2006. London, UK.

Goldstone, A.J. & Mudder, T.I. 1988. Cyanide Regeneration Process. U.S. Patent No. 4,994,243.

Habashi, F. 1987. One Hundred Years of Cyanidation. CIM Bulletin 80 (905): 108–114.

Hamilton, E.M. 1920. Manual of Cyanidation. McGraw-Hill Book Company, Inc. New York.

Harris, D. 1984. The Knelson Concentrator—Applications in Australia. The Australasian Institute of Mining & Metallurgy, Perth and Kalgoorlie Branches, regional Conference on Gold Mining, Metallurgy and Geology. October 1984.

Heinen, H.J., Peterson, D.G., and Lindstrom, R.E. 1978. Processing gold ores using heap leach adsorption methods. U.S. Department of the Interior, Bureau of Mines.

Julian, H.F. & Smart, E. 1904. Cyaniding Gold and Silver Ores. 1st Edition. Charles Griffin and Company, Limited, London.

Kappes, R. 2007. N2TEC Flotation Technology—A Review and Update. Precious Metals Processing—Advances in Primary and Secondary Operations Symposium, October 3–6, 2007. SME, Littleton, CO.

Kappes, R., Fortin, C. & Dunne, R. 2011. The Current Status of the Chemistry of Gold Flotation in Industry. Proc. World Gold 2011. 50th Conference of Metallurgists, CIM, Montreal, Quebec, Canada. pp. 385–396.

Keane, J. 2007. Engineering and Construction of a 20,000 TPD Gold Recovery Process Plant in North China. Precious Metals Processing—Advances in Primary and Secondary Operations Symposium, October 3–6, 2007. SME, Littleton, CO.

King, A. 1949. Gold Metallurgy on the Witwatersrand. Transvaal Chamber of Mines, Johannesburg, South Africa.

Kinross Fort Knox Mine. 2012. Fort Knox Mine Annual Meeting Presentation, Thursday March 29. www.dnr.alaska.gov accessed on September 30th, 2012.

Kinross Gold 2012. www.kinross.com Accessed on September 30th, 2012 and Article: Paracatu Mine, Brazil published in Crusher News July 21st, 2012.

Knelson, B. & Edwards, R. 1990. Development and Economic Application of Knelson Concentrators in Low Grade Alluvial Gold Deposits. The AusIMM Annual Conference—The Mineral Industry in New Zealand. Rororua, New Zealand, 18–21 March 1990. pp. 123–128.

Laplante, A.R. 2000. Testing Requirements and Insight for Gravity Gold Circuit Design, Randol Gold and Silver Forum, 2000, pp. 73–83.

Laplante, A.R., Liu, L., & Cauchon, A. 1990. Gold Recovery at the Mill of Les Mines Camchib Inc. 22nd Annual Meeting of the Canadian Mineral Processors, CIM, Ottawa, Canada. pp. 397–414.

Laxen, P.A. 1982. Carbon-in-Pulp in South Africa. AusIMM Perth and Kalgoorlie Branches and Murdoch University Carbon-in-Pulp Seminar. AusIMM, Parkville, VIC, Australia.

Longley, R.J., McCallum, A. & Katsikaros, N. 2003. Intensive Cyanidation: Onsite Application of the InLine Leach Reactor to Gravity Gold Concentrates. Minerals Engineering 16, pp. 411–419. Elsevier Science Ltd.

MacPhail, P.K., Fleming, C.A. & Sarbutt, K.W. 1998. Cyanide Recovery by the SART Process for the Lobo-Marte Project, Chile. Randol Gold & Silver Forum '98. Randol International, Golden, CO. pp. 319–323.

Mansanti, J.G., Arnold, J.R., Gourdie, J.H. & Marsden, J.O. 1987. Double-Thickener Circuit at Gold Fields' Chimney Creek Mine. Min. & Met. Proc. J., November. SME, Littleton, CO.

Marsden, J.O. 1996. How low can you go? Processing of low grade gold ores in North America. Proc. Hidden Wealth Conference, Ed. Glen, H.W.S. African Inst. Min. & Met., Johannesburg, South Africa. pp. 115–123.

Marsden, J.O. 2006. Overview of Gold Processing Techniques Around the World. Minerals & Metallurgical Processing J., Society for Mining, Metallurgy and Exploration Inc., Littleton, CO.

Marsden, J.O. 2009. Lessons learned from the copper industry applied to gold extraction. Proc. World Gold 2009. Johannesburg, South Africa. October 2009. S. Afr. Inst. Min. & Met., Marshalltown, Guateng, South Africa.

Marsden, J.O. & Fuerstenau, M.C. 1993. Comparison of Merrill-Crowe Precipitation and Carbon Adsorption for Precious Metals Recovery. Proc. XVIII International Mineral Processing Congress, Sydney, Australia. Aust. Inst. Min. & Met., Sydney, Australia.

Marsden, J.O. & House, C.I. 2006. *The Chemistry of Gold Extraction.* Second Edition. Society for Mining, Metallurgy & Exploration, Inc., Littleton, CO.

Marsden, J.O. & Wilmot, J.C. 2007. Sulfate-based flowsheet options for hydrometallurgical treatment of copper sulfide concentrates. Proc. Sixth International Copper-Cobre 2007 Conference, Vol. IV (Book 1), Eds. Riveros, P.A., Dixon, D.G., Dreisinger, D.B. & Collins M.J., Canadian Inst. Min., Met. & Pet., Montreal, Quebec, Canada. pp. 77–100.

Marsden, J.O., Mansanti, J.G. & Sass, S.A. 1993. Innovations Methods for Precious Metals Recovery in North America. SME/AIME Annual Meeting, Reno, NV. SME, Littleton, CO.

Marsden, J.O., Todd, L.R. & Moritz, R. 1995. The effect of lift height, overall heap height and climate on heap leaching efficiency. AIME/SME Annual Meeting Preprint Number 95-229. Society for Mining, Metallurgy & Exploration Inc., Littleton, Colorado.

McNulty, T. 1989. A Metallurgical History of Gold. Paper presented at American Mining Congress, San Francisco, CA.

McQuiston, F.W. and Shoemaker, R.S. 1975. Gold & Silver Cyanidation Plant Practice. Vol. I. AIME Monograph. American Inst. of Min., Met. & Petroleum Eng., New York.

Pekrul, J. 2007. Update on LIXKill at the Autoclave. Precious Metals Processing—Advances in Primary and Secondary Operations Symposium, October 3–6, 2007. SME, Littleton, CO.

Potter, G.M., Bergmann, A., Haidlen, U. 1986. Process of Recovering Copper and of Optionally Recovering Silver and Gold by Leaching of Oxide- and Sulfide-Containing Materials with Water-Soluble Cyanides. U.S. Patent 4,587,110.

Rickard, T.A. 1907. Recent Cyanide Practice. Mining & Scientific Press, San Francisco.

Robinson, T., Mayhew, K., Jones, D. & Murray, K. 2011. The CESL Gold Process. Proc. World Gold 2011, 50th Conference of Metallurgists, CIM, Montreal, Quebec, Canada. pp. 183–196.

Rose, T.K. 1896. The Metallurgy of Gold. Second Edition. Charles Griffin & Co. Ltd., London.

Rose, T.K. 1935. The Metallurgy of Gold. Sixth Edition. Charles Griffin & Co. Ltd., London.

Schnabel, C. 1921. Handbook of Metallurgy. Third edition. Volume 1. Macmillan, London. pp. 936–1134.

Seal, T. 2007. HYDRO-JEX: Heap Leach Pad Stimulation Technology—Ready for World Wide Industrial Adoption? Presented at SME Annual Meeting, Feb, 2007. SME, Littleton, CO.

Shoemaker, R.S. 1984. Gold: Quid Non Mort Alia Pectora Cogis, Auri Sacra Fames. Proc. Precious Metals: Mining, Extraction and Processing. Ed. V. Kudryk, D.A. Corrigan, and W.W. Liang. SME-AIME, Littleton, CO.

Simmons, G.L. and Gathje, J.C. 2003. High Temperature Pressure Oxidation of Ores and Ore Concentrates Containing Silver Using Controlled Precipitation of Sulfate Species. U.S. Patent No. 6,641,642.

Stanley, G.G. 1987. The Extractive Metallurgy of Gold in South Africa. Monograph M7. South African Inst. Min. & Met., Johannesburg.

Thompson Reuters. 2012. Gold Survey 2012. Thompson Reuters Gold Fields Mineral Services, London, UK.

Watson, B. and Steward, G. 2002. Gravity Leaching—The ACACIA Reactor. Proc. Symp. Metallurgical Plant Design and Operating Strategies. Sydney, NSW, Australia. pp. 383–390.

Whitlock, J.L. and Mudder, T.I. 1986. The Homestake Wastewater Treatemnt Process: Biological Removal of Toxic Parameters from Cyanidation Wastewaters and Bioassay Effluent Evaluation. Fundamental and Applied Biohydrometallurgy. Edited by R.W. Lawrence et al. Elsevier, Amsterdam.

Whyte, R.M., Dempsey, P. & Stange, W. 1990. The Development and Testing of the AAC Pumpcell at Vaal Reefs Exploration and Mining Company Limited. International Reef Mining Conference: Innovations in Metallurgical Plants. S. Afr. Inst. Min. & Metall., Johannesburg, South Africa.

Yen, W.T., Deschenes, G., & Aylmore, M. 2001. Thiosulfate Leaching as an Alternative to Cyanidation: A Review of the Latest Developments. Paper 24, 33rd Annual Operators Conference of Canadian Mineral Processors. January 2001, Ottawa, ON, Canada.

ABOUT THE AUTHORS

John O. Marsden has over 30 years of broad international experience in the development and management of major, innovative copper, gold, silver, molybdenum, and cobalt operations and projects in the USA, Chile, Peru, Brazil, Mexico, South Africa, and the Congo (DRC). Marsden is currently president of Metallurgium, established in 2009 to provide consulting services to the mining industry worldwide. Prior to that he held various operations and technical management positions during 18 years with Freeport-McMoRan Copper & Gold Inc. (formerly Phelps Dodge), most recently serving as president of Freeport-McMoRan Mining Company. He also spent 9 years with Consolidated Gold Fields plc in the USA and South Africa. He is a director of Romarco Minerals Inc. and serves on a number of industry and university advisory boards. He has published 34 technical papers, holds 15 U.S. patents, and is the author of *The Chemistry of Gold Extraction*, now in its second edition. He is a member of the National Academy of Engineering, a distinguished member of the Society for Mining, Metallurgy & Exploration (SME), a fellow of the Australasian Institute of Mining and Metallurgy, and a registered professional engineer (Nevada). He is the recipient of the AIME Robert H. Richards Award, and the Milton E. Wadsworth and Arthur F. Taggart awards of SME.

S. Andrew Sass has more than 40 years of experience in the design and operation of process plants for gold, silver, copper, molybdenum, aluminum, and evaporates spanning six continents. Sass is currently president of Sass Engineering, Inc., established in 2009 to provide process engineering and project management consulting services to the mining industry. Prior to that he spent 30 years in design and construction of metallurgical plants, mostly for Davy McKee and its successors, including Aker Kvaerner, leaving at the vice-presidential level. He has also worked for Kaiser Aluminum, FMC, and Freeport-McMoRan Copper & Gold Inc. He holds a BS in chemical engineering from UC Davis and an MBA from Cal Poly. He is a member of the Society for Mining, Metallurgy & Exploration (SME), having published numerous technical papers. He is a registered professional engineer in California, Nevada, Utah, and Arizona.

Innovations in Zinc and Lead Hydrometallurgy

Tim Robinson
Republic Alternative Technologies, Strongsville, Ohio, United States

Corby G. Anderson
Kroll Institute for Extractive Metallurgy, Colorado School of Mines, Golden, Colorado, United States

ABSTRACT: This paper covers the history and implementation of key commercial hydrometallurgical technologies that the authors humbly view as innovative in zinc and lead hydrometallurgy. Breakthrough leaching, purification, separation, concentration and final recovery unit operations are considered. This includes discussions of zinc Roast Leach Electrowinning, pressure oxidation, iron removal and solvent extraction. The Betts process for lead electro refining is also discussed in detail.

INTRODUCTION

The vast majority of commercial hydrometallurgical application and innovation in these two metals has been for zinc rather than lead. Hence, the bulk of this paper focuses on those key industrial areas of innovation and application. Ironically, the advent of and longest standing applied hydrometallurgical lead innovation still exists and operates today over a century after implementation in the form of the Betts refining process.

ZINC HYDROMETALLURGY

Roast Leach Electrowin Process

Frenchman Léon Letrange in 1881 first developed and patented (Letrange 1883) on an experimental scale the reduction of zinc ores in a sulfate system. This method used roasting, dilute sulfuric acid leaching and electrowinning (EW) with zinc starter sheets as per copper electrorefining (ER) and carbon anodes. Zinc sulfate solution purification was not described in Letrange's patent but was simplistically known at the time in the chemical preparation of the paint pigment, lithopone. This was the basis of the industrial innovative Roast-Leach-Electrowin (RLE) process. A combined zinc concentrate processing flowsheet is shown in Figure 1. The traditional historic pyrometallurgical smelting route follows the top path while the hydrometallurgical RLE path follows the bottom path. Of note, in both cases, pyrometallurgical roasting precedes both established process routes. This key roasting step is best expressed as

$$ZnS + O_2 \rightarrow ZnO + SO_2 \quad (1)$$

The sulfur dioxide generated is scrubbed to produce sulfuric and the oxidized zinc sulfide becomes a zinc oxide calcine that is progressively leached with the sulfuric acid produced. While of historic importance for zinc production, RLE has been utilized for example copper and cobalt

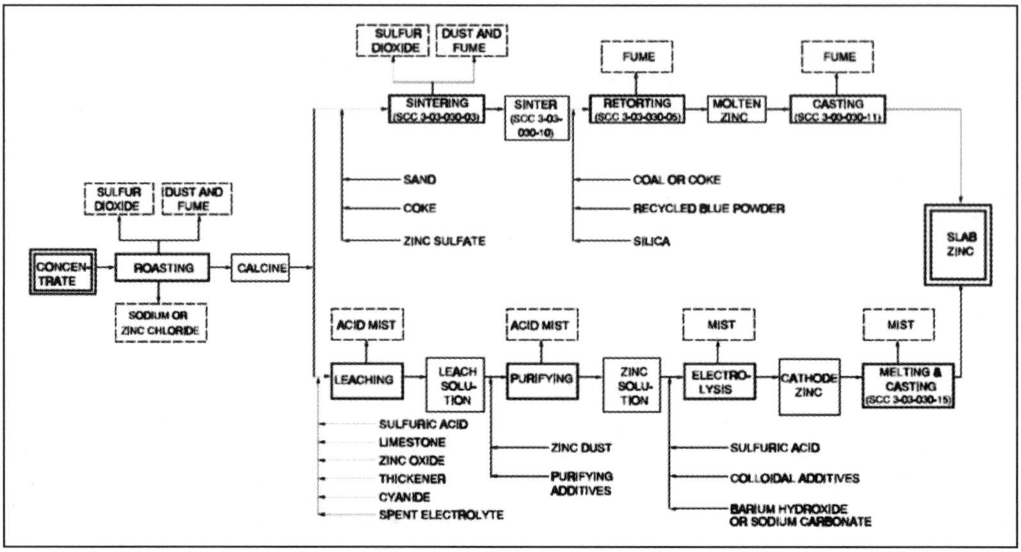

Figure 1. Traditional zinc concentrate processing summary flowsheet illustrating roasting followed by calcine smelting or by calcine leaching and electrowinning (USEPA 1986)

concentrates (Aird et al. 1980). While other zinc extraction process development at the time focused on chloride systems, due to simplicity and materials of construction, the sulfate system advanced to large scale commercial zinc production. Numerous attempts had been made, without success, to scale up and it was soon found that purification requirements were specific to the type of ore treated.

Driven by the World War I demand for pure EW zinc for brass (Ralston 1921), pilot plants and eventually larger production plants were built at

- Anaconda, Montana: 25 TPD (1915) increasing to 150 TPD (1916)
- Trail, BC, Canada: 50–75 TPD (1914)
- Risdon, Tasmania: 15–100 TPD (1917–1919)

In electrowinning, lead anodes were used to replace the carbon anodes and it was found that pure electrowon zinc metal could be easily stripped from an aluminum cathode, thereby replacing zinc starter sheet technology and producing the first permanent cathode harvesting system. It is interesting to note that two large copper electrowinning plants were first built at this period also at Chuquicamata, Chile, and New Cornelia, Ajo, Arizona (Robinson 2003). These large scale copper EW plants would still use copper starter sheets and both carbon and lead anodes.

It was the removal and control of iron that further propelled the RLE process through the application of the Jarosite, Goethite and Hematite processes. This is addressed later in the paper.

It became apparent that on a large scale, the purification step for the zinc electrolyte became specific to the plant and ore treated. Copper and cadmium could be removed by air atomized zinc dust. At Anaconda with the high antimony and arsenic levels in the ore, copper sulfate and zinc dust could be used. At Risdon, cobalt impurities in the Broken Hill zinc ore were very difficult to remove and additives such as glue, nitroso beta napthol, and zinc dust (in warmer conditions) assisted with cobalt removal.

Since these first plants, a number of innovations have been made in impurity removal:

1. Addition of activators such as oxides of antimony and arsenic to speed up purification kinetics in cobalt cementation.
2. Minimization of recycled zinc dust requirements by controlling purification Eh to regulate zinc dust addition (Gonzalez-Dominguez 1997).
3. Use of fluidized bed reactors for efficient and pure cadmium removal (Free 2011).
4. Use of water atomized zinc dust with larger surface area to reduce zinc dust consumption compared to air atomized zinc dust (Free 2011).

The first three pioneering commercial zinc sites built smaller cellhouses before expanding to commercial production levels. All sites had ready and reliable sources of cheap hydropower which was essential for the new energy intensive zinc EW technology. It is interesting to note that Trail's first zinc EW plant was the largest at 50 TPD compared to Anaconda and Risdon, as Trail had already had a strong electrometallurgical technical team and had been operating a commercial lead refinery since 1902 on the site.

The first manual zinc cellhouses used a cascade flow system (like lead ER) in lead lined wooden cells. These cells were small compared to today with only 24 cathodes. As current density was gradually increased, the electrolyte had to be cooled by water circulation in lead coils in the cell. Trail used a zinc EW electrode size of approximately $1.2m^2$ ($0.6m^2$ each side) which was the same size as the commercial lead ER size of 2 by $0.6m^2$.

This manual type of cellhouse remained largely the same, only slightly increasing length of cell and size of electrodes and increasing current density. The Japanese had made some improvements in automation, but it was in the late 1960s when productivity began to drive innovation with construction of the jumbo design cellhouse at Union Miniere Zinc in Belgium (Caufriez 1998). This jumbo cellhouse was built with 2.6 m^2 electrodes, 375 ASM, 44 cathodes per cell, parallel cell feeding electrolyte cooling with air cooling towers and automated electrode handling machines.

Since then, new zinc EW cellhouse designs have included increased electrode sizes up to 3.8 m^2, cells with over 110 cathodes, totally automated driverless cranes and further automation that has removed operators from cellhouse acid mist.

Future innovation will be driven by energy savings and include

1. New amorphous coated titanium anodes like those pioneered by Freeport and Republic Alternative/Doshisha University in copper which lower the anode oxygen overpotential (Robinson 2011),
2. Water cooled double intercell contact designs,
3. Real time cell voltage or current monitoring, and
4. Large monolithic polymer concrete cells that use no liners.

The Iron Precipitation Purification Processes

The advent of elegantly simple processes to leach zinc oxide calcine from roasting processes made the zinc electrolytic process dominant. This made it possible to recover zinc and other metals that were lost in zinc ferrite residues and to remove high concentrations of iron from metal sulfate solutions in readily separated crystalline forms. The three main innovative technologies for iron

Table 1. Comparison of iron precipitation purification processes (Gordon and Pickering 1975)

	Goethite	Jarosite	Hematite
pH	2–3.5	1.5	Up to 2% H_2SO_4
Temperature (°C)	70–90	90–100	~200
Anion	Any	SO_4^{2-} only	SO_4^{2-} only
Added cation required	0	Na^+, K^+, NH_4^+	0
Compound formed	α,βFeO·OH Fe_2O_3	R $Fe_3(SO_4)_2\ (OH)_6$	Fe_2O_3
Cationic impurities	Medium	Low	Low
Anionic impurities	Medium	High	Medium
Filterability	Very good	Very good	Very good
Fe left infiltrate after precipitation (g/L)	<0.05	1–5	3

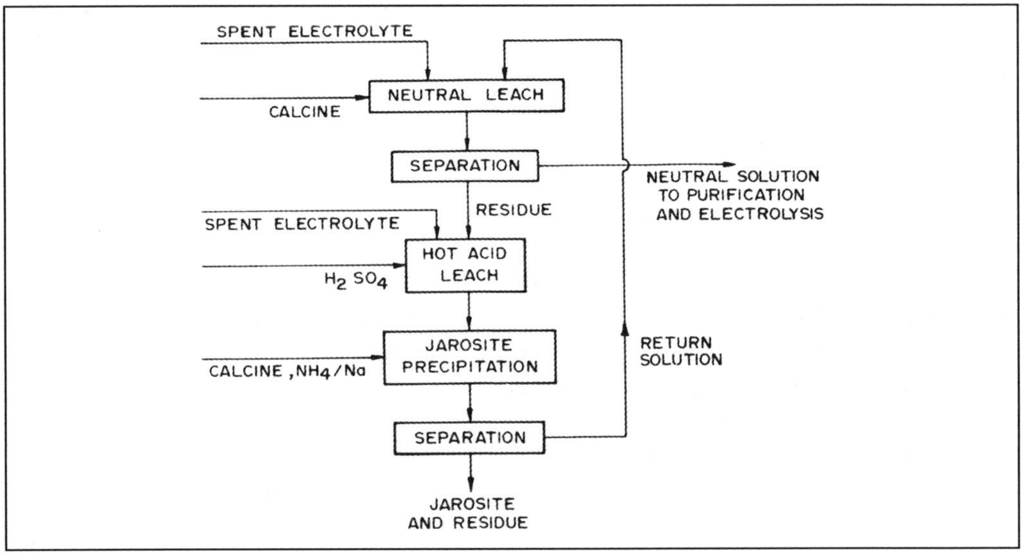

Figure 2. Simplified jarosite process flowsheet (Gordon and Pickering 1975)

precipitation are the Jarosite Process, the Goethite Process and the Hematite Process and removal (Gordon 1975). Table 1 illustrates the operating parameters for each process (Davey 1976).

Of these, the Jarosite Process, as illustrated in Figures 2 and 3, became dominant. The Jarosite Process became a practical reality after it was developed independently by the Electolytic Zinc Co. of Australasia Ltd., Det Norske Zink-kompani A/S of Norway, and Asturiana de Zinc S.A. of Spain (Haigh 1967, 1970; Wood 1973; Steinvelt 1968, 1970a, 1970b, 1970c; Arreguiet et al. 1979; Dutrizac and Kaiman1976; Dutrizac et al. 1980; Dutrizac 1980, 1982, 1983; McAndrew et al. 1975; Limpo et al. 1976; Menendez, and Fernandez 1967; Pammenter, and Haigh 1981; Zapuskalova and Margolis 1977).

The Jarosite Process basically consists of the precipitation of iron in the form of (NH_4/Na) $Fe(SO_4)_2(OH)_6$ from an acidic solution of less than pH 1.5 at a temperature of 90–100°C in the presence of cations like NH_4^+ or Na^+. After jarosite precipitates, the iron content of the solution is typically brought down to 1 to 5 g/L. The jarosite precipitation reaction generates free acid and this must be neutralized as the reaction progresses. This is illustrated in Equation 2.

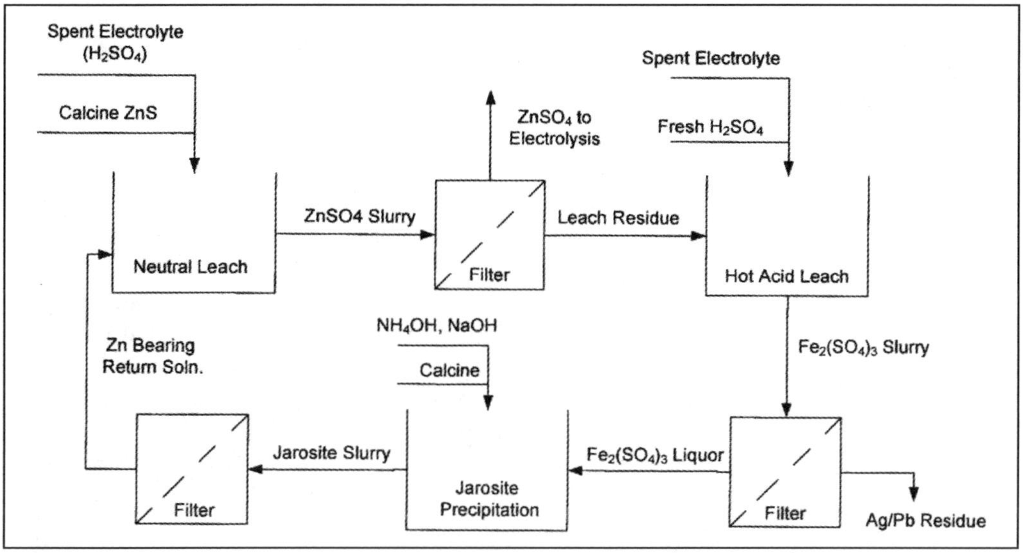

Figure 3. Detailed jarosite process flowsheet (Gordon and Pickering 1975)

$$3Fe_2(SO_4)_3 + 2(NH_4, Na)OH + 10H_2O \rightarrow 2(NH_4, Na)Fe_3(SO_4)_2(OH)_6 + 5H_2SO_4 \quad \textbf{(2)}$$

The Jarosite Process depends on temperature, pH, alkali concentration, iron concentration, seeding and presence of impurities. The rate of formation improves with an increase of temperature and can be complete in several hours at 100°C (Brown 1970, Brown 1971).

Two variations of the process called the VM and EZ processes exist for iron removal from zinc sulfate (Allen 1970, Andre 1973, Davey 1976).

The VM goethite process was introduced by Balen Plant, Vielle Montague S.A., Belgium. In this process, the leach liquor should have iron in the ferrous state and it should be oxidized with air at a controlled rate. If leach liquors contain appreciable quantities of Fe^{+3}, it should first be reduced to Fe^{+2} in a separate step, by using a replacement agent like ZnS itself.

Goethite precipitates at pH of 2 to 3.5 and at a temperature of 70 to 90°C, according to the following reaction:

$$FeSO_4 + ½O_2 + H_2O \rightarrow FeO{\cdot}OH \text{ (a-form)} + H_2SO_4 \quad \textbf{(3)}$$

Figure 4 illustrates the goethite process flowsheet.

The EZ process was introduced by the Electrolytic Zinc Co of Australasia Ltd. In this process, the concentration of Fe^{+3} in the solution is precipitated directly without prior reduction and re-oxidation of the iron.

Akita Zinc Smelter of Japan practices the Hematite Process for the removal of soluble iron as Fe_2O_3 (Mealey 1982). In this process, the solution generated after hot acid leaching of the ferrite residue is first treated with H_2S to separate copper and then neutralized with limestone to produce $CaSO_4$. The solution remaining after separating $CaSO_4$ contains zinc, cadmium and ferrous iron. Iron is removed from the solution as hematite by heating it at 200°C in titanium-lined autoclaves in the presence of O_2. Oxygen is necessary to convert Fe^{+2} to Fe_2O_3. The following reaction takes place:

$$Fe_2(SO_4)_3 + 3H_2O \rightarrow Fe_2O_3 + 3H_2SO_4 \quad \textbf{(4)}$$

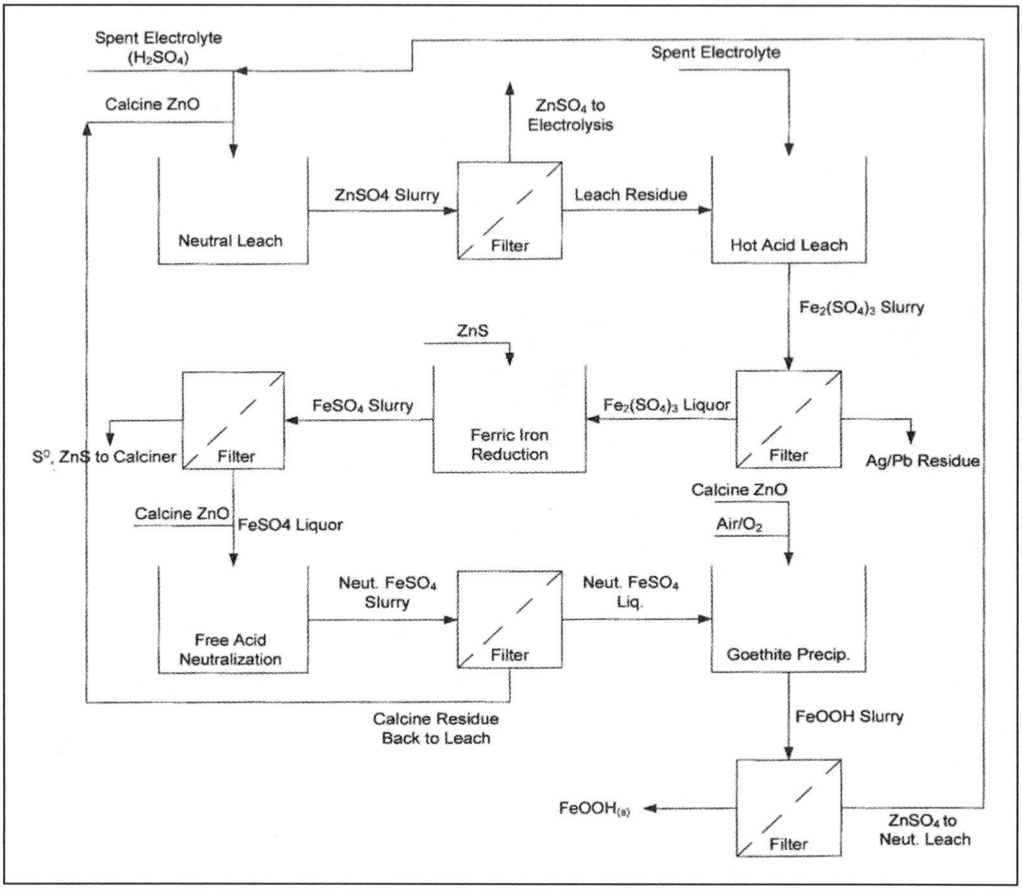

Figure 4. Goethite process flowsheet (Gordon and Pickering 1975)

Figure 5 illustrates the hematite process.

Zinc Solvent Extraction

Because of large scale commercial success of the solvent extraction (SX) purification and concentration process in hydrometallurgical copper extraction, zinc operators began to investigate zinc SX to produce zinc electrolyte (Cole 2003, Gnoinski 2007, Gnoinski and Sole 2007).

Technicas Reunidas of Madrid, Spain pioneered this zinc SX with their ZINCEX process installed at a number of small plants in Spain and Portugal that focused on leaching of secondary zinc feeds. The experience gained from these smaller plants led to design and start-up of larger commercial zinc SX plant installations at Skorpion, Namibia (150 ktpa), and more recently Akita, Japan (20 ktpa).

This SX purification process is particularly suited to zinc silicate/oxide ores and secondary feeds such as from electric arc furnace (EAF) dust that contain halides and where low grade zinc leach solutions need to concentrated to EW electrolyte concentrations. SX was also used to make zinc EW electrolyte at the Teck bioleach pilot plant at Sullivan, BC, and at the Metmex, Penoles, Mexico, lead smelter zinc flue dust leach project (5 ktpa).

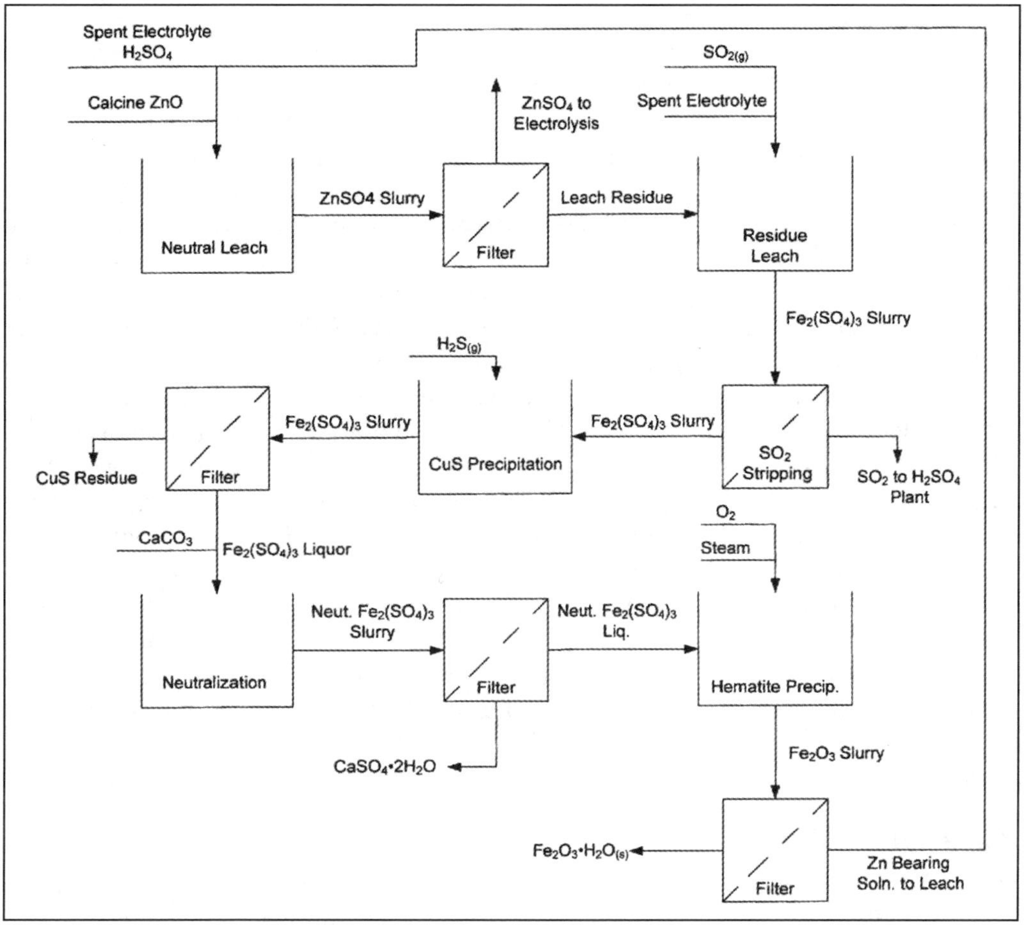

Figure 5. Hematite process flowsheet (Gordon and Pickering 1975)

Zinc Concentrate Pressure Leaching

The concept of direct oxidative pressure leaching of zinc sulfide concentrate in H_2SO_4 media was demonstrated on a laboratory scale by Sherritt Gordon Mines Ltd. in 1957 (Forward 1959). The outstanding features of the process were high zinc extraction, rejection of iron in an insoluble form and the production of sulfur in a non-polluting elemental form. In this process, the maximum leach temperature was restricted to below the melting point of the elemental sulfur (i.e., 119°C) because molten sulfur formed in the leach coated the partially reacted metal sulfides and limited zinc extraction. In order to obtain high zinc extraction at the low temperature employed, retention times of 6 to 8 hours were required. This in fact was major driver for implementation of the process at the Trail smelter about two decades later in January 1981 which already had a Roast Leach Electrowin plant in place. The adoption of pressure oxidation to produce elemental sulfur in part alleviated the need to expand either the acid plant or fertilizer plant processes. Cominco Limited and Sherritt Gordon Mines Ltd. (Parker 1981) adopted the above-mentioned process after suitable modification and optimization to run a 3 t/d pilot plant in Fort Saskatchewan, Alberta for processing a zinc sulfide concentrate containing 47.6% zinc, 11.3% iron, 6.96% lead and 30.6% total sulfur. The essential steps involved in the process were grinding of the concentrate to 98.5% less than 44 microns, pressure leaching of the ground concentrate with dilute sulfuric acid from electrowinning return electrolyte and separation of sulfur from the leached residue.

Pressure leaching was conducted in a four compartment brick lined autoclave with 1.27 m^3 operating volume. Spent electrowinning electrolyte was mixed with sufficient concentrated sulfuric acid to raise the free acid concentration to about 165 g/L, preheated and pumped into the first compartment along with oxygen, which was added to maintain and oxygen partial pressure of 620 kPa. The addition of surface active compounds such as Quebracho and/or lignin sulfonate slat inhibited the molten sulfur that was formed from occluding unreacted sulfide particles and allowed leaching to more rapidly go to completion. About 80 to 85% of the total exothermic reaction takes place in the first compartment, raising the temperature to 146°C. About 97% of the zinc is extracted into the leach liquor after total residence time of 90 minutes. At the elevated temperature and pressures utilized, ZnS, PbS and FeS mineralization present in the concentrate react with oxygen and sulfuric acid to form simple metal sulfates, water and elemental sulfur. A key aspect is that at the elevated temperatures and with diminishing acid levels, ferric iron in solution eventually precipitated in the form of basic jarosites. Recovery of the elemental sulfur from the leach residue was accomplished by using either decantation or a flash-flotation technique. The decantation method was preferred and resulted in about 91% of the sulfur being removed as a product containing 94 to 98% sulfur. The success of the original pilot plant resulted in a subsequent successful first commercial acid-pressure leaching plant in 1981. It was designed to treat about 25% of Cominco's planned zinc capacity of 270,000 t/year at the Trail operation. Figure 6 illustrates the zinc pressure leach process at Cominco's Trail operations (Parker 1981, Chalkley et al. 1982, Parker 1983, Maschmeyer et al. 1978.

LEAD HYDROMETALLURGY

Applied hydrometallurgical processes for lead are very limited, but the Betts Electrolytic Process stands as a monument and testament to innovation in both lead the hydrometallurgical community.

Betts Electrolytic Process (BEP)

Electricity had been discovered in the 1800s and in the period of 1865 to 1870 the first patents (English) for electrolytic refining of copper from smelter anodes were issued to James Elkington. Development also began on an electrolytic lead refining process using an acetate electrolyte and a lead bullion anode by N.S. Keith (1878) and Tommasi (1871/1891), but the lead cathode produced was spongy and difficult to handle.

Using starter sheet cathode technology and glue as an additive, commercial scale copper refineries were able to be built in Hamburg and Belgium and the first commercial scale refinery built in the USA was Balbach Smelting and Refining Company in Newark, New Jersey, in 1883. Within ten years, eleven more copper refineries were constructed in the USA, particularly in the New York/New Jersey area, including Laurel Hill (Phelps Dodge), Raritan (Anaconda), Perth Amboy (ASARCO), and Carteret (Amax) producing a quarter of the world's refined copper.

The lead cathode produced in an acetate electrolyte could not use starter sheets, and glue (gelatine) did not improve the lead cathode deposit. In the late 1800s Anson Betts (Betts 1908) began a series of experiments with different electrolytes in a small pilot cell arrangement consisting of four cells (10 in. W, 16 in. D and 30 in. L) that contained nine anodes each. The first cathodes

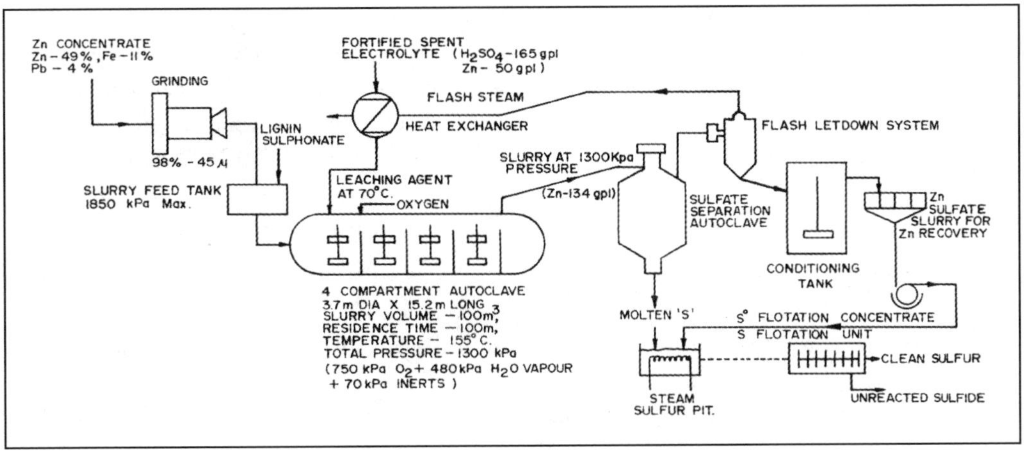

Figure 6. The zinc pressure leach process at Cominco's Trail operation (Parker 1981)

were sheet iron dipped in lead. Current density was 7–8.8 ASF and typical cell voltage was 0.175 V. Electrolytes tested included acetate, fluoboric acid and hydrofluosilicic acid. Hydrofluosilicic acid was ultimately chosen due to high lead solubility (in excess of 400 gpl), low cost, higher conductivity and that like copper refining with the addition of glue (gelatine), made a dense deposit that could make a refined lead starter sheet from a lead bullion anode. Both hydrofluosilicic acid electrolyte and the glue addition were key innovations that led to commercial application and led to the Betts Electrolytic Process or BEP being patented in 1901.

The main difference between the BEP process and copper refining was the lead bullion impurities such as silver, gold, antimony, arsenic and silver remain attached to the lead bullion anode as a thick slime rather than fall to the bottom of the cell in copper. The main advantages of the BEP process over other lead refining processes was that it could produce a purer lead product in a single step from lead bullion that included impurities such as bismuth, antimony, arsenic silver and gold. Hygiene in the BEP was generally improved over the pyro-metallurgical refining processes and drosses and slag recycle streams were minimized. Lead fumes and dust also were not produced as this was a hydro-metallurgical process. The only loss of current efficiency in the BEP process is short circuits between the anode and cathode. Theoretically it can be 100%. The BEP process today remains the most efficient method to separate lead from bismuth.

Commercial scale up (electrode size) in electrolytic processes can be faster than in pyrometallurgical applications, so from Bett's small pilot plant, a small commercial lead refining plant was built at Cominco's existing lead and zinc plant at Trail, BC, in Canada in 1902 (Thomas 1980). The Trail location had a ready source of continuous hydropower supply, an existing lead-zinc processing plant and was close to several large lead ore deposits.

Trail has pioneered the commercial application of the BEP process with continuous improvement and innovation focussed largely on productivity since 1902. The first cellhouse built in 1902 had an anode size of $0.6m^2$ each side and used a monorail system for electrode change out and asphalt coated wooden cells containing 25 cathode starter sheets and 24 lead bullion anodes. By 1922 the plant had expanded to 135 TPD with cranes and asphalt covered concrete cells. In 1926, the refinery was expanded to 1116 cells (350 TPD), the present footprint today.

A major innovation in improving lead cathode quality was the discovery and patenting of aloe in 1953 as a strong and improved lead refining smoothing agent to replace glue (Mather 1953). In 1954 Trail implemented aloe and this improved the lead cathode deposit at higher current densities and enabled the cathode cycle to be increased to 7 days (from 3 days). At Trail, current density was able to be raised to 240 ASM and in turn increasing refining capacity of 550 TPD with the existing refinery footprint from 1926.

Innovations and developments since 1954 at Trail have included the following

1. Continuous drossing of lead bullion (first developed at BHAS in Pt. Pirie, South Australia) to remove copper before the refining process (BHAS < 1966)
2. Improvements in anode casting with controlled lead bullion composition for optimum slimes adherence to anode
3. Improvements in starter sheet manufacture to make straight electrodes and minimize short circuiting
4. Development in the mid-1970s of an in-plant instrument that continuously measures cathode polarization of electrolyte and the active concentration of the additives in the electrolyte for high current density (Krauss 1976a)
5. Implementation in the mid-1970s of programmed rectifier current to keep cell voltage below 0.2 V anode overpotential so that bismuth does not dissolve from anode slime layer back into electrolyte (Krauss 1976b)
6. Polymer concrete cells that require no liners and minimal maintenance replaced lined concrete cells
7. Slimes treatment processes

Since 1926, with continuous innovation, Trail has effectively increased production by over 50% using the same refinery footprint, cell size (24 anodes) and electrode size (0.6 m^2).

Over the last several decades BEP refineries have since been mainly built in Japan and China. Developments at those BEP refineries include (Kubota 1998, Hiraki 1998, and Gonzales 1991) the following

1. Implementation of real-time cell voltage monitoring to detect and correct short circuits
2. Periodic current reversal rectifiers that smooth the lead cathode
3. Improving electrode handling automation leading to increased productivity, reduction in labor cost and improving occupational health
4. Improved electrode straightening for higher current efficiency

Future innovation and implementation opportunities could be as follows:

1. Bipolar lead refining technology tested at Trail in the 1980s that enables electrode size to be larger leading to smaller BEP plant footprint, lower energy consumption and more automated permanent electrode system to be installed.
2. Wireless cell voltage monitoring technology used in copper refining
3. Mettop high current density technology (used in copper ER) that introduces electrolyte flow to cathode and does not disturb slimes layer on anode. At present, the BEP cell flow limited to 30–40 L/min

4. Amorphous coatings pioneered by Freeport and Doshisha/Republic in copper EW that do not precipitate lead oxide that could be used for lead ER liberators

In summary, the innovative success of the BEP process can be due to several factors:

1. Successful commercial application of copper refining using glue addition as a smoothing agent and starter sheet cathode technology before the BEP was developed
2. Relative ease of construction of a pilot plant to test the many different electrolytes specifically for lead refining
3. Relatively fast scale up and technology implementation time from Betts's pilot cells (patent in 1901) to Cominco Trail's first commercial plant in 1902
4. Strong technical development and innovation team continuity at the first commercial site at Trail, BC, since 1902 to present day

SUMMARY

This paper has summarized what the authors humbly deem as the significant developments to date in the zinc and lead hydrometallurgical field. As with zinc concentrate pressure oxidation, it is likely that the pace of lead hydrometallurgical development and implementation will proceed more rapidly due to increasing restrictions on the established and dominant pyrometallurgical smelting process.

REFERENCES

Aird, J., Celmer, R.S., and May, A.V., 1980. New Cobalt production From Chambishi's Roast Leach Electrowon process, paper presented at the 10th annual Hydrometallurgical meeting (CIM) Edmonton.

Allen, R.W., Haigh, C.J., and Hamdorf, C.J., 1970. An improved method of removing dissolved ferric iron from iron-bearing solutions, Australian Patent 4242,095.

Andre, J.A. and Masson, N.J.A., 1973. The goethite process in re-treating zinc leaching residues, in AIME 102nd Annual Meeting AIME, Chicago.

Arregui, V., Gordon, A.R., and Steinvelt, G., 1979. The Jarosite Process—Past Present and Future, in Lead-Zinc-Tin, '80, Cigan, J.M., Mackey, T.S., and O'Keefe T.J., Eds, AIME, New York, 97.

Betts, AG. 1908. Lead Electrorefining by Electrolysis, John Wiley and Sons, New York.

BHAS, 1966. Can. Patent 733437, 3 May 1966, Can. patent 736884 21 June.

Brown, J.B., 1970. A chemical study of some synthetic potassium-hydronium jarosites, Can. Mineral, 10(4), 696.

Brown, J.B., 1971. Jarosite-goethite stabilities at 25°C, 1 atm, Miner. Deposita, 245.

Caufriez, A., Dubois, M., Lejay, E. Zinc Cellhouses, 1998. Concept and practice at Union Mineire, pp. 337–352, Zinc and Lead Processing, Dutrizac, J.E., Gonzales, J.A., Bolton, G.L., Hancock, H., ed. Aug 1998, Calgary, Canada, The Metallurgical Society of CIM, August.

Chalkley, M.E., Collins, M.J., Masters, I.M. and Ozberk, E., 1982. The Integration of the Sherritt Zinc pressure Leach Process with Commercial Iron Precipitation Processes, Hydrometallurgy.

Cole, P.M., Zinc Solvent Extraction in the Process Industries, Mineral Processing and Extractive Metallurgy Review,Volume 24, Number 2, April–June 2003, pp. 91–137(47).

Davey, P.T., and Scott, T.R., 1976. Removal of iron from leach liquors by the Goethite process, Hydrometallurgy, 2, 25.

Dutrizac, J., 1982. The behavior of impurities during jarosite precipitation, in Proceedings NATO Adv. Res. Inst. Hydrometall., Cambridge, July 25 to 31.

Dutrizac, J.D., 1983. Jarosite type compounds and their application in the metallurgical industry, in Hydrometallurgy—Research Development and Plant Practice, Osseo-Asare, K. and Miller, J.D., Eds. The Metalllurgical Society of the AIME, New York, 513.

Dutrizac, J.E., 1980. The physical chemistry of iron precipitation in the zinc industry, in Lead-Zinc-Tin, '80, Cigan, J.M., Mackey, T.S. and O'Keefe, T.J., Eds., AIME, New York, 1979, 532.

Dutrizac, J.E., Dinardo, O., and Kaiman, S., 1980. Factors affecting lead jarosite formation, Hydrometallurgy, 5, 305.

Dutrizac, J.E., and Kaiman, S., 1976. Synthesis and properties of jarosite type compounds, Can. Mineral, 14, 151, 1976.

Forward, F.A., and Veltman, H., 1959. Direct Leaching Zinc Sulfide Concentrates by Sherritt Gordon, J. Met., 12, 836.

Free, M., Moats, M., Robinson, T.G., Neelameggham, N., Houlachi, G., Ginatta, M., Creber, D., Holywell, G. 2011. Electrometallurgy–Now and in the Future, Electrometallurgy 2011, TMS, Orlando.

Gnoinski, J., 2007, Skorpion Zinc: Optimisation and Innovation, The South African Institute of Mining and Metallurgy, Fourth Southern African Conference on Base Metals.

Gnoinski, J., and Sole, K. 2007, The Influence and Benefits of an Upstream Solvent Extraction Circuit on the Electrowinning of Zinc in Sulfate Media, Electrometallurgy.

Gonzales, J.A., 1991. Electrochemical Processes Within the Slimes Layer of Lead Anodes During Betts Electrorefining, PhD Thesis UBC March.

Gonzalez-Dominguez, J.A., Dreisinger, D.B., 1997. Identifying Research Opportunities in Zinc Electrowinning, 1997 Review of Extraction and Processing, Journal of Metals, April.

Gordon, A.R. and Pickering, R.W., 1975. Improved Leaching Technologies in the Electolytic Zinc Industry, Metallurgical Transactions B, Volume 6B, March, 43.

Haigh, C.J., 1967. The hydrolysis of iron in acid solutions, Proc. Aust. Inst. Metall., September, 49.

Haigh, C.J. and Pickering, R.W., 1970. The treatment of zinc plant residue at the Risdon Works of the Electrolytic Zinc Company of Australasia Ltd., AIME World Symposium. Min. Metall. Lead Zinc, 2, 423.

Hiraki, T., Ogata, M., 1998. An Outline of the lead electrolysis operation at the Chigirishima Refinery pp. 379–388, Zinc and Lead Processing, Dutrizac, J.E., Gonzales, J.A., Bolton, G.L., Hancock, H., ed. Aug, Calgary, Canada, 37th Annual Conference of Metallurgists of CIM.

Krauss, C.J. and Shaw G., 1976a, Can. Patent 988879, May 11.

Krauss, C.J., 1976b, Can. Patent 1020491, Nov. 8.

Kubota, H., Kusakabe, T., Takei, K., Takewaki, M., 1998. Current operation of Sumitomo Metal Mining's (SMM) Betts Lead electrorefinery. Zinc and Lead from steel mill dusts pp. 353–366, Zinc and Lead Processing, Dutrizac, J.E., Gonzales, J.A., Bolton, G.L., Hancock, H., ed. Aug, Calgary, Canada, 37th Annual Conference of Metallurgists of CIM.

Letrange, L.,1883. US. Patent 286208.

Limpo, J.L., Luis, A., Sigwin, D., and Hernandez, A., 1976. Kinetics and mechanism of the precipitation of iron as Jaorosite, Rev. Metal. Cerium, 12(3), 123.

Maschmeyer, D.E.G., Kawulka, P., Milner, E.F.G., and Swinkels, G.M., 1978. Application of the Sherritt Cominco process to Arizona copper concentrates, J. Met., 30, 27.

Mather, F.C., Lantz, G.B. 1953. US Patent 2664393, Dec 29.

McAndrew, R.T., Wang, S.S., and Brown, W.R., 1975. Precipitation of iron compounds from sulfuric acid leaching solutions, CIM Bull, January, 101.

Mealey, M., 1973. Hydrometallurgy plays a big role in japan's new Zn smelter, Eng, Min. J., January.

Menendez, F.J.S., and Fernandez, V.A., 1967. Process for the recovery of zinc from ferrites, Canadian Patent, 770,555.

Pammenter, R.V., and Haigh, C.J. 1981. Improved metal recovery with the low contaminated jarosite process, in Proc. Extraction Metall. '81, Institute of Mining and Metallurgy, London, 379.

Parker, E.G., 1981. Oxidative Pressure Leaching of Zinc Concentrates, CIM Bull., 74, 145.

Parker, E.G., McKay, D.R., and Salomon De-Friedberg, 1983. Zinc Pressure Leaching at Cominco's Trail operation, in Proc. 3rd Int. Symp. Hydromet. Research, Development and Plant Practice, Osseo-Asare, K. and J.D. Miller, Eds., TMS, AIME, New York, 927.

Ralston O.C., 1921. Electrolytic Deposition and Hydrometallurgy of Zinc, McGraw-Hill Book Co, New York.

Robinson T.G., 2003. Copper Elecrowinning—A Historical Perspective, Hydrometallurgy 2003, TMS, Vancouver, August.

.Robinson T.G., Sole, K.C., Moats, M.C., Crundwell F.K., Morimitsu, M., Palmu, L. 2011. Developments in Base Metal Electrowinning Cellhouse Design, Electrometallurgy 2011, TMS, Orlando.

Steinvelt, G. 1968. Treatment of zinc containing acid sulfate solutions to separate iron therefrom, Canadian Patent, 793,766.

Steinvelt, G. 1970a. Electrolytic Zinc Plant and Residue Recovery, Det Norske Zink KOmpani A/S, AIME World Symposium Min. Metall. Lead Zinc, 2, 223.

Steinvelt, G., 1970b. Precipitation of iron as Jarosite and its applications in the wet metallurgy of zinc. Erzmetall, 23, 523.

Steinvelt, G. 1970c. Treatment of Zinc Plant residues by the Jarosite process, in Advances in Extractive Metallurgy and Refining, Jones, M.J. Ed. Institute of Mining and Metallurgy London.

Thomas, D.L., Krauss, C.J., Kerby R.C. 1980. Betts Lead Electrorefining at Cominco, Lead Zinc TMS.

USEPA, 1986. Metallurgical Industry Emissions Report, Section 12-7.

Wood, J.T., 1973. Treatment of electrolytic zinc plant residues by the jarosite process, Aust. Min., 65, 23.

Zapuskalova, N.A. and Margolis, E.V., 1977. Study of hydrolytic precipitation of iron (III) from zinc sulfate solutions if the presence of monovalent cations, Izv. Vyssh. Uchebn. Zaved. Tsvetn. Metall. 4, 38.

ABOUT THE AUTHORS

Tim Robinson graduated with a BE Chem (Hons) from the University of Melbourne in 1986. His work for MIM Holdings, Simons Mining, Global Industrial Technologies, Freeport-McMoRan, and currently with Republic Alternative Experience has spanned four countries: United States, Canada, Chile, and Australia. Robinson has been involved with copper, nickel, zinc, and lead production as well as in the fields of crushing, milling, flotation, hydrometallurgy, smelting, refining, and fabrication of metals. His career focus has been on developing sustainable mineral processing and hydrometallurgical process technologies. He was responsible for process design and commissioning of more than a dozen copper tank houses around the world from 1993 to 1997. Robinson has written more than 30 technical and operational benchmarking papers for the copper, nickel, and zinc industries from 1995 to present. He also co-authored *Extractive Metallurgy of Nickel, Cobalt and Platinum Group Metals* (Elsevier). He led the development of low-energy non-lead-anode technology at Freeport-McMoRan and holds more than 20 patents in electrode design and in electrometallurgical processes.

Corby G. Anderson has more than 34 years of global experience in industrial operations, management, engineering, design, consulting, teaching, research, and professional service. His career includes positions with Morton Thiokol, Key Tronic Corporation, Sunshine Mining and Refining Company, H.A. Simons Ltd., and Montana Tech. Anderson holds a BSc in chemical engineering and an MSc and PhD in metallurgical engineering. He is a fellow of the Institution of Chemical Engineers and the Institute of Materials, Minerals and Mining. He currently serves as the Harrison western professor in the Kroll Institute for Extractive Metallurgy as part of the George S. Ansell Department of Metallurgical and Materials Engineering at the Colorado School of Mines. In 2009 he was honored by SME with the Milton E. Wadsworth Extractive Metallurgy Award for his contributions in hydrometallurgical research.

Innovations in Nickel and Cobalt Hydrometallurgy

Corby G. Anderson
Kroll Institute for Extractive Metallurgy, Colorado School of Mines, Golden, Colorado, United States

ABSTRACT: This paper covers the history and implementation of key technologies in nickel and cobalt hydrometallurgy. Breakthrough leaching, purification, separation, concentration and recovery unit operations are considered. The author has humbly chosen illustrative technologies which have innovative drivers. The key driver has been the enhanced separations offered by hydrometallurgy albeit global politics has also played a significant role in the case of Cuban developments. This includes industrial examples such as the Sherritt Gordon Process, the Falconbridge Process, the Roast Reduction Ammoniacal Leach Process, and the High Pressure Acid Leaching Process along with a treatise on the advent of Solvent Extraction separations in these endeavors. To that end, the proposed and imminent Voisey's Bay sulfide concentrate industrial process, relying heavily on modern hydrometallurgical innovations such as continuous pressure oxidation and multiple solvent extraction separations is briefly elucidated.

INTRODUCTION

In the processing of nickel and cobalt ores and concentrates a parallel path has been taken developmentally for both sulfides and lateritic oxides. The application of hydrometallurgy to nickel and cobalt has clearly delineated and developed any and all process unit operations now common in the field. The following sections elucidate this.

SULFIDE CONCENTRATE PROCESSES

The innovations in treatment of nickel and cobalt sulfide concentrates are best demonstrated by the Sherritt Gordon and Falconbridge Matte Leach processes.

Sherritt Gordon Ammoniacal Pressure Leach

The Sherritt process is a hydrometallurgical process named for Sherritt Gordon Mines Ltd. (now Sherritt International) of Sherridon and Lynn Lake, Manitoba, Canada, based on the older Forward process developed by Frank Forward for the recovery of copper and nickel from those same mines (Boldt, 1967, Forward & Mackiw, 1955, Mackiw and Benz 1961). The key driver was the ability to separate nickel and cobalt from the high iron bearing sulfide concentrates form Lynn Lake, Manitoba. Nickel sulfide concentrates can be treated by either roasting or flash smelting to produce matte from which nickel and cobalt can be recovered hydrometallurgically, or they may be treated by an ammonia solution pressure leach. The residue is removed. A feed of matte and sulfide concentrate containing approximately 0.4% cobalt and 30% sulfur is pressure leached at elevated temperature and pressure in an ammoniacal solution to produce a solution of nickel, copper and cobalt.

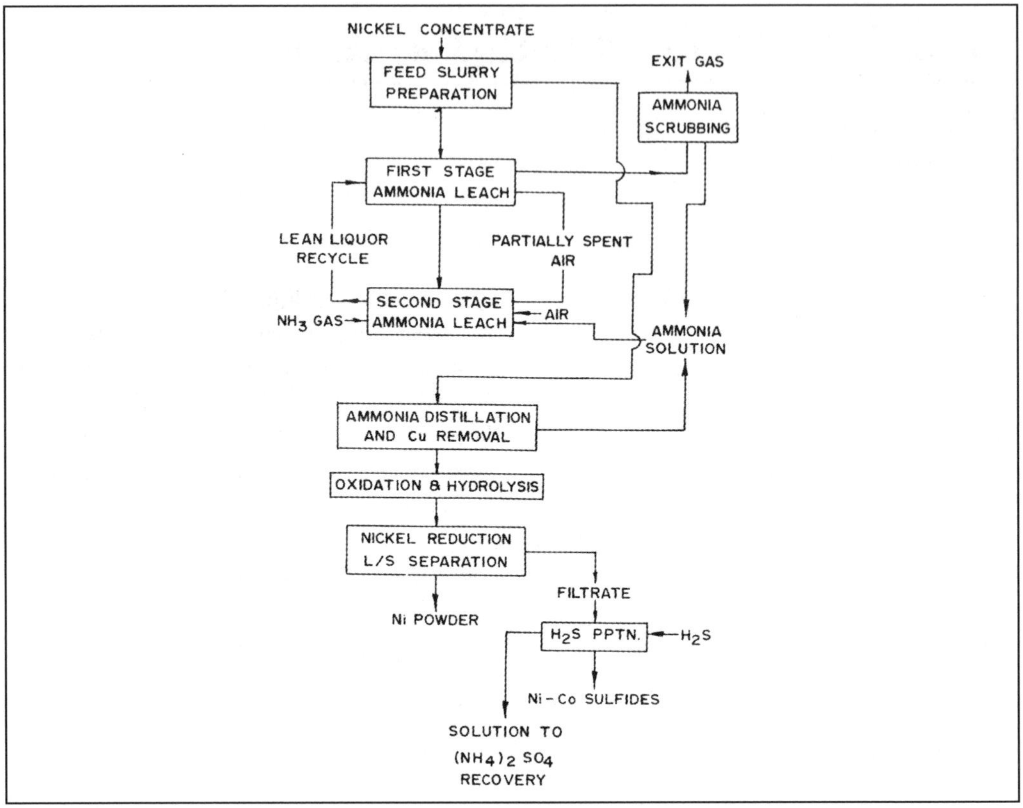

Figure 1. Sherritt Gordon ammoniacal leach process for nickel sulfide concentrate (Gupta and Mukherjee 1990)

By boiling away the ammonia; copper is precipitated as a sulfide and sent to a smelter. Hydrogen sulfide is added to the autoclave to remove nickel sulfide and copper sulfide which is fed back to the leaching process. Air is then passed through the solution in the autoclave for oxyhydrolysis. The solution is then reduced with hydrogen, again at high temperature and pressure, to precipitate nickel powder (>99%). The remaining solution (containing approximately equal proportions of nickel and cobalt sulfides), is then adjusted (to a lower temperature and pressure) to precipitate the mixed sulfides and the fluid is concentrated and crystallized into ammonium sulfate ($(NH_4)_2SO_4$). The mixed sulfides are pressure leached with air and sulfuric acid. Ammonia is then added to remove potassium and iron as jarosite ($KFe^{3+}{}_3(OH)_6(SO_4)_2$).

More ammonia and air is added for oxidation. The solution is removed from the autoclave and sulfuric acid added to remove nickel as nickel(II) sulfate-ammonium sulfate hexahydrate ($(NiSO_4)\cdot((NH_4)_2SO_4)\cdot 6H_2O$) which is then sent to have its nickel recovered. The solution is then further reduced with more sulfuric acid and cobalt metal powder is added to aid in the nucleation of precipitants (seeding). Addition of hydrogen gas to saturation precipitates cobalt powder with a purity of approximately 99.6%. Figures 1, 2, and 3 outline the leaching and hydrogen reduction and overall flowsheets for this innovative process.

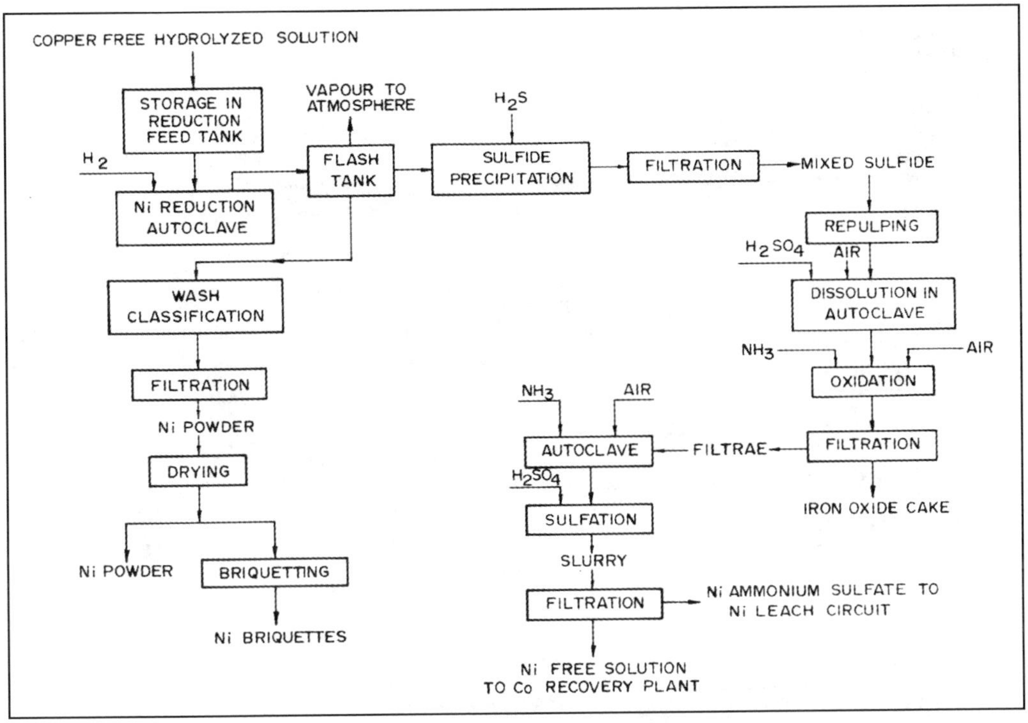

Figure 2. Flowsheet for recovery of nickel by hydrogen reduction as practiced at Fort Saskatchewan (Gupta and Mukherjee 1990)

Falconbridge Matte Leach Process

The refinery at Kristiansand, Norway has a long history of nickel production (Boldt 1967; Stensholt 1986, 1988, 2001). It was erected in 1910 to process matte from the smelter at Evje, Norway, where ore from the Flaat mine was treated. One of the earliest mines opened for nickel, the Flaat was opened in 1872 and did not close down until 1945. The Kristiansand plant from the outset employed the pioneer electrolytic method pioneered by N.V. Hybinette. Falconbridge acquired the plant in 1928.

The crushed converter matte produced at the Falconbridge nickel smelter was shipped to Falconbridge Nikkelverk Aktieselskap at Kristiansand, Norway. This refinery produces cathode nickel, nickel plating anodes, nickel sulfate, as well as refined copper, cobalt, gold, silver and platinum metals. Liquid sulfur dioxide and selenium are also produced. Since Falconbridge makes a bulk nickel. copper mineral concentrate in its beneficiation practice, the matte contains virtually all the copper in the original ore. Nickel and copper are separated by roasting the matte to sulfide and acid leaching the calcine to remove copper. Then smelting the leached calcine to anode metal. Figure 4 shows the plant in Norway.

This matte is ground to 95% passing a 10 mesh screen. To remove its sulfur content it is fed to a multi-hearth roaster, mixed with recirculated material and cement copper, and roasted to oxide at about 1450°F maximum in a virtually autogenous operation. The cooled product is air elutriated,

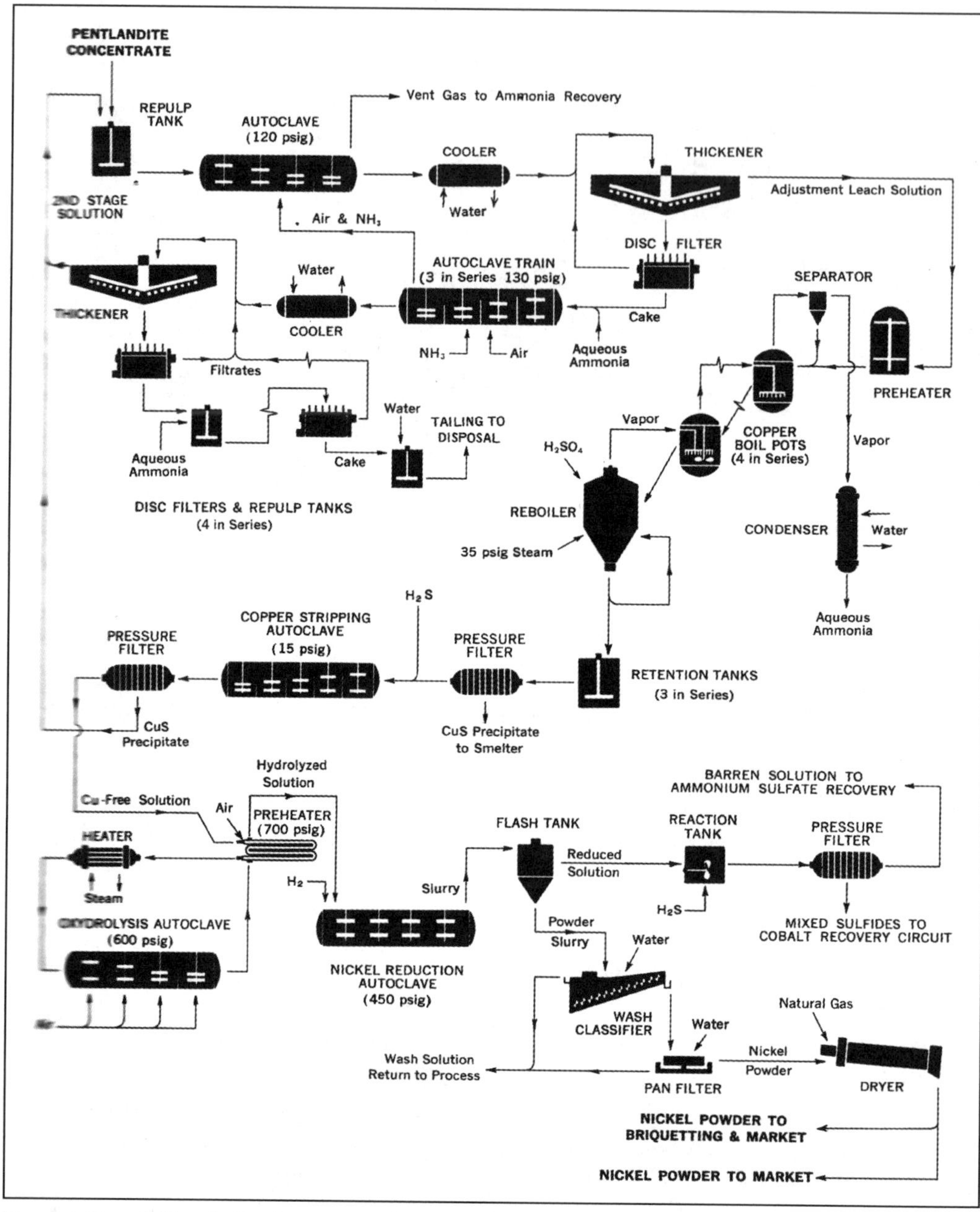

Figure 3. Overall Sherritt Gordon process flowsheet (Boldt 1967)

and the undersize fine calcine low in sulfur is sent to the leaching section. The oversize is reground in separate mills and recirculated to the roasting furnaces. Dust in the roaster gas is collected in cyclones and returned to the roasters. The gas is cleaned in wet electrostatic precipitators to remove residual dust and sulfuric acid mist.

Part of the slimes so obtained is treated to extract selenium, and part is returned to the roasters. The sulfuric acid, containing nickel and copper, is used for electrolyte make-up in the nickel

Figure 4. Falconbridge refinery at Kristiansand, Norway (Boldt 1967)

or copper electrolytic tank houses. The cleaned gas goes to an absorption plant where liquid sulfur dioxide is produced for market.

The fine calcine is continuously leached in a cascade of pachuca tanks with spent electrolyte from the copper tank house. The free sulfuric acid in the electrolyte dissolves most of the calcine's copper content. The coarse fraction of the leach residue pulp is removed by cyclone classifiers, filtered, and washed on continuous horizontal vacuum filters, This cake is then hydrogen-reduced to impure active nickel, for use in cementing out copper from nickel electrolyte. Most of the pachuca residue escapes removal in the cyclones and is collected by stainless steel filter presses. It is repulped with water, filtered, dried, and reduction melted.

The filtrate from the filter presses goes to copper electrowinning tanks equipped with copper cathode starting sheets and insoluble anodes of antimonial lead. Since the copper-bearing solution also contains some nickel dissolved in the leaching, a portion of this spent copper electrolyte is bled off and further decopperized electrolytically, then partially evaporated. Crude nickel sulfate crystals form as the concentrated solution cools. The solution is returned to the leaching tanks. The crude nickel salts are redissolved, the solution is purified, and pure nickel sulfate crystals are produced.

Melting is done using coke in a three-phase electric furnace. Revert material to the furnace includes anode scrap form the nickel tankhouse and off grade nickel and copper. After slag skimming, the metal is tapped into a forehearth heated by an electric arc, from which it is poured into anode molds on a train of flat cars on tracks. After cleaning, and grinding the contact points to bright metal, the anodes are transferred to the nickel tank house. In the tank house at Kristiansand each deposition tank takes up to 39 anodes and 38 cathodes. Cathode dimensions are normally 3.3 by 2 ft. A flow of purified electrolyte of about 5 gallons per hour to each cathode compartment maintains a hydrostatic head inside the compartment of about 2 inches. The current density is normally 17 amperes per square foot.

In its purification process Falconbridge first oxidizes the solution by bubbling with air to precipitate iron and substantially all of the arsenic. The filtered ferric hydroxide precipitate is treated with acid to dissolve the nickel and copper content, washed, and discarded. Copper is cemented out with the sponge nickel-copper alloy produced by low-temperature hydrogen reduction of the coarse fraction of the leach residue. Cementation takes place in mechanically agitated tanks, an

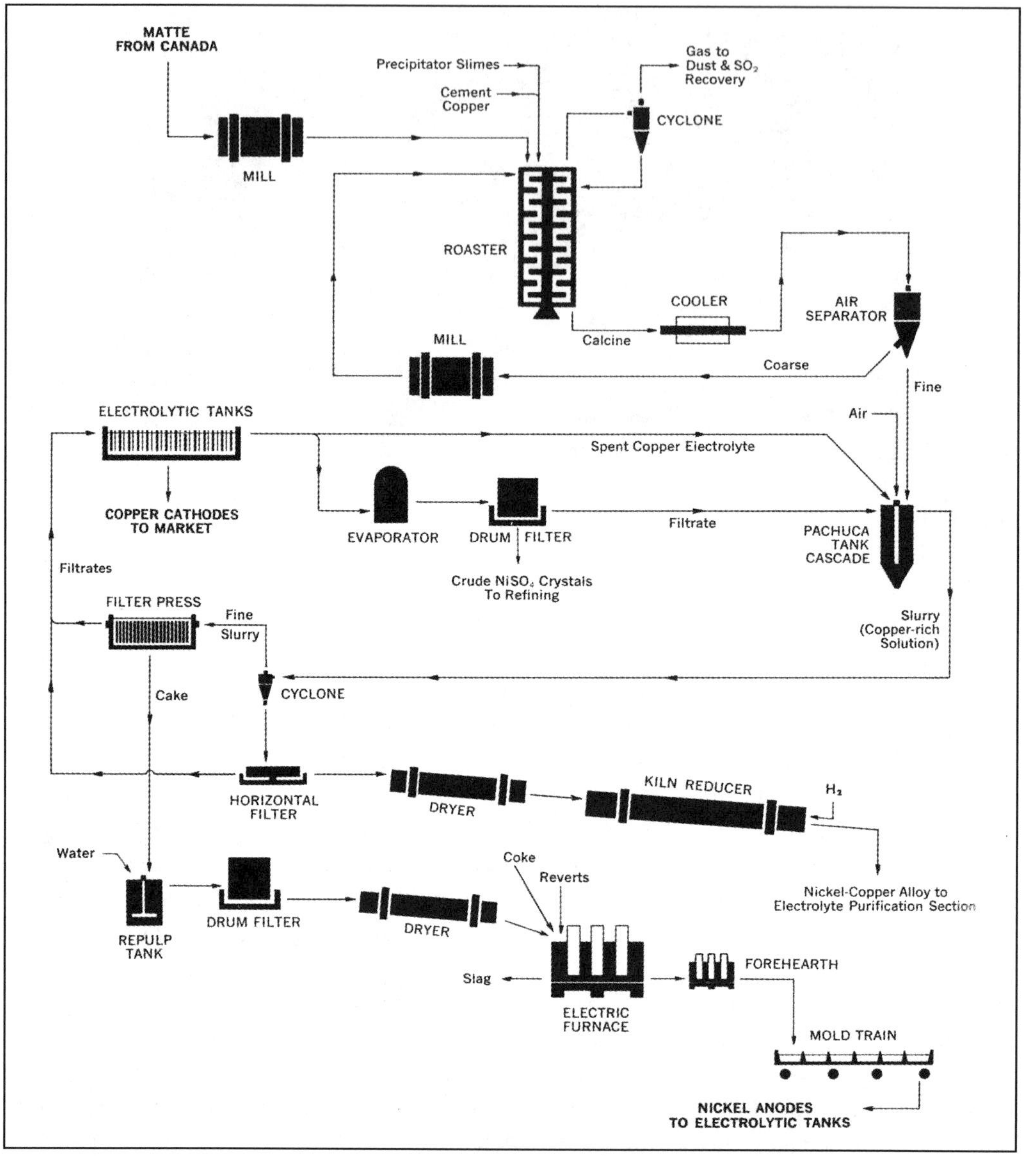

Figure 5. Flowsheet of nickel anode production at Kristiansand (Boldt 1967)

equivalent amount of nickel dissolving. In addition, aeration oxidizes and solubilizes some nickel, and its dissolution results in a desirable rise in the pH. The cement copper is filtered, washed, dried, and returned to the roasters. The solution is splashed through stoneware towers in the presence of gaseous chlorine and discharged into reaction tanks, where cobaltic hydroxide precipitates. Nickel carbonate prepared at the refinery is added to neutralize the acidifying effect that occurs as cobalt is removed from the solution. The precipitate is treated with acid to lower nickel content, then goes to the cobalt refinery where pure electrolytic cobalt is produced by electrowinning. Acid is added to the purified nickel electrolyte to adjust the pH, following which it is pumped back to the

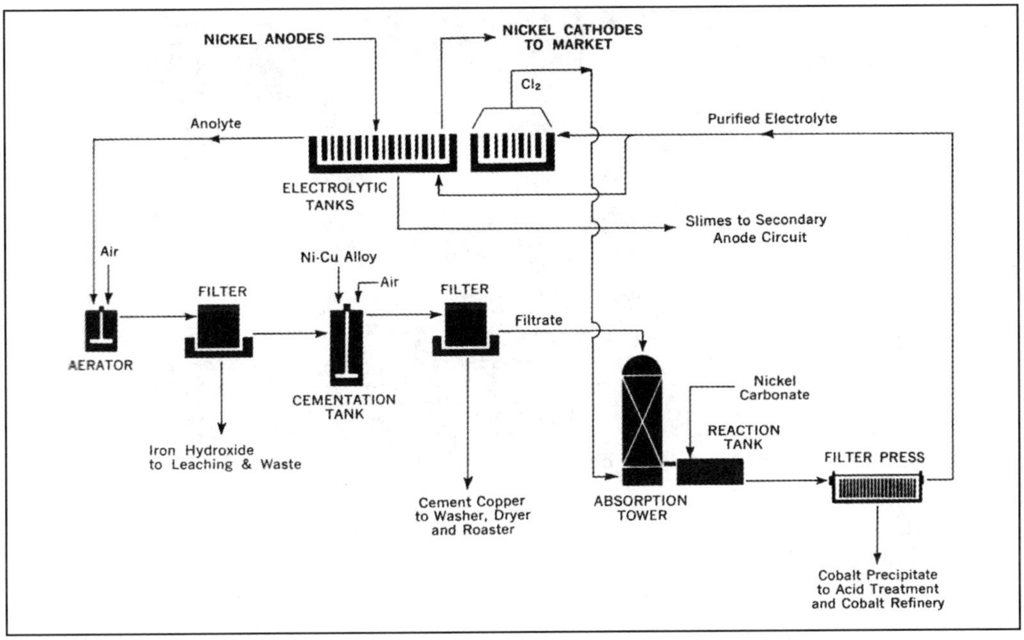

Figure 6. Flowsheet of Kristiansand electrolyte purification (Boldt 1967)

electrodeposition tanks. Chlorine gas used to precipitate the cobalt is obtained by electrolysis of part of the purified electrolyte in a group of tanks employing insoluble anodes. As nickel from the solution deposits, the chlorine evolves at the anode and is collected.

Since pure electrolyte is used in these tanks, the cathodes are of the same quality as those produced in the regular tanks. Finished cathodes are taken out of the tanks after about 8 days when they weigh about 115 pounds. The slime accumulating in tank bottoms contains residual nickel, copper and precious metals. It is subsequently treated for recovery of these metals. Figures 5 and 6 illustrate the nickel anode production and electrolytic treatment flowsheets. In summary, the key driver for implementation of this technology was both enhanced hydrometallurgical separations as well as an existing facility commensurate with the proposed technology and a less restrictive intellectual property climate allowing implementation by Falconbridge outside of North America without overriding repercussions.

OXIDE LATERITE ORE PROCESSES

Roast Reduction Ammoniacal Leaching (RRAL)

Cuba's nickel industry developed as a result of World War II, when the demand for nickel increased dramatically due to the need for high-strength stainless steel and other nickel alloys for use in the production of guns, ammunition and vehicles (Boldt 1967; Baragwanath 1936, 1961). U.S. Patent 1,487,145 on ammonia leaching of nickel ores was issued to M.H. Caron of Delft University March 18, 1924 (Caron 1924, 1942, 1946, 1950a, 1950b). Equivalent patents in other countries were obtained later. The Dutch Syndicate Brikcarbo became interested in the process in 1935; and a couple of years later the late H. Foster Bain sent samples of Surigao laterite iron ore to Delft and witnessed tests made there on this material for the Philippines Commonwealth. As a

Figure 7. Nicaro, Cuba, RRAL plant (Boldt 1967)

result, Caron was asked in April 1940 by Bain to go to the Philippines to erect and take charge of a pilot plant for the process. The war prevented carrying out this proposal. The Dorr Co. also became interested in the process at about the same period. The investigations by the Freeport Sulfur Co. on the process as applied to the Cuban nickeliferous laterites, which resulted in the Nicaro enterprise, and the results of this operation have been well described elsewhere. This Nicaro plant was, in operation for almost two years, and during this period produced about 10% of the world nickel production from laterites containing 1.35% nickel. This plant and its operation were war measures and, in view of this, activities were suspended in April 1947. The results obtained fully demonstrated the technical feasibility of the process and its economical aspects on a commercial scale. In this respect, it should be understood that it is probable that improvements may be made by further development, and that there are possibilities for advantageous application of the process to garnierite and similar ores with higher values in nickel than the laterite iron ores at Nicaro. All nickel and cobalt ores which originate from weathering of peridotites or similar basic rocks having sufficient values are suitable for treatment by the ammonia leaching process after a preliminary reduction under proper conditions. The formation of these deposits was probably as follows: In the course of time the basic rocks were attacked by atmospheric agencies; MgO and SiO_2 were gradually leached out, and secondary nickel minerals formed, such as garnierite, a hydrated silicate of nickel and magnesia. These secondary products are, however, not stable. They decompose in a further stage of weathering and ultimately only a relatively small residue of insoluble oxides remains, known as laterite iron ore with a small nickel content and very little cobalt. Under these mantles of laterite, richer nickel values may be found, usually indicated by the occurrence of garnierite. The more the ore is disintegrated by nature, the higher the iron content and the better the nickel extractions that may be expected therefrom.

As before, on December 31, 1941, the U.S. government financed the creation of Nícaro Nickel Company, a Cuban venture under the management of the Freeport Sulfur Company. Under contract with the U.S. government, Freeport's plant, located east of the Bay of Nipe, processed ore mined from the Piñares de Mayarí deposits until it closed on March 31, 1947.

The Nicaro RRAL plant is illustrated in Figures 7 and 8. Flowsheets of the Nicaro RRAL process are shown in Figures 9 and 10.

Figure 8. Leaching tanks and thickeners at Nicaro (Boldt 1967)

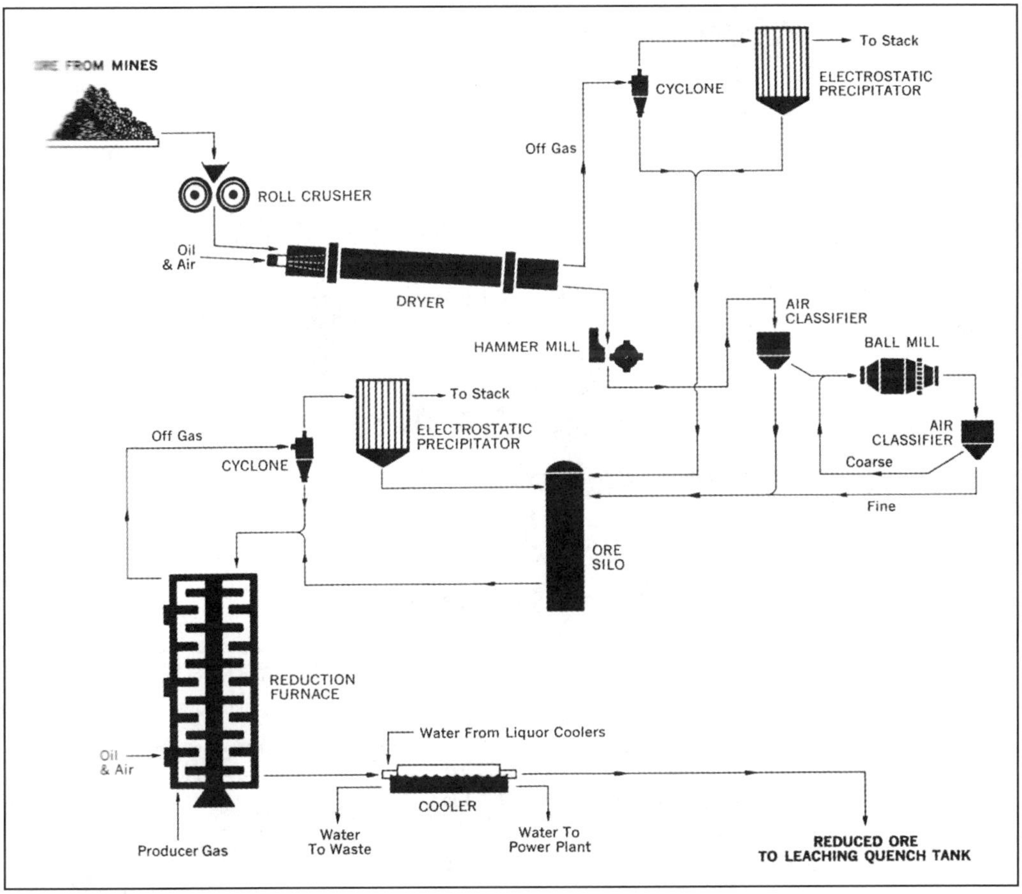

Figure 9. Nicaro, Cuba, RRAL laterite ore preparation flowsheet (Boldt 1967)

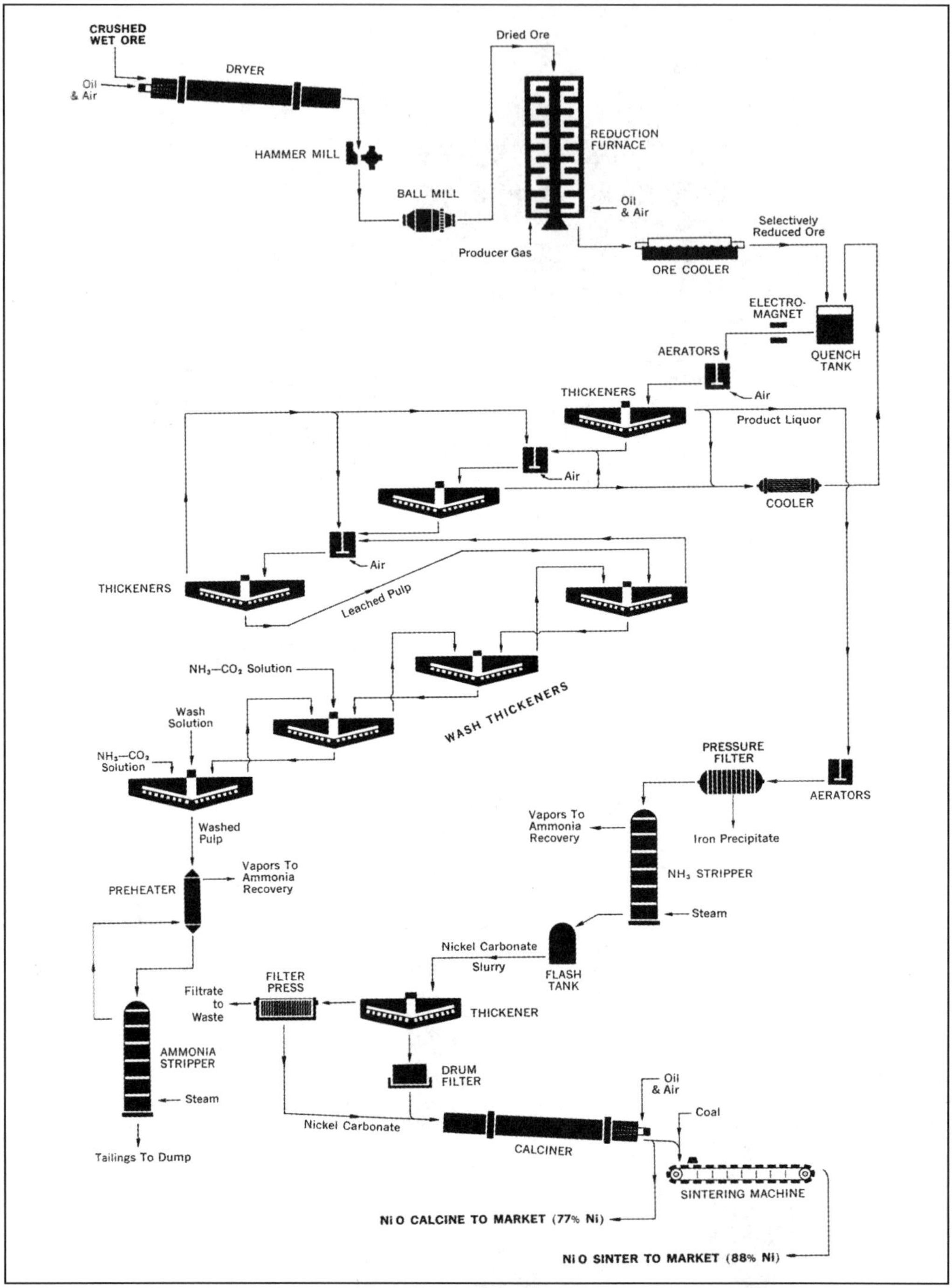

Figure 10. Nicaro, Cuba, RRAL process flowsheet (Boldt 1967)

The Caron process has been utilized for the production of nickel since 1944 and there are currently commercial operations at Nicaro, Punta-Gorda, and Yabulu Nickel. In general the process now consists of: ore beneficiation and blending, reduction roasting, leaching of roaster product in ammoniacal solvent, separation of Ni and Co using solvent extraction, precipitation from solution of basic nickel carbonate (BNC) intermediate product, and thermal decomposition of BNC to NiO and gaseous reduction of NiO to Ni. The main feed for the process is a mixture of limonitic (Fe-rich oxides/hydroxides) and saprolitic (Mg-Fe silicates) nickel laterites. The mixture is dried to reduce moisture from 35% to approximately 8%. The dried laterites are then ground to size less than 850 µm before they are fed into reduction- roasters. The reduction-roasting is carried out in Herreshoff multiple hearth furnaces. Heavy fuel oil is mixed with the laterites to provide the high temperature inside the furnace through its partial combustion, and provide reducing agent for the reduction process. The maximum temperature of the reduction roasting process is 740–750°C. This process is the critical stage that determines the recovery of nickel from the laterite ores. The reduced ores discharged from the roasters and cooled under a neutral atmosphere to temperature below 250°C. The cooled roaster product is subsequently leached in an ammonia-ammonium carbonate solution. In this process, the nickel is selectively leached from the solid. The simplified reactions of the leaching process are described as follows:

$$Ni(s) + \tfrac{1}{2}O_2\ (g) + 8NH3\ (g) + 2CO_2\ (g) + H_2O \leftrightarrow Ni(NH_3)6^{\ 2+} + 2NH^{\ 4+} + 2CO_3^{\ 2-} \quad (1)$$

$$FeNi(s) + O_2\ (g) + 12NH_3\ (g) + 3CO_2\ (g) + H_2O \leftrightarrow Ni(NH_3)_6^{\ 2+} + Fe(NH_3)_4^{\ 2+} + 2NH_4^{\ +} + 3CO_3^{\ 2-} \quad (2)$$

The pregnant liquor is then separated from the undissolved solid. Iron and cobalt are also dissolved into the ammonia-ammonium carbonate liquor. The Fe metal complex is unstable and reacts with the oxygen from the atmosphere and precipitates as solid iron hydroxide. Cobalt, on the other hand, is separated from the nickel by ammonium solvent extraction process and is extracted by hydrometallurgical refining in a separate cobalt processing circuit. The purified liquor is then steam stripped to remove ammonia-ammonium carbonate and to recover the nickel as basic nickel carbonate (BNC), i.e., a mixed carbonate hydroxide solid with approximately 52% nickel content (dry basis). **Figure 11** illustrates the ammonia stripping towers.

The remaining liquid is separated from the solid by filtering process. The BNC has various compositions and is present in one of the following forms: $NiCO_3{\cdot}xH_2O$, $NiCO_3{\cdot}xNi(OH)_2{\cdot}yH_2O$, $NiCO_3{\cdot}xNiO{\cdot}yH_2O$. The BNC is further processed through a series of high temperature processes involving various gas solid reactions. The BNC at the Yabulu refinery is fed to a rotary kiln operated in slightly reducing atmosphere conditions, fired with naphta or coal seam gas. This process is aimed at partially reducing the BNC to Ni-NiO mixtures. Dehydration and decomposition of BNC in the kiln occurs through the following reactions:

$$90\text{–}200°C:\ xNiCO_3{\cdot}yNi(OH)_2{\cdot}zH_2O \leftrightarrow xNiCO_3{\cdot}yNi(OH)_2 + zH_2O \quad (3)$$

$$270\text{–}420°C:\ xNiCO_3{\cdot}yNi(OH)_2 \leftrightarrow (x + y)NiO + xCO_2 + yH_2O \quad (4)$$

Partial reduction of nickel oxide takes place subsequent to the decomposition process:

$$NiO(s) + H_2/CO(g) \leftrightarrow Ni(s) + H_2O/CO_2 \quad (5)$$

Figure 11. Ammonia stripping towers at Nicaro (Boldt 1967)

The kiln produces a mixture of metal-metal oxide containing 92% nickel. The second stage of the reduction process is carried out in a reduction furnace. The kiln product is crushed and mixed with saw dust to provide sufficient porosity in the bed during the reduction process. The residence time in the furnace is approximately one hour. The furnace carries out reduction at an average temperature of 900°C under an atmosphere containing cracked ammonia, 75% H_2 and 25% N_2. The reduction furnace product contains approximately 97.5% Ni. The reduction furnace product is then crushed, blended with stearic acid and pressed into compacts. The compacts are then fed into sintering furnace, which has a similar operation as the reduction furnace. In the furnace, final stage of reduction and sintering of the nickel take place simultaneously. The final product of this process contains 99% Ni. The flowsheet for this calcination process is illustrated in Figure 12.

In summary, the innovative driver for his innovation was the initial critical need for nickel supply for the World War II efforts by the United States.

High Pressure Acid Leaching of Nickel Laterites

Cuba's Nickel industry developed as a result of World War II, when the demand for nickel increased dramatically due to the need for high-strength stainless steel and other nickel alloys for use in the production of guns, ammunition and vehicles (Boldt 1967). Founded in 1912, Freeport Sulfur

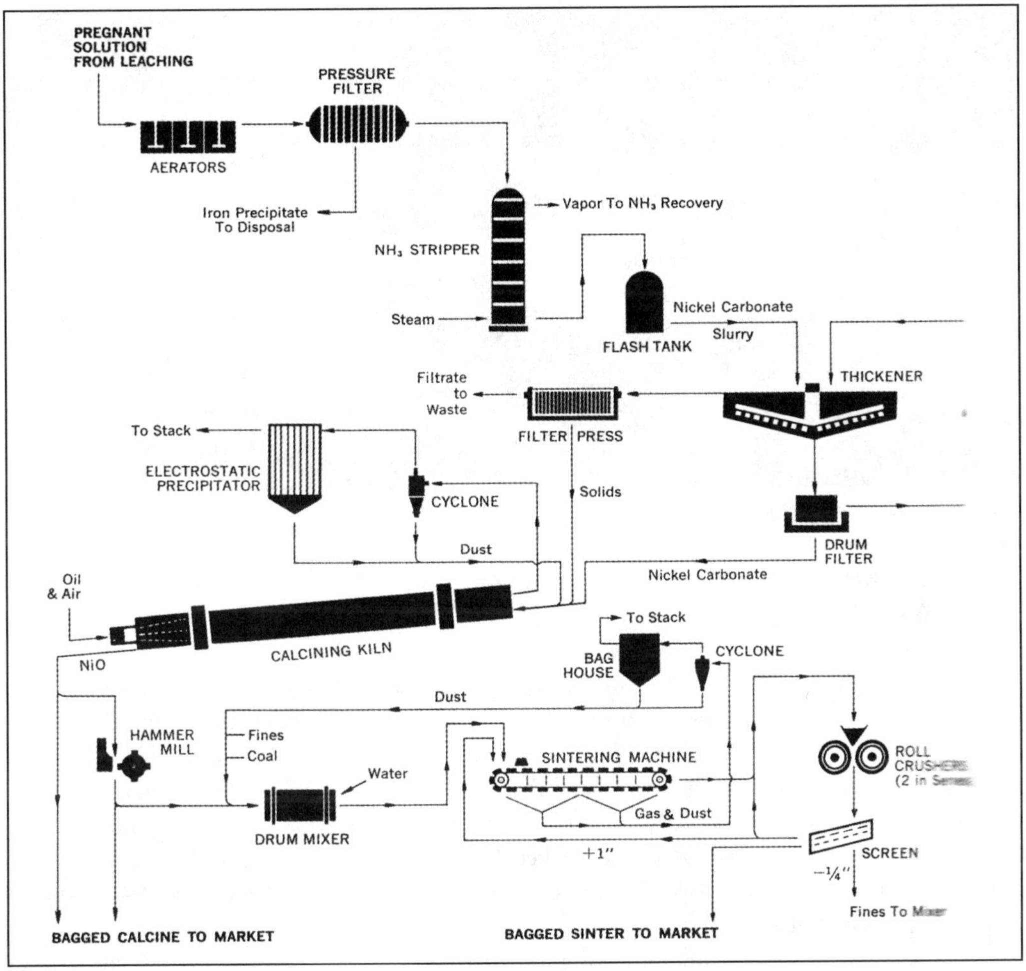

Figure 12. Nickel carbonate recovery and nickel oxide recovery at Nicaro (Boldt 1967)

Company played a complex and critical role for over fifty years in Cuba's nickel mining and processing business. Formed to develop the huge sulfur deposits found in Freeport, Texas, the company began to diversify in 1931, purchasing manganese deposits in the Oriente province in Cuba. In 1955, Freeport invested over $100 million in the construction of a second nickel-cobalt mine and processing facility at Moa Bay, Cuba, and a refinery at Port Nickel, Louisiana. On March 11, 1957, the U.S. government announced a contract for the purchase from Cuba of Freeport's production of nickel as strategic commodities. The contract would last until June 30, 1965.

In 1960, the Cuban government nationalized and failed to provide financial compensation to both the Moa and Nícaro facility. In 1971, The U.S. Federal Claims Settlement Commission certified the claims against the government of Cuba for improved real property and corporate assets by Moa Mining Company and Nicaro Nickel Company for $88,349,000 and $33,014,08. In December 1994, Canada's Sherritt Inc., which had been processing Cuban nickel-cobalt mixed sulfides since 1991, announced the formation of a 50/50 vertically integrated joint venture with Cuba's General Nickel Co. S.A., The joint venture, known as Metals Enterprise S.A., includes Moa Nickel

Figure 13. Moa Bay plant at commissioning (Boldt 1967)

S.A., which owns and operates the mining and processing facilities at Moa; the Cobalt Refinery Company, which owns and operates the nickel and cobalt refinery located in Fort Saskatchewan, Alberta, Canada; and the International Cobalt Company. The mixed sulfides produced at Moa are shipped to the east coast of Canada, where they are then transported by train to the Fort Saskatchewan facility and refined into finished nickel or cobalt. Nickel Oxides from the wholly owned General Nickel's Nicaro and Punta Gorda processing plants are exported to China and other countries. The Bahamas based International Cobalt Company (ICC) Inc., purchases mixed sulfides from Moa Nickel and other third parties, and contracts (tolling) with the Fort Saskatchewan refinery to process the raw materials. ICC also markets Metal Enterprise S.A.'s refined nickel and cobalt products internationally. The Moa Bay process at start up is shown in Figure 13 and the shown flowsheet is in Figure 14.

At Moa Bay leaching of limonitic laterites that typically contains 0.1 to 0.3% Ni, 0.05 to 0.257% Co, 35% Fe is undertaken. Efficient and selective extraction of nickel from such high iron-bearing material is not possible by direct atmospheric leaching with H_2SO_4. It requires leaching at a temperature above 200°C in autoclaves. Under these conditions, both iron and aluminum restrictively enter the leach liquor due to poor solubilities of their sulfates. The plant became operational in 1959. The flowsheet depicting the acid-leaching system is shown in Figure 10. The ore as a 45 V% solid slurry is heated to the operating temperature (230 to 250°C) and pressure (4.3 MPa) by means of a number of preheaters and heaters before going into the first of four autoclaves placed in series. The autoclaves are fabricated from welded steel cylinders with spherical ends and are lined internally first with 6-mm lead, then 114-mm acid-proof brick, and finally 76-mm-thick carbon bricks. The carbon brick serves to protect the acid- resistant bricks from corrosion and erosion. All other intimae parts of the autoclaves which come in direct contact with the superheated acidic slurry are made of titanium. The slurry is fed by means of specially designed high-pressure slurry pumps and is agitated by injecting steam at a pressure of 4.24 MPa into the bottom of the autoclave. Concentrated H_2SO_4 (98%) is added to the first autoclave from the top at an acid to ore ratio of approximately 0.25. The slurry moves from one autoclave to another under gravity at such

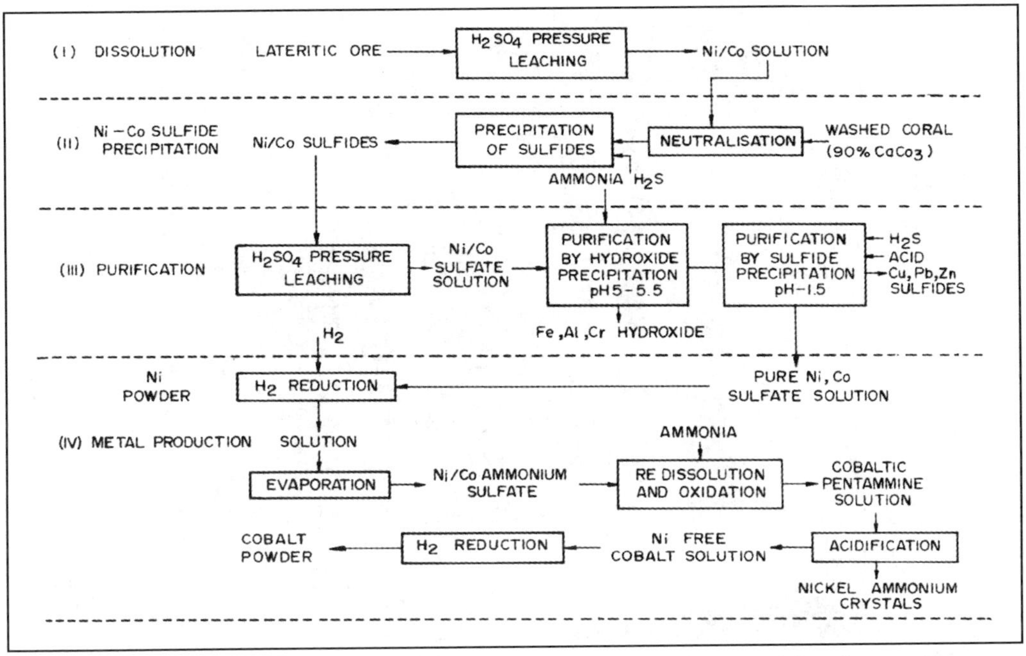

Figure 14. Moa Bay process for high pressure acid leaching of nickel and cobalt laterite ores (Gupta and Mukherjee 1990)

a rate that a total residence time of 1 to 2 hr is achieved. Figures 15 and 16 illustrate the HPAL system at Moa Bay.

The slurry coming out of the autoclaves is cooled and subjected to solid/liquid separation and washing to generate a liquor containing 5.95 g/L Ni, 0.64 g/L Co, 0.8 g/L Fe, and 2.3 g/L Al representing 96% Ni and 95% Co extractions, The liquor is treated with H2S which results in the recovery of Ni and Co in the form of their sulfides. The sulfide products were historically shipped to Port Nickel, Louisiana where they are further treated for the production of nickel and cobalt metals. Now they are shipped to Fort Saskatchewan, Alberta, Canada. So, the main driver for the HPAL process implementation was market driven albeit the eventual development was truncated by international political tensions.

MODERN HYDROMETALLURGICAL INNOVATIONS

All nickel ores contain cobalt. Both metals have similar chemistries and hence their separations are very difficult. Early on the Sherritt Gordon process utilized an incomplete separation via hydrogen gas reduction. For more definitive separations precipitation of cobaltic hexamine salts, precipitation of cobaltic hydroxide using nickelic hydroxide and solvent extraction are used. Currently, solvent extraction is the favored route with growing application in traditional processes such as RRAL, HPAL and ammoniacal based leaching flowsheets.

Nickel Cobalt Solvent Extraction Separation

Solvent extraction of nickel and cobalt is practiced in three different solution environments: (i) chloride (ii) sulfate and (iii) ammoniacal solutions (Crundwell et al. 2011). Generally cobalt

Figure 15. HPAL leaching system at Moa Bay, Cuba (Boldt 1967)

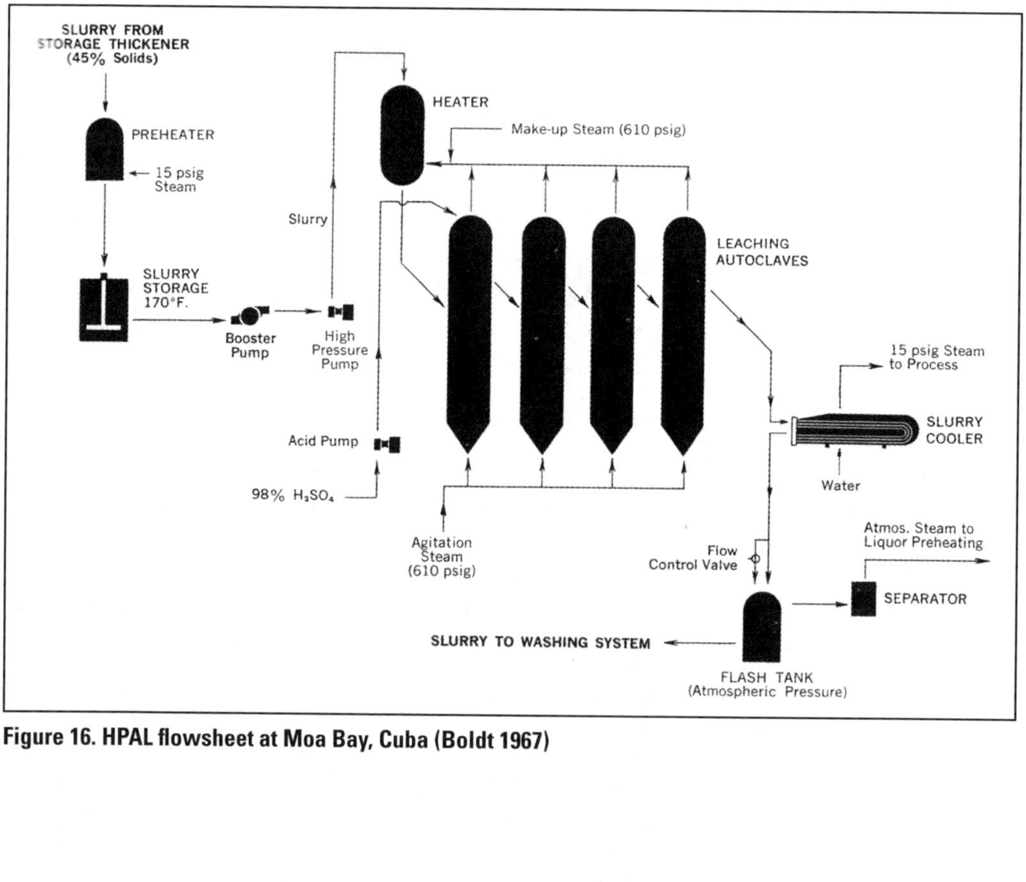

Figure 16. HPAL flowsheet at Moa Bay, Cuba (Boldt 1967)

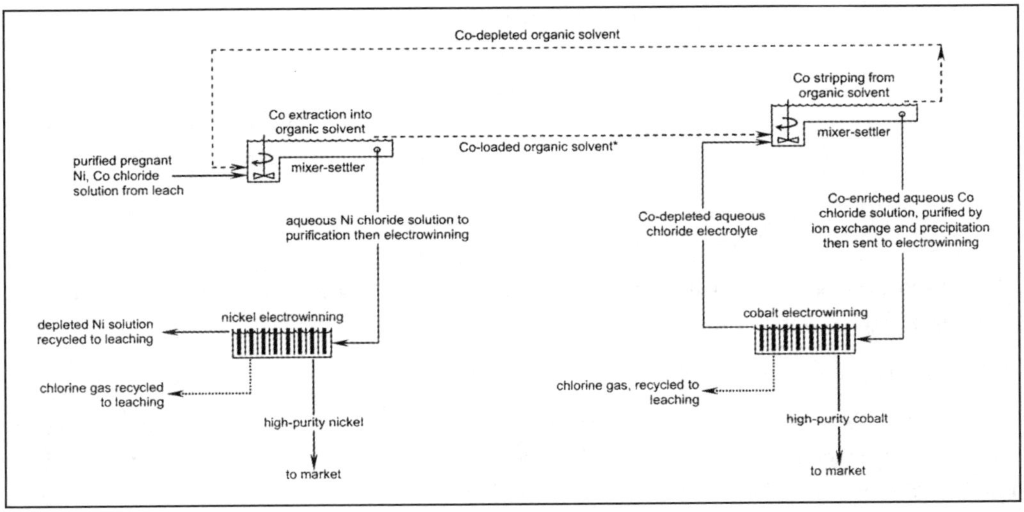

Figure 17. Solvent extraction nickel and cobalt separation flowsheet for a chloride-based system (Crundwell 2011)

is extracted selectively leaving nickel in solution. In ammonical solutions nickel is extracted selectively, leaving cobalt in solution. In rare cases, nickel and cobalt are co extracted and then separated from each other.

Solvent extraction in chloride solutions is applied at the following operations (Dotterud et al. 2009, Eramet 2010, Higuchi et al. 2006):

i. Xstrata Nikkelverk, Kristiansand, Norway
ii. Sumitomo Metal Mining, Niihama, Japan
iii. Eramet, Le-Havre-Sandouville, France

Chloride solvent extraction accounts for about 150,000 tonnes per year of nickel and 10,000 tonnes per year of cobalt. This section is based on the Nikkelverk plant, which is described in several publications (Dotterud et al. 2009, Stensholt et al. 1986, Stensholt et al. 1988, 2001). In the Nikkelverk operation, cobalt is extracted from solution using tri-isooctyl amine. The extractant is available commercially as Alamine 308 (Cognis 2010) and several other brands. The concentration of the tri-isooctylarnine is 17% by volume in an aromatic diluent. Figure 17 illustrates a simplified solvent extraction of cobalt/nickel separation flowsheet including metal electrowinning as applied to chloride solutions.

Solvent extraction of cobalt from sulfate solutions is practiced at the following operations:

i. Minara Resources, Murrin Murrin, Australia
ii. Norilsk Nickel, Harjavalta, Finland
iii. Sherritt Toamasina, Madagascar

Solvent extraction in sulfate accounts for about 140,000 tonnes per year of nickel and 15,000 tonnes per year of cobalt.

The extractant used is bis(2,4,4-trimethyltpentyl)phosphonic acid which is commercially available as Cyanex 272, LIX 272 and Ionquest 290 (Sole 2008). The concentration of the extractant is typically 10–25% by volume in an aliphatic dilutent. The aqueous solution is purified to remove zinc and copper impurities prior to solvent extraction so as to prevent their co-extraction when using Cyanex 272 or LIX 272 (Rickelton & Nucciarone 1997). The extraction reaction at 50°C is given as follows:

$$2H_2A_2\text{ (l)} + CoSO_4\text{ (aq)} \rightarrow CoA_2{:}H_2A_2\text{ (l)} + H_2SO_4\text{ (aq)}$$

where A represents the 2,4,4-trimethylpentyr phosphonic conjugate base. The stripping reaction is the reverse of the extraction reaction:

$$CoA_2{:}H_2A_2\text{ (l)} + H_2SO_4\text{ (aq)} \rightarrow 2H_2A_2\text{ (l)} + CoSO_4\text{ (aq)}$$

Extraction is favored at the low concentrations of sulfuric acid about 0.001 g/L, in the aqueous solution to the extraction stage. Stripping is favored by high concentration of about 180 g/L in the aqueous solution to the strip. Extraction and stripping are each carried out in several mixer-settlers. Murrin Murrin, for example uses four stages of extraction and three stages of stripping.

The maximum separation factor of Cyanex 272 for Co:Ni is 6700:l (Cytec 2010). This means that up to 6,700 times more cobalt than nickel will be extracted into Cyanex 272 from an aqueous sulfate solution containing equal amounts of nickel and cobalt. Sole (2008b) indicates that preferential extraction of cobalt is due to the simple structure and hydrophobicity of the organic complex of cobalt.

Solvent extraction of cobalt from ammoniacal solutions is practiced at Queensland Nickel, Australia. This refinery at Yabulu, near Townsville, Australia, uses LIX-84-I to extract nickel from an ammoniacal solution (MacKenzie & Virnig 2004; Sole & Cole 2002). The reaction is

$$Ni(NH_3)_4^{2+}\text{(aq)} + 2RH\text{(l)} \rightarrow R_2Ni\text{(l)} + 2NH_3\text{ (aq) } 2NH4{+}\text{(aq)}$$

Solvent extraction from ammoniacal solutions is different from chloride and sulfate solutions because nickel is loaded preferentially, not cobalt. However, in order for this extraction to be effective, cobalt must be present as Co^{3+} prior to nickel solvent extraction. The extraction occurs in three stages of counter-current mixer-settlers with advance organic:aqueous ratio of 1.5:1. The loaded organic is stripped with a solution of 270 g/L ammonia in three stages using an organic:aqueous ratio of 1.5:1. The strip solution is contains 75–80 g/L Ni, with low levels of impurities. The cobalt is precipitated from the nickel extraction raffinate as a sulfide. This sulfide is redissolved. The impurities, mainly iron and zinc, are removed from the solution using Cyanex 272. The cobalt and nickel are extracted from this solution using D2EPHA and stripped into a solution containing ammonia and ammonium carbonate. The cobalt is oxidized to Co^{3+}, and nickel, manganese and copper are removed from the strip solution using LIX 84-I. The raffinate from this extraction is the final purified cobalt solution. There are two disadvantages to the process chemistry used at Queensland Nickel. The first disadvantage is that the Co^{3+} oxidizes the extractant to form a ketone, which means that the costs of replenishing extractant are higher than at other facilities. The second disadvantage is that the high ammonia concentrations cause the extractant to form an imine. The extractant is regenerated using hydroxylamine at Queensland Nickel.

Vale has implemented a novel process at Goro in New Caledonia (Bacon et al. 2005; Sole & Cole 2002). Laterite ore is leached at 270°C and 56 bar. The residence time in the leaching autoclave is about 30 min. The discharge slurry is washed in the counter-current decantation circuit. The solution from leaching is partially neutralized to remove aluminum, chromium, silicon, copper and iron. Copper is further removed using ion exchange. This purified solution is transferred to solvent extraction. Nickel and cobalt are extracted from the purified solution, along with zinc, using bis(2,4,4-trimethylpentyl)dithiophosphinic acid, available as Cyanex 301. Calcium, magnesium and manganese are not extracted. The loaded organic is stripped from solution using concentrated hydrochloric acid. Pulsed columns are used. The stripped solution contains nickel, cobalt and zinc. Zinc is removed from solution using ion exchange, and the purified solution is pumped to cobalt solvent extraction. Cobalt is separated from the nickel using Alamine 336. Cobalt is stripped from the loaded organic using a dilute solution. The raffinate contains mainly nickel chloride and the strip solution contains cobalt chloride. The nickel chloride solution is treated by pyrohydrolysis to form nickel oxide powder and to regenerate the hydrochloric acid needed for stripping the nickel and cobalt from the Cyanex 301.The cobalt is recovered from solution by precipitation with sodium carbonate to product cobalt carbonate.

Voisey's Bay Pressure Oxidation Process

A schematic flowsheet for the Voisey's Bay process is shown in **Figure 18** (Crundwell et al. 2011). This emerging technology relies on continuous horizontal pressure oxidation along with a mixed lixiviant of sulfate and chlorine gas. It also relies heavily on the use of multi stages of solvent extraction using multiple reagents (Stevens et al. 2009). The feed to this process is ground concentrate, which is contacted with a mixture of oxygen and chlorine gases from the nickel tankhouse in a series of stirred tank reactors. The step is called the chlorine preleach. The slurry is pumped into the pressure-leaching autoclave where the concentrate is dissolved using nickel spent electrolyte (sometimes called nickel anolyte). The anolyte is acidic and provides the acid required for leaching. The discharge slurry from the autoclave goes through the counter-current decantation section, where undissolved and precipitated solids are separated from the solution. The solids are neutralized and transferred to disposal. The solution goes to impurity removal where iron and other contaminants are removed. After impurity removal, the solution is transferred to copper solvent extraction. Copper is removed from the solution by solvent extraction using LIX 84. The stripped solution from solvent extraction is the feed to copper electro-winning, where high-purity copper is produced. Cobalt is extracted from the solution by solvent extraction using Cyanex 272. Cobalt is electrowon from the strip solution from solvent extraction as cobalt rounds. The raffinate from cobalt extraction is the feed to nickel electrowinning. The nickel electrowinning occurs in a divided cell, with anode bags for the collection of chlorine and oxygen gases. These gases are used in the leaching section of the process. The nickel spent, or anolyte, is recycled to leaching. The raffinate from the copper solvent extraction is passed through further purification steps such as cadmium removal by sulfide precipitation, and a second iron removal step. Then the solution goes on to impurity solvent extraction where D2EPHA is used to remove calcium, copper, lead, iron, manganese and zinc. In summation, full production is anticipated in 2014. The driver here has been a combination of superior hydrometallurgical separations, the required local production of value added metals and the perception of a choice of a more environmentally controlled process over conventional sulfide concentrate smelting.

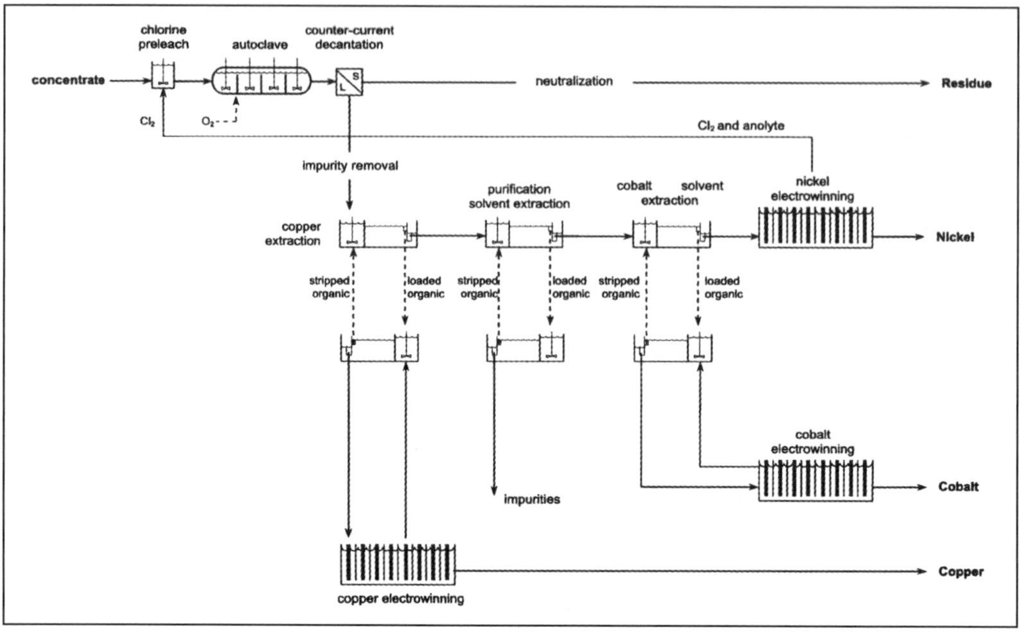

Figure 18. Schematic diagram of the Voisey's Bay process (Crundwell et al. 2011)

SUMMARY

The innovations in nickel and cobalt sulfide concentrates and lateritic oxide ores has been extensive and all encompassing. More than any other metals, the comprehensive development and application of any and all unit operations is clearly elucidated in processing of nickel and cobalt. These key developments long after the advent of the Bayer and cyanidation processes served to further broaden the acceptance and implementation of hydrometallurgy as an advanced and accepted technology. Again, the main driver has been the enhanced separations inherent with hydrometallurgy. However, as in the case of Cuba, the global political dynamic shaped the development of technology In particular, the advent of the Voisey's Bay industrial process now encompassing mixed lixiviants, pressure oxidation and multiple solvent extraction separations may hail a new era for hydrometallurgy. The key driver here again is the superior separations, the required local production of value added pure metal products and the growing global perception of hydrometallurgy as more environmentally acceptable than conventional smelting.

REFERENCES

Adams, M. van der Meulen, D., Czerny. C., et al. 2004. Piloting of the beneficiation and EPAL circuits for Ravensthorpe Nickel Operations, In W.P. Imrie, D.M. Lane, S.C.C.Barnett, R. Berezowsky, E.J.M. Jahnsen & P.G. Mason (Eds.). International Laterite Symposium—2004, pp. 79–99, TMS.

Bacon, W.G., Colton, D.F., Krause, E. Mihaylov, I.O., Singhai. A., & Duterque, J.P., 2005. Development of the Goro nickel process. In ALTA 2005 Nickel/Cobalt Forum. ALTA Metallurgical Services.

Baily, T.F., 1945. Production of Nickel and Chrome Steels from Laterite Ores, The Electrochemical Society, vol. 88, 186–193.

Baragwanath, J., 1936. *Pay Streak*, Garden City, N.Y., Doubleday, Doran & Co., Inc.

Baragwanath, J.,1961. *A Good Time Was Had*, New York, N.Y., Appleton-Century-Crofts, Inc.

Battle, T., Moats, M., Cocalia, V. et al. Eds, 2013. Ni Co 2013, Proceedings of a Ni Co Symposium, TMS, Wiley, March.

Boldt, J.R., Jr., 1967. *The Winning of Nickel*, ed. By P. Queneau, Princeton, N.J., D. Van Nostrand Co., Inc.

Brooks, J. (Chairman), 1958. "Disposal problems of government owned nickel plant at Nicaro, Cuba," *Hearings before a subcommittee of committee on government operations, House of Representatives*, 85th Congress, second session (Washington, D.C.).

Budac, J.J. Fraser, R. & Mihaylov, I. (Eds.)., 2009. Hydrometallurgy of Nickel and Cobalt 2009. Metallurgical Society of CIM.

Burkin, A.R., 1987. Extractive metallurgy of nickel, Society of (British) Chemical Industry.

Caron, M.H., 1924. *U.S. Patent 1,487,145*, 8 March.

Caron, M.H., 1942. *U.S. Patent 2,290,313*, 21 July.

Caron, M.H., 1946. *Mining and Metallurgy*, Dec. 1945, 613; Pawel, *M&M*, Feb. 1946, 143; Baragwanath, *M&M*, March, 181.

Caron, M.H., 1950. "Fundamental and Practical actors in Ammonia Leaching of Nickel and Cobalt Ores," *Transactions AIME*, vol. 188, January, 67–90.

Caron, M.H., 1950a. "Separation of Nickel and Cobalt," *Transaction AIME*, vol. 188, January, 91–103, 1387.

Caron, M.H., 1950b. "Discussion: Fundamental and Practical Factors in Ammonia Leaching of Nickel and Cobalt Ores and Separation of Nickel and Cobalt," *Transactions AIME*, vol. 188, November, 1387.

Chalkley, M.E. & Toirac, I.L., 1997. The acid pressure leach process for nickel and cobalt laterite. In W.C. Cooper & I. Mihaylov (Eds.), Nickel/cobalt 97, Vol. I, Hydrometallurgy and refining of nickel and cobalt, pp. 341–353, CIM.

Cognis, 2010. Alamine 348 Technical brochure. Cognis Corporation.

Collins, M.J., Barta, L.A. Buban. K.R., et al. 2005. Process development by Dynatec for the Ambatovy nickel project. CIM Bulletin, 98,90.

Collins, M.J., Buban, K.R., Holloway, P.C., et al. 2009. Ambatovy laterite ore preparation plant and high pressure acid leach pilot plant operation. In J.J. Budac, R. Fraser & I. Mihaylov. et al. (Eds.). Hydrometallurgy of nickel and cobalt 2009, pp. 499–510, CIM.

Cook, W.R., Quenett, R.T., and Shefford, D.M., 1973. Current practice at the Thompson Nickel Refinery: AIME Annual Meeting, Chicago, TMS.

Cooper, W.C. and Mihaylov, I., Eds., Nickel Cobalt 97, Vol. 1. Hydrometallurgy and Refining of Nickel and Cobalt, The Metallurgical Society of CIM, Montreal, Canada.

Cordingley, P.D., & Krentz, R., 2005. Corefco refinery–Review of operations, In J. Donald & R. Schonewille (Eds.), Nickel and Cobalt 2005 Challenges In Extraction and Production, pp. 407–425, CIM.

Cox, J.S., 1911. The Iron-Ore Deposits of the Moa District, Oriente Province, Island of Cuba, *AIME Transactions 42*, 73–90.

Crundwell, F.K., Moats, M.S., Ramachandran, V., Robinson, T.G. and Davenport, W.G. 2011. Extractive Metallurgy of Nickel, Cobalt and Platinum-Group Metals, Elsevier, Oxford, UK.

Cyanex 272 Extraction Technical Brochure, Cytec Corporation.

De Graff, J.E., 1979. The Treatment of Lateritic Nickel Ores—A Further Study of the Caron Process and Other Possible Improvements, *Hydrometallurgy*, 5 47–65.

Donald, J. & Schonewille, R. (Eds.), 2005. Nickel and cobalt 2005, challenges in extraction and production, Metallurgical Society of CIM.

Donegan, S., 2006. Direct solvent extraction of nickel and Bulong operations. Minerals Engineering, I9, 1234–1245.

Dotterud, O.M., Peek, E.M.L., Stenstad, O., & Ramsdal. P.O. 2009. Iron control and tailings disposal in the Xstrata chlorine leach process. In J.J. Budac. R. Fraser. I. Mihaylov. V.G. Papangelakis & D.J. Robinson (Eds.), Hydrometallurgy of nickel and cobalt 2009, pp. 321–333, CIM.

Dufour, M.F. and R.C. Hills, 1945. "Nickel from Cuba: Nicaro Produces Nickel Oxide by Leaching with Ammonia," *Chemical Industries*, October, 621–627.

Duyvensteyn, W.P., Wicker. G.R. & Doane. R.E., 1979.. An omnivorous process for laterite deposits. In D.J. Evans. R.S., Shoemaker & H. Veltman (Eds.). International Laterite Nickel.

Editor, 1944. Nicaro Nickel's New Cuban Plant Begins Production, *Mining & Metallurgy*, May, 254–256.

Editor, 1945. The Building of Nicaro, *Fortune*, July, 146–149, 178, 180–183.

Eramet, 2010. Le Havre-Sandouville Refinery. Company brochure.

Evans, D.J. 1979. International Laterite Symposium: Proceedings, New Orleans, Louisiana, February 19 to 21, 1979 Society of Mining Engineers of the American Institute of Mining, Metallurgical, and Petroleum Engineers, 688 pages.

Flett, D.S. 2005. Solvent extraction in hydrometallurgy: The role of organophosphorus extractants, Journal of Organometallic Chemistry, 690, 2426–2438.

Forward, F.A., and Mackiw, V.N., 1955. Chemistry of the Ammonia Pressure Process for Leaching Nickel Copper and Cobalt from Sherritt Gordon Sulfide Concentrates, J. of Metals, 7, pp. 457–462.

Gupta, C.K. and Mukherjeee, T.K., 1990. Hydrometallurgy in Extraction Processes, Vol. I & II, CRC Press, Boca Raton Florida Cytec, 2010.

Habashi, F., 2009. A history of nickel. In J. Liu, J. Peacey & M. Barati, et al. (Eds.), Pvrometallurgy of nickel and cobalt 2009, proceedings of the international symposium (Taylor A., ed. (pp. 77–98). Metallurgical Society of CIM.

Hayward, K., 2008. Murrin Murrin leads the way. Sulfuric Acid Today, 14.7–10, Kofluk, R.P., & Freeman. G.K.W. 2006, Iron control in the Moa Bay laterite operation. In J.E. Dutrizac & P.A. Riveros (Eds.). Iron Control Technologies, pp. 573–589, CIM.

Higuchi, H. Ozaki, Y., Sugiura, T. & Kemori. N., 2006. Iron removal from the MCLE [matte chlorine leach electrowin circuit]. In J.E. Dutrizac & P.A. Riveros (Eds.). Iron control technologies, pp. 403–413, CIM.

Imrie, W.P. Lane, D.M. & Barnett. S.C.C. (Eds.), 2004. International Laterite Nickel Symposium, TMS.

Kerfoot, D.G.E., 1989. The Development of the Sherritt ammonia pressure leach process:, CIM Bulletin, V. 82, No.926, June, p. 136–141.

Kerfoot, D.G.E., and Cordingley, P.D., 1997. The acid pressure leach process for nickel and cobalt late rites. Part III: Review of operations at Fort Saskatchewan, in Cooper, W.C. and Mihaylov, I., Eds., Nickel Cobalt 97, Vol. 1. Hydrometallurgy and Refining of Nickel and Cobalt, The Metallurgical Society of CIM, Montreal, Canada, pp. 355–369.

Kerfoot, D.G. E., Collins, M.J., Holloway, P.C and Schonewille, R.H., 2012. The Nickel Industry In Canada, The Canadian Metallurgical and Materials Landscape 1960 to 2011, CIM.

Kofluk. R.P. & Freeman. G.K.W., 2006. Iron control in the Moa Bay operation. In J.E. Dutrizac & P.A. Riveros (Eds.), Iron control technologies, pp. 573–589, CIM.

Lewis, A.E., 2010. Review of metal sulfide precipitation. Hydrometallurgy, 104,222–234.

Little, J.E., 1911. The Mayari Iron-Mines, Oriente Province, Island of Cuba, as Developed by the Spanish American Iron Co., *AIME Transactions 42*, 152–169.

MacKenzie, J.M.W., & Virnig, M.J., 2004. Solvent extraction technology for the extraction of nickel using LIX 84-INS. An update and circuit comparisons. In W.P. Imrie & D.M. Lane (Eds.),International laterite symposium, pp. 457–475, TMS.

Mackiw, V.N. and Benz, T.W., 1961. Application of Pressure Hydrometallurgy to the Production of Metallic Cobalt, Extractive Metallurgy of Copper Nickel and Cobalt, Interscience Publishers, New York, AIME, 503.

Makinen, T. Fagerlund. K. Anjala. Y. & Rosenback. L. 2005. Outokumpu's technologies for efficient and environmentally sound nickel production. In J. Donald & R. Schonerville (Eds.).Nickel and Cobalt 2005 Challenges in Extraction and Production, pp. 71–89). CIN'I.

Mansumoto, N. Matsumoto, S., Nakagawa. H. & Sugita. I., 2009. Solvent extraction technology for nickel and cobalt separation crowding reaction. In J.J., Budac, R. Fraser. I. Mihaylov, I.V.G.

Matos, R., R., 1997. Industrial experience with the Ni/Co sulphide precipitation process. In W.C. Cooper & I. Mihaylov (Eds.), Nickel/cobalt 97, Vol. 1, Hvdrometallurgy and refining of nickel and cobalt, pp. 371–378, CIM.

Meddings, B. and Mackiw, V., 1963. The Gaseous Reducation of Metals From Aqueous Solutions, Unit Processes in Hydrometallurgy, AIME, Dallas, pp. 345–384.

Mihaylov, I., Grinbaum, B., Ilan, Y., & Efraim, A. 2009. Opportunities for nickel-cobalt extraction and separation using Cyanex 301. In J.J. Budac. R. Fraser. I. Mihavylov. V.G. Papangelakis & D.J. Robinson (Eds.), Hydrometallurgy of nickel and cobalt 2009 (pp. 383–391). CIM.

Molina, N., 2009. Nickel and cobalt sulfide precipitation, a proven method of selective metal precipitation in laterite process foresheets. In J.J. Budac. R. Fraser & I. Mihaylov. et al. (Eds.), Hydrometallurgy of nickel and cobalt 2009, pp. 271–281, CIM.

Okita, Y., Singhal. A. & Peraud, J.J., 2006. Iron control in the Goro nickel process. In J.E. Dutrizac & P.A. Riveros (Eds.), Iron control technologies (pp. 635–651). CIM.

Paatero, E., Nymar, B., Laital, H., et al. 2009. Extraction of nickel with pre-neutralized organic acids using Outotec mixer settler technology. In J.J. Budac. R. Fraser, L. Mihavylov, V.C. Papangelakis & D.J. Robinson (Eds.), Hydrometallurgy of nickel and Cobalt 2009, pp. 211–220, CIM.

Papangelakis & D.J. Robinson et al. (Eds.). Hydrometallurgy of nickel and cobalt 2009. pp. 283–293, CIM.

Prado, F.G. and J.P. Dempsey, 1986. Indigenous Solids Reductants in Caron type Nickel Plants, 25th Annual Conference of Metallurgists –1986, Toronto, Ontario, Canada, August 17–20.

Reddy, R.G. and Weizenbach, R.N., 1993. Extractive Metallurgy of Copper, Nickel and Cobalt, Proceedings of the Paul E. Queneau International Symposium, TMS, Vol. I & 2.

Reid, J.G., & Price, M.J., 1993. Ammoniacal Solvent Extraction at Queensland Nickel Process installation and operation, Solvent Extraction in the process industries. Vol. I. Society of chemical industries.

Reid, J.G. and J.B. Fittock, 2004. "Yabulu 25 Years On," *International Laterite Nickel Symposium*, ed. W.P. Imrie and D.M. Lane, Warrendale, Pa., TMS, 599–618.

Rickelton, W.A., & Nucciarone. D., 1997. The treatment of cobalt/nickel solutions using Cyanex extractants. In W C. Cooper & I. Mihaylov (Eds.). Nickel Cobalt 97, Vol. 1. Hydrometallurgy and refining of nickel and cobalt (pp. 275–292). CIM.

Sakai, O., Kawata, M. Kyoda. Y., & Tsuchida, N., 2006. Commissioning of Coral Bay nickel project. In ALTA 2006Nickel/Cobalt Conference Proceedings, ALTA Metallurgical Services.

Smith, R., 2004. Kwinana nickel refinery, October 28.

Sole, K. 2008a. Solvent extraction in base metal hydrometallurgy, Johannesburg: Short course at Anglo Research. July 9. 2006.

Sole, K., 2008b. Solvent extraction in hydrometallurgical processing and purification of metals. In M. Aguilar & J.L. Cortina (Eds.), Solvent extraction and liquid membranes fundamentals and applications in new materials (pp. 141–200). CRC Press.

Sole, K.C., & Cole, P.M., 2002. Purification of nickel by solvent extraction. In Y. Marcus, A.K. SenGupta & J.A. Marinsky (Eds.), Ion exchange and solvent extraction (pp. 143–195). Marcel Dekker Inc.

Southwick, L.M., 2013. When laboratory work and operating plant don't agree (Part 1): Commercializing the Caron ammonia-ammonia carbonate Ni process, Proceedings of Ni-Co 2013, TMS, Wiley.

Stensholt, E.O., Zachariasen. H. & Lund J.H., 1986. Falconbridge chlorine leach process, Trans. Inst. Min. Met. C,5. C10—C16.

Stensholt. E.O. Zachariansen. H. Lund. J.H. & Thornhill, P.G., 1988. Recent improvements in the Falconbridge Nikkelverk nickel refinery. In G.P. Tyroler & C.A. Landolt (Eds.), Extractive Metallurgy of Nickel and Cobalt (pp. 403–413). TMS.

Stensholt, E.O., Dotterud. O.M. Henriksen. E.E., et al. 2001. Development and plant practice of the Falconbridge chlorine leach process, CIM Bulletin, 94. 101–104.

Stevens, D., Bishop, G. Singhai, et al. 2009. Operation of the pressure oxidative leach process for Voisey's Bay nickel concentrate at Vale Inco's hydromet demonstration plant. In J.J. Budac, R. Fraser, and I. Mihaylov, et al. Eds. Hydrometallurgy of nickel and cobalt 2009, pp. 3–16, CIM.

Tatebe, T., 1930. Paper 351, *International Engineering Conference*, Tokyo.

Tobelmann, H.A., and H.J. Morgan, 1948. *Review of the Nicaro Nickel Project, Nicaro, Oriente, Cuba, Plancor 690*, Washington, D.C., Reconstruction Finance Corp., GSA Report 97271.

Tsuchida, N., Ozaki, Y., Nakai, O., & Kobayashi, H., 2004. Development of process design for Coral Bay nickel project. In W.P. Imrie. D.M. Lane & S.C. C Barnett et al. (Eds.), international laterite nickel symposium 2004, Process development for prospective projects, pp. 152–160,TMS.

Weld, C.M., "The Residual Brown Iron Ores of Cuba," *Iron Age, Vol. 84* (August 26, 1909), 614–617.

Woodward. T.M. & Bahri, P.A., 2007. Steady-state optimisation of the leaching process at Kwinana nickel refinery. In V. Plesu & P.S. Agachi (Eds.). 17th European symposium on computer aided process engineering—ESCAPE17, pp. l–6, Elsevier.

ABOUT THE AUTHOR

Corby G. Anderson has more than 34 years of global experience in industrial operations, management, engineering, design, consulting, teaching, research, and professional service. His career includes positions with Morton Thiokol, Key Tronic Corporation, Sunshine Mining and Refining Company, H.A. Simons Ltd., and Montana Tech. Anderson holds a BSc in chemical engineering and an MSc and PhD in metallurgical engineering. He is a fellow of the Institution of Chemical Engineers and the Institute of Materials, Minerals and Mining. He currently serves as the Harrison western professor in the Kroll Institute for Extractive Metallurgy as part of the George S. Ansell Department of Metallurgical and Materials Engineering at the Colorado School of Mines. In 2009 he was honored by SME with the Milton E. Wadsworth Extractive Metallurgy Award for his contributions in hydrometallurgical research.

The Bayer Hydrometallurgical Alumina Process Evolution and Innovations

Joseph L. Anjier
JA Engineering Services LLC, Baton Rouge, Louisiana, United States

Corby G. Anderson
Kroll Institute for Extractive Metallurgy, Colorado School of Mines, Golden, Colorado, United States

ABSTRACT: The Bayer process was developed in 1888 and has been continuously improved through the years by process breakthroughs to become an efficient, economical part of the aluminum industry. This paper describes the history of the process, the mineralogy of bauxite, the Bayer Process development, design and operational innovations, and energy supply and waste streams.

BAUXITE AND ITS MINERALOGY

Bauxite is the only raw material used in the production of alumina on a commercial scale in the United States (Cundiff and Reese 1955, Cundiff 1972, Hose 1963). Bauxite was discovered in 1821 by the French chemist P. Berthier who called it alumina hydratee des Beaux after the village of Les Beaux near Arles in southern France. These are illustrated in **Figures 1 and 2**. The name was changed to beauxite in 1847 by A. Dufrenoy and to bauxite in 1861 by Henri Sainte-Claire Deville (Habashi 1988).

Vast resources of clay are also technically feasible sources of alumina. Residual rocks, in which alumina trihydrate and monohydrate minerals gibbsite, boehmite, and diaspore predominate, are classified as bauxites. Some of the residual minerals which are developed in the unique type of rock weathering which produces bauxites, are, kaolinite, halloysite, goethite, hematite, magnetite, anatase, quartz and some phosphatic and manganiferous minerals may form the lesser constituents of bauxites. Traces of the elements arsenic, barium, boron, carbon, bismuth, beryllium, calcium, cesium, cerium chromium, cobalt, copper, hafnium, gallium, sulfur, tin, uranium, zinc and others have been found in bauxites. The reactions required to produce these bauxites take place in high rainfall areas of the tropics and subtropics where most of the known bauxite deposits are found.

There are many different minerals found in bauxite. The composition varies within different bauxite deposits. Probably the most important minerals for alumina plants are gibbsite, boehmite, iron, zinc and limestone. The organic compound contamination in the bauxite is also very important as the organics dissolve and produce sodium oxalate. Sodium oxalate interferes with the size control in the precipitation process. The undissolved minerals are separated from the process streams and are stored in waste solid deposits or discharged when in solution. The quality of these bauxites varies widely as does the delivered cost. Therefore the source of the bauxite greatly influences the cost of a facility and has supported process development through the years.

Figure 1. Bauxite deposit in Les Beaux, France (Anjier 2013)

Figure 2. The village of Les Beaux, France (Anjier 2013)

To utilize the aluminum values in bauxite, the alumina minerals must be extracted and purified. Depending upon its location and sources of bauxite, alumina refineries are designed to process a variety of bauxites and produce a finished product of calcined alumina. According to the U.S Geological Survey, Mineral Commodity Summaries, January 2021, bauxite resources are estimated to be 55 to 75 billion tons in Africa, Oceania, South America and the Caribbean. Domestic sources will not meet the long term U.S. demand. Table 1 shows the 2011 production by country, listing the world bauxite production and reserves as published by the USGS Mineral Commodity Summaries in January 2012 (USGS 2012).

BAYER ALUMINA PROCESS

The Bayer process as we know it was developed by Karl Josef Bayer in 1888 (Bayer, 1888 & 1892; Figure 3).

Table 1. World Bauxite Production and Reserves, kilotonnes (USGS 2012)

Country	2011 Mine Production, KT	Reserves, KT
United States	NA	20,000
Australia	67,000	6,200,000
Brazil	31,000	3,600,000
China	46,000	830,000
Greece	2,000	600,000
Guinea	18,000	7,400,000
Guyana	2,000	850,000
India	20,000	900,000
Jamaica	10,200	2,000,000
Kazakhstan	5,400	160,000
Russia	5,800	200,000
Sierra Leone	1,700	180,000
Suriname	5,000	580,000
Venezuela	4,500	320,000
Vietnam	80	2,100,000
Others	2,600	3,300,000
World total (rounded)	220,000	29,000,000

Figure 3. Karl Josef Bayer 1847–1904 (Habashi 1988)

He found that the process developed by Louis Le Chatelier in 1855 which involved mixing sodium carbonate with a red mineral (bauxite) found at Chateau Des Beaux, France, could be greatly improved using sodium hydroxide or (NaOH, or caustic). By heating with a solution of NaOH, aluminum compounds in the bauxite will dissolve and produce sodium aluminate ($NaAlO_2$). The compounds, such as iron, which do not dissolve can be filtered from the solution leaving a pure sodium aluminate solution. This simple discovery is the basis for the digestion section of alumina plants.

In the case of (1) gibbsite and (2) boehmite/diaspore, the dissolution is described by the following equations:

$$Al(OH)_3 + Na^+ + OH^- \rightarrow Al(OH)_4^- + Na^+ \quad (1)$$

$$AlO(OH) + Na^+ + OH^- + H_2O \rightarrow Al(OH)_4^- + Na^+ \quad (2)$$

This leaching process is processed at elevated temperature and pressure, depending on the composition of the bauxite. Ores that are high in gibbsite are leached at 130°C and ores that are higher in boehmite/diaspore are leached at temperatures closer to 200°C.

Bayer also found that if the sodium aluminate is allowed to cool and it is diluted, aluminum trihydrate ($Al_2O_3{}^*{}_3H_2O$) crystals will precipitate as a hydrolytic reaction with water, shown in Equation 3.

$$AlO(OH)_2^- + H_2O \rightarrow Al(OH)_3 + OH^- \quad (3)$$

If these alumina trihydrate crystals are removed from the clarified sodium aluminate solution, and added (seeded) to a newly digested and clarified sodium aluminate solution, additional alumina trihydrate crystals will precipitate without extensive cooling. Precipitation initiated by seeding rather than cooling allows control of the particle size of the alumina trihydrate and reduces energy consumption. The precipitation has to be done under precise conditions to generate a precipitate that can be handled easily. Flash precipitation would result in a colloidal precipitate that would be difficult to separate and wash. In order to make a crystalline precipitate with large particle size, the solution is seeded with a large excess of $Al(OH)_3$, carefully diluted with water and slowly cooled under agitation. The precipitation temperature is maintained at 25–35°C. The crystallization process is slow with each batch taking about 4 days. This second discovery is the basis for the Precipitation section of the alumina plant. Figures 4 and 5 illustrate the precipitation vessels and the final precipitated aluminum hydroxide product.

Figure 4. Precipitating tanks for aluminum hydroxide (Habashi 1999)

Figure 5. Electron micrograph showing the crystalline nature of aluminum hydroxide precipitated from the Bayer process. The edge of the hexagon is about 8 microns and the thickness is about 4 microns. (Habashi 1999)

Further testing by Bayer allowed him to develop a multistep process for producing purified alumina (Al_2O_3):

1. Bauxite digestion in heated sodium hydroxide to produce sodium aluminate.
2. Clarification of the sodium aluminate slurry by removal of undissolved solids with settling vessels or filters.
3. Cooling of the solution prior to placing in holding vessels called precipitators.
4. Adding previously precipitated alumina trihydrate solids (seed) to the cooled sodium aluminate liquor to precipitate additional alumina trihydrate crystals.
5. Classifying the alumina trihydrate crystals into fine seed, coarse seed and product.
6. Heating the product sized alumina trihydrate crystals to drive off the water of crystallization and produce alumina, the raw material for producing aluminum.

Figure 6 illustrates these simple steps which make up the Bayer process while Figure 7 further elucidates the unit operations. Figure 8 provide more process detail and along with Table 2 illustrates the typical Bayer process products and compound deportment from ore to alumina final product and to red mud wastes. For reference, about 1 dry tonne and 5 wet tonnes of red mud are produced per tonne of alumina produced. Annually about 120 MT of red mud are produced globally.

At the time of his discovery, aluminum was one of the most expensive metals on earth. Conversely, it is aloes one of the most common elements in the earth's crust trailing only oxygen and silicon in relative abundance. These findings laid the basis for an economical process to produce purified alumina. This could be used as the raw material for the Hall/Héroult electrolytic process for making aluminum metal, also discovered in 1888. Bayer built many alumina refineries in Russia, France and Austria in the late 1800s to produce alumina. This innovative process is still in use throughout the world. Figure 9 illustrates the Queensland alumina plant in Australia where one of the authors, Anjier, spent a good portion of his career developing, designing and operating this facility.

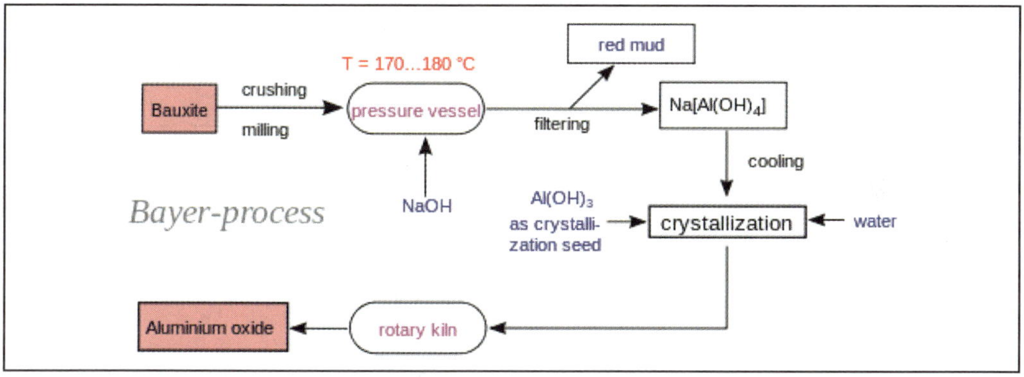

Figure 6. The Bayer process simplified flowsheet (Wikipedia 2013)

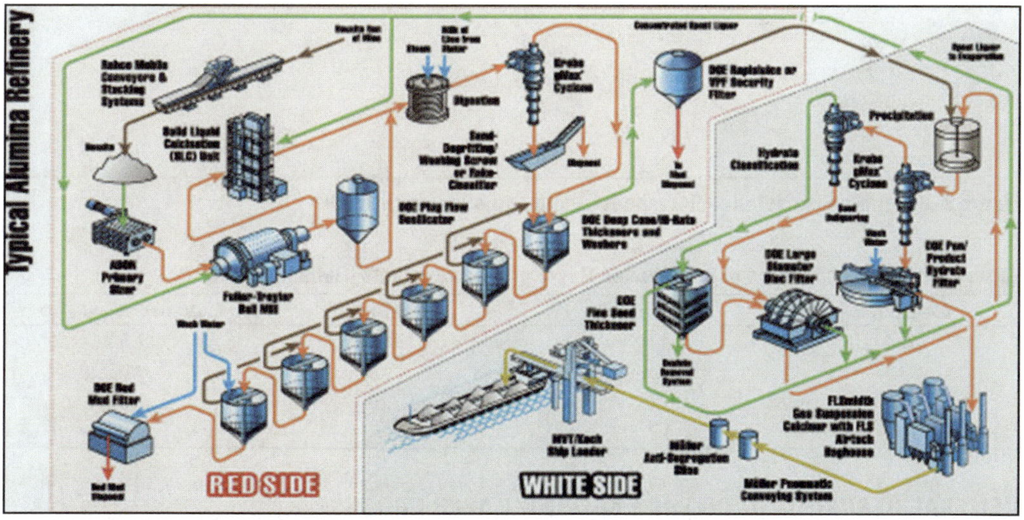

Figure 7. The Bayer process unit operations with both the alumina (white side) and red mud (red side) process streams shown (FLSmidth 2013)

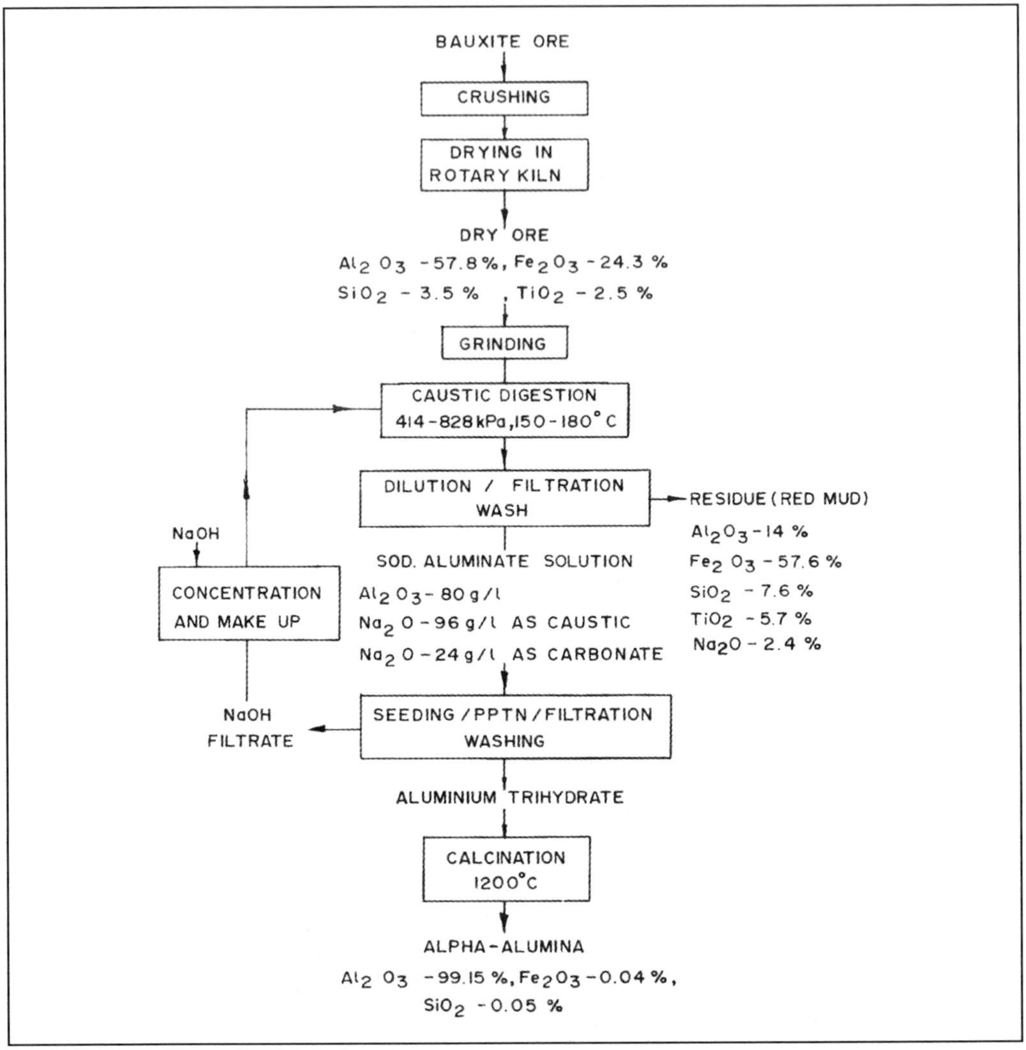

Figure 8. Bayer process detailed flowsheet (Gupta and Mukherjee 1990)

Table 2. Typical composition of bauxite and Bayer process products (Habashi 1999)

Compound	Bauxite, %	Red Mud, %	Calcined Al_2O_3, %
Al_2O_3	57.8	14.0	99.55
SiO_2	3.5	7.6	0.05
Fe_2O_3	24.3	57.6	0.04
TiO_2	2.5	5.7	Nil

GENERAL DESCRIPTION OF THE EVOLVING BAYER PROCESS

Bauxite is unloaded and stored in large buildings until required. Bauxite is removed from storage and if required undergoes a two stage grinding operation before it is fed to a high temperature digest in sodium hydroxide(spent liquor) solution to yield a slurry containing sodium aluminate

Figure 9. Queensland Bayer process alumina plant (Anjier 2013)

Figure 10. Bayer process bauxite digestor system equipment (Anjier 2013)

$NaAlO_2$ and suspended waste solids called red mud as well as sand. To allow efficient recovery and reduction of waste, the sand is removed and red mud is settled from the liquor using synthetic flocculants to produce a clear sodium aluminate solution. Figure 10 illustrates Bayer plant equipment and facilities (Adamson 1956, Adamson et al. 1963, Gerard 1963).

The red mud is thickened in a settler, the mud underflow is reslurried and pumped to the washing circuit trains. The mud travels down the multi stage washing circuit to recover a portion of the alumina and alkali values. The red mud waste slurry is finally discharged to a mud lake or other storage facility.

The overflow is pregnant liquor. After filtering to remove residual solids, the pregnant liquor is cooled through heat exchangers prior to entry into precipitation tanks. Seed crystals of previously precipitated alumina trihydrate are added and the tanks are agitated. This hydrolyzes the aluminate to yield alumina trihydrate, which grows on the seed material. This process is continued until about half of the alumina is precipitated. The alumina trihydrate is then sent to classification where it is separated into fine seed, coarse seed and product sized hydrate before thickening tanks to separate the trihydrate from the spent liquor. The trihydrate is filtered washed and calcined for aluminum manufacture. Following the removal of the hydrate crystals, spent liquor is concentrated

Figure 11. Typical Bayer process plant facility layout (Anjier 2013)

by evaporators and returned to the digestion circuit. Figure 11 illustrates an aerial view of the size and layout of a typical Bayer process plant facility.

Power House and Utilities

Significant quantities of steam, electricity and water are required for the successful operation of an alumina refinery. The steam may be produced in coal fired or oil fired high-pressure boilers. These boilers supply steam to the digestion and evaporation area and the electrical power plant, assuming that they produce their own electricity. Ash from the coal is usually deposited in the red mud lake and is a potential source of metals in the discharge from the ponds. Fresh water is a requirement for mud washing, clean up and alumina washing. In many plants river water is the source, others use domestic water or wells. Large plants can use up to 2 million U.S. gallons of water per day. Figure 12 shows the Bayer heat exchanger system.

Waste Streams

Alumina plants have a variety of waste streams with potential environmental hazards. The Australian Government in the Department of the Environment, Water, Heritage and the Arts has published *Emission Estimation Technique Manual for Alumina Refining*, Version 2.0. This is a manual for estimating environmental pollutants (Australian 1992). The manual does not include plant data from the five alumina plants in Australia but does describe the source and potential pollutants found in each plant process.

The streams, which either leave or are accumulated in the typical alumina plant, are as follows:

- Bauxite screening sand. Bauxites must be screened to remove limestone rocks, oversize from the grinding operation and sand from the digestion discharge slurry. Based on the production rate of the plant this can be many tons per day.

Figure 12. Bayer process plant heat exchange system (Anjier 2013)

Figure 13. Red mud in an impoundment facility (Anjier 2013)

- Bauxite mud (red mud). After digestion and primary mud settler separation, the bauxite mud passes through a counter current decantation washing circuit where fresh water reduces the soda content of the mud before it is discharged into the mud impoundments and the concentrated water from the washing circuit is used to adjust the concentration of the pregnant liquor feeding the mud settler. This mud will contain all of the dangerous compounds found in the various bauxites including radioactive particles (Ryle 2010). Figure 13 illustrates red mud in an impoundment area.
- Acidic streams from acid cleaning piping and heat exchangers.
- Sodium oxalate streams. The organics, which are in the bauxite, form sodium oxalate during the digestion process. Sodium oxalate impedes the agglomeration and growth of alumina trihydrate particles resulting in and out of particle size specification product. To solve this problem the sodium oxalate is washed off of the seed used to initiate the precipitation reaction. This dissolved sodium oxalate stream is discharged into the mud lake (Mallinckrodt Baker 1990).

- Equipment scaling is a plant wide problem because of the scaling nature of the supersaturated sodium aluminate solution. This scale must be manually removed from the piping and vessels or in some cases acid cleaned. This produces additional solid and liquid waste streams.

MAJOR EQUIPMENT AND PROCESS INNOVATIONS

Experimentation and process development is a continuous undertaking to improve the quality of the manufacturing process, increase the yield of the alumina and reduce capital and operating cost. There have been some major innovations which changed the design of alumina plants.

Conversion from Batch to Continuous Digestion—Alcoa

Digestion was originally a batch process. Alcoa designed and built their first plant at East St. Louis as a continuous process from digestion to calcinations. This design improved production capacity and was used to build plants during World War II.

Conversion from Batch to Continuous Precipitation—Pechiney

The first precipitators mixed pregnant liquor and alumina trihydrate seed in single vessels which were agitated with belt powered mechanical agitators and held until the alumina trihydrate was precipitated. Pechiney was a leader in converting the batch precipitators to continuous precipitators (Porter 1961). This simplified the precipitation operation and allowed development of new techniques to control the particle size of the precipitated alumina trihydrate.

Development of Air Lift Agitators for Precipitators

Original precipitation agitators were belt operated and propeller driven. Since a line of precipitators may be 10–12 precipitators long the plant was large and ungainly. The development of air lifts allowed a simplification of design and operation. Figure 14 illustrates this.

Figure 14. Aluminum hydroxide precipitator (Anjier 2013)

Development of Rotary Kilns

The Hall–Héroult electrolytic reduction process specifies particle size distribution, loss on ignition as well as other quality parameters. Rotary kilns aided in achieving these specifications.

Development of High Temperature Digestion—Kaiser

As was mentioned earlier, bauxites contain monohydrate and alumina trihydrate. The trihydrate alumina can be dissolved at a lower digest temperature than the monohydrate. As the supply of trihydrate rich bauxite decreased, Kaiser Aluminum undertook the development of a high temperature digest system to dissolve the monohydrate alumina. This was installed in 1958 at the Gramercy alumina plant and has been used in all plants which have been built since then unless they have a supply of trihydrate bauxite.

Improvements in the Red Mud Circuit

Improvements include equipment design and operation, mud settling, chemical development and application. The red mud from the digesters is cooled in a series of flash tanks with the steam used to heat the caustic to be used to digest the alumina. The discharge from the digestion train is a slurry of pregnant liquor, red mud, sand and all of the undissolved solids, both good and bad. The first step is the removal of the sand in a variety of sand trap designs, and then the desanded slurry is sent to a mud thickener. The pregnant liquor overflows and the mud underflow is transferred to the 4–10 vessel train of mud washers. In the washing train fresh water is brought into the last washer while the thickened mud is sent to the first washer where it is mixed with the overflow from the previous washer. In this way the caustic is removed from the mud and transferred to the digestion discharge for caustic concentration control.

The development flocculating chemicals by Nalco (Nalco 1988) and several other chemical companies has significantly improved the operation of these vessels. The previous flocculent was starch slurry. The new flocculants are easy to use significantly improve the operation of the washing and settling circuit.

Development of Individual Coolers for Precipitators

The temperature of the liquor in the first precipitator has a well-defined effect on the growth and agglomeration of fine seed. The use of individual precipitator heat exchangers allows high fill temperatures but allows staged cooling so that particle size can be controlled and precipitation yield can be increased.

SUMMARY

This paper has detailed the history, development, implementation, and improvements to the Bayer process for production of purified alumina. This landmark innovation created a viable global industry providing a source of inexpensive aluminum to propel our society forward. It also helped to place hydrometallurgy as a viable industrial technology throughout the world.

REFERENCES

Anjier, J., 2013. Personal communication.

Australian Government Department of the Environment, Water, Heritage and the Arts. 1992. NOPI (National Pollutant Inventory) Emission Estimation Technique Manual for Alumina Refining, Version 2.0.

Bayer, K.J., 1888. Darsteeling von Thonerhydrat and Alkalialuminat, German patent 43,977.

Bayer, K.J., 1892. Darsteeling von Thonerhydrat and Alkalialuminat, German patent 65,604.

Cundiff, W.H., and Reese, K.M., 1955. In Aluminum Production the First Stage is Alumina, Industrial and Engineering Chemistry, Sept. pp. 1673–1675.

FLSmidth, F. 2013. http://www.flsmidth.com/en-Us/Products/Light+Metals/Alumina+and+Bauxite.

Gupta, C.K., and Mukherjee, T.K., 1990. Hydrometallurgy In Extraction Processes, Volume I, CRC press.

Habashi, F.H., 1988. A Hundred Years of the Bayer Process for Alumina Production, CIM Bulletin, January, Volume 81, No 909, pgs 70–74.

Habashi, 1999. Textbook of Hydrometallurgy, Second Edition, Laval University, Quebec, City, Canada.

Hose, H.R., 1963. Bauxite Mineralogy, Extractive Metallurgy of Alminum, Vol. 1, Interscience, p. 6.

Porter, J.L., 1961. Process of purifying caustic aluminate liquors, US Patent 2981600A.

Ryle, G. 2010. The Great Red Mud Experiment That Went Radioactive, *Saving Iceland.* 7 May 2002. Web. 17 August. http://www.savingiceland.org/2004/05/the-great-red-mud-experiment-that-went-radioactive/.

USGS, 2012. Bauxite and Alumina in the 4th Quarter, Mineral Information Services.

Vereinigte Aluminum Werke, A.G., German Patents 960,811 1957, 963,602 (1957); British Patent 799,243 (1958).

Wikkipedia, 2013. The Bayer Process, http://en.wikipedia.org/wiki/Bayer_process.

BIBLIOGRAPHY

Adamson, A.N., 1956. Alumina Production: Principles and Practice, The Chemical Engineer, June pp. 156–164, 171.

Adamson, A.N., Bloore, E.J., and Carr, A.R., 1963. Basic Principles of Bayer Process Design, Extractive Metallurgy of Aluminum Vol. 1, Alumina, Interscience, pp. 23–58.

Bielfeldt, K., 1968. Practical Experiences with the Tube Digester, Journal of Metals, Sept., pp. 48–54.

Breuer, R.C., Barsotti, L.R., and Kelly, A.C., 1953. "Behavior of Silica in Sodium Aluminate Solutions," Extractive Metallugy of Aluminum, Vol. 1, Alumina, Interscience, pp. 133–157.

Chaplin, N.T., 1971. Reaction of Lime in Sodium Aluminate Liquors, Light Metals, AIME, New York, pp. 47–61.

Chemical Engineering, 1954. Nov., pp. 112–112, pp. 334–337.

Cundiff, W.H., 1972. Influence of Bauxite Composition on Alumina Plant Technology, Proceedings, 3rd International Symposium of ICSOBA, Banska Bystrica, CSSR, Research Institute for Non-Ferrous Metals, Budapest.

Cundiff, W.H., 1985. Alumina, SME Mineral Processing Handbook, Chapter 19, 20 pages.

Edwards, J.D., 1949. "The Combination Process for Alumina," Trans. AIME, Vol. 182, pp. 9–13.

Energy Conservation and Recycling in the Aluminum Industry, 1974. Conference cosponsored by the Center for Materials Science and Engineering, MIT, and by the US Bureau of Mines, April.

Evans, H.J., 1960. High Alumina Laterites of North Queensland, Proceedings, Laterites and Assorted Minerals Symposium, Brisbane, Australia, p. 11.

Gerard, G. 1963. Extractive Metallurgy of Aluminum, Vol. 2, Interscience.

Harder, E.C., 1952, Examples of Bauxite Deposits Illustrating Variations in Origin, Problems of Clay and Laterite Genesis, AIME, New York, pp. 35–64.

Hill, V.J., 1963. "Factors Influencing Angle of Repose of Calcined Alumina for Electrolytic Reduction, Extractive Metallurgy; of Alumina Vol. 1, Alumina, Interscience, pp. 219–278.

lkkatai, T., and Okada, N., 1963. Viscosity, Specific Gravity, and Equilibrium Concentration of Sodium Aluminate Solutions, Extractive Metallurgy of Aluminum, Vol. 1, Alumina, Interscience, pp. 159–173.

Jamey, M., 1956. Contributions to the Kinetics of Alumina Trihydrate Precipitation, AIME Meeting, February, New York.

Kemmer, F.N., 1988. The Nalco Water Handbook, Nalco Chemical Company, Second Edition.

King, W.R., 1971. The Iron Minerals in Jamaican Bauxites, Light Metals, AIME, New York, pp. 3–18.

Lanyi, B., 1956. Continuous Digestion of Bauxite, Femipari Kotato Intezet Kazlemenyei, Vol. 1, pp. 5–28.

Loos, H., 1971. Red Mud Filtration with Rotary Filters, Bauxite- Alumina-Aluminum, Vol. 3, Proceedings 2nd International Symposium of ICSOBA, Research Institute for Non-Ferrous Metals, Budapest, pp. 175–180.

Mallinckrodt Baker, 1990. MSDS 4586, Sodium Oxalate.

Pearson, T.G., 1955, The Chemical Background of the Aluminum Industry, The Royal Institute of Chemistry, London, Lecturer, Monographs, Reports, No. 3.

Perry, K.W., 1970. Modifications to Bauxite Digestion, Bauxite Alumina, Aluminum, Vol. 1, Proceedings, 2nd International Symposium of ICSOBA, Research Institute for Non-Ferrous Metals, Budapest, pp. 43–57.

Plaetschke, H., 1971. Separation of Red Mud Wash Rotary Filters, Bauxite-Alumina-Aluminum, Vol. 3, Proceedings. 2nd International Symposium of ICSOBA, Research Institute for Non-Ferrous Metals, Budapest, pp. 181–183.

Regnier, J., 1963. Continuous Digestion of Monohydrate Bauxites, Extractive Metallurgy of Aluminum, Alumina, Interscience, Vol. 1, pp. 251–264.

Reh, R.L., et al. 1971. Experience with the Calcination of Alumina Trihydrate in a Circulating Fluid Bed, Light Metals, AIME New York, pp. 109–120.

Sato, C., and Furukawa, A., 1971. Comparative Evaluation of the Economics of Floury Type and Sandy Type Bayer Alumina, Bauxite-Alumina-Aluminum, Vol. 3, Proceedings, 2nd International Symposium of ICSOBA, Research Institute for Non-Ferrous Metals, Budapest, pp. 227–235.

Scandrett, H.F., 1971. Bayer Heat Cycle via McCabe-Thiele Diagrams, Bauxite-Alumina-Aluminum, Vol. 3, Proceedings 2nd International Symposium of ICSODA, Research Institute for Non-Ferrous Metals, Budapest.

Schwalbach, V., 1971. The History of Continuous Red Mud Filtration and Methods of Aluminate Liquor Clarification and Hydrate Separation, Bauxite-Alumina-Aluminum, Vol. 3, Proceedings 2nd International Symposium of ICSOBA, Research Institute for Non-Ferrous Metals, Budapest, pp. 187–195.

Scott, I., 1963. Effect of Seed and Temperature on the Particle Size of Bayer Hydrate, Extractive Metallurgy of Aluminum, Vol. I, Alumina, Interscience, pp. 203–218.

Shimosato, J., Kogyo Kagaka Zanski, 1963. Vol. 66, No. 2, pp. 172–177.

Soymar, K., 1970. Bauxite-Alumina-Aluminum, Vol. 3, Alumina, Proceedings, 2nd International Symposium of ICSOBA, Research Institute for Non-Ferrous Metals, Budapest, pp. 45–65.

Staudinger, J., 1971. Some Problems in Underground Bauxite Mining, Proceedings, 2nd International Symposium of ICSOBA, Vol. 2, Research Institute for Non-Ferrous Metals, Budapest, pp. 253–261.

Szepcshegyi, I., 1971. Up-to-Date Mining and Transport System in the Mines of the Bakony, Bauxite-Alumina-Aluminum, Vol. 2, Proceedings, 2nd International Symposium of ICSOBA, Vol. 2, Research Institute for Non-Ferrous Metals, Budapest, 1971, pp. 239–251.

Torngren, S., 1971. Some Up to Date Heat Exchangers for Use in the Bayer Process, Bauxite-Alumina-Aluminum, Vol. 3, Proceedings, 3rd International Symposium of ICSOBA, Research Institute for Non-Ferrous Metals, Budapest, pp. 243–252.

Wefers, K., and Bell, G.M., 1972. Technical Paper No. 19, Oxides and Hydroxides of Aluminum, A Revision of Technical Paper No. 10, "Alumina Properties," Alcoa Research Laboratories, 512 pp.

Weipa Bauxite, 1970. Comalco Ltd., Melbourne, Australia, p. 32.

ABOUT THE AUTHORS

Joseph L. Anjier has more than 52 years of experience in chemical engineering, which has varied from the building of rockets to the technology, operation, and startup of major alumina plants in the United States and Australia. He is a petroleum refining engineer from the Colorado School of Mines and earned an MBA from Louisiana State University. His career included positions with Aerojet General Corporation, Kaiser Aluminum and Chemical Corporation, Comalco Aluminum Company, and JA Engineering Services, LLC. He is a member of AICHE, TMS, and is a professional chemical engineer in the state of Louisiana. He holds three patents and has presented two technical papers to the TMS and one to the AICHE. After retirement from Kaiser, he established JA Engineering Services, LLC, in which he consults on process and environmental problems with a variety of customers.

Corby G. Anderson has more than 34 years of global experience in industrial operations, management, engineering, design, consulting, teaching, research, and professional service. His career includes positions with Morton Thiokol, Key Tronic Corporation, Sunshine Mining and Refining Company, H.A. Simons Ltd., and Montana Tech. Anderson holds a BSc in chemical engineering and an MSc and PhD in metallurgical engineering. He is a fellow of the Institution of Chemical Engineers and the Institute of Materials, Minerals and Mining. He currently serves as the Harrison western professor in the Kroll Institute for Extractive Metallurgy as part of the George S. Ansell Department of Metallurgical and Materials Engineering at the Colorado School of Mines. In 2009 he was honored by SME with the Milton E. Wadsworth Extractive Metallurgy Award for his contributions in hydrometallurgical research.

Rare Earth Hydro- and Pyrometallurgy

Karl A. Gschneidner Jr.
Ames Laboratory (U.S. Department of Energy) and Department of Materials Science and Engineering, Iowa State University, Ames, Iowa, United States

ABSTRACT: The processes for separating the rare earth elements were born as a part of the World War II Manhattan Project to develop the atomic bomb. Ion exchange was the initial successful technique, and the liquid-liquid solvent extraction process soon evolved after World War II as the primary technique for separating the rare earths on a commercial basis. The preparation of the rare earth metals was also initially started in the War years, but most of the development occurred in the late 1940s and early 1950s. This research has led to many important and critical applications of the rare earths, which most persons are completely unaware of the impact these elements have on their lives. For example, when a person buys an automobile it is the largest purchase of rare earth materials they will make until they purchase the next automobile.

PRE–WORLD WAR II

Before World War II the individual trivalent rare earth elements, other than Ce, were separated by either fractional crystallization or fractional precipitation (Powell 1979). Scientists and their technicians might have made 50,000 crystallizations (or precipitations) in their life time to obtain a rare earth element that was only 99% pure with respect to its neighboring lanthanide elements. Cerium was easily separated and purified from La, Pr, and Nd by making use of its 3+ and 4+ oxidation states, a process still used today commercially to remove Ce from the other rare earth elements (Callow 1967).

However, this level of purity was insufficient for scientists who were studying the optical properties of individual rare earth elements in the late 1930s, one needs at least four nines purity to obtain good spectroscopic data; today commercial optical devices, e.g., color television, requires five to six nines purity. F.H. Spedding, a physical chemist who was carrying out research on the spectra of rare earth compounds in the 1930s and early 1940s, realized that a more efficient and speedier process was critically needed to obtain high purity rare earths to measure the wave lengths of the intrinsic spectral lines of each lanthanide element and thus determine their electronic nature. Much of this early research was on Sm compounds because Sm could be easily purified by making use of its dual 2+/3+ valence states. He also studied a few other lanthanide elements (i.e., Gd, Tm, and Nd) which had only one valence state (3+), and noted the big difference in obtaining a high purity compound for optical studies. Spedding realized that in order to get a pure lanthanide in a reasonable time frame he needed to develop an automatic system which would repeat the separation process over a thousand times. He and his students worked on chromatographic columns with various absorbers (e.g., Al_2O_3 or SiO_2 gel) and paper chromatography but none were successful. His research was abruptly halted in December 1941, when he was asked to head up the chemistry effort to make uranium metal for the first atomic reactor under the stands of Stagg Field in Chicago (Svec 1988).

WORLD WAR II RESEARCH EFFORTS

During the atomic fission of a U^{235} atom lanthanide atoms are formed, several of which have extremely high neutron cross sections, and these highly radioactive lanthanides needed to be removed from unspent U^{235} fuel in order to reuse the U^{235}. Thus studies were undertaken at both Iowa State College (Ames) and Oak Ridge to develop chemical processes for separating the rare earths. Both came up with an ion-exchange process to separate moderate amounts of the rare earths, and Ames was awarded the patent for the ion exchange process by an independent panel of lawyers. Oak Ridge worked with fusion products, and was rewarded by the discovery of the missing lanthanide element, atomic number 61, which lies between Nd and Sm. They named this element appropriately promethium (Svec 1988). This research showed that promethium is unlikely to be found in nature because its longest half-life is 14 years, and definitely ruled out claims in the 1930s by several scientific research groups of having discovered element "61" in rare earth ores.

The initial ion exchange process which was a big improvement over fractional crystallization or precipitation was far from perfect, since the elutation curves had a "bell-shape" and there was considerable overlap of adjacent lanthanides (Powell 1979). In time stronger complexing compounds were discovered and their utilization led to displacement ion chromatography which had rectangular shaped bands and very little overlap between adjacent rare earth elements in the columns. This allowed scientists and engineers to use large columns (up to one meter diameter and 8 meters high) to obtain kilogram quantities of the individual elements with only a few parts parts million of other rare earth impurities. The main problem with ion exchange is the long time required to obtain the given rare earth or rare earths, although the individual elements are very pure (5 nines or greater) (Powell 1979).

At the same time research was being carried out on solvent or liquid-liquid extraction at Ames, Argonne and Oak Ridge to separate the rare earth elements. The advantage of solvent extraction is that it is faster than the other separation technique, but a large number of mixer settlers, which are quite expensive, is required. The number needed depends on the purity required and the number of rare earth elements in the start solution. Usually where high purities (>4 nines) are required the given rare earth element is purified by ion exchange (Bautista 1995).

CURRENT TECHNOLOGY

A summary of the commercial processes developed after World War II based on the research carried out by the U.S. Atomic Energy Commission during the war years (above) were summarized by Gschneidner (1981). These included the Molycorp and Rhone-Poulenc liquid-liquid solvent extraction processes. More detailed information is briefly discussed below.

ION EXCHANGE

Initially a column is loaded with a retaining ion, such as Cu^{2+} or Zn^{2+}. A retaining ion is used instead of H^+ because of the insolubility of H_4(EDTA). Then the rare earth solution is loaded onto the column and then eluted by a complexing agent such as EDTA (ethylenediamine tetraacetate). Separation occurs because of the slight difference in the stability of the R(EDTA) complexes, the most stable complex remains in solution while least stable is retained on the ion exchange resin. The smallest rare earth ion forms the most stable complex, i.e., Lu, and it is the first lanthanide eluted from the column while La, the largest rare earth ion, is the last element eluted from the column. The retaining ion and complexing agent are recycled. The eluted rare earth is precipitated as the oxalate, which is generally converted to the oxide by firing (heating in air) the rare earth oxalate at 800 to 1200°C in a muffle furnace (Gschneidner 1981 and Powell 1979). There are a variety of complexing agents whtch can be utilized in the separation process: some are better for the light lanthanides and others for the heavy rare earths. More details, including an extensive list of complexing agents can be found in a review by Powell (1979).

SOLVENT EXTRACTION

The solvent extraction process involves mixing an aqueous solution containing a mixture of rare earths with an organic phase, allowing the aqueous and organic phases to separate and to decant one phase from the other. Separation occurs because one rare earth ion is more soluble in the aqueous phase, while the other rare earth is more soluble in the organic phase when they are mixed and then settled. This process is continued many times and the aqueous phase contain a given lanthanide at a three to four nines purity, while the organic phase becomes richer in the other lanthanide also reaching three to four nines purity. Of course when separating the rare earths from a given ore there may be up to twelve elements to be separated and purified making the process much more challenging and difficult. There are a variety of mineral acids used as aqueous phase (i.e., HCl, HNO_3), and likewise many different organic based solvents (i.e., kerosene, tributyl phosphate). Generally the given combination of aqueous/non-aqueous reagents may be used to concentrate the light lanthanides from the original mixture, then a different aqueous/non-aqueous combination is used to separate the lights from one another, and finally a third set for separating the heavies. In the solvent extraction process La remains in the aqueous phase and is the first element separated from the mixture and it is then followed by Pr, Nd, etc. The Ce, of course, as noted above, is initially removed by an oxidation/reduction step and is not involved in the solvent extraction process. More details about the solvent extraction process are given in the reviews by Gschneidner (1981) and Bautista (1995).

PREPARATION OF THE METALS

Electrolytic Process

The rare earth metals are basically prepared by two processes: electrolytic and metallothermic reduction. The electrolytic method is limited to the metals which melt below 1050°C, because at higher temperatures the metals and fused salt electrolytes readily react with the crucible and electrodes. Thus the only metals prepared by this process are the first four light lanthanides (La, Ce, Pr and Nd) and mischmetal (a mixture of the naturally occurring unseparated rare earth metals). Today, commercially, the corresponding oxide is reduced to the metal in a fused salt eutectic solution of the respective rare earth fluoride and LiF. It is an environmentally hazardous process because of the low solubility of the oxide in the eutectic electrolyte and it is difficult to maintain a uniform concentration of the oxide in the electrolytic cell. If the oxide is consumed F_2 gas is released at the anode instead of O_2. Before the oxide process was developed (i.e., ~1995) the metals were prepared by the electrolytic reduction of the chloride or fluoride (Gschneidner 1980, and Beaudry and Gschneidner 1978).

METALLOTHERMIC REDUCTION PROCESS

The low vapor pressure rare earths La→Nd, Gd→Er, Lu, Y and Sc are prepared by metallothermic reduction of the chloride or fluoride usually using Ca metal as the reductant. The Ca + RX_3 are heated to ~100°C above the melting point of the metal and the metal sinks to the bottom of the Ta crucible and the immiscible halide slag floats on top of the molten metal. The two phases are readily separated physically. The high vapor pressure metals Sm, Eu, Tm and Yb are prepared by heating the respective oxide with La (sometimes Ce) metal. The La (Ce) forms $La_2O_3(CeO_2)$ and the high vapor pressure metal sublimes out of the mixture and is collected at the cold end of the Ta crucible. There are significant differences in the processes when preparing commercial grade rare earth metals (purities vary from 95 to 98 at.%–the major impurities are O, H, C and N), and high purity (Ames Laboratory) research grade metals (99.5+ at.%); typical impurity levels of both grades have been summarized by Gschneidner (1980). Special handling procedures are used for the preparation of the Ames research grade metal, to limit the amounts of the interstitial impurities—O, H, C and N in the final product. One significant difference is that a graphite or MgO crucible is used in industrial processes instead of a Ta one, which accounts for the higher O and C contents in the commercial grade product. Furthermore, industrially, the halide is used as initially prepared and not further purified, while the high purity Ames metals the fluoride is processed to lower the residual oxide content from ~300 ppm by wt. to <10 ppm by wt. (see below). Details can be found in the reviews by Gschneidner (1980) and Beaudry and Gschneidner (1978).

PREPARATION OF RCL_3 AND RF_3

The trihalides are prepared from the oxide. The chlorides are prepared by dissolving the R_2O_3 (CeO_2 or Pr_6O_{11}, or Tb_4O_7) in HCl and evaporating to form $RCl_3{\cdot}6H_2O$. Further removal of H_2O is done under a vacuum at low temperatures to prevent the ROCl from forming. In addition to the slow drying process, the RCl_3 is hygroscopic and must be handled in an inert dry atmosphere. The chlorides may also be prepared by reacting the oxide, or hydroxide, or carbonate with anhydrous HCl. Another process is to react phosgene gas, $COCl_2$, with the oxide at high temperature to obtain the corresponding anhydrous rare earth trichloride and carbon dioxide. Similarly one can react the rare earth oxide with chlorine gas in the presence of graphite to yield the same products as with phosgene (Callow 1967).

The RF_3 phases are preferred because they are stable in air and can be more easily handled. There are three processes used to prepare the RF_3 compounds (Gschneidner 1980): (1) the precipitation of RF_3 when liquid HF acid is added to a RCl_3 or $R(NO_3)$ solution. The RF_3 compound needs to be dried at ~100°C in air to remove any water. (2) Instead of adding HF(l) to the aqueous chloride or nitrate solution, the rare earth oxide is mixed and heated with NH_4HF_2 at ~400°C to directly form the respective RF_3 compound (Gschneidner 1966). (3) The reaction of HF gas with the oxide to yield RF_3 and H_2O at ~650°C. Processes 1 and 2 are used commercially and the resultant fluoride may contain a few tenths of a percent Fe, which is present in the starting HF(l) and NH_4F_2 chemicals. This is especially a problem for process 2. Process 3 yields the highest purity fluoride, especially if the RF_3 is "topped," i.e., heated to above its melting point under HF(g); this process removes the last residual oxygen in RF_3. The last technique process 3 is the one used to prepare the high purity Ames metals (Gschneidner 1980 and Beaudry and Gschneidner (1978).

APPLICATIONS

The availability of reasonably pure to very high purity rare earths has greatly impacted our lives. Many of the items we utilize in our daily life include: cell phones, i-pods, televisions, automobiles, computers, electric motors, rechargeable nickel metal hydride batteries, petroleum cracking catalysts, MRI units, wind turbines, energy efficient compact fluorescent lamps (CFL) and LED lights. The optical displays, and CFL and LED lamps utilize 99.999+% pure Eu (red color), Tb (green color), and the Y host material; electric motors (automobiles and aircraft), speakers (cell phones, automobiles), wind turbine engines, MRI units (magnets), and computers (spindles) all utilize 98–99% pure Nd in $Nd_2Fe_{14}B$ permanent magnets; nickel metal hydride rechargeable batteries contain 98–99% pure La and mischmetal; cracking catalysts for petroleum refining utilize 98% pure mixed rare earth oxides, La and Ce oxides; automotive windshields and semiconductor chips are polished by 98–99% pure Ce oxides; automotive exhaust three-way catalysts contain 98–99% pure Ce oxides; sensors to control lean/rich gas mixture for automotive gasoline engines utilize 99% pure Y_2O_3. In addition to our daily lives, *all* of our country's weapons systems used to defend the United States critically depend upon the rare earths—computers, electric motors, optical displays, refined gasoline and diesel fuel, communications, etc. Yes, we really need the rare earths, and *we cannot get along without them in our modern world.*

ACKNOWLEDGMENTS

The author thanks Yaroslav Mudryk for his constructive comments regarding this manuscript. This work was supported by the U.S. Department of Energy, Office of Basic Energy Science, Division of Materials Sciences and Engineering. The research was performed at the Ames Laboratory. Ames Laboratory is operated for the U.S. Department of Energy by Iowa State University under contract No. DE-AC02-07CH11358.

REFERENCES

Bautista, R.G. 1995. Separation chemistry. In *Handbook on the Physics and Chemistry of Rare Earths*, Vol. 21, ch. 139, p. 1. Edited by K.A. Gschneidner, Jr. and L. Eyring. The Netherlands: Elsevier Science Publisher.

Beaudry, B.J. and Gschneidner, K.A., Jr. 1978. Preparation and basic properties of the rare earth metals. In *Handbook on the Physics and Chemistry of Rare Earths*, Vol. 1, ch. 2, p. 173. Edited by K.A. Gschneidner, Jr. and L. Eyring. The Netherlands: Elsevier Science Publisher.

Callow, R.J. 1967. *The Industrial Chemistry of the Lanthanons, Yttrium, Thorium and Uranium*. Oxford: Pergamon Press.

Gschneidner, K.A., Jr. 1966. The application of vacuum metallurgy in the purification of rare earth metals. In *Transactions of the Vacuum Metallurgy Conference*, p. 98. Edited by L.M. Bianchi. Boston: American Vacuum Society.

Gschneidner, K.A., Jr. 1980. Preparation and purification of the rare earth metals and the effect of impurities on their properties. In *Science and Technology of Rare Earth Materials*, p. 25. Edited by E.C. Subbarao and W.E. Wallace. New York: Academic Press.

Gschneidner, K.A., Jr. 1981. Rare earth speciality inorganic chemicals. In *Speciality Inorganic Compounds*, p. 403. Edited by R. Thompson. London: The Royal Society of Chemistry.

Powell, J.E. 1979. Separation chemistry. In *Handbook on the Physics and Chemistry of Rare Earths*, Vol. 31, ch. 22, p. 81. Edited by K.A. Gschneidner, Jr. and L. Eyring. The Netherlands: Elsevier Science Publisher.

Svec, H.J. 1988. Prologue. In *Handbook on the Physics and Chemistry of Rare Earths*, Vol. 11, p. 1. Edited by K.A. Gschneidner, Jr. and L. Eyring. The Netherlands: Elsevier Science Publisher.

ABOUT THE AUTHOR

Karl A. Gschneidner Jr. is an Anson Marston distinguished professor in the Department of Materials Science and Engineering at Iowa State University, and a senior metallurgist of the Ames Laboratory, U.S. Department of Energy. He was the founding director of the Rare-earth Information Center from 1966 to 1996. Gschneidner received a B.S. degree from the University of Detroit in 1952 and his Ph.D. from Iowa State University in 1957. He was at the Los Alamos National Laboratory from 1957 to 1963 and joined Iowa State University in 1963. He has published more than 500 papers in peer-reviewed journals, holds 15 patents (plus 4 pending), and has given 316 invited presentations. He retired in 2011 as the founding and senior editor of the 41-volume series of the *Handbook on the Physics and Chemistry of Rare Earths*. He was elected to the National Academy of Engineering in 2007 for "contributions to the science and technology of rare-earth materials," and received over 25 awards and honors, including fellowships in five professional societies. Gschneidner was named the 2000 University of Detroit Mercy's Science Alumnus.

Biohydrometallurgy Innovations—Discovery and Advances

James A. Brierley
Brierley Consultancy LLC, Highlands Ranch, Colorado, United States

ABSTRACT: The author has selected noteworthy events in the historical development of biohydrometallurgy. The events are categorized as discovery of the microbial process for biooxidation of pyrite; demonstration that sulfide mineral biooxidation and bioleaching are conducted by numerous types of microbes composing a consortium; definition of how microbes oxidize sulfide minerals; and development of commercial plants for economic use of biohydrometallurgy. The latter topic is considered for heap and stirred-tank reactor systems. Possible future developments are contemplated.

INTRODUCTION

Innovations in biohydrometallurgy have historically followed an evolutionary rather than a revolutionary research and development path. Innovations can be traced to the fundamental, biochemical characteristics of the microbes with advancements in practical development of the technology occurring in a stepwise manner from academic understanding of the catalysts (the microbes) to implementation of applied processes based on experience and expertise.

Bioleaching is the practice of using microbes for solubilization of base metal values, which are subsequently recovered from PLS using conventional metallurgical processes. Biooxidation pretreatment refers to use of microbes for oxidation of minerals, such as pyrite, that occlude precious metal values, which can then be recovered by standard metallurgical processes for dissolution of the precious metals from the oxidized residue.

This paper focuses on the role of microbes in the hydrometallurgical extraction of precious and base metals from sulfide ores and concentrates, beginning with what is believed to be the first commercial application of microbes for metal extraction. Additional background reading on developments leading to applications of biotechnology for metallurgical processes are Brierley (2008), Gericke et al. (2009), and the textbook, *Biohydrometallurgy*, by Rossi (1990); the latter is a comprehensive collection of references and a presentation of the science and engineering of biohydrometallurgy to that date.

THE MICROORGANISMS

The microbiology community has long known about bacteria, called chemolithotrophic autotrophs that gain their energy for growth and reproduction by oxidation of inorganic substrates such as elemental sulfur and ferrous iron. These bacteria are also capable of carbon fixation, which is the assimilation of carbon from atmospheric CO_2 and synthesis of organic carbon compounds required for life. These characteristics impart the chemolithotrophic bacteria with the ability to effectively perform biohydrometallurgical applications. This section addresses the role and types of microorganisms involved in biohydrometallurgical processes.

Bacterial Generation of Acidity by Oxidation of Pyrite

The first observations of bacteria interacting with a mineral substrate, pyrite, were reported by Colmer et al. (1950). The authors' interests centered on biogenic acid generation, and they speculated that pyrite biooxidation of pyrite could be a source of the acid in drainages from some bituminous coal mines. A bacterium, *Thiobacillus ferrooxidans,* was isolated from an acid drainage and named by Temple and Colmer (1951). This organism was renamed *Acidithiobacillus ferrooxidans* by Kelly and Wood (2000) after additional biochemical study. The reactions by which the bacterium causes acid generation were subsequently reported (Temple and Delchamps 1953). Definition of these microbially-catalyzed reactions was fundamental to the understanding and subsequent development of applied minerals biooxidation and bioleaching. Initiation of the process was postulated to be abiotic oxidation of pyrite or marcasite to ferrous sulfate:

$$FeS_2 + H_2O + 3\tfrac{1}{2}O_2 \rightarrow FeSO_4 + H_2SO_4 \quad \textbf{(1)}$$

This chemical oxidation is followed immediately by bacterial oxidation of the ferrous sulfate (Equation 2) providing energy for growth of the bacteria. The ferric sulfate reacts rapidly with pyrite; ferric sulfate is chemically reduced in oxidizing the pyrite (Equation 3).

$$2FeSO_4 + \tfrac{1}{2}O_2 + H_2SO_4 \rightarrow Fe_2(SO_4)_3 + H_2O \quad \textbf{(2)}$$

$$Fe_2(SO_4)_3 + FeS_2 \rightarrow 3FeSO_4 + 2S \quad \textbf{(3)}$$

Sulfur formed during the ferric oxidation of the pyrite is biooxidized by the bacteria:

$$S + 1\tfrac{1}{2}O_2 + H_2O \rightarrow 2H_2SO_4 \quad \textbf{(4)}$$

The discovery of the acidophilic iron-oxidizing bacteria opened the door to research into the role of bacteria in oxidizing sulfide minerals; this resulted in the new discipline that became known as bioleaching and minerals biooxidation. Subsequent research and development focused on the few bacteria identified at the time, primarily *At. ferrooxidans* and the sulfur-oxidizing *Acidithiobacillus thiooxidans.* This changed as the numbers and diversity of microbial types involved in bioleaching and minerals biooxidation became increasingly recognized.

An early indication of the diversity of mineral oxidizing bacteria occurred with the discovery of an unnamed moderately-thermophilic (oxidizes ferrous iron at 55°C) bacterium isolated from a copper leach dump (Brierley 1978) and further characterized in 1977 (Brierley and LeRoux 1977). The first named microbe in this moderately-thermophilic, bacterial group was *Sulfobacillus thermosulfidooxidans* (Golovacheva and Karavaiko 1978). There have since been many more genera and species identified as moderately-thermophilic bacteria (Schippers 2007).

The Archaea

In the mid-1960s (Brierley 1965) a very unusual microorganisms was discovered in an acidic thermal spring in Yellowstone National Park. A study of the isolate demonstrated its ability to oxidize both sulfur and ferrous iron while growing at 70°C and pH 2.0 (Brierley and Brierley 1973). This organism belonged to an unusual group of microorganisms, generally designated as "extremophiles," which were recognized in the late 1970s as uniquely different from bacteria. This group was further defined by Woese (1982) and designated as Archaea. The Archaea have an independent evolutionary history and have many differences in their biochemistry from other forms of life. The discovery and characterization of the Archaea impacted biohydrometallurgy as the Archaea include the extreme thermophiles (growth at 70–80°C), which oxidize iron and sulfur in acid conditions and utilize carbon dioxide as their carbon source—all characteristics of value for biohydrometallurgy. Their ability to bioleach primary sulfide minerals such as chalcopyrite and molybdenite was reported in the 1970s (Brierley and Brierley 1978; Brierley 1974).

Marsh and Norris (1983) discovered an archaean, which they named *Sulfolobus metallicus*, in an exothermic coal spoil. Acid generation in the spoil was associated with this microbe's ability to oxidize pyrite. The significance of this discovery was the demonstration that an archaean can exist in a man-made heap.

With the recent emphasis on heap leaching of low-grade primary ore, principally chalcopyrite and enargite (Roberston et al. 2011; Lee et al. 2011), research and development using the Archaea have accelerated (Du Plessis et al. 2007; Plumb et al. 2007).

A Community of Microorganisms

During the late 1990s and the first decade of the 2000s R&D showed that biooxidation of sulfide minerals was accomplished, not by just a few genera and/or species of bacteria as earlier believed, but rather by a consortium of different genera of bacteria and archaea spanning a range of temperatures and physical conditions. This finding exemplifies the evolutionary process of understanding mineral biooxidation. The challenge now is how to optimize the application of these consortia of microbes operating in hydrometallurgical environments (Rawlings 1997; Norris 2007; Schippers 2007).

Advances in understanding the complexity of microbial communities in biomining processes must be facilitated by developments in modern molecular biology techniques that allow for analyzing complex microbial populations in bioleaching operations. Methods to accomplish this have been described by Chávez et al. (2011). These protocols can lead to a comprehensive definition of the diversity of the microbial populations present in mineral biooxidation and bioleaching operations.

MECHANISM FOR MICROBIAL BIOLEACHING OF SULFIDE MINERALS

Processes by which microbes oxidize sulfide minerals have long been described as "direct" and "indirect" biooxidation. This characterization has resulted in some misunderstandings during the last quarter century. The "indirect" biooxidation mechanism was defined as biochemical oxidation of ferrous iron with chemical oxidation of the sulfide minerals by the ferric iron. The "direct" mechanism was described as some unidentified metabolic product, such as an enzyme, that was responsible for the oxidation of sulfide minerals.

In the last 15+ years the processes that microorganisms employ for sulfide biooxidation have been resolved and better defined (Sand et al. 1995, 1999, 2001; Schippers and Sand 1999). These authors define a mechanism of "contact" biooxidation. Bacteria possess an extracellular-polymeric-substance (EPS) on the cell exterior that facilitates attachment of the bacteria to sulfide mineral surfaces. The EPS is also the location for the oxidation of ferrous to ferric iron. This brings the cyclic process of bacterial oxidation of ferrous iron and chemical ferric iron oxidation of the sulfide mineral in direct contact. The "contact" mechanism is essentially what was previously referred to as the "direct" process of sulfide biooxidation. While many microorganisms are attached to the mineral surface via EPS, there are also a number of microorganisms that are suspended in the leach solution; these are known as planktonic microbes, and they, too, oxidize ferrous iron and soluble sulfur compounds.

APPLICATIONS OF BIOHYDROMETALLURGY

Commercial applications of biohydrometallurgy are divided into two processes: bioleaching and minerals biooxidation. Although bioleaching and minerals biooxidation have principally been applied for copper, nickel, gold and silver, the potential exists for the biohydrometallurgical processing of other metals and metalloids.

Silverman and Ehrlich (1964) described the oxidation and reduction reactions that would be involved in the leaching of many of these constituents. This early publication provides a listing of metal values and reactions that could be applied for commercial benefit.

Commercial application of biohydrometallurgy includes heap bioleaching, which is primarily employed by the copper industry. The microbes have been more recently used for bioleaching nickel. The second successful application is use of stirred-tank reactor systems for biooxidation pretreatment of refractory sulfidic-gold concentrates. In addition a stirred-tank reactor system is also in use for bioleaching cobalt from tailings. This section discusses these applications of biohydrometallurgy.

Bioleaching—Heap Reactors

The role of iron-oxidizing bacteria in solubilization of copper in sulfide ores was first reported in 1954 (Bryner et al. 1954). Industrial interest in bioleaching was demonstrated by a patent issued to what was then Kennecott Copper Corporation (Zimmerley et al. 1958). This is the first known industrial patent for a biohydrometallurgical process. This patent claimed the use of bacteria for a cyclic process of ferric sulfate oxidation of sulfide minerals wherein the ferric sulfate served as the lixiviant for oxidation of the sulfides and the bacteria regenerated the ferric sulfate from the oxidation of reduced ferrous sulfate. The patent recognized the need for aeration of the lixiviant to promote the biooxidation of the ferrous sulfate. The role of bacteria in the Kennecott waste dumps was investigated by Beck (1967). This study focused on *At. ferrooxidans* as the microbial catalyst. Of interest, Beck concluded that copper leaching was a major function of bioleaching, but high temperatures (60–80°C) within the dump facilitated a chemical leach process. At that time there was no knowledge of thermophilic microbes for bioleaching.

Improvement of bioleach heaps, although controversial to some, was brought about by use of active aeration systems installed beneath the ore bed. Microbial processes for bioleaching and biooxidation of sulfide minerals require aeration as a source of oxygen and carbon dioxide. The debate about active (or forced aeration) centers not on the need for air as a source of O_2 and CO_2 but rather on whether natural ventilation is sufficient to supply those components. The first reported use of an active aeration system for a copper bioleach heap was at Girilambone, Australia (Lancaster and Walsh 1997). Subsequently, aeration has been in use at many heap bioleaching operations.

The early developments in heap bioleaching have led to worldwide use of the process for bioleaching copper. It is estimated that 23 copper heap bioleaching plants have been constructed since 1980 with 16 in currently in operation (Anonymous 2010). The majority of these plants are in South America (Table 1) with others occurring throughout the world (Table 2). The tabulated plants are not comprehensive including all possible examples, these serve as examples for the development of bioleaching.

Bioleaching using a heap reactor has been developed and is in operation for recovery of nickel, zinc, cobalt and copper from a black schist (Puhakka et al. 2007; Saari and Riekkola-Vanhanen 2011) at the Talvivaara Sotkamo Mine located in eastern Finland. The internal heap temperatures range from 10–90°C; heating is attributed to rapidly oxidizing pyrrhotite in the ore. A complex

Table 1. Copper heap leaching plants in South America

Plant/Operator/Commissioned	Cathode Copper Production, t/y
Cerro Colorado, Chile/BHP Billiton/1993	115,000
Ivan-Zar, Chile/Compania Minera Milpro/1994	10,000–12,000
Punta del Cobre, Chile/Sociedad Punta del Cobre/1994	7,000–8,000
Quebrada Blanca, Chile/Teck Resources/1994	75,000
Andacollo Cobre, Chile/Teck Resources/1996	21,000
Dos Amiogos, Chile/CEMIN/1996	10,000
Cerro Verde, Peru/Freeport-McMoran/1997	54,200
Zaldivar, Chile/Barrick Gold/1998	150,000
Los Bronces/AngloAmerican plc	50,000
Lomas Bayas, Chile/Xstrata/1998	60,000
Escondida, Chile/BHP Billiton/2006	180,000
Radomiro Tomic/CODELCO/2012	Not reported
Spence, Chile/BHP Billiton/2007	200,000
Tres Valles/Vale SA/2010	18,500

Table 2. Copper heap leaching in other countries

Plant/Operator/Commissioned	Cathode Copper Production, t/y
Skouriotissa Copper, Cyprus/Hellenic Copper/1996	8,000
Monywa, Myanmar/Myanmar Mining Enterprise/1998	40,000
Nifty Copper, Australia/Fadityz Birla/1998	16,000
Morenci, Arizona, USA/Freeport-McMoran/2001	380,000
Jinchuan Copper, China/Zijin Mining Group/2006	10,000
Whim Creek, Australia/Straits Resources/2006	17,000

consortium of bioleaching microorganisms is present in the heap. The reported microbes span the range of operating temperatures with the mesophilic and moderately thermophilic bacteria and archaea, and the extremely thermophilic archaea.

An expansion of the engineered heap concept for biohydrometallurgy was done by Newmont Mining Corporation (Logan et al. 2006). Heaps were used for biooxidation pretreatment of refractory sulfidic gold ores. Heaps were stacked, aerated and following biooxidation removed and processed by milling for gold recovery through cyanidation.

Biooxidation Pretreatment—Tank Reactors

Gencor pioneered research and development for use of stirred-tank reactors for biooxidation of sulfidic-refractory gold concentrates for enhanced recovery of gold (van Aswegen et al. 1991). The first commercial engineered bioreactor system for commercial biooxidation pretreatment of refractory sulfidic-gold concentrate was implemented by Gold Fields in 1986 (Gericke et al. 2009) at the Fairview Mine which remains in operation today in South Africa.

The success of the Gold Fields BIOX® technology is reflected by the number of plants built and those in operation listed in Table 3 (Gericke et al. 2009; BIOX® 2012). In addition to the Gold Fields BIOX® system, other stirred-tank reactor systems have been built and listed in Table 4 (Gericke et al. 2009; Anonymous 2010).

Table 3. Gold Fields BIOX® plants for biooxidation of concentrates

Mine	Country	Concentrate Capacity, tpd	Date of Commissioning	Status
Fairview	South Africa	62	1986	Operating
Sao Bento	Brazil	160	1990	Mine under care and maintenance
Harbour Lights	Australia	40	1991	Mine depleted 1994
Wiluna	Australia	158	1993	Operating
Ashanti	Ghana	960	1994	Operating
Coricancha	Peru	60	1998	Operating
Fosterville	Australia	211	2005	Operating
Suzdai	Kazakhstan	520	2006	Operating
Bogoso	Ghana	820	2007	Operating
Jinfeng	China	790	2007	Operating
Kokpatas	Uzbekistan	2,138	2008	Operating
Agnes	South Africa	20	2010	Demonstration

Table 4. Stirred tank reactor plants for biooxidation of concentrates

Mine	Country	Concentrate Capacity, tpd	Date of Commissioning	Status
Youanmi	Australia	120	1994	Mine depleted 1998
Beaconsfield	Tasmania	70	2000	Operating
Laizhou	China	100	2001	Operating
Olimpiada	Russia	8,220	2003	Operating

The number of tank reactor plants built and those in operation demonstrate the acceptance by industry of biohydrometallurgy. Not only are the numbers of plants increasing but the amount of sulfide concentrate processed is also increasing.

Stirred tank reactor technology has expanded beyond initial use as a pretreatment for refractory sulfidic gold concentrates. A tank reactor plant was developed by BRGM and is in operation at the Kasese Cobalt Company, Uganda, since 1988 for bioleaching cobalt from pyritic tailings (Morin and D'Hugues 2007).

FUTURE DEVELOPMENTS

Prophesying future developments in hydrometallurgy is a challenging and perhaps fruitless task. Current reports do allow for speculation on what the future may bring. There will be other developments not yet foreseen given the evolutionary nature of advancements in biohydrometallurgy.

The use of extremely thermophilic archaea will be dependent upon the mineralogy of target sulfidic ores and concentrates. The pyrite must be in sufficient quantity and amenable to rapid oxidation to supply the heat required for the microbes' growth. A reactive sulfide-sulfur content of 1.58% with 22% oxidation was found sufficient to increase a heap temperature to 81°C (Logan et al. 2007). Research and development for thermophilic bioleaching of copper sulfide ores is in progress. The extremely thermophilic archaea demonstrated effective bioleaching of enargite in column tests using crushed ore (Lee et al. 2011). Extraction of copper was high with a secondary advantage of arsenic deposition as scorodite in the columns operated at 60–65°C. This portends the development of the process for on-site heap reactors. Robertson et al. (2011) reported testing of primary chalcopyrite ores in columns for prediction and simulation of temperature behavior in pyrite-containing heaps. Dew et al. (2011) also present column test methods for simulation of thermophilic bioleaching of chalcopyrite. Their column simulation showed that self-generation of heat from bioleaching of pyrite within the ore could be managed to sustain average temperatures of 70°C in a heap. The microbial population transitioned from mesophilic microbes at the start of a leach cycle (15–20°C) to a population dominated by thermophiles at 70°C.

The complex interaction of thermophilic archaea that leads to the mechanism for bioleaching of primary sulfide minerals chalcopyrite, enargite and covellite (Acar et al. 2005) has yet to be defined. Also, not yet demonstrated for the archaea is the presence of extracellular polymeric compounds that enable these microbes to attach to sulfide surfaces. Understanding the mechanisms by which this group of microbes biooxidize and bioleach sulfide minerals could lead to improvement in their use in commercial operations.

Understanding the complexity of microbial populations in minerals biooxidation and bioleaching processes and knowing how these microbial populations interact with sulfide minerals could lead to the design of appropriate consortia to treat specific ore types. Control of the physical/chemical conditions for bioleaching and biooxidation could be used with a specific microbial consortium to optimize leaching and biooxidation pretreatment. This would require producing a set of select microbes for inoculation of an ore or a concentrate. A process for microbial inoculation of ore for a heap reactor has been described (Brierley and Hill 1993). Other processes for inoculating heaps were recently described and referenced (Gericke 2011).

Continued biohydrometallurgical developments could be problematic. This is attributable to a decline in research capacity and numbers of experienced individuals, especially in North America. The decreasing number of government, industrial and academic research and development institutions that have programs in the bioleaching field is a concern. The ageing population of those involved in the historic development of biohydrometallurgical fundamentals, practice and service signifies a loss of expertise and experience. However, this latter also presents an opportunity for those who are dedicated and eager to contribute significantly to a field that is ripe for basic and applied research discoveries and innovative engineering advances.

ACKNOWLEDGMENTS

The author is greatly appreciative of the suggestions and review by Corale L. Brierley and Francisco Roberto in preparation of this manuscript.

REFERENCES

Anonymous. 2010 Biomining beckons. *Mining Magazine.* March, p. 36.

Acar, S., Brierley, J.A. and Wan Rong Y. 2005. Conditions for bioleaching a covellite-bearing ore. *Hydromet.* 77: 239.

Beck, J.V. The role of bacteria in copper mining operations. *Biotechnol. Bioengineer.* 9: 487.

BIOX®. 2012. The BIOX® Brochure 2012. Gold Fields, South Africa.

Brierley, C.L. 1974. Molybdenite-leaching: use of a high-temperature microbe. *J. Less-Common Metals* 36: 237.

Brierley, J.A. 1978. Thermophilic iron-oxidizing bacteria found in copper leaching dumps. *Appl. Environ. Microbiol.* 36:323.

Brierley, J.A. 2008. A perspective on developments in biohydrometallurgy. *Hydromet.* 94: 2.

Brierley, C.L. and Brierley, J.A. 1973. A chemoautotrophic and thermophilic microorganism isolated from an acid hot spring. *Can. J. Microbiol.* 19: 183.

Brierley, J.A. and Brierley, C.L. 1978. Microbial leaching of copper at ambient and elevated temperatures. In *Metallurgical Applications of Bacterial Leaching and Related Microbiological Phenomena*. Edited by L.E. Murr, A.E. Torma and J.A. Brierley. New York, Academic Press Inc., p. 477.

Brierley, J.A. and Hill, D.L. 1993. Biooxidation Process for Recovery of Gold from Heaps of Low Grade Sulfidic and Carbonaceous Sulfidic Ore Materials. U.S. Patent No. 5,246,486.

Brierley, J.A. and LeRoux, N.W. 1977. A facultative thermophilic Thiobacillus-like bacterium: oxidation of iron and pyrite. In *Conference Bacterial Leaching 1977*. Edited by Schwartz, W. Weinheim, Verlag Chemie, p. 55.

Bryner, L.C., Beck, J.V., Davis, D.B. and Wilson, D.G. 1954. Microorganisms in leaching sulfide minerals. *Ind. Eng. Chem.* 46: 2587.

Chávez, P., Obreque, J., Vera, J., Quiroga, D. and Castro, J. 2011. Microorganisms counting techniques in hydrometallurgy. In *Biohydrometallurgical Processes: A Practical Approach*. Edited by L.G.S. Sobral, D.M. de Oliveira and C.E.G. de Souza. Rio de Janeiro, Brazil, Centre for Mineral Technology, p. 73.

Colmer, A.R., Temple, K.L. and Hinkle, J.E. 1950. An iron-oxidizing bacterium from the acid drainage of some bituminous coal mines. *J. Bacteriol.* 59: 317.

Dew, D.W., Rautenbach, G.F., van Hille, R.P., Davis-Belmar, C.S., Harvey, I.J. and Truelove, J.S. 2011. High temperature heap leaching of chalcopyrite: method of evaluation and process model validation. In *Percolation Leaching: the Status Globally and in Southern Africa*. Johannesburg, SAIMM. p. 201.

Du Plessis, C.A., Batty, J.D. and Dew, D. 2007. Commercial applications of thermophilic bioleaching. In *Biomining*. Edited by D.E. Rawlings and D.B. Johnson. Heidelberg, Germany, Springer-Verlag, p. 57.

Gericke, M. 2011. Review on the role of microbiology in the design and operation of heap bioleaching processes. In *Percolation Leaching: the Status Globally and in Southern Africa*. Johannesburg, SAIMM. p. 165.

Gericke, M., Neale, JW and van Staden PG. 2009. A Mintek perspective of the past 25 years in minerals bioleaching. *J. SAIMM*. 108: 567.

Golovacheva, R.S. and Karavaiko, G.I. 1978. A new genus of thermophilic spore-forming bacteria, *Sulfobacillus*. *Mikrobiologiya* 47: 815.

Kelly, D.P. and Wood, A.P. 2000. Reclassification of some species of *Thiobacillus* to the newly designated genera *Acidithiobacillus* gen. nov., *Halothiobacillus* gen. nov. and *Thermithiobacillus* gen. nov. *International J. Syst. Evol. Microbiol.* 50: 511.

Lancaster, R. and Walsh, D. 1997. The development of the aeration of a copper sulphide ore at Girilambone. In *IBS/BIOMINE 97*, Proceedings of the International Biohydrometallurgy Symposium, Glenside, South Australia, Australian Mineral Foundation, p. M5.4.1.

Lee, J., S. Acar, D.L. Doerr and J.A. Brierley. 2011. Comparative Bioleaching and Mineralogy of Composited sulfide Ores Containing Enargite, Covellite and Chalcocite by Mesophilic and Thermophilic Microorganisms. *Hydromet.* 105: 213.

Logan, T.C., T. Seal and J.A. Brierley. 2006. Whole-Ore Heap Biooxidation of Sulfidic Gold-Bearing Ores. In *Biomining*. Edited by D.E. Rawlings and D.B. Johnson. Heidelberg, Germany, Springer-Verlag, p. 113.

Marsh, R.M. and Norris, P.R., 1983. The isolation of some thermophilic, autotrophic, iron- and sulphur-oxidizing bacteria. *FEMS Microbiol. Letts.* 17: 311.

Morin, D.H.R. and D'Hugues, P. 2007. Bioleaching of a cobalt-containing pyrite in stirred reactors: a case study from laboratory scale to industrial application. In *Biomining*. Edited by D.E. Rawlings and D.B. Johnson. Heidelberg, Germany, Springer-Verlag, p. 35.

Norris, P. 2007. Acidophile diversity in mineral sulfide oxidation. In *Biomining*. Edited by D.E. Rawlings and D.B. Johnson. Heidelberg, Germany, Springer-Verlag, p. 199.

Plumb, J.J., Hawkes, R.B. and Franzmann, P.D. 2007. The microbiology of moderately thermophilic and transiently thermophilic ore heaps. In *Biomining*. Edited by D.E. Rawlings and D.B. Johnson. Heidelberg, Germany, Springer-Verlag, p. 217.

Puhakka, J.A., Kaksonen, A.H. and Riekkola-Vahnen, M. 2007. In *Biomining*. Edited by D.E. Rawlings and D.B. Johnson. Heidelberg, Germany, Springer-Verlag, p. 139.

Rawlings, D.E. 1997. Mesophilic, autotrophic bioleaching bacteria: description, physiology and role. In *Biomining: Theory, Microbes and Industrial Processes*. Edited by D.E. Rawlings. Berlin, Springer-Verlag and Landes Bioscience, p. 229.

Robertson, S.W., van Staden, P.J. and Seyedbagheri, S. 2011. Advances in high temperature heap leaching of refractory copper sulphide ores. In *Percolation Leaching: the Status Globally and in Southern Africa.* Johannesburg, SAIMM. p. 97.

Rossi, G. 1990. *Biohydrometallurgy.* New York, McGraw-Hill Book Company.

Sand, W., Gehrke, T., Hallmann, R. and Schippers, A. 1995. Sulfur chemistry, biofilm, and the (in) direct attack mechanism—a critical evaluation of bacterial leaching. *Appl. Microbiol. Biotechnol.* 43: 961.

Sand, W., Gehrke, T., Jozsa, P-G. and Schippers, A. 1999. Direct versus indirect bioleaching. *Process Metall.* 9: 27.

Sand, W., Gehrke, T., Jozsa, P-G. and Schippers, A. 2001. (Bio)chemistry of bacterial leaching—direct vs. indirect bioleaching. *Hydromet.* 59: 159.

Saari, P. and Riekkola-Vanhanen. 2011. Talvivaara bio-heap leaching process. In *Percolation Leaching: the Status Globally and in Southern Africa.* Johannesburg, SAIMM. p. 53.

Schippers, A. 2007. Microorganisms involved in bioleaching and nucleic acid-based molecular methods for their identification and quantification. In *Microbial Processing of Metal Sulfides.* Edited by E.R. Donati and W. Sand. Springer, The Netherlands. p. 3.

Schippers, A. and Sand, W. 1999. Bacterial leaching of metal sulfides proceeds by two indirect mechanisms via thiosulfate or via polysulfides and sulfur. *Appl. Environ. Microbiol.* 65: 319.

Silverman, M.P. and Ehrlich, H.L. 1964. Microbial formation and degradation of minerals. *Adv. Appl. Microbiol.* 6: 158.

Temple, K.L. and Colmer A.R. 1951. The autotrophic oxidation of iron by a new bacterium: *Thiobacillus ferrooxidans. Jr. Bacteriol.* 62: 605.

Temple, K.L. and Delchamps, E.W. 1953. Autotrophic bacteria and formation of acid in bituminous coal mines. *Appl. Microbiol.* 1: 255.

Woese, C.R. 1982. Archaebacteria and cellular origins: an overview. In *Archaebacteria.* Edited by O. Kandler. *Zbl. Bakt. Hyg., I. Abt. Orig. C.* 3:1.

van Aswegen, P.C., Godfrey, M.W., Miller, D.M. and Haines, A.K. 1991. Developments and innovations in bacterial oxidation of refractory ores. *Min. Metall. Proc.* November, p. 168.

Zimmerley, S.R., Wilson, D.G. and Prater, J.D. 1958. Cyclic leaching process employing iron oxidizing bacteria. U.S. Patent 2,829,964.

ABOUT THE AUTHOR

James A. Brierley has 40 years of experience as a professional microbiologist with a focus on microbe/mineral interactions and process development. Initially he conducted research, and was involved in teaching and department management as professor of biology at the New Mexico Institute of Mining and Technology (School of Mines) for a period of 17 years. For five years he served as research director for Advanced Mineral Technologies developing microbial processes for the metals industries. He then worked for Newmont Mining Corporation as chief research scientist–biohydrometallurgy for a period of 12 years testing and developing use of microbes for processing low-grade sulfidic refractory ores. This work resulted in Newmont's commercialization of a biooxidation-heap pretreatment technology to process low-grade refractory gold ores. Currently he offers consulting services as a principal of Brierley Consultancy LLC. His professional recognitions include Fellow of the American Association for the Advancement of Science, 1985; Recipient of the Colorado State University Honor Alumnus Award, College of Veterinary Medicine & Biomedical Sciences, 2001; Society for Mining, Metallurgy and Exploration Recipient, 2000 Milton E. Wadsworth Extractive Metallurgy Award; and election to the U.S. National Academy of Engineering, 2002.

Uranium Processing Practices, Innovations, and Trends

Henry Schnell
HA Schnell Consulting Inc., British Columbia, Canada

ABSTRACT: Uranium is naturally found in a wide range of concentrations and consequently mining tonnages and grades vary greatly. Uranium production methods have adapted to this wide range of grades, evolving from conventional operations to in-situ leach recovery and agglomerated heap leach.

This paper describes the current practices and trends from a variety of operations worldwide with a review of the individual unit operations required to produce a uranium concentrate.

INTRODUCTION

Uranium is a relatively common element in the crust of the earth (very much more than in the mantle). It is a metal approximately as common as tin or zinc, and it is a constituent of most rocks and even of the sea. Some typical concentrations are as shown in Table 1.

Historically, uranium production has been dominated by underground or open pit mining and conventional milling. However, in-situ leaching (ISL or ISR) of uranium began as early as 1961 in the Ukraine and has today made ISL, at over 40% of uranium production, the cornerstone of the uranium production, most notably in the USA, Kazakhstan, and Uzbekistan. In addition, several other countries have either begun ISL operations or are assessing the amenability of their uranium resources to ISL extraction (WNA 2011).

Although uranium recovery is moving away from the "conventional" type of a flowsheet to ISL and heap leach, many of the unit operations still apply and this paper will look at the trends within these basic unit operations as they apply.

URANIUM PROCESSING OVERVIEW

Historically flowsheets based on a tank leach process tended to be very similar and are referred to as "conventional." There is a trending away from such a conventional flowsheet or the technology within each unit operation, as high grade deposits are exploited in Canada or as extreme low grades are considered in places such as Namibia. Today we see a much wider range of unit operation adaptations. The basic flowsheet in Figure 1 is provided as a reference for the basic uranium processes, including in-situ recovery.

URANIUM PROCESSING UNIT OPERATIONS

Comminution

Uranium recovery is a hydro-metallurgical process using either acid or carbonate to leach the uranium from ore. The front end of the flowsheet, comminution is generally limited to exposing the uranium mineral within the ore matrix.

Table 1. Common uranium concentrations (ppm = parts per million)

Very high-grade ore (Canada)—20% U	200,000 ppm U
High-grade ore—2% U	20,000 ppm U
Low-grade ore—0.1% U	1,000 ppm U
Very low-grade ore (Namibia)—0.01% U	100 ppm U
Granite	4–5 ppm U
Sedimentary rock	2 ppm U
Earth's continental crust (average)	2.8 ppm U
Seawater	0.003 ppm U

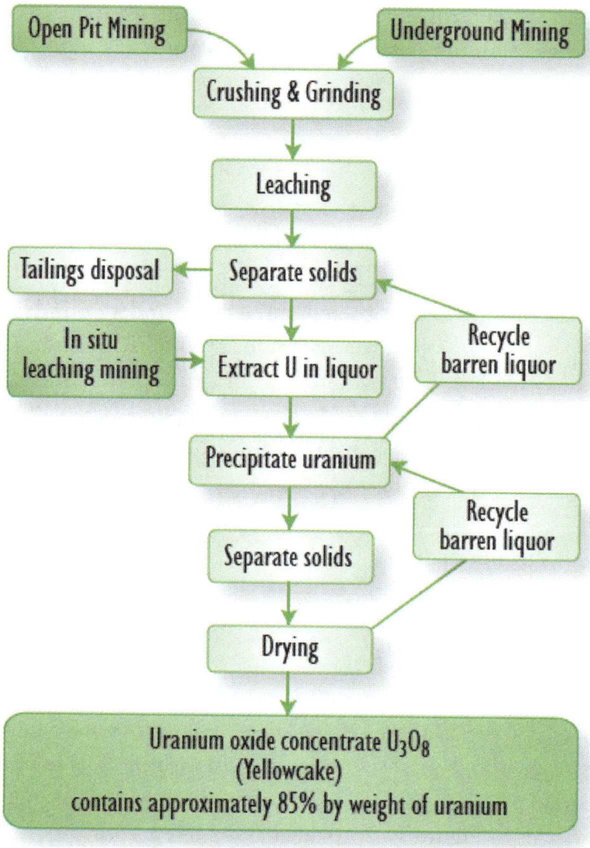

Figure 1. General unit processes for uranium production (Lunt et al. 2007)

Traditionally, ore was first crushed in jaw or gyratory crushers. Gyratory crusher installations are no longer the norm due to costs and the fact that uranium tonnages for high grade ores are small and many low grade ores tend to be within softer rock matrixes. Jaw crushers are still applied, as per the Langer Heinrich operation, but there is a trend towards the use of mineral sizers for the low grade operations such as Trekkopje in Nambia or Kaylekera in Malawi.

An interesting alternative are the mining methods used for the very high grade ores in Canada. McArthur River uses raise bore mining, eliminating the need for crushing and Cigar Lake will use water jet boring to recover the ore directly into a coarse slurry form.

Grinding for high grade deposits generally use smaller autogenous mills, as in the case of McClean Lake in Canada. Most interesting are the grinding circuits at the McArthur River and the planned Cigar Lake operations in Canada, where small autogenous mills have been placed underground to avoid high grade ore hoisting. In these cases the ore is transported to surface with positive displacement pumps. Both these operations then transport the ground slurry by truck in special dual containment containers to the mills over 80 km away.

Most grinding circuits use closed circuit operations with cyclones. Cyclones are preferred so that the uranium minerals with a specific gravity higher than the associated gangue will tend to be ground finer, improving leaching.

Pre-Concentration

There have been renewed attempts to use radiometric ore sorting to reduce treatment tonnages. Several pilot ore sorters have been installed; at the Ranger operation in Australia (ERA 2008) and at Rossing in Namibia (Rossing 2007). However, neither of these operations have installed commercial ore sorters.

An interesting pre-concentration step is operating at the Langer Heinrich mine in Namibia where the uranium as carnotite is hosted in gypcrete ore (Paladin 2012). This operation does an attrition grind followed by sizing separation of the various ore fractions to reduce ore tonnage.

Gravity separation has been applied to uranium ores in the past with some success, but this was associated with radiation issues (Schnell 2010). This is currently not used with one exception where a Knelson concentrator has been installed in the ore receiving circuit of Key Lake to help reduce cement carry over coming from the McArthur River underground mine backfill (Cameco 2009).

Leaching

Uranium leaching is most commonly carried out with the use of sulfuric acid and an oxidant. There is a trend towards the use of oxygen due to the development of low cost oxygen production using vapor swing technology. The McClean lake operation has installed 40 tonnes of oxygen production capacity and Key Lake is currently also commissioning a new vapor swing oxygen plant (Cameco 2012). In the case of the McClean lake operation, hydrogen peroxide was used for the Sue ores, but for the Cigar Lake ore to be processed at McClean mill, oxygen will be the primary oxidant, but supplemented with hydrogen peroxide.

There is a general move away from the use of chlorates due to environmental and corrosion issues. Pyrolusite still is considered where it is economically available, but often the ores need good iron content or ferric sulfate must be added.

Carbonate leaching has re-emerged with the start-up of the Langer Heinrich operation in Namibia and Trekkopje will be the first application using carbonate solutions in a heap leach operation (Schnell 2010).

Leaching equipment ranges from pressure leach (Key Lake and McClean Lake) to agitated tank leach (Langer Heinrich, Rossing, etc.) to a return to pachuca leach (Cameco 2002).

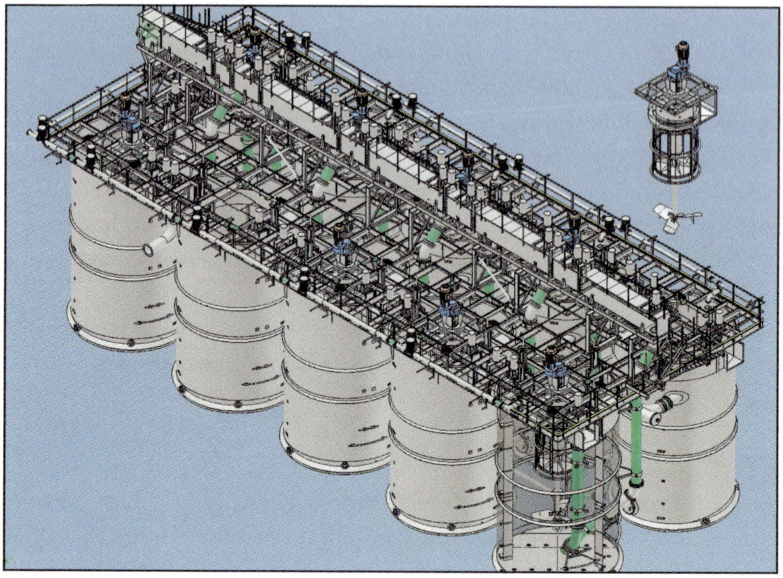

Figure 2. Schematic of resin-in-pulp at Kayelekera (Source: KEMIX 2012)

Solid-Liquid Separation

Filtration and Counter-Current Decantation (CCD) are the primary methods used to separate the uranium bearing solutions from the leached ore. Horizontal belt filters are still used in the Niger operations and are being considered for some projects. CCD also is considered were applicable, such as in the expansion projects at Langer Heinrich in Namibia. A notable development for CCD is the use of high rate thickeners at the McClean Lake operation (Schnell 2000).

Interesting is the trend towards the use of resin-in-pulp (RIP) associated with the development of more durable resins and improved equipment. Most notable is the Kayelekera operation in Malawi is using RIP with the application of the AAC Pumpcell as shown in **Figure 2** (KEMIX 2012).

Solution Clarification

The leach solutions will commonly contain fine solids that need to be clarified prior to solution purification. Conventional clarifiers and sand filtration are the norm. Some continuous sand filtration types have shown poor availability, but the use of pinned bed clarifiers is being considered for some heap leach projects which have high PLS flows (Baxter 2010).

Solution Purification

Various other metals are dissolved during leaching, requiring purification of the uranium bearing solutions. The purity requirements of the final uranium precipitate is becoming stricter and most uranium conversion plants require a precipitate quality better than the ASTM standard. Most commonly molybdenum and vanadium are associated with uranium, but care must be taken to also consider other metals, most notably zirconium, arsenic, copper and nickel.

Solvent extraction is the norm for uranium, but little progress has been made for improved extractants and modifiers. The tertiary amines are most predominant, but there is some minor usage of DEPHA. Solvent extraction circuit equipment is usually consists of conventional mixer-settlers or Krebs mixer-settlers for the lower solution flows in the Canadian operations of McClean Lake and Key Lake. At Olympic Dam Bateman pulse columns have been applied (Grinbaum 2006). Future projects will also consider Vertical Smooth Flow technology (VSF) (Paatero 2010).

Uranium stripping in solvent extraction is carried out with ammonium sulfate (Key Lake, McClean Lake), sodium carbonate (Somair), or strong acid (Rabbit Lake). One interesting alternative exists where the uranium is precipitated directly in the organic extractant.

Molybdenum, vanadium or other metals such as zirconium values may still be high after solvent extraction or ion exchange. The two main options for molybdenum removal are solvent extraction with a LIX extractant, as used in the past at Key Lake in Canada or the use of activated carbon absorption as used at the McClean Lake operation in Canada. Zirconium removal by precipitation as practiced at the Somair operation in Niger. Also in Niger, vanadium removal is carried out within the solvent extraction circuit at the Cominak operation.

Normally IX is chosen for uranium solution grades at the less than 0.5 g/L range; however slightly lower solution grades will at times use the SX option. Ion exchange (IX), as mentioned has progressed both in the resin quality and equipment design.

The recovery of uranium from lower grade deposits and in-situ leach operations produces lower grade solutions making IX the preferred choice. In the case of IX, there may also be a need for additional metal contaminant removal, which usually adds a small SX circuit, this is known as an Eluex. The most notable example of an Eluex circuit is the Rossing operation in Namibia (ERA 2008).

There are three basic technologies used for uranium ion-exchange, fixed bed columns, pinned bed columns and fluid bed columns (NIMCIX). Fixed bed columns have been applied in the Langer Heinrich operation, with pinned bed columns predominant in the Kazakhstan in-situ leach operations. NIMCIX columns have been installed at Trekkopje (Figure 3) and at the Langer Heinrich operations in Namibia (Paladin 2012). NIMCIX offers the advantage of handling high suspended solids, making them preferred for heap leach of poor settling solids.

Figure 3. Trekkopje NIMCIX columns (Source: ENVIRO 2012)

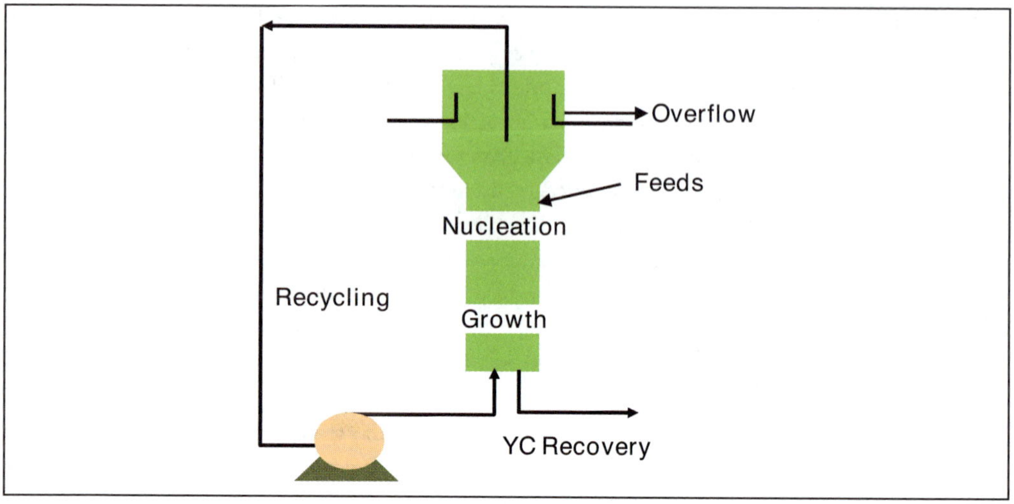

Figure 4. Schematic of fluid bed precipitator (Source: Courtaud 2011)

Precipitation

Uranium precipitation has typically used a series of agitated tanks using magnesium oxide (MgO), sodium hydroxide (NaOH), ammonium hydroxide (NH_3OH) or hydrogen peroxide (H_2O_2). The main trend is to move away from alkali metal precipitation towards use of hydrogen peroxide or ammonia. Most new projects are using hydrogen peroxide, such as Langer Heinrich, Imouraren, etc. In Canada at Key Lake and McClean, ammonia is used but with the additional complication of an ammonium sulfate crystallization plant to remove ammonia from process water to avoid ammonia releases into the environment.

A two stage precipitation circuit is used in the case of alkaline leach operations such as Langer Heinrich or Trekkopje. These operations use first stage sodium hydroxide precipitation, followed by re-dissolution of the sodium diuranate and final hydrogen peroxide precipitation.

A recent equipment development is a fluid bed precipitator (Courtaud 2011) that reduces impurity occlusion within the precipitate as shown in Figure 4. This technology has been used in the Somair operation and is included in the Trekkopje and Imouraren projects.

Drying and Packaging

The final moisture removal and drying of the uranium precipitate varies among operations. There is a process preference to dewater the precipitate with a horizontal belt filter, as is the case at Somair and Trekkopje. This allows for better washing of the precipitate. However, since belt filters are at a higher cost, centrifuge dewatering is commonly used as is the case at the operation of Key Lake, McClean Lake and Langer Heinrich.

Final product drying and/or calcining equipment ranges from multi-hearth furnaces (Key Lake and McClean Lake) to rotary tube dryers and calciners (Kazakhstan), to Hollow-Flyte driers (Rabbit lake and Langer Heinrich) to spray drier (Cominak). The choice of calcining or drying to a lower temperature is dependent upon economics and destination of the final product.

In all cases the final precipitate is packed into standard drums for shipping. Packaging equipment and installations continue to be upgraded to reduce personnel exposures. These packaging

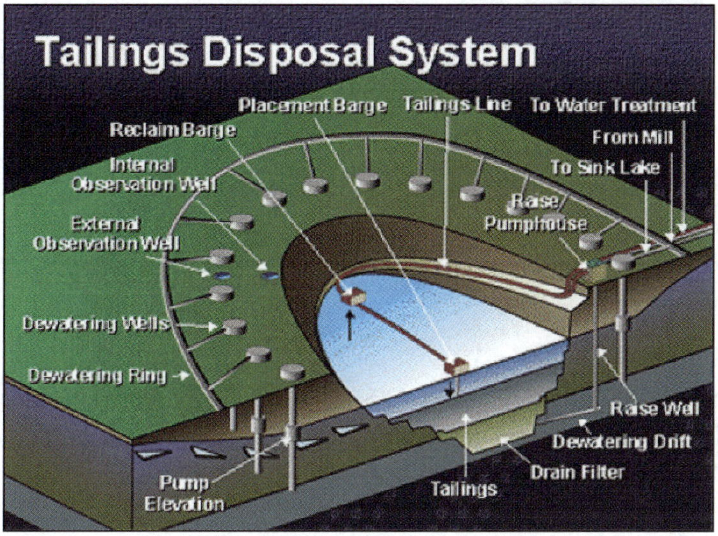

Figure 5. Schematic of in-pit tailings disposal system (Source: Infomine)

facilities are generally in a separate enclosed space with remote operation and even the drum lids are sometimes added with glove box type of equipment.

Tailings

The characteristics of tailings impoundments have undergone many changes in recent decades. Historically, tailings were generally deposited in above ground dam impoundments or in natural ground low points, with minimal treatment. In most cases, tailings are now impounded in purpose-built lined cells, placed in a mined out pit, or sent to an engineered facility. Modern mines have tailings neutralization systems that use lime, together with other additives such as barium chloride, to stabilize radium content and prevent metal contaminants from creating an environmental issue.

The purpose-built lined pit or system of tailings cells has been adopted as the current practice in the United States. This is combined with a final cover to stabilize the tailings and prevent future contamination. For acid leach plants, all tailings need to be neutralized before disposal.

An alternative to the tailings cell design, as used in Canada, is to use in-pit disposal, where the tailings are placed in a designed open pit that allows the tailings to become less permeable than the surrounding rock, and a 'French' drain prevents ground water from entering the tailings mass (Figure 5). The tailings are placed sub-aqueous to prevent dust and to protect workers from potential radiation exposure. For final closure, the tailings mass is required to be below the surrounding ground level, and an engineered cover is installed to prevent contamination and to stabilize the area.

HEAP LEACH

Heap leach, based on the crushed agglomerated ore technology originating in Chile for copper is emerging for uranium processing. It should be noted that the recent emergence of heap leach for uranium ores is very different from historical uranium heap operations now referred to as "Dump Leach." This adaptation from copper with acid was pioneered by the Caetite operation in Brazil, but has now been adopted to supplement production at Somair in Niger and will be used at the future Imourarn project in Niger.

A novel adaptation of this technology is the application of alkaline heap leach at the Trekkopje project in Namibia (Schnell 2010). At Trekkopje the ore contains chlorides and therefore the process has added a pre-leach ore wash cycle which is then followed by a three or four stage counter-current alkaline heap leach using sodium carbonate/bi-carbonate solutions. Solution recovery uses the NIMCIX uranium recovery technology.

Many other projects currently are considering the heap leach option and it will be interesting to see this technology evolve.

BY-PRODUCT RECOVERY

By-product recovery of uranium has previously been carried out from phosphoric acid production. The future will see the re-emergence of this uranium source depending on environmental and economic conditions.

Uranium has been produced from the gold operations at AngloGold Ashanti for many years, and many South African gold operations and gold tailings recovery projects continue to review the possibilities of uranium recovery.

Olympic Dam is the largest resource of uranium and continues to be a significant co-product uranium producer, combined with copper and precious metal production. More recently the Talvivaara project is constructing a uranium recovery plant as part of its multi-metal mining operation (Paatero 2010).

As simple uranium ore resources become scarce, we must learn to produce uranium as a by-product, or produce by-products from uranium resources. These projects are emerging and by-product uranium recovery will become more significant from phosphoric acid, rare earth element recovery and multi-metal projects.

CONCLUSION

The most significant trend in uranium processing is the move toward in-situ leach, with agglomerated crushed ore heap leach now also gaining momentum. Uranium unit processes continue to advance and adapt the technology and equipment from other metal processing facilities.

With good easy to process uranium resources diminishing, production will increase from uranium as a by-product or metal by-products from uranium processing.

The requirements of the final uranium precipitate are increasingly more stringent for acceptance by a converter. Further process improvements will need to continue due to environmental considerations, worker health and safety. These challenges will require improved reagents and equipment combined with more stringent operations and management practices.

REFERENCES

Ausenco 2010. *Imouraren Uranium Heap Leach Pad & Tailings Project*, Ausenco.com.

Baxter et al., 2010. *Design and Operation of a Pinned Bed Clarifier in Solvent Extraction*—Bateman.

Cameco 2009. *Annual Report 2009.*

Cameco 2012. *Quarterly Report 2012—Revitalization Plan.*

Cameco 2002. *Uranium Recycle Project Proposal, November 2002.*

Courtaud B. et al., 2011–Method for Preparing Uranium Concentrates by Fluidized Bed Precipitation, US Patent Application, Sept 1, 2011.

Durupt, N. 2009. *Heap Leaching Of Low Grade Uranium Ores At Somair*, ALTA Conference.
ENVIRO 2012. environment & société magazine—Oct 18, 2012.
ERA 2008. *Sustainable Development Report.*
Gomero et al. *Evaluation of a New Milling Process for the Caetite-Brazil Uranium Ore*, Uranium 2010 conference, Saskatoon.
Grinbaum et al. 2006. *Comparison of SX ofUranium Using Mixer Settler and Columns*, Bateman.
IAEA 1993. *Uranium Extraction Technology*, Technical Publication 359.
Infomine 2010. *Malawi RIP Technology*. http://www.mining-technology.com/projects/kayelekerauraniummin/.
KEMIX 2012. *AAC Pumpcell*–KEMIX Corporate Brochure.
Lunt et al., 2007. *Uranium Extraction: the key process drivers*, The Journal of The Southern African Institute of Mining and Metallurgy, #107.
Murray, R. 2007. *Desert Riches—Langer Heinrich*, Mining Magazine.
National Research Council, Academies of Science, 2011. *Uranium Mining in Virginia: Scientific, Technical, Environmental, Human Health and Safety, and Regulatory Aspects of Uranium Mining and Processing in Virginia, Committee on Uranium Mining in Virginia.*
Paatero E. et al., 2010–*Recovery of Molybdenum and Uranium as By-products with Solvent Extraction*, Metals and Materials Recycling Workshop, Santiago.
Paladin 2012. *Financial Report*, Feb 14, 2012.
Public Information Document 2007. *Rössing Uranium Mine Expansion Project–Social and Environmental Impact Assessment.*
ROYMEC 2012. http://www.mining-technology.com/contractors/.
Schnell, H. 2009. *Uranium from unconventional sources. Presentation from the IAEA Technical Meeting on Uranium from Unconventional Resources, Vienna, Austria, November 4–6.*
Schnell, H. 2010. *Cluff Lake Mill Operations—a historical review*, Uranium 2000 Conference, Saskatoon.
Schnell, H. 2010. *The Trekkopje Project*–IAEA Workshop 2010.
Schnell, H. 2000. *Processing of Cigar Lake Ore in the McClean Lake Mill*, Uranium 2000 Conference, Saskatoon.
WNA, 2010. *Uranium: From Mine to Mill.*

ABOUT THE AUTHOR

Henry Schnell has more than 40 years of experience in management, plant operations, plant design, engineering, and research and development in mining and ore treatment that includes several gold projects, the Quebrada Blanca Bio-leach copper operation in Chile, and over 20 years of this experience specializing in uranium metallurgy, operations, and mining projects. Schnell has recently retired from the position of technical authority (senior expert) in the Expertise & Technical Department, Mining Business Unit, with AREVA NC Inc. In his role as technical authority for uranium and gold, he was responsible for review and support of existing operations and new projects worldwide, and for final technical authorization of plant design and modifications.

Innovations in Hydrometallurgy

Corby G. Anderson
Kroll Institute for Extractive Metallurgy, Colorado School of Mines, Golden, Colorado, United States

ABSTRACT: This paper covers the history and implementation of key technologies in the relatively new but emerging field of hydrometallurgy. Breakthrough leaching, purification, separation, concentration and recovery unit operations are considered that the author humbly deems most significant and best illustrative of primary enabling hydrometallurgical innovations. Comparisons and contrasts to pyrometallurgy are also drawn. In all cases, the key industrial implementation aspect of inherent superior hydrometallurgical separations, oftentimes from both low grade materials and dilute solutions as an enabling technical characteristic is elucidated. This is the basis of hydrometallurgy's lasting and now growing application and development. As well, the growing global perception that hydrometallurgy is an environmentally favorable technology has hastened its development, adoption and application.

INTRODUCTION

Hydrometallurgy is a relatively new discipline with recorded evidence of copper production via cementation at least four hundred years ago. However, its significant industrial emergence coincided with the Bayer process and the Cyanidation process both about occurring 1887. Since that time significant application and innovation has occurred to allow a wide range of function applications in this growing field. Table 1 illustrates a timeline of some significant developments in this field (Gupta and Mukherjee 1990).

Habashi (Habashi 2005) further classifies historical hydrometallurgical developments by separate eras of advancement. These are the Early Period, the Modern Era and World War II, and Recent Advances. These are shown in Tables 2, 3, and 4 (Habashi 2005).

In most cases of industrial application, it is the distinct defining separations advantage that has propelled hydrometallurgy to the forefront. Recently, the increasing global perception of less environmental impacts has also propelled the adoption of hydrometallurgical technologies. For reference, a comparison of the merits of hydrometallurgy with pyrometallurgy is listed in Table 5 (Gupta and Mukherjee 1990).

Hence the rest of this paper focuses on the history as well as the author-chosen key leaching, separation, concentration and recovery innovations that have now founded this technology worldwide.

THE BEGINNING OF HYDROMETALLURGY: INTRODUCTION OF THE BAYER ALUMINA AND CYANIDATION PROCESSES

Evidence of hydrometallurgical recovery of copper by cementation goes back at least 400 years to Rio Tinto Spain. Others have noted that copper cementation may have been practiced in China

Table 1. Landmarks in the history of hydrometallurgy

1700	Heap leaching of copper from weathered pyrite at Rio Tinto, Spain. Copper precipitated from solution with iron.
1900	Bayer process for aluminum. Pure alumina from caustic soda leach of bauxite followed by crystallization.
1900	Cyanide leaching of gold and silver ores. Precipitation of gold with zinc powder.
1920	Electrolytic zinc process. Sulfide ore roasted then leached in H_2SO_4 to produce electrolyte.
1926	Large-scale (6,800 t/d of ore) H_2SO_4 vat leaching of copper ore with electrowinning (Inspiration Consolidated Copper Co., United States).
1940	Caron process for nickel from oxide ores. Selective reduction followed by ammoniacal leach.
1953	Uranium recovery from low-grade ore. Acid leaching-ion exchange.
1954	Sherritt Gordon process for nickel from sulfide concentrates. Ammoniacal pressure leach followed by pressure hydrogen reduction to produce nickel powder (Canada).
1956	Introduction of amine solvent extraction for uranium recovery from leach liquor.
1956	J.J. Aman, British Patent 793700 on the high temperature spray hydrolysis of metal chloride solutions to recovery of hydrochloric acid.
1958	Pressure sulfuric acid leach of oxide nickel ore, Moa Bay, Cuba.
1960	Expansion of copper production by bacteria leaching of low-grade ore dumps in United States.
1964	First application of the Jarosite process for iron removed from zinc sulfate leach liquors (Asturiana de Zinc, Spain).
1968	Dump leaching of oxide copper ore with H_2SO_4 followed by solvent extraction and electrowinning (Ranchers: Exploration & Development Co. Ltd. copper production, 18,200 kg/d).
1968	Falconbridge matte leach process for copper, nickel, cobalt, HCl leach, and processing of chloride liquors by solvent extraction.
1974	Pachucal each of oxide copper ore with H_2SO_4 followed by solvent extraction and electrowinning (Nchanga Consolidated Copper Mines Ltd., Zambia); copper production, 182,800 kg/d.
1975	Arbiter process-ammoniacal leach of copper sulfide concentrate followed by solvent extraction and electrowinning; plant at Montawa closed 1979.
1977	Successful piloting of direct pressure leaching of zinc concentrates by Cominco, Canada.
1981	Commercialization of Cominco's Trail operation for the processing of zinc concentrate.

Table 2. Early period of hydrometallurgical development

7th century	The apparent transmutation of iron into copper by alchemists ($Cu^2 + Fe^- \rightarrow Cu + Fe^2$)
8th century	The discovery of aqua regia by the Arab alchemist Jabir Ibn Hayyan (720–813 AD). This was the only known solvent for gold. Still used today in gold refining.
16th century	Heap leaching of copper-containing pyrite in the Harz mountains in Germany and in Rio Tinto in Spain, and the precipitation of copper from the solutions by iron.
18th century	Production of potash for soap and glass industries by leaching ashes left after burning wood, e.g., in Quebec, where the clearing of forests was in progress on a large scale.

well before the birth of Christ. But, it was almost the simultaneous invention and application of the Bayer process for alumina and the cyanidation process for gold that propelled hydrometallurgy to a universal industrial status (Habashi 2005). The year 1987 also marked the centennial of the beginning of modern hydrometallurgy. Karl Joseph Bayer was the inventor of the Bayer alumina process while John Stewart MacArthur along with the Forrest brothers were the inventors of cyanidation.

Commercial production of smelter-grade Al_2O_3 from bauxite ores by caustic leaching is popularly known as the Bayer Process, named after its inventor Karl Joseph Bayer. The process, since introduced over a century ago, has continued to remain as the principal process practiced by all

Table 3. Modern era of hydrometallurgical development and during Wordl War II

1887	The invention of the cyanidation process, i.e., dissolution of gold from ores by a dilute sodium cyanide solution and the precipitation of gold from the solutions by zinc. The invention of Bayer's process: precipitation of crystalline $Al(OH)_3$ from sodium aluminate solution by seeding, followed in 1892 by his invention of pressure leaching of bauxite by NaOH solution.
1912	Recovery of copper from leach solution in Chile by electrolysis.
1916	The use of ammonium hydroxide for leaching native copper ore in Lake Superior District, and for malachite-azurite ore in Alaska. Development of the hydrometallurgical electrowinning zinc process at Trail and Anaconda. The recovery of cadmium as a by-product of the zinc hydrometallurgical process.
1924	Caron process for ammonia leaching of metallic nickel produced by reduction of laterites.
1927	Henglein process for pressure leaching of ZnS at high temperature in the presence of oxygen .
1930	Sullivan process for ambient leaching of copper sulfides by ferric chloride solution.
1940s	Development of the uranium technology in connection with the U.S. Manhattan Project aimed at producing an atomic bomb. Introduction of sodium carbonate as a leaching agent for uranium, the widespread use of ion exchange and solvent extraction for uranium recovery, and the separation of the lanthanides by ion exchange.

Table 4. Recent advances in hydrometallurgical development

1950s	The application of pressure hydrometallurgy for leaching nickel sulfide ores and the precipitation of pure nickel from solution by hydrogen under pressure.
1960s	Discovery of the role played by microorganisms in leaching processes and the widespread use of heap and in situ leaching for extracting copper from low-grade material. The application of pressure leaching to a variety of raw materials, e.g., laterites, tungsten ores, uranium ores. The application of solvent extraction for copper.
1970s	Discovery of galvanic action in leaching sulfide minerals. Recovery of traces of uranium from waste leach solutions after copper precipitation with scrap iron. Pressure leaching of zinc sulfide concentrate in dilute H_2SO_4 at Trail and Timmins in Canada.
1980s	The hydrometallurgy of gold greatly advanced: widespread application of activated carbon technology, and aqueous oxidation of gold refractory ores.

the aluminum-producing countries in the world. On August 3, 1888, German Patent No. 43977 entitled "A Process for the Production of Aluminum Hydroxide" was issued. The discovery which led to the patent was made by the Austrian chemist Karl Josef Bayer (1847–1904) (Figure 1) who was at that time in Russia, and the process became known as the Bayer Process in his honor. The process immediately achieved industrial success, displacing the pyrometallurgical process that had been used until that time to produce alumina. Coincidentally and fortunately, the advent of the Bayer process coincided with the Hall–Héroult process introduction in which effectively and efficiently reduced pure alumina to aluminum metal.

The Hall–Héroult process was invented independently and almost simultaneously in 1886 by the American chemist Charles Martin Hall and the Frenchman Paul Héroult (Figure 2). In 1888, Hall opened the first large-scale aluminum production plant in Pittsburgh. It later became the Alcoa corporation.

Aluminum cannot be produced by the electrolysis of an aluminum salt dissolved in water because of the high reactivity of aluminum with the protons of water and the subsequent formation of hydrogen. As, in an aqueous solution, protons (H^+) are preferentially reduced to atomic

Table 5. Key characteristics of hydrometallurgical and pyrometallurgical processes

	Description	
Terms of Reference	**Pyrometallurgy**	**Hydrometallurgy**
Treatment of high-grade ores	More economical	Less economical
Treatment of low-grade ores	Unsuitable because large amounts of energy are required to melt associated gangues	Suitable provided a selective leachant can be used
Treatment of complex ores	Unsuitable because separation is difficult	Possesses the flexibility for treatment of complex ores and for production of a variety of by-products
Process economics	Best suited for large-scale operations requiring a large capital investment	Can be used for small-scale operations; it has a lower capital investment sensitivity to plant size in lower plant capacity ranges
Treatment of secondary resources	Unsuitable in most cases	Suitable
Separation of chemically similar metals	Not possible	Possible
Treatment of sulfide ores	Sulfur dioxide is generated which in high concentrations must be converted to sulfuric acid and in low concentrations must be disposed of in other ways. For the former, a market has to be found, and for the latter, methods are available but are expensive.	Can be treated without generating sulfur dioxide; the need to make and market sulfuric acid is eliminated; possibility exists to recover sulfur in elemental form
Reaction rate	Rapid as high temperatures are involved	Generally slow as all operations are carried out at relatively low temps
Throughput	Very high unit throughput	Small throughput for production unit
Materials handling	Handling of molten metals, slags, mattes, etc., appears problematic	Solutions and slurries can be easily transferred in closed pipe lines
Environmental pollution	Problems of waste gas and noise; many processes emit large amounts of dusts which must be recovered to abate pollution or because the dust itself contains valuable metals	No atmospheric pollution problems. Problems of waste disposal, problems of wastewater; no dust problem because materials handled usually are wet
Solid residues	Many residues, such as slags, are coarse and harmless, so that they can be stored in exposed piles without danger of dissolution; piles may be unsightly and esthetically unacceptable	Most solid residues are in finely divided states: in their dried conditions create dust problems, in their wet conditions release metal ions in solutions; this may contaminate the environment
Toxic gases	Many processes generate toxic gases; expensive systems are required to combat their nonrelease to the atmosphere	Many processes do not generate gases, and if they do, reactors can be made gas-tight easily
Process feed	Suitability for inhomogeneous feed	Sensitivity to major variations instead; the process feed, in general, has to be more uniform and more attention to composition and mineralogy is required
Operational feature	Represents uniquely metallurgical plant features; engineering is not considered complex	In general, more sophisticated control schemes are required to maintain satisfactory operation; plants operate more like chemical plants with control philosophy associated with chemical plants; plant engineering is more complex

Figure 1. Founders of modern hydrometallurgy: (left) Karl Joseph Bayer and (right) John Stewart MacArthur (Habashi 1993a)

Paul-Louis-Toussaint Héroult
(April 10, 1863–May 9 1914)
Patent: April 23rd, 1886

Charles Martin Hall
(Dec. 6, 1863–Dec. 27, 1914)
Patent: July 9th, 1886

Figure 2. Héroult and Hall, co-inventors of the alumina reduction process (Entner 2013)

hydrogen before Al^{3+} ions, the reduction of Al^{3+} is done by electrolysis of a molten aluminum salt. This is a water free medium, and hence, H^+ reduction is avoided.

In the Hall–Héroult process, alumina, Al_2O_3, is dissolved in an industrial carbon-lined vat of molten cryolite, Na_3AlF_6 (sodium hexafluoroaluminate), called a "cell." Aluminum oxide has a melting point of over 2,000°C (3,630°F) while pure cryolite has a melting point of 1,012°C (1,854°F). With a small percentage of alumina dissolved in it, cryolite has a melting point of about 1,000°C (1,830°F). Some aluminum fluoride, AlF_3 is also added into the process to reduce the melting point of the cryolite-alumina mixture.

The molten mixture of cryolite, alumina, aluminum fluoride is then electrolyzed by passing a direct electric current through it. The electrochemical reaction causes liquid aluminum metal to be deposited at the cathode as a precipitate, while the oxygen from the alumina combines with carbon from the anode to produce carbon dioxide, CO_2. An electric potential of three to five volts is needed to drive the reaction, and the rate of production is proportional to the electric current. An industrial-scale smelter typically consumes hundreds of thousands of amperes for each cell. Figure 3 illustrates this innovative recovery and reduction methodology.

The Bayer Process involved the pressure leaching of bauxite with NaOH solution to obtain sodium aluminate solution from which aluminum hydroxide was precipitated by seeding. The Bayer process steps are (1) digestion of dried bauxite ore with NaOH solution under enhanced pressure and temperature conditions to dissolve aluminum as sodium aluminate, (2) separation of the aluminum-laden solution from the residue (it is carried out by running the liquor through thickeners and filter presses for removing suspended matter), and (3) precipitation of aluminum hydroxide, which is carried out by introducing the clear liquid, obtained after filtration from the previous step, into precipitation tanks made of sheet-iron cylinders provided with agitators. A small amount of freshly prepared $Al(OH)_3$ is added to the tank to act as seeds, and the whole is then agitated. Precipitation of $Al(OH)_3$, takes place over a period of time that may run up to 60 h. About 70% of the aluminum hydroxide precipitates in 36 h and the remainder on further standing. The precipitated $Al(OH)_3$,settles to the bottom of the tank, where it is drawn off, washed, and filtered before calcination to alumina. The NaOH liquor from the tank is evaporated and concentrated for reuse in the treatment of raw bauxite. The flowsheet of the Bayer process depicting these essential steps is shown in Figure 4. The residue remaining after alkali digestion contains iron oxide, titania, silica, and other impurities from bauxite as well as some alumina not dissolved. This residue is called redmud and is discarded. It is normally washed to recover the sodium hydroxide, and it has found a limited use in the purification of artificial gas. In Germany, red mud is being used in the manufacture of iron oxide pigment and has been used for painting freight cars.

Aluminum is present in bauxite ores either as one or a combination of three hydroxides, namely, gibbsite, boehmite, and diaspore. Gibbsite, chemically represented by the formula $Al(OH)_3$, exhibits a monoclinic structure and dissolves readily even in relatively dilute NaOH solution as given in the following reaction:

$$Al(OH)_3 + NaOH \rightarrow NaAl(OH)_3 \tag{1}$$

The other two hydroxides, boehmite and diaspore, possessing the same chemical formula (AlO:OH) and crystal structure (orthorhombic) dissolve significantly only in hot, concentrated NaOH solution according to the following reaction:

$$AlO{:}OH + NaOH + H_2O \rightarrow Na[Al(OH_4)] \tag{2}$$

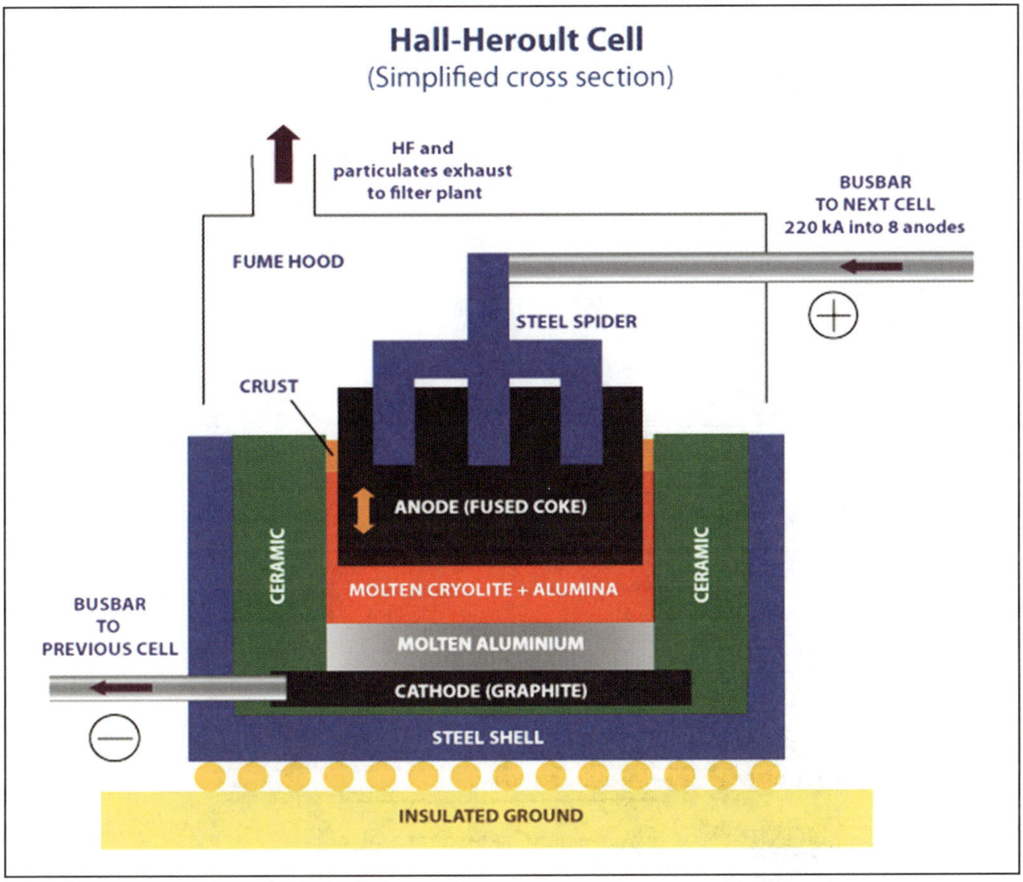

Figure 3. The Hall–Héroult alumina reduction process (Wikipedia 2013)

Figure 5 clearly indicates the dramatic impact the introduction of the Bayer process along with the Hall–Héroult reduction process had on production of aluminum. Because of its high reactivity and difficulty in reduction, before 1887, aluminum was thought to be a precious metal prized for its rarity, strength and low density.

On October 19, 1887, British Patent No. 14174 entitled Process of Obtaining Gold and Silver from Ores was issued. The discovery was made by John Steward MacArthur (Figure 1), who was a chemist-metallurgist, supported and assisted by Robert and William Forrest, who were two brothers practicing medicine in Glasgow. The discovery was made in the basement of their clinic. The process, which became known as the cyanidation process in reference to the alkali cyanide reagent used, immediately received industrial success in New Zealand in 1889 and in South Africa in 1890. It replaced the chlorination method which had been in use. In 1903, MacArthur was awarded the first medal of The Institution of Mining and Metallurgy in London. The cyanidation process was responsible for the doubling of gold production in the world in the two decades which followed its first application (Figure 6).

To the modern metallurgist, it may be surprising to think of MacArthur coming across an obscure reagent such as an alkali cyanide as a leaching agent for gold ores.

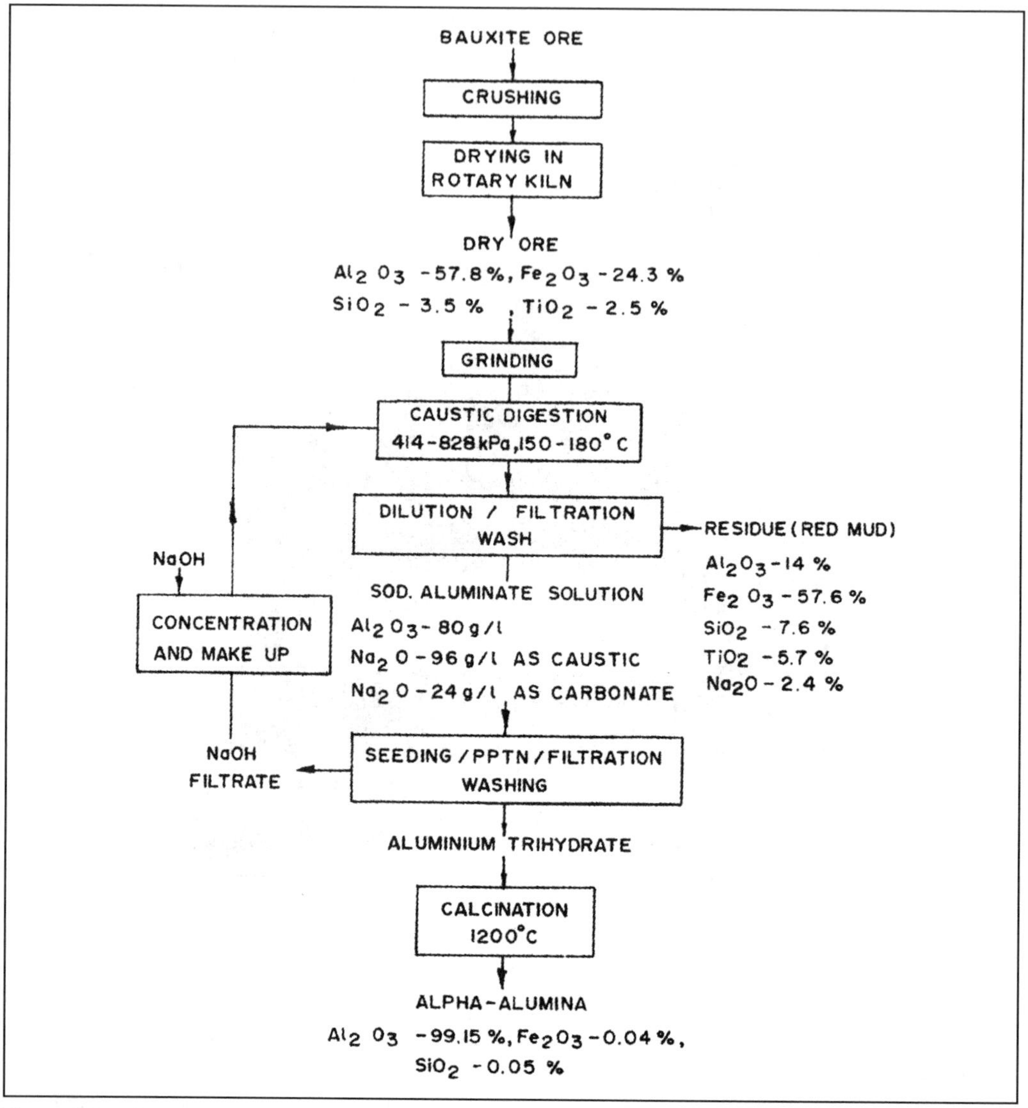

Figure 4. The Bayer Process flowsheet (Gupta & Mukherjee 1990)

The dissolution of gold in cyanide solution provides a fine example of selective complexing and has provided the basis, worldwide, for extracting gold from its ore. In this case, oxygen, is absorbed on the gold surface and is reduced according to the following reaction:

$$O_2 + 2H_2O + 2e^- \rightarrow 2OH^- + H_2O_2 \tag{3}$$

In an anodic zone, gold is oxidized to Au^+ ions which are immediately complexed by cyanide ion as shown:

$$Au \rightarrow Au^+ + e^- \tag{4}$$

$$Au^+ + 2CN^- \rightarrow [Au(CN)_2]^- \tag{5}$$

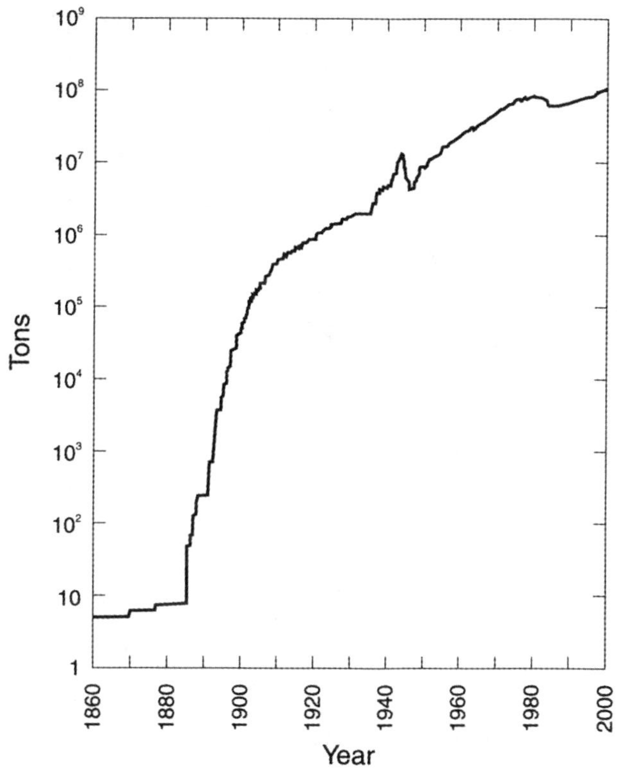

Figure 5. Tonnage of bauxite treated annually (Habashi 1993a)

The overall reaction, therefore, can be expressed simply as

$$2Au + 4CN^- + O_2 + 2H_2O \rightarrow 2[Au(CN)_2]^- + 2OH^- + H_2O_2 \quad \textbf{(6)}$$

Several findings are worth noting regarding leaching of gold in cyanide solution. A dilute solution of NaCN is as good as a concentrated solution. At low cyanide concentration, oxygen pressure has no effect on the rate of reaction, while at high cyanide concentration where the rate of leaching is independent of cyanide concentration the reaction rate shows dependency on oxygen pressure. In summary, Figure 6 clearly illustrates the impact that the invention and adoption of cyanidation had on gold and silver production.

Now however, as seen in Figure 7, cyanidation along with many other hydrometallurgical unit operations such as pressure oxidation has allowed effective production of gold from challenging ore bodies on a global scale.

INNOVATIONS IN LEACHING

Arguably the two most dramatic advances in leaching are pressure oxidation and dump or heap leaching. However, dump and heap leaching as applied to either gold or copper are largely beholding to the development of their key selective enabling technologies of the Zadra process with activated carbon and solvent extraction capable of treating the resultant dilute solutions. These aspects

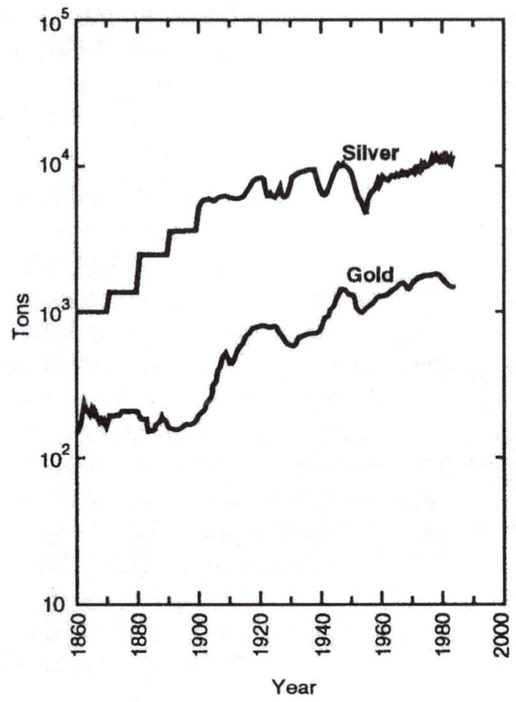

Figure 6. Annual production of gold and silver (Habashi 1993)

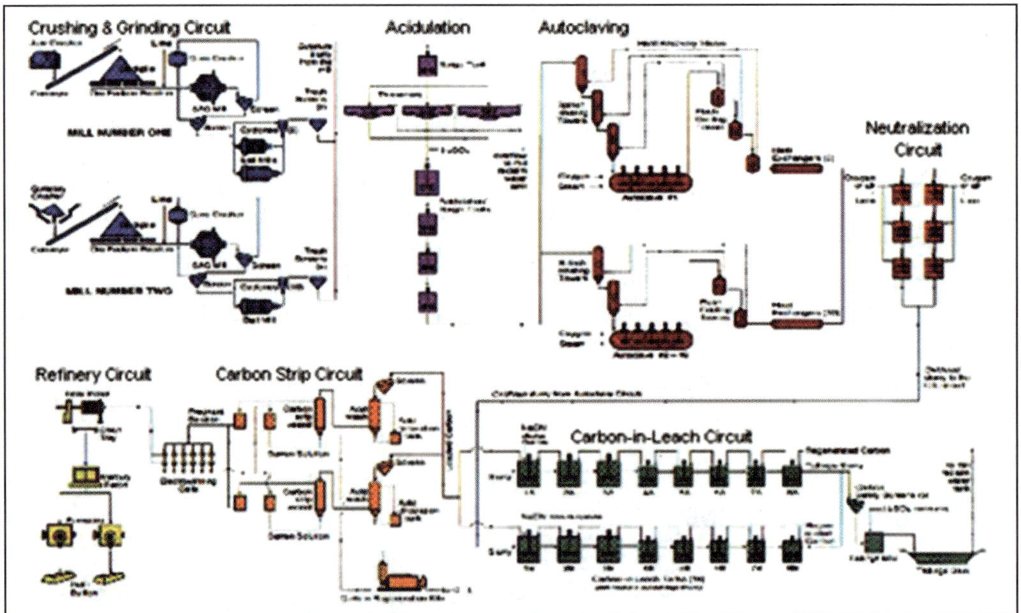

Figure 7. Barrick Goldstrike refractory gold ore treatment flowsheet (CIM METSOC 2013)

have been or will be considered to some degree in this paper. Hence, the focus in this part of the paper will be on pressure leaching. Outside of the Bayer process the three events humbly deemed most significant by the author are the commercialization of the pressure leaching of zinc concentrates at Trail by Cominco, the advent of the Sherritt Gordon selective ammoniacal pressure leach and the high pressure acid leaching of nickel laterites at Moa Bay.

Zinc Concentrate Pressure Leaching

The concept of direct oxidative pressure leaching of zinc sulfide concentrate in H_2SO_4 media was demonstrated on a laboratory scale by Sherritt Gordon Mines Ltd. in 1957. The outstanding features of the process were high Zn extraction, rejection of iron in an insoluble form and the production of sulfur in a non-polluting elemental form. In this process, the maximum leach temperature was restricted to below the melting point of the elemental sulfur (i.e., 119°C) because molten sulfur, formed in the leach coated the partially reacted metal sulfides and limited zinc extraction. In order to obtain high zinc extraction at the low temperature employed, retention times of 6 to 8 hours were required. This in fact was major driver for implementation of the process at the Trail smelter about two decades later in January 1981 which already had a Roast Leach Electrowin plant in place. The adoption of pressure oxidation to produce elemental sulfur in part alleviated the need to expand either the acid plant or fertilizer plant processes. Cominco Limited and Sherritt Gordon Mines Ltd. adopted the above-mentioned process after suitable modification and optimization to run a 3 t/d pilot plant in Fort Saskatchewan, Alberta, for processing a zinc sulfide concentrate containing 47.6% zinc, 11.3% iron, 6.96% lead and 30.6% total sulfur. The essential steps involved in the process were grinding of the concentrate to 98.5% less than 44 microns, pressure leaching of the ground concentrate with dilute sulfuric acid from electrowinning return electrolyte and separation of sulfur from the leached residue.

Pressure leaching was conducted in a four compartment brick lined autoclave with 1.27 m^3 operating volume. Spent electrowinning electrolyte was mixed with sufficient concentrated sulfuric acid to raise the free acid concentration to about 165 g/L, preheated and pumped into the first compartment along with oxygen, which was added to maintain and oxygen partial pressure of 620 kPa. The addition of surface active compounds such as quebracho and/or lignin sulfonate salt inhibited the molten sulfur that was formed from occluding unreacted sulfide particles and allowed leaching to more rapidly go to completion. About 80 to 85% of the total exothermic reaction takes place in the first compartment, raising the temperature to 146°C. About 97% of the zinc is extracted into the leach liquor after. total residence time of 90 minutes. At elevated temperature and pressures utilized, ZnS, PbS and FeS mineralization present in the concentrate react with oxygen and sulfuric acid to form simple metal sulfates, water and elemental sulfur. A key aspect is that at the elevated temperatures and with diminishing acid levels, ferric iron in solution eventually precipitated in the form of basic jarosites. Recovery of the elemental sulfur from the leach residue was accomplished by using either decantation or a flash-flotation technique. The decantation method was preferred and resulted in about 91% of the sulfur being removed as a product containing 94 to 98% sulfur. The success of the original pilot plant resulted in a subsequent successful first commercial acid-pressure leaching plant in 1981. It was designed to treat about 25% of Cominco's planned zinc capacity of 270,000 t/year at the Trail operation. **Figure 8** illustrates the zinc pressure leach process at Cominco's Trail operations. **Figure 9** illustrates the Cominco Trail smelter zinc concentrate pressure leaching autoclave.

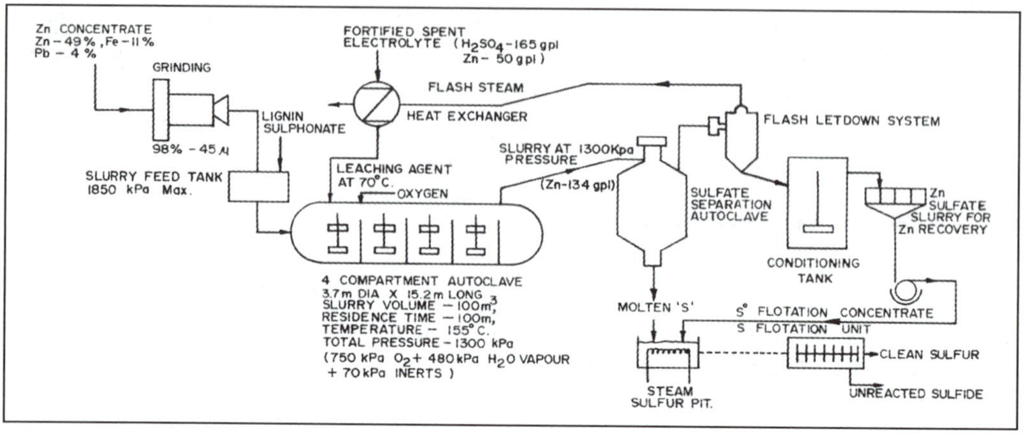

Figure 8. The zinc pressure leach process flowsheet (Gupta and Mukherjee 1990)

Figure 9. The zinc concentrate pressure leaching process at Trail, British Columbia (TECK 2013)

Sherritt Gordon Ammoniacal Pressure Leach

The Sherritt process is a hydrometallurgical process named for Sherritt Gordon Mines Ltd. (now Sherritt International) of Sherridon and Lynn Lake, Manitoba, Canada, based on the older Forward process developed by Frank Forward for the recovery of copper and nickel from those same mines. Nickel sulfide concentrates can be treated by either roasting or flash smelting to produce matte from which nickel and cobalt can be recovered hydrometallurgically, or they may be treated by an ammonia solution pressure leach. The residue is removed. A feed of matte and sulfide concentrate containing approximately 0.4% cobalt and 30% sulfur is pressure leached at elevated temperature and pressure in an ammoniacal solution to produce a solution of nickel, copper and cobalt. By boiling away the ammonia; copper is precipitated as a sulfide and sent to a smelter. Hydrogen sulfide is added to the autoclave to remove nickel sulfide and copper sulfide which is fed back to the leaching process. Air is then passed through the solution in the autoclave for oxyhydrolysis. The solution is then reduced with hydrogen, again at high temperature and pressure, to precipitate nickel powder (>99%). The remaining solution (containing approximately equal proportions of nickel and cobalt sulfides), is then adjusted (to a lower temperature and pressure) to precipitate the mixed sulfides and the fluid is concentrated and crystallized into ammonium sulfate ($(NH_4)_2SO_4$). The mixed sulfides

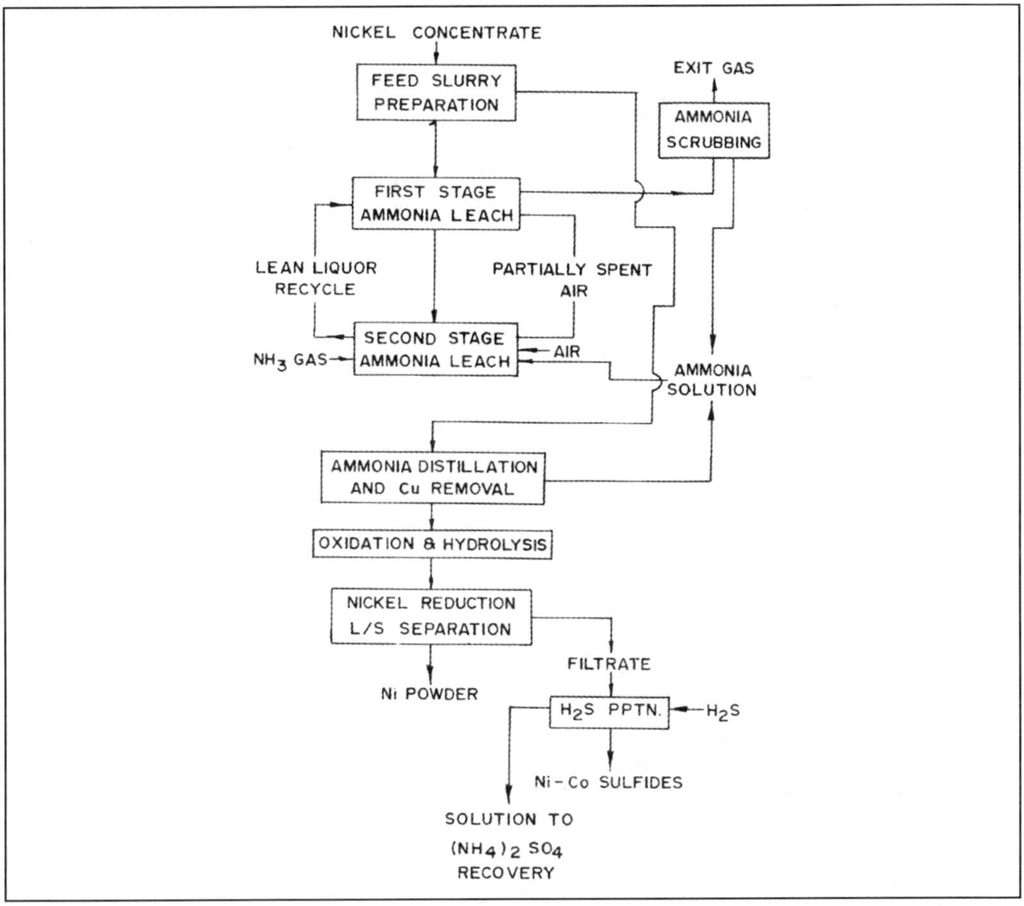

Figure 10. The Sherritt Gordon ammoniacal leach process for nickel sulfide concentrate (Gupta & Mukherjee 1990)

are pressure leached with air and sulfuric acid. Ammonia is then added to remove potassium and iron as jarosite ($KFe^{3+}{}_3(OH)_6(SO_4)_2$). More ammonia and air is added for oxidation. The solution is removed from the autoclave and sulfuric acid added to remove nickel as nickel(II) sulfate-ammonium sulfate hexahydrate ($(NiSO_4)\cdot((NH_4)_2SO_4)\cdot 6H_2O$) which is then sent to have its nickel recovered. The solution is then further reduced with more sulfuric acid and cobalt metal powder is added to aid in the nucleation of precipitants (seeding). Addition of hydrogen gas to saturation precipitates cobalt powder with a purity of approximately 99.6%. Figures 10, 11, and 12 outline the leaching and hydrogen reduction and overall flowsheets for this innovative process.

High Pressure Acid Leaching of Nickel Laterites

Cuba's Nickel industry developed as a result of World War II, when the demand for nickel in-creased dramatically due to the need for high-strength stainless steel and other nickel alloys for use in the production of guns, ammunition and vehicles (Boldt 1967). On December 31, 1941, the U.S. government financed the creation of Nícaro Nickel Company, a Cuban venture under the management of the Freeport Sulfur Company. Under contract with the U.S. government, Freeport's plant,

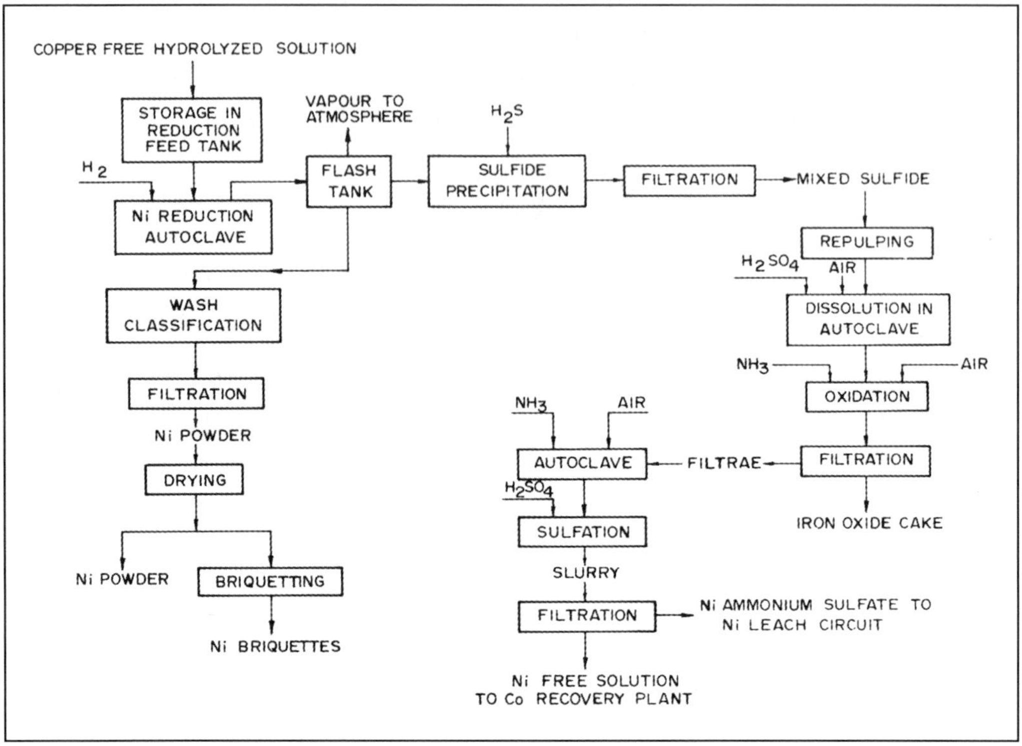

Figure 11. Flowsheet for recovery of nickel by hydrogen reduction as practiced at Fort Saskatchewan (Gupta & Mukherjee 1990)

located east of the Bay of Nipe, processed ore mined from the Piñares de Mayarí deposits until it closed on March 31, 1947. Founded in 1912, Freeport Sulphur Company played a complex and critical role for over fifty years in Cuba's nickel mining and processing business. Formed to develop the huge sulfur deposits found in Freeport, Texas, the company began to diversify in 1931, purchasing manganese deposits in the Oriente province in Cuba. In 1955, Freeport invested over $100 million in the construction of a second nickel-cobalt mine and processing facility at Moa Bay, Cuba, and a refinery at Port Nickel, Louisiana. On March 11, 1957, the U.S. government announced a contract for the purchase from Cuba of Freeport's production of nickel as strategic commodities. The contract would last until June 30, 1965.

In 1960, the Cuban government nationalized and failed to provide financial compensation to both the Moa and Nícaro facility. In 1971, The U.S. Federal Claims Settlement Commission certified the claims against the government of Cuba for improved real property and corporate assets by Moa Mining Company and Nicaro Nickel Company for $88,349,000 and $33,014,08. In December 1994, Canada's Sherritt Inc., which had been processing Cuban nickel-cobalt mixed sulfides since 1991, announced the formation of a 50/50 vertically integrated joint venture with Cuba's General Nickel Co. S.A. The joint venture, known as Metals Enterprise S.A., includes Moa Nickel S.A., which owns and operates the mining and processing facilities at Moa; the Cobalt Refinery Company, which owns and operates the nickel and cobalt refinery located in Fort Saskatchewan, Alberta, Canada; and the International Cobalt Company. The mixed sulfides produced at Moa

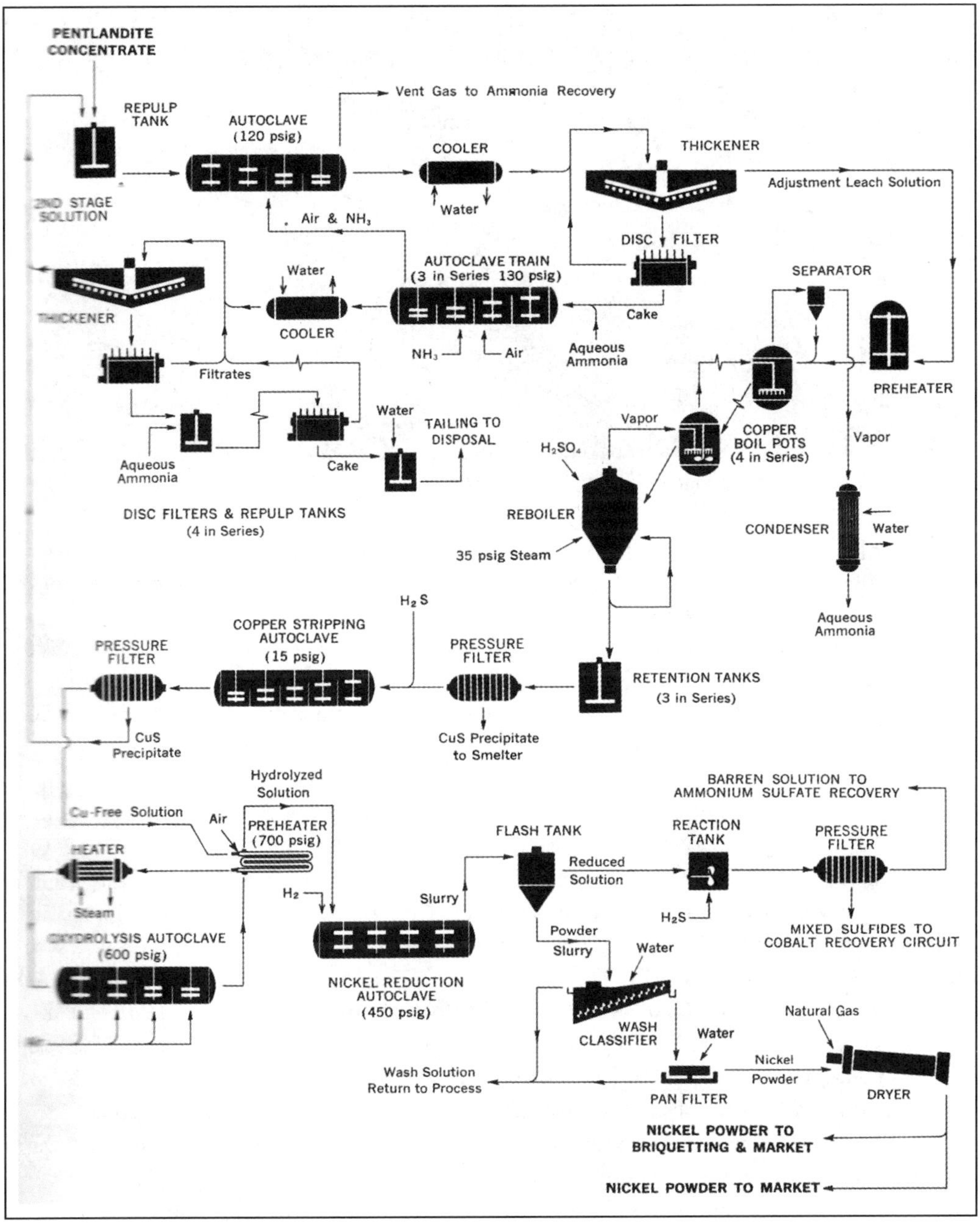

Figure 12. Overall Sherritt Gordon process flowsheet (Boldt 1967)

are shipped to the east coast of Canada, where they are then transported by train to the Fort Saskatchewan facility and refined into finished nickel or cobalt. Nickel Oxides from the wholly owned General Nickel's Nicaro and Punta Gorda processing plants are exported to China and other countries. The Bahamas based International Cobalt Company (ICC) Inc., purchases mixed sulfides from Moa Nickel and other third parties, and contracts (tolling) with the Fort Saskatchewan

Figure 13. Moa Bay plant at commissioning (Boldt 1967)

refinery to process the raw materials. ICC also markets Metal Enterprise S.A.'s refined nickel and cobalt products internationally. The Moa Bay process at start up is shown in Figure 13 and the detailed flowsheet is in Figure 14.

At Moa Bay leaching of limonitic nickel and cobalt bearing laterites that typically contains 0.1 to 0.3% Ni, 0.05 to 0.257% Co, 35% Fe is undertaken. Efficient and selective extraction of nickel from such high iron-bearing material is not possible by direct atmospheric leaching with H_2SO_4. It requires leaching at a temperature above 200°C in autoclaves. Under these conditions, both iron and aluminum restrictively enter the leach liquor due to the very poor solubilities of their sulfates. The plant became operational in 1959. The flowsheet depicting the acid-leaching system is shown in Figure 10. The ore as a 45 V% solid slurry is heated to the operating temperature (230 to 250°C) and pressure (4.3 MPa) by means of a number of preheaters and heaters before going into the first of four autoclaves placed in series. The autoclaves are fabricated from welded steel cylinders with spherical ends and are lined internally first with 6-mm lead, then 114-mm acid-proof brick, and finally 76-mm-thick carbon bricks. The carbon brick serves to protect the acid- resistant bricks from corrosion and erosion. All other intimae parts of the autoclaves which come in direct contact with the superheated acidic slurry are made of titanium. The slurry is fed by means of specially designed high-pressure slurry pumps and is agitated by injecting steam at a pressure of 4.24 MPa into the bottom of the autoclave. Concentrated H_2SO_4 (98%) is added to the first autoclave from the top at an acid to ore ratio of approximately 0.25. The slurry moves from one autoclave to another under gravity at such a rate that a total residence time of 1 to 2 h is achieved. The slurry coming out of the autoclaves is cooled and subjected to solid/liquid separation and washing to generate a liquor containing 5.95 g/L Ni, 0.64 g/L Co, 0.8 g/L Fe, and 2.3 g/L Al representing 96% Ni and 95% Co extractions, The liquor is treated with H_2S gas which results in the recovery of Ni and Co in the form of their mixed sulfides. The sulfide products were historically shipped to Port Nickel, Louisiana where they were further treated for the production of nickel and cobalt metals. After the Cuban nationalization of Moa Bay, they now go to Fort Saskatchewan, Alberta, Canada.

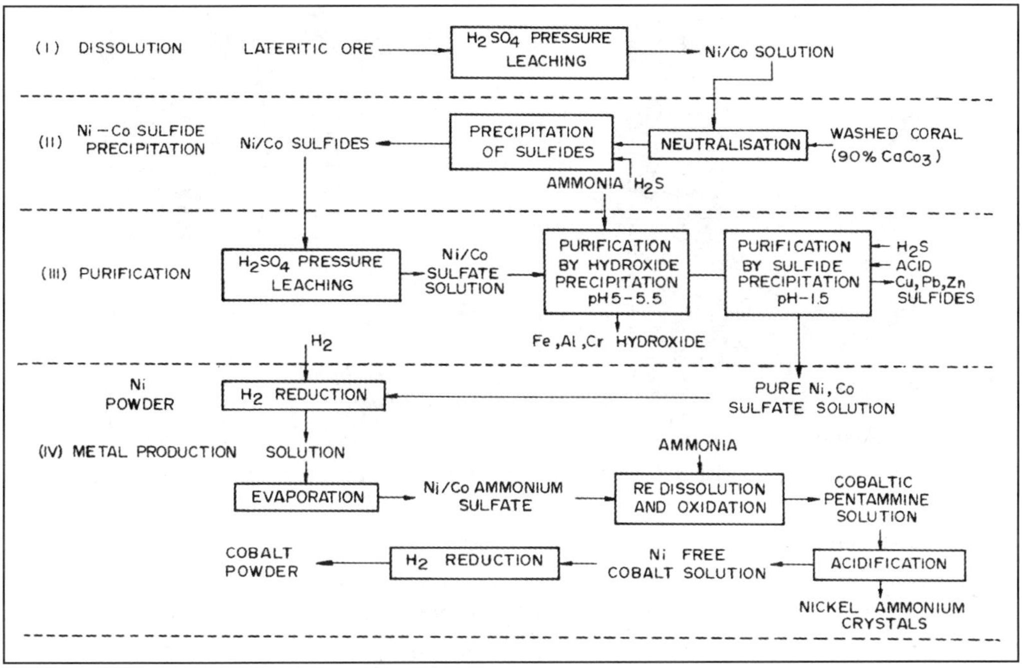

Figure 14. Moa Bay process flowsheet for high pressure acid leaching of nickel and cobalt laterite ores (Gupta and Mukherjee 1990)

INNOVATIONS IN SEPARATIONS AND SELECTIVE CONCENTRATION

The key separations utilized in hydrometallurgy involve derivations of ion exchange technology largely derived from uranium separations during the World War II Manhattan Project. This is a relatively new development and it exponentially enabled the use of hydrometallurgy by allowing selective recovery and concentrations from dilute solutions sometimes resulting in pure final value added products on site. Uranium, gold and copper hydrometallurgy stand as successful testaments to this technical development. The Manhattan Project also directly resulted in the creation of the Rare Earths (i.e., RE, REE or TREO) industry which furthered the development and application of this enabling technology. A summary of this key development utilizing the superior separations available with hydrometallurgy follows (Kronholm et al. 2013). This work lead directly to the wider industrial application of ion exchange, activated carbon for gold and eventually copper solvent extraction

When approaching the use of ion-exchange techniques to separate rare earth elements there are three paths to be considered. The first of these uses a chelating eluent which is selective between individual lanthanides, combined with resins which show little selectivity. The second and third are extraction with a selective stationary phase; one case where the resin has been chemically modified at a molecular level and one where a resin's surface is coated with an existing extractant. These methods are discussed at length in literature and have advantages and drawbacks.

The chelating eluent technique was the first method to achieve separation of rare earths on the kilogram and pound scale. Spedding describes the use of ammonium citrate/citric acid to selectively elute RE's loaded onto Amberlite IR-100 sulfonic acid cation exchange resin (Spedding 1947). He

Figure 15. Frank Spedding (left) and Jack Powell (right) (courtesy of K. Gschneidner, Ames Laboratory)

and Powell (Figure 15). were instrumental in this and subsequent ion exchange developments and applications. This led to industrial adoption of IX and SX as we know it.

The principle behind this method is that RE's are loaded onto cation exchange resin with little order of preference; then as a dilute solution of ammonium citrate or citric acid is pumped through the column an exchange of the RE takes place between the resin and the citrate anion. The citrate anion can be represented as Y^{3-} and the citrate-RE complex as YM where M represents the trivalent RE cation. The metal, when coordinated to citrate in the mobile phase is able to flow through the column without interacting with the resin. The RE coordinates and dissociates from the resin many times as it moves down the column and because the formation of some citrate-RE complexes are favored over others, bands of RE begin to develop in the column. Much like RE separations performed by solvent extraction these separations take advantage of the changing stability of coordinated REs based on the change in ionic radius due to the lanthanide contraction. As one progresses from left to right on the periodic table the ionic radii of elements get smaller and the complexes formed become more stable due to better overlap of molecular orbitals (Spedding et al. 1953). As a result of this, RE's elute from right to left as they appear on the periodic table, the heavier elements first. Yttrium tends to be an exception to this; the position where it elutes is dependent on the mobile phase used. Though citric acid was the first eluent used it was replaced by acids that are stronger chelators. Aminopolyacetic acids have shown increased heavy RE selectivity and also can be used to elute higher concentrations of RE (Spedding et al. 1960). An illustration of separation using elution chromatography techniques can be seen in Figure 16, where conceptually A and B represent individual REs that are separated.

Several parameters have been found to affect the degree and ease of separation of REEs using complexing eluents. Adjusting pH has been shown to change the stability of the complexes that the RE forms with the eluent. A low pH weakens the complex formed by the eluent with the RE giving good separation due to more exchanges of RE between resin and eluent, but at the cost of long elution times and dilute RE effluent streams. Higher pH causes more stable complexes to be formed allowing for more concentrated bands of RE to be generated in the effluent but requires the use of "retaining ions" (Spedding et al. 1954). Retaining ions are loaded onto the resin at the leading edge of RE bands and form more stable complexes with the eluent than REs; this acts to hold back RE

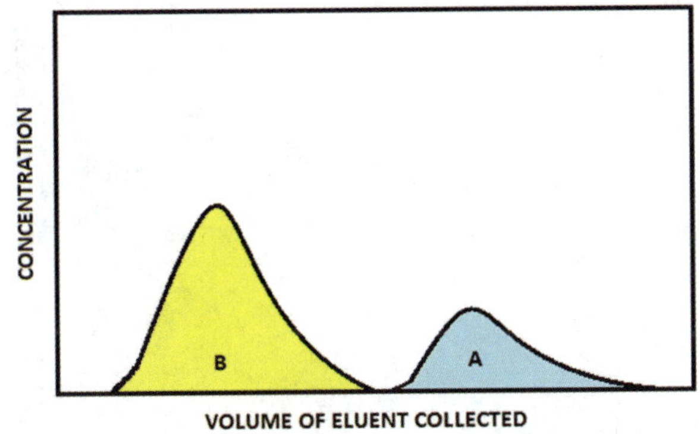

Figure 16. Rare earth elution chromatography (Kronholm et al. 2013)

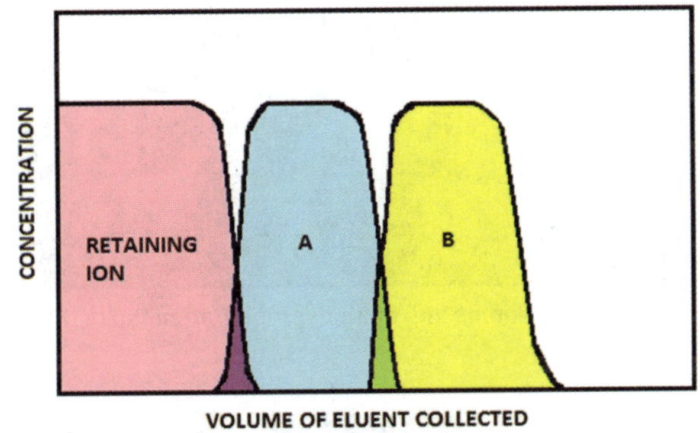

Figure 17. Displacement chromatography (Kronholm et al. 2013)

bands and prevents them from eluting too quickly before separation is achieved. This is known as "displacement chromatography," an illustration of this can be seen in Figure 17 where again conceptually A and B represent individual REs that are separated. Displacement chromatography was the initial basis of and was widely used for commercial scale preparation of REs for almost 30 years following its development in the early 1950s and is still used in certain high purity applications. Figure 18 illustrates an early RE ion exchange column configuration. The IX columns are in the back and the oxalate precipitation takes place in the tanks in the foreground. Figure 19 illustrates ion exchange columns for the industrial separation of the lanthanides.

There are many choices of eluent (Choppin 1993) but the most current available literature suggests that EDTA and HEDTA are still the most widely used eluents for macro scale cation exchange separations (Gupta and Krishnamurthy 2005) HEDTA allows for H^+ to be used as the retaining ion which is a benefit over EDTA as it forms precipitates on the resin when completely protonated and is used with a cupric ion to retain the RE (Asher 1962) The eluent used affects the process conditions and efficacy of the separation. In addition to pH, temperature can improve characteristics of RE separations. Increased temperature has shown to improve production rates due to increased solubility of RE/aminopolyacetic acid complexes (Powell and Burkholder 1968).

Figure 18. Rare earth ion exchange columns and oxalate precipitation tanks (Kronholm et al. 2013)

Figure 19. Ion exchange columns for the industrial separation of the lanthanides at Michigan Chemical Corporation, St. Louis, Michigan (Habashi 1993)

Increasing solubility of these complexes allows more concentrated RE solutions to be separated which reduces process water use and elution times. Columns run at increased pressure have exhibited improved kinetics, resulting in faster elution times of RE bands (Campbell 1973) Though the original selective eluent research was done on separation using cation exchange resin, later work explored the use of anion exchange resin.

Subsequently, the use of activated carbon, as a regenerable anionic adsorption media for gold cyanide in the Zadra process further established these types of selective ion exchange type concentrating technologies. However, in the author's humble opinion no hydrometallurgical separation technology has been as important or significant than the eventual use of liquid ion exchange, or solvent extraction, for copper production. Coupled with electrowinning about 30% of the world's copper is now derived from this technology. The perceived chemistry is noted in the following equation.

$$2RH^{+} + Cu^{+2} + SO_4^{-2} \leftrightarrow R_2Cu + H^{+} + SO_4 \qquad (7)$$

As well **Figure 20** denotes the proposed solvent extraction chemical chelation mechanism in graphic form.

When the sulfuric acid copper L/SX/EW flowsheet (**Figure 21**) was put forth by the Minerals Development Group of General Mills in 1960 solvent extraction had been known for over 100 years. It was used extensively on a very small scale in analytical chemistry and on a large scale for the recovery of uranium from sulfuric acid leach solutions (Kordosky 2002). General Mills had already developed and commercialized Alamine® 336 as an SX reagent for the recovery of uranium from sulfuric acid leach liquor and believed that a similar technology for copper recovery would be welcome. However, an extensive market survey showed that the industry reception for copper recovery by L/SX/EW technology was almost hostile.

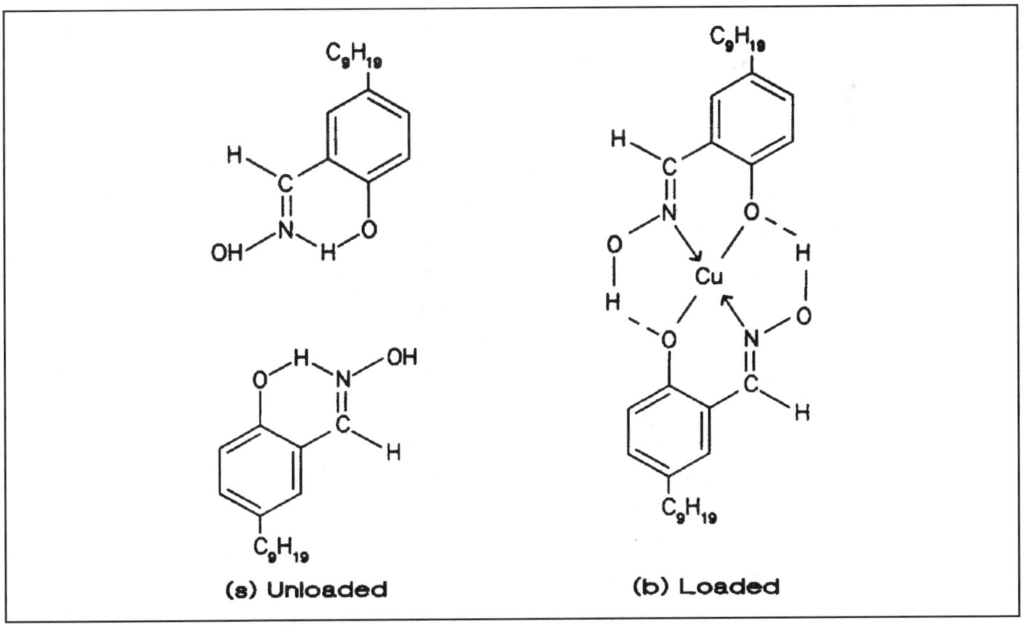

Figure 20. Solvent extraction oxime chelation chemistry (Biswas and Davenport 1994)

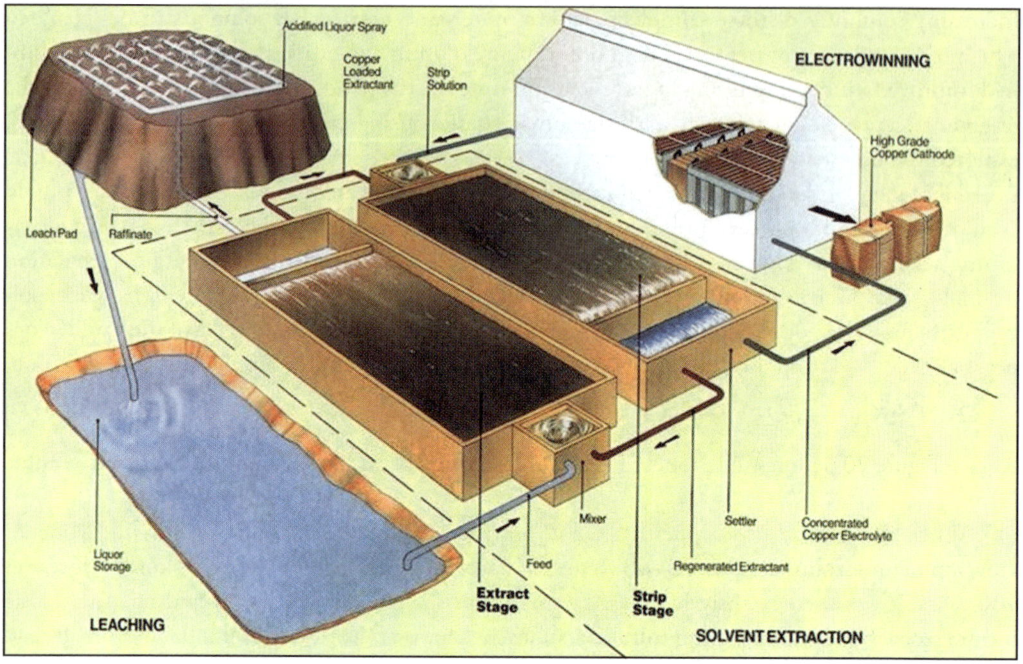

Figure 21. Leach SX EW flowsheet (AGORGA 1990)

The R&D director of a large copper producer predicted at an AIME annual meeting that there would never be a pound of copper recovered using solvent extraction and his comment prompted applause. Fortunately, the Minerals Development Group of General Mills, in the person of Joe House, Don Agers and Ronald Swanson, believed so much in the copper L/SX/EW process that they kept the development of this technology alive as a "bootleg" project, that is, a project where the three individuals had other primary work duties and could only work on this project when they found time or on their own time. By late 1962 this group had identified and formulated an organic soluble molecule containing a hydroxy-oxime functionality as the reagent LIX®63. While LIX 63 had many of the properties required for a reagent to be successful in the proposed L/SX/EW flow sheet, LIX 63 was not compatible with the copper leaching process because it did not extract copper below pH~3. Technically this problem could be solved by neutralizing the acid leach liquor or by leaching oxide copper with ammonia, but neither solution was economical. a molecule that extracted copper well from typical dump leach liquors (pH ~1.8) had to be built. The reagent LIX 64, containing the ketoxime LIX and a catalytic amount of LIX 63, was introduced in 19657 and in March 1968 the first commercial copper L/SX/EW operation, the Bluebird plan of Ranchers Exploration and Development Corporation, came on line. In 1968 there were only two widely practiced copper leaching processes using dilute sulfuric acid. The first process, vat leaching of high-grade copper oxide ore followed by EW of copper from the leach solution, produced low quality copper cathode at relatively high cost. In 1968 the tonnage of high grade oxide ores was decreasing and vat leaching was on the decline. The second process, heap and dump leaching of low grade oxide and/or sulfide ore followed by precipitation of low quality copper from the leach solution on scrap iron, was practiced on oxide ore that was too low-grade for vat leaching, or low grade sulfide ore that had to be mined in order to expose the underlying high grade sulfide ore. Copper recovered from leaching low grade copper ores was considered a bonus and little effort had been made to fully understand the leaching process or to maximize copper recovery. While some fundamental leaching studies had taken place on the whole leaching was a poorly practiced art and little had been done to make it a well-practiced science.

Solvent extraction for copper was not yet proven commercially so most copper companies were taking a wait and see attitude. Furthermore LIX 64, the only reagent available, had significant limitations with respect to extractive strength, metal transfer kinetics and copper/iron selectivity and it could only be used up to a maximum copper loading of about 3 g/L because of entrainment problems. These properties restricted copper leach solutions which could be effectively treated by SX using LIX 64 to $\leq$ 3 g/L Cu at a pH $\geq$ 1.8. In 1968 the copper industry as a whole did not believe large quantities of high quality copper could be produced by hydrometallurgy and at the time the Ranchers plant came on line, expectations for the eventual success of copper L/SX/EW technology were rather modest.

Ranchers leached run of mine copper oxide ore (>0.5% Cu) placed in a sealed canyon in 20-foot layers (heaps) by distributing dilute sulfuric acid over the fresh ore via spray emitted from needle valves with pipes and needle valves placed to ensure complete coverage of the ore. New heaps were built over old heaps. Pregnant leach solution flowing from the bottom of the heap was collected in a pond created by placing a dam downstream from the leach area. The Bluebird copper SX plant was built similar to existing uranium SX plants with tall, single baffled mixers and long narrow settlers. Dispersion exiting the mixer was pumped to the opposite end of the settler and the phases flowed back toward the mixers. In electrowinning rich electrolyte with ~36 g/L Cu, 3 g/L Fe and 145 g/L H_2SO_4 entered one end of the EW cells and spent electrolyte with ~32 g/L Cu and 151 g/L H_2SO_4 exited the other end. Cathodes weighing 60 kilograms were grown on copper starter sheets at a current density of about 180 A/m^2 with a current efficiency of 80 to 85%. The anode was 6% Sb in Pb. Design production at Ranchers was 30,000 pounds of copper daily. In its first fiscal year of operation, July 1968–June 1969, the Bluebird plant produced 9 million pounds of copper (82% of design), a remarkable achievement for a metallurgical plant using new technology. By later adding only rectifier capacity the plant often exceeded its design production by 50%. Ranchers' Bluebird plant proved that L/SX/EW technology could produce large quantities of good quality cathode copper on a consistent day to day basis at a profit. This raised the awareness and sparked the interest of the industry in copper L/SX/EW. Improvements in the technology quickly followed from many sources including consultants, copper producers, engineering companies, chemical companies and metallurgical testing laboratories.

INNOVATIONS IN PRECIPITATION, REDUCTION, AND RECOVERY

Outside of selective crystallization as used in the Bayer process on a massive scale globally, hydrogen reduction as used in the Sherritt Gordon process or longstanding applications of copper cementation with iron can be considered as historic breakthrough innovations. However, in the opinion of the author, the selective electrolytic refining and reduction technologies have arguably had the most impact on hydrometallurgical advancement. These provide high purity final metal products.

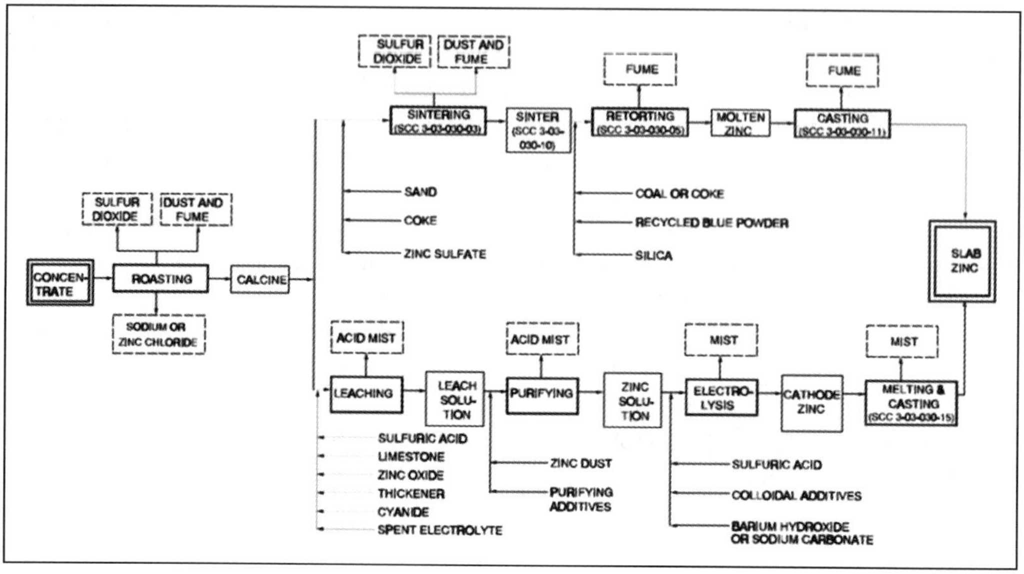

Figure 22. Traditional zinc concentrate processing summary flowsheet illustrating roasting followed by calcine smelting or by calcine leaching and electrowinning (USEPA 1986)

Of these the innovative and long standing highly selective Betts electrorefining process as well as selective zinc electrowinning as part of the Roast Leach Electrowin process for zinc concentrates stand out most dramatically.

Roast Leach Electrowin Process

Frenchman Léon Letrange in 1881 first developed and patented (Letrange 1883) on an experimental scale the reduction of zinc ores in a sulfate system. This method used roasting, dilute sulfuric acid leaching and electrowinning (EW) with zinc starter sheets as per copper electrorefining (ER) and carbon anodes. Zinc sulfate solution purification was not described in Letrange's patent but was simplistically known at the time in the chemical preparation of the paint pigment, lithopone. This was the basis of the industrial innovative Roast-Leach-Electrowin (RLE) process. A combined zinc concentrate processing flowsheet is shown in Figure 22. The traditional historic pyrometallurgical smelting route follows the top path while the hydrometallurgical RLE path follows the bottom path. Of note, in both cases, pyrometallurgical roasting precedes both established process routes. This key roasting step is best expressed as

$$ZnS + O_2 \rightarrow ZnO + SO_2 \quad \textbf{(8)}$$

The sulfur dioxide generated is scrubbed to produce sulfuric and the oxidized zinc sulfide becomes a zinc oxide calcine that is progressively leached with the sulfuric acid produced. While of historic importance for zinc production, RLE has been utilized for example copper and cobalt concentrates (Aird et al. 1980). While other zinc extraction process development at the time focused on chloride systems, due to simplicity and materials of construction, the sulfate system advanced to large scale commercial zinc production. Numerous attempts had been made, without success, to scale up and it was soon found that purification requirements were specific to the type of ore treated.

Driven by the World War I demand for pure EW zinc for brass (Ralston 1921), pilot plants and eventually larger production plants were built at

- Anaconda, Montana: 25 TPD (1915) increasing to 150 TPD (1916)
- Trail, BC, Canada: 50–75 TPD (1914)
- Risdon, Tasmania: 15–100 TPD (1917–1919)

In electrowinning, lead anodes were used to replace the carbon anodes and it was found that pure electrowon zinc metal could be easily stripped from an aluminum cathode, thereby replacing zinc starter sheet technology and producing the first permanent cathode harvesting system. It is interesting to note that two large copper electrowinning plants were first built at this period also at Chuquicamata, Chile, and New Cornelia, Ajo, Arizona (Robinson 2003). These large scale copper EW plants would still use copper starter sheets and both carbon and lead anodes.

It was the removal and control of iron that further propelled the RLE process through the application of the Jarosite, Goethite and Hematite processes. This is addressed later in the paper.

It became apparent that on a large scale, the purification step for the zinc electrolyte became specific to the plant and ore treated. Copper and cadmium could be removed by air atomized zinc dust. At Anaconda with the high antimony and arsenic levels in the ore, copper sulfate and zinc dust could be used. At Risdon, cobalt impurities in the Broken Hill zinc ore were very difficult to remove and additives such as glue, nitroso beta napthol, and zinc dust (in warmer conditions) assisted with cobalt removal.

Since these first plants, a number of innovations have been made in impurity removal:

1. Addition of activators such as oxides of antimony and arsenic to speed up purification kinetics in cobalt cementation.
2. Minimization of recycled zinc dust requirements by controlling purification Eh to regulate zinc dust addition (Gonzalez-Dominguez 1997).
3. Use of fluidized bed reactors for efficient and pure cadmium removal (Free 2011).
4. Use of water atomized zinc dust with larger surface area to reduce zinc dust consumption compared to air atomized zinc dust (Free 2011).

The first three pioneering commercial zinc sites built smaller cellhouses before expanding to commercial production levels. All sites had ready and reliable sources of cheap hydropower which was essential for the new energy intensive zinc EW technology. It is interesting to note that Trail's first zinc EW plant was the largest at 50 TPD compared to Anaconda and Risdon, as Trail had already had a strong electrometallurgical technical team and had been operating a commercial lead refinery since 1902 on the site.

The first manual zinc cellhouses used a cascade flow system (like lead ER) in lead lined wooden cells. These cells were small compared to today with only 24 cathodes. As current density was gradually increased, the electrolyte had to be cooled by water circulation in lead coils in the cell. Trail used a zinc EW electrode size of approximately 1.2m^2 (0.6m^2 each side) which was the same size as the commercial lead ER size of 2 by 0.6m^2.

This manual type of cellhouse remained largely the same, only slightly increasing length of cell and size of electrodes and increasing current density. The Japanese had made some improvements in automation, but it was in the late 1960s when productivity began to drive innovation with construction of the jumbo design cellhouse at Union Miniere Zinc in Belgium (Caufriez 1998). This jumbo cellhouse was built with 2.6 m^2 electrodes, 375 ASM, 44 cathodes per cell, parallel cell feeding electrolyte cooling with air cooling towers and automated electrode handling machines.

Since then, new zinc EW cellhouse designs have included increased electrode sizes up to 3.8 m^2, cells with over 110 cathodes, totally automated driverless cranes and further automation that has removed operators from cellhouse acid mist.

Future innovation will be driven by energy savings and include

1. New amorphous coated titanium anodes like those pioneered by Freeport and Republic Alternative/Doshisha University in copper which lower the anode oxygen overpotential (Robinson 2011),
2. Water cooled double intercell contact designs,
3. Real time cell voltage or current monitoring, and
4. Large monolithic polymer concrete cells that use no liners.

Betts Electrolytic Process (BEP)

Electricity had been discovered in the 1800s and in the period of 1865 to 1870 the first patents (English) for electrolytic refining of copper from smelter anodes were issued to James Elkington. Development also began on an electrolytic lead refining process using an acetate electrolyte and a lead bullion anode by N.S. Keith (1878) and Tommasi (1871/1891), but the lead cathode produced was spongy and difficult to handle.

Using starter sheet cathode technology and glue as an additive, commercial scale copper refineries were able to be built in Hamburg and Belgium and the first commercial scale refinery built in the USA was Balbach Smelting and Refining Company in Newark, New Jersey, in 1883. Within ten years, eleven more copper refineries were constructed in the USA, particularly in the New York/New Jersey area, including Laurel Hill (Phelps Dodge), Raritan (Anaconda), Perth Amboy (ASARCO), and Carteret (Amax) producing a quarter of the world's refined copper.

The lead cathode produced in an acetate electrolyte could not use starter sheets, and glue (gelatine) did not improve the lead cathode deposit. In the late 1800s Anson Betts (Betts 1908) began a series of experiments with different electrolytes in a small pilot cell arrangement consisting of four cells (10 in. W, 16 in. D and 30 in. L) that contained nine anodes each. The first cathodes were sheet iron dipped in lead. Current density was 7–8.8 ASF and typical cell voltage was 0.175 V. Electrolytes tested included acetate, fluoboric acid and hydrofluosilicic acid. Hydrofluosilicic acid was ultimately chosen due to high lead solubility (in excess of 400 gpl), low cost, higher conductivity and that like copper refining with the addition of glue (gelatine), made a dense deposit that could make a refined lead starter sheet from a lead bullion anode. Both hydrofluosilicic acid electrolyte and the glue addition were key innovations that led to commercial application and led to the Betts Electrolytic Process or BEP being patented in 1901.

The main difference between the BEP process and copper refining was the lead bullion impurities such as silver, gold, antimony, arsenic and silver remain attached to the lead bullion anode as a thick slime rather than fall to the bottom of the cell in copper. The main advantages of the BEP process over other lead refining processes was that it could produce a purer lead product in a single step from lead bullion that included impurities such as bismuth, antimony, arsenic silver and gold. Hygiene in the BEP was generally improved over the pyro-metallurgical refining processes and drosses and slag recycle streams were minimized. Lead fumes and dust also were not produced as this was a hydro-metallurgical process. The only loss of current efficiency in the BEP process is short circuits between the anode and cathode. Theoretically it can be 100%. The BEP process today remains the most efficient method to separate lead from bismuth.

Commercial scale up (electrode size) in electrolytic processes can be faster than in pyro-metallurgical applications, so from Bett's small pilot plant, a small commercial lead refining plant was built at Cominco's existing lead and zinc plant at Trail, BC, in Canada in 1902 (Thomas 1980). The Trail location had a ready source of continuous hydropower supply, an existing lead-zinc processing plant and was close to several large lead ore deposits.

Trail has pioneered the commercial application of the BEP process with continuous improvement and innovation focussed largely on productivity since 1902. The first cellhouse built in 1902 had an anode size of $0.6m^2$ each side and used a monorail system for electrode change out and asphalt coated wooden cells containing 25 cathode starter sheets and 24 lead bullion anodes. By 1922 the plant had expanded to 135 TPD with cranes and asphalt covered concrete cells. In 1926, the refinery was expanded to 1116 cells (350 TPD), the present footprint today.

A major innovation in improving lead cathode quality was the discovery and patenting of aloe in 1953 as a strong and improved lead refining smoothing agent to replace glue (Mather 1953). In 1954 Trail implemented aloe and this improved the lead cathode deposit at higher current densities and enabled the cathode cycle to be increased to 7 days (from 3 days). At Trail, current density was able to be raised to 240 ASM and in turn increasing refining capacity of 550 TPD with the existing refinery footprint from 1926.

Innovations and developments since 1954 at Trail have included the following:

1. Continuous drossing of lead bullion (first developed at BHAS in Pt. Pirie, South Australia) to remove copper before the refining process (BHAS < 1966)
2. Improvements in anode casting with controlled lead bullion composition for optimum slimes adherence to anode
3. Improvements in starter sheet manufacture to make straight electrodes and minimize short circuiting
4. Development in the mid-1970s of an in-plant instrument that continuously measures cathode polarization of electrolyte and the active concentration of the additives in the electrolyte for high current density (Krauss 1976a)
5. Implementation in the mid-1970s of programmed rectifier current to keep cell voltage below 0.2 V anode overpotential so that bismuth does not dissolve from anode slime layer back into electrolyte (Krauss 1976b)
6. Polymer concrete cells that require no liners and minimal maintenance replaced lined concrete cells
7. Slimes treatment processes

Since 1926, with continuous innovation, Trail has effectively increased production by over 50% using the same refinery footprint, cell size (24 anodes) and electrode size ($0.6\ m^2$).

Over the last several decades BEP refineries have since been mainly built in Japan and China. Developments at those BEP refineries include (Kubota 1998, Hiraki 1998, and Gonzales 1991) the following:

1. Implementation of real-time cell voltage monitoring to detect and correct short circuits
2. Periodic current reversal rectifiers that smooth the lead cathode
3. Improving electrode handling automation leading to increased productivity, reduction in labor cost and improving occupational health
4. Improved electrode straightening for higher current efficiency

Future innovation and implementation opportunities could be as follows:

1. Bipolar lead refining technology tested at Trail in the 1980s that enables electrode size to be larger leading to smaller BEP plant footprint, lower energy consumption and more automated permanent electrode system to be installed.
2. Wireless cell voltage monitoring technology used in copper refining
3. Mettop high current density technology (used in copper ER) that introduces electrolyte flow to cathode and does not disturb slimes layer on anode. At present, the BEP cell flow limited to 30–40 L/min
4. Amorphous coatings pioneered by Freeport and Doshisha/Republic in copper EW that do not precipitate lead oxide that could be used for lead ER liberators

In summary, the innovative success of the BEP process can be due to several factors:

1. Successful commercial application of copper refining using glue addition as a smoothing agent and starter sheet cathode technology before the BEP was developed
2. Relative ease of construction of a pilot plant to test the many different electrolytes specifically for lead refining
3. Relatively fast scale up and technology implementation time from Betts's pilot cells (patent in 1901) to Cominco Trail's first commercial plant in 1902
4. Strong technical development and innovation team continuity at the first commercial site at Trail, BC, since 1902 to present day

SUMMARY

The advent of integration of hydrometallurgical industrial technology globally has been recent and dramatic as evidenced by the Bayer and Cyanidation processes starting circa 1887. Further innovation predicated on definitive separations technologies such as pressure leaching, heap leaching, ion exchange, solvent extraction, and electrolytic refining and reduction has further propelled the usage and growth of hydrometallurgical technology. Normally, it is the high degree of selectivity that distinguishes hydrometallurgy as a viable technology over pyrometallurgy. Hence, the Bayer process, the Cyanidation process, the Sherritt Gordon process, the HPAL laterite process, copper Solvent Extraction, the Zinc electrowinning process, and the Betts electro refining process as described in this paper all still stand today as viable competitive hydrometallurgical technologies predicated on very distinct separation chemistries. Another author may choose different illustrative hydrometallurgical technologies other than those shown in this paper, but the underlying quality and high degree of separations will still be the key enabling aspect. Further, hydrometallurgy stands as the only technology suitable to treat low grade complex resources of metals. Hence, the selective separations from dilute solutions and then ready recovery to a final value added product are distinct technology leveraging capabilities of current and future hydrometallurgical technology. As well, as was part of the case for the zinc concentrate pressure oxidation implementation by Cominco at

Trail, British Columbia, the global perception, and often reality, of lessened environmental impacts and complications with hydrometallurgy relative to traditional pyrometallurgy will further drive the development and implementation of industrial hydrometallurgy innovations.

REFERENCES

Hydrometallurgical History and Background

Gupta, C.K and Mukherjee, T.K., 1990. Hydrometallurgy In Extraction Processes, Volume I & II, CRC press.

Habashi, F., 1987. One hundred years of cyanidation. Bulletin, Can. Inst. Min. Metall. 80 (905), 108–114, ed. M.L. Wayman, pp. 78–85, Canadian Institute of Mining and Metallurgy, Montreal.

Habashi, F., 1993a. A Textbook of Hydrometallurgy, Metallurgie Extractive Quebec, Sainte Foy, Quebec 1993, second edition, 1999, distributed by Laval University Bookstore 'Zone.'

Habashi, F., 1993b. The discovery and industrialization of the rare earths. Bull. Can. Inst. Min. Metall. 87 (976), 80–87.

Habashi, F., 1995. Bayer's process for alumina production: a historical perspective. Bull. Hist. Chem. (17118), 15–19.

Habashi, F., 1998. Principles of Extractive Metallurgy- Amalgam and Electrometallurgy, vol. 4. Metallurgie Extractive Quebec, Laval University Bookstore Zone, Quebec City, pp. 15–16.

Habashi, F., 1999. Solvent extraction in hydrometallurgy. A historical perspective. Bull. Can. Inst. Min. Metall. 92 (1033), I 03–106.

Habashi, F., 1999. Textbook of Hydrometallurgy, Second Edition, Laval University, Quebec, City, Canada.

Habashi, F., 2005. A Short History of Hydrometallurgy, Hydrometallurgy 79 Elsevier, pp. 15–22.

Habashi, F., 2009. A history of nickel. In J. Liu, J. Peacey & M. Barati et al. (Eds.), Pvrometallurgy of nickel and cobalt 2009, proceedings of the international symposium (Taylor A., ed. pp. 77–98). Metallurgical Society of CIM.

Ingallis, V.R., 1936. History of the metallurgy of zinc, Metallurgy of Lead and Zinc. AIME, New York, pp. 339–373.

The Bayer Process

Adamson, A.N., 1956. Alumina Production: Principles and Practice, The Chemical Engineer, June pp. 156–164,171.

Adamson, A.N., Bloore, E.J., and Carr, A.R., 1963. Basic Principles of Bayer Process Design, Extractive Metallurgy of Aluminum Vol. 1, Alumina, Interscience, pp. 23–58.

Australian Government Department of the Environment, Water, Heritage and the Arts. 1992. NOPI (National Pollutant Inventory) Emission Estimation Technique Manual for Alumina Refining, Version 2.0.

Bayer, K.J., 1888. Darsteeling von Thonerhydrat and Alkalialuminat, German patent 43,977.

Bayer, K.J., 1892. Darsteeling von Thonerhydrat and Alkalialuminat, German patent 65,604.

Cundiff, W.H., and Reese, K.M., 1955. In Aluminum Production the First Stage is Alumina, Industrial and Engineering Chemistry, Sept. pp. 1673–1675.

Cundiff, W.H., 1972. Influence of Bauxite Composition on Alumina Plant Technology, Proceedings, 3rd International Symposium of ICSOBA, Banska Bystrica, CSSR, Research Institute for Non-Ferrous Metals, Budapest.

Cundiff, W.H., 1985. Alumina, SME Mineral Processing Handbook, Chapter 19, 20 pages.

Entner, P. 2013. http://www.peter-entner.com/e/theory/PrincHH/PrincHH.aspx.

Gerard, G. 1963. Extractive Metallurgy of Aluminum, Vol. 2, Interscience.

Habashi, F.H., 1988. A Hundred Years of the Bayer Process for Alumina Production, CIM Bulletin, January, Volume 81, No. 909, pp. 70–74.

Hose, H.R., 1963. Bauxite Mineralogy, Extractive Metallurgy of Aluminum, Vol. 1, Interscience, p. 6.

Jamey, M., 1956. Contributions to the Kinetics of Alumina Trihydrate Precipitation, AIME Meeting, February, New York.

Pearson, T.G., 1955. The Chemical Background of the Aluminum Industry, The Royal Institute of Chemistry, London, Lecturer, Monographs, Reports, No. 3.

Perry, K.W., 1970. Modifications to Bauxite Digestion, Bauxite Alumina, Aluminum, Vol. 1, Proceedings, 2nd International Symposium of ICSOBA, Research Institute for Non-Ferrous Metals, Budapest.

Porter, J.L., 1961. Process of purifying caustic aluminate liquors, US Patent 2981600A.

Schwalbach, V., 1971. The History of Continuous Red Mud Filtration and Methods of Aluminate Liquor Clarification and Hydrate Separation, Bauxite-Alumina-Aluminum, Vol. 3, Proceedings 2nd International Symposium of ICSOBA, Research Institute for Non-Ferrous Metals, Budapest, pp. 187–195.

Scott, I., 1963. Effect of Seed and Temperature on the Particle Size of Bayer Hydrate, Extractive Metallurgy of Aluminum, Vol. I, Alumina, Interscience, pp. 203–218.

Shimosato, J., Kogyo Kagaka Zanski, 1963. Vol. 66, No. 2, pp. 172–177.

Soymar, K., 1970. Bauxite-Alumina-Aluminum, Vol. 3, Alumina, Proceedings, 2nd International Symposium of ICSOBA, Research Institute for Non-Ferrous Metals, Budapest, pp. 45–65.

Staudinger, J., 1971. Some Problems in Underground Bauxite Mining, Proceedings, 2nd International Symposium of ICSOBA, Vol. 2, Research Institute for Non-Ferrous Metals, Budapest, pp. 253–261.

Cyanidation Process

Adams, M.D. 1990. The chemical behavior of cyanide in the extraction of gold. 1. Kinetics of cyanide loss in the presence and absence of activated carbon. Journal of South African Institute of Mining and Metallurgy 90(2):37–44.

Barsky, G., S.J. Swainson, and N. Hedley. 1934. Dissolution of gold and silver in cyanide solution. Transactions A of the Institution of Mining and Metallurgy 712: 66A-677.

Capone, S., A. Robertis, and S. Sammartano, 1986. Studies on hexacyano-ferrate(ll) complexes: Formation constants for alkali metals. Thennochemica Acta 772(3):7–74.

Cathro, K.J. 1963. The effect of oxygen in the cyanide process for gold recovery. Pages 181–205 in Proceedings Australian Institute of Mining & Metallurgy. Carlton, Victoria, Australia: Australasian Institute of Mining & Metallurgy.

Cathro, K.J., and D.F.A. Koch. 1964. The anodic dissolution of gold in cyanide solutions. Journal of Electochemical Society 117:L416–1420.

Cathro, K.J., and A. Walkley. 1961. The cyanidation of Gold. CSIRO publication Melbourne, Australia: CSIRO.

Finkelstein, N.P. 1972. The chemistry of the extraction of gold from its ores. Pages 284–351 in Gold Metallurgry on the Witwatersrand. Edited by R.J. Adamson. Cape Town, South Africa: Cape & Transvaal Printers Ltd.

Fivas, C.E., 1988. Presidential Address: How the MacArthur-Forrest cyanidation process ensured South Africa's golden future, Journal of the South African Institute of Mining and Metallurgy, September.

Habashi, F. 1966. The theory of Cyanidation. Transactions of the Mineralogical society of NME 235: 236–239.

Heath, A.R., and J.A. Rumball. 1998. Optimizing cyanide:oxygen ratios in gold CIP/CIL circuits. Minerals Engineering 1 1 (1 1): 999–1010.

Hiskey J.B., and V.M. Sanchez. 1990. Mechanistic and kinetic aspects of silver dissolution in cyanide solutions. Journal of Applied Electrochemistry.

Hogfeldt, E. 7982. Stability Constants of Metal-Ion Complexes. Port A: Inorganic Ligands. No. 21. New York: Pergamon Press.

Jeffrey, M., and I. Ritchie. 2000. Electrochemical aspects of gold cyanidation. Pages 176–186 in Proceedings of Electrochemistry in Mineral & Metal Processing V. Edited by R. Woods and F.M. Doyle. Pennington, NJ: The Electrochemical Society.

Jeffrey, M.I., I. Chandra, I.M. Ritchie, G.A. Hope, K. Waling, and R. Woods. 2005. Innovations in gold leaching research and development. Pages 207–221 in lnnovations in Natural Resource Processing: Proceedings of Jan D. Miller Symposium. Edited by C.A. Young, J.J. Kellar, M.L. Free, J. Drelich, and R.P. King. Littleton, CO: SME.

Kirk, D.W., F.R. Foulkes, and W.F. Graydon. 1978. A study of anodic dissolution of gold in aqueous alkaline cyanide. Journal of Electrochemical Society 7257436–1443.

Kudryk, V., and H.H. Kellogg. 1954. The mechanism and rate controlling factors in the dissolution of gold in cyanide solution. Journal of Metals 6(5):541–548.

Lorenzen, L., and J.S.J. van Deventer. 1992. Electrochemical interactions between gold and its associated minerals during cyanidation. Hydrometallurgy 3A:777–704.

MacArthur, J.S. et. al, 1888. Improvements in obtaining gold and silver from ores and other compounds—British patent # 14174, August 10.

Marsden, J.O., and House, C.I., 2006. The Chemistry of Gold Extraction, Second Edition, SME.

Nicol, M.J. 1980. The anodic behavior of gold. Part II. Oxidation in alkaline solutions. Gold Bulletin 13: 105–111.

Nicol, M.J., C.A. Fleming, and R.L. Paul. 1987. The chemistry of the extraction of gold. Pages 831–905 in The Extractive Metallurgy of Gold. Monograph M7. Edited by G.G. Stanley. Johannesburg: South African Institute of Mining and Metallurgy.

Sharpe, A.G. 1976. The Chemistry of Cyano-Complexes of the Transition Metals. London: Academic Press.

Sillen, L.G., and A.E. Martell. 1964. Stability Constants of Metal-lon Complexes. Publication No. 17. London: The Chemical Society.

Sillen, L.G., and A.E. Martell. 1970. Stability Constants of Metal-Ion Complexes. Supplement No. 1. Publication No. 25. London. The Chemical Society.

Wang, X., and K.S.E. Forssberg. 1990. The chemistry of cyanide-metal complexes in relation to hydrometallurgical processes of precious metals. Mineral Processing & Extractive Metallurgy Reviews 6:81–725.

Zurilla, R.W., R.K. Sen, and E.Yeager. 1978. The kinetics of oxygen reduction reaction on gold in alkaline solution. Journal of Electrochemical Society 125:1103–1109.

Sherritt Gordon Process and HPAL Laterite Process

Baragwanath, J., 1936. *Pay Streak*, Garden City, N.Y., Doubleday, Doran & Co., Inc.

Baragwanath, J., 1961. *A Good Time Was Had*, New York, N.Y., Appleton-Century-Crofts, Inc.

Battle, T. Moats, M., Cocalia, V. et al. Eds, 2013. Ni Co 2013, Proceedings of a Ni Co Symposium, TMS, Wiley, March.

Boldt, J.R., Jr., 1967. *The Winning of Nickel*, ed. By P. Queneau, Princeton, N.J., D. Van Nostrand Co., Inc.

Budac, J.J. Fraser, R. & Mihaylov, I. (Eds.), 2009. Hydrometallurgy of Nickel and Cobalt 2009. Metallurgical Society of CIM.

Burkin, A.R., 1987. Extractive metallurgy of nickel, Society of (British) Chemical Industry.

Chalkley, M.E. & Toirac, I.L., 1997. The acid pressure leach process for nickel and cobalt laterite. In W.C. Cooper & I. Mihaylov (Eds.), Nickel/cobalt 97, Vol. I, Hydrometallurgy and refining of nickel and cobalt, pp. 341–353, CIM.

Cooper, W.C. and Mihaylov, I., Eds., Nickel Cobalt 97, Vol. 1. Hydrometallurgy and Refining of Nickel and Cobalt, The Metallurgical Society of CIM, Montreal, Canada.

Cox, J.S., 1911. The Iron-Ore Deposits of the Moa District, Oriente Province, Island of Cuba, *AIME Transactions 42*, 73–90.

Crundwell, F.K., Moats, M.S., Ramachandran, V., Robinson, T.G. and Davenport, W.G. 2011. Extractive Metallurgy of Nickel, Cobalt and Platinum-Group Metals, Elsevier, Oxford, UK.

Donald, J. & Schonewille, R. (Eds.), 2005. Nickel and cobalt 2005, challenges in extraction and production, Metallurgical Society of CIM.

Evans, D.J. 1979. International Laterite Symposium: Proceedings, New Orleans, Louisiana, February 19 to 21, 1979 Society of Mining Engineers of the American Institute of Mining, Metallurgical, and Petroleum Engineers, 688 pages.

Forward, F.A., and Mackiw, V.N., 1955. Chemistry of the Ammonia Pressure Process for Leaching Nickel Copper and Cobalt from Sherritt Gordon Sulfide Concentrates, J. of Metals, 7, pp. 457–462.

Gupta, C.K and Mukherjee, T.K., 1990. Hydrometallurgy In Extraction Processes, Volume I, CRC press.

Habashi, F., 2009. A history of nickel. In J. Liu, J. Peacey & M. Barati, et al. (Eds.), Pvrometallurgy of nickel and cobalt 2009, proceedings of the international symposium (Taylor A., ed. (pp. 77–98). Metallurgical Society of CIM.

Imrie, W.P. Lane, D.M. & Barnett. S.C.C. (Eds.), 2004. International Laterite Nickel Symposium, TMS.

Kerfoot, D.G.E., 1989. The Development of the Sherritt ammonia pressure leach process:, CIM Bulletin, V. 82, No. 926, June, p. 136–141.

Kerfoot, D.G.E., and Cordingley, P.D., 1997. The acid pressure leach process for nickel and cobalt late rites. Part III: Review of operations at Fort Saskatchewan, in Cooper, W.C. and Mihaylov, I., Eds., Nickel Cobalt 97, Vol. 1. Hydrometallurgy and Refining of Nickel and Cobalt, The Metallurgical Society of CIM, Montreal, Canada, pp. 355–369.

Kerfoot, D.G. E., Collins, M.J., Holloway, P.C. and Schonewille, R.H., 2012. The Nickel Industry In Canada, The Canadian Metallurgical and Materials Landscape 1960 to 2011, CIM.

Kofluk, R.P., & Freeman. G.K.W. 2006. Iron control in the Moa Bay laterite operation. In J.E. Dutrizac & P.A. Riveros (Eds.). Iron Control Technologies, pp. 573–589, CIM.

Lewis, A.E., 2010. Review of metal sulfide precipitation. Hydrometallurgy, 104,222–234.

Little, J.E., 1911. The Mayari Iron-Mines, Oriente Province, Island of Cuba, as Developed by the Spanish American Iron Co., *AIME Transactions 42*, 152–169.

Mackiw, V.N. and Benz, T.W., 1961. Application of Pressure Hydrometallurgy to the Production of Metallic Cobalt, Extractive Metallurgy of Copper Nickel and Cobalt, Interscience Publishers, New York, AIME, 503.

Matos, R.R., 1997. Industrial experience with the Ni/Co sulphide precipitation process. In W.C. Cooper & I. Mihaylov (Eds.), Nickel/cobalt 97, Vol. 1, Hvdrometallurgy and refining of nickel and cobalt, pp. 371–378, CIM.

Meddings, B. and Mackiw, V., 1963. The Gaseous Reduction of Metals From Aqueous Solutions, Unit Processes in Hydrometallurgy, AIME, Dallas, pp. 345–384.

Molina, N., 2009. Nickel and cobalt sulphide precipitation, a proven method of selective metal precipitation in laterite process flowsheets. In J.J. Budac, R. Fraser, I. Mihaylov. V.C. Papangelakis & D.J. Robinson (Eds.). Hydrometallurgy of nickel and cobalt 2009 pp. 271–281, CIM.

Reddy, R.G. and Weizenbach, R.N., 1993. Extractive Metallurgy of Copper, Nickel and Cobalt, Proceedings of the Paul E. Queneau International Symposium, TMS, Vol. 1 & 2.

Rare Earths Ion Exchange

Asher, D.R. et al. 1962. *Yttrium Oxide of High Purity; DTPA–Ion Exchange Process.* I&EC Process Design and Development, Vol. 1, pp. 52–56.

Atanasyants, A.G. and Seryogin, A.N. 1995. *The Reaction of the Electrochemical Reduction Eu(III)+e = Eu(II) in Hydrochloric Solution.* Hydrometallurgy, Vol. 37, pp. 367–374.

Campbell, D.O. 1973. *Rapid Rare Earth Separation by Pressurized Ion Exchange Chromatography.* Journal of Inorganic Nuclear Chemistry, Vol. 35, pp. 3911–3919.

Choppin, G.RA. 1993. *A Half Century of Lanthanide Aminopolycarboxylates.* Journal of Alloys and Compounds, Vol. 192, pp. 256–261.

Cotton, F.A. et al. 1999. *Advanced Inorganic Chemistry.* New York: Wiley, 0-471-19957-5.

Demets, M., Morino, L. and Fernando, L. 1993. *Procede Pour L'extraction de Cerium d'une Solution Aqueuse d'un Melange D'elements des Terres Rares. 92203314.7.*

Gschneidner, K 2013. Personal communication, Ames Laboratory.

Gupta, C.K. and Krishnamurthy, N. 2005. *Extractive Metallurgy of Rare Earths.* Boca Raton: CRC Press, 0-415-33340-7.

McCoy, H.N. 1936. *The Extraction of Europium from Monazite Residues and the Preparation of Pure Europium Compounds,* Journal of the American Chemical Society, Vol. 58, pp. 2279–2281.

Powell, J E and Burkholder, H.R. 1968. *Augmenting the Separation of Adjacent Pairs of Rare Earths in Elutions With Hydroxyethylenediamineacetate (HEDTA).* Journal of Chromatography, Vol. 36, pp. 99–104.

Sayed, S.A., Rabie, K.A. and Salama, I.E., 2005. *Studies on Europium Separation From a Middle Rare Earth Concentrate by In Situ Reduction Technique.* Separation and Purification Technology, Vol. 46.

Spedding, F.H. 1947, *The Separation of Rare Earths by Ion Exchange. Pilot Plant Scale Operations.* J. Am. Chem. Soc, pp. 2812–2818.

Spedding, F.H., Wheelright, E.J. and Schwarzenbach, G. 1953. *The Stability of the Rare Earth Complexes With Ethylenediamenetetraacetic Acid.*, Journal of the American Chemical Society, Vol. 75, pp. 4196–4201.

Spedding, F.H., Powell, J.E. and Wheelwright, E.J. 1954. *The Use of Copper as the Retaining Ion in the Elution of Rare Earths With Ammonium Ethylenediamine Tetraacetate Solutions*, Journal of the American Chemical Society, Vol. 76, pp. 2557–2560.

Spedding, F.H., Powell, J.E. and James, D.B., 1960. *Cation-Exchange Elution Sequences–Divalent and Rare Earth Cations With EDTA, HEDTA and Citrate.* Journal of Inorganic Nuclear Chemistry, Vol. 19, pp. 133–141.

Vasudevan, S. et al. 2005. *An Electrochemical Process for the Separation of Cerium From Rare Earths.* Hydrometallurgy, Vol. 76, pp. 115–121.

Vickery, R.C. 1948. *The Influence of Basicity Upon the Efficiency of Oxidation/Hydrolysis Procedures For the Separation and Purification of Cerium.* Journal of the Society of Chemical Industry, Vol. 67, pp. 333–336.

Vickery, R.C. 1950. *Some Reactions of Cerium and Other Rare Earths with Chlorine and Hypochlorite.* Journal of the Society of Chemical Industry, Vol. 69, pp. 122–125.

Kronholm, B., Anderson, C.G. and Taylor, P.R. 2013. A Primer on Rare Earths Separations, Journal of Metals, Vol. 68, No. 10, TMS.

Copper Solvent Extraction

Acorga, 1990. Personal communication.

Agers, D.W. and DeMent, E.R., 1972. The Evaluation of New LIX Reagents for the Extraction of copper and Suggestions for the Design of commercial Mixer-settler Plants, TMS preprint 72–87, Warrendale, Pennsylvania.

Biswas, A.K. and Davenport, W.G., 1994. Extractive Metallurgy of Copper, 3rd Edition, Pergamon Press.

Dalton, R.F., Severs, K.J. and Stephans, G., 1936. Advances in Solvent Extraction for Copper by Optimized Use of Modifiers, paper presented at Mining Latin America, Chile, November 1986.

Henke, 1990,The Chemistry of Metals Recovery Using LIX Reagents, 1990–1991 edition, Henkel Corporation, Tucson, Arizona.

House, J.E., 1989. The Development of the LIX Reagents, Minerals and Metallurgical Processing, 6(1), 1–6.

Jenkins, J.G., 1991. Magma San Manuel SX-EW Improvements in Phase Disengagement Time of circuit organic and Recovery of organic from the Raffinate Pond through clay Treatment, paper presented at AIME Arizona conference Annual Meeting, Tucson, Arizona, December.

Jenkins, J.G. and Eamon, M.A., 1990. Plant Practices and Innovations at Magma Copper Company's San Manuel SX-EW Plant, in Electrometallurgical Plant Practice edited by Claessens, P.L. and Harris G.B., Pergamon press, New York, 41–56.

Kordosky, G.A.,1992. Copper Solvent Extraction: The State of the Art, Journal of Metals, 44(5), 40–45.

Kordosky, G., 2002. Copper recovery using leach/solvent extraction/electrowinning technology: Forty years of innovation, 2.2 million tonnes of copper annually, Journal of the South African Institute of Mining and Metallurgy, November-December, pp. 445–450.

Kordosky, G.A., Olafson, S.M., Lewis, R.G., Deffner, V.L. and House, J.E., 1987. A State-of the-Art Discussion on the Solvent Extraction Reagents used for the Recovery of copper from Dilute sulfuric Acid Leach solutions, Separation Science and Technology 22 (2 and 3), Marcel Dekker Inc., New York, 215–233.

Phelps Dodge, 1993. handout from AIME Annual Hydrometallurgy Meeting, hosted by Phelps Dodge Mining Company, Hurley, New Mexico, May 22.

Phillips, 1992. Product Information Sheet, Phillips petroleum company, Bartlesville, Oklahoma.

Readett, D. and Clayton, B. 1993. Cleaning Hydrometallurgical Liquor Using Jameson Cells, in Flotation Plants: Are They Optimized? edited by Malhotra, D., SME, Littleton, Colorado, 165–170.

Ritcey, G.M. and Ashbrook, A.W., 1984. Solvent Extraction: Principles and Applications to Process Metallurgy, Part I, Elsevier-Science Publishing Company, New York.

Townson, B., 1989. Advances in Copper Solvent Extraction, paper presented at Non-ferrous Smelting Symposium, Port Pirie, South Australia, September.

Zinc Concentrate Pressure Oxidation, Zinc Electrowinning and Betts Lead Electrorefining

Aird, J., Celmer, R.S., and May, A.V., 1980. New Cobalt production From Chambishi's Roast Leach Electrowon process, paper presented at the 10th annual Hydrometallurgical meeting (CIM) Edmonton.

Betts, AG. 1908. Lead Electrorefining by Electrolysis, John Wiley and Sons, New York.

Caufriez, A., Dubois, M., Lejay, E. Zinc Cellhouses, 1998. Concept and practice at Union. Mineire, pp. 337–352, Zinc and Lead Processing, Dutrizac, J.E., Gonzales, J.A., Bolton, G.L., Hancock, H., ed Aug 1998. Calgary, Canada, The Metallurgical Society of CIM, August.

Chalkley, M.E., Collins, M.J., Masters, I.M. and Ozberk, E., 1982. The Integration of the Sherritt Zinc pressure Leach Process with Commercial Iron Precipitation Processes, Hydrometallurgy.

Forward, F.A., and Veltman, H., 1959. Direct Leaching Zinc Sulfide Concentrates by Sherritt Gordon, J. Met., 12, 836.

Free, M., Moats, M., Robinson T.G., Neelameggham, N., Houlachi, G., Ginatta M., Creber D., Holywell G. 2011. Electrometallurgy–Now and in the Future, Electrometallurgy 2011,TMS, Orlando.

Gonzales, J.A., 1991. Electrochemical Processes Within the Slimes Layer of Lead Anodes During Betts Electrorefining, PhD Thesis UBC March.

Gonzalez-Dominguez, J.A., Dreisinger, D.B., 1997. Identifying Research Opportunities in Zinc Electrowinning, 1997 Review of Extraction and Processing, Journal of Metals, April.

Gordon, A.R. and Pickering, R.W., 1975. Improved Leaching Technologies in the Electolytic Zinc Industry, Metallurgical Transactions B, Volume 6B, March, 43.

Hiraki, T., Ogata, M., M. 1998. An Outline of the lead electrolysis operation at the Chigirishima Refinery pp. 379–388, Zinc and Lead Processing, Dutrizac, J.E., Gonzales, J.A., Bolton, G.L., Hancock, H., ed. Aug, Calgary, Canada, 37th Annual Conference of Metallurgists of CIM.

Krauss, C.J. and Shaw G., 1976a, Can. Patent 988879, May 11.

Krauss, C.J., 1976b, Can. Patent 1020491, Nov 8.

Kubota, H., Kusakabe, T., Takei, K., Takewaki, M., 1998. Current operation of Sumitomo Metal Mining's (SMM) Betts Lead electrorefinery. Zinc and Lead from steel mill dusts pp. 353–366, Zinc and Lead Processing, Dutrizac, J.E., Gonzales, J.A., Bolton, G.L., Hancock, H., ed Aug, Calgary, Canada, 37th Annual Conference of Metallurgists of CIM.

Letrange, L.,1883. US. Patent 286208.

Mather, FC, Lantz GB. 1953. US Patent 2664393, Dec 29.

Parker, E.G., 1981. Oxidative Pressure Leaching of Zinc Concentrates, CIM Bull., 74, 145.

Parker, E.G., McKay, D.R., and Salomon De-Friedberg, 1983. Zinc Pressure Leaching at Cominco's Trail operation, in Proc. 3rd Int. Symp. Hydromet. Research, Development and Plant Practice, Osseo-Asare, K. and J.D. Miller, Eds., TMS, AIME, New York, 927.

Ralston O.C., 1921. Electrolytic Deposition and Hydrometallurgy of Zinc, McGraw-Hill Book Co Inc. New York.

Robinson T.G., 2003. Copper Elecrowinning—a Historical Perspective, Hydrometallurgy 2003, TMS, Vancouver, August.

Robinson T.G., Sole, K.C., Moats, M.C., Crundwell FK, Morimitsu, M, Palmu, L. 2011. Developments in Base Metal Electrowinning Cellhouse Design, Electrometallurgy 2011, TMS, Orlando.

Thomas, D.L., Krauss, C.J., Kerby R.C. 1980. Betts Lead Electrorefining at Cominco, Lead Zinc TMS.

ABOUT THE AUTHOR

Corby G. Anderson has more than 34 years of global experience in industrial operations, management, engineering, design, consulting, teaching, research, and professional service. His career includes positions with Morton Thiokol, Key Tronic Corporation, Sunshine Mining and Refining Company, H.A. Simons Ltd., and Montana Tech. Anderson holds a BSc in chemical engineering and an MSc and PhD in metallurgical engineering. He is a fellow of the Institution of Chemical Engineers and the Institute of Materials, Minerals and Mining. He currently serves as the Harrison western professor in the Kroll Institute for Extractive Metallurgy as part of the George S. Ansell Department of Metallurgical and Materials Engineering at the Colorado School of Mines. In 2009 he was honored by SME with the Milton E. Wadsworth Extractive Metallurgy Award for his contributions in hydrometallurgical research.

Pyrometallurgy

Brief Overview of Some Innovations in Pyrometallurgy

Patrick R. Taylor
Kroll Institute for Extractive Metallurgy, Colorado School of Mines, Golden, Colorado, United States

ABSTRACT: A brief overview of some of the innovations in pyrometallurgy is presented. It is a very large subject as much of the metals that are used are produced by pyrometallurigcal methods and there have been significant advances in nearly all aspects of the field. Most of the innovations can be attributed to the need for reduced operating costs through improved productivity; the need to treat more complex and difficult feed materials; the need to meet increasing environmental standards; and to an increased understanding of the fundamental physical chemistry and process metallurgy applied to pyrometallurgy. An attempt has been made to illustrate innovations through several examples and to point out where additional innovations might occur.

INTRODUCTION

Any discussion regarding innovations in pyrometallurgy should include some mention of the numerous engineers that have contributed to these efforts and the numerous technical publications that document these innovations. This is clearly impossible in a brief overview, so the author has chosen to highlight only a few of those technologies. This discussion neglects much of the work that might have been discussed. The author apologizes to everyone who believes that they, or someone else they know, should have been mentioned as part of this review.

Pyrometallurgy has made many advances over the past several decades, but the start of a significant amount of the current innovations really began right after World War II when many countries were rebuilt and many economies expanded. At the same time, several Universities began to develop new educational and research programs to address fundamental topics in pyrometallurgy and to train engineering students with new tools to be utilized in industry. The driving force for innovation has come primarily from the mining, minerals and metals companies in their efforts to improve productivity, increase profitability and meet ever increasingly stringent environmental regulations. These innovative changes were reflected in several university programs through the increased understanding that was incorporated into the classes and research laboratories and in the many company research facilities.

This overview will discuss several aspects of pyrometallurgy: fundamentals; ore preparation; and smelting and reduction. The author has selected examples from each topic to illustrate innovations; but there are many more that might have been mentioned.

PYROMETALLURGY FUNDAMENTALS

Beginning after World War II, several leading groups emerged that were involved in the increased understanding of the physical chemistry of process metallurgy. These efforts provided the foundation, and many of the engineers, that have helped develop many of the innovations in pyrometallurgy. One was located at Imperial College in London, England and another was at MIT in Cambridge, Massachusetts, USA. One of the first really fundamental research updates that gave an overview of the status of our fundamental understanding of pyrometallurgy was published in a general discussion of The Faraday Society in 1948. The group at Imperial College was particularly productive for a significant period of time up until the late 1960s. One book from 1967 included many familiar names discussing heat and mass transfer in process metallurgy (Hills 1967). Students from these programs went on to lead efforts at other universities and in industry. One could compile a long list of notable metallurgists that were trained through these efforts, the efforts at MIT and the many other universities that adopted this curriculum.

In addition, many mining, minerals and metals companies developed their own research facilities to evaluate new technology and applications for metal extraction, refining and product development. There are numerous examples of innovations that came out of these laboratories. Unfortunately, very few of these company research laboratories currently exist. Recently, much of the innovations in pyrometallurgy have come from equipment vendors, many times in conjunction with mining, minerals or metals companies.

Fundamentals in pyrometallurgy may be separated into the separate categories: thermodynamics; energy and materials balances; kinetics and reactor design; heat, mass and momentum transfer; and unit operations. Good overviews may be found in text books by Rosenqvist (2004), Alcock (1976), and Vignes (2011a, 2011b, 2011c).

The development of advanced modeling of various aspects of pyrometallurgy has also led to innovations. Models such as: Gibbs Energy Minimization, computational fluid dynamics and discrete element modeling have given the modern engineer very powerful tools to evaluate changes and improved design. Among the first leaders in these modeling efforts were Szekely, Themelis, Evans and Brimacombe (Szekely and Themelis 1971, Szekely 1979, Szekely et al. 1989).

ORE PREPARATION

Innovations in ore preparation are illustrated using two examples: iron ore induration and roasting technology.

Iron Ore Induration

The development of pelletization and induration to turn iron ore fines into consistent size and property pellets was a significant innovation that greatly affected two industries: the iron ore mining companies and the iron and steel making companies. A schematic of a grate-kiln process is shown in Figure 1. This technology supplanted sintering to prepare fine feed for the iron blast furnace. A significant amount of fossil fuel energy is used in the induration process and with ever increasing fossil fuel costs and some governmental concerns regarding carbon emissions, innovation is going to take place in this aspect of ore preparation.

The future of methods to treat fine iron ore concentrates will require new technological innovations. If certain countries decide to impose a tax on carbon emissions, then induration of iron ore pellets could become much more expensive. The innovative approaches that may be taken might include

- Use of alterative heat sources during induration (electricity in place of fossil fuels),
- Development of cold bonded pellets,

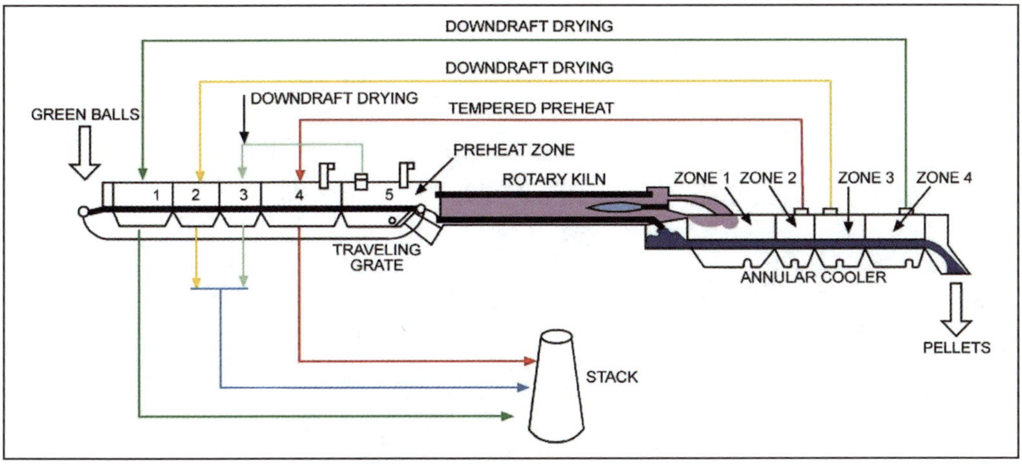

Figure 1. Schematic of a grate-kiln induration system (Trescot et al. 2004)

- Development of iron making technology that would use the iron ore fines directly, and
- Use of hydrogen as a reluctant to convert iron ore fines.

These topics are being actively pursued worldwide.

Roasting Technology

Much of the innovations in roasting technology have been related to an increased understanding of the fundamental behavior of fluidized beds. Two stage fluidized beds with oxygen enrichment have been developed for refractory precious metal ores. An example of such a two stage fluidized bed reactor is shown in Figure 2. The advantage of multiple stages is related to several factors, including: the ability to change conditions from one stage to another to affect specific chemical reactions; and, to enhance the residence time distributions of the solid particles in the bed, as two beds give a better overall probability of particles being in the reactor for a fixed period of time. In addition, the use of oxygen enriched air in roasting has enhanced the ability to operate thermally autogenously and providing a richer off gas for sulfuric acid production. Some of the areas where innovation is occurring in roasting might include

- Evaluating methods to treat ores with higher amounts of problematic elements, either by capture in the calcine during roasting (lime roasting), selective volatilization (Oudenne 2006) or by improved gas treatment systems;
- Continued design and scale up improvements; and
- Continued development and application of recirculating fluidized beds. The higher operating gas velocities provide for a greater difference in slip velocity and enhance heat and mass transfer rates and thus the kinetics of heterogeneous reactions.

SMELTING AND REDUCTION

Smelting

Flash smelting represents a significant innovation that was developed initially by INCO (Queneau 1996) and Outokumpu (Kojo 2000). The recognition that fine sulfide particles, reacted in an

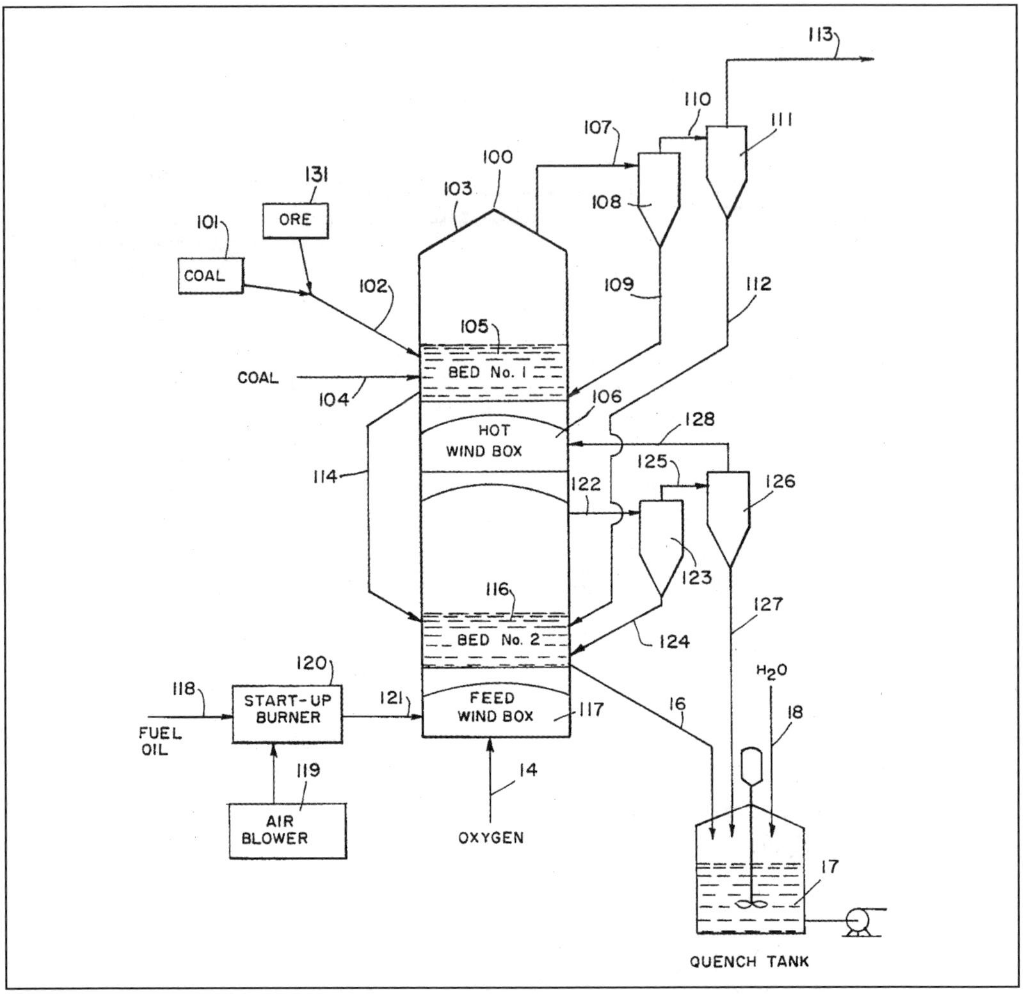

Figure 2. Schematic for a two-stage fluidized bed for refractory gold ores (Smith et al. 1990)

oxygen rich environment, in flight, would lead to a faster process that also could be thermally autogenesis was a significant breakthrough. Outokumpu's flash smelting process was developed after World War II in response to electric energy shortages in Finland. Another major advantage of using oxygen, or oxygen enriched air, was that the SO_2 in the produced gases was more economically recovered in the form of sulfuric acid. A schematic for the Outokumpu's flash smelting furnace is shown in Figure 3. A very good review of the various alternative copper pyrometallurgy technologies is given by Moskaluk and Alfantazi (2003).

Another innovation in smelting was the Noranda bath smelting process (Figure 4). The process is also energy efficient. The reactor is a refractory-lined cylindrical vessel and is used to smelt a broad range of copper-bearing materials (concentrates, electronic scrap, and industrial wastes). Fluxes, fossil fuel, and feedstock are injected into one end of the reactor via a high-speed belt, while oxygen-enriched air is forced into the liquid melt by means of submerged tuyeres. Additional process heat may be provided by supplementary fuel consisting of oil, natural gas, coal, or coke. Wet coal may be added with the solid charge without pulverizing or sizing the carbonaceous material. A high SO_2 concentration in furnace off-gases is suitable for the production of metallurgical grade sulfuric acid. The process allows a high recovery of copper and associated precious metals than some of the alternatives. This is enhanced by the stirred slag and capability of maintaining a low silica slag.

Possible innovations in smelting might include

- Development of methods to treat more complex concentrates,
- Development of more continuous processes, and
- Development of additional smelting systems to treat complex recycled electronic components.

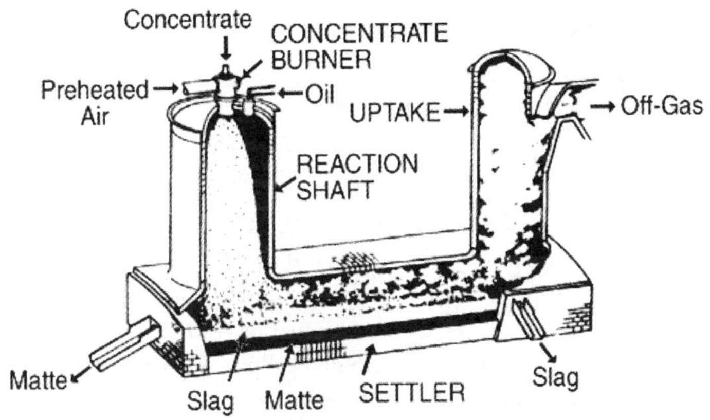

Figure 3. Outokumpu's flash smelting furnace (King 2007)

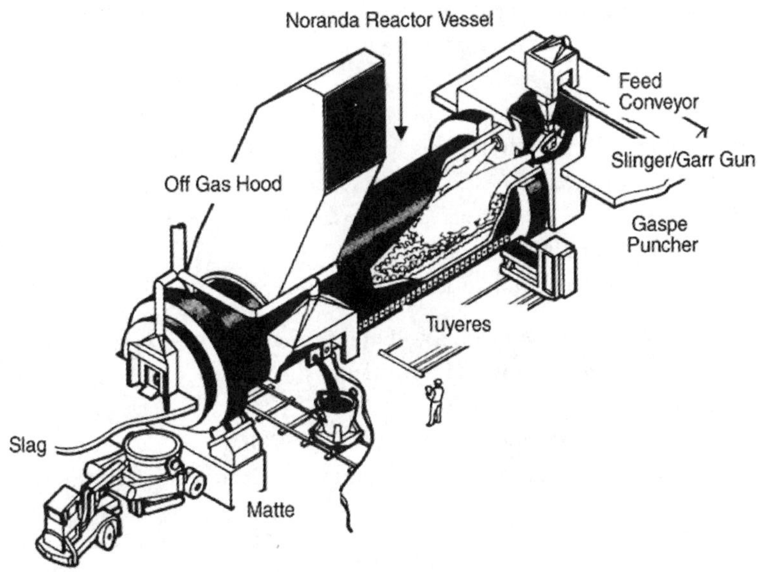

Figure 4. Noranda reactor (Moskaluk and Alfantazi 2003)

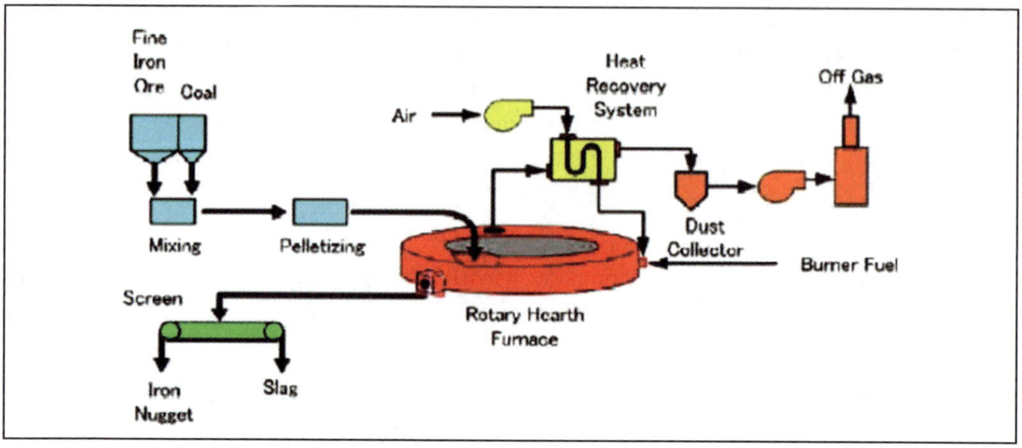

Figure 5. Midrex "nugget" process (Ishikawa et al. 2009)

Reduction

New iron making technology is being developed that will allow the use of iron ore fines, coal instead of coke, and feed materials with higher amounts of impurities. There is a long list of these alternative processes and only two are discussed here. First, Midrex (Ishokawa et al. 2009) has developed a number of alternative processes, but one that has received quite a bit of attention lately is the Nugget process. This process is shown schematically in Figure 5. This process uses a rotary hearth to reduce un-sintered pellets at temperatures below the operating range of blast furnaces. The discovery that the reduced iron tends to migrate and form nuggets during the process led to this effort. The process uses coal instead of coke and is said to allow for higher impurity levels in the feed. The process has been piloted and a commercial scale process is being developed.

Another approach to treating fine iron ores is the HIsmelt process (Figure 6). This process uses iron ore fines, coal instead of coke, is air based and is said to produce flexibility and premium grade products. The thermal efficiency is enhanced by post-combustion. Coal is injected through bottom tuyers into the molten bath. Carbon is dissolved and reacts with the oxygen from incoming iron ore to from carbon monoxide and iron. This reaction is endothermic and additional heat has to be supplied. To provide this needed heat, carbon monoxide released from the bath is reacted with oxygen from top injection of air. The reacted hot gasses exit the vessel and are used in a fluidized bed to pre-heat and pre-reduce incoming ore.

Some of the innovations that will be addressed in iron reduction might include:

- Further development of iron making technology that utilizes iron ores fines, coal instead of coke, allows for greater amounts of impurities in the feed and provides an acceptable quality pig iron for steel production;
- The use of hydrogen as a reductant for iron ore fines; and
- Development of co-generation facilities as part of the iron reduction process.

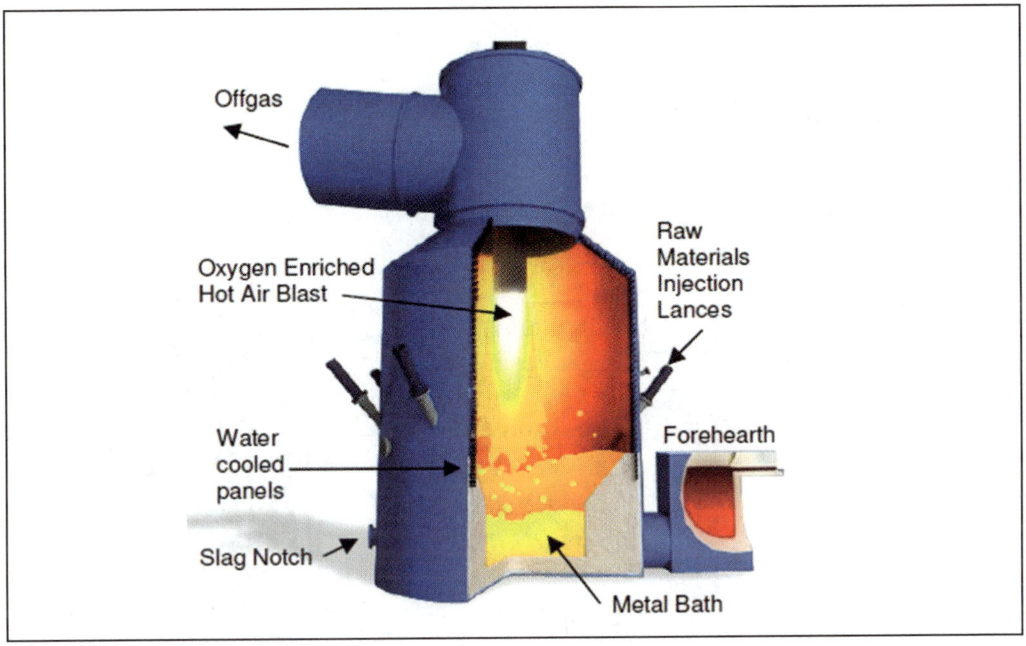

Figure 6. HIsmelt process (Goodman 2007)

DC Arc Furnace

A number of significant innovative contributions in pyrometallurgy have been developed at Mintek in South Africa. A good review of much of this work may be found in the paper by Jones and Curr (2006). One example is shown in Figure 7 for the Mintek Thermal Magnesium Process (Abellatif 2011). This process uses a DC arc furnace to convert calcined dolemite with aluminum and ferrosilicon to produce magnesium at 1,650–1,750°C.

SUMMARY

Several examples of innovations in pyrometallurgy are given. Each is briefly discussed and possible additional innovations are identified. The field of pyrometallurgy is an area of active improvements and innovations. Some of the specific advantages of pyrometallurgical processing might be illustrated by looking at the number of smelters that now take used circuit boards and other high value used electronics as feed to their operations. Smelters such as Umicore (Hageluken 2006) and Boliden (Lehner and Vikdahl 1998) have adapted their processes to take more recycled materials. The versatility of high temperature, molten metal operations are ideally suited to recover many of the valuable and critical metals being used in society today. Continued efforts on the part of the equipment vendors, the mining minerals and metals companies, and the process metallurgy oriented universities will lead to additional innovations. These innovations will be driven by: the need to treat more difficult and complex feeds; to increase productivity and reduce costs: and by ever increasing environmental requirements.

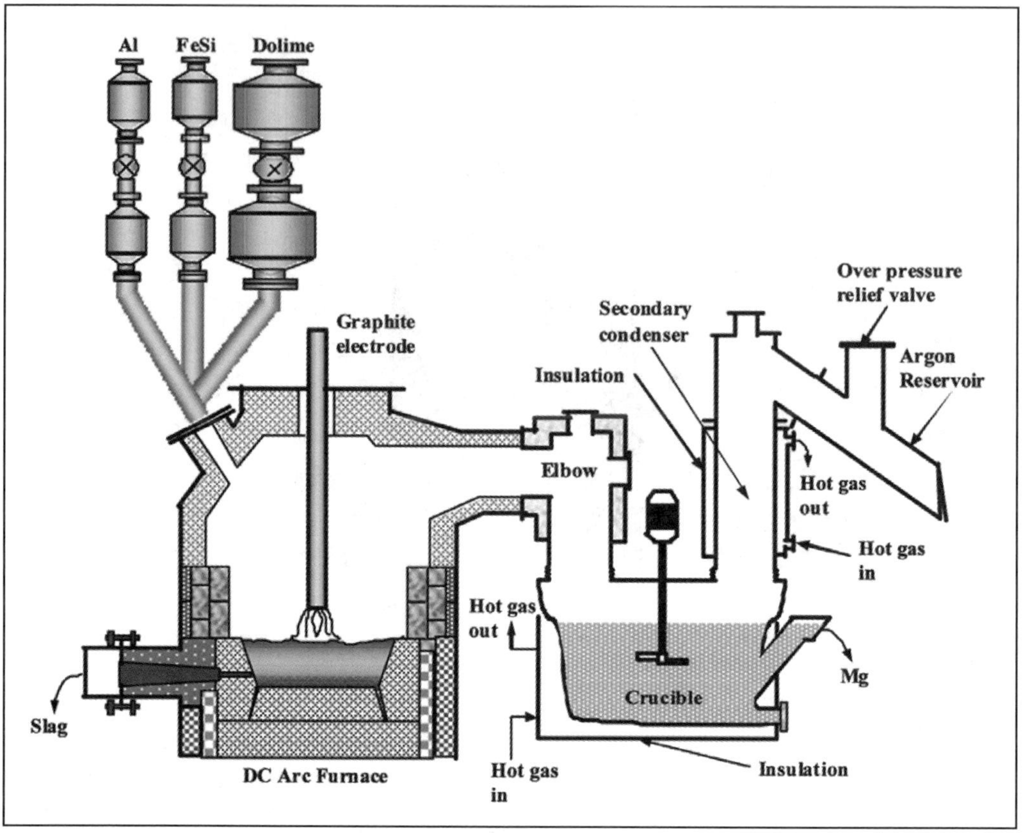

Figure 7. Schematic of MTMP pilot plant (Abdellatif 2011)

REFERENCES

Abdellatif, M., 2011. Review of the development work on the Mintek Thermal Magnesium Process (MTMP), *SAIMM*, Vol. 111(6), pp. 398–403.

Alcock, C.B., 1976. *Principles of Pyrometallurgy*, Academic Press, New York.

Goodman, N. 2007. *Operations at the HIsmelt Kwinana Plant*. HIsmelt Corporation. Available at: www.hismelt.com/EN/Documents/TechnicalPapers/AIST2007.PDF.

Hagelucken, C. 2006. Recycling of Electronic Scrap at Umicore Precious Metals Refining, *Acta Metallurgica Slovaca*, 12, pp. 111–120.

Hills, A.W.D. 1967. *Heat and Mass Transfer in Process Metallurgy*, The Institution of Mining and Metallurgy, London.

HIsmelt, 2005. http://www.hismelt.com/technology/.

Ishikawa, H., Kopfle, J., McClelland, J. and Ripke, J. 2009. Rotary Hearth Furnace Technologies for Iron Ore and Recycling Applications, *Archives of Metallurgy and Materials*, 53 (2) pp. 541–545.

Jones, R.T. and Curr, T.R., 2006. Pyrometallurgy at Mintek, *South African Pyrometallurgy*, Edited by R.T. Jones, SAIMM, Johannesburg, pp. 127–150.

King, M.G., 2007. The Evolution of Technology for Extractive Metallurgy over the Last 50 Years—Is the Best Yet to Come?, *JOM*, pp. 21–27, Apr.

Kojo, I.V., Jokilaakso,A. and Hanniala, P., 2000. Flash Smelting and Converting Furnaces: A 50 Year Retrospect, *JOM*, pp. 57–61, Feb.

Lehner, T. and Vikdahl, A. 1998. Integrated recycling of non-ferrous metals at Boliden Ltd Ronnskar smelter. *Sulfide Smelting '98: Current and Future Practices*; San Antonio, TX; USA; pp. 353–362.

Moskaluk, R.R. and Alfantazi, A.M., 2003. Review of copper pyrometallurgical practice: today and tomorrow, *Minerals Engineering* 16, pp. 893–919.

Oudenne, P.D., 2006. The Stage Selective Roasting, an Efficient Way for the Treatment of Arsenical and Antimonical Bearing Concentrates and Other Complex Metallurgical By-products, *Sohn International Symposium*, Edited by F. Kongoli and R.G. Reddy. TMS.

Queneau, P.B. and Marcuson, S.W., 1996. Oxygen Pyrometallurgy at Copper Cliff—a Half Century of Progress, *JOM*, pp. 12–21. Jan.

Rosenqvist, T., 2004. *Principles of Extractive Metallurgy*, Tapir Academic Press, Trondhiem.

Smith, J.C., McCord, T.H. and O'Neil, G.R., 1990, *Treating Refractory Gold Ores via Oxygen-Enriched Roasting*, US Patent 5,919,715, April 24.

Szekely J., 1979. *Fluid Flow Phenomena in Metals Processing*, Academic Press, New York.

Szekely J. and Themelis N.J., 1971. *Rate Phenomena in Process Metallurgy*, John Wiley & Sons, Inc., New York.

Szekely J., Evans J.W. and Brimacombe J.K., 1989, *The Mathematical and Physical Modelling of Primary Metals Processing Operations*, John Wiley & Sons, Inc., New York.

The Faraday Society, 1948. The Physical Chemistry of Process Metallurgy, *Discussions of the Faraday Society, No. 4*, Gurney and Jackson, London.

Themelis N.J., 1995. *Transport and Chemical Rate Phenomena*, Gordon and Breach Science Publishers SA.

Trescot, J., Connor, D., Faulkner B., Beloy, F. and Vogely S., 2004. Application of Ported Kiln Technology to the Grate-Kiln Process at Minntac, *SME preprint 04–27.*

Vignes, A., 2011a. *Extractive Metallurgy 1—Basic Thermodynamics and Kinetics*, John Wiley and Sons, Hoboken.

Vignes, A., 2011b. *Extractive Metallurgy 2—Metallurgical Reaction Processes*, John Wiley and Sons, Hoboken.

Vignes, A., 2011c. *Extractive Metallurgy 3—Processing Operations and Routes*, John Wiley and Sons, Hoboken.

ABOUT THE AUTHOR

Patrick R. Taylor is a registered professional engineer with more than 38 years of experience in mineral processing and extractive metallurgy engineering, research, teaching, and consulting. He has authored or co-authored approximately 200 papers and presentations, and holds 9 patents. He is currently the George S. Ansell Distinguished Professor of Chemical Metallurgy at the Colorado School of Mines. He is the director of the Kroll Institute for Extractive Metallurgy. Taylor has supervised research for numerous graduate students and post-doctoral researchers. In 1990 he was named Distinguished Faculty Member at the University of Idaho and received the Research Excellence Award in 1994. Other honors and awards include Fellow of ASM International (1996), SME's Milton E. Wadsworth Award (2003), TMS Extraction and Processing Division Distinguished Lecturer (2006), Distinguished Member of SME (2008), TMS EPD Distinguished Service Award (2010), and AIME James Douglas Gold Medal (2013).

Innovations in Copper Smelting—Keys to Productivity and Efficiency

Eric H. Partelpoeg
EHP Consulting, Inc., Tucson, Arizona, United States

ABSTRACT: Recent decades have experienced prolific improvements in smelter productivity and efficiencies due to the copper smelting innovations of numerous researchers around the world. This review presents several of the significant innovations advanced by dedicated smelter teams collaborating across smelters, universities, private technology service companies, and technology equipment providers. Open communication between smelter organizations are often facilitated from technical papers and communications at national and international conferences and has led to continually evolving innovations and commercialization. Subject areas focused on in this review include (1) smelting fundamentals and metallurgy, (2) energy efficiency, (3) furnace feed systems, (4) furnace and off-gas system cooling, (5) converting technologies, (6) anode refining innovations, (7) emissions control improvements; and, in general, innovations in (8) bath and (9) flash smelting technologies.

INTRODUCTION

In recent decades, the global copper smelter industry has shifted towards fewer, but larger capacity smelters (Ramachandran et al. 2003). Copper smelter production intensity has increased while emissions have decreased. Davenport and co-authors' Extractive Metallurgy of Copper text (Schlesinger et al. 2011) provides a comprehensive overview of smelting technologies while Caballero (Caballero et al. 2009) provides a focused review of innovations in Chilean smelters. These reviews as well as numerous symposiums (Kapusta et al. 2009; Ross et al. 2005; Diaz et al. 2003, 1999, 1991, 1987; Stephans et al. 2002; George et al. 1999; Asteljoki et al. 1997; McCain et al. 1994; Landolt 1993; Smith et al. 1992; Sohn et al. 1983) provide background information for this paper. More comprehensive explanations and citations of the numerous advancements and innovations extending beyond this brief review are available in this paper's reference section, and on the SME onemine.org database.

There are numerous examples in the references listed above that demonstrate how innovations have led to increased smelter productivity, particularly since the 1970s, which was a catalyzing decade for copper smelters with new incentive for developments due to increased energy costs and programs to decrease emissions. One example of how a smelter has applied innovations to improve productivity while decreasing emissions is the Norddeutsche Affinerie (now Aurubis) Smelter (Willbrandt 1993, Edens et al. 2006, Lossin et al. 1999). Figure 1 shows the dramatic improvements experienced through the one smelting furnace as a direct result of smelter innovations and

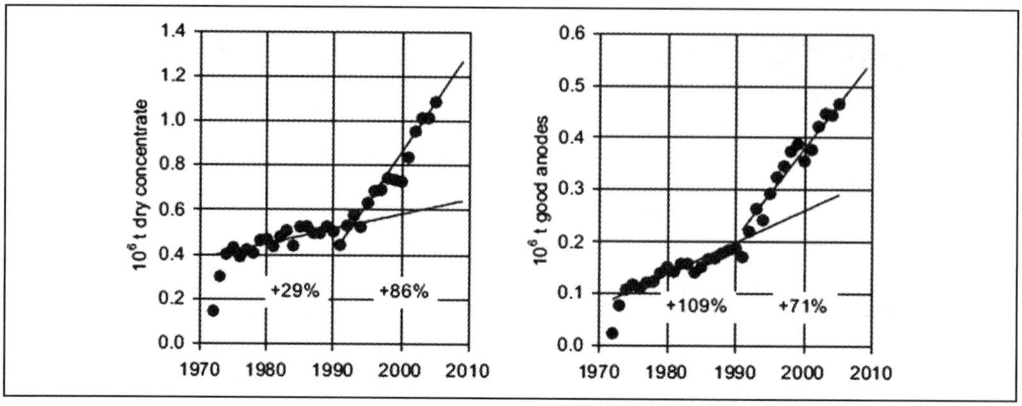

Figure 1. Example increase in smelter productivity from innovations

dedicated operations. It is the goal of this paper to highlight these key innovations from an industry wide perspective.

SMELTING FUNDAMENTALS, METALLURGY, AND ENERGY STUDIES

A review of smelting fundamentals and metallurgy provides a useful foundation for further examination of copper smelting innovations. This brief synopsis of the efforts of the researchers and educators associated with smelter basics reviews the following areas: (1) slag chemistry, (2) impurity distribution, (3) mathematical modeling, and comments regarding smelter energy assessments.

Research leaders Yazawa, Mackey, Toguri, and Sohn have advanced developments in smelter slag chemistry and slag physical properties relevant to the successful operation of a copper smelter. The extensive work by Yazawa and his students (including Professor Kimio Itagaki and Meguru Nagamori) was instrumental to many successful innovations. An important example is the research and development of the calcium ferrite copper oxide slag that has led to its implementation in industry.

See Tanaka's paper (Tanaka 2006) for details including an excellent list of references to Yazawa's work and the work of his colleagues. Mackey's efforts to review the physical chemistry of copper smelting slags (Mackey 1982) have helped smelter operators to better control and optimize their smelting and converting furnaces. Toguri (with help from his colleagues) has left a legacy of efforts that study and explain slag chemistry in a clear manner (including Sridhar 1997 and Toguri et al. 1976). Sohn's efforts include slag chemistry research, impurity distribution studies, and important contributions in mathematical modeling of smelting reactions. Samples of Sohn's contributions include Kim et al. (1996), Hahn et al. (1990), and Perez et al. (2001). Any review of smelting reaction studies would be remiss to not point out the efforts at CSIRO led by Jorgensen, which include Jorgensen et al. (2001) and Solnordal et al. (2003). Ahokainen et al. (2006) describe recent advances in burner modeling and include a comprehensive reference list of related work.

The importance of impurity distribution control in copper smelting is increasing for two compounding reasons: (1) smelters are reducing their emissions, which historically may have been outlets for some impurity elements, and (2) impurity levels in copper concentrates are increasing. Impurity distribution studies were reviewed by George et al. in 1976 and summarized by Bigley in 1978. These efforts were followed in 1982 by the U.S. Bureau of Mines Report of Investigations

8655 (Johnson et al. 1982). Current research in Chile is directed on impurity distributions, with a focus on arsenic, which has increased in levels in several Chilean concentrates (Valenzuela et al. 2006). Larouche (2005) is commended for his Master of Engineering thesis (McGill 2001) that provides a comprehensive review of minor elements in copper smelting and electrorefining that includes an exhaustive list of references and data from smelters.

Innovative contributions of dozens of other key researchers from all continents extend far beyond those aforementioned in this brief review of smelter fundamentals. Energy use in smelting has always been scrutinized by smelter operators. The energy crisis in the 1970s underscored the importance of smelter energy efficiency. Lewis et al. (1977) document how smelter costs increased in the inflationary years of 1972–1976 by 85%, with the energy component increasing by 183% over the same period. McNulty (1983) followed up with a review in 1983 that indicates that energy costs to produce cathode copper increased from \$0.07/pound to \$0.17/pound during the period 1970 to 1983. These 1970s market forces led to the 1976 effort by Kellogg et al. (1976) to estimate energy use in sulfide smelting of copper. Riekkola-Vanhanen (1999) provides a more recent (1999) but less comprehensive review of smelter energy consumption. The Chilean Copper Commission published a 2008 review of copper industry-wide energy consumption. This review provides a high level assessment of copper industry energy consumption with a focus on CO_2 equivalent emissions for each phase of copper production. Interestingly, the study concludes that the total energy consumption of concentrating, smelting, and refining is nearly equivalent to the SX-EW option (utilized primarily for oxide copper deposits).

There have been significant innovations in copper smelting since Kellogg's energy review in 1976. Following are some of the energy related technological changes since Kellogg's review:

1. The development of flash converting. Energy studies of flash converting should include granulation water pumping energy (associated with the upstream flash furnace matte and the flash converter slag) as well as the energy requirements associated with the copper containing streams.
2. The successful development and commercialization of TSL (top submerged lance) technologies of which Isasmelt is the predominant brand of TSL technology. Kellogg summarily concludes that processing wet concentrate is a disadvantage. A detailed assessment may well prove this assertion, but TSL energy consumption should be evaluated completely.
3. Flash furnace concentrate drying technology as shifted from direct fuel-fired drying systems towards steam drying. The source of this medium pressure steam (15–20 bar) is typically from a let-down of smelting boiler steam (typically 60 bar). An updated energy review should reflect this loss of high pressure steam credits (as well as the benefits of reduced fuel consumption).
4. Flash furnace cooling intensity has increased in many cases to significantly increase furnace campaign life. This improvement (particularly with flash smelting) results in increased energy costs associated with cooling water pumping. In many instances, two cooling water circuits are employed (a recirculating loop between the furnace and heat exchangers that are serviced from a cooling tower water system).
5. Oxygen consumption has increased per unit of copper while oxygen plant energy efficiency has also improved.

6. The extent of smelter fugitive gas capture and treatment has increased significantly since the 1970s.
7. Acid plant technologies have improved to accept an inlet gas with an O_2/SO_2 ratio that can be significantly lower than the 1.0 value used by Kellogg in his review. Acid plants can accept inlet gas concentrations of 14% SO_2 without special technology and up to 18% SO_2 with the adaptation of new technologies or pre-converter designs. Acid plants can now produce medium pressure steam (which can be considered for concentrate drying).
8. Water consumption in a smelter is an increasingly important issue. An updated energy study may want to assign an energy-equivalence to water consumption.

The importance of Kellogg's 1976 effort and advancements since his review demonstrate the need for an objective re-evaluation of the energy costs associated with copper smelting in the 21st century.

COPPER SMELTER INNOVATIONS

The remainder of this paper focuses on key innovations from the world's leading smelters which have contributed to significant production increases.

Furnace Feed and Burner System Innovations

Flash furnaces have enjoyed substantial increases to their original rated production limits. The Magma Metals furnace was no exception. This furnace was installed in 1988, originally designed for a rate of 2,700 tpd (approximately 115 tph concentrate). The original design capacity of just less than 900,000 tons per year concentrate was an ambitious goal at the time. Due to the efforts of a dedicated Magma Smelter team led by J.D. McCain, this smelter was the first (or among the first) smelters where a single smelting vessel processed more than one million tons of new concentrate in a year (accomplished in 1994 [Jones et al. 1999]). A key factor that contributed to this accomplishment was a series of improvements to the concentrate burner (the final component of the furnace feed system. The original Outokumpu burner did not have the capability to meet the Magma goal of one million tons per year of concentrate. They first worked with Outokumpu to modify the burner (Gonzales et al. 1993) and decrease its diameter and increase the process air velocity by 42%. The goal of this modification was to improve the distribution of process air at the discharge of the burner. Results did not meet expectations and the Magma team elected to continue development independently and build a ¼ scale exact functioning model of the existing process air duct and windbox.

The result of this testing confirmed that the installed burner was not distributing the process air evenly. The Magma engineers proceeded to modify the model—they ran over one hundred tests to develop a burner with a significantly improved distribution of process air through the burner. Magma continued their development efforts to improve the distribution of concentrate to the burner. The Magma-developed burner significantly improved flash furnace performance. Table 1 summarizes the before and after data.

In addition to these early Magma efforts, other smelters were also improving their flash furnace concentrate burner design. Two examples of progress with flash furnace burner improvements include the work at the Sumitomo Toyo Smelter (Hattori et al. 2003) and at the Saganoseki Smelter (Suzuki et al. 2006). The Toyo development resulted in decreased dust carry-over and decreased slag magnetite levels. The Saganoseki development work focused on improving the burner performance

Table 1. Comparison of original and magma-modified burner (Gonzales et al. 1993) (Heat loss units converted from original MMBtu units)

	Original Burner—1988	Modified Burner—1992
%Fe_3O_4 in slag	20–25%	6–7%
Matte grade deviation	±4–5%	±1%
Flash furnace dust recirculating load	6–8%	4–4.5%
Uptake shaft throat accretions	Excessive, requiring frequent blasting for removal	Virtually none
Reaction shaft heat loss	7,500–9,500 MJ/hour with frequent excursions over 10,500 MJ/hour	5,300–6,300 MJ/hour, very stable

in order to achieve flash furnace feed rates (from 160 tph in 1996 to 195 tph in 2005). The improved burner performance allowed for increased matte grade (61% to 68%) without an increase of the copper content in slag.

The success of efforts to improve the combustion of concentrate in a flash furnace is dependent on an even flow of concentrate to the burner. The standard delivery conveyance mechanism of dry concentrate to a flash furnace burner has historically been a drag chain conveyor. Noranda is believed to be among the first to use air slides (in this case to convey concentrate to the inlet of the concentrate injection system). Phelps Dodge (Hidalgo Smelter) installed air slides to convey concentrate from the concentrate lift system discharge to the concentrate dry bin immediately upstream of the furnace feed system. These successes led the Magma Smelter to install air slides to feed the concentrate burner (Jones et al. 1999); the smelter was permanently shut down prior to the start-up of this feed system. The Chagres Smelter is believed to be among the first to implement air slide technology to feed the flash furnace burner. These early efforts to incorporate air slide technology have led to this technology becoming the preferred feed mechanism to a flash furnace burner (Peuraniemi et al. 2006). The even flow achieved with the air slide is important—but equally important is accurately and consistently metered flow that can be achieved with the use of mass flow bins and loss in weight (LIW) feeding systems. The BCL Smelter in Botswana was the first flash furnace system to install a LIW feeding system (Malema et al. 2006). The importance of mass flow bins and flow control are described by Goodwill et al. (1999).

Smelting vessel feed system innovations have not been limited to flash furnaces. An example of a significant innovation with bath smelting technology is the development by Codelco and Clyde Materials Handling of a controlled, accurate injection method for concentrate into a Teniente Converter (Caballero 2009, Coleman et al. 2008).

Furnace and Off-Gas Cooling Innovations

An industry-wide effort by smelter operators, specialized cooling system design firms, and smelter technology providers was paramount to the smelting intensity rise in flash furnace and the associated increased productivity and emission controls demands of downstream converters.

Water-cooling of smelting furnaces and water-cooled tapping plates were common in the early 1900s (Laist 1934), often fabricated from cast iron. Newton and Wilson's 1942 "Metallurgy of Copper" describes the water cooling technology of the early 1900s in some detail. By the early 1990s, copper smelter cooling elements had evolved to improved materials and quality of manufacturing. In 1992, the Phelps Dodge Hidalgo Smelter team replaced vertical plate cooling jackets

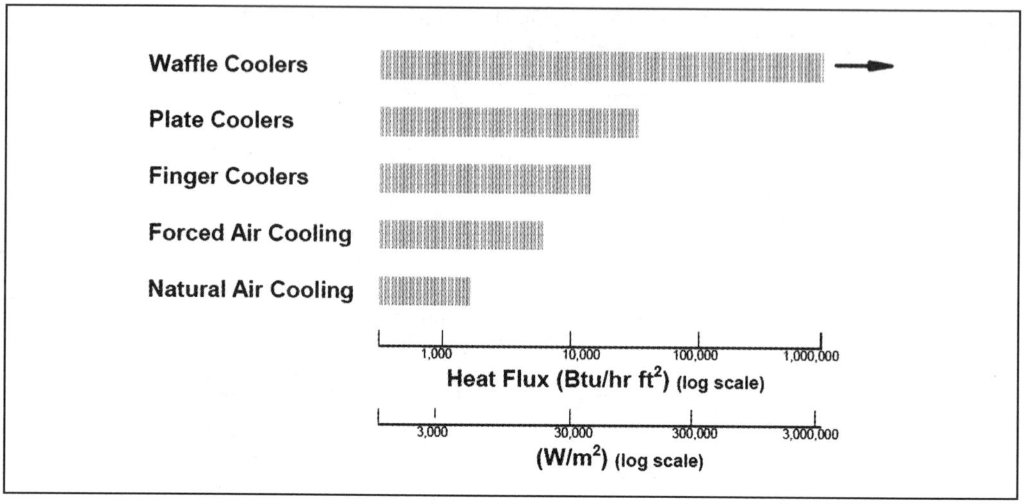

Figure 2. Effectiveness of various cooling methods (Voermann 1999)

(that had been cast with copper with embedded copper pipes in the casting mold) with fabricated cathode billet jackets. The water passages for these new jackets were drilled and plug welded as necessary (Mitchell et al. 1992). One of the challenges with early design castings with embedded copper pipe was that the copper pipe required cooling and/or the casting temperature had to be low to avoid melting the pipe during casting. There was often a gap between the copper pipe and the casting and this gap reduced heat transfer efficiency of the unit. Copper castings have benefited from the availability of copper-nickel pipe alloy material (MacRae 2001) to achieve a good metallurgical bond between the cast metal and the pipe coil.

The Ashio Smelter team was a leader in flash furnace reaction shaft cooling innovations (Shima et al. 1997). Their efforts to replace reaction shaft spray cooling with jackets led the industry to a technology that is now a standard offering.

The design of copper cooling elements for pyrometallurgical furnaces has had interesting developments in recent years. Drilled billet cooling blocks have improved in reliability by incorporating hot working to refine the grain size, and micro alloying of the copper to achieve solid welds at the plugs.

Many designers now incorporate a pattern (often a "waffle" shape) on the hot face of the copper for retention of slag and to form a stable accretion. Typically the pattern is first installed with refractory inserts to avoid over-heating before the slag is retained on the cooler (Voermann 1999). These types of cooling elements can be exposed to higher heat loads (**Figure 2**). The design of composite systems (copper and refractory) has been proven to increase furnace life (Kyllo et al. 2005).

One example of the results of innovations in furnace cooling is shown in Figure 3 that shows the improvements to flash converter cooling elements at Kennecott (Walton et al. 2005).

Kojo et al. (2000) provide a general overview (to complement the Hatch paper Voermann et al. 1999) of the features of the different types of water-cooling technologies used in copper smelting.

In addition to improved furnace cooling, the downstream converting process, typically of Peirce-Smith converters, has responded to increased furnace productivity with commensurate increased converter productivity. The increased converting productivities (e.g., as described by

Figure 3. Flash converter reaction shaft/settler cooling elements (Walton et al. 2005)

Enrico et al. have to a great degree been due to improvements in converter hood design (Enrico et al. 2005). While water-cooled converter hoods have been used for decades, the more recent efforts by engineers and designers have greatly enhanced their reliability and safety (Drummond et al. 1999, Safe et al. 2000, Le Roux et al.). The focus of much of this work has been related to the development of safer cooling jackets to minimize the risk of water leaks. Consideration to the skin temperature of the jacket is critical to reduce the propensity of acid dew point condensation on the surface of the jacket.

Converting Technology Innovations

Peirce-Smith converting, first implemented in the early 1900s (Kapusta et al. 2009), continues to dominate the copper industry. This dominance is due to the hard work of numerous operators and engineers. The efforts of Cardona et al. (2012) and Tan and Vix (2005) are two examples of research and development that have improved the knowledge base and performance of Peirce-Smith converting. Efforts such as these, as well as efforts by smelter operators have let to industry wide improvements that are similar to those shown in Figure 4 (Tanaka et al. 2005).

One of the limiting factors of Peirce-Smith converting has been tuyere area refractory life and limitations of oxygen enrichment to the converter. Kapusta et al. (2005) describe the success of high oxygen shrouded injection that enables oxygen enrichment of up to 40% while eliminating the requirement of mechanical punching of the tuyere. This improvement is off-set by the heat balance challenges associated with operating a Peirce-Smith as even with lower oxygen enrichment levels the converter requires significant levels of cold material (reverts) addition to avoid overheating.

While Peirce-Smith converting innovations have been instrumental to its continued dominance, the development of continuous converting technology has proven to be a breakthrough in copper smelting. The efforts of D.B. George (and the associated support of the Kennecott Utah Copper Corporation) were instrumental to the development of flash converting. A perspective of continuous converting is summarized by George (2002) who suggests that the continued implementation of flash converting will be influenced by the following factors:

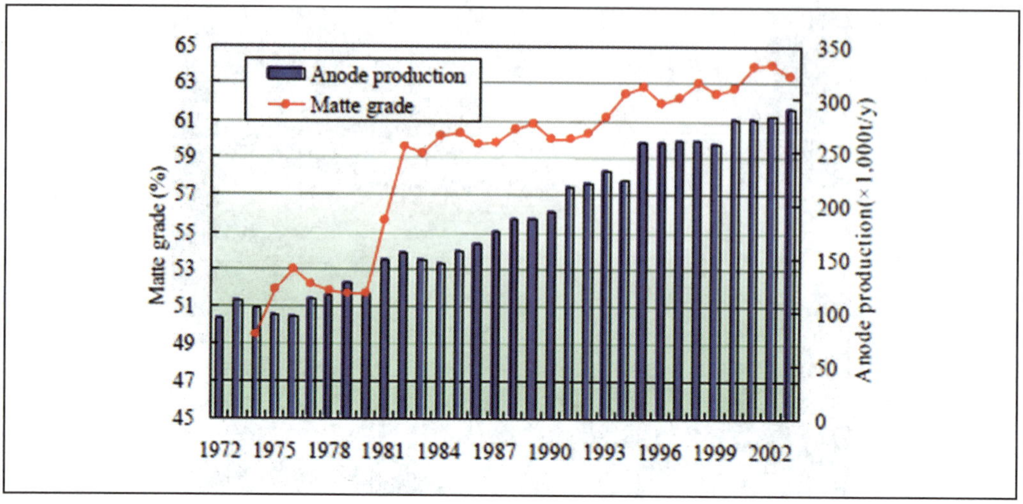

Figure 4. Productivity Improvements at the Tamano smelter (Tanaka et al. 2005)

- The promulgation of more effective and stringent emission control regulations in the main copper smelting centers.
- The recognition that economies of scale favor large fully integrated smelters.
- The recognition that many smelters can be utilized to produce only granulated matte that can be shipped to large, centrally located smelters, thus increasing sulfur capture, reducing operating complexity and costs, and realizing increased value from sulfuric acid.

Ojima (2003) compares converting options; he estimates that flash converting operating costs are more than 50% higher than Peirce-Smith converting costs (Table 2). Overall smelter costs are indicated by Ojima at approximately 10% higher than with Peirce-Smith converting. It is unclear from Ojima's work if both cost estimates were based on identical emission control criteria and if the maximum amount of acid plant operating cost reductions were included. It is also unclear what the unit energy costs were assumed for this comparison. The conclusion that Peirce-Smith converting is a lower operating cost option should be reviewed with a more "open book" approach—similar to how Kellogg assessed energy requirements in the 1970s.

Alternative continuous converting options to flash converting include Mitsubishi converting (Goto et al. 2003) and Codelco continuous converting (Moyano et al. 2006). Xstrata Technology has been developing Isaconvert™ technology in recent years. Based on recent publications, it appears that this continuous converting technology may soon be installed in a commercial application (Nikolic et al. 2009). One of the significant challenges with these alternates to flash converting is to design a system with a campaign life that matches the performance currently achieved at the Kennecott Smelter.

Anode Refining Innovations

Anode furnace and casting systems have quietly kept pace with the recent productivity improvements of smelting and converting systems. The implementation of porous plug technology has been an integral component to anode area improvements. The porous plugs are used predominantly to inject low amounts of nitrogen in the bath to improve mixing in the bath with resultant

Table 2. Ojima's comparison of Peirce-Smith and flash converting smelter overall operating costs (from Figure 6 in Ojima 2003)

	Peirce-Smith Converting	Flash Converting	Difference
Drying	6.2%	6.2%	0.0%
Flash smelting	39.4%	42.3%	7.2%
Electric furnace slag cleaning	5.4%	7.2%	33.3%
Converting	12.1%	19.1%	57.4%
Anode furnace	5.2%	7.0%	35.0%
Acid plant	25.8%	18.8%	–27.0%
Others	5.9%	8.8%	47.8%
	100.0%	109.3%	9.3%

improvements to oxidation and reduction efficiencies. Landolt et al. (at Inco) were among the first to document the benefits of porous plug technology in the smelting industry (Landolt et al. 1993). The success of porous plug technology moved quickly from Inco to Arizona and New Mexico smelters in the early 1990s. The benefits of this technology are documented by Acuna and Sherrington (2003). Porous plug technology is now a main-stream component of the efficient anode copper processing operations that is often utilized to support high production smelters (Rigby et al. 2003). Porous plug technology is not, however, an absolute requirement to achieve high levels of anode furnace productivity. The team at Norddeutsche (now Aurubis) has achieved world-class anode operation without the use of porous plugs (Edens et al. 2006).

Another important development with anode furnace operation is the increased use of oxy-fuel burners. An example of an effective oxy-fuel burner is Praxair's Dilute Oxygen Combustion (DOC) burner (Riley et al. 2001). DOC technology injects the fuel and oxygen into the furnace through separate ports rather than through a single burner. The reactants mix with the hot furnace gases before reacting with each other. This design results in a lower flame temperature and lower NO_x levels. The successful implementation of DOC technology at the Kennecott Smelter (U.S. Department of Energy 2001) has led to its implementation in several copper smelters.

The improvements of anode refining and burner technology have led to increased anode production rates which have resulted in increased demands of the anode casting process and the anode molds (both in term of quality and mold life). Wenzl et al. (2007) discuss the factors that affect physical anode quality; Edens et al. (2005) describe the development of improved anode molds at the Norddeutsche (now Aurubis) smelter. Initial mold life was 800–2,200 tons with molds cast into a permanent steel mold. The second development was to cast blocks and press the anode shape into the block. Anode molds are now produced by milling continuously cast copper cakes. With this approach, mold life has increased to more than 4,000 tons.

Emission Control Innovations

Copper smelter emissions have been on a steady decrease for decades. The main source of smelter emissions has been the converter aisle. **Figure 5** shows typical emissions from Peirce-Smith converter activities. The Anaconda Butte Smelter photo shows a view of the converter aisle, with emissions evident from a converter that had recently rolled out. The Hudson Bay Mining Flin Flon Smelter photo shows converter aisle emissions during the addition of matte to the converter.

Figure 5. Smelter converter aisle photos (left: Anaconda Smelter 1979; right: Flin Flon Smelter circa 2000)

Before 1990 | Between 1990 and 2002 | After 2003

Figure 6. Chuquicamata smelter photos (Caballero et al. 2009)

Figure 6 provides a view of emissions from the Chuquicamata Smelter from the 1980s through to the early 2000s. The Chilean smelter industry is continuing to focus on a variety of strategies to reduce emissions (Caballero et al. 2009).

Willbrandt (1993) provides a detailed review of the efforts at the Norddeutsche Smelter from the mid-1970s until the early 1990s to reduce emissions. Vikdahl et al. (1992) provide a similar chronology of emission control improvements carried out at the Rönnskär Smelter. Similar efforts have occurred at smelters throughout the world.

More recently (particularly since the 1990s), new acid plant technologies have been developed to accommodate the increased demands of the acid plants to accommodate higher inlet SO_2 concentrations while reducing acid plant tail-gas emissions. Innovations have been developed by acid plant catalyst technology companies (notable Haldor Topsoe and MECS). They have each developed catalysts with the following improvements from acid plant catalyst available 20 years ago:

1. Acid plant energy savings (lower pressure drop); realized by improving the shape and size of catalyst.
2. Lower emissions—with new catalysts acid plant tail-gas with less than 100 ppm SO_2 can be designed and operated.
3. Longer operating campaigns (by developing catalyst shapes that can accommodate twice the amount of fouling (dust) before pressure drops force a plant shutdown to screen catalyst).

Acid plant technology developers have also supporting copper smelters with innovative approaches to processing high concentration SO_2:

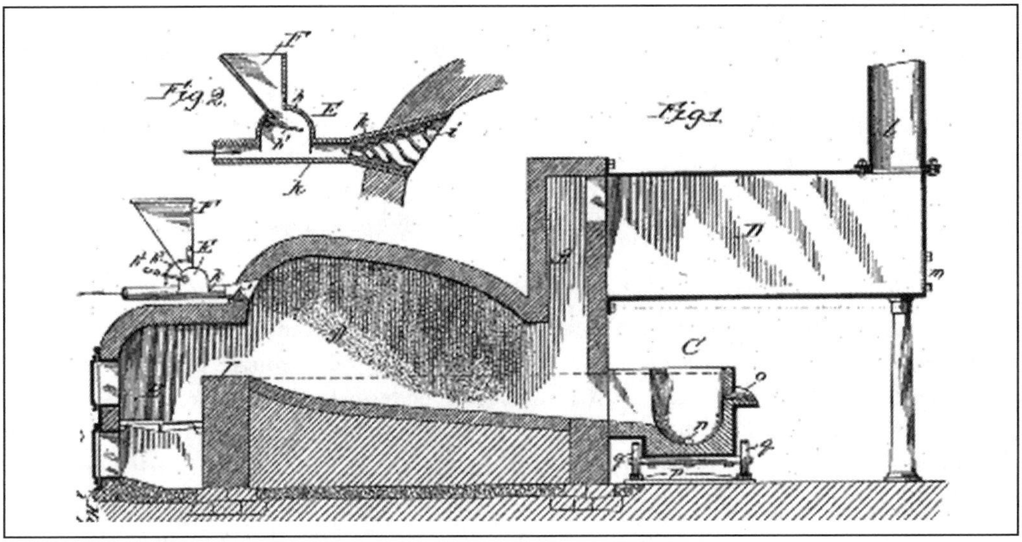

Figure 7. First concept of flash smelting (U.S. Patent 578,812, March 1897)

1. Outotec has developed the Lurec™ process. This design accommodates high strength SO_2 by shifting the SO_2 to SO_3 equilibrium of the first pass by recycling a portion of converted gas back to the first pass. This shift is required to limit the first pass outlet temperature to approximately 630°C (close to the catalyst upper temperature limit). This technology has been proven at an inlet concentration of approximately 18% SO_2 and the Xiangguang Smelter in China implemented this technology in 2007.
2. MECS has proven plant configurations that operate with a portion of the gases flowing to a pre-converter. The SO_2/SO_3 gas stream from the pre-converter is blended with the main process gas to limit the conversion (again to control the converter outlet temperature).
3. Bayer has developed the Bayqik® isothermal pre-converter process. This technology provides yet another option on how to control the conversion temperature of the main converter's first pass outlet temperature.

Flash Smelting Innovations

Flash smelting as a proven and industry leading process has made substantial gains in productivity and efficiency since the concept of introducing fine concentrate feed into the furnace freeboard space was disclosed in Henry Bridgman's 1897 patent (Figure 7).

Many of the copper smelting innovations discussed in the previous pages were specifically developed to improve flash furnace performance. These include the improvements discussed regarding furnace feed/burner systems and improvements to furnace cooling.

Flash smelting was developed nearly simultaneously at two fronts, Inco developed oxygen flash smelting and implemented it at its Copper Cliff smelter while Outokumpu developed air flash smelting at the Harjavalta Smelter. Both technologies are major innovations that have served the copper industry well. Inco technology was implemented successfully at the Chino Smelter (Kennecott owned at the time of implementation) and continues to operate well at the ASARCO Hayden Smelter. The Outokumpu (now Outotec) flash furnace has become the world leader in smelter technologies. While the goal of this paper has been to highlight the efforts of small teams around the world to develop innovations, the efforts and successes of Outotec have been instrumental to the current state of copper smelting. A review of the innovations that Outotec has delivered to the copper smelter industry merits an extensive document to themselves. Kojo and Storch (2006) provide a good synopsis of Outotec's successes in copper smelting including a description of how their Lurec® acid plant technology matches the low volume, high SO_2 concentration that is typical of flash smelters and particularly flash smelting/flash converting smelters.

Bath Smelting Innovations

Innovations associated with Teniente Converter technology have been discussed earlier and this technology continues to process a significant amount of Chile's concentrate. Xstrata's Isasmelt™ technology has quickly assumed a leadership role among bath smelting technologies. From a modest start at Mount Isa in the 1980s, the technology now accounts for approximately ten million tons a year of feed material (Alvear et al. 2010); four million of which has occurred since 2007. One feature of this technology that is attractive to operators with complex concentrates is the ability for Isasmelt vessels to deal with impurity elements (Alvear et al. 2006). Based on their success rate with smaller capacity smelters (<1 million tons per year concentrate), it appears that, at least in this production range, the technology is competitively priced compared to alternative smelting options.

DISCUSSION

The 1970s was a decade of change for the copper smelting industry. Increased energy costs coupled with increased emphasis to reduce smelter emissions were the catalyst to many of innovations discussed in this paper. In some cases the innovations were designed to address emission reductions or energy efficiency directly. In other cases, motivation to offset high energy costs and reduce maintenance costs spurred smelters to increase productivity and initiate process changes.

Since the 1970s there has been a significant geographical shift of smelter production as summarized in Table 3. Production in North America and Africa has significantly decreased in both absolute and relative terms. South American production has lost some ground as well (from 14% of world production to 13%). Europe's gains have been significant but overshadowed by the tremendous growth in Asia, due to new smelter projects in China and India.

Figure 8 shows a circa-2003 estimate of the relative operating costs of a Chinese smelter compared to a European smelter, providing a general view of how operating costs are affected by smelter location (Tuominen et al. 2005).

It is generally believed that Chinese and Indian copper smelter production will be consumed internally or exported as a component of a finished manufactured good (motors, vehicles, etc.). Though geo-political projections are beyond the scope of this paper, shifts of global influence are is likely to affect the trajectory of the copper smelting industry, similarly to the energy-related catalysts for change of the 1970s. In recent years, the mining industry as a whole has experienced several challenges associated with capital projects. KPMG has published a white paper reviewing how mining industry leaders rank various business risks. Table 4 summarizes KPMG's survey of mining industry leaders.

The risks of mining industry projects may have an impact on future trends in smelter innovations. Cost escalation concerns and the recent trends of low concentrate treatment and refining charges may give pause to the development of new smelters. Government involvement can be

Table 3. Copper smelter production distribution (thousand metric tons)

	1976*	% of 1976 Total	2011†	% of 2011 Total	Absolute % Change 1976–2011
North America	2,578	36%	891	6%	–65%
Africa	1,432	20%	514	3%	–64%
Asia	1,186	17%	7,334	50%	518%
South America	978	14%	1,873	13%	92%
Europe	740	10%	3,376	23%	356%
Oceania	218	3%	733	5%	237%
Totals	7,132	100%	14,721	100%	106%

(Kazakhstan and Iran data included with Europe totals; only top 20 countries included in 2011 data estimate).
*Agarwal et al. 1976.
†International Copper Study group 2012.

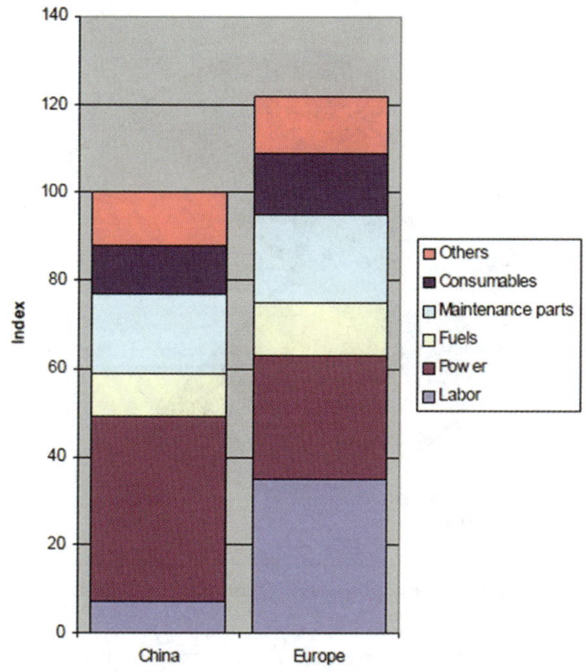

Figure 8. Global differences in smelter operating costs, 2003 (Tuominen et al. 2005)

interpreted as risk of more stringent environmental, health, and safety regulations. Labor shortages of smelter operators, engineers and smelter managers also influence how smelters approach in-house innovation projects. All of these macro issues affect which smelter innovations are to be developed and the level of urgency.

Smelter executives may balance project risk categories across the areas of the smelter that promise the most likely reduction of project risks (for example continuous converting to address government intervention risk associated with more stringent regulations associated with batch converting).

Table 4. Mining project business risk rankings

Business Risk Category	% of Mining Industry Leaders Who Ranked Category as a Major Risk				
	2008	2009	2010	2011	2012
Cost escalation	78%	58%	36%	55%	55%
Government involvement in the industry	38%	53%	30%	40%	40%
Access to new projects	36%	57%	52%	38%	38%
Ability to raise capital	53%	42%	41%	35%	35%
Labor shortages	73%	53%	22%	33%	34%
Year average	56%	53%	36%	40%	40%

Source: http://www.kpmg.com/PE/es/IssuesAndInsights/ArticlesPublications/Documents/Business-Risks-Facing-The-Mining-Industry.pdf

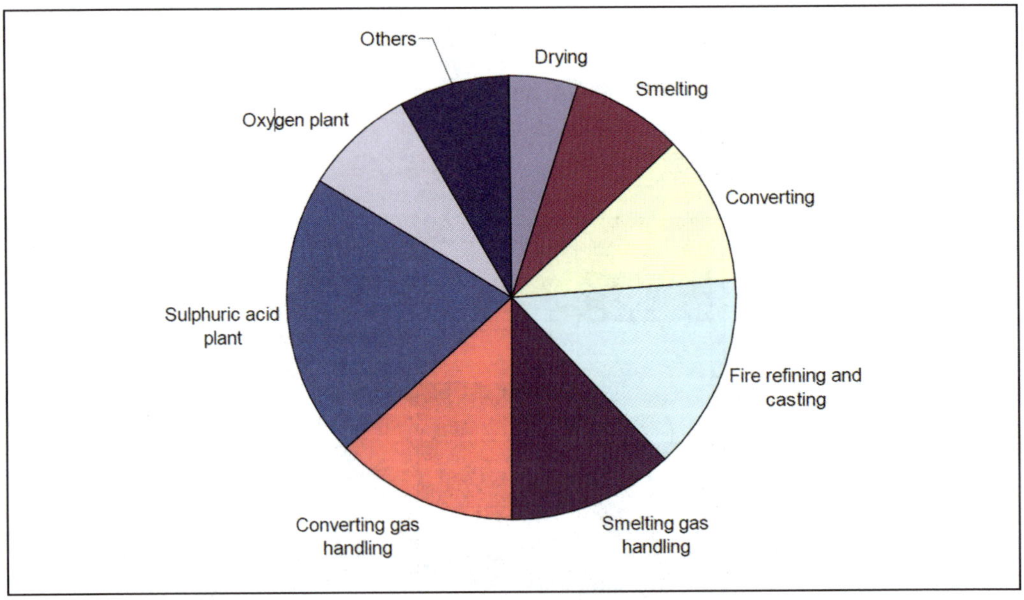

Figure 9. Relative allocations of smelter capital costs (Tuominen et al. 2005)

Figure 9 shows the relative allocation of smelter capital costs as projected by Outotec (Tuominen et al. 2005). Their conclusion is that continuous converting creates a lower smelter off-gas volume lowering gas handling and acid plant capital costs.

Macro-economic issues aside, the following areas of smelter processes could be developed in the next 10–20 years:

1. Continued shifts from batch to continuous converting. This will primarily be in response to environmental issues, though capital and operating costs may also encourage this trend. Top submerged lance (TSL) continuous converting will likely offer an alternative to flash converting. This technology will likely face "growing pains" and its first application should be selected in a smelter with an experienced smelter team.
2. Due to the importance of even distribution of concentrate to a flash furnace, pneumatic feed systems may displace mechanical (e.g., air slide) feed systems to the furnace. This

development may progress to the next generation of concentrate burner, to further improve smelting efficiency by reducing dust carry-over and slag magnetite levels.

3. TSL smelting vessels (Isasmelt™ and Ausmelt™) may implement intense cooling to achieve multi-year campaign lives.
4. The use of boiler tube panels to replace furnace freeboard areas may become more common (to increase energy recovery).

The continued climate of depressed concentrate treatment charges remains a challenge for further development in the smelting industry. Operating revenues are low and smelter budgets may not sustain the efforts require to develop significant innovations.

A review of the references cited in this paper show that significant innovations were developed in all corners of the copper smelting world: the Americas, Europe, Asia, Africa, and Oceania including Australia. This global participation and progress is in part due to open communications between smelters shared at national and international conferences and symposiums, and in publications. There is some evidence that the amount of detail forthcoming in technical papers is decreasing in recent years. In order for innovations to proceed at the pace experienced in the last 25–35 years, the industry must maintain its long-standing "open-door policy" between smelting researchers and experts.

REFERENCES

Acuna, C., Sherrington, M., 2003. "Efficiency of Porous Plugs in Fire Refining of Crude Copper." *Metallurgical and Materials Processing: Principles and Technologies Vol. 2*. pp. 265–279. TMS ISBN 0-87339-534-4 onemine.org reference. http://www.onemine.org/view/?d=DCC5BDC3689E918727349CF9CDB723A78688912534F395381A64BFFC23461B49167623.

Agarwal, J.C., Yannopoulus, J.C., 1976. "Copper Introduction and Overview." *Extractive Metallurgy of Copper Vol. 1*. pp. 15–23. TMS onemneorg reference. http://www.onemine.org/view/?d=466A9FBA4054ACFA7B6A4FC183BD8F44A528E67D014E42F30EE964F4C2E53261167892.

Ahokainen, T., Jokilaakso, A., Taskinen, P., Kytö, M.,. "A New Advanced CFD Model for Flash Smelting and Converting Processes. In." *Advanced Processing of Metals and Materials Vol. 8*. Edited by Kongoli and Reddy. pp. 529–543. TMS 2006 onemine.org reference. http://www.onemine.org/view/?d=6ECB2157EE225F9EDEEB74A482FD89A043882DEA2E1286D94D39C75E7BB83249202525.

Alvear, G.R., Hunt, S.P., Zhang, B., 1976. "Copper Isasmelt—Dealing with Impurities." *International Symposium on Sulfide Smelting Vol. 8*. pp. 673–685. Edited by Kongoli and Reddy. TMS.

Alvear, G.R., Arthur, P., Partington, P., 2010. "Feasibility to Profitability with Copper Isasmelt™." *Pyrometallurgy I GDMB Vol. 2*. pp. 615–630. Clausthal-Zellerfeld, Germany: Copper 2010.

Asteljoki, J.A., Stephens, R.L., editors, 1997. "Sulfide Smelting '98 Current and Future Practices." *TMS Annual Meeting*. ISBN 0-87339-397-X.

Bigley, A.C., 1978. "Impurity Distribution and Control in Copper Smelters and Refineries." Presented at the *107th AIME Annual Meeting*. Denver, CO, March 1. A78-67.

Cardona, N., Mackey, P.J., Coursal, P., Parada, R., Parra, R., 2012. "Optimizing Peirce-Smith Converters Using Thermodynamic Modeling and Plant Sampling TMS." *JOM* 64 (8).

"Chilean Copper Commission, Energy Consumption and Greenhouse Gas Emissions in the Chilean Copper Mining Industry, Events of 2008, Intellectual Property Registration No. 181718; Internet Reference (valid September 2012)." http://www.cochilco.cl/english/productos/doc/energy_consumption_and_greenhouse.pdf.

Coleman, M., Money, G., 2008. "Increasing Capacity and Productivity in the Metals Markets Through Pneumatic Conveying and Process Injection Technologies." *The Southern African Institute of Mining and Metallurgy.* onemine.org reference. http://www.onemine.org/view/?d=F3C2D1E001963C70C0A28E5D0DFB77FAD5305F0950A07FDB0039F31021F40AE1185008.

Díaz, C., Landolt C., Luraschi, A., editors, 1987. *Pyrometallurgy of Copper*. Copper 87.

Díaz, C., Landolt, C., Utigard, T., editors, 1999. *Smelting Technology Development Process Modeling and Fundamentals*. Copper 99 ISBN 0-87339-440-2.

Díaz, C., Landolt, C., Utigard, T., editors, 2003. *Pyrometallurgy of Copper*. Copper 2003, ISBN 1-894475-38-0.

Díaz, O.C., Landolt C., Luraschi, A., Newman, C.J., editors, 1991. *Pyrometallurgy of Copper*. Copper 91 ISBN 0-08-041432-X.

Drummond, W., Deakin, J., 2000. *A Water-Cooled Hood System for Peirce-Smith Converters and Similar Furnace Vessel*. pp. 40–41 JOM, May.

Edens, T., Hannemann, D., 2005. "New Casting Moulds for Anode copper." *Converter and Fire Refining Practices*. pp. 159–165. Edited by Ross, Warner, and Scholey. TMS.

Edens, T.L., Willbrandt, P., Specht, A., 2006. "Optimisation of the Anode Sector at Norddeutsche Affinerie, Volume 7 of Advanced Processing of Metals and Materials (Sohn International Symposium)." *International Symposium on Sulfide Smelting*. onemine.org reference. http://www.onemine.org/view/?d=FCA123673E2113525D1404A44CCC324BA9262B56700F440E318236D853EC6FBD202438.

Enrico, W.A., Vasilev, I., Naidenov, N., Marinov, E., Kirilov, D., 2005. "Converter Modernization and Operational Improvements at the Pirdop Smelter." *Converter and Fire Refining Practices*. pp. 109–117. TMS onemine.org reference. http://www.onemine.org/view/?d=8C4E85193D72835334CE77FDC9D4203DD8A2D1098F68FCD06E0367C8A4D5B632151601.

George, D.B. 2002. "Continuous Copper Converting—A perspective and View of the Future." *Sulfide Smelting 2002*. pp. 3–13. Edited by Stephens, R.L. and Sohn, H.Y. Seattle, WS: TMS.

George, D.B., Donaldson, J.W., and, Johnson, R.E., 1976. "Minor Element Behavior in Copper Smelting and Converting." *World Mining and Metals Technology*. AIME onemine.org reference. http://www.onemine.org/search/summary.cfm/Minor-Element-Behavior-In-Copper-Smelting-And-Converting?d=059AB59DD9B33BB637F33730AFA17F51A12147329A554BD588BFE3041194DF3339229&fullText=Minor%20Element%20Behavior%20George%20Donaldson%20Johnson.

George, D.B., Chen, W.J., Mackey, P.J., Weddick, A.J., editors, 1999. *Smelting Operations and Advances*. Copper 99 ISBN-10: 0873394399.

Gonzales, T.W., Jones, D.M., 1993. "Flash Smelting at the Magma Metals Company San Manuel Smelter." *Extractive Metallurgy of Copper, Nickel, and Cobalt Vol. 2*. pp. 1395–1407. TMS, ISBN 0-87339-219-1 onemine.org reference. http://www.onemine.org/search/summary.cfm/Flash-Smelting-at-Magma-Metals-Company-San-Manuel-Smelter?d=C7F9120D1289C122ECD7AFFB7E45FE8597DD74E7A81060EADAD8668BB384FBAF167868&fullText=Magma%20Smelter%20Flash.

Goodwill, D.J., Jones, D.M., Royal, T.A., 1999. "Redesigning the Flash Furnace Feed System at BHP Copper." *Smelting Operations and Advances Vol. 5*. pp. 517–531. Edited by George, Chen, Mackey, and Weddick. Copper 99 TMS.

Goto, M., Hayashi, M., 2003. "Recent Advances in Modern Continuous Converting." *Metallurgical and Materials Processing: Principles and Technologies, Vol.2: High Temperature Metals Productions*. pp. 179–187. Edited by Kongoli, Itagaki, Yamanchi, and Sohn. TMS ISBN 0-87339-5547-6 onemine.org reference. http://www.onemine.org/view/?d=7D5AF509E4F7A6F0E184A71D99EFDF9B69F2EA58932FB6FE99B11CF7BC215C91167558.

Hahn, Y.B., and Sohn, H.Y., 1990. "Mathematical Modeling of Sulfide Flash Smelting Process: Part I. Model Development and Verification with Laboratory and Pilot–Plant Measurements for Chalcopyrite Concentrate Smelting." *Transactions of AIME B*. 21B: 945–958.

Hahn, Y.B., and Sohn H.Y., 1990. "Mathematical Modeling of Sulfide Flash Smelting Process: Part II. Quantitative Analysis of Radiative Heat Transfer." *Transactions of AIME B*. 21B: 959–966.

Hattori, Y., Mori, Y., Kondo, Y., Sasaki, Y., Okubu, T., Baba, K., 2003. "Development of Sumitomo Premixed Concentrate Burner for Copper Flash Smelting." *High-Temperature Metals Production Vol. 2*. pp. 293–304. Edited by Kongol, Yamauchi, Sohn. TMS onemine.org reference. http://www.onemine.org/view/?d=6C457D7A2DF77E9B17BC499DEF807A66D273D514DCC21C890C5B29A03962C8A3167567.

"International Copper Study Group—The World Copper Factbook." 2012. http://www.icsg.org/.

Johnson, E.A., Sanker, P.E., Oden, L.L., and See, J.B., 1982. "Copper Losses and the Distribution of Impurity Elements Between Matte and Silica-Saturated Iron Silicate Slags at 1,250°C." US BOM, ROI 8655 onemine.org reference. http://www.onemine.org/search/summary.cfm?d=DB2AAC4160D9D9D4B755 61EBD6DC05E788A6739794E04C6D14EE4119C350EA91164120&fromView.

Jones, D.M., Cardoza, R., Baus, A., 1999. "Rebuild of the BHP San Manuel Outokumpu Flash Furnace." *TMS Copper 99 Smelting Operations and Advances Vol. 5.* TMS.

Jorgensen, F.R.A. and Koh, P.T.L., 2001. "Combustion in Flash Smelting Furnaces." *Journal of Metals* (May): 16–20.

Kapusta, J.P.T., Stickling, H., and Tai, W., 2005. "High Oxygen Shrouded Injection at Falconbridge: Five Years of Operation." *Converting and Fire Refining Practices.* pp. 47–60. Edited by Ross, Warner, and Scholey. TMS onemine.org reference. http://www.onemine.org/view/?d=67024F90AB72E953467AAC 6B04CBF2B19FC21A876D69B919913F542BE1EADB61151606.

Kellogg, H.H., Henderson, J.M., 1976. "Energy Use in Sulfide Smelting of Copper." *Extractive Metallurgy of Copper.* ASIN: B000HB0T8K onemone.org reference. http://www.onemine.org/search/summary .cfm/Energy-Use-in-Sulfide-Smelting-of-Copper?d=3518921B343E3D18DCA424B6DC1366F51A BEBA93546F281A6CEBF047EDD867FF167911&fullText=Energy%20use%20in%20sulfide%20 smelting%20of%20copper%20Kellogg%20Henderson.

Kim, H.G. Sohn, H.Y., 1996. "Thermodynamic Modelling of Minor-Element Behavior in In-Bath Copper Smelting and Converting with Calcium Ferrite Slag." *Transactions of AIME Sec C.* C151-C163.

Kojo, I.V., Storch, H., 2006. "Copper Production with Outokumpu Flash Smelting: An Update." *International Symposium of Sulfide Smelting Vol. 8.* pp. 225–238. Edited by Kongoli and Reddy.

Kojo, I.V., Jokilaakso, A., Hanniala, P., 2000. *Flash Smelting and Converting Furnaces: A 50 Year Retrospect.* JOM,. pp. 57–61. February.

Kyllo, A.K., Gray, N.B., 2005. "Composite Furnace Module Cooling Systems in the Electric Slag Cleaning Furnace." *Proceedings of EMC.* pp. 1027–1034.

Laist, F., 1934. "Reverberatory Smelting Practice." *History of Reverberatory Smelting in Montana.* pp. 1879–1933. AIME Transactions onemine.org reference. http://www.onemine.org/view/?d=9DC83BF3589C1 498270D7AA33484A7B2E8CED4C8A204258064DFFD39B8B4C49117036.

Landolt, C., Dutton, A., Fritz, A., and Segsworth, S., 1993. "Nickel & Copper Smelting at Inco's Copper Cliff Smelter." *Extractive Metallurgy of Copper, Nickel, and Cobalt Vol. 2.* pp. 1497–1527. Edited by Landolt, C. TMS ISBN 0-87339-219-1 onemine.org reference. http://www.onemine.org/view/?d=72576925B6 CB58BAED6DA2DB5EA7B2EAC81ABCBA0A1D23DE2A2DFE673464C80D167875.

Landolt, C.A., editor, 1993. *Extractive Metallurgy of Copper, Nickel and Cobalt.* TMS, ISBN 0-87339-219-1.

Larouche, P., 2001. "Minor Elements in Copper Smelting and Electrorefining.Thesis submittal McGill University." November.

Le Roux, H.J., Drummond, W.B. "Reducing/Eliminating Gas Emissions from Peirce [sic] Smith Converters." *The Third Southern African Conference on Base Metals.* pp. 71–86 onemine.org reference. http://www .onemine.org/view/?d=CFFBDD876D935A15D1FA778EA4EBC15741B6DCC1CDF0468D6FBAB 09CD1F671FE47321.

Lewis, F.M., Chase, C.K., and Bhappu, R.B., 1977. "Copper Production Costs Update." *SME Fall Meeting.* ST Louis, MO Preprint 77-K-379: AIME onemine.org reference. http://www.onemine.org/view/?d= DB734B7AE4A652B10DA64C43FF4770A73E7A0BC99C36BAAB418C9C3DD03C0A9B166652.

Lossin, A., Windhager, H., 1999. "Improving the Quality of Smelter Acid: The Example of Norddeutsche Affinerie's " Premium Quality."" *British Sulphur* 99: 209–220.

Mackey, P.J., 1982. "The Physical Chemistry of Copper Smelting Slags—A Review." *Canadian Metallurgical Quarterly.* 21(3):221–260.

MacRae, A., (Kvaerner E&C), 2002. "Pipe Coil Selection for Cast Copper Cooling Blocks." *Sulfide Smelting 2002.* Seattle, WA: TMS ISBN-10: 0873395255.

Malema, M.T., Legg, A.C., 2006. "Recent Improvements at the BCL Smelter." *Southern African Pyrometallurgy.* pp. 215–231. Edited by Jones, R.T. onemine.org reference. http://www.onemine.org/view/?d=198BA22 F536BB7FDDA9ED46AE2DFE79F70B5D33AB28F95E0E46B1836AF3B5F8347394.

McCain, J.D., Floyd, J.M., editors, 1994. "Converting, Fire Refining and Casting." *TMS Annual Meeting.* ISBN 0-87339-263-9.

McNulty, T.P., 1983. "Changing Energy Economics in Extractive Metallurgy." *SME Meeting. Preprint 83-432.* Salt Lake City, UT: AIME onemine.org reference. http://www.onemine.org/search/summary.cfm/Changing-Energy-Economics-In-Extractive-Metallurgy?d=453572A7C5465F7919E986BA2BED846BACD22EC654DB306AEBF755FDD9504F60174376&fullText=Changing%20Energy%20Economics%20McNulty.

Mitchell, W.J., Reed, R.A., 1992. "Phelps Dodge Hidalgo Smelter—Major Turnaround." *Extractive Metallurgy of Copper, Nickel and Cobalt Vol. 2.* pp. 1675–1680. Edited by Landolt, C. Warrendale, PA: TMS onemine.org reference. http://www.onemine.org/view/?d=E34FFA5032212C9A86AACFE5F110C7453F742BBB7D5AF14822B93C09B1127B5B167888.

Moyano, A., Caballero, C., Mackay, R., Morales, P., Cordero, D., Font, J., 2006. "The Development of the Codelco-Chile Continuous Converting Process." *International Symposium of Sulfide Smelting Vol. 8.* pp. 239–250. TMS.

Nikolic, S., Eddwards, J.S., Burrows, A.S., Alvear, R.F., 2009. "Isaconvert™—TSL Continuous Copper Converting Update." *International Peirce-Smith Converting Centennial.* pp. 407–414. TMS ISBN 978-0-87339-736-0.

Ojima, Y., 2003. "Future of Copper Converting Process." *Metallurgical and Materials Processing: Principles and Technologies Vol. 2: High Temperature Metals Productions.* pp. 255–264. Edited by Kongoli, Itagaki, Yamanchi, and Sohn. TMS ISBN 0-87339-534-4 onemine.org reference. http://www.onemine.org/view/?d=84EE8F44369CE66978CA01E82C99B81BF89C129408D02C5909F823E11DBA54EF167622.

Perez-Tello, M., Sohn, H.Y., St. Marie, K., and Jokilaakso, A., 2001. "Experimental Investigation and 3-D Computational Fluid Dynamics Modeling of the Flash Converting Furnace Shaft: Part I. Experimental Observation of Copper Converting Reactions in Terms of Converting Rate, Converting Quality, Changes in Particle Size." *Morphology, and Mineralogy Transactions of AIME B.* 32B: 847–868.

Peuraniemi, E.J., Lahtinen, M., 2006. "Outokumpu Blister Flash Smelting Processes." *Advanced Processing of Metals and Materials Vol. 8.* pp. 303–312. Edited by Kongoli and Reddy. TMS onemine.org reference. http://www.onemine.org/view/?d=0E761ECD3DCD72351622B78EFF160A77CC6A9A4EDB860316C960DB3F2C8C0CC2202503.

Riekkola-Vanhanen, M. 1999. *Best Available Technique in Copper Producrion and By-production of Precious Metals.* Finish Expert Report. Helsinki, Australia: Finish Environment Institute ISBN 952-11-0506-2.

Rigby, A.J., Filzwieser, A., Wallner, S., and Caulfield, K., 2003. "The COP KIN® System Part II: Performance and benefits—a world wide overview." *Proceedings of Copper 2003.* Santiago, Chile.

Riley, M.F., Kobayashi, H., Deneys, A.C., 2001. "Praxair's Dilute Oxygen Combustion Technology for Pyrometallurgical Applications." *JOM* (May): 21–24.

Ross, A.G., Warner, A., Scholey, K., editors, 2005. *Converter and Fire Refining Practices.* TMS ISBN 978-0-87339-586-1.

Safe, P., Stephens, R.L., 2000. "Peirce-Smith Converter Hood Design Analysis Using Computational Fluid Dynamics Modeling." *EPD Congress.* pp. 51–61. TMS, ISBN 0-87339-459-3 onemine.org reference. http://www.onemine.org/view/?d=097037C0D77DB091C8512B070688D75D795C26908DFBBCF5B18E252C2081853C164452.

Shima, M., Kohra, A., Shoji, S., 1997. "Flash Smelting with 95% Oxygen Process Air in Ashio." *Adelaide Branch, Research and Development in Extractive Metallurgy.* pp. 57–60. AusIMM onemine.org reference, May. http://www.onemine.org/view/?d=A783AA236E3BDC5784C55755A0DCB412F503219128BBF0A16E05C9D831C32A44195479.

Slaven, G., MacRae, A. and Valentas, L., 2003. "The Implementation of Ultralife™ Copper Casting." *Technology in the EAF.* Pittsburg, PA, May.

Smith, J.A., Newman, C.J., editors, 1992. *Smelter Gas Handling and Treatment.* TMS ISBN 0-87339-141-1.

Sohn, H.Y., George, D.B., Zunkel, A.D., editors, 1983. *Advances in Sulfide Smelting.* TMS ISBN 0-89520-464-9.

Solnordal, C.B., Jorgensen, F.R.A., Koh, P.T.L. and Hunt, A., 2003. "CFD Modeling of the Flow and Reactions in a Flash Smelter Reaction Shaft." *Third International Conference of CFD in the Minerals and Process Industries.* pp. 161–166. CSIRO, December.

Sridhar, R., Toguri, J.M., and Simenov, S., 1997. "Copper Losses and Thermodynamic Considerations in Copper Smelting, Metallurgical and Materials Transactions B, Vol. 28B." pp. 191–200. April.

Stephans, R.L., Sohn, H.Y., editors, 2002. "Sulfide smelting 2002: Proceedings of a symposium sponsored by the Extraction and Processing Division (EPD) of TMS (the Minerals, Metals & Materials Society): held during the." *2002 TMS Annual Meeting.* Seattle, WA, February. pp. 17–21.

Suzuki, Y., Hashiuchi, F., Yasuda, Y., 2006. "Recent Operational Improvements at Saganoseki Smelter." *International Symposium of Sulfide Smelting Vol. 8.* pp. 211–222. Edited by Kongoli and Reddy. TMS onemine.org reference. http://www.onemine.org/view/?d=2F2A154A9F89580DE2C79058EBBEC096 3ED592C23459ED95C2784DF10D5794E4202496.

Tan, P., Vix, P., 2005. "Modelling of Slag Blow in Copper Peirce-Smith Converters." *First Extractive Metallurgy Operators' Conference.* pp. 110–115. Brisbane, Australia: QLD ISBN 978-1-920806-39-2 onemine.org reference. http://www.onemine.org/view/?d=66C734BFFB07EF733FBB5BE88C94B0092B613A7E38 78984C79221A6B9072DD3B189131.

Tanaka, F., 2006. "Metallurgy for Continuous Converting in the Mitsubishi Process, pp. 357–366 of Advanced Processing of Metals and Materials (Sohn International Symposium), International Symposium on Sulfide Smelting." *Advanced Processing of Metals and Materials Vol.8.* ISBN 0873396332, 9780873396332 onemine.org reference. http://www.onemine.org/search/summary.cfm?d=94292DAE886D6B006FA92 9B9BD18FA636273E01A59E643EB719155BEE3CD0A92202509&fromView.

Tanaka, S., Hamamoto, M., Hashimoto, M., Udo, S., 2005. "Operation and Improvements on Peirce-Smith Converters at the Tamano Smelter." *Converter and Fire Refining Practices.* pp. 79–88. TMS ISBN 978-0-87339-586-1.

Toguri, J.M., Kaiura, G.H., and Marchant, G., 1976. "The Viscosity of the Molten FeO-Fe2O3-SiO2 System." *Extractive Metallurgy of Copper, Vol.1.* TMS onemine.org reference. http://www.onemine.org/search/summary.cfm?d=185D06CF957317D05E7FCADD6F4DA15EAA121E96A26CA1112B1C34 B62361FC8D167905&fromView.

Tuominen, J., Kojo, I.V., 2005. "Blister Flash Smelting—Efficient and Flexible Low-Cost Continuous Copper Process." *Converter and Fire Refining Practices.* pp. 271–282. Edited by Warner and Scholey. TMS onemine.org reference. http://www.onemine.org/view/?d=022C0DBA217A4E7FFCAAF4058DDE376 19606D925473830543720D0FF93B8702B151597.

United States Department of Energy, Anonymous, "Kennecott Utah Copper Retrofits. n.d. www.oit.doe.gov.

Valenzuela, A., Balladares, E., Cordero, D., Sánchez, M., 2006. "Arsenic Management in the Metallurgical Industry: The Chilean Experience." *Advanced Processing of Metals and Materials Vol.* TMS onemine .org reference, n.d. Edited by Kongoli and Reddy. http://www.onemine.org/search/summary.cfm/ Arsenic-Management-In-The-Metallurgical-Industry-The-Chilean-Experience?d=E90683124D3096 BFE7AD55E198483300C3C434069451C49A5EE229474882D733202578&fullText=Arsenic%20 Management%20in%20the%20Metallurgical%20Industry%.

Vikdahl, A., Lehner, T., 1992. "Productivity, Quality and Smelter Operations, Rönnskär 1992." *Extractive Metallurgy of Copper, Nickel, and Cobalt.* pp. 1529–1547. Edited by C. Landolt. TMS, 1993.

Voermann, N., Ham, F., Merry, J., Veenstra, R., Hutchison, K., 1999. "Furnace Cooling Design for Modern, High-Intensity Pyrometallurgical Processes "4th Int. Conference Copper 99—Cobre 99."" *Smelting Operations and Advances Vol. 5.* pp. 573–582. Edited by George, D.B. et al., Warrendale, PA: TMS.

Walton, R., Foster, R., George-Kennedy, D., 2005. "An Update of Flash Converting at Kennecott Utah Copper Corporation." *EPD Congress Edited by Schlesinge, M.E.R.* TMS ISBN-10: 0873395816.

Wenzl, C., Filzwieser, A., Antrekowitsch, H., 2007. "Review of Anode Casting—Part II: Physical Anode Quality." *World of Metallurgy, Erzmetall.* 60 (2): 83–88.

Willbrandt, P., 1993. "Operational Results of Norddeutsche Affinerie Copper Smelter." *Extractive Metallurgy of Copper, Nickel, and Cobalt Vol. II.* TMS ISBN 0-87339-219-1. onemine.org reference. http://www .onemine.org/search/summary.cfm/Operational-Results-of-Norddeutsche-Affinerie-Copper-Smelter?d= B9366E07EBAE26F9D626B196303C5E15870A831F0359BCDAE82AC50CFA4142CF167865&ful lText=norddeutsche.

ABOUT THE AUTHOR

Eric H. Partelpoeg has worked more than 30 years in the mining industry with a focus on copper smelter, acid plant, roasting, and off-gas system operations and projects. His career has included technical services and operations management during his 15 years with Phelps Dodge. He has provided consulting services to the mining industry since 1996 and established his own consulting firm in 2002. His consulting practice has leveraged his operating experience with a focus on pyrometallurgical projects and sulfuric acid plant projects. Most recently he has provided process leadership for a prefeasibility study of a greenfield copper smelter.

Innovations in Zinc and Lead Pyrometallurgy

Len Harris
Company Director and Consultant, Denver, Colorado, United States

ABSTRACT: In the "early years," zinc was universally produced pyrometallurgically by retorting after roasting of its ores and concentrates. In the early 1950s Cerro de Pasco Corporation installed the ill-fated Sterling thermal electric arc furnace process in La Oroya, Peru, employing the fuming of zinc metal and recovery in splash condensers. Today hydrometallurgical processing is mostly used but the Imperial Smelting Process continues to be used for smelting mixed zinc/lead concentrates in Canada and England. The fuming of zinc from residues and zinc silicate and zinc oxide ores is still practiced throughout the world employing the Waelz kiln process. The production of zinc dust is mainly carried out by retorting. In the case of lead, pyrometallurgical methods are universally used today for production; Important changes made in this technology over the years and finishes with the latest in direct lead smelting including the development and installation of the NFC SKS Process in China with their bottom blown oxygen furnace.

ZINC

Unlike copper, iron and lead which were known from ancient times, zinc is a relatively new comer having not been discovered and produced until about 1000 AD. in China and India The first slab of zinc (known then as spelter) was imported to Europe in about 1740 and a zinc smelter was erected in Bristol. England. A patent was granted in 1758 for smelting zinc from zinc sulfide. In 1836 the horizontal retort zinc process began treating calamine (zinc oxide) ores mixed with coal. In the United States natural gas was introduced to fire the horizontal retorts in 1878 and was used continuously until 1925. Elsewhere in the world, retorting was used until the 1960s (Cotterill 1970; Lehto 1970). During that time, the main innovations were made in the roasting area especially with the introduction of fluidized bed reactors.

Roasting

In the early days of zinc production, "oxide" ores were used and treated in retorts with coal to distill off the zinc metal and collect in condensers. The next big step was to treat sulfide ores (sphalerite) which required pretreatment by roasting. Roasting was already known in the middle-ages by burning in heaps and later by roasting in Edwards single hearth rabble roasters (Edwards 1902) or multi-hearth Herreshoff roasters (Habashi 2004). Later sinter machines were used and finally fluidized bed roasters by Dorr-Oliver and Lurgi came on the scene This was a great breakthrough in roasting technology as not only was roaster throughput greatly increased but it produced higher strength sulfur dioxide gas ideally suited for sulfuric acid production. It also allowed for waste heat boilers to be placed in the off gas stream and thus reduce costs. Lurgi's very first roaster treating zinc concentrates was installed in the late 1950s at La Oroya, Peru (Harris 1965). The trouble-free start up and ease of operation provided Lurgi with the basis to quickly take over the market throughout the world. It is interesting to note that it was originally thought that the high altitude at La Oroya (12,300 feet)) would have a negative effect on throughput which caused Lurgi to down rate the reactor. This turned out to not be so as the thin air at this altitude (lower oxygen content) had to be increased in volume (increased space rate) to maintain fluidity in the reactor bringing it into line with conditions at sea level.

Extraction

For decades, zinc was extracted from ores and concentrates (roasted) by distillation in retorts and collection of the zinc metal in condensers. Horizontal retorts were initially used and later changed to vertical retorts for ease of loading and unloading at much reduced costs (Mathewson 1959).

In the 1950s, New Jersey Zinc Company developed an electro-thermal furnace heated by graphite electrodes (Bunce and Peirce 1959). Pelletized feed (zinc calcine) was charged to the furnace with coal and the distilled zinc was collected in molten zinc splash condensers. The process was called the Sterling Process which appeared to offer many benefits especially increased throughput and of great benefit where power was cheap (Handwerck et al. 1952). Such a furnace was installed at Cerro de Pasco's operations in La Oroya in the mid-1950s (Harris and Wadia 1965). Two things plagued the operation. One was the relatively low recovery of the volatilized zinc in the condenser and the other was unexpected and random explosions in the furnace caused by a buildup of carbon monoxide.

It was thought that the condenser problem could be overcome by switching from zinc metal in the condenser to lead metal as used in the Imperial Smelting furnaces. The problem of the explosions in the furnace could not be solved and the furnace was shut down. The chairman of Cerro de Pasco at that time was very keen on this process as the company operated hydroelectric power plants at that time producing very cheap power. He was very unhappy that the process was shut down and declared that the reason that it did not work was because the operators did not want it to work as they only believed in the electrolytic process for producing zinc.

Meanwhile, St. Joe continued to operate their version of electro-thermic process at their Monaca, Pennsylvania (Weaton et al. 1944). This process was first installed in 1930 (following piloting) and continues going strong today (Lund 1970). It consists of a vertical furnace with graphite electrodes inserted in the sides and is fed with zinc calcine and coke. The zinc is volatilized and collected in a bath of molten zinc in a large U-tube arrangement. This plant also produces zinc oxide which together with the zinc metal, amounts to something in excess of 200,000 tons per year of zinc equivalent. It is interesting to note that in their early years, roasting was carried out in Herreshoff multi hearth furnaces and they were able to produce a high sulfur dioxide gas from these furnaces to make sulfuric acid.

In 1968 the Imperial Smelting Corporation revealed to the world from their home base in Avonmouth, England, their very innovative Imperial Smelting furnace for the treatment of mixed lead/zinc concentrates which could not be separated by flotation (Lee 2006). It turned out to be a great success and such furnaces were quickly installed in Canada, France, Australia and elsewhere where they continue to operate today. The operation is similar to that of a lead blast furnace except that it is operated with a "hot top." Blast air through the tuyeres is preheated to 950°C. Zinc lead hot sinter is charged to the furnace with preheated coke and the zinc is fumed off and collected in molten lead splash condensers and transferred to the zinc refinery. The lead bullion is tapped off in the bottom of the furnace with slag to a fore hearth for separation and subsequent treatment.

Slag Fuming and Treatment of Residues

Slag from lead smelters often contains a high percentage of zinc in economical amounts for recovery. These slags are commonly treated in slag fuming furnaces to produce zinc oxide for selling to the market if sufficiently pure or sent to electrolytic zinc plants as feed to this process for recovery as zinc metal. The slag furnace is a fully water jacketed shaft type furnace fitted with tuyers somewhat like the lead blast furnace. The slag is charged to the furnace in the molten state in a batch wise fashion (about 50-ton charges) and pulverized coal is blown through the tuyeres with air. Some lead bullion is produced and tapped off at the bottom of the furnace. The zinc is fumed off in the furnace gases and after passing through a waste heat boiler and on to a bag house where the zinc oxide fume is collected. Such practice has been conducted at Trail in B.C., Canada, since 1928.

After World War II in 1947, Preussag Al Metall located in Oker, Germany, opened their first "half shaft" furnace for treating retort residues and slag (Dumont and Mueller 1970). Zinc leach residues are treated in Spain and Italy by the same method. The plant at Oker grew to 15 furnaces and was later introduced to Mitsui's electrolytic zinc plant in Japan for treating zinc leach residues. Unlike the classical lead blast furnace, the half shaft furnace has tuyers on one side of the furnace. The opposite side is a bricked slope. The zinc fume and combustion gases pass through a boiler and the zinc and lead oxides are collected in a bag house and sent to a de-leading kiln. Premium grade zinc oxide is produced and sold. The feed to the furnaces is briquetted with pitch which forms part of the reductant augmented with coke. If copper is present in the furnace charge, it is converted to matte and tapped off at the bottom of the furnace.

Another process which has become widely used throughout the world involves the use of a Waelz rotary kiln (Hoffman 1928). The process was originally developed in Germany to fume zinc from electric arc furnace off gases. The feed to the furnace consists of residue pelletized with pulverized coal. Within the furnace, the charge makes a backward and forward motion like waltzing. Thus it gets the name in German Waelz. As with other fuming furnace processes, the zinc is fumed off and the zinc oxide product is collected in a bag house. The author visited Germany to observe the operation as a possibility for treating zinc leach residue and as a result, a plant was installed there in the 1960s. The leach residue feed to this furnace assayed 26.0% Zn, 8.0% Pb, 29,8% Fe, 10.2 oz/t Ag, 17 oz/t Ag and 0.12% Cd. The idea was to volatilize the zinc, lead and cadmium, indium and part of the silver and collect in a bag house. The rest of the silver and the copper would be collected in a sponge iron product for use in precipitating copper from acid mine waters. The process worked well except that the sponge iron was now fully reduced. There was a need for iron at that time in the copper reverbatory furnaces so the copper and the copper and silver recovered in the matte. The zinc fume was subjected to a two stage leach in which zinc and cadmium was dissolved in the first stage and sent to the electrolytic zinc plant and the indium recovered in the second stage leach at lower pH.

Zinc Dust

Zinc dust, or finely dispersed zinc metal, is used in many places in the Merrill Crowe process to precipitate gold and silver from cyanide solutions. It is also used in electrolytic zinc plants to purify the electrolyte by removing copper and cadmium from leach solutions. The dust can be produced in two ways: volatilization or atomization. Very fine dust is produced by volatilization by distilling molten zinc metal in a retort and collection in a condenser. Atomization is carried out by pouring molten zinc metal into a pot which has a small hole in the bottom. The zinc flowing through the hole is met by a stream of high pressure air and blown into a tunnel and collected from the floor of the tunnel. The volatilization procedure produces a finer product which is much favored by gold plants whereas the atomized product is cheaper and suitable for purifying zinc electrolyte.

LEAD

Lead was probably one of the first metals produced by man. It is known to have been used prior to 3500 BC from discoveries made in Egypt. The oldest piece of lead in existence is found in the British Museum and predates 3800 BC

The manner in which prehistoric people extracted lead from its ores is largely unknown but it can be supposed that that primitive furnaces consisting of small pits dug in the ground and enclosed by stones to form a shaft, and lead was recovered from oxide ores by heating with wood or charcoal. In the Third Century BC the Romans operated the lead mines of Spain which are still operating today. During the period 700 AD to 1000 AD, many mines were worked in Germany and these are described in some detail in *De Re Metallica* (Agricola 1556).

Extraction

To date, the extraction of lead from its ores and concentrates is by pyrometallurgical means only and is quite complex, involving the treatment mainly of lead sulfide (galena) (Hoffman 1901; Fuller 1968). This is not because the reduction of galena is difficult but because other metals such as copper, tin, antimony, arsenic, silver, gold and bismuth are commonly associated with the lead sulfide which subsequently to the production of lead bullion must be separated by a series of refining steps. In the early days, lead ores were treated in reverberatory or hearth furnace but the use of blast furnaces has now been in use for decades (Cotterill 1970). Roasting or sintering is used prior to adding to the furnace with coarse coke.

Sintering

Prior to the use of sintering machines in the lead smelting business, the ores and concentrates were roasted in single hearth rabble roasters or multi hearth roaster similar to that used to oxidize zinc sulfide materials for retorting.

The sintering process was a great leap forward in the roasting of lead sulfides and serves a two fold purpose, namely to remove most of the sulfur present in the feed and to produce a material with suitable physical characteristics for smelting in the lead blast furnace. The material should be in the form of coarse hard lumps with sufficient strength and porosity to provide a permeable charge inside the furnaces. High sulfur in the blast furnace charge tends to throw an undue proportion of the lead into the matte thus resulting in poor recovery and form zinc and copper crusts inside the furnace shaft. The sinter produced is screened to remove fines from the furnace charge which are returned to the sinter machines with the new feed. In the early days of the sintering process, the charge was fed to the machines where it met the flame from a fire box which ignited the charge and as the grates moved along on a conveyor like grated moving surface (known as pallets), the ignition gases were drawn through the burning charge into wind boxes and discharger to atmosphere. The gases were generally too weak for feed to a sulfuric acid plant and the wind boxes below the grates build up with molten lead trickling down from the burning charge material. Someone made the great discovery of reversing the sintering operation to the so called "up draft" sintering mode were a thin layer of charge was fist ignited under down draft conditions and then received the bulk of the charge as the grates moved along and received an up draft blast of air as it continued its way along the machine. Thus, throughput of the machines was increased, high strength sulfur dioxide gas was produced for making sulfuric acid and the presence of molten lead in the wind boxes was eliminated. The author is not sure of where up draft sintering began but he thinks it was at St. Joe's smelter in Missouri. In any event, the change over from down draft to up draft sintering quickly spread throughout the industry. The plant in Peru consisted of 10 small sinter machines measuring 22 feet by 3½ feet and one larger machine measuring 40 feet by 6 feet. The author led a team to convert all the machines from down draft to up draft.

Blast Furnace

The lead blast furnace was and still is, the "work horse" of lead pyrometallugical plants. It consists essentially of a rectangular refractory brick and water jacketed shaft mounted above a crucible and fitted tuyeres along the lower sides for the introduction of blast air, Suitable tapping arrangements are made at the crucible for the removal of molten lead, matte, spice and slag. Following the invention of continuous tapping of the blast furnaces by ASARCO in the 1960s (Roy and Stone 1963), the tendency nowadays is to fill the furnaces with continuous tappers for removal of all molten products continuously to a fore hearth for separation of the lead, matte, spice and slag outside the furnace itself. This invention was probably the most important innovation in lead smelting procedures ever. Following a visit to the Tacoma lead smelter of ASARCO to observe the patented tapper in operation (known as the Roy tapper named after the inventor Mr. Roy), such tappers were placed on all three blast furnaces in operation in La Oroya. In an attempt to increase the throughput of any given lead blast furnace, operators (including those in La Oroya) have experimented with adding substances to the furnace blast air such as oxygen, coke oven byproduct gas and oil. Very little improvements in furnace operations have been reported.

Lead Bullion Refining

The lead bullion produced from the blast furnace often contains other metals which have to be removed to produce a high quality refined product for marketing. This can be done electrolytically by the Betts process (Fingland 1930) or pyrometallurgically in a series of refining steps. The metals to be removed are normally copper, silver, gold, antimony and arsenic. In pyrometallurgical refining, the first step is to remove the copper which is achieved simply by cooling the bullion in pots to about 450°C (just above the freezing point of lead) At this temperature, the copper compounds become insoluble which separates out as a dross and can be scraped off from the surface of the molten bullion. Elemental sulfur is often added to this drossing step to help the removal of copper. The next step is to remove the arsenic, antimony and tin by two methods known as "softening" and the Harris process (Harris 1970). In the "softening" process the molten lead bullion at about 750°C is contacted with air by means of pipe lances and the metals alloyed with lead are converted to their oxides and slagged off. The operation is performed continuously in some places and was first introduced at the Trail smelter in British Colombia in Canada, In the Harris process, the molten lead at close to its melting point is pumped through molten sodium hydrate. Arsenic, antimony and tin are thereby removed completely from the lead and maybe recovered by leaching from the caustic mixture. Continuing with the pyrometallurgical process, the next step is to remove silver and gold from the bullion, if present. Again, this can be done by two processes know as the Pattinson process (Schnabel 1898) and the Parkes process (Eurich 1912). To a large extent, the Pattinson process has been eliminated by the Parkes process but still used in a few plants. The Pattinson process is rather complex and is based on the fact that molten lead containing silver is cooled to near its freezing point, crystals of lead containing less than 0.0015% Ag separate. The crystals are removed and more lead is added to the vessel and the procedure is repeated. Finally the molten lead contains up to 1.5% Ag are cupelled to recover the silver. In the Parkes process, molten zinc is used to remove and recover the silver and gold. The zinc alloy of silver and gold rises to the top of the molten lead as a crust. This crust is scraped off and distilled in retorts to recover the zinc in condensers for reuse in the process and enriched bullion for cupellation and recovery of the silver and gold, The Parkes process was another innovative process in the pyrometallurgical recovery of lead. It too, can be carried out in a continuous mode. The last remaining metal to be recovered from the lead bullion is bismuth. Bismuth is removed by the Kroll–Betterson process (Evers 1949) which involves the use of calcium and magnesium (Betterton 1932). When introduced into the lead bullion, they form the compounds calcium bismuth and magnesium bismuth which are removed as dross and then treated for the recovery of refined bismuth.

THE NEW PROCESSES

There have been several good reviews of new lead smelting processes, including one by Rob Stephens from Teck's Trail operations in his fine paper at the Lead-Zinc 2010 meeting held in Vancouver, Canada (Stephens 2000). He used the term "new smelters" to describe the direct lead smelting process (no sintering) operating throughout the world. I am sure that he doesn't mind me using his definition of these smelters.

Among the newer processes (Siegmund 2003) are QSL (Meurer and Ambroz 2011), Kivcet (Chaudhuri et al. 1980), Isasmelt (Errington et al. 2010), and Ausmelt (Creedy et al. 2010)

Lead metallurgists have long desired to do away with the sintering step used in the classical sintering/blast furnace technology (Lee and Park 2003; Errington et al. 2010; Dongba and Xinya 2011; Gaoa et al. 2012).

REFERENCES

Agricola, Georgius 1556, De Re Metallica.

Betterton, J.O., 1932, Metal Refining and Removing Bismuth from Lead: United States Patents 1,853,534; 1,853,535; 1,853,536; 1,853,537; 1,853,538; 1,853,539; 1,853,540; 1,853,541; Apr. 12.

Bunce, E.H. and Peirce W.M., 1959, New Jersey Zinc develops a new condenser, Eng. Min. J., 150, No. 3, March, p. 59.

Chaudhuri, K.B., Koch, M. and Patino, J.L., 1980, The technical-scale realization of the Kivcet process for lead, Non-Ferrous Metallurgy, Bull CIM, p. 146–150.

Cotterill, C.H., 1970, Extractive Metallurgy of Lead and Zinc, AIME World Symposium on Mining & Metallurgy of Lead and Zinc, AIME.

Creedy, S., Reuter, M., Hughes, S., Swayn, G., Andrews, R. and Matusewicz, R., 2010, The Versatility of Outotec's Ausmelt Process for Lead Production, PbZn 2010 Conference, pp. 439–450.

Dongbo, L. and Xinyu, W., 2011, Oxygen bottom blowing lead melting technology (SKS) Presented in Lisbon, Portugal, pp. 185–215.

Dumont, H. and Mueller, E.A., 1970, Processing of zinc and lead bearing residues in the Half-Shaft Furnace Process by Preussag Ag Metall, World Symposium on Mining and Metallurgy of Lead and Zinc, AIME, 2, pp. 389–408.

Edwards, T., 1902, Furnace for Roasting Ores, US patent No. 714,464.

Errington, B., Hawkins, P. and Lim, A., 2010, ISASMELT for Lead Recycling, Lead-Zinc 2010 Symposium, pp. 709–720.

Eurich, E., 1912, The Development of the Parkes Process in the United States, Bull AIME, 72, p. 1531–1540.

Evers, D., 1949, Removal of bismuth by the Kroll-Betterton Methods: Ztschr. Erzbergbau u. Metallhuttenw, 2, p. 129.

Fingland, J.J., 1930, The Betts Electrolytic Lead-Refining Process, Trans Am Electrochem Soc, 57, p. 177.

Gaoa, W., Wang, C., Yinc, F., Chend, Y. and Yang, W., 2012, Situation and technology progress of lead smelting in China, Advanced Materials Research, 581–582. pp. 904–911.

Habashi, F., 2004, Fire and the art of metals—A short history of pyrometallurgy, International Conference on Molten Slags, Fluxes and Salts, VII, SAIMM.

Handwerk, E.C. Mahler, G.T and Fetterolf, L.D., 1952, Electric furnace smelting of zinc ores—The Sterling Process. J Met, 4, pp. 581–586.

Harris, L., 1965, Lead smelting improvements at La Oroy, Pyrometallurgical Processes in Non-Ferrous Metallurgy Symposium, Pittsburg, PA.

Harris, L., 1970, Lead, The Encyclopedia of the Chemical Elements (ed. C.A. Hampel), Reinhold Book Corporation.

Harris, L. and Wadia, B.M., 1965, The Cerro de Pasco Process compared with other methods for zinc leach residue treatment, CIM Annual Meeting. Ottawa, Canada.

Hoffman, H.O., 1901, The Metallurgy of Lead, Scientific Publishing Company.

Hoffman, H.O., 1928, The Waelz Process, AIME, pp. 537–553.

Fuller. F.T., 1968, Process for Direct Smelting of lead concentrates, J Metals, 20, 12, pp. 26–30.

Lee R.W., 2006, Developments with the Imperial Smelting Process, Non-ferrous Materials Extraction and Processing (ed. F. Kongoli and R.G. Reddy), TMS pp. 129–138.

Lee, Y.H. and Park Y.M., 2003, The experience of lead direct smelting in Korea Zinc's Onsan refinery, Metallurgical and Materials Processing: Principles and Technologies, High Temperature Metals Production (ed. F. Kongoli, K. ltogaki, C. Yamauchi and H.Y. Sohn), TMS, 2, pp. 91–98.

Lehto, R.S., 1970, Zinc, The Encyclopedia of the Chemical Elements (ed. C.A. Hampel), Reinhold Book Corporation.

Lund, R.E., 1970, Josephtown Electrothermic Zinc Smelter of St. Joe Minerals Corporation, World Symposium on Mining and Metallurgy of Lead and Zinc, AIME, 2, 20, pp. 549–580.

Mathewson, C.H., 1959, Zinc The Science and Technology of the Metal. Its Alloys and Compounds, American Chemical Society, Reinhold Publishing Corporation.

Meurer, U. and Ambroz, H., 2011. Primary Lead Production with the QSL Process—an Ecological and Economical Advanced Technology. European Metallurgical Conference, Duesseldorf, Germany, Berzelius Stolberg GmbH, 2, pp. 481–488.

Roy, J.T. and Stone, J.R., 1963, Continuous tapping of a lead blast furnace, Trans TMS-AIME, 227, pp. 172–177.

Schnabel, C., 1898, Copper, Lead, Silver, Gold, Handbook of Metallurgy, MacMillan and Co, New York, pp. 503–511.

Siegmund, A., 2003, Modern applies technologies for primary lead smelting at the beginning of the 21st century, Yazawa International Symposium, Metallurgical and Materials Processing: Principles and Technologies, High Temperature Metals Production (ed F. Kongoli, K. ltagaki, C. Yamauchi and H.Y. Som), TMS, 2, pp. 43–62.

Stephens,R., 2000, Advances in primary lead Smelting, Lead-Zinc 2000 (ed. J.E. Dutrizac), TMS, pp. 45–71.

Weaton, G., Najarian, H.K. and Long, C.C., 1944, Production of Electrothermic Zinc at Josephtown Smelter, Trans AIMME, 159, pp. 141–160.

ABOUT THE AUTHOR

Len Harris currently serves as a director for several mining companies and is chairman emeritus, Mining Energy and Petroleum Task Force Chamber of the Americas. Previously he worked for Newmont for 21 years in New York, Denver, and Peru. He held numerous positions including general manager of Minera Yanacocha, president and general manager of Newmont Peru, vice president and general manager of Newmont Latin America, and many other appointments. He also spent 18 years with Cerro de Pasco Corporation in Peru and New York and 2 years as mine manager at Texada Mines in Australia. Harris is a graduate of the Mount Morgan School of Mines (Australia). He has received numerous awards, including the MMSA Gold Medal, Western Mining Hall of Fame Medal of Merit, Newmont Chairman Award, the Peruvian Society of Engineers Engineer of the Year Award, and the Peruvian College of Journalism Award. He is a Legion of Honor Member of SME and has received from SME/AIME the Saunders Gold Medal, Krumb Lecturer, President's Citation, Distinguished Member, and Arthur C. Daman Lifetime Achievement awards.

Technology Advancement in the Refining of Platinum Group Metals

Neville Plint
Anglo American Platinum, South Africa

Stephen Woollam
Anglo American Technical Solutions, South Africa

ABSTRACT: Early Refining methods for Platinum Group Metals typically involved translation of laboratory techniques for assaying of concentrates. This led to highly inefficient processes which required multiple re-working to obtain the required degree of impurity rejection and to separate the Platinum metals from each other. This paper gives a high level review of comparison of "classical refining" flowsheets, largely based on selective dissolution techniques and precipitation against modern refining approaches. In the case of the latter, the approach taken was to apply a series of sequential hydrometallurgical extraction steps for removal of each metal, which formed a core separation. Secondary refining of the extracted metals was aimed at further impurity rejection and then final production of metal. The advantages of this approach are discussed.

INTRODUCTION

The world-wide production of Platinum Group Metals (PGMs) has increased exponentially over the past century, driven by the growth in demand across a range of applications; this is depicted in Figure 1. PGMs are used in a number of applications for reducing harmful emissions, most notably from automobiles. This, coupled with the potential of fuel cells to deliver zero emissions from vehicles, is likely to result in continued growth in demand for PGMs in the long-term.

Natural PGM deposits are almost always associated with copper, nickel and iron sulfides. In South African deposits, the concentration of the PGMs is typically less than 5 g/ton. Sequential physical separation by flotation and pyrometallurgical enrichment, increases this concentration by about 100-fold.

Hydrometallurgical processes are applied to recover base metals and in so doing, enrich the remaining PGMs to a concentrate of approximately 50%. This concentrate still contains residual base metals and a range of amphoteric elements, such as Se, Te, As, Sb, Sn, Pb, etc. The refining of this concentrate requires separation of each of the PGMs from each other and from the residual impurities. This requires a multitude of processing steps and application of a wide range of metallurgical processes.

There are some 200 hydrometallurgical steps in the current Precious Metals Refinery (PMR) of Anglo American Platinum (Amplats). Production facilities in comparable industries such as nuclear, petrochemical and pharmaceutical industries typically have plants having 300 to 400 "main plant items"; those being for unit operations. The Amplats refinery, while small in comparison to other

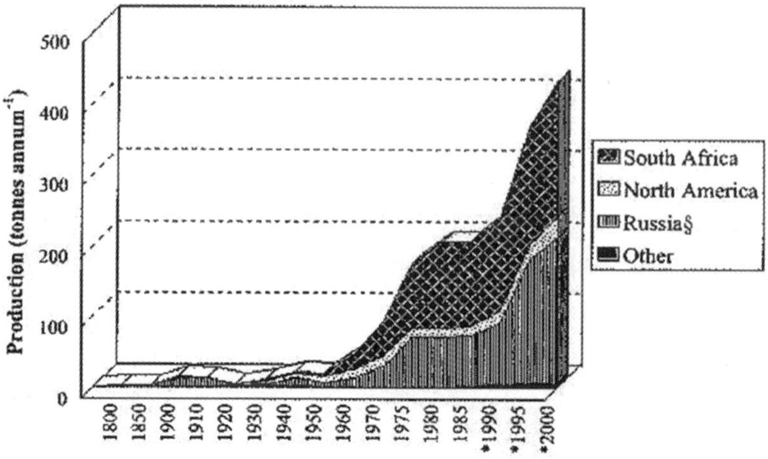

Figure 1. World-wide production of PGMs (tones per annum) since 1800

processing operations or production facilities, consists of approximately 1,500 "main plant items." This gives a sense of the number of unit operations and pieces of equipment needed to perform those operations; in many cases there are convoluted recycles and a web of cross-linking interactions.

In the earlier years, PGM concentrates were shipped to UK/Europe to be refined. The first refinery in South Africa was commissioned in 1975 by Impala Platinum and was based on a "Classical Process" which was, in essence, based on the chemical analysis procedure for determining the PGM content of ores, etc. This then spawned the Lonmin refinery.

This paper will shed some light on how PGM Refining technology has evolved over the years and will offer thought on what the future drivers will be for the refining industry.

DISSOLUTION OF THE PGM CONCENTRATE

PGMs are noble metals and as such offer high inertia to corrosion (and dissolution). This property is exploited during selective leaching of the base metals in the upstream processing. However, the highly oxidizing conditions required effect dissolution of the PGMs from the high-grade concentrate typically mean that all base metals present will also dissolve. Early methods of dissolution of the PGM concentrate exploited a difference in rate of dissolution between Platinum and Palladium and the other PGMs (also referred to as the Insoluble Metals—Rhodium, Ruthenium, Osmium and Iridium). Therefore digestion of the PGM concentrate (which had been pre-roasted) in aqua regia (mixture of hydrochloric and nitric acids) typically gave good dissolution of Platinum (Pt) and Palladium (Pd) and left the Insoluble Metals in a residue; the insoluble metals readily forming an oxide layer which offered resistance to leach. This practice has largely been phased out over the past several decades, with the exception of refineries at Lonmin and Krastsvetmet, where classical methods still form the basis of their flowsheets.

Since then, the common practice for dissolution of the PGM concentrate involves use of chlorine gas and hydrochloric acid at atmospheric pressure and elevated temperatures. Under these conditions, and provided the Insoluble Metals are sufficiently finely divided or alloyed with Pt and/or Pd, near-complete dissolution of the concentrate is possible. A major advantage in moving away from use of aqua regia for selective dissolution of Pt and Pd is elimination of NO_x emissions and the challenges associated with dealing with NO_x levels with narrowing legislative allowances.

More recently, in the case of Amplats, a further enhancement of the dissolution process has been adopted. This has involved use of higher temperatures (and consequently pressures) during the dissolution process. There are numerous advantages to this, including a reduction in the consumption of chlorine (since the reactor is merely pressurized under chlorine and not sparged), improved dissolved yields and better control over Osmium deportment. Any chlorine not consumed in the dissolution will find its ultimate fate as sodium chloride and this is a costly path to follow. Under the conditions employed, Osmium is readily converted to its tetroxide species, OsO_4, which is volatile and can therefore subsequently be efficiently removed from the dissolve liquor, ahead of the core separation processes.

CLASSICAL REFINING OF THE "SOLUBLE PGMs"

Early refining flowsheets involved selective leach of Platinum and Palladium from the roasted PGM concentrate. Platinum and Palladium from the filtered solutions were then typically precipitated as insoluble salts in sequential steps. In the case of Platinum, the ammonium hexachloro platinate salt ($(NH_4)_2PtCl_6$) was usually used, which in the case of Palladium, the diammine salt ($(NH_3)_2PdCl_2$) was typically used.

The fall-back with the above approach was that separation factors were not sufficiently high to achieve the required purity in a single step. As a consequence, repeated precipitation of the PGM salt with intermediate calcination and re-dissolution was needed to reject impurities to low enough levels. The consequence of this was that the classical precipitation processes inherently had very poor first pass yields and a high degree of re-work. Of significant importance, was that manual handling of PGM salts gave rise to a high rate of incidence for Platinum Salt Sensitivity (PSS). This is an allergenic condition, brought upon by exposure to platinum salts (not as metal) and requires the removal of affected the individual from the source of exposure.

CLASSICAL REFINING OF THE "INSOLUBLE PGMs"

The insoluble metals from the dissolution process, described above, were dissolved by high-temperature fusion with alkaline and oxidizing fluxes, followed by digestion in water. Ruthenium and Osmium were then present as volatile oxides (MO_4) and therefore could be removed by distillation. The remaining Rhodium and Iridium were recovered by precipitation of insoluble ammonium salts and then calcined. Control of oxidation potential allowed for separation of the two metals. As with the example described for final recovery of Platinum and Palladium, repeated precipitation, calcination and re-dissolution steps were needed to obtain sufficient rejection of impurities.

In most cases, the final product for each PGM is a sponge, which is a form of metal produced by calcination of a metal salt. It has a high surface area and so is a useful precursor to digestion to produce salts for autocatalyst applications or other catalyst compounds for industrial use. In some cases, the metal is melted in an induction furnace and then either poured into water (to produce metal grains) or into ingots (to produce metal bars).

The plant for the classical separation process, was typically arranged in a cascade system, allowing for movement of liquids by gravity. A typical "classical" refining flowsheet is shown in Figure 2.

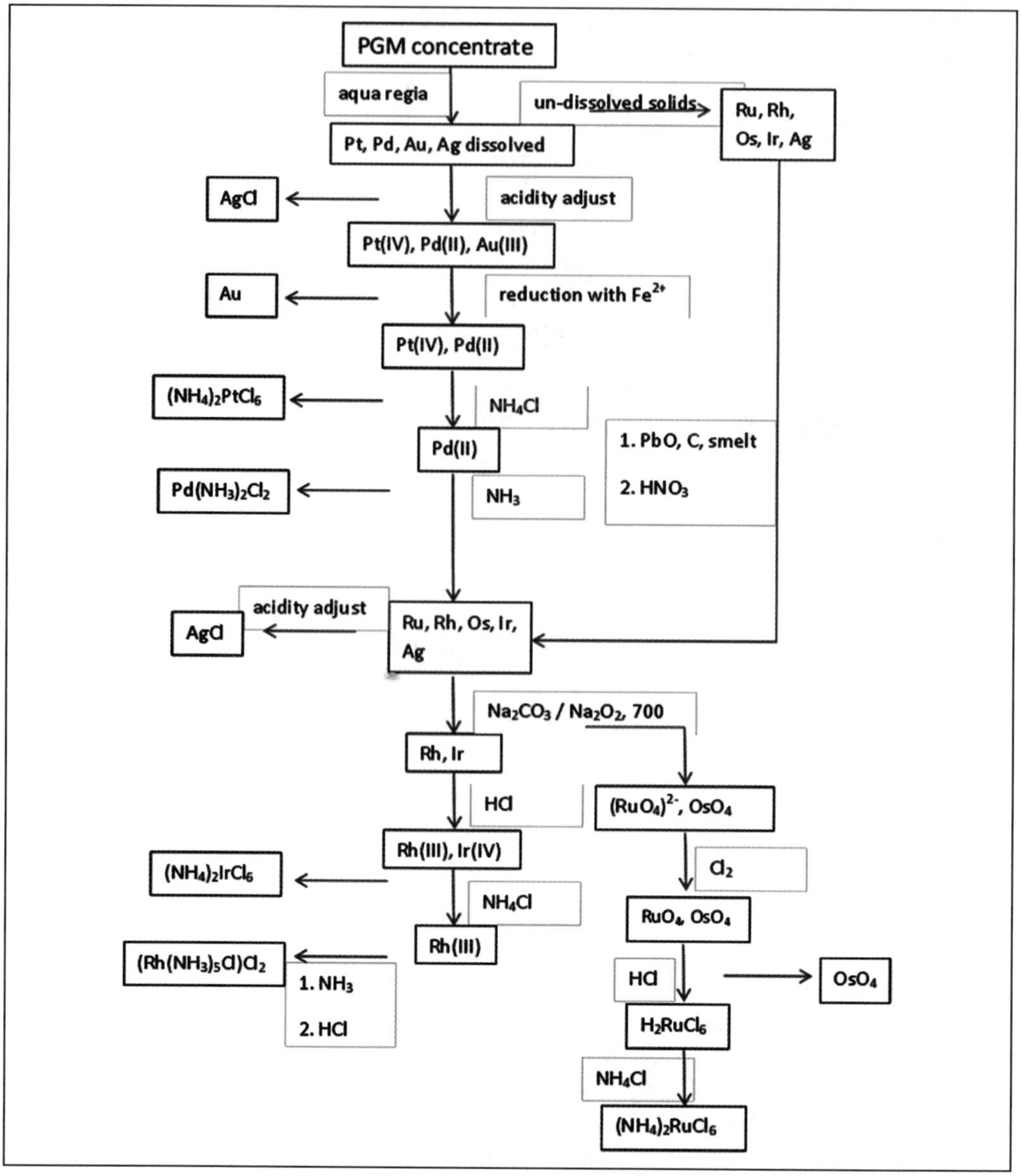

Figure 2. Typical "classical" PGM refining flowsheet

ADVANCED TECHNOLOGIES FOR REFINING OF PGMs

In the 1970s work commenced on developing a modern flowsheet for the separation and purification of PGMs for Amplats. This continued into the 1980s, with commissioning of the plant in 1989, near Rustenburg. A number of fundamental changes to the process have since been made and so the refinery is now based on a flowsheet very different to that of the classical refining processes. Despite this fundamental shift, many of the process steps or chemistries are still applied, but in a very different manner or put together in a very different way or coupled with novel separation techniques.

A fundamental shift in approach was taken, moving away from repeated processing to achieve the overall separation factors required. The new approach for separation, after dissolution, was based on three main areas:

1. A core primary separation set of processes,
2. Secondary purification, and
3. Final reduction to metal.

The core separation set of processes involved sequential extraction of each PGM (primary separation), using reagents which were designed to exploit subtle differences in chemistries between the PGMs. The extracted PGM was then put through a series of final purification processes (secondary purification) and then converted to metal sponge. In the case of Amplats, the core separation processes were based on Solvent Extraction (SX) and similarly for Vale's Action refinery and Johnson Matthey's refinery (both in the UK) and the Heraeus refinery (in Germany). In the case of other refiners, such as Impala, Ion Exchange (IX) is integral to its flowsheet. The principle of the new approach was based on achieve an extraction of each PGM in high yield and with good selectivity over other metals. The consequence was that the secondary purification processes typically only needed to increase purity from 99 to 99.99%.

This approach delivered on a number of objectives:

- Better containment of PGM salts, therefore significantly reducing the number of cases of PSS
- Significantly improved first pass yields and consequence reduction in inventory
- Significant improvement in overall yield
- Significantly less manual operations, leading to opportunities for automation and advanced process control. A further benefit here is improved ability to secure valuable metal in process.
- Increased process robustness and flexibility

A high-level flowsheet for a refinery using solvent extraction as a basis for primary separations, is shown in Figure 3.

By way of example of illustrating the benefits of conversion to a modern flowsheet, the quoted first pass yield, as percentage, for Lonmin versus Amplats is given in Table 1. While improvement have been made to first pass yields at Lonmin, the technology will limit the extent to which this can be done.

The quoted overall recoveries, as percentage, for Lonmin versus Amplats is given in Table 2. It should also be borne in mind that additional upfront leaching steps are needed in the case of the Lonmin PGM concentrate, which are not incorporated into these calculations).

Every so often, a reagent will come along which could reverse the direction of the above-described trend away from precipitation processes as a core separation technology. One such example is the use of diethylenetriamine (DETA) which forms of salt with the chloro-rhodium species, of extremely low solubility. Moreover, the selectivity of precipitation over base metals and other PGMs is extremely high. On this basis, separation and recovery of Rhodium is based on this reagent in many refining operations. It is interesting to note that application of this reagent was discovered, developed and patented by Amplats and yet were one last in PGM refining to apply this process commercially. With that said, Amplats undertook a detailed fundamental research programme on this process and this led to a far superior process design than any of its competitors and which offered a safer operation and a more robust and efficient process. In this instance, the exceptional performance of this reagent drove refining back towards the classical precipitation approach. However, the methodology is still aligned with the principles of a modern flowsheet in that it still involves a sequential series of extractions of each PGM.

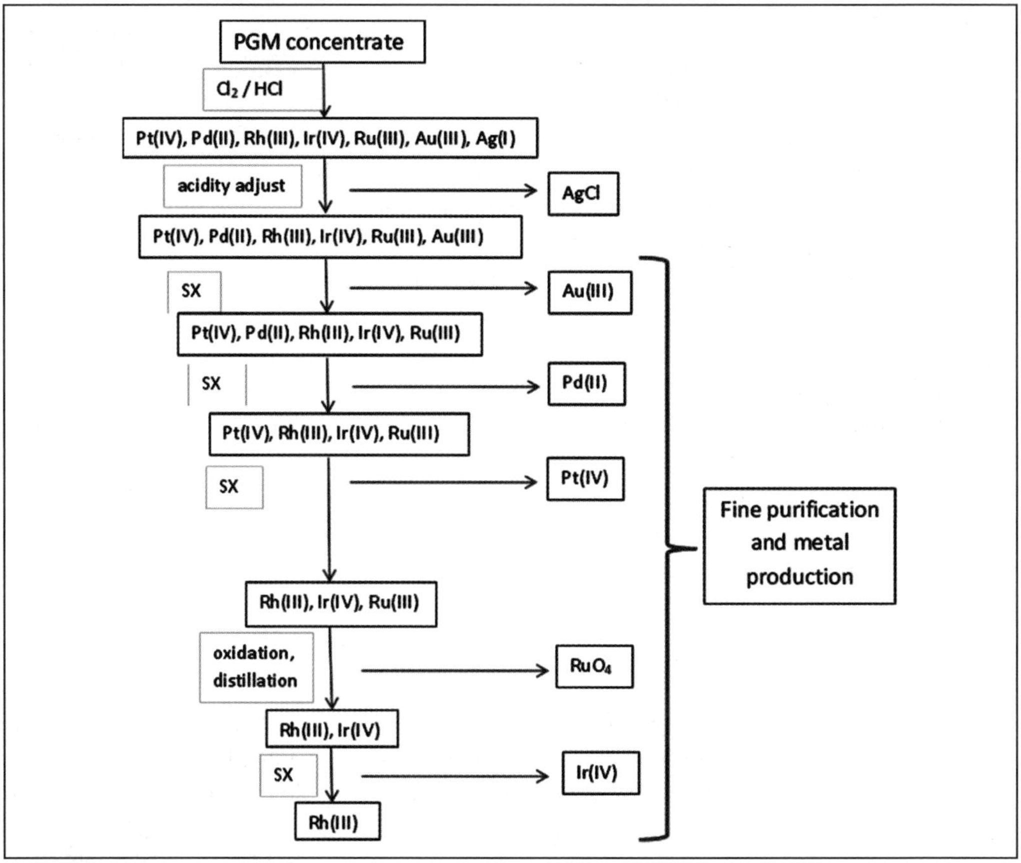

Figure 3. A modern PGM refining flowsheet, using solvent extraction for primary separations

Table 1. First pass yield of the main PGMs for flowsheets based on classical and modern flowsheets

	Lonmin*	Amplats†
Pt	86	94
Pd	90	92
Rh	77	82
Ru	86	90

*Commissioned in 1974.
†Commissioned in 1989.

Table 2. Overall recoveries of PGMs for refiners adopting differing flowsheets

	Lonmin*	Amplats†
Pt	99	99.5
Pd	98	99.7
Rh	93	99.5
Ru	98	99.5

*Commissioned in 1974.
†Commissioned in 1989.

REFINERY OF THE FUTURE

The capacity at some refineries is being reached. Furthermore equipment and technology is becoming outdated. This presents an opportunity to make a step-change to refining flowsheets or to make incremental modification and improvements to existing refineries.

A number of important factors should form the basis of a future refinery, summarized as follows:

- Safety. It is likely that current legislative exposure levels to platinum salts in the working environment (2 $\mu g/m^3$ on TWA) will be further reduced in future. A point may be reached where it just becomes impractical to contain existing processes and a step-change in technology or flowsheet is needed to meet these limits. Furthermore, there is an ever-tightening control over use of harmful chemicals in processes and this may force a move to processes which are inherently safer, require lower usage of harmful reagents and which guarantee curtailing of environmental risk.
- Recycling of reagents
- Reduction in inventory; this still presents a significant opportunity
- Processes with lower OPEX and CAPEX
- Processes which are more amenable to automation and advanced process control
- Water and energy conservation
- Process robustness and ability to accommodate varying inputs
- Security of value in process
- Multiple separations in a single process

SUMMARY

In summary, a modern refinery is likely to employ a wide range of technologies:

- Leach
- Distillation
- Precipitation and crystallisation
- Solvent extraction
- Ion exchange
- Complexation with organic reagents
- Hydrolysis
- Oxidation/reduction reactions
- Thermal processing under various gases

Successful operation of a PGM Refinery in a modern world depends on having a strong technical support team and highly skilled production management. Examples of specific skills sets required include the following:

- Analytical chemistry and sampling
- Process chemistry
- Process engineering
- Advanced Process Control and analytics
- Materials engineering
- Mechanical and electrical engineering

- Medical
- Security

REFERENCES

Beath, C.B., Westwood, R.J., and Cousins, C.A. 1961. Platinum Metals Review, 5(3):102.
Bernadis, F.L., Grant, R.A., and Sherrington, D.C. 2005. Reactive and Functional Polymers 65:205.
Black, W.H., Izatt, S.R., Dale, J.B., and Bruening, R.L. 2006. IPMI Conference, Nevada, June 10–13.
Cleare, M.J., Charlesworth, P., and D.J. Bryson, D.J. 1979. J. Chem. Tech. Biotechnol. 29:210.
Crundwell, F.K., Moats, M.S., Ramachandran, V., Robinson, T.G., and Delport, W.G. 2011. Extractive Metallurgy of Nickel, Cobalt and Platinum Group Metals, Elsevier.
Gouldsmith, A.F.S., and Wilson, B. 1963. Platinum Metals Review, 7(4):136.

ABOUT THE AUTHORS

Neville Plint joined Anglo Platinum in 1996 and has held a variety of research and management positions in metallurgical research, process control, and analytical chemistry. He has a PhD and an MBA (distinction) from the University of the Witwatersrand and is currently head of research and development for Anglo American Platinum. He is also the chairman on AMIRA International and sits on the advisory boards of Ian Wark Research Institute, ACCESS (University of South Australia), JKMRC, and SMI—University of Queensland, University of Cape Town Chemical Engineering and Catalysis, and University of Pretoria Chemical Engineering. His personal research interests include mineral processing, geometallurgy, resource to market optimization, material stewardship, water management, and advanced instrumentation for plant control and optimization.

Stephen Woollam has worked 20 years in the mining industry focusing on process chemistry related to refining of the Platinum Group Metals (PGMs). His career has been based entirely with the world's largest producer of platinum, Anglo American. He currently leads a team of research scientists and engineers at Anglo American's Technical Solutions–Research group and which offers a center of expertise for refining of PGMs and base metals (Ni, Cu and Co) for Anglo American Platinum's processing operations. A significant number of process improvements to the refining operations have been implemented as a consequence of this group's R&D activities over the years. A major part of their current role is to develop a next-generation PGM refinery.

Metallurgical Innovations in the Iron and Steel Industry

Glenn E. Hoffman
Cardero Iron Ore Management Company, Lancaster, South Carolina, United States

ABSTRACT: Every successful company starts with a business plan that propels an idea, invention or technology into commercial practice with the intention of providing a product or service to the marketplace that meets an unfulfilled niche demand. Often the road to success is riddled with trials and tribulations that entail much risk taking and consumption of precious resources—mainly man-hours and capital. For those who exercise proper due diligence and are willing to assume the risk, the end result can be a reward that leads to a profitable business and a healthy return on investment that more than justifies the assumed risk. A secondary reward is that the new business or venture has the potential to become a lasting entity that offers sustainable income to both investors and a host of workers dedicated to and dependent on the original venture.

In order to achieve a higher level of long range sustainability, the original business plan, idea or technology will naturally mature to the point where it is challenged by competing technologies. Thus the key for sustainable success is to adopt a policy that promotes the creation of new ideas, inventions or technologies that are able to adapt and adjust to whatever changes and challenges the marketplace imposes on the venture. The mechanism for adaptation is *innovation*.

The intent of this paper is to present some basic information on the state of the iron resource condition that the iron and steel industry is currently facing, as well as the current state of energy environment where site specific costs can make or break the success of a metallurgical technology application. One can appreciate that perhaps the greatest competitive challenge for the modern day iron processing facility manager is to obtain low cost feed materials for value added processing step while minimizing exposure to various logistical influences (transportation costs for both feed materials and products) and implementing the most cost and energy efficient technology for production of the final product. If one can understand the nature of these factors, and safely navigate through the mine field created by the interaction among these factors, then the end result will be a successful and sustainable business operation.

Two case studies are presented in the paper highlighting the innovation theme from two different perspectives. The first study considers the challenges and innovative responses taken to sustain a maturing technology, while the second study outlines the approach needed to accommodate a paradigm shift in thinking for metallurgical processing. That is, illustrating what innovative steps can be taken so that a low grade resource can be transformed into high margin commodity product. Both case studies employ innovation as the key to success.

INTRODUCTION

Every successful company starts with a business plan that propels an idea, invention or technology into commercial practice with the intention of providing a product or service to the marketplace that meets an unfulfilled niche demand. Often the road to success is riddled with trials and tribulations that entail much risk taking and consumption of precious resources—mainly man hours and capital. For those who exercise proper due diligence and are willing to assume the risk, the end result can be a reward that leads to a profitable business and a healthy return on investment that more than justifies the assumed risk. A secondary reward is that the new business or venture has the potential to become a lasting entity that offers sustainable income to both investors and a host of workers dedicated to and dependent on the original venture. That is, it is not a flash in the pan. Creative accounting and necessary cost control measures guard against the declining cash flow bumps in the road of success, but these are temporary fixes and not a wise strategy to employ when it comes to crafting a plan for long term sustainability. In order to achieve a higher level of long range sustainability, the original business plan, idea or technology will naturally mature to the point where it is challenged by competing technologies. Thus the key for sustainable success is to adopt a policy that promotes the creation of new ideas, inventions or technologies that are able to adapt and adjust to whatever changes and challenges the marketplace imposes on the venture.

The mechanism for adaptation is *innovation*. Innovation is a critical component to the longevity and success of a company, and it applies equally to the iron and steel industry. But innovation is not something that comes naturally or easily to people. One would like to be able to turn on a switch and be able to 'create' at will, but the reality of the situation is that innovation needs a proper environment to get the creative juices flowing. It requires money, patience, more money, perseverance, and did I say money? Most managers and executives liken the notion of innovation to R&D, and R&D budgets can range from two to perhaps twenty percent of gross earnings YOY, with the higher figure reflecting the requirements for 'high tech' industries. Unfortunately, research and development budgets are usually the first to get a hair-cut when times get tough for a company. Nevertheless, a great company promotes innovation and actively seeks innovators. It's not so much the budget as it is the people who create and use the budget. Great innovators are somewhat of a rare breed—like great managers. But a great company can recognize those in their organization who have the potential to become good innovators, and create the proper environment to nurture grow these candidates into great innovators so that their powers and abilities can constructively contribute to the sustainability of the business over time.

WHAT MAKES A GREAT INNOVATOR?

Most good innovators share some common characteristic traits and discovery skills. These traits and skills can be viewed as the first steps towards developing an effective prescription or methodology towards creating or cultivating innovation. Among these discovery skills are an innate ability for associating, questioning, observing, experimenting and networking. These skills will be expanded in the following paragraphs. First, associating relates to the ability of a successful innovator to connect seemingly unrelated questions, problems or ideas from different fields to the task at hand. The more diverse the experience and knowledge level, the more potential for connections to be used for innovative purposes.

As any parent can attest, the first and second words a toddler learns to say are "Mama" and "Dada" or "Papa." But the third most popular word is "Why?" The repeated question ("Why something...?"), can test the limit of the parent's ability to think of creative responses until at long last the final response is "Because"! But questioning leads to knowledge, and knowledge to understanding, and ultimately understanding to success. Innovators often find that the real difficulty in the innovation process is not getting to the end game or finding the right answer, but formulating the right questions. A good innovator asks "Why," "Why not," and "What if?" Another trait of questioning is to imagine opposites or a completely different alternative, and then perform a synthesis on these findings. The good innovator, through repeated questioning, forces constraints on the analysis process which leads to a new or better understanding of the subject or task at hand.

Observing is an innovation trait that entails scrutinizing the common phenomena. The power of observation takes time and patience to properly develop. Good observation skills enable the innovator to make the necessary connections between facts and findings, and this leads to creativity which is the lifeblood of the innovator. Steve Jobs said, "Creativity is the ability to connect things" (Quinn and Tayal 2012).

Experimenting can take the form of either loosely organized experiments that test the envelope (which resembles piddling), or carefully structured laboratory programs. Experiments are constructions in interactive experiences that are used to provoke unorthodox responses to see what insights might emerge. Often experiments are exercises in repetition and seemingly futile attempts towards success that are perceived as failures. But failure is also a learning experience. Edison once said, "I haven't failed, I simply found 10,000 ways that do not work" (Runes 1968).

The last trait of the successful innovator is networking. The good innovator likes to increase his or her knowledge base, and one good way to do that is to interact with other people who have different viewpoints or ideas and perspectives from one's own. Variety is the spice of life and the solo 'not invented here' attitude has no place in the stewpot of constructive innovation.

The aforementioned traits and discovery skills are the necessary ingredients that are found in a good innovator. But how does one acquire these traits and skills? Are innovator's naturally born with these skills, or can they be learned and developed over the course of a career? One can contend that many of the required traits and skills can be acquired through a conscious effort to trial and hone through dedicated effort and constant application. Surely there are exceptions who are 'naturals,' but the majority of good innovators are ordinary folk who are unafraid to rise to the occasion of a challenge, and like an athlete, practices every day until a level of confidence and perfection is attained. It takes a proper attitude, a willingness to learn, and a lot of practice. But how can this technique apply to the iron and steel industry?

One way the innovation technique can be learned is to become exposed to various case studies or programs that navigate through the complexities of a problem and ultimately result in a solution. Think of it as initiating the innovation process by first establishing a statement of what the problem is and then assessing the current state of understanding or listing the available information at hand on the problem. This is the *truth* or reality of the situation. A healthy response to the *truth* will result in an *innovation* that is a positive adaptation to the challenge presented. An unhealthy response to the *truth* can also result, which is an undesirable *consequence*. A *consequence* then is the result if no innovation or positive action that arises to the challenge. The *approach* then is the plan that is developed and executed so as to arrive at the desired innovation that solves the problem or resolves the challenge. One challenge that faces the Iron and Steel Industry is meeting the demands of a marketplace when the quality of mineral resources decreases in a volatile energy environment.

THE CHALLENGE TO THE METALLURGICAL INDUSTRY

Processing mineral resources into saleable commodities is a substantial capital intensive activity. It requires a commitment to a particular technology for the long term, which means promoters and investors have to have a clear vision to the future as well as courage and conviction. However, the market is always right, and in an environment where energy types and prices are volatile, it may mean that a particular technology that makes economic sense one day can be a loser the next.

In any case, relax, at least for the moment. The world today is not running short of many mineral commodities even though prices are spiraling upward. Einstein's observation still holds true, everything is still relative, as long as one can pass on the incremental costs of production to the consumer. However, a strong consensus exists among many of the companies comprising the minerals processing industry; that is, it is becoming increasingly more difficult to find the easy-to-mine, high quality resources that make juicy profit margins which shareholders like. For the most part, the resource cherries have all been picked, and established mineral processing companies who are exhausting their reserves, or in the process of looking to add reserves so as to create longevity for their company, have no recourse but to look long and hard at what were once considered the dregs of the earth. Even end users, such as steel companies, are now looking to become 'miners' so as to establish captive iron ore resources. Owning a captive resource facilitates a dependable supply of feed materials which translates into effective cost control and a guarantee for a thriving sustainable business in the future.

The problem is how can these low grade marginal resources become economically viable? Don't despair; even orphan properties can turn out to be gems in the rough with proper attention to doing your homework, applying innovative skills, developing the right technology, and possessing proper site specific conditions. The right technology application may require considerable development effort, both bench and pilot scale. As for site specific requirements that can tip the scales, these include availability of cheap power or fuel, a paradigm shift in the mining method and resultant mining costs, proximity of the main processing facility to required infrastructure, a local community that is hungry for jobs and willing to embrace industry, local/state governments offering subsidies for industry to establish greenfield processing sites, and even developing market acumen in the form of knowing what to target for production. The list goes on, but the attention to detail and the ability to adapt to changes can spell the difference between long term success and failure.

There needs to be dynamic interaction of economical routes with the decision making process for flow sheeting. Although 'Grade is King,' 'Recovery is Queen' and the Queen almost always gets her way (or loses her head)! When we focus on the recovery, we can usually get the grade as well. It's just a matter of re-arranging the puzzle pieces to balance grade and recovery. But don't forget the throughput; it's like a three legged bar-chair that can't stand on two legs.

The intention of this paper is to shed some light on the proposed methodology that can be employed to target development of a marginal resource into a viable resource, or a the shifts in an existing technology that transforms a fading cash cow back to a rising star. There is no magic or secrets to the method, just application of a healthy dose of common sense along with sweat equity, perseverance and a little luck. The focus of this paper will be to review the state of iron ore resource and reserve grades, the state of the energy environment, and then examine the methodology used (or being considered) for some technologies selected as case studies to meet the challenges. These technology and resource case studies are characterized as being either hard rock or iron sand assets.

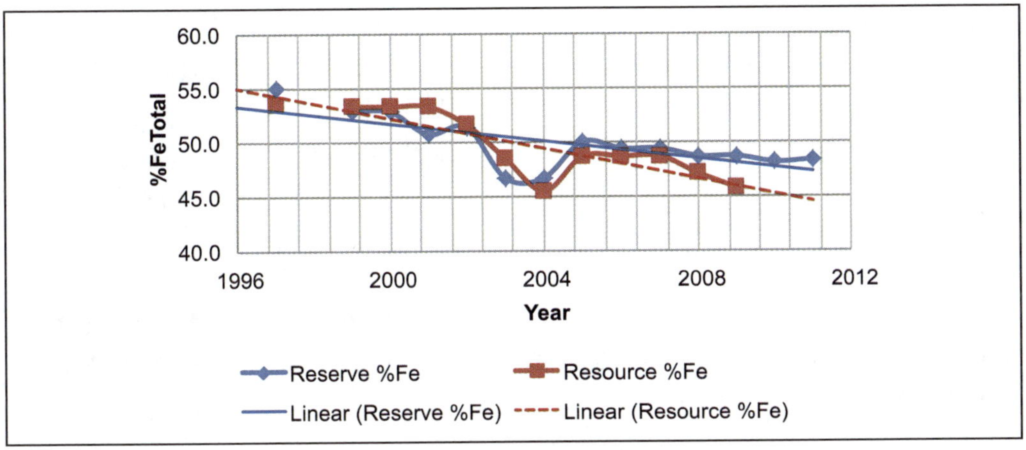

Figure 1. World mine reserve and resource average iron content (Data Source: USGS 2012)

STATE OF THE IRON ORE AND ENERGY INDUSTRIES

Many iron ore producing mines have been in production for 20 to 30 years, or more. The highest quality iron ore was the initial focus of mine operations when many of the mines were first brought on line. Over the years, the quality of iron ore mine reserves and resources have been declining. This claim can be supported by examining world-wide data for the quantity of ore reserves and resources and their respective grades. Figure 1 illustrates this trend on a world average basis for the period 1997–2011 for both reserves and resources (data extracted by USGS commodity reports). A 'resource' is considered a natural occurring concentration of solids in or on the Earth's crust in such a form and amount that economic extraction of a commodity from the concentration is currently or potentially feasible. A 'reserve base' is that part of a resource that meets specified minimum physical and chemical criteria related to current mining and production practices, including those for grade, quality, thickness, and depth. The reserve base is the in-place demonstrated (measure plus indicated) resource from which reserves are estimated. The reserve base includes those resources that are currently economic (reserves), marginally economic (marginal reserves) and some of those that are considered sub-economic (sub-economic resources). The 'reserve' then is that part of the reserve base which could be economically extracted or produced at the time of determination, and include only recoverable materials.

Energy is the life blood for metallurgical processes due to the energy intensive nature of the various technologies employed to produce saleable commodities. Figure 2 illustrates the pricing for Brent petroleum ($/barrel) during the period from May 1987 to present, and for natural gas ($/MM BTU) during the period 1997 to Q3 2012. The Brent pricing trend is upwards with the exception of a price spike in 2008, followed by a precipitous fall in 2009, and then a resumption of the upward trend from 2009 to the present fall A general rising trend in price can be seen for natural gas from 1999 through 2009, with occasional spikes to $10 to $15/MM BTU in 2001, 2003, 2005 and mid-2008. However, a recent low of $1.82/MM BTU was reached in April of 2012. This precipitous fall in natural gas prices in the US from 2008 to present is primarily due to the advent of more supply of natural gas due to cheap 'frac sand' wells.

A recent compilation for electricity prices (US cents per kWh) is shown in Table 1. The table contains an alphabetical listing of select worldwide countries. Unfortunately, prices for India and

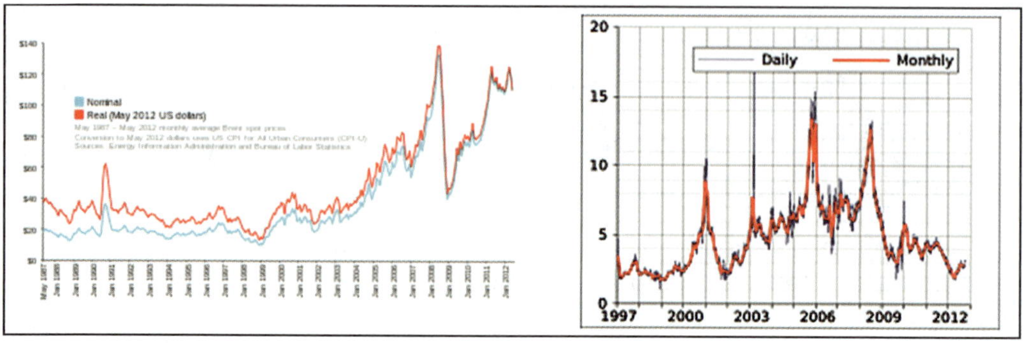

Figure 2. Brent crude oil prices ($/barrel) (Henry Hub Brent, 2012a) and US natural gas prices $/MM BTU (Henry Hub Brent, 2012b)

Japan are not included. According to this list, relatively low electricity prices exist (<US $0.12 kWh) in developed countries that are consumers of raw material commodities; e.g., USA, China, certain European countries, Turkey, Canada, etc. One would think that the BRIC countries would all have reasonable electricity costs as these are relatively strong, industrialized economies. However, electricity prices in third world countries that generally possess the mineral resources are typically greater than US $0.15/kWh. The USA and China are to some extent an exception to this generalization in that these countries have both mineral resources and relatively cheap electricity costs. In addition, the USA has cheap natural gas, which can be used to produce electricity via gas turbines or combined cycle power plant production methods.

In light of this basic information on the iron resource condition, and the costs associated for variable types of energy, one can appreciate that one of the greatest competitive challenges for the modern day iron processing facility is to obtain low cost feed materials for value added processing step while minimizing exposure to various logistical influences (transportation costs for both feed materials and products) and implementing the most cost and energy efficient technology for production of the final product. If one can successfully juggle these factors, then the end result is a successful and sustainable business operation. The remainder of this paper will illustrate how innovation can influence the outcome for a sustainable technology by presenting two case studies for illustrative purposes.

CASE 1: THE ASCENT OF NATURAL GAS BASED DIRECT REDUCTION

The ascent of the MIDREX natural gas based technology (Atsushi et al. 2010) for the production of direct reduced iron is an excellent example of how a mature 'cash cow' technology can be reinvented a number of times through innovation to lead to continued sustainability. Pioneered in the early 1970s, the original MIDREX technology was a novel melding of technologies whereby the reforming natural gas for the production of CO and H_2 reductants, was 'connected' with a moving bed shaft furnace reactor to produce cold direct reduced iron (DRI). The unique feature of the original process was that the gas reformer was capable of delivering a reducing gas at elevated temperature to the moving bed reactor such that the endothermic and exothermic reduction reactions required for converting FeO to metallic iron and with H_2 and CO respectively were essentially thermodynamically balanced. Thus the CO and H_2 ratio in the reducing gas as well the gas quality (ratio of reductants to oxidants) was optimized with respect to inclusion of spent gas from

Table 1. Worldwide electricity prices US cents/kWh (Source: http://en.wikipedia.org/wiki/Price_of_electricity; sources as cited in the Wikipedia article)

Country/Territory	US cents/kWh	As of	Sources
Argentina	5.74*	2006	[4][2]
Australia	22 to 46.56	August 23, 2012	[5],[6]
Belgium	29.06	November 1, 2011	EEP[5]
Bhutan	1.88 to 4.40	March 23, 2012	BPC[6]
Bulgaria	16.33	July 1, 2012	DKER
Brazil	34.18	January 1, 2011	ANEEL
Canada	10.78	January 1, 2011	PEI
China	7.5 to 10.7	May 17, 2012	[7]
Chile	23.11	January 1, 2011	Chilectra[8]
Croatia	17.55	July 1, 2008	HEP
Denmark	40.38	November 1, 2011	EEP[5]
Dubai	7.62	2011	DEWA[9]
Finland	20.65	November 1, 2011	EEP[5]
France	19.39	November 1, 2011	EEP[5]
Germany	27.81	November 1, 2011	EEP
Guyana	26.80	April 1, 2012	GPL
Hungary	23.44	November 1, 2011	EEP[5]
Hong Kong (HK Is.)	12.04	January 1, 2012	HEC[10]
Iceland	9 to 10	June 1, 2012	OR[11]
Iran	2 to 19	July 1, 2011	
Ireland	28.36	November 1, 2011	EEP[5]
Israel	12.34*	January 1, 2012	IEC[12]
Italy	28.39	November 1, 2011	EEP[5]
Jamaica	7.35 to 16.80	June 1, 2011	JPSCo[13]
Latvia	18.25	June 1, 2012	Latvenergo[14]
Malaysia	7.42	December 1, 2007	ST[15]
Mexico	19.28**	August 22, 2012	CFE[16]
Moldova	11.11	April 1, 2011	RUF[citation needed]
Netherlands	28.89	November 1, 2011	EEP[5]
New Zealand	19.15	April 19, 2012	
Pakistan	2.00 to 15.070	May 16, 2012	LESCOFESCO
Perú	10.44	2007	PTL[17]
Philippines	30.46	March 1, 2010	[7]
Portugal	25.25	November 1, 2011	EEP[5]
Russia	9.58	January 1, 2012	Mosenergosbyt
Singapore	22.24	July 4, 2012	[18]
Spain	22.73	July 1, 2012	Iberdrola,energy.eu
Solomon Islands	83 to 89	2012	SIEA, (CBSI)
South Africa	7.60 to 14	July 1, 2012	Eskom
Sweden	27.10	November 1, 2011	EEP
Taiwan	7 to 17	October 1, 2008	[8]

the reactor (top gas). This process feature, in combination with the high heat transfer efficiency afforded by a counter current moving bed of iron oxide feed, led to an overall high energy efficiency for the overall process and thereby optimized the natural gas requirement to 2.1~2.3 Gcal/mt DRI. Further innovations such incorporation of the remaining top gas fuel for firing the reformer, process gas pre-heat, and combustion air preheat resulted in improved energy efficiency. The basic flow sheet including heat recovery is shown in Figure 3.

Other technology 'twists' and innovations which independently contributed to extending the life of the MIDREX gas based direct reduction technology include hot discharge of direct reduced iron (HDRI), hot DRI transport, briquetting of hot DRI (HBI), the HOTLINK concept where the DRI shaft furnace is 'linked' to an electric arc furnace to take advantage of hot charging DRI so as to reduce both the power requirement to produce liquid steel and the tap-to-tap time. In parallel to these innovative steps was an effort to increase the throughput of the shaft furnace by increasing the reactor inlet gas, or 'bustle gas,' temperature. Figure 4 details the various methods of temperature increase and the relative effect on process performance. As one can see from the figure, Case 6 which is the feature of combined oxygen injection with OXY+ offers the potential of increasing the plant productivity some 50% greater than the original plant process design.

One can see that continued sustainability of the MIDREX natural gas based technology has been ensured through successful development and implementation of novel technology innovations and improvements. But ever increasing bustle gas temperatures puts a constraint on the quality of the feed iron oxide to the process. The original MIDREX raw material specification for natural gas based direct reduction in the shaft furnace is rather stringent. High quality DRI grade pellets usually sell for a 5–10% premium over normal blast furnace grade pellets.

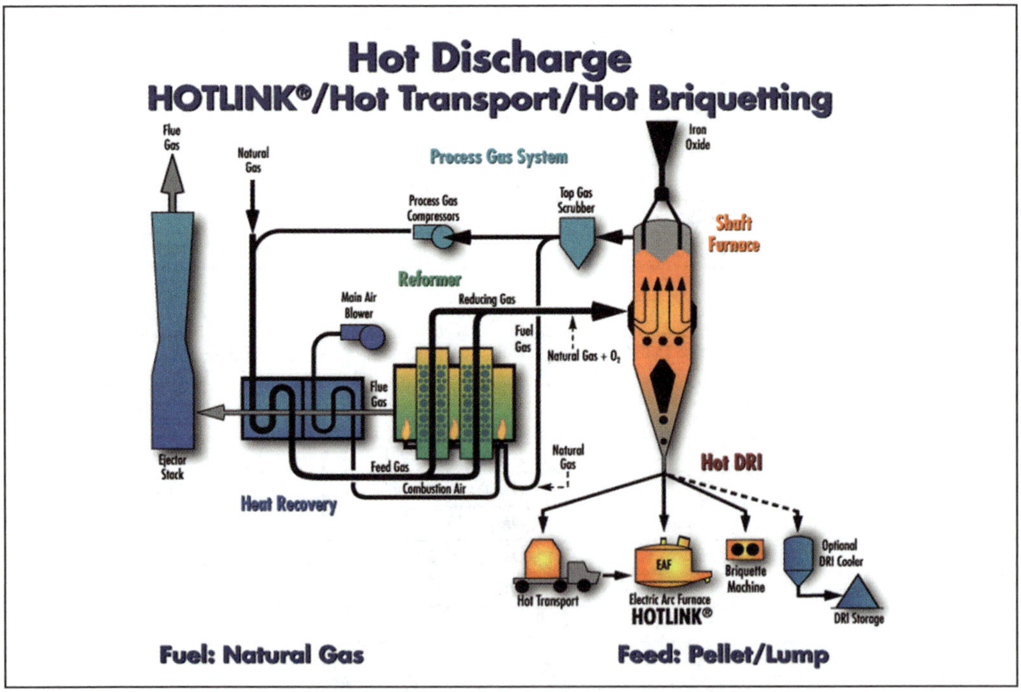

Figure 3. The MIDREX natural gas based flowsheet (Courtesy of MIDREX Technologies, Inc.)

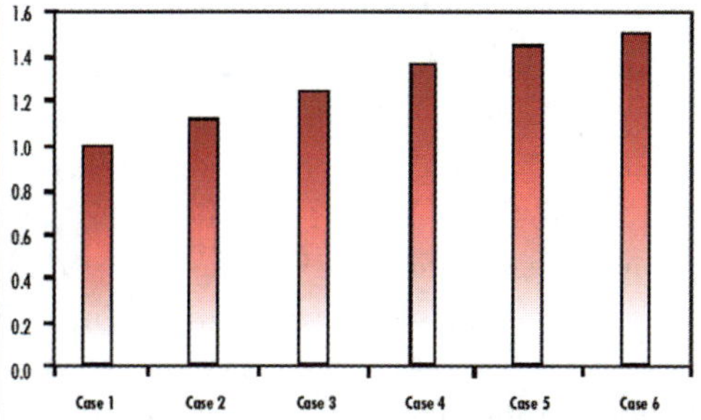

Key
Case 1: Original practice, 1970s
Case 2: Practice using lump ore, 1980s
Case 3: Practice using coating of oxide feed materials, 1990s
Case 4: Oxygen injection practice, late 1990s
Case 5: OXY+ practice, 2000
Case 6: Combined oxygen injection and OXY+, Future

Figure 4. Productivity comparison—MIDREX natural gas based flowsheet (Courtesy of MIDREX Technologies, Inc.)

Too much silica in the gangue in the iron ore pellets can result in sticking among the hot DRI particles, which promotes formation of clusters in the burden of the shaft furnace. Excess clustering can result in serious material and gas flow problems within the shaft furnace, even to the extent that the entire burden can be blocked. One of the remaining challenges for the natural gas based direct reduction technology is the use of high gangue (or blast furnace grade) iron ore pellets in the shaft furnace. If the MIDREX natural gas based technology is incapable of processing pellets produced from low grade iron ore concentrate into high grade DRI, then perhaps this creates an opportunity for a paradigm shift in technology that favors production of an alternative iron product of high quality from an 'inferior' or black sheep type resource. One such alternative is hot metal production for pig iron from iron sands using Cardero smelting technology.

CASE 2: THE ASCENT OF SMELTING TECHNOLOGY AIMED AT PROCESSING LOW GRADE IRON ORE CONCENTRATE

Cardero Iron Ore Solutions

Cardero Iron Ore is focused on the next level of opportunity in iron ores around the world. These deposits tend to have the disadvantage of lower grade and/or high titanium levels but have other advantages that result lower operating and capital costs. Such constraints make these projects ideal candidates for the application innovative solutions. These advantages include

1. Iron sands that are already at or near the liberation size, so they do not require blasting, crushing, and grinding, which can add significant operating and capital costs;
2. Simple flow sheets that require only two to four stages of magnetic separation (Hoffman et al. 2012);
3. Close proximity to infrastructure and ports; and
4. Employing the patented Cardero Technology to reject the titanium co-product to a slag while producing blast furnace quality pig iron in an electric melting furnace (Hoffman et al. 2007).

Cardero Iron Ore has developed its own in-house capabilities to move these projects forward including a metallurgical laboratory to aid in flow sheet development and design, and Cardero is capable of bringing the project through the basic engineering stage and supervise the detailed engineering through project start-up and beyond.

CARDERO MATERIAL TESTING LABORATORY

In 2011, Cardero Iron Ore built their own million-dollar, state-of-the-art research and development facility. The lab features the latest laboratory- and pilot-scale equipment for mineral processing and smelting/reduction. The lab was specifically designed for processing iron ores, rare earths, and other non-ferrous ores and is being tailored to suit Cardero's projects. The equipment was carefully selected so that it encompasses every commercial iron ore beneficiation unit operation available throughout the world. In the laboratory, iron ore can be crushed and ground and concentrated by rejecting gangue constituents by their differences in gravity/density, magnetic susceptibility, surface chemical properties (froth flotation), and electrical conductivity differences (high tension/electrostatic). Highlighted equipment includes: A Thermo-Niton handheld xrf for elemental analysis (including optional rare earth element calibration), gravity spiral, shaking table (gravity/density separator), wet and dry low-intensity magnetic separators and a dry high intensity magnetic separator, froth flotation, electrostatic separator, rotary kilns, an electric arc furnace. The laboratory also includes a variety of equipment for crushing/grinding, jigging, hydro-cycloning, screening, drying and firing, sample preparation and splitting, and pressure filtering. While the Cardero Iron Ore's laboratory can handle a wide variety of minerals, it has been specifically tailored for processing iron ore. This laboratory allows for rapid flow sheet development in-house by Cardero Iron Ore Ltd. and reduces delays associated with waiting for an outside laboratory to schedule and complete its work.

HYDROELECTRIC IRON AND STEELMAKING—THE NORTH ATLANTIC IRON CORPORATION (NAIC) PROJECT IN GOOSE BAY, LABRADOR, CANADA

Cardero has formed a joint venture partnership with Grand River Ironsands (GRI) Incorporated and Petmin Limited (South Africa) to form North Atlantic Iron Corporation (NAIC). The ironsands of Labrador were first discovered in 1896 and are located near the Town of Happy Valley-Goose Bay, NL. GRI has 1,800 staked mineral claims that cover an area of 450 km^2 and include sands in the Churchill River east of Muskrat Falls and lands to the south and west. The innovative steps for this project are to efficiently make a high margin end product (pig iron) from a low grade iron resource (iron sands) by way of first developing and implementing an optimized sands process flow sheet that facilitates production of a high grade iron ore concentrate. The second innovative step focuses on the development and implementation of proprietary pyrometallurgy technology that allows conversion of the iron sand concentrate (which contains elevated levels of TiO_2 and V_2O_5) into high quality pig iron.

The NAIC iron sands and pig iron project will complete a Resource Estimate in Q4 of 2012 with the related drilling already completed in 2011 through 2012. Cardero Iron has engineered a demonstration-scale (approximately 20 tons per hour raw sand throughput capacity) concentrating facility that is being built in Goose Bay by the NAIC joint venture. The NAIC demonstration concentrating facility will process a ~7,000 ton bulk sample of sands into iron ore concentrate. The plant was commissioned in August, 2012, and approximately 150 mt of finished concentrate at ~54% Fe(Total) has been produced through October 2012. Figure 5 shows the operational pilot plant.

This bulk concentrate sample is to be utilized in a demonstration level smelting test scheduled for Q4 of 2012 or Q1 2013. The magnetite and titano-magnetite in the sands will be concentrated and the vast majority (approximately 90%) of the sands will be returned to the land. Blasting, crushing and the use of chemicals are all avoided. The iron sands concentrate will be treated by

Figure 5. Operational NAIC demonstration concentrating plant (Courtesy of NAIC)

a demonstration test of the Cardero Process that will reduce and melt the feed materials with an electric furnace into pig iron, the main ingredient in steelmaking.

The NAIC project has all of the key elements in one location—the resource, a full service community, low cost hydroelectric power, and a port. The first commercial plant is planned to produce ~500,000 tons of high purity pig iron. The pig iron product is expected to have a market in the United States and Europe where most ports are less than seven days by sea from Happy-Valley-Goose Bay. NAIC will continue to work closely with the key stakeholders including the local community, regulatory and permitting authorities, and the First Nation's populations.

PAMPA EL TORO (PET) IN PERU

Pampa el Toro is another Iron Ore Project situated in the desert coastal region of southern Peru (Hoffman et al. 2010). The property is located near the city of Nazca in the desert coastal region of southern approximately 45 km northeast of the port of San Juan, close to the large Marcona Iron Mine (approximate resource of 1.4 billion tons at 54% iron) and to Cardero's recently sold Pampa de Pongo Iron Project. Cardero initiated a feasibility study by starting a sampling program of the iron ore dune sand resource along with a trenching and percussion drilling program managed by Cardero's own geologists in 2006.

PeT Resource Estimate

In mid-2009, Cardero received an initial mineral resource estimate for the Pampa el Toro deposit from SRK Consulting (South Africa) Inc. (SRK). Indicated and inferred resources have been defined over an approximate area of 1,595 ha (approximately 15% of the total 10,300-ha Pampa el Toro concession area and only to a nominal depth of 30 m from the surface.

Cardero's PeT Property Background

The project was originally investigated by National Steel & United Nations in 1970s. In 1974, the Peruvian Ministry of Energy and Mines originally highlighted the iron sand potential of the area as part of a nationwide iron inventory study. Geologically, the large property comprises magnetite (iron)-bearing sands consisting of active dune fields and a large Quaternary basin containing a thick sequence of friable sands with visible magnetite throughout many of the exposures.

Cardero obtained the property in 2003. SRK was hired to manage the resource estimate drilling program that was completed in 2009 and resulted in sufficient reserves for greater than 30 years life of mine (LOM) at 2M mtpa iron ore concentrate production. PeT has an indicated resource of 241.8 million tons at 6.66% Fe_2O_3, 0.72% TiO_2 with an inferred resource at 629.9 million tons at 6.48% Fe_2O_3, 0.70% TiO_2. The SRK Resource Estimate is based on 1,595 ha out of 12,200 ha at 30-m depth; can easily be expanded to greater area and 50-m depth to accommodate 5 million or 10 million mtpa of merchant iron ore concentrate production. PeT is amenable for both merchant iron ore concentrate and merchant pig iron project scenarios.

Preliminary sampling by Cardero indicates that the magnetic fraction for active dune material ranges between 0.7 and 21.24 weight percent (wt. %) magnetite and averages approximately 11.0 wt. %. Samples of Quaternary sand returned magnetic fractions ranging between 1.62 and 13.59 wt. %, averaging about 7.7 wt. % magnetite. Assays of the magnetic concentrates showed fairly uniform values of approximately: 61.5% iron (Fe), 4.3–6.41% titanium oxide (TiO_2), 0.2–0.28% phosphate (P_2O_5), 2.41–4.34% silica (SiO_2), 1.22–1.82% alumina (Al_2O_3), 0.61–1.14% magnesium (MgO), 0.4–0.55% manganese (MnO), and 0.01% sulphur (S).

Cardero's PeT Pilot Plant

Cardero engineered and built a simple beneficiation pilot plant on-site (Figure 6) that was used to upgrade the sand and produce 60 metric tons of iron ore concentrate at 64% Fe. This concentrate is ready for testing by off-takers or partners. Commercial scenarios include 2, 5, or 10 million tpa iron ore concentrate for a minimum of 20-year life of mine.

Figure 6. Cardero's PeT pilot plant (Courtesy of Cardero)

Table 2. PeT pig iron compares favorably with commercial grade pig iron

Element	Commercial Grade Pig Iron (PI), %	[NETL #s] Average PeT 300 Kw-fce PI with High S Coal, %; ±STDEV		PeT 50 Kw-fce PI with High S Coal, %	PeT 50 Kw-fce PI with Lower S Graphite, %
Iron	>92	94.92	0.53	95.26	92.79
Carbon	3.5–4.5	3.59	0.24	3.02	4.64
Silicon	<1.5	0.18	0.01	0.17	0.31
Manganese	<0.5–1.0	<0.01	0.00	<0.01	0.17
Sulphur	0.05 max	0.21	0.01	0.20	0.05
Phosphorus	<0.05	0.03	0.01	0.03	0.04
Additional Elements					
Copper		0.01	0.005	0.01	0.02
Chromium		0.07	0.01	0.07	0.08
Vanadium		0.26	0.01	0.45	0.92
Titanium		0.01	0.01	0.01	0.40

Cardero's PeT Metallurgical Processing

Cardero completed its planned three-phase pilot-scale pig iron production program. The Pampa el Toro concentrate produced from the pilot-scale magnetic beneficiation plant in Peru was agglomerated and processed in the US into liquid hot metal (molten iron) in a total of three 10-hour-long, pilot-scale electric arc-based melting furnace tests and cast into pigs. A total of one-half metric ton of pig iron has been produced.

These pilot-scale melting tests have now confirmed that scale-up from the previous bench scale tests was successful. High quality pig iron can be produced from PeT concentrate. High quality pig iron, such as blast furnace grade, contains at least 92% iron and 3.5–4.5% carbon with a maximum of 0.05% sulfur and 0.05% phosphorus. The pig iron was analyzed by the U.S. Department of Energy's National Energy Technology Lab (NETL) and verified by an independent certified lab specializing in pig iron analyses. Results from the independent certified laboratory are shown in Table 2. The pig iron produced during the pilot-scale tests showed a relatively high concentration of 0.21% sulfur, this was due to the fact that a relatively low quality high sulfur reductant (petroleum coke) was selected so as to demonstrate processing flexibility with respect to raw material feed selection. This result compares with a previous 50 Kw furnace test that used the higher sulfur reductant and resulted in 0.20% sulfur. An additional bench-scale test utilizing lower sulfur (0.02%) graphite produced pig iron with only 0.05% sulfur.

Highlights from these pilot-scale melting tests are demonstrating that extremely low titanium and silicon levels were achieved resulting in a high quality pig iron. Carbon levels of 3.59% in the pig iron, which is desirable, shows that the melting environment was very reducing. Highly reducing environments tend to also reduce elements such as silicon and titanium and increase their contamination in the pig iron. However, the Cardero process shows its superiority in rejecting these pig iron contaminants to the slag. Theoretically it is possible to utilize this liquid hot metal to produce steel without the need for scrap; so for instances like the NAIC project where cheap hydroelectric power is abundant, this can be deemed hydro-electric steelmaking.

This pilot-scale demonstration was the last of three phases. Phase I was design and testing of agglomeration procedures for the melter feed materials. Phase II was bench-scale testing with a 50 kW electric furnace which determined the optimum agglomerate feed formulation along with operating control parameters. The following three goals of the 300 kW pilot-scale melting test were found to be achievable over an extended period of time:

1. Consistent melter operation using a range of feed reductants
2. Uniform hot metal quality
3. Uniform slag quality (possessing enriched titanium levels of >18% for titanium recovery)

Cardero continues to advance its PeT Iron Sands Deposit in southern Peru with ongoing discussions and technical reviews with potential strategic partners. Strengthening iron ore markets have triggered an increased interest in this iron deposit, which is strategically situated in close proximity to substantial infrastructure, including highways, power, a deep-water port, and an operating iron ore mine.

The U.S. Department of Energy's National Energy Technology Lab (NETL) successfully demonstrated the Cardero Process at the Bench- and Pilot-Scale by producing pig iron from the Pampa el Toro iron concentrate which reduces the amount of carbon dioxide produced during iron making.

In an initial series of tests in the cooperative work under a funds-in agreement with Cardero Iron, NETL researchers briquetted 500 pounds of a PeT concentrate, and then performed direct smelting tests on the agglomerates in an electric arc furnace.

Since the iron sand is essentially already liberated and does not need grinding it avoids the energy consumed by size reduction—which is about 10 kWh per ton of ore. In addition, the sintering of the iron ore, which oxidizes the magnetite to hematite (Fe_2O_3), is eliminated thereby reducing the production of CO_2 by approximately 11%.

Titanium and Vanadium Recovery

Another innovative step associated with the processing of iron sands concerns dealing with the elevated levels of TiO_2 and V_2O_5 byproducts contained in the slag from the smelting step. Most steelmakers view TiO_2 and V_2O_5 components in iron ore concentrate as being detrimental to the blast furnace operation for the production of hot metal. The blast furnace hot metal will contain traces of Ti and V, which may affect downstream steel quality. And the addition of TiO_2 to the blast furnace slag affects its melting point and viscosity. However, processing of this concentrate material using the Cardero smelting technology promotes the production a pig iron that has minimal Ti and V tramp components, and moreover, concentrates these oxide species in the slag. The concentrating feature offers an opportunity to the potential plant owner as both the TiO_2 and V_2O_5 can be potentially recovered. Cardero commissioned a research program that was conducted by the Colorado School of Mines (CSM) Kroll Institute for Extractive Metallurgy (KEIM). The institute is named after William J. Kroll who developed the Kroll Process which is the pyrometallurgical industrial process used to produce metallic titanium. Most vanadium is used as ferrovanadium or as a steel additive to improve its strength and is common in hand tools. Titanium is used in paints to provide opacity and as a white pigment that is a safer alternative to lead-based paints.

Slag samples produced during Cardero's successful pilot scale pig iron production program held at the U.S. Department of Energy's (DOE) National Energy Technology Lab (NETL) were sent to CSM for investigation. The slag exhibited 0.7–0.8% V and 18% TiO_2.

Vanadium

Three methods were investigated to recover the vanadium out of the slag co-product from pig iron production:

1. Salt roasting and hot water leaching,
2. Calcified roasting and acidic leaching, and
3. Acidic and alkaline leaching.

Calcified roasting gave a recovery of 45% of the vanadium. This is an excellent recovery considering the mode of occurrence and fine liberation size. In a commercial operation, this leaching process would be followed by a solvent extraction or ion exchange purification process and then the V_2O_5 would be precipitated from solution to produce a final marketable product.

Titanium

Two methods were investigated:

1. Hydrochloric acid (HCl), and
2. Sulphuric acid (H_2SO_4).

The HCl exhibited superior recovery of 50.2% of the titanium in solution. In a commercial operation, the titanium will then be precipitated from the solution by hydrolysis and collected as TiO_2. Application of this approach to TiO_2 recovery enhances the economics of the overall PeT as well as NAIC projects by way of a secondary revenue stream.

SUMMARY AND CONCLUSIONS

The intent of this paper has been to first present some basic information on the state of the iron resource condition that the Iron & Steel Industry is currently facing, as well as the current state of energy environment where site specific costs can make or break the success of a metallurgical technology application. These basic inputs have a significant impact on the shaping of the response that the guiding manager must make so as to answer to the market needs and the sustainability of the commercial operation to which he or she is entrusted to guide (whether existing or conceptual). One can appreciate that perhaps the greatest competitive challenge for the modern day iron processing facility manager is to obtain low cost feed materials for value added processing step while minimizing exposure to various logistical influences (transportation costs for both feed materials and products) and implementing the most cost and energy efficient technology for production of the final product. If one can understand the nature of these factors, and safely navigate through the mine field created by the interaction among these factors, then the end result will be a successful and sustainable business operation.

Innovation is not something that the individual can easily produce on command. While it is true that necessity breeds invention, the wise manager knows how to spot the innovative individual, instill a desire in him or her to create, and establish the healthy environment to ultimately reap the benefits this new found creativity. Innovation can be managed and cultivated, but it takes time, patience and money. It also requires the manager to constantly be aware of the pulse of the market, and the ever changing influences and constraints that the market is under.

Two case studies were presented in this paper that highlighted the innovation theme from two perspectives. The first study considered the challenges and innovative responses taken to sustain a maturing technology, while the second study outlined the approach needed to accommodate a paradigm shift in thinking for metallurgical processing. That is, illustrating what innovative steps can be taken so that a low grade resource can be transformed into high margin commodity product. Both case studies employed innovation as the key to success. The rest is up to the individual.

REFERENCES

Atsushi, M., Uemura, H. and Sakaguchi, T. 2010. Midrex® Process, Kobelco technology review, 29, Dec, p. 50–57.

Henry Hub Brent. 2012a. Crude Oil Spot Prices. http://ycharts.com/indicators/brent_crude_oil_spot_price.

Henry Hub Brent. 2012b, Natural Gas Spot Prices http://ycharts.com/register?next=/indicators/natural_gas_spot_price.

Hoffman, G., McClelland, J. and Ripke, J. 2007. Techniques for utilization of low quality iron ore feeds for shaft furnace direct reduction, Iron and Steel Technology Conference, AISTech 2007.

Hoffman, G., and Ripke, S.J. 2012. New approach to developing the optimal mineral processing flow sheet. Mining Engineering, March, pp. 23–30.

Quinn, J. and Tayal, A. 2012. Steve Jobs Genius by Design, Random House Inc.

Runes, D.D. 1968. Diary and Sundry Observations of Thomas Alva Edison, Abbey Publishing.

USGS. 2012. "Iron ore statistics and information. http://minerals.usgs.gov/minerals/pubs/commodity/iron_ore/.

Wikipedia. 2012. Worldwide electricity prices. http://en.wikipedia.org/wiki/Price_of_electricity.

ABOUT THE AUTHOR

Glenn E. Hoffman is president and CEO of Cardero Iron Ore Management. His education includes a B.S. in metallurgical & materials engineering from Illinois Institute of Technology and an M.S. in metallurgical engineering from Michigan Technological University. Hoffman's work experience includes 30 years at MIDREX Technologies as a process development engineer and manager of research & development, which covered a variety of development projects including exotic combustion technologies, natural gas and coal based direct reduction technologies for iron ore, and new iron making technologies including HIsmelt, FASTMET, FASTMELT, and ITmk3. During the last six years, Hoffman has been directing work related to developing the various iron ore assets owned by the Cardero Group, as well as directing the technology effort for the North Atlantic Iron Corporation JV iron sands project. Hoffman is a Metallurgical QP, and past winner of SME's Henry Krumb lecturer award and A.M. Gaudin award. He also holds 16 U.S. patents.

Mineralogy and Analytical Chemistry

Innovations in Process Mineralogy and Laboratory Automation

Wolfgang Baum
FLSmidth, Salt Lake City, Inc., Midvale, Utah, United States

ABSTRACT: Changes inherent in lower grade, finer grained, deeper, harder and more complex ores in concert with mushrooming costs for power, water, steel and environmental aspects have transformed the view and use of mineralogy during the last 20 years. Robotics sample preparation, automated mineral analyzers, high volume X-ray diffraction (XRD) analyses, and large-scale use of near infrared (NIR) are only some examples of massive innovations which have changed the laboratory and plant operation in mining. These changes have initiated the robust application of quantitative and automated mineralogy as a pivotal planning, optimizing and production control tool. Mineralogy has become equally as critical as analytical chemistry and metallurgical testing. Step-change technology with fully automated central labs has emerged at various mining companies. The pioneering installation of fully integrated partially automated mineralogy mine site labs was advanced in the copper and gold industry. Additional laboratory technology such as laser ablation inductively coupled mass spectrometry (LA-ICP-MS), laser induced breakdown spectroscopy (LIBS), X-ray computed tomography (X-ray CT), nuclear magnetic resonance (NMR), or cross belt neutron analysis (CNA) and the increasing use of other cross belt analyzers are initiating more innovative transformation in ore characterization and plant control. This paper will highlight the extraordinary innovations which have been implemented by process mineralogists during the last two decades.

INTRODUCTION

Geo-metallurgy, better flow-sheet design and equipment use, plant and process optimization, precision ore treatment and better production forecasts have become synonymous for the extended and innovative use of mineralogical analyses and data (Baum 2007). Although the hand-lens continues to be an integral part of field work, the mining industry could not be farther from the long-standing perception as a low-tech, dirty and rock-moving business. Our industry has made giant leaps into lab automation and as stated by Follink (2007) the future competitive mining will require in-situ ore characterization in real time during the operation. That will permit continuous process adjustments via smart feedback controls to achieve maximum recoveries at the lowest cost and minimal environmental impact.

Based on these new lab technologies, ore body profiling, modeling, simulation, process diagnostics and plant optimization are now fundamental standard operating procedures in many mining companies. The engineering firms and equipment suppliers have taken notice and seismic changes are also occurring in their areas.

SHIFTS AND CHANGES IN MINING AND MINERALS PROCESSING

The implementation of process mineralogy as an integral tool for ore body development, process design, troubleshooting and optimization occurred in the 1980s when many mining companies realized the important contribution of quantitative mineralogy in minerals processing and hydrometallurgy (Lotter et al. 2011). However, the advance of computerization during the 1990s and since 2000 with extremely high data output at very low cost was the leading factor of the initiating events. Today, routine automated mineralogy easily yields several gigabytes of data collected daily. Chemometrics calculations for NIR or FT-NIR analyses and their models also require significant computational power.

Not all other initiating events on the operational site can be listed here; but some examples will illustrate the contributing factors for the changes:

- The introduction of automated mineral analyzers and XRD Rietveld analysis enabling quantitative 24/7 mineralogy.
- Many of the early gold heap leach operations in the Western United States were initially troubled or failures.
- Operators starting to use mineralogy as a process improvement tool, e.g., El Indio (Baum et al. 1987).
- The increasing number of capital-intensive projects.
- The large-scale development of heap leaching for oxide copper ores in Chile and Peru.
- Rising water, steel, power and environmental costs.
- The recognition that chemistry and visual ore characterization simply do not suffice for process design.

Substantial credit needs to be given to the pioneers (first movers) in automated mineralogy who applied the new computer power, hardware and software to the critical process-related mineralogical challenges in copper, platinum, nickel, gold and other mining operations. These innovation drivers challenged the status quo and the traditional resistance and skeptical thinking in many operations. This author (and his lab team) were told as recently as 2002: *"What the XRD does, I can do out of the window of my truck."* Their foresight, dedication and tenacity enabled Mineralogical Plant Surveys, Heap Leach Audits, Acid Consumption Forecasting, and reliable Permeability Classifications to name a few.

THE INNOVATION YEARS FOR MINERALOGY

The major innovation breakthrough for mineralogy was the capability to obtain rapid quantitative data with XRD Rietveld, automated mineral analyzers (QEMSCAN, MLA, ASPEX: FEI, Hillsboro; OR USA TIMA: Tescan USA, Cranberry Township/PA USA; RoqSCAN: Fugro Robertson/Carl Zeiss, Houston, TX USA; Particle SCAN VP: Carl Zeiss, Oberkochen, Germany) and the use of NIR for clay and alteration minerals. These analytical technologies made mineralogy as powerful as analytical chemistry. In addition, the large scale application of robotics sample preparation has allowed mining companies to handle very heavy sample loads and make them available to ore profiling, daily production analyses, or other applications.

Although QEMSCANs were operating routinely in the 1990s, the mineralogists faced with the application challenges, contributed significantly (with the manufacturers) to optimizing software, hardware and service. For the copper industry, great breakthroughs were achieved by taking on unknown territories (such as Heap Leach Audits) and establishing quantitative, material-balanced copper deportments in leach residues. This had to be accomplished against the continued doubts of operators and long-tenured individuals who were afraid of the new technology.

In the copper industry, one of the first highly successful QEMSCAN concentrator surveys (Kendrick et al. 2003) and the first comprehensive Heap Leach Audit in a secondary sulfide leach (Fennel et al. 2005) were breakthrough events. They certainly contributed to the pioneering decision of Phelps Dodge to establish to first fully integrated Process Mineralogy Lab at a porphyry copper mine site in Peru.

RE-TOOLING ORE CHARACTERIZATION AND PROCESS MINERALOGY

Ore variability in terms of mineralogy continues to be one of the neglected issues of modern mining. Although modeling efforts are targeting some of the variability, their past focus has been primarily on comminution using numerous parameters (SPI = SAG Power Index, SAGDesign = SAG Design Test, Ai = Bond Abrasion Index, CWi = crushing work index, JKDWT = JK Drop Weight Test) for mill sizing. Hardness/abrasion/breakage parameter testing can be considered robust, but it is not fail-free. The SPI, for example can have failures in predicting certain rock types. This can result in unexpected mill-throughput problems. Virtually all of the standard comminution tests (i.e., the interpretation of their data) can be optimized with mineralogical data such as XRD, NIR and/or Automated Mineral Analysis for better design and forecasting.

Two large impact factors for metallurgy, in both heap leaching and concentrators, are clay and reagent consumptions, specifically acid control in hydrometallurgy. Today's innovations such as NIR and XRD Rietveld analysis permit reliable clay data base construction and continuous alert for the operators from ore control via blasting to crusher/agglomeration, optimization and comminution/flotation circuit control (Brandt et al. 2011). The use of quantitative clay mineralogy can all but eliminate permeability failures in heaps, assist in optimal agglomeration and can be connected with Acid Consumer Data Bases. All of this can dramatically lower the cost of copper leaching (Baum 1998, Allen et al. 2007).

LAB AUTOMATION—THE STEP TOWARD 24/7 MINERALOGY AND CHEMISTRY

Lab automation efforts in metals mining started with isolated activities in the late 1980s (e.g., Newmont's automation of sample preparation and chemical labs in Nevada). Smaller automated sample preparation islands were installed in various other operations around the world and include lab modules in aluminum production (Mozal, Hillside, Bayside), mineral sands (Richards Bay), multiple iron ore labs and zinc (Skorpion), to name a few.

The innovation potential received large-scale attention with break-through robotics labs such as the Anglo EBRL (Eastern Bushveld Regional Lab) for fire assays and the Phelps Dodge/Freeport Central Analytical Service Center CASC (Best et al. 2007). The construction and full operational use of the Automated XRD-NIR Lab AXN at Freeport in Arizona constitutes an additional large innovation step toward modern plant mineralogy (Baum 2009). Central automated labs at BHP in Australia followed together with a worldwide rise of smaller and larger lab automation modules in iron ore, copper, gold and other industries.

MINERALOGY INNOVATIONS IN PROJECT DEVELOPMENT AND PLANT OPTIMIZATION

Scoping, pre-feasibility and feasibility studies should make more use of process-related mineralogical ore characterization including thorough and large-scale analysis of the bench or pilot test products. Mineralogical work (more than just chemistry) has the powerful tools to identify "red flags" during these assessments and suggest metallurgical remedies.

In addition, mineralogical characterization methods have been pivotal in resolving process problems such as plugging of drip lines, salt buildup and reduction thereof, minimizing ponding, channeling and dry zones as well as laboratory assistance in the SXEW section of the plant (crud analysis). The recent developments by CSIRO researchers on the in-situ use of XRD for electrowinning cells (McGregor & Madsen 2011) indicates the future potential for in-plant mineralogy. Early efforts by a few copper mining companies for using automated mineral analyzers, XRD and NIR in "*Heap Audit/Survey*" applications similar to concentrator surveys were extremely successful and assisted in reducing expensive and time-consuming column leach tests (Fennel et al. 2005). This effort can be considered as the first true implementation of quantitative heap leach mineralogy in copper.

Anglo American Platinum (Rule & Schouwstra 2011) has demonstrated the economic value of using process mineralogy for better PGM plant design and optimization.

The courageous first implementation of a neutron cross belt analyzer at a Ni laterite operation in New Caledonia serves as another example of the step-changes experienced in the last decade.

With the increasing improvements of cross belt analyzers, mineralogical analyses are now also being implemented "on-the-belt." The cross belt neutron analyzer is making slow in-roads to metals mining (being already present in coal and cement). Several mining operations have implemented the use of Blue Cube systems. On-line analysis via LIBS has been spearheaded since 2000 by the Florida Industrial and Phosphate Research Institute (FPIR) leading to the first LIBS use on a conveyor belt (Gaft et al. 2007). The innovative use of NMR analysis, also in phosphate applications, in the early 1990s (Cooper 1991) and substantial technology improvements will result in increasing use of this technology for fast on-line analysis (Davis 2006).

OPERATIONAL MINERALOGY—THE NEW PARTNER AT THE MINE SITE

During the time of preparation of this paper, numerous plants have been outfitted with automated mineral analyzers (Cerro Verde, Kennecott, Climax, Barrick Goldstrike, etc.). They exemplify the rising acceptance and innovative use of Process Mineralogy. The use of automated mineral analyzers by mining contract labs and universities around the world and the startup use by newcomers in iron ore, nickel, platinum, and oil and gas document the growth of process mineralogy lab technologies as a complementary segment supporting chemistry and metallurgy. Overall, the introduction of the automated minerals analyzers and, especially, their daily production use represent one of the most significant innovation breakthroughs in mining.

Further, many mining operations have been installing modular mineralogical lab technology, especially XRD, Fourier-Transform (FT-NIR), or NIR at their productions facilities. Daily XRD and NIR mineralogy is now being used in several copper operations in the US, Chile and Peru and they have been used via FTIR systems in the aluminum industry for years. The German potash company Kali & Salz was most likely the innovation leader in using cross belt FT-NIR in their salt operations for production control (Rest et al. 2002).

This illustrates a point of critical importance: Too many mining operations continue to miss the opportunity of exploring innovation in other areas. Too often we become too focused on just the commodities with which we are working. Adaptable breakthrough innovation may have already occurred in a different industrial field or operation. However, how many metallurgists from copper or gold operations benchmark what is happening in potash mineral processing? The *"look over the fence"* is more important than ever. We do not have the time and resources to develop everything ourselves or in our own industrial sector.

GEO-METALLURGY—THE NEW CUSTOMER OF PROCESS MINERALOGY

During the mid-1990s into the 2000–2011 periods, the emergence of Geo-Metallurgy documents the efforts toward better ore characterization and plant operation. Part of Geo-Metallurgy is the integration of well known Process Mineralogy into mine site geology, project development, metallurgical testing, flow sheet design and better plant start-up plus faster troubleshooting. Geo-Metallurgy has taken a key role in the age of modeling and ever increasing needs for fast-track development—often without detailed and time-consuming metallurgical evaluations.

Analytical down-hole tools from the oil industry are (finally) receiving some pilot testing for monitoring heap leach operations. This will represent another major break-through innovation in regard to operating heaps and stockpiles like plants. The future potential for bio leaching low-grade or waste rock materials through "engineered heaps" such as practiced by the Morenci copper operation (Ekenes and Caro 2012) will be enhanced by the use of down-hole analysis tools.

In conclusion, the last 10 years of using innovative mineralogical analysis techniques in concert with powerful computer development have shown that a large step-change has occurred by applying daily, quantitative mineralogical data in the following ore control practices:

- Mine geology and resource control
- Clay modeling
- Hardness monitoring
- Blast index improvements
- Alteration mineral models
- Refined agglomeration
- Acid consumption improvements
- Permeability models, forecasts and optimization
- Optimized reagent use in flotation

More opportunities exist with the emerging lab technologies such as x-ray micro-tomography, or a fully automated mineral analysis module which provides high turnaround mineral data, down-hole probes, and considerably more use of cross belt analyzers.

ECONOMIC IMPACT

In the past 12-year period (2000–2012), it is my estimation that economic improvements related to mineralogical and lab automation innovations, have provided more than \$2.3 billion of benefits to the metals mining industry. In the environmental area, the use of metal speciation, tailings analysis, efforts to avoid multi-party litigation, beneficial reuse and residual value recovery indicate economic impacts in the multi-billion dollar range.

THE PATH FORWARD

The next two decades will be governed by the mining industry's current efforts in autonomous mining, remote control centers, automated process equipment and more step changes in robotics lab technology. The increasing significance of underground mining will mandate even more innovations in 24/7 mineralogy and down hole logging techniques. Ore body characteristics require continuous updates and revisions and the complex ores plus lower grades and larger equipment are unforgiving in the event of insufficient quantitative data related to mineralogical variance.

In regard to Process Mineralogy's and Lab Automation's future, this author foresees a significant potential in Automated Mineralogy Modules, i.e., containerized or otherwise compacted lab modules which provide automated sample preparation, sizing, potting/polishing and mineralogy via automated mineral analyzers. These systems will ultimately provide 8–15 hour turnaround time for quantitative mineralogy and be linked to modern process control. Down-hole technology from the oil industry and innovative use of mineralogy lab technology transferred to cross-belt applications will constitute the large re-tooling fields of the future. Specifically the down-hole analytical technologies will permit us to operate heaps and stockpiles like plants and cross belt mineralogy will enable real-time fine tuning of metallurgical plants.

REFERENCES

Allen, D., Baber, B., Baum, W., and Eady, S. 2007. Ore characterization, alteration coding and plant feed control in copper operations as directed by routine semi-automated mineralogical analyses. Proceedings, Copper 2007, Vol. IV, Book 2, pp. 247–258.

Baum, W. 1998. Copper mineralogy and extraction problems—how to reduce your losses. Proceedings, Randol Gold Forum '92, Vancouver, pp. 151–154.

Baum, W. 2007. Material characterization on the cutting edge. CSIRO Process, June 2007, p. 12.

Baum, W. 2009. The new Automated XRD-NIR Mineralogy Lab—High capacity tool for short and long-range ore characterization. SME 2009 Arizona Conference, Presentation & abstract.

Baum, W., Sanhueza, J., Smith, E.H. and Tufar, W. 1987. The use of process mineralogy for plant optimization at the El Indio gold-silver-copper operations (Chile). Erzmetall, 42, No. 9, pp. 373–378.

Best, E., Baum, W., Gilbert, R., Hohenstein, B. and Balt, A. 2007. The Phelps Dodge Central Analytical Service Center: Step change technology implementing robotics systems and lab automation in the copper industry. Proceedings, Copper 2007, Toronto, Vol. VII, pp. 119–126.

Brandt, D., Martinez, I. French, T. Slattery, L. and Baker, G. 2011. Better blast fragmentation through teamwork at the Dos Pobres Mine in Safford, AS. Mining Engineering, October 2011, pp. 30–35.

Cooper, H. 1991. An investigation of potential for improved efficiencies in phosphate rougher flotation through on-line BPL measurements. FPIR Publication # 04-032-091.

Davis, V. 2006. Process nuclear magnetic resonance (NMR) analyzers—benefits for phosphate beneficiation plants. Beneficiation of Phosphates: Technology and Sustainability, P. Zhang, K. Swager, L.L. Filho, and H. El-Shall, eds. SME, Littleton, CO.

Ekenes, J.M. and Caro, C.A. 2012. Improving recovery of copper from low-grade chalcopyrite ores. SME Annual Meeting, Seattle, 2012, Pre-Print 12-099.

Fennel, M., Guevara, J., Canchaya, S., Velarde, G., Baum, W. and Gottlieb, P. 2005. QEMSCAN mineral analysis for ore characterization and plant support at Cerro Verde. 27th Mining Convention, Arequipa, Peru, 2005, Technical Papers, pp. 1–11.

Follink, B. 2007. The gateway to precision processing. CSIRO Process, June 2007, p. 2.

Gaft, M., Stana, R., and Sapir-Sofer, I. 2007. LIBS module for dolomite content evaluation on a conveyor. FIPR Publication # 04-067-224.

Lotter, N. 2010. Modern Process Mineralogy: an integrated multi-disciplined approach to flowsheeting. Keynote lecture "Process Mineralogy '10" MEI, Cape Town, November, Proceedings, pp. 1–15.

Lotter, N., Kormos, L.J., Oliveira, J., Fragomeni, D., and Whiteman, E. 2011. Modern Process Mineralogy: Two case studies. Minerals Engineering, 24 (2011), pp. 638–650.

McGregor, K., and Madsen, I. 2011. Analysis in extreme environments. Process, October, 2011.

Rest, T., Fuss, J., Behling, J., and Stahl, J. 2002. On-line element determination using near infrared (NIR) spectroscopy for process control and limit value monitoring of solutions with high salt content in the potash industries (in German). Kali & Steinsalz, 3, 2002, pp. 6–13.

Rule, C. and Schouwstra, R.P. 2011. Process mineralogy delivering significant value at Anglo Platinum concentrator operations. Proceedings of 10th ICAM, pp. 613–620.

ABOUT THE AUTHOR

Wolfgang Baum has worked for 35 years in the mining industry focusing on process mineralogy related to ore control, metallurgy, flotation, heap leaching and environmental technology. His career includes assignments with Hazen Research, Texasgulf, St. Joe Minerals Corporation, PMET, Phelps Dodge Mining Company, and Freeport-McMoRan Copper & Gold. At Phelps Dodge and Freeport, he initiated the large-scale use of automated mineral analyzers, near infrared, and robotics technology in laboratories in support of geo-metallurgy and production. In 2011, he joined FLSmidth Salt Lake City where is managing the ore characterization labs.

Innovations in Measurement of Mineral Structure and Surface Chemistry in Flotation: Past, Present, and Future

Roger St.C. Smart and Andrea R. Gerson
Minerals and Materials Science and Technology, Mawson Institute, University of South Australia, Mawson Lakes, South Australia

Brian R. Hart
Surface Science Western, University of Western Ontario Research and Technology Park, London, Ontario, Canada

David A. Beattie
Ian Wark Research Institute, University of South Australia, Mawson Lakes, South Australia

Courtney Young
Metallurgical and Materials Engineering, Montana Tech, Butte, Montana, United States

ABSTRACT: This paper surveys the development of surface analytical methods that have contributed fundamental understanding of the flotation process and the ability to measure changes in mineral structure, surface chemistry and reagent uptake in single minerals and now between flotation feeds, concentrates and tails in synthetic and real plant samples. The early applications and past development of new techniques, i.e., XPS (or ESCA), Auger spectroscopy, scanned probe microscopies (STM, AFM), ion beam techniques (ToF-LIMS, ToF-SIMS) and vibrational spectroscopies (FTIR, Raman) to mineral processing are briefly surveyed. In current practice, further innovations in methodologies and principal component analysis applied to ToF-SIMS, contact angle estimation and vibrational spectroscopies (ATR FTIR, TIR Raman, SFG) with examples are reported including cases where the techniques have been combined with electrokinetic and electrochemical analyses (i.e., spectroelectrochemistry). Differences in hydrophobic/hydrophilic surface species on specific mineral phases, which can control recovery and grade, can now be analysed statistically using ToF-SIMS to diagnose problems in grinding, conditioning (physical and reagent) and interfering species on surfaces of value minerals. In future, new methods based on synchrotron techniques offering information on surface structure, amorphous phases and analysis of diffraction and composition on single particles (<2 μm) are being developed for location of toxic and penalty impurities.

INTRODUCTION

In froth flotation, the chemistry of the top few monolayers of different mineral surfaces determines the recovery and grade in operation. Attachment of a collector-conditioned mineral particle to a bubble and the stability of this attachment in both pulp and froth phases is dependent on the hydrophobic/hydrophilic ratio of surface species in the outermost molecular layers (1–2 nm) on individual mineral particle surfaces (Piantadosi & Smart 2002). In practice, adsorption of the hydrophobic collector molecules to surfaces of the valuable mineral phase particles can be compromised by the presence of many hydrophilic oxidation and other surface species with preference for water attachment. In reality, the surface of each individual value mineral particle is a complex, distinctly non-uniform array of oxidation products (i.e., oxyhydroxides, oxy-sulphur species), adsorbed ions, precipitates, collector molecules and attached fine and ultra-fine particles of other mineral phases. In order to understand and improve poor flotation recovery, it is first necessary to know whether this is being caused by poor hydrophobic/hydrophilic conditioning of the value mineral surfaces (or inadvertent hydrophobic conditioning of gangue minerals). It is then possible to define the reasons for this before changes to mechanical (e.g., grinding, agitation, fines separation) or chemical (e.g., redox control, reagent doses, addition points, water quality) conditioning can be usefully made. Hence, ideally, we would like to know the hydrophobic/hydrophilic ratio by particle and as a statistical distribution between different mineral phases across a flotation circuit from roughers, scavengers, cleaners to tail. This is a difficult proposition but surface analysis has come a long way towards this goal. This paper will briefly review the development of these measurements, their current applications and future prospects.

These techniques are equally applicable to recovery of metals by hydrometallurgical leaching, where leach kinetics and selectivity can be cost-limiting, rate-determining surface species and surface layers forming during leaching may limit access of reactants to the surface and of reaction products to the solution. The limiting actions of complexants, oxidation products, fine gangue interference, precipitates as colloidal products and thick surface layers (amorphous and crystalline) need to be defined before the sub-processes requiring change can be usefully addressed. We will focus on the flotation applications in this paper.

THE PAST: DEVELOPMENT OF METHODS

Until the introduction of direct surface analytical methods, studies of the actions of collectors, activators, depressants and other chemical reagents used in the flotation process were largely based on indirect methods. Fuerstenau and Chander (1982) noted in their early review "Industrial Applications of Surface Analysis" that "mineral processing research has extensively involved determination of electrokinetic behavior, voltammetry, gas adsorption, aqueous-phase adsorption, infrared and UV-VIS spectroscopy, measurement of contact angles, and, more recently, ESCA analysis to delineate surface composition. Lack of techniques to study solid/liquid interfaces *in-situ* has led to the development of many ex-situ and indirect methods for surface characterization. The results of such measurements are often interpreted through use of models requiring various underlying assumptions. It has been often necessary to use a combination of two or more techniques to delineate the physico-chemical interfacial phenomena."

The introduction and evolution of direct surface analytical techniques in their application to fundamental aspects of mineral structure and surface chemistry has provided the mineral processing industry with valuable tools to better understand factors involved in process control. Advances in these techniques and instrumentation have produced new tools and procedures for the comprehensive statistical analysis of ore samples and process stream products. These techniques, together with the mineral structure techniques, and the information they have given on the behaviour of different minerals in the flotation process have been extensively reviewed by experts in each technique in the SME Centenary of Flotation volume (Smart et al. 2007). The significance of these techniques is that they provide not only a compositional analysis of the surface layers but also information on chemical states (e.g., oxidation, bonding), spatial and statistical distribution of adsorbed species between individual particles and mineral phases in complex mixtures as a function of depth through the surface layers. Nevertheless, Fuerstenau and Chander (1982) remain correct in that the combination of these ex-situ techniques with adsorption and solution modelling provides the most complete information on surface chemical conditioning and interferences in flotation. It is recognised that, since they are ex-situ techniques that operate in ultra high vacuum, validation of the relationship between the measured surface compositions or chemical states and those prevailing in the original pulp solution in the flotation cell or circuit, is required. Sampling methodologies have been developed and tested in extensive sets of correlated surface analysis/solution analysis/flotation response testing in major projects over more than 20 years (Smart et al. 2003a, 2007).

Two of the complementary techniques now widely used with the direct surface analysis techniques are EDTA extraction and solution speciation modeling.

EDTA Extraction

This technically much-simpler chemical-based analysis dissolves oxidised metal ions (i.e., Cu^{2+}, Fe^{3+}, Zn^{2+}, Ni^{2+}, Pb^{2+}) as EDTA (ethylene diamine tetraacetic acid) complexes from in situ surface reaction layers, adsorbed colloids, precipitates and ions to give solution assays that provide a bulk estimate of the extent of oxidation of all minerals. Comparison with the chemical information (i.e species and composition) from specific minerals in the surface sensitive analyses of the top 1–10 nm can then provide a more complete picture of the hydrophilic species likely to be interacting with the bubbles and potentially interfering with flotation. Examples of the methodology and its use in this way can be found in early work (e.g., Grano et al. 1990) and more recently in Gerson and Jasieniak (2008).

Solution Speciation Modelling

Basic determination of solution speciation (i.e., predicted relative abundance of specific dissolved and precipitated chemical species) is readily accessible from solution assays but is often neglected. While there are numerous examples of the use of such simulation programs for environmental studies, particularly for adsorption and examination of oxidation states for toxic elements (e.g., Halim et al. 2005), there are surprisingly few applications for mineral flotation studies. Speciation simulation programs, e.g., GEOCHEM-EZ (Shaff et al. 2010), PHREEQC (Toran and Grandstaff 2002), MINTEQ (Allison et al. 1990) and others are available, often free to download. Solution concentrations of all cation and anions, pH, temperature and particularly E_H or ORP (often not measured) are required input. Gas partial pressures and the presence of specific solids phases can be specified in the estimates. The programs come with their own databases of equilibrium constants, solubility products and redox couples, including collector and complexant species, but data on more obscure species may need to be added to the data base.

The value of these simulation programs for minerals processing is to take a 'snap-shot' of the system, assuming equilibrium, so that under-saturated but potentially adsorbing species and precipitating supersaturated solids may be identified. This can again be compared with the surface analyses to extend the correlation of indirect and direct information on hydrophobic and hydrophilic species. An example of the value of this correlation of solution modelling and surface analysis can be found in defining correct procedures for copper activation of sulphide flotation (e.g., pentlandite) (Malysiak et al. 2002) avoiding precipitation of hydrophilic $Cu(OH)_2$. A recent example of combined information from solution modelling with zeta potential, X-ray photoelectron spectroscopy (XPS) and EDTA extraction is in the flotation study on Cu and Pb speciation as a function of contrasting Eh and pH conditions during grinding (Peng et al. 2012).

XPS or ESCA

The technique of electron spectroscopy for chemical analysis (ESCA), later renamed more specifically X-ray photoelectron spectroscopy (XPS), was developed by Kai Siegbahn for study of atomic, molecular and solid state structure (Siegbahn et al. 1967) for which he received the Nobel Prize in 1981. In XPS, the sample surface is irradiated with mono-energetic x-rays producing photoelectrons analysed to determine their binding energy. From the binding energy and intensity of a photoelectron peak, the elemental identity, chemical state, and percentage of an element in the top 2–10 nm are determined. Two-dimensional imaging can now provide species mapping of surface chemistry with a practical spatial resolution of 10–20 µm but the data collection is relatively slow. In minerals processing research, again for practical collection times, it has mostly been used in larger area-average mode (100 µm–1 mm) applied to single mineral or pure mineral mixture studies with limited application to statistical analysis of plant samples. The first studies of copper and copper oxide were made in 1958 (Siegbahn et al. 1967). The technique appears to have been first used in studies applied to mineral processing by Clifford, Purdy and Miller (1975) and Brion (1980). A series of seminal studies by Buckley and Woods across 1984–1994, combining XPS and cyclic voltammetry, reviewed in Buckley (1994), initially defined many of the surface oxidation products, including oxidised sulphur species (from sulphide, disulphide, polysulphide, elemental sulphur, to thiosulphate, sulphite and sulphate) and adsorbed xanthate species on the major base metal sulphide surfaces as single minerals. Later work on oxidation, collector, activator and other adsorption mechanisms as well as interfering species, reviewed by Smart et al. (2003a), extended the studies of single minerals and laboratory flotation systems giving more direct insight to the factors affecting recovery and grade. Most of the hydrophobic, including both reagent and reaction (e.g., polysulphide) products, and hydrophilic species affecting flotation were defined in these studies. The mechanisms and products of in-pulp oxidation and the actions of collectors in partial removal of oxidised species and hydrophobisation of the surfaces were studied using XPS combined with other surface analytical methods summarised in the two reviews (Smart et al. 2003a, 2007). The mechanisms of copper sulphate as flotation activator in sphalerite, in ion exchange with Zn^{2+}, reduction in situ to Cu^{+} with collector attachment, and adjacent oxidation of sulphide to polysulphide-like species were revealed by these species seen in XPS (Gerson et al. 1999). Inadvertent activation by species such as Pb^{+} and Ag^{+} were first seen in these studies. Actions of depressants, such as cyanide and bisulphite (e.g., Khemleva et al. 2006), on this surface chemistry were also defined in XPS studies of single minerals. XPS has probably contributed more understanding of mechanisms of surface chemical actions in flotation than any other surface analytical technique.

Auger and Scanning Auger Spectroscopy

Auger electron spectroscopy, first suggested for surface analysis by Lander (1953), was also introduced to analysis of mineral surfaces in the same period as XPS (e.g., Eadington 1974, 1977). This is an analytical technique that uses a nm-focussed primary electron beam, similar to scanning electron microscopy (SEM), to probe the surface of a solid material. Electrons emitted as a result of the quantum-specific Auger process are analysed and the identity and quantity of the elements determined from the kinetic energy and intensity of the Auger peaks. Like X-ray photoelectrons, Auger electrons can only escape from the outer 2–10 nm of a solid surface at their characteristic energy. The finely focused electron beam can be scanned to create secondary electron and Auger images, or the beam can be positioned to perform microanalysis of specific particles but surface charging (i.e., low-conductivity minerals) can limit applicability of this technique. In mineral mixtures after grinding and conditioning, scanning Auger spectroscopic images and analysis first illustrated the complexity of hydrophilic oxidised layers, adsorbed species and attached fine particles on mineral surfaces that can interfere with bubble-particle attachment (e.g., Smart et al. 2003a).

FTIR Spectroscopy

Just as XPS and Auger spectroscopy contributed to the greatest increase in our understanding of mineral surface chemistry of relevance for flotation, Fourier-transform infrared (FTIR) spectroscopy has contributed the most to our understanding of the adsorption mechanisms of reagents on those mineral surfaces. FTIR spectroscopy, with its ability to record high resolution spectra rapidly and with high signal to noise, revolutionised the structural determination and identification of organic molecules when introduced in the late 1960s. However, the application of infrared spectroscopy to surface adsorbed species pre-dates the advent of the FTIR spectrometer, with dispersive instruments being used to obtain spectra of monolayers of material on reflective surfaces through the sampling method of reflectance-absorbance infrared spectroscopy (RAIRS—also known as external reflectance infrared) (Francis and Ellison 1961; Greenler, 1966, 1969). In addition, dispersive instruments were used to study the adsorption of flotation collectors to metal sulphide surfaces (Greenler 1962). That being said, it was the introduction of FTIR instruments that resulted in a widespread uptake of infrared spectrometers in labs world-wide, and thus facilitated a step-change in our understanding of adsorption phenomena.

As mentioned above, external reflection infrared spectroscopy, which relies on the alteration of surface optical properties by adsorbed layers of molecules (and more importantly, specifically at energies that correspond to vibrations of the adsorbed molecules), has the sensitivity to detect monolayer and sub-monolayer quantities of adsorbing molecules. It has been used to great effect to elucidate binding mechanisms and orientation of flotation collectors on metallic reflective surfaces (gold, silver, copper, etc.) and on non-metallic surfaces (Mielczarski 1993; Mielczarski and Leppinen 1987). However, the general requirement to analyse samples dry (i.e., outside of the flotation-relevant solution environment) limits the relevance of the methodology for flotation studies, in spite of it being applied to real mineral surfaces (Mielczarski et al. 1993). RAIRS has been applied to metal surfaces in electrochemical cells whilst the electrode surface has been immersed in liquid, in studies of relevance to flotation (such as xanthate adsorption on copper metal (Bozkurt et al. 1996). The addition of surface potential control increases the usefulness of the methodology, but it is still one step removed from the analysis of real mineral surfaces.

A second infrared sampling methodology that is particularly useful for analysing real mineral particles is diffuse reflectance FTIR (termed DRIFT). The methodology requires that conditioned mineral particles be dried and mixed with a non-absorbing medium (potassium bromide, KBr), placed in a sample cup, irradiated with an IR beam, and diffuse (i.e., multiple non-specular reflections) are collected. DRIFT was shown to be useful in gaining some insight into binding mechanisms for collectors and polymers (Gärd et al. 1997; Khemleva et al. 2006; Raju et al. 1997), although the need to dry samples prior to analysis means that its use for analysing polymer adsorption is not without complications, and studies of adsorption kinetics are not possible.

A more routine FTIR sampling methodology that is suited to analysing adsorbates on surfaces immersed in liquid (in addition to dry surfaces) is ATR (attenuated total reflection) in which IR radiation is internally reflected in a reactive IR-transparent crystal and adsorbing molecules are probed by interactions with an IR evanescent wave at the solid-liquid interface (Harrick 1967). ATR combines the sensitivity and structure determination characteristics of RAIRS while enabling the study of reagents adsorbing from solution directly. The one caveat is that the routine application of the methodology is limited to adsorption substrates that are IR transparent (e.g., ZnS, CaF_2, NaCl, KCl, Al_2O_3, $CaCO_3$, germanium, silicon). The technique has been used to study collector adsorption, including studies of orientation (Larsson et al. 2000) and co-adsorption (Larsson et al. 2001). The in situ nature of the techniques also allows for the determination of collector adsorption kinetics (Fredricksson and Holmgren 2007). ATR using crystals as substrates suffers from the same limitations as RAIRS in terms of connecting directly with real flotation systems and real mineral particles. As will be discussed later, studies have also been done *in-situ* on mineral particles by pressing them as a bed against the ATR crystal, and then controlling the bed potential as needed. In addition, D_2O is usually used in all of these ATR/FTIR techniques to minimize the interfering signals from normal water (H_2O). However, a series of studies at The University of Utah with Jan Miller using reactive ATR crystals are particularly noteworthy and discussed in the following.

Kellar et al. (1990, 1991) examined fluorite (CaF_2)/oleate system and confirmed that oleate not only chemisorbed at the fluorite surface but also formed a near complete monolayer. As the monolayer formed, they found that the CH-stretching vibrations would change depending on the amount of surface coverage as well as temperature. Noting that it was similar to conformational/structural changes observed in the literature for liquid crystals, they likened it to the formation of coagel/gel phases as opposed to hemi-micelles. At high oleate concentrations, a small amount of calcium dioleate surface precipitation was also observed but would decrease when the solution concentration exceeded the critical micelle concentration (CMC).

Similarly, Young and Miller (2002a,b) investigated the calcite ($CaCO_3$)/oleate system. In their case, the ATR calcite crystal was fashioned from a natural specimen and required the study to be done using the Near-IR (NIR) as opposed to Mid-IR (MIR). They observed the same behaviour as for fluorite but found that chemisorbed oleate only gave submonolayer coverage equivalent to no more than approximately 25% of the maximum observed at fluorite surfaces. Surprisingly, thermodynamic analysis showed that the free energy for chemisorbed oleate formation was the same at both surfaces but enthalpic and entropic contributions varied significantly and were attributed to differences in hydration of the two surfaces prior to adsorption and therefore the desorption of water that had to occur in order for oleate adsorption to take place. By likening the chemisorbed oleate to a "protective coating" and comparing it to fluorite, they noted the lower amount of chemisorbed oleate on calcite allowed significantly more calcium dioleate surface precipitation to occur. Calculations showed that both surface precipitation events correlated well to the mineral solubility and Free and Miller (1996) suggested earlier that the mechanism was caused by the formation of calcium dioleate in solution followed by its transport back to the surface and not nucleation and growth at the surface. Likewise, the amount of chemisorption was related to the spacing between calcium adsorption sites with fluorite being able to accommodate the size of the oleate molecule at every site whereas calcite could not.

By comparison, Cross et al. (1993) used the ATR/FTIR technique to examine the alumina (Al_2O_3)/sodium dodecyl sulphate (SDS) system. They found similar conformational changes and concluded that the adsorbed species went through four stages of adsorption until maximum adsorption occurred. During the initial stages, the surface species resembled cylindrical micelles in solution but transitioned into a coagel-like phase at higher adsorption densities. At increased temperatures, the coagel phase developed conformations similar to hexagonal liquid crystals. In this case and in agreement with the literature, adsorption was physisorption and therefore dependent on surface charge. Finally, Free and Miller (1994) used many of the same synthetic ATR crystals and observed the same characteristics when the collectors were transferred as Langmuir-Blodgett films. Their work revised the ATR adsorption density calculation and helped improve our understanding of the ATR techniques and flotation.

Laser Doppler Electrophoresis

On the other hand, Miller et al. (1992) and Yalamanchili et al. (1993) extended the ATR/FTIR studies at The University of Utah to halite (NaCl) and sylvite (KCl) in the presence of dodecylamine hydrochloride (DAH). As expected and in accordance with the literature (Fuerstenau and Fuerstenau 1956), they found DAH had minimal adsorption on halite and maximum adsorption on sylvite. However, through the use of laser-Doppler electrophoresis, they were able to surprisingly show that sylvite particles exhibited a negative surface charge as they dissolved in a near-saturated solution and halite yielded a positive charge, neither of which should occur according to classical DLVO theory via compression of the double layer. With this technique, particles with a surface charge are set in motion in a conductive fluid by applying a voltage across the fluid. Particle velocity and direction are instantaneously measured by laser Doppler and used to calculate the electrophoretic mobility and zeta potential based on traditional electrokinetic equations. By comparison, the cumbersome microscopic technique used previously by numerous authors for a variety of mineral/collector flotation systems would take several minutes to complete and simply would be impossible to use for studying soluble salt minerals. To explain how sylvite can be floated from halite, they found that DAH would form a positively charged colloid that adsorbed at the negatively charged surface of sylvite. Furthermore, they observed the opposite phenomenon when the solution pH was adjusted above the pzc of the colloid as well as when negatively charged laurate was used as the colloid/collector. Although they confirmed this surface-charging phenomenon by simultaneously studying other alkali halides, there were a few exceptions which led to subsequent studies that revealed how the interfacial structure of water also plays a role. In this regard, Hancer et al. (2000) showed that, depending on the soluble salt, the ions in the brine would either facilitate hydrogen bonding and thereby promote water structuring or disrupt it and thereby promote water breaking. They concluded that flotation would only be possible if the soluble salt was a structure breaker. Overall, these studies showed that there was significantly more to soluble-salt flotation than the size of the amine head group being able to accommodate the sylvite structure/lattice spacing as opposed to that of halite.

Perhaps the most significant contribution from this research campaign at the University of Utah was the colloid/coagel/gel extensions to the Hemi-Micelle Theory of Adsorption (Gaudin and Fuerstenau 1955). In essence, because it was understood that liquid crystals like coagel/gel phases would transform into micelles at temperatures above the Krafft point in bulk solution (Cross et al. 1992), the same must be true for adsorbed species, depending on the type of collector.

Laser Raman Spectroscopy

Young and Miller (1999) extended their studies on fluorite and calcite with oleate using Laser Raman Spectroscopy (LRS) and found that the carbon-carbon double bond would react between neighbouring chemisorbed oleate molecules on fluorite but not on calcite. This phenomenon was attributed to epoxide formation that polymerized the surface by cross-linking the chemisorbed oleate helping to explain fluorite's enhanced hydrophobicity under certain conditions. The spacing between chemisorbed oleate on fluorite was close enough to allow the cross-linking to occur but not on calcite. This work was done *in-situ* by reflecting the laser beam off a bed of mineral particles and was among the first to use LRS for studying flotation systems.

Spectroelectrochemistry

Since then, most LRS studies have concentrated on examining sulphide minerals in the absence and presence of collector and often in conjunction with ever-improving sensitivity of electrical measurement and control associated with advancements in galvanostats/potentiostats. These "spectroelectrochemical" studies with voltammetric, galvanostatic and impedance techniques required the mineral to be prepared as the working electrode with a fresh surface being created by polishing in between tests. Greg Hope's research groups at Griffith University in Brisbane have been leading the charge in this area having studied a variety of sulphides ranging from simple to complex minerals including galena (PbS), pyrite (FeS_2), covellite (CuS), chalcocite (Cu_2S), chalcopyrite ($CuFeS_2$), bornite (Cu_5FeS_6), tetrahedrite ($Cu_{12}Sb_4S_{13}$) and tennantite ($Cu_{12}As_4S_{13}$). Woods et al. (1998, 2000), Asbjornsson et al. (2003a,b), Parker et al. (2003) and Jeffrey et al. (2005), for example, examined a variety of collector types and lengths (xanthate, thiourea, mercaptan and dithiophosphinate) as well as collectorless flotation (often with additional application to hydrometallurgical leaching). Results typically showed direct evidence of collector bonding with the metal of interest such as Pb and Cu in a chemisorbed state (as opposed to inferred evidence). Results were often compared to E_H-pH (Pourbaix) Diagrams. Their latest efforts have included researchers at Montana Tech (Gow et al. 2012 and 2013) involving, for example, enargite (Cu_3AsS_4) electrochemistry and the use of mass-balanced diagrams determined with StabCal software (Huang et al. 2005 and 2012). In these studies, the utility of LRS for determining reaction products has been critical and even forced particular species to be excluded from and others included in the calculations. They stress the importance of mass-balanced calculations, particularly in systems involving at least three elements (e.g., Cu, As and S).

As alluded to earlier, spectroelectrochemical studies have also been conducted using ATR/FTIR by packing a bed of mineral particles against an inert ATR crystal, often with potential-control. Jaako Leppinen, Jerzy Mielczarski, and Roe-Hoan Yoon have been leaders with this technology, often with their own research groups but periodically in collaboration with one another (Basilio et al. 1988 and 1992; Laajalehto et al. 1999; Leppinen et al. 1988, 1989 and 1990; Mielczarski et al. 1979, 1986, 1987 and 1993; Talonen et al. 1991). They commonly used cyclic voltammetry and constant-potential control to study sulphide mineral flotation systems including the interaction of xanthate, thionocarbamate, mercaptobenzothiazole and dithiophosphate with galena (PbS), chalcocite (Cu_2S), chalcopyrite ($CuFeS_2$), sphalerite (ZnS), marmatite [(Zn,Fe)S], pyrrhotite ($Fe_{1-x}S$) and pyrite (FeS_2) with and without copper activation. Depending on the system being examined and often with polarized IR, their results showed the collectors oriented at the sulphide surface depending on whether it chemisorbed as a monodentate or bidentate structure or, in the case of xanthate, additionally formed dixanthogen or a surface-precipitated metal xanthate. Different results were observed if contaminants were present in varying amounts, particularly if oxidation products were present prior to collector adsorption. In an innovative electrochemical study, Woods et al. (1990) used cyclic voltammetry to determine xanthate chemisorption isotherms as a function of potential, plotted them on E_H-pH diagrams, and fit the charge-transfer isotherms to a Frumkin adsorption model. By comparing the chemisorption isotherms to flotation recoveries of chalcocite measured with a Hallimond tube, they respectively showed that 50% and 90% recoveries were obtained with surface coverages of 20% and 50%, specifically showing that flotation could be achieved with submonolayers.

Clearly, the amount and presence of hydrophobic-inducing species can vary with potential control. Research like this has ultimately lead to methods of controlling potential in industrial flotation practices including, for examples, using nitrogen gas as a flotation carrier to minimize oxidation, redox reagents to control pulp potentials, and specific types of steel as grinding media to control galvanic interactions. These concepts have even led to successful sulphide flotation operations in the absence of collector and, in this regard, arguments about whether the sulphide mineral surfaces become metal-deficient or sulphur-enriched (Buckley and Woods 1994).

The increasing hydrophobicity of some sulphide surfaces in initial oxidation in solution has been attributed to polysulphide S_n species in which the loss of metal ions to solution (metal-deficiency) results in some in situ S-S bonding (Buckley and Riley 1991). Using the correlation of XPS S2p signals from the oxidized sulphide surfaces with ToF-SIMS mass markers reflecting increased S-S bonding, has given confidence in the XPS assignment to hydrophobic polysulphide species. For instance, freshly-cleaved galena (PbS) surfaces reacted in pH 8 aqueous solution for increasing periods of time have shown a systematic increase in S_n/S ratios with increasing% of oxidised S2p species from XPS spectra (Smart et al. 2000). The identification of S2p species in oxidized sulphide surfaces was summarized in Smart et al. (1999).

Scanned Probe Microscopies: STM and AFM

The introduction of the scanned probe microscopy techniques of scanning tunnelling microscopy (STM) and atomic force microscopy (AFM) to study surface structure at the nm scale in solution has revealed sometimes astonishing detail on oxidation and adsorption reactions. STM was invented by Binnig et al. in 1982 for which Binnig and Röhrer were subsequently awarded the Nobel Prize for physics. In STM, an atomically sharp tip is moved towards a reasonably conductive surface in sub-nm steps using piezoelectric crystal drivers until a tunnelling current is first observed between the tip and surface atoms. It can then be moved across the surface to produce atomic images and some identification of chemical structure in ST electrochemical mode. The AFM, was first demonstrated in 1986 by Binnig et al. (1986). AFM can image insulating surfaces, cannot chemically identify atoms, but can give information on particle–particle and even particle–bubble interactions and changes with reaction and adsorption in force–distance (approach–retract) mode (Ralston 1994). In AFM, the force between a piezo cantilever with attached tip or bubble and the surface is detected by deflection of the cantilever using a laser reflection device. For example, Eggleston and Hochella (1992) observed STM images of galena surfaces in which individual atoms react on contact with air-equlibrated water for 1 min. On oxidation, the tunnelling intensity maximum from single surface sulphur atoms disappeaed due to alteration of the local chemical structure. The apparent vacancies are in fact the sites of individual oxidised sulphur atoms. From their work, and from another study described below, it is clear that, when a surface S atom is oxidised, S atoms in adjacent sites are activated, thus promoting further reaction rather than initiating new oxidation reaction at S sites further away leading to development with time of pits of sub-nm dimensions. The z-dimensions of these pits corresponded to half or full unit cell parameters of the PbS crystal structure (Smart et al. 2003a, 2007). XPS and AFM studies of the same surfaces with closely similar development of pits confirmed that congruent dissolution of Pb and S occurs with unaltered Pb 4f and S 2p spectra releasing lead and sulphate ions to solution. The rate of formation of dissolution pits was strongly dependent on the concentration of impurities, on pH, and on the gas (air, O_2, N_2) used to purge the aqueous solution. *In-situ* STM images of a freshly cleaved galena in air-equilibrated 10^{-4} M ethyl xanthate solution show colloidal particles of lead ethyl xanthate, confirmed by XPS, formed at the surface corresponding to multilayer surface coverage. In-situ STM studies of ethyl xanthate treated preoxidized galena surfaces have also shown the removal of oxidized lead species and the formation of colloidal lead ethyl xanthate particles as flattened spheres with diameters of 10–20 nm and average heights of 6 nm (Kim et al. 1995). Other examples of the fundamental understanding generated from these atomic-level techniques are reviewed in Smart et al. (2003a).

Ion Beam Mass Spectrometry: D-SIMS, ToF-LIMS, and ToF-SIMS

But the development of the ion micro-beam techniques based on secondary ion mass spectrometry (SIMS), and time of flight laser ionisation mass spectrometry (ToF-LIMS) and time of flight secondary ion mass spectrometry (ToF-SIMS) in particular, have provided the statistical basis for identification of surface species determining the hydrophobic/hydrophilic ratio and interfering with flotation performance. The development and applications of these techniques has been recently reviewed by Nessett (2012). Early applications of dynamic SIMS in quantitative trace element analysis in geological materials (e.g., Lovering 1975, Shimizu et al. 1979) required complex correction factors inherent to the technique. Calibration using selected ion-implanted standards was developed by Chryssoulis et al. (1986). Using this method, the concentration of Ag in various sulphides from Brunswick base metal ore, Au in common sulphides and As bearing sulphides from refractory ores were measured (Chryssoulis et al. 1987). The capacity of dynamic SIMS to image the distribution of sub-microscopic Au within the sulphides was also demonstrated. An analytical program developed in the AMTEL labs has become an industry standard for quantification of sub-microscopic (invisible) gold and other precious metals and their deportment in mineral phases (Chryssoulis and Cabri 1990).

The advent of the highly focussed (90 nm), highly sensitive (7 decade intensity range), highly mass resolved ($m/\Delta m$ up to 10,000 times) time of flight mass analysers heralded the ability to determine all surface species on a specific mineral phase comparatively between feed, concentrates and tails in flotation. The use of the ToF-LIMS technique, using laser pulses to produce secondary ions for mass analysis, in minerals process analysis was pioneered by Clarke et al. (1986) and extensively developed by Chryssoulis (e.g., Chryssoulis et al. 1995) producing valuable new insight to flotation separation processes in plant operation. This work is reviewed in Smart et al. (2007) and Nesset (2011). This technique produces surface species in the first two laser pulses but has the disadvantage that the laser energy may destroy information from some organic reagents. In an early review of spectroscopic techniques to study the interactions between minerals and reagents in froth flotation by Giesekke (1983), it is noted that static SIMS using a focussed ion beam (i.e., ToF-SIMS) operates in a non-destructive mode analysing only the top-most monolayers of the sample surface and that this is the most appropriate tool for ion beam surface analysis.

In this short paper, we will focus on ToF-SIMS because it is now the most advanced technique for statistical analysis of the hydrophobic/hydrophilic surface chemistry directing flotation response. In ToF-SIMS, a focussed (90 nm), pulsed primary beam of Ga, Ar, Cs or, more recently, Bi ions and Au ion clusters, is directed at the mineral surfaces in fixed, rastered or pixellated mode. A fixed beam will analyse a chosen area in the practical range from 90 nm to 1 mm diameter. A rastered beam produces lateral images for secondary ion products selected from the total mass spectra collected in the scan. In the pixellated mode, a full mass spectrum is recorded in each pixel (up to 256×256 pixels) in a chosen area giving more than 10^7 data points. Images for any ion can then be selected from this stored data but, importantly, the data can be statistically analysed for correlation of surface species within and between mineral phases. In static analysis mode, a very low flux of the heavy ions impacts surface layers so that, in the time of routine measurement, less than 5 surface atoms in 1,000 are impacted. The secondary ions, both positive and negative, ejected from these impacts are mass analysed by charge (m/z) with reversed polarities using their time of flight (t) to the detector in the relationship $m/z = at^{1/2} + b$. The very high mass resolution $m/\Delta m$ in the range 7,000–10,000 is able to easily separate almost all closely similar ion masses, e.g., O_2^- from S^-. Secondary elemental ions and molecular fragment ions detached from the first two molecular layers of the surface provide a very detailed set of positive and negative mass fragments from simple ions, e.g., Na^+, OH^- through to large molecular ions of specific reagents, e.g., isobutyl xanthate $(CH_3)_2CHOCS_2^-$. This data collection generally takes less than one hour.

There is a strict protocol for sample collection, storage, transfer and instrument introduction from either laboratory or operating plant flotations that preserves the surface chemistry at the time of sampling (Smart 1991). Validation tests of the method are described in the Centenary of Flotation volume (Smart et al. 2007).

Some examples of the first use of ToF-SIMS in mineral processing help to explain the current development. Early studies by Brinen and co-workers at Cytec (Brinen & Reich 1992, Brinen et al. 1993) and by Chryssoulis and his AMTEL group (e.g., Stowe et al. 1995, Chryssoulis et al. 1995) confirmed that this technique can identify parent and fragment ions from collectors and map their distribution on single mineral grains. Brinen and co-workers found the distribution of dithiophosphinate collector on galena particle surfaces to be distinctly non-uniform, an important observation in understanding the mechanisms of bubble-particle attachment.

Mineral Phase Imaging: Hydrophobic/Hydrophilic Ratios

A method developed at the Ian Wark Research Institute first provided a statistical comparison of differences in surface species between streams by mineral phase (Piantadosi et al. 2000). Imaging for specific signals and combinations of signals (e.g., Pb, Zn, Cu, Fe) is used to identify the particles of a specific mineral phase (e.g., galena, sphalerite, covellite, pyrite, chalcopyrite) for specific analysis. Sufficient particles of each mineral phase are selected for reliable statistics and the full mass spectrum from each particle recorded and stored. The statistical analysis (Piantadosi et al. 2000) then determines a mean value for each atomic and molecular species with 95% confidence intervals for each signal. This analysis was first applied to the effects of calcium ion depression on a galena flotation (Piantadosi et al. 2000) using collector adsorption of both isobutyl xanthate (IBX) and diisobutyl dithiophosphinate (DBPhos). Linear regression and mean analyses of surface species on galena particles in the first concentrate and tail (26 particle sets) were used to estimate hydrophobic (e.g., $DBPhos^-/SO_3^-$) and hydrophilic (e.g., Ca/Pb, Al/Pb, PbOH/Pb, SO_3^-/S_2^-) indices. The results correlated closely with the flotation response. Hydrophilic species indices such as calcium (~2×), but also aluminum and metal hydroxides, were statistically higher on tail than on concentrate particle surfaces. The method suggested that it was possible to quantitatively assess conditioning of sulphide surfaces for optimum selectivity. A second statistical study (Piantadosi and Smart 2002) of the effect of iron hydroxide oxidation products and collector, IBX, on galena and pyrite flotation separation at pH 9 confirmed a clear separation of IBX normalised intensities with galena in concentrate 5±1.5 times higher than those in feed or and tail. A hydrophobicity/hydrophilicity index, based on a ratio of signals from the hydrophobic collector (IBX) to ions from hydrophilic oxy-sulphur products (SO_3) and iron hydroxide (FeOH), gave a value for the concentrate of 44.7 ± 13.7 compared with 7.1 ± 2.4 for the tail. A similar differentiation was found using DBPhos collector. These hydrophobic/hydrophilic indices systematically decreases across the flotation sequence between concentrates 1, 2, 3 and tail (Smart et al. 2003b) and can be used to diagnose parts of the circuit in which this ratio is not effective for bubble-particle attachment therefore requiring changes to conditioning procedures.

THE PRESENT: NEW METHODS

ToF-SIMS: Statistical Analysis of Real Ores

ToF-SIMS analysis is now applied to ore samples from lab floats and operating plants. Currently, this statistical diagnosis of surface chemical factors affecting flotation recovery and grade is used in two forms: direct comparison of intensities of surface species on selected mineral phases through feed, concentrates and tail; and principal component analysis (PCA) of all surface species correlated with specific mineral phases. An example of each mode is presented.

ToF-SIMS: Comparison of Intensities

In a Brunswick Mines plant survey, sphalerite was reporting to the copper–lead concentrate providing an opportunity for improvement of Cu-Pb-Zn separation. Plant samples were taken in the copper-lead circuit from the rougher feed (RF), rougher bank A concentrate (Con A) and tail (Tl A) and rougher bank F concentrate (Con F) and tail (Tl F). Sphalerite particles were selected using mineral phase imaging to give reliable statistics. TOF-SIMS normalized intensities (**Figure 1**) show the range and distribution of both Cu and Pb on the surface of sphalerite grains. Both Cu and Pb intensities in the con A/tail A pair and Con F relative to sphalerite in the feed (RF) suggesting dissolution/adsorption in conditioning. The data suggest that for the con/tail pairs at A and F, Cu on the surface of sphalerite grains is discriminatory. Pb on the other hand does not show any significant discrimination between Con A and Tail A, and intensities are only slightly higher on sphalerites from Con F relative to the tail but there is extreme variation in the Pb content on the surface of the sphalerite grains reporting to Tail A. The down-bank increase in Pb activation may indicate a decrease in solution Cu. These data fit the early observations of Ralston and Healy (1980) who showed in adsorption studies that sphalerite surface uptake for the metal ion Cu occurs before that of Pb and the later observations that Cu activation is favoured 3–4 times over Pb at pH >9 (Sui et al. 1999).

This kind of detail from surface analysis can suggest changes to conditioning, residence time and Cu deactivation reagents. For instance, Zn activation control in an industrial Cu flotation circuit treating a complex Cu-Pb-Zn-Au-Ag sulphide ore at the LaRonde Division of Agnico-Eagle Mines Limited (Quebec, Canada) was investigated through COREM (Olsen et al. 2012). The action mechanisms of $ZnSO_4$, TETA (triethylenetetramine) and $NaHSO_3$ used as depressants to

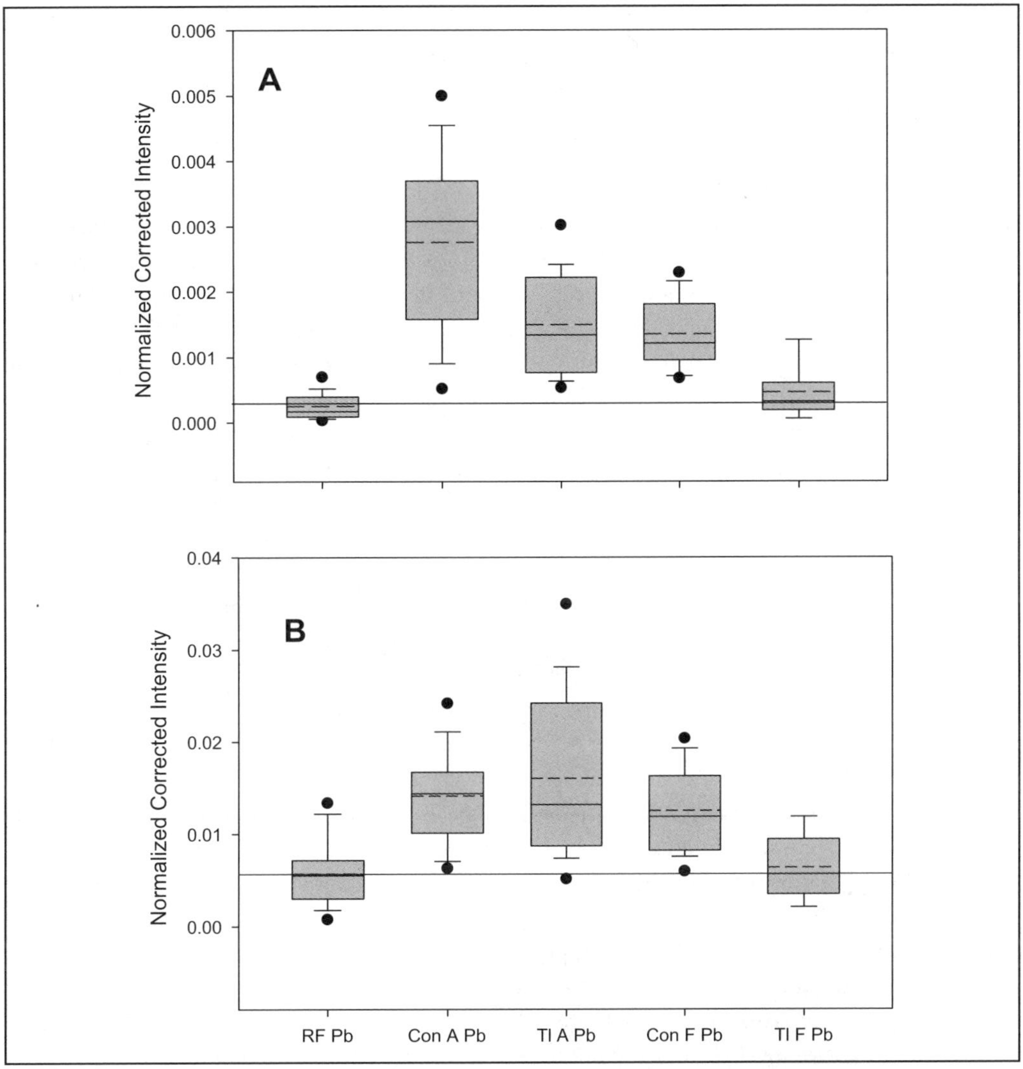

Figure 1. Vertical box plots for ToF-SIMS analyses of sphalerite surfaces from Con and Tail samples Brunswick Mines. (A) Cu distribution and (B) Pb distribution. Solid horizontal line in each box identifies median Cu or Pb content of the sample. The solid horizontal line across the entire plot identifies the mean Cu or Pb content in the sample group and is used as a base line.

limit ZnS activation were examined. A combined ToF-SIMS/XPS examination of sphalerite surfaces in copper plant samples with Cu^{2+} addition was performed to evaluate the response to the three depressants. Although copper transfer occurred under all test conditions, both the $ZnSO_4$ and TETA test samples indicate that Cu was partially inhibited from attaching to the surface of the sphalerite grains. The operational mechanisms however were probably different, the former likely related to the development of oxidative species on sphalerite grains (Khmeleva et al. 2006), the latter to the chelating capacity of TETA (Tukel et al. 2010). Sphalerite surfaces in both the $ZnSO_4$ and $NaHSO_3$ tests reported the highest proportion of species indicative of oxidation such as Zn

$(OH)_2$ and SO_3. It is possible that a greater development of sulfoxyl and hydroxide species on the surface of the sphalerites, as identified by ToF-SIMS, may inhibit collector attachment, produce hydrophilic surfaces, and, in combination, result in reduced sphalerite flotation (Khmeleva et al. 2006, Chandra & Gerson 2006).

ToF-SIMS: Principal Component Analysis (PCA)

In the most recent innovation, multivariate PCA an identify statistical relationships between secondary ions from species that contribute to variation in surface chemistry between minerals and between flotation streams, i.e., feed, concentrates and tail (Hart et al. 2006; Brito e Abreu & Skinner 2011). This form of analysis can be applied to an entire mapped area of ToF-SIMS data without the need to manually define regions of interest. As each set of correlated components are identified, this correlated data is removed from further analysis to enable further correlations to be defined. Each set of correlations is termed a principal component (PC) while the degree of correlation of each species within a given PC is termed the 'factor loading' (in both positive and negative sets in the PC). The factor loading is not a measure of the amount or intensity of that species present but of its correlation with other species in that PC. Many of these PCs immediately identify specific minerals with their associated, correlated surface species. An indication of the significance of a species identified in a given PC may be obtained by examining maps of the PC as compared to images of the actual mass fragment distribution (e.g., **Figures 2a**, b and c). In practice, the first PC 1 is found to be associated with topographic effects (like an SEM image) and does not hold any chemical information. This appears, after removal of the topography, in PC 2 onwards. This analysis can be processed in much shorter time (<1 h) compared with other techniques. Examples from PCA applied to ToF-SIMS data to provide chemical diagnosis capabilities illustrates the value of the method, The full set of data, combining mineralogy, liberation, solution speciation, microscopy (SEM), surface analysis and flotation testing, can be found in Gerson et al. (2012).

The initial flotation stage in the Kennecott Utah Copper Corporation (KUCC, Rio Tinto) concentrator flotation circuit is the bulk recovery (maximised) of copper and molybdenum sulphides, which are then separated via subsequent flotation steps. The Bingham Canyon porphyry copper deposit, the source of the ore, is geologically complex (Triffett and Bradshaw 2008) but can be simplified into limestone skarn (LSN) ore, containing economic concentrations of Cu sulphide minerals, and monzonite (MZ) ore, containing economic concentrations of both Mo and Cu sulphide minerals. It had been proposed, as a result of plant-based flotation observations that that blending of these two ore types leads to 'poisoning' of the flotation response. Laboratory flotation demonstrated that chalcopyrite and bornite recovery was near pro-rata in the blend but molybdenite recovery was substantially adversely affected (Gerson et al. 2012). PCA ToF-SIMS on the laboratory concentrates and tails from MZ, LSN and blended (70:30 respectively) feeds was used to explain these differences in behaviour.

PCA factor loadings from the MZ float experiments found a very strong correlation of all three Mo isotopes in PC 2 from Con 1 specifically identifying molybdenite surfaces (**Figure** 2d). The lack of significant correlation to other species, except a minor correlation to hydrophobic Cu-collector fragments (dicresyldithiophosphate), shows very clean molybdenite surfaces. In contrast, significant correlation of Cu isotopes with Fe, suggesting identification of bornite and chalcopyrite, is only obtained in PC 4 where this is correlated to both hydrophobic collector fragments and hydrophilic FeO, FeOH, K and Mg species. MZ Con 4 still displayed a very strong correlation of the

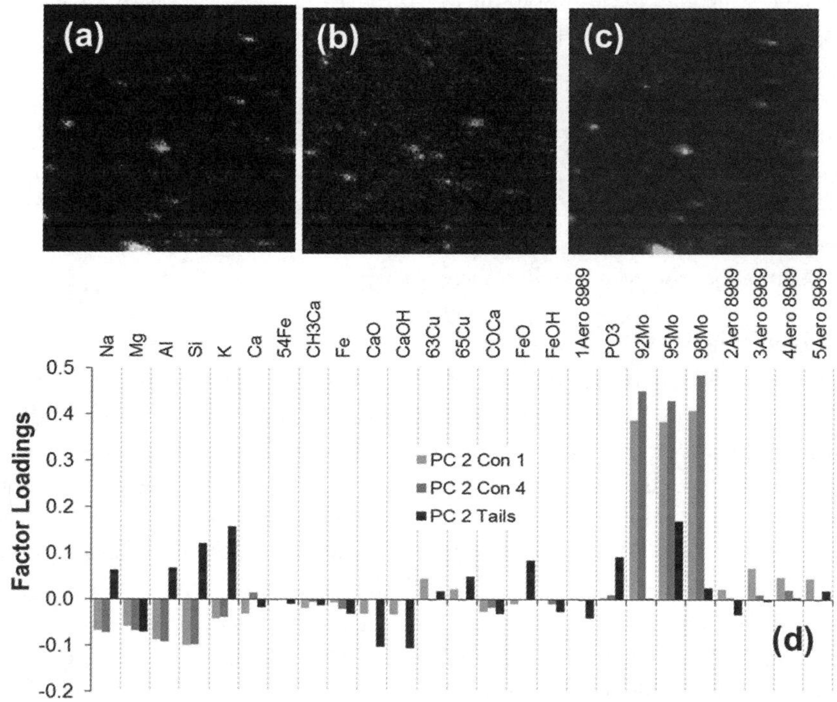

Figure 2. 500 × 500 mm images of MZ Con 1: (a) ToF-SIMS ^{98}Mo data; (b) ToF-SIMS ^{63}Cu data; and (c) PC 2 showing areas of positive factor loadings as pale. It is clear that there is a high degree of correspondence between PC 2 and ^{98}Mo. (d) PC 2 factor loadings for MZ float Cons 1 and 4 and tails.

Mo isotopes in PC2 (Figure 2d). In this later concentrate, no PCs identifying strong factor loadings for Cu and Fe were found suggesting that the surfaces of the copper containing minerals were largely obscured by oxidation and/or attached particles. For the MZ tail, all Mo isotopes were not found in PC2 (Figure 2d) and were only weakly selected in PC 4 but no Cu and Fe association were observed. These statistical correlations show that, while the surfaces of molybdenite were relatively clean in Cons 1 and 4, the surfaces of the Cu containing minerals carried a much higher hydrophilic loading and were substantially obscured.

Table 1. Absolute factor loadings for PC 4 MZ flotation experiment

Species	Con 1	Con 4	Tails
^{63}Cu	0.37	0.17	0.07
Fe	0.20	0.04	0.01
Ca	0.10	0.13	0.28
CH_3Ca	0.01	0.07	0.13
CaO	0.06	0.07	0.21
CaOH	0.01	0.14	0.21

A similar trend was found for the Cu-containing components of the LSN ore. An increasing surface loading of Ca-containing and other hydrophilic components was observed as demonstrated in Table 1. For the blended ore, PCA ToF-SIMS from copper containing minerals in concentrates and tails was similar to that for the MZ and LSN ores with a relatively high hydrophilic species surface loading which increased through the flotation process, i.e., from Con 1 to Con 4 to tails. However, the ToF-SIMS PCA analysis for the blended ore after Con 1 did not select Mo in any PC for either Con 4 or the tails, in contrast to the MZ ore analyses, showing either very low concentrations or high surface coverage of molybdenite.

These PCA ToF-SIMS analyses suggest that the copper-containing components within both the MZ and LSN ores were already associated with significant surface contamination so that, on blending, their flotation response was not significantly affected. However, the surface of the molybdenite component of the MZ ore was largely clean. On blending, partial transfer of the hydrophilic load in the LSN ore took place on to the MZ molybdenite resulting in apparent 'poisoning' of the flotation response of this component. Unless other factors apply, e.g., stock piling, throughput, there is no flotation benefit from blending these ore types.

In Situ Vibrational Spectroscopy

The combined limitations of traditional sampling methodologies (dry samples, or limited substates) in vibrational spectroscopy led current flotation researchers to adopt variations of FTIR sampling techniques and other forms of vibrational spectroscopy to enable information to be obtained from real mineral systems under flotation-related conditions (i.e., adsorption at the solid-liquid interface with control of solution conditions). The most widespread vibrational spectroscopy variant in this area is particle film ATR FTIR (McQuillan 2001). Rather than study the adsorption directly to the surface of an IR transparent crystal, layers of nano-sized mineral particles are deposited on the ATR crystal, forming a layer than is still thin enough to be probed by the IR evanescent wave (Cuba-Chiem et al. 2008; Beaussart et al. 2012). This procedure not only increases the range of substrates for analysis (including real mineral particles), it also increases the available surface area for adsorption, enabling detailed studies of surfactant (Khmeleva et al. 2006) and polymer adsorption (Kirwan et al. 2003) onto minerals.

In addition to particle film ATR FTIR, a similar variant of Raman spectroscopy, total internal reflection Raman (Woods and Bain 2012) has also been adopted for studies of flotation reagent adsorption (Beattie et al. 2006). The ability to minimize spectral interference due to water (water vibrations are much less intense in Raman than FTIR), and the electric field enhancements possible due to control of incident light angles, provides the technique with the ability to study complex systems such as collector mixtures (Beattie et al. 2006) and co-adsorption of surfactants (Woods et al. 2011). Surface enhanced Raman spectroscopy (SERS) has also been used to study collector adsorption on noble metal surfaces (with the same substrate limitations as RAIRS discussed above) (Woods et al. 2000), in addition to some novel studies using deposited metal nanoparticles to allow the study of collector adsorption on real mineral surfaces (Hope et al. 2007).

In addition to Raman spectroscopy, non-linear optical vibrational spectroscopic techniques (sum frequency generation spectroscopy—SFG) (Williams and Beattie 2002) are being applied to the study of adsorption on mineral surfaces (Wang et al. 2008). Although more complicated than Raman or FTIR (due to the need for advanced laser systems to generate tunable or broadband infrared light), the technique allows in situ analysis of adsorption to mineral-solution interfaces, is surface specific (i.e., no signal is generated from molecules in the bulk solution above the surface), and is highly sensitive to the presence of small amounts of adsorbates (detection limits of 1% of a monolayer). Given the low dosages/surface coverage of collectors in flotation circuits, the ability to probe adsorption and structure with higher surface sensitivity than FTIR and Raman is a distinct advantage. However, the power of the methodology is limited by a general requirement for analyzing flat surfaces (like RAIRS or traditional ATR FTIR).

Contact Angle Determination

The hydrophobic/hydrophilic ratio of species on each mineral particle determines the effectiveness of capture by the rising bubbles and, on average for a specific mineral phase, defines the three-phase (air/solution/gas) contact angle which is generally used in advanced flotation models to define the attachment and stability efficiencies (Ralston, Fornasiero & Grano 2007, Ralston et al. 2007, Eriksson & Yoon 2007). In the first use of hydrophobic/hydrophilic ratios to estimate contact angles on particular mineral phases (Piantadosi et al. 2001), contact angles on single mineral chalcopyrite, measured using the equilibrium capillary pressure technique across a packed bed of particles, were correlated with the amount of collector (dialkyl dithiophosphate, DTP) adsorbed from solution, the ToF-SIMS index DTP/S and flotation recovery at different collector dosages. These correlations were then used to estimate the contact angle of chalcopyrite particles in concentrates and tails from flotation of Ok Tedi ore at different collector dosages based on the statistical analysis of DTP/S ratios from 26 particles in the feed.

PCA ToF-SIMS has been used most recently by Brito e Abreu et al. (2010) and Brito e Abreu & Skinner (2011) to extend and simplify the estimates of contact angle in single mineral studies of chalcopyrite flotation. They have used geometrical representation of the observations in 2D scatterplots relating successive PCs for identifying patterns within the data and potential outliers. This method has shown that the contact angle of chalcopyrite can be predicted from just three secondary ions: oxygen, sulphur, and a thiol collector fragment. In these studies (Brito e Abreu et al. 2010), differences in surface species and contact angles with particle size (20–38, 75–105, and 150–210 μm) were also determined with a considerable spread of contact angles within each size range from 20° to 90°. For low contact angles, no apparent difference in surface chemistry between particle sizes was found but particles with high contact angles show decreasing S and collector signals with increasing hydrophilic gangue signals (e.g., Al, Na, K) as the particle size increases. Estimates of contact angles using these statistical ToF-SIMS approaches have been incorporated into flotation model predictions of rate constants for mineral particles in operating plants (Ralston et al. 2007a, b; Muganda et al. 2012).

THE FUTURE: NEW TECHNIQUES AND ANALYSIS

The structure and composition of minerals in froth flotation can also compromise both grade and recovery in the process. The excellent primary tools for mineral structure and liberation analysis, QEMSCAN and MLA, are well known and widely used (Smart et al. 2007). They are based on compositional identification of minerals from EDS databases. They do not, however, separate crystalline from amorphous phases, identify different structures with the same composition (e.g., sphalerite, wurtzite) or different elemental substitution and are practically limited to >5 μm particles. Recent developments, to be described, have extended this structural analysis.

There are a number of emerging technologies and analytical approaches which are poised to play an important role in the understanding of the chemical processes occurring during mineral processing. We discuss here a few that are founded on the application of synchrotron radiation. Synchrotron X-ray radiation offers a number of technical advantages over laboratory based X-ray sources. For general purposes these can be roughly placed into three categories. The much greater brilliance, often by a factor of 10^{10} or more; enables the use of in situ or time resolved measurements not possible in a laboratory. The high degree of collimation, which enables focusing and imaging with spatial resolution at times now reaching less than 10 nm (Ice et al. 2011), offers the opportunity to probe surfaces and structures with a resolution not before possible. The wide energy range offered by synchrotron radiation enables energy resolved or X-ray scanning experimentation thus offering new forms of analysis not practicably possible in a laboratory.

Mineralogy can now be probed using synchrotron radiation with a spatial resolution routinely of 1–2 μm. By using synchrotron microprobe analysis it is possible to measure three different but complementary types of information using the same instrumentation. X-ray fluorescence (XRF) maps can be measured to enable location of elements of interest. This is particularly useful where high value elements at relatively low concentrations (low ppm or even ppb) are the focus of interest in tailings or leach residue samples as a means for process optimisation and improved recovery.

These areas may then be subject to micro-diffraction or entire micro-diffraction maps may be collected. This enables correlation of mineral phases to value elements that may be present as solid solution. Such elemental forms are often missed by techniques such as QEMSCAN and MLA. Furthermore application of these SEM-based techniques can result in error due to the assumption of crystal phase based on elemental quantification, the basis for these analyses, rather than on actual measurement of structure. Where crystallite size is larger than incident X-ray beam size, as is becoming increasingly the case, single crystal diffraction may arise rather than powder diffraction, resulting in very limited diffraction information from any particular voxel. However, this may be overcome by the application of Laue (broad bandpass) diffraction which enables sampling of many diffraction planes simultaneously. While the analysis of such data is more complicated than for monochromatic X-ray diffraction data, considerable effort is currently being directed into developing this approach into a rapid and accurate tool which will in the near future be able to be applied to mineral processing and earth sciences analyses (e.g., the development of the FOXMAS analytical software http://sciencestudioproject.com/foxmas.php).

These micro-XRF and –XRD measurements may also be coupled to micro-X-ray absorption spectroscopy. This enables the local spatial and/or electronic environment of an element to be examined. XAS measurements are carried out by scanning the incident X-ray energy across a specific electron excitation energy of a specific element. The photoelectron emitted then interacts with the local surrounding atoms, up to 5 Å away, giving rise to an interference fringe. The region of up to 50 eV above the edge energy is sensitive to the oxidation state of the element of interest and highly localised electronic environment. This region of measurement is commonly referred to as X-ray absorption near edge spectroscopy (XANES). The interference fringe arising from more than 50 eV above the edge energy, called extended X-ray absorption fine structure (EXAFS), is more sensitive to local spatial environment. These types of measurements enable the geochemical location of elements to be probed which is of particular application for examining surface adsorption mechanisms, and interstitial and substitutional environments of trace value elements. XAS can be measured in a number of different ways, e.g., total absorbance, fluorescence, partial or total electron yield. Each of these gives rise to varying probe depths and are described in detail elsewhere (Gerson et al. 2009). A recent example of the application of these three techniques applied to the mineralogy of nickel laterites can be found at Fan and Gerson (2011).

Two other emerging synchrotron-based techniques are worthy of mention. These are the related techniques of photoemission electron microscopy (PEEM) and scanning photoemission microscopy (SPEM). These are highly spatially resolved surface sensitive techniques that may be applied to the examination of the distribution of surface species on mineral samples. However, in their current incarnation both techniques require flat mineral surfaces and in contrast to the previous techniques described in this section, both PEEM and SPEM require UHV environments. For both techniques the surface is subjected to photo-ionising radiation. For PEEM an entire surface area is irradiated simultaneously and the resulting excitation processes are imaged, whereas for SPEM the surface is raster scanned by a highly focused X-ray beam. Two approaches may be taken, with either the incident energy being varied with the intensity of the resulting emission intensity being measured, i.e., as for XAS, or the incident energy remains the same and a range of the kinetic energy of the excited electrons is measured as for X-ray photoelectron spectroscopy (XPS). Spatial resolution of less than 10 nm is now possible on the newest PEEM instrumentation. A recent example of the application of PEEM, coupled to synchrotron IR analysis deals with the examination of Cu activation of, and adsorption of xanthate onto pyrite (Chandra and Gerson 2012). In addition, PEEM has been used to study the evolution of surface chemistry on composite sulphide mineral surfaces, with direct evidence seen for galvanic interactions (Acres et al. 2010).

The improved sensitivity to structure and speciation coupled to high spatial resolution offered by synchrotron radiation based techniques offers potential for improved understanding of mineral flotation systems. Moreover with increasing accessibility, there are now more than 50 synchrotrons worldwide and costs (for IP restricted data collection only) that are competitive with top-end lab based analyses these approaches provide an as yet relatively untapped resource for minerals processing science and technology development.

SUMMARY

Poor flotation recovery or grade and misreporting losses can be caused by poor hydrophobic conditioning of the value mineral surfaces or inadvertent hydrophobic conditioning of gangue mineral surfaces. To define this surface chemistry, the hydrophobic/hydrophilic ratio by particle and as a statistical distribution between different mineral phases across a flotation circuit from roughers, scavengers, cleaners to tail is needed before changes to mechanical or chemical operation can be effective. Surface analysis using statistical methods in ToF-SIMS can now provide this information. Direct comparison of intensities and principal component analysis from hydrophobic (principally collectors) and hydrophilic species between feeds, concentrates and tails can identify surface species discriminating in bubble-particle attachment. In addition to MLA and QEMSCAN analysis, losses due to mineralogy (e.g., minor phases and solid solutions) can now be probed using both compositional and diffraction analysis in synchrotron radiation with a spatial resolution of 1–2 μm. These techniques open new opportunities for improvements in recovery and in unrealized value if they are used in plant surveys.

ACKNOWLEDGMENTS

The considerable assistance from Robert Dunne in the content, structure, and editing of this paper is sincerely appreciated.

REFERENCES

Acres, R.G., Harmer, S.L., and Beattie, D.A. 2010. Synchrotron PEEM and ToF-SIMS study of oxidized heterogeneous pentlandite, pyrrhotite and chalcopyrite, *J. Synch. Rad.* 17:606–615.

Allison, J.D.; Brown, D.S.; Novo-Gradac, K.J. 1990. MINTEQA2/PRODEFA2—A geochemical assessment model for environmental systems—version 3.0 user's manual: *Environmental Research Laboratory, Office of Research and Development, U.S. Environmental Protection Agency*, Athens, Georgia.

Asbjornsson, J., Kelsall. G.H., Pattrick, R.A.D., Vaughan, D.J., Wincott, P.L. and Hope, G.A. 2002a. Electrochemical behavior and surface characterization of enargite (Cu_3AsS_4) in 0.1M HCl. In: *Electrochemistry in Mineral and Metal Processing VI*, Doyle et al. (Editors), The Electrochemical Society, Pennington, NJ, 15–26.

Asbjornsson, J., Kelsall, G.H., Vaughan, D.J., Pattrick, R.A.D., Wincott, P.L. and Hope, G.A. 2002b. Electrochemical behavior and surface characterization of tennantite in 0.1M HCl. In: *Electrochemistry in Mineral and Metal Processing VI*, Doyle et al. (Editors), The Electrochemical Society, Pennington, NJ, 37–47.

Basilio, C.I., Leppinen, J.O., Yoon, R.H., Nagaraj, D.R. and Wang, S.S. 1988. Flotation and adsorption studies of modified thionocarbamates on sulfide minerals. Preprint 88-156, SME, Littleton, CO.

Basilio, C.I., Kim, D.S., Yoon, R.H., Leppinen, J.O., and Nagaraj, D.R. 1992. Interaction of thiophosphinate collectors with precious metals. Preprint 92-174, SME, Littleton, CO.

Beattie, D.A., Larsson, M.L., and Holmgren, A.R. 2006. In situ total internal reflection Raman spectroscopy of surfactant adsorption at a mineral surface, *Vib. Spectrosc.* 41:198–204.

Beaussart, A., Petrone, L., Mierczynska-Vasilev, A., McQuillan, A.J. and Beattie, D.A. 2012. In situ ATR FTIR study of dextrin adsorption on anatase TiO_2, *Langmuir*, 28:4233–4240.

Binnig, G., and Röhrer, H. 1982. *Helv. Phys. Acta* 55:726– .

Binnig, G., Quate C.F., and Gerber, C. 1986. *Phys. Rev. Lett.* 56:930-.

Bozkurt, V., Brienne, S.H.R., Xu, Z., Rao, S.R., Butler, I.S., and Finch, J.A. 1996. Development of in situ external reflection spectroscopy technique for adsorbed films, *Miner. Eng.* 9:351–355.

Brinen, J.S., and Reich, F. 1992. Static SIMS imaging of the diisobutyl dithiophosphinate on galena surfaces, *Surf. Interface Anal.* 18:448–452.

Brinen, J.S., Greenhouse, S., Nagaraj, D.R., and Lee, J. 1993. SIMS and SIMS imaging studies of adsorbed dialkyl dithiophosphinates on PbS crystal surfaces. *Int. J. Miner. Process.* 38:93–109.

Brion D. 1980. Etude par spectroscopie de photoelectrons de la degradation superficielle de FeS2, ZnS et PbS a l'air et dans l'eau. *Appl. Surf. Sci.* 5: 133–152.

Brito e Abreu, S., Brien, C., and Skinner, W. 2010. ToF-SIMS as a new method to determine the contact angle of mineral surfaces. *Langmuir* 26:8122–8130.

Brito e Abreu, S., and Skinner, W. 2011. ToF-SIMS-derived hydrophobicity in DTP flotation of chalcopyrite: contact angle distributions in flotation streams. *Int. J.Miner. Process.* 98:35–41.

Buckley A.N. 1994. A survey of the application of X-ray photoelectron spectroscopy to flotation research, *Colloids Surf.* 93:159–172.

Buckley A.N., and Riley K.W. 1991. The self-induced floatability of sulfide minerals: examination of recent evidence for elemental sulfur as being the hydrophobic entity. *Surf. Interf. Anal.* 17: 655–659.

Buckley A.N., and Woods R. 1994. On the characterisation of sulfur species on sulfide mineral surfaces by XPS and Raman spectroscopy. *J. Electroanal. Chem.*, 370: 295–296.

Chandra, A.P., and Gerson, A.R. 2006. A review of the fundamental studies of the copper activation mechanisms for selective flotation of the sulfide minerals, sphalerite and pyrite. *Adv. Colloid Interf, Sci.* 145: 97–110.

Chandra, A.P., Puskar, L., Simpson, D.J., and Gerson A.R. 2012. Copper and xanthate adsorption onto pyrite surfaces: Implications for mineral processing. *Int. J. Miner. Process.* 114–117:16–26.

Chryssoulis, S.L., and Cabri, L.J., 1990. The significance of gold mineralogical balances in mineral processing. *Inst. Min. Metall. (London), Trans. Sec.* 99:C1-C10.

Chryssoulis, S.L., Chauvin, W.J., and Surges, L.J. 1986. Trace element analysis by secondary ion mass spectrometry with particular reference to the Brunswick sphalerite. *Can. Metall. Quart.* 25: 233–239.

Chryssoulis, S.L., Cabri, L.J., and Salter, R.S. 1987. Direct determination of refractory gold in sulphide ores. In *Proc. Int. Symp. Gold Metallurgy 1*, Edited by R.S. Salter, D.M. Wyslouzil and G.W. McDonald. 235–244.

Chryssoulis, S.L., Stowe, K.G., Niehuis, E., Cramer, H.G., Bendel, C., and Kim J.Y. 1995. Detection of collectors on concentrator mineral grains by time of flight secondary ion mass spectrometry TOF-SIMS. *Trans. Inst. Min. Metall., Sect. C* 104:141–150.

Chryssoulis, S.L., Weisener, C.G., and Dimov, S. 1995. Detection of mineral collectors by TOF-LIMS. In *Proceedings, Secondary Ion Mass Spectrometry SIMS X.* Edited by A. Benninghoven, B. Hagenhoff, H.W. Werner, John Wiley & Sons, 899–902.

Clarke, N.S., Ruckman, J.C., and Davey, A.R. 1986. Laser microprobe mass spectrometry of solid surfaces. In *Proc. Third Int. Laser Microprobe Mass Spectrometry Workshop*. Eds. F. Adams and L. van Vaeck (University of Antwerp, Belgium).

Clifford, R.K., Purdy, K.L. and Miller, J.D. 1975. Characterization of sulfide mineral surfaces in froth flotation systems using electron spectroscopy for chemical analysis. *AIChe, Symp. Ser. 71*, 150:138–147.

Cross, W.M. and Miller, J.D. 1993, Adsorbed hydrocarbon chain conformation and its effect on hydrophobicity. Preprint 93–213, SME, Littleton, CO.

Cross, W.M., Kellar, J.J. and Miller, J.D. 1992, Fourier transform near-infrared examination of the coagel-to-micelle transition for sodium laurate. *Appl. Spectrosc.* 46:701–704.

Cuba-Chiem, L.T., Huynh, L., Ralston, J., and Beattie, D.A. 2008.In situ particle film ATR FTIR spectroscopy of carboxymethyl cellulose adsorption on talc: Binding mechanism, pH effects, and adsorption kinetics, *Langmuir*, 24:8036–8044.

Eadington, P. 1974. Use of Auger-electron spectroscopy to determine the composition of fracture surfaces of minerals. *Trans. Inst. Min, Metall.* 83:C223–C227.

Eadington P. 1977. Study of the oxidation layers on surfaces of chalcopyrite by use of Auger electron spectroscopy. *Trans. Inst. Min. Metall.* 86:C186–189.

Eggleston, C.M. and Hochella M.F.Jr. 1992. Tunneling spectroscopy applied to PbS(001) surfaces: fresh surfaces, oxidation and sorption of aqueous Au. *Amer. Mineral.* 78:877–883.

Eriksson, J.C. and Yoon, R-H. 2007. The nature of hydrophobic attraction forces, In *Froth Flotation: a Century of Innovation*, Edited by Fuerstenau, M.C., Jameson, G., Yoon, R.-H., SME Publications, Littleton, Colorado pp. 133–178.

Fan R., and Gerson, A.R. 2011. Nickel geochemistry of a Phillipine laterite examined by bulk and microprobe synchrotron analyses. *Geochim. Cosmochim. Acta* 75:6400–6415.

Francis, S.A., and Ellison, A.H. 1961. Reflection infrared studies of zinc dialkyl dithiophosphate films adsorbed on metal surfaces, *J. Chem. Eng. Data*, 6:83–86.

Fredriksson, A., and Holmgren, A. 2007. An in situ ATR-FTIR study of the adsorption kinetics of xanthate on germanium, *Colloids Surf. A.* 302:96–101.

Free, M.L. and Miller, J.D. 1994. Effect of sample and incident beam areas on quantitative spectroscopy, *Appl. Spectrosc.* 48:891–893.

Free, M.L. and Miller, J.D. 1996. The significance of collector colloid adsorption phenomena in the fluorite/oleate flotation system as revealed by FTIR/IRS and solution chemistry analysis. *Int. J. Miner. Process.*, 48:197–216.

Fuerstenau, D.W., and Chander, S. 1982. Industrial applications of surface analysis, *ACS Symp. Series*, 199:283–312.

Fuerstenau, D.W. and Fuerstenau, M.C. 1956. Ionic size in flotation collection of alkali halides. *Mining Engineering* 8:302–306.

Gärd, R., Holmgren, A., and Forsling, W. 1997.Spectroscopic studies of dextrin adsorption onto colloidal ZnS, *J. Colloid Interf. Sci.* 194:319–325.

Gaudin, A.M. and Fuerstenau, D.W. 1955. Quartz flotation with cationic collectors. *Trans. AIME*, 202:958–962.

Gerson, A.R.; Jasieniak, M. 2008. The effect of surface oxidation on the Cu activation of pentlandite and pyrrhotite. In *Proceedings XXIV International Minerals Processing Congress*, Edited by W.D. Guo, S.C. Yao, W.F. Liang, Z.L. Cheng, H. Long; Science Press Beijing, 1054–1063.

Gerson, A.R., Lange, A.G., Prince, K.P., and Smart. R.St.C. 1999. The mechanism of copper activation of sphalerite. *Appl. Surface Sci.* 137:207–223.

Gerson, A.R., Smart R.St.C., Li, J., Kawashima, N., Weedon, D., Triffett, B., and Bradshaw D. 2012. Diagnosis of the surface chemical influences on flotation performance: Copper sulfides and molybdenite. *Int J Miner Process.* 16:106–109.

Gerson, A.R., Cookson D.J., and Prince, K. 2009. Synchrotron based techniques. In *Handbook of Surface and Interface Analysis Methods for Problem-Solving*. Edited by J.C. Rivière and S. Myhra. CRC Press ISBN: 978-0-8493-7558-3, pp. 193–222.

Gow, R., Huang, H., Young, C., Takasaki, Y., and Hope. G. 2012. Electrochemistry of enargite I: reactivity in alkaline solutions. *Electrometallurgy 2012*, M. Free et al. Editors, TMS, Warrendale, PA, 217–225.

Gow, R., Huang, H., Young, C., and Hope. G. 2013. Electrochemistry of enargite II: reactivity in acidic solutions. *Min. & Met. Process.* (submitted).

Grano, S. Ralston J. and Smart R.St.C. 1990. Influence of electrochemical environment on the flotation behaviour of Mt. Isa copper and lead-zinc ore., *Int. J. Min. Proc.* 30:69–97.

Greenler, R.G. 1962. An infrared investigation of xanthate adsorption by lead sulfide, *J. Phys. Chem.* 66:879–883.

Greenler, R.G. 1969. Reflection method for obtaining the infrared spectrum of a thin layer on a metal surface, *J. Chem. Phys.* 50:1963–1968.

Greenler, R.G. 1996. Infrared study of adsorbed molecules on metal surfaces by reflection techniques, *J. Chem. Phys.* 44:310–315.

Halim, C.E.; Short, S.A.; Scott, J.A; Amal, R.; Low, G. 2005. Modelling the leaching of Pb, Cd, As, and Cr from cementitious waste using PHREEQC. *J. Hazard. Mat.*125:45–61.

Hancer, M., Celik, M.S., and Miller, J.D. 2000. The significance of interfacial water in soluble salt flotation systems. *J. Coll. Interf. Sci.*, 235:150–161.

Harrick, N.J. 1967. Internal reflection spectroscopy, Harrick Scientific Corp, Ossining, N.Y.

Hart, B., Biesinger, M.C. and Smart. R. St.C. 2006. Improved statistical methods applied to surface chemistry in minerals flotation, *Miner. Eng.* 19:790–798.

Hope, G.A., Buckley, A.M., Munce, C.G., and Woods, R. 2007. Gold enhanced spectroelectrochemical investigation of 2-mercaptobenzothiazole, isopropyl xanthate and butylethoxycarbonylthiourea adsorption on minerals, *Miner. Eng.*, 20:964–969.

Huang, H. 2012. *StabCal, Software for calculating stability diagrams from thermodynamics.* Metallurgical & Materials Engineering, Montana Tech, Butte, MT.

Huang, H., Twidwell, L. and Young, C. 2005. Speciation for aqueous systems—an equilibrium calculation approach. In: Computational Analysis in Hydrometallurgy, D.G. Dixon et al. (Editors), CIM, Westmount, QC, 295–310.

Ice, G.E., Budai, J.D., and Pang J.W.L. 2011. The race to X-ray microbeam and nanobeam science. *Science* 334(6060):1234.

Jeffrey, M.I., Chandra, I., Ritchie, I.M., Hope, G.A., Watling, K. and Woods, R. 2005. Innovations in gold leahing research and development. In: *Innovations in Natural Resource Processing*, C.A. Young et al. (Editors), SME, Littleton, CO, 207–222.

Kellar, J.J., Cross, W.M., and Miller, J.D., 1990, In-situ internal reflection spectroscopy of surfactant adsorption reactions using reactive internal reflection elements, *Sep. Sci. & Tech.*, 25: 2133–2155.

Kellar, J.J., Young, C.A., Knutson, K., and Miller, J.D., 1991, Thermotropic phase transitions of adsorbed oleate species at a fluorite surface by in-situ FI –IR/IRS spectroscopy. *J. Coll. Interf. Sci.*, 144:381–389.

Khmeleva, T.N., Chapelet, J.K., Skinner, W.M., and Beattie, D.A. 2006. Depression mechanisms of sodium bisulfite in the xanthate-induced flotation of copper activated sphalerite, *Int. J. Miner. Process.* 79(1):61–75.

Kim, B.S., Hayes, R.A., Prestidge, C.A., Ralston, J., and Smart, R.St.C. 1995. Scanning tunneling microscopy studies of galena: The mechanisms of oxidation in aqueous solution. *Langmuir* 11:2554–2562.

Kirwan, L.J., Fawell, P.D., and van Bronswijk, W. 2003. In situ FTIR-ATR examination of poly(acrylic acid) adsorbed onto hematite at low pH, *Langmuir*, 19:5802–5807.

Laajalehto, K., Leppinen, J., Kartio, I., and Laiho, T. 1999. XPS and FTIR study of the influence of electrode potential on activation of pyrite by copper or lead. *Coll. Surf. A.* 154:193–199.

Lander J.J. 1953. Auger peaks in the energy spectra of secondary electrons from various materials. *Phys.Rev.* 91:1382–1387.

Larsson, M.L., Holmgren, A., and Forsling, W. 2000. Xanthate adsorbed on ZnS studied by polarized FTIR-ATR spectroscopy, *Langmuir* 16:8129–8133.

Larsson, M.L., A. Holmgren, A., and W. Forsling, W. 2001. Structure and orientation of collectors adsorbed at the ZnS/water interface, *J. Colloid Interf. Sci.* 242:25–30.

Leppinen, J.O. 1990. FTIR and flotation investigation of the adsorption of ethylxanthate on activated and non-activated sulfide minerals. *Int. J. Min. Proc.* 30:245–263.

Leppinen, J.O., Basilio, C.I. and Yoon, R.H. 1988. In-situ FTIR spectroscopic study of ethylxanthate electrosorption on sulfide minerals. In: *Electrochemistry in Mineral and Metal Processing*, P.E. Richardson and R. Woods (Editors), The Electrochemical Society, Pennington, NJ, 49–65.

Leppinen, J.O., Basilio, C.I. and Yoon, R.H. 1989. *Int. J. Min. Proc.* 26:259.

Lovering, J.F. 1975. Applications of SIMS microanalysis techniques to trace element and isotopic studies in geochemistry and cosmochemistry. In: *Workshop on secondary ion mass spectrometry (SIMS) and ion microprobe analysis:* Washington D.C. Natl. Bur. Standards Spec. Publ. 427, 135–178.

Malysiak, V., O'Connor, C.T., Ralston, J., Gerson, A.R., Coetzera, L.P. and Bradshaw, D.J. 2002. Pentlandite–feldspar interaction and its effect on separation by flotation. *Int. J. Miner. Process.* 66:89–106.

McQuillan, A.J. 2001. Probing solid-solution interfacial chemistry with ATR-IR spectroscopy of particle films, *Adv. Materials*, 13:1034–1038.

Mielczarski, J.A. 1986. In-situ ATR-IR spectroscopic study of xanthate adsorption on marcasite, *Coll. Surf.* 17: 251–271.

Mielczarski, J.A. 1993. External reflection infrared spectroscopy at metallic, semiconductor, and nonmetallic substrates. 1. Monolayer films, *J. Phys. Chem.* 97:2649–2663.

Mielczarski, J.A., 1994. Ex-situ and in-situ characterization of surfactants monolayers adsorbed on nonmetallic substrates by infrared reflection spectroscopy. *Surf, Interface Anal.* 22:162.

Mielczarski, J., and Leppinen, J. 1987. Infrared reflection-absorption spectroscopic study of adsorption of xanthates on copper, *Surf. Sci.* 187:526–538.

Mielczarski, J., Nowak, P., Strojek, J.W., and Pomianowski, A. 1979. Infrered internal reflection spectroscopic investigations of potassium ethyl-xanthate sorption on sulphide minerals. In: J.S. Laskowski (Editor), *Proceedings of 13th IMPC*, 35–53.

Mielczarski, J.A., Cases, J.M., Bouquet, E., Barres, O., and Delon, J.F. 1993. Nature and structure of adsorption layer on apatite contacted with oleate solution. 1. Adsorption and Fourier transform infrared reflection studies, *Langmuir*, 9:2370–2382.

Mielczarski, J., Cases, J.M., Mielczarski, E., Alnot, M., Ehrhardt, J.J. and Franco, A. 1995. Surface products of water and xanthate interaction with copper sulfide minerals: selective separation. In: *Proceedings of XIX IMPC* 3:29–34.

Miller, J.D., Yalamanchili, M.R., and Kellar, J.J., 1992. Surface charge of alkali halide particles as determined by laser-Doppler electrophoresis. *Langmuir*. 8:1464–1469.

Muganda, S. Zanin, M. and Grano, S.R. 2012. Benchmarking the flotation performance of ores, *Miner. Eng.* 26:70–79.

Nesset, J.E., 2011. Metallurgical and material processing, In: *The Canadian Metallurgical and Materials Landscape 1960–2011*, Edited by J. Kapusta, P. Mackey and N. Stubina. The Canadian Institute of Mining, Metallurgy and Petroleum, p. 497.

Olsen, C., Makni, S., Hart B., Laliberty, M., Pratt, A., Blatter, P., Lanouette M. 2012. Application of surface chemical analysis to the industrial flotation process of a complex sulphide ore, *Proceedings of the 26th. International Mineral Processing Congress (IMPC 12)*, September 24–28, 2012, New Delhi, Paper 771.

Parker, G., Woods, R., and Hope, G., 2003. Raman investigation of sulfide leaching. In: *Hydrometallurgy 2003*, C.A. Young et al. (Editors), TMS, Warrendale, PA, 447–460.

Peng, Y., Bo. W.; and Gerson, A. 2012. The effect of electrochemical potential on the activation of pyrite by copper and lead ions during grinding. *Int. J. Miner. Process.*102:141–149.

Piantadosi, C., Jasieniak, M., Skinner W.M., and Smart. R.St.C. 2000. Statistical comparison of surface species in flotation concentrates and tails from ToF-SIMS evidence. *Miner. Eng.* 13:1377–1394.

Piantadosi, C., Pyke, B.L. and Smart. R.St.C. 2001. TOF-SIMS statistical analysis of surface hydrophobic-hydrophilic species ratios and contact angle estimation in real ore systems. In *Interactions in Mineral Processing*. Edited by Finch, J.A. Rao, S.R. & Huang, L. Toronto Metallurgical Society, pp. 305–323.

Piantadosi, C. and Smart, R.St.C. 2002. Statistical comparison of hydrophobic and hydrophilic species on galena and pyrite particles in flotation concentrates and tails from TOF-SIMS evidence. *Int. J. Miner. Process.* 64:43–54.

Raju, G.B., Holmgren, A., and Forsling, W. 1997. Adsorption of dextrin at mineral/water interface, *J. Colloid Interf. Sci.* 193:215–222.

Ralston, J. 1994. Bubble-particle capture, In *Flotation II.* Edited by S. Castro and J. Alvarez. Chile: Publ. Andros, 2:1464.

Ralston, J., Fornasiero, D. and Grano, S. 2007a. Pulp and solution chemistry, In: *Froth Flotation: A Century of Innovation*, Edited by Fuerstenau, M.C., Jameson, G., Yoon, R.-H., SME Publications, Littleton, Colorado, pp. 227–258.

Ralston, J., Fornasiero, D., Grano, S., Duan, J., and Akroyd, T. 2007b. Reducing uncertainty in mineral flotation-flotation rate constant prediction for particles in an operating plant ore. *Int. J. Miner. Process.* 84:89–98.

Ralston, J. and Healy, T.W. 1980. Activation of zinc sulphide with CuII, CdII and PbII: II. Activation in neutral and weakly alkaline media, *Int. J. Miner. Process.* 7(3):203–217.

Shaff, J.O., Schultz, B.A., Craft., E.J., Clark, R.T., and Kochian, 2010. L.V. GEOCHEM-EZ: a chemical speciation program with greater power and flexibility. *Plant Soil.* 330:207–214.

Shimizu, N., Semet, M.O., and Allegre, C.J. 1978. Geochemical applications of quantitative ion-microprobe analysis. *Geochim. Cosmochim. Acta*, 42:1321–1334.

Siegbahn, K., Nordling, C., Fahlman, A., Nordberg, R., Hamrin, A., Hedman, J., Johansson, G., Bergman, T., Karlson, S., Lindgren, I., & Lindberg, B., 1967. *Electron Spectroscopy for Chemical Analysis—Atomic, Molecular and Solid State Structure by Means of Electron Spectroscopy.* Almquist and Wiksells, Stockholm.

Smart, R.St.C., 1991. Surface layers in base metal sulphide flotation. *Miner. Eng.* 4:891–909.

Smart, R.St.C., W.M. Skinner, and A.R. Gerson. 1999. XPS of sulfide mineral surfaces: Metal-deficient, polysulfides, defects and elemental sulfur. *Surf. Interface Anal.* 28:101–105.

Smart, R.St.C., M. Jasieniak, K.E. Prince, and W.M. Skinner. 2000. SIMS studies of oxidation mechanisms and polysulfide formation in reacted sulfide surfaces. *Miner. Eng.* 13:857–870.

Smart, R.St.C., Amarantidis, J., Skinner, W.M., Prestidge, C.A., LaVanier, L., and Grano, S.G. 2003a. Surface analytical studies of oxidation and collector adsorption in sulfide mineral flotation. In: *Topics in Applied Physics. Vol. 85. Solid-Liquid Interfaces.* Edited by Wandelt K., Thurgate S., Springer-Verlag, Berlin, pp. 3–60.

Smart, R.St.C., Jasieniak, M., Piantadosi, C., and Skinner, W.M. 2003b. Diagnostic surface analysis in sulfide flotation. In *Flotation and Flocculation: From Fundamentals to Applications*, Edited by Ralston, J., Miller, J.D., Rubio J., Publ. Ian Wark Research Institute, University of South Australia, pp. 241–248.

Smart, R.St.C., Skinner, W.M., Gerson, A.R., Mielczarski, J., Chryssoulis, S., Pratt, A.R., Lastra, R., Hope, G.A., Wang, X., Fa, K., and Miller, J.D. 2007. Surface characterisation and new tools for research. In *Froth Flotation: A Century of Innovation.* Edited by Fuerstenau, M.C., Jameson, G., Yoon, R.-H., SME Publications, Littleton, Colorado, pp. 283–338.

Stowe, K.G., Chryssoulis, S.L., Kim, J.Y. 1995. Mapping of composition of mineral surfaces by TOF-SIMS. *Miner. Eng.* 8:421–430.

Sui, C.C., Lee, D., Casuge, A., and Finch, J.A. 1999. Comparison of the activation of sphalerite by copper and lead ions. *Miner. Metall. Process.* 16(2):53–61.

Talonen, P., Rastas, J., and Leppinen, J. 1991. In-situ FTIR study of ethyl xanthate adsorption on gold, silver and copper electrodes under controlled potential. *Coll. Surf. A.* 17:669–674.

Toran, L. and Grandstaff, D. 2002.PHREEQC and PHREEQCI: Geochemical modeling with an interactive interface. *Ground Water* 40:462–464.

Triffett, B. and Bradshaw, D. 2008. The role of morphology and host rock lithology on the flotation behaviour of molybdenite at Kennecott Utah Copper. *Proc. 9th Int. Congress Appl. Mineral. (ICAM 2008)*, AusIMM Publication Series, pp. 465–473.

Tukel, C., Kelebek, S., and Yalcin, E. 2010. Eh-pH stability diagrams for analysis of polyamine interaction with chalcopyrite and deactivation of Cu-activated pyrrhotite. *Canad. Metall. Quart.* 49:411–418.

Wang, X., Liu, J. and Miller, J.D. 2008. Adsorption and self-assembly of octyl hydroxamic acid at a fluorite surface as revealed by sum-frequency vibrational spectroscopy, *J. Colloid Interf. Sci.* 325:398–403.

Williams, C.T., and Beattie, D.A. 2002. Probing buried interfaces with non-linear optical spectroscopy, *Surf. Sci.* 500:545–576.

Woods, D.A., and Bain, C.D. 2012. Total internal reflection Raman spectroscopy, *Analyst*, 137:35–48.

Woods, D.A., Petkov, J., and Bain, C.D. 2011. Surfactant adsorption kinetics by total internal reflection raman spectroscopy. 2. CTAB and triton X-100 mixtures on silica, *J. Phys. Chem. B* 115:7353–7363.

Woods, R., Hope, G.A., and Brown, G.M. 1998. Spectroelectrochemical investigations of the interaction of ethyl xanthate with copper, silver and gold: III SERS of xanthate adsorbed on gold surfaces. *Coll & Surf. A.* 137:329–337.

Woods, R. Hope, G.A., and Watling, K. 2000. SERS spectroelectrochemical investigation of the interaction of 2-mercaptobenzothiazole with copper, silver and gold surfaces, *J. Appl. Electrochem.* 30:1209–1222.

Woods, R. Hope, G.A., and Watling, K. 2002. SERS investigation of the interaction of the flotation collectors—diisobutyldithiophosphinate and butylethoxycarbonylthiourea—with coinage metal surfaces. In: *Electrochemistry in Mineral and Metal Processing VI*, Doyle et al. (Editors), The Electrochemical Society, Pennington, NJ, 15–26.

Woods, R., Young, C.A. and Yoon, R.H. 1990. Ethyl xanthate chemisorption isotherms and E_H-pH diagrams for the copper/water/xanthate and chalcocite/water/xanthate systems. *Inter. J. Miner. Proc.* 30:17–33.

Yalamanchili, M.R., Kellar, J.J. and Miller, J.D. 1993. Adsorption of collector colloids in the flotation of alkali halide particles. *Int. J. Miner. Proc.* 39:137–153.

Young, C.A. and Miller, J.D. 1999. In-situ FT-IR/IRS and MLRS examination of adsorbed oleate at fluorite and calcite surfaces. In: L.J. Cabri, C.H. Bucknam, E.B. Milosavljevic, S.L. Chryssoulis and R.A. Miller (Editors), *Analytical Technology in the Mineral Industries*, TMS, Warrendale, PA, 119–136.

Young, C.A. and Miller, J.D. 2002a. Conformation of chemisorbed oleate adsorbed at a calcite surface, Preprint 02-28, SME, Littleton, CO.

Young, C.A. and Miller, J.D. 2002b. Thermodynamic Evaluation of Oleate Chemisorption at calcium semi-soluble salt surfaces, Preprint 02-29, SME, Littleton, CO.

ABOUT THE AUTHORS

Roger St.C. Smart is emeritus professor in minerals and materials science and technology at Mawson Institute, University of South Australia, where he founded the SA Surface Technology Centre in 1987 and was deputy director of the Ian Wark Research Institute from 1995 to 2002. Smart currently leads AMIRA where company projects focus on geochemical and surface chemical control of minerals separation by flotation, reactions, and particle aggregation in tailings and waste rock systems with particular expertise in techniques for surface compositional and structural analysis.

Andrea R. Gerson is the leader of the research concentration minerals and materials science & technology within the Mawson Institute. Gerson has published 135 peer-reviewed book chapters, reviews, journal articles, and conference papers. She was a member of the Australian Research Council College of Experts from 2008 to 2010. Gerson's scientific expertise lies in the influence of structure, both spatial and electronic, and surface chemistry on reactivity of minerals and materials. Specific areas of research are in synchrotron science and structure, reactivity, hydrometallurgy, flotation, precipitation, and crystallisation of minerals.

Brian R. Hart is an adjunct faculty member in the Department of Earth Sciences at Western University (UW). Hart is director of the mineral chemistry and mineral processing group at Surface Science Western, WU. His expertise lies in contract analytical services, process optimization, and industry sponsored as well as independent research. Hart has written more than 40 refereed publications, 65+ conference proceedings, and 3 book chapters.

David A Beattie obtained his BSc (Hons) degree in chemical physics and PhD in chemistry from the University of Edinburgh. After completing his studies, he undertook postdoctoral research at the University of Oxford, during which time he was awarded a British Ramsay research fellowship. He took up the position of research fellow at the Ian Wark Research Instituteat the University of South Australia in November 2000. He was promoted to the level of senior research fellow in January 2007, and to associate research professor in January 2011. Beattie's research spans a range of areas including nano- and biolubrication, the surface chemistry of mineral processing and waste management, and novel stimulus-responsive coatings. The main aim of his work is to link the adsorption and structure of molecules at surfaces with the ability of those molecules to alter the macroscopic interfacial properties of materials (wettability, reactivity, friction). The objective of his research is to gain the ability to control and manipulate surface properties for a range of scientific and industrial outcomes.

Courtney Young is department head and Lewis S. Prater distinguished professor of metallurgical & materials engineering at Montana Tech of The University of Montana. He is a graduate of three premiere mineral/coal processing and extractive metallurgy institutions, having obtained his BS in mineral processing engineering from then Montana College of Mineral Science and Technology in 1984, his MS in mining and minerals engineering from Virginia Polytechnic Institute and State University in 1987, and his PhD in metallurgical engineering from the University of Utah College of Mines and Earth Sciences in 1995. He has been teaching, conducting research, and serving the community and profession at Montana Tech for 20 years. Young is a member of SME, TMS, AIST, NWMA, MMA, and MMSA. He is an honorary member of MMA and registered as a qualified professional with MMSA. In 2009, he was honored by AIME with the Frank F. Aplan Award for his dedication to mineral and coal processing education and research as well as SME with the President's Citation for his leadership and contributions to the society and profession.

Innovations in Analytical Chemistry—The Key to Metallurgical Advancement

Meg Dietrich
Newmont Mining Corporation, Englewood, Colorado, United States

ABSTRACT: Innovation is the development of new solutions to meet the demands of the times. In the mining industry innovations have been driven by government regulatory changes, economic influences and the demand for metallurgical process improvement. Analytical chemistry is the force behind metallurgical advancement and discovery. Measurement is the key to understanding what you have achieved.

The analysis of metals was among the earliest applications of analytical chemistry. While fire assay for analysis of precious metals has changed very little in the past 500 years, the majority of analytical equipment in the laboratory has changed dramatically in the past 60 years. As a result assays and analyses which took days to complete, changed to hours, hours improved to minutes, and minutes to seconds in some cases. These improvements came through the discovery and optimization of electronic instrumentation; particularly flame Atomic Absorption Spectrometer, Inductively Coupled Plasma–Optical Emission Spectrometer, Inductively Coupled Plasma-Mass Spectrometer, X-Ray Fluorescence and quick combustion gas analyzers.

INTRODUCTION

While the early alchemists may not have succeeded in turning base metals into gold or discovering the elixir of life, they did develop a framework of theory, terminology, experimental process and basic laboratory techniques that are still recognizable today. Robert Boyle, an alchemist considered to be the father of modern chemistry, published *The Sceptical Chymist* in 1661. This was an important early reference text in which he hypothesizes that matter consists of atoms in motion and that every phenomenon was the result of collisions of particles in motion. His innovative thinking and experimentation proved that there were more chemical elements than the classic four, earth, fire, air and water. He challenged Aristotle's theory which was being taught at universities.

The practice of alchemy declined throughout the 18th century as modern chemistry began to form, using more rational principles but the early alchemists with their mystical notions contributed to the industries of today, including dyes, ceramics, glass, blasting reagents and especially to metalworking, ore testing and refining.

De Re Metallica, a Latin text published in 1556 by Georgius Agricola, was an important chemistry text for the period and was used for mining, refining and smelting for nearly 200 years. It was translated to English in 1912 by former U.S. President Herbert Hoover, a mining engineer, and his wife Lou Henry Hoover, a geologist and Latinist. The text describes in detail the assaying of ore using flux with litharge in a crucible, fusing in a muffle followed by scorifying. The metal is transferred to a preheated cupel. Silver is parted from gold using nitric acid. This all sounds very familiar and after nearly 500 years an assayer today follows similar steps (Agricola 1556).

In 1908, one of the earliest texts with assaying instructions was written by Richard W. Lodge, professor of mining and metallurgy at MIT. The text included many current techniques. He compiled years of notes from his staff and solicited contributions from other professionals at the time and wrote "Notes on Assaying and Metallurgical Experiments." The student conducting metallurgical tests was expected to learn proper assay methods for their metallurgical products. An assay balance was used and metallurgical samples related to gold and silver were weighed against an "assay ton" system of weights. This was a system devised by C.F. Chandler of Columbia College to avoid calculation of the conversion from percent into troy ounces per ton avoirdupois of 2000 lbs. Weights could be made using a whole assay ton or fractions of an assay ton. Despite modern electronic balances now used in the 21st century which are capable of accurately weighing micrograms and other mathematical devices, assayers still use the terminology of assay ton when referring to the starting weights of ores for precious metal determination (Lodge 1908).

Lodge gives instructions for assay of metals by either fire or wet methods. He details fire methods for Ag, Au, Cu, Sn, Pb Platinum Group Metals and bullion, one element at a time. The method taught to metallurgical students for Cu is described as a simplification of a Cornish method. "The ore is smelted into a matte and sieved through 40 mesh. It's placed in an iron roasting dish and covered with ruddle (Fe_2O_3 + water) or chalk and placed in a red muffle and stirred constantly to obtain a dead roast (a roast in which sulfates and sulfides are both gone). The roasted ore is now all oxide and is sieved through 40 mesh again. The ore is then placed in a crucible and fluxed with cream of tartar, soda, charcoal, silica and borax glass. The crucible is fused and allowed to cool. The mass is broken and the copper button weighed. The impure copper button is refined (Lodge 1908)."

Edward Bugbee taught fire assay at MIT. He continued adding to the mimeographed notes from Richard Lodge and wrote his first fire assay text in 1915 with 150 pages. Bugbee thought a course in fire assay was the logical place to introduce the study of metallurgy. Many of the principles of metallurgy are used in one stage or another of fire assay. This piqued the students' interest and through fire assay the students increased their understanding of thermochemistry of metals and their oxide and sulfide forms, physical constants of slags and refractories, principles of sampling, cooling of alloys and chemistry of oxidation and reduction reactions. Bugbee's 1940 third edition of *A Textbook of Fire Assaying* continues to be used today as a primary fire assay reference (Bugbee 1940).

Education for the Metallurgical Chemist in 1903 involved significantly more chemistry hours than that of a student receiving a bachelor degree in chemistry today, where much of their education includes liberal arts electives. The student in 1903 was expected to blow their own glassware and make their own gasses and acids, Table 1.

By 1917, Wilfred Scott and a host of chief chemists from various mining companies and research scientists from all over the United States compiled analytical methods that would become the primary reference for metallurgical assaying for the next 50 years (Scott 1907).

The methods were complex and time consuming compared to modern methods. One example is the measurement of arsenic in iron and copper ores. "The sample is brought into solution by fusion with 1:1:2 Na_2CO_3, potassium nitrate, and zinc oxide in a platinum dish. The solution is added to a 5-liter flask filled with strong hydrochloric acid and attached to an elaborate distillation apparatus. Arsenic gas is distilled off and determined volumetrically" (Scott 1907) as shown in Figure 1.

Table 1. Curriculum for the metallurgical chemist, 1903, Harvard School of Mines, 3-year course (Macleod and Walker 1903)

Year	Course	Hours/Week	Weeks	Total Hours
1	Qualitative analysis and preparation of gases	9	11	
		6	22	231
2	Advanced qualitative analysis	3	33	
	Quantitative analysis	12	33	494
3	Assaying	16	33	
	Technical analysis	8	33	792
			Total Hours	1,518

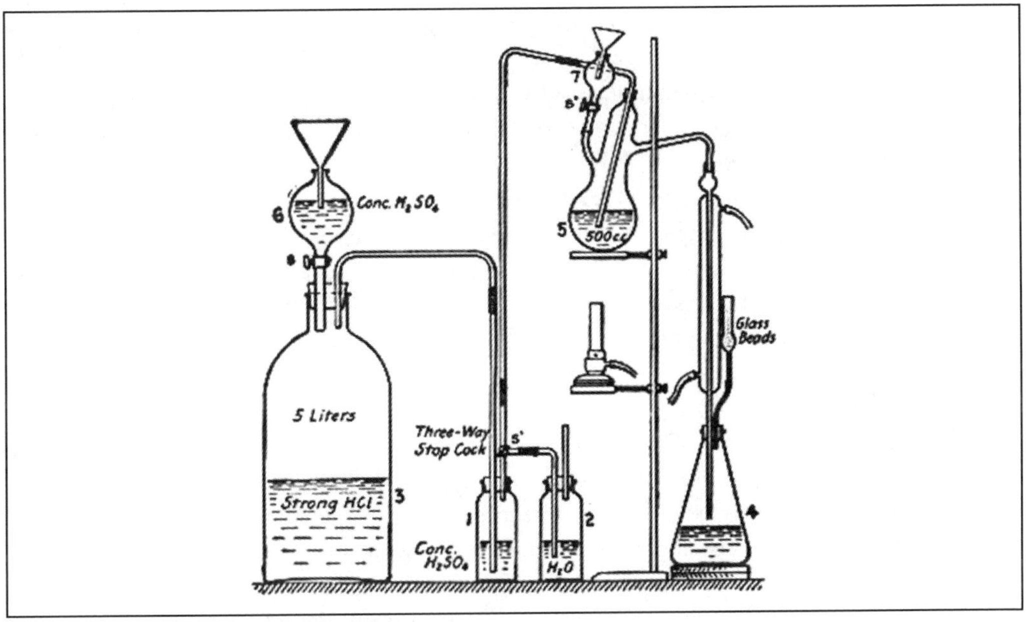

Figure 1. Apparatus for distillation of arsenious acid

While modern analytical chemistry is dominated by instrumental analysis, classical methods (also known as Wet Chemistry) use separations such as precipitation, extraction, and distillation. Gravimetric, titrimetric and color comparison techniques were used for quantitative and qualitative analysis by measurement of weight or volume. Some of these traditional techniques are still used today for special purposes.

Laboratories in the early 1900s to 1960s were stocked with blown glassware, beakers, Erlenmeyer flasks, glass pipets, and burets, items rarely used in today's modern laboratory where costly glassware has been replaced with disposable plasticware. Analyses were slow and tedious. Synthetic color indicators used in acid-base and redox titrations were one of the most notable advancements of the previous 50 years. Another advancement was the pH meter developed in 1936 by what is now the Beckman Corp. in response to the need for fast and accurate acidity from lemons for the California Fruit Growers (Szabadváry 1960).

ADVANCES IN SPECTROSCOPY

In 1666, Newton showed that the white light from the sun could be dispersed into a continuous series of colors and introduced the word "spectrum" to describe this phenomenon. By 1859 Kirchoff and Bunsen recognized that each atom and molecule has its own characteristic spectrum and established spectroscopy as a scientific tool for studying atomic and molecular structure, and founded the field of spectrochemical analysis for analyzing the composition of materials. The study of Quantum Mechanics in the latter half of the 19th century helped refine measurements (Szabadváry 1960).

MIT built an early design spectrometer in 1931 which included a Rowland circle of 10-m diameter and a large 21-ft vacuum spectrograph. The large dimensions were necessary to resolve spectra. The newly designed instrument used 35-mm motion picture film containing intensity traces as a function of wavelength to accuracy of seven figures. Data interpretation was slow and tedious (MIT 2012).

MIT Physics Professor Harrison designed and constructed the first automatic comparator for measuring spectral line wavelengths with high speed and precision. The instrument could automatically read 20 in. of spectrum in 2 minutes. This data resulted in the collection of the MIT. Wavelength tables containing 110,000 wavelengths assigned to elements. These tables are still in use today and are used to develop modern spectrometers. National Bureau of Standards (now National Institute of Standards and Technology) republished the table in 1969. Visitors came from all over the world for the annual summer conference. Harrison recognized the value of cooperative work with other fields including metallurgy (MIT 2012).

POST-WAR BOOM

Starting in the middle decades of the 20th century, the speed, scope, and nature of the chemical practice were transformed through researchers' use of new analytical instrumentation. Many of these instruments combined electronics with chemical measurement. Measurements that had taken days became the task of hours, and hours to minutes. New questions could be asked of chemical systems, and old questions could be answered in new ways.

These developments from the 1940s to the 1970s formed a profound change in the practice of analytical chemistry. The chemistry laboratory changed more between 1950 and 2000 than from 1600 to 1950.

The opening decades of this Second Chemical Revolution—the 1940s and 1950s—saw a flurry of activities surrounding analytical instrumentation and chemical research associated with the needs and demands of the projects of World War II. The Manhattan Project cost $2 billion (equivalent to $25.8 billion today) and employed 130,000 people from 1939 to 1946 (Consumer Price Index 2012). New methods, techniques and approaches to problems were explored for uranium and rare earths. Analytical Chemistry of the Manhattan Project was published in 1950.

The period from the end of World War II to the early 1970s was a golden era of American capitalism. $200 billion in war bonds matured, and the G.I Bill financed a well-educated work force. The middle class swelled, as did GDP and productivity (French 1997).

End to depression and wartime scarcities, and a pent up demand for consumer goods fueled a steady industrial expansion. Shopping centers multiplied, rising from eight at the end of World War II to 3,840 in 1960. U.S. Steel's production peaked at more than 35 million tons of steel in

1953. Its employment was greatest in 1943 (during World War II) when it had more than 340,000 employees. Personal savings grew, so people could afford to buy more. In 1950, Americans bought over 6 million cars and GNP reached $318 billion. Baby boom and suburban flight helped increase consumer demand. Oil was abundant and gasoline cheap. The electronics industry boomed and TV was invented as was the first generation of computers. Construction, cars, and electronics all did well (French 1997).

In 1948, from radar technology learned from World War II, American physicists William Shockley, John Bardeen, and Walter Brattain invented the transistor. All three share the Nobel Prize (Blakemore 1985).

With these technological innovations, the need for natural resources and metals had never been greater. The production demands were unprecedented. Not only did military and industrial expansion prompt the development of the new laboratory instrumentation to keep up with metal production, they also provided the funding for the widespread adoption of these tools by chemists. Both *innovation* and *adoption* provided significant momentum to the expansion of the instrumentation industry in both domestic and foreign markets.

Instrumentation companies explored the capabilities of integrating microprocessors and microcomputers. New instruments incorporating microcomputers led approaches that accelerated the speed of data collection.

CARBON AND SULFUR ANALYSIS

Many corporations evolved with innovations and demands of the times. The Leco Corporation, St. Joseph, MI, is a good example of that evolution and how it has played a large supportive role to the mining and process plant laboratories throughout the world.

Lower and lower carbon specs were being placed on steel and the laboratory was pressured for quicker carbon measurements. The on-going friction between the metallurgist and the analytical laboratory for faster results has been around for the past century! As new metallurgical processes evolved, the methods of analysis had to be revamped to keep up with the demands. With the Bessemer process steel could be made in 30–40 minutes while carbon analysis of samples from the "heat" still took 7–8 hours for colorimetric carbon analysis. It was impossible to maintain carbon control. The wet combustion carbon method was developed and took carbon analysis requirements down to 2 hours. However, the Open Hearth, Electric furnace and BOF steel making processes each resulted in quicker steel making and the carbon analysis again had to be modified to match the needs of the process (Anderson 1967).

Leco Corporation, starting in 1936, had a role in the carbon and sulfur analysis evolution with the introduction of the first rapid carbon analyzer to the steel industry. WWII prevented importation of ceramic ware so Leco began manufacturing ceramics allowing Leco to grow at a tremendous rate through the war years, supplying instrumentation for rapid analysis (Figure 2). In the 1950s Leco instruments benefited from the newly created transistor and instruments changed from glassware and chemical based to electronic. In the 1960s electronics continue to improve and analysis times go from minutes to seconds with increased accuracy and automation. The 1970s brought solid state electronics and digital instrumentation as well as rolling out the first automated sulfur analyzer that replaced the classic wet chemistry Eschka method (Ostwald 1917) again cutting hours off of the required analysis time. The 1980s saw internal micro-processors capable of self-calibration

Figure 2. 1940s chemist measures carbon on one of the first Leco carbon analyzers (Leco Corp Archives)

and self-diagnostics and faster analysis. In 1990s and 2000s new instruments are developed for a wider variety of markets and the internet is used for on-line diagnostics, eliminating the need for service calls. Leco has been stable throughout the recession years and continues to be a valuable partner in the mining industry (Leco Corporation 2011).

METALS ANALYSIS—THE FIRST DIRECT READING SPECTROMETER

Because of the demands from a robust steel industry, the Arc/Spark DC plasma was the first direct reading instrument used commercially in 1950 for quantitative multi-element analysis. These instruments reduced the multi-element analysis of metals from hours to minutes (and later, to seconds). It worked on conductive material and was ideal for steel and metallics. It could be used with nonconductive ores when mixed with graphite, packed into a carbon electrode and subjected to a DC arc. Liquids had to be dried or plated onto electrodes. Spectral interferences were immense and results had large errors even with the best analyst. It gained popularity with alloy producers and foundries and is still used in that area today. It never gained wide acceptance because of its complexity.

FLAME ATOMIC ABSORPTION SPECTROMETER

The first Flame Atomic Absorption Spectrometer (AAS) was built by CSIRO scientist Alan Walsh in 1954 (Figure 3). He worked on improving spectrometers. It was known that when a metal was

Figure 3. Alan Walsh developed the technique of atomic absorption (CSIRO Archives)

heated in a flame, it emitted unique specific signatures of light. This quality is used by spectrometers to detect the presence of metals in samples by detecting these signatures (McKay 1976).

Alan Walsh had the realization that by shooting a beam of light at a chosen frequency into a heated sample, the amount of light absorbed would be an indication of how much metal was present. It had never occurred to him to use absorbance instead of emission. This realization led to the first flame AAS being built. The first applications solved by flame AAS were ones in medical treatment, agriculture and mining (McKay 1976).

Atomic absorption provided quick, easy, accurate and highly sensitive means of determining the concentrations of over sixty-five of the elements. The method has found important application world-wide in mineral exploration and metallurgy, and environmental monitoring. It has been described as *the most significant advance in chemical analysis* in the twentieth century. Walsh's work has been well recognized. In 1977, he received a knighthood (McKay 1976).

Arc/spark technique declined in popularity in the 1960s and 1970s while atomic absorption gained in popularity. Since absorption of light by ground state atoms was used as the mode of detection, the need for very high temperatures to populate excited states of atoms was no longer a limitation. The instabilities and spectral interferences which plagued arc/spark emission techniques were also greatly reduced by atomic absorption techniques.

By the time the original patents of atomic absorption had expired around 1969, twenty licenses had been issued, and there were also several manufacturers in countries such as Japan in which patents had not been sought. During 1963–1967 sales of atomic absorption instruments experienced exponential-like growth. By 1969 there were more than 10,000 atomic absorption spectrophotometers in use in laboratories around the world, and by 1977 this number had grown to around 40,000.

The cost-benefit analysis of the atomic absorption project was studied and conservatively assessed the value of the net benefits to the Australian economy at around $22 million (in 1968 Australian dollars), compared with $1.3 million originally spent on the research. Laboratories all over the world benefitted from productivity gains, especially the ability to perform large numbers of assays very rapidly and with a high order of accuracy (McKay 1976).

GRAPHITE FURNACE

In 1961, Russian scientist Boris V. L'vov developed graphite furnace atomic absorption spectroscopy (GFAAS). Not all elements responded well using flame AAS. The GFAAS is either a separate instrument or an accessory to the standard flame AAS. Small sample sizes (5 µL) are placed on an L'vov platform inside a graphite tube and the sample is vaporized and atomized. Very low detection compared to flame AAS could be obtained and it served as a nice compliment to standard flame AAS for trace level gold, penalty elements in concentrates and environmental elements of interest (Robinson et al. 2005).

INDUCTIVELY COUPLED PLASMA–OPTICAL EMISSION SPECTROMETER

British chemist Stanley Greenfield is credited with developing the plasma-based Inductively Coupled Plasma–Optical Emission Spectrometer (ICP-OES) instrument in the mid-1960s about the same time flame-based AAS became prominent. This instrument would reduce the multi-element analysis of metals from hours to minutes (and later, to seconds). American chemist Velmar Fassel and his team at Iowa State University made improvements. These first plasma-based instruments used direct current (DC) and microwave-induced (MI) systems to generate the plasma. Interference effects and plasma instability limited the usefulness of plasma instruments; consequently flame-based AAS dominated the analytical market for metals analysis. The limitations of the first plasma instruments were overcome by utilizing an argon ICP instead of DC or MI generated plasma. Argon ICP-OES became popular in the 1980s due to their decreased cost, multi-element capability, and labor saving advantages. Flame AAS instruments require a radiation source (lamp) per element for the approximately 35 elements they can measure. Because the lamp must be changed between each element of interest, flame AAS techniques analyze a single element at a time and are unable to easily analyze metalloids. ICP optical systems, by contrast, were now able to analyze about 60 different elements at the same time with a single source (the plasma)(Boss and Fredeen 1997; Robinson et al. 2005).

Early ICP-OESs were designed with a monochrometer and photo-multiplier tube which either limited the wavelength selection or required sequentially stepping to each wavelength. Advances to ICP-OESs include echelle grating and charge-coupled device (CCD) which allowed the ICP-OES to measure 1000s of wavelengths simultaneously using a smaller footprint. Improvements were made to spray chambers, nebulizers, and torch orientation to improve detection and stability. The solid state generator replaced the RF tube in the late 1990s and simplified software made the ICP-OES of today virtually maintenance free.

X-RAY FLUORESCENCE

The X-ray Fluorescence Spectrometer (XRF) is another instrument developed as a result of the robust steel industry of the 1950s.

X-rays were first discovered by Wilhelm K. Roentgen (German physicist, 1845–1923) who won the Nobel Prize in 1901 for the discovery of X-rays. In 1922, A. Hadding first applied X-ray spectra to chemical analysis of minerals and in 1955 the first XRF became commercially available for industrial use. From the 1950s to 1960s nearly all the X-ray spectrometers were wavelength dispersive (WD) spectrometers. John W. Criss and LaVerne Stanley Birks in 1968 produced the first computer program for fundamental parameter calculations in X-ray fluorescence which simplified method development (Thomsen 2006). The XRF provided quick multi-element analysis of major elements. While precision is very good, accuracy due to error in matrix changes such as shift from copper oxide to copper sulfide caused skepticism among plant operators. Enhanced sample preparation with binders and fusion improved accuracy but often analysis by flame AAS was required for accurate metallurgical balance in a plant. Light elements are not very sensitive. However, WD-XRF continues to be a workhorse in many major mining laboratories sacrificing some accuracy for speed since they are capable of measuring millions of analytes per year.

The first benchtop Energy dispersive XRF (ED-XRF) surfaced in the early 70s as an alternative to large bulky, maintenance intensive and expensive WD-XRF. It has made gains in popularity in recent years with the addition of silicon drift detectors and low powered tubes allowing improved detection and resolution. The hand-held portable XRF is popular among exploration geologists in the field.

REGULATORY DEMANDS

In the mid-20th century an environmental movement pointed out there were environmental issues associated with the material benefits that were being enjoyed from the industrial boom.

The USEPA created in 1970 by Richard M. Nixon drove legislation pertaining to protection of air, water, land, and hazardous wastes. This opened up a new branch of laboratory analyses, Environmental Chemistry.

The Clean Water Act (CWA) established the basic structure for regulating discharges of pollutants into the waters of the United States and regulating quality standards for surface waters. The basis of the CWA was enacted in 1948 and finalized in 1972. In 1963, the Clean Air Act was introduced. The Resource Conservation and Recovery Act (RCRA) became effective in 1976 and gave direction for management of hazardous wastes.

Environmental laboratories sprouted up all over the country in the 80s to meet the regulatory needs. When meeting the legislative requirements laboratories were, and still are, required to follow very prescriptive methods with strict quality control/quality assurance and accreditation programs.

The USEPA publication SW-846, entitled *Test Methods for Evaluating Solid Waste, Physical/Chemical Methods*, is the official compendium of analytical and sampling methods that have been evaluated and approved for use in complying with the RCRA regulations. It has been issued by USEPA since 1980 and is currently in its third edition. Advances in analytical instrumentation and techniques are occasionally reviewed. Another text, *Standard Methods for the Examination of Water and Wastewater* originally written in 1905 is in its 22nd edition and provides standard methods for water analyses.

The rigid environmental methods with extensive quality control, although necessary, have hindered the innovative spirit of the analytical chemists. The cookbook methods leave little room for creativity in productivity or use of alternative chemicals with superior qualities. Deviations from standard methods require documentation and the process to incorporate improved methods into standards is complex.

INDUCTIVELY COUPLED PLASMA–MASS SPECTROMETER

Regulatory requirements drove the need for quick trace element analysis. AAS and ICP-OES did not have low enough detection limits for all elements. GFAAS was single element, consumables were expensive and methods were too slow for productive use.

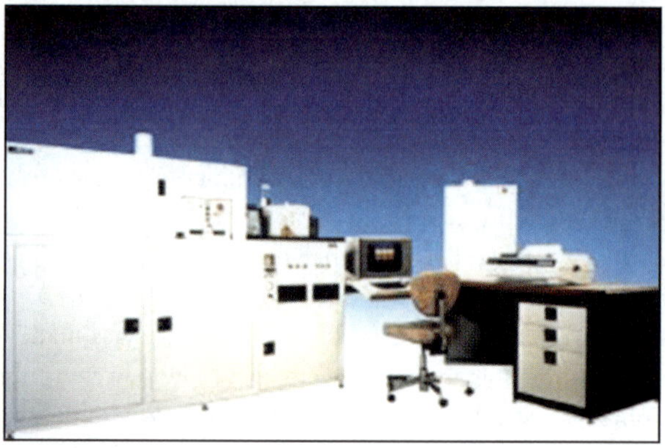

Figure 4. First commercially available ICP-MS, Elan 250 (PerkinElmer/Sciex)

Over 25 years ago, mass spectrometry was rarely used for the routine determination of trace elements. There were several issues with the technique that made routine trace analysis impractical at the time. Therefore, researchers began the search for a new technique for the determination of trace elements that could address problems experienced by other techniques. Much of the research on Inductively Coupled Plasma–Mass Spectrometer (ICP-MS) was performed by researchers Robert S. Houk, Velmer A. Fassel, Gerald D. Fiesch, and Harry J. Svec from Iowa State University (ISU, Ames, IA) and Allen Gray from the University of Surrey (Surrey, U.K.). The objective of their work was to demonstrate the feasibility of ICP-MS for the routine determination of elemental concentrations and isotopic abundance ratios in solution (Robinson et al. 2005; Thomsen 2006). Figure 4 shows the first commercially available ICP-MS in 1984.

Today's ICP-MS has virtually displaced the use of graphite furnaces and types of emission spectroscopy. When compared to other techniques, ICP-MS is a more accurate and powerful technique with many more advantages. It is 1,000–10,000 times more sensitive than it was 30 years ago. ICP-MS offers a large dynamic range as it can analyze elements at parts per trillion (ppt) to parts per million (ppm) and provide both quantitative and qualitative isotope ratio analyses at high speeds with high sample throughput. ICP-MS is also capable of determining the individual isotopes of each element. Advancements in software have simplified or automated the tuning procedures and interference corrections.

Laser-ablation accessories for ICP-MS are recent developments and are used to measure trace elements in solids. They are becoming commonplace in geological laboratories and are useful in the gold industry for microscopic gold mapping studies.

DATA MANAGEMENT

Prior to the mid-1970s before electronic calculators became commonplace, laboratory calculations were tedious and methods were designed to eliminate mathematical calculations as much as possible. An "assay ton" was measured as an exact weight, so final bead weight would report in miligrams but was equal to ounces per ton without any calculations. Titrants were made from weights that conveniently included the molecular weight of the analyte so measurements read directly from the burette were already in g/L eliminating calculation. The AAS was calibrated with standards that already included the dilutions and sample weight so direct readout was in percent. The results could be reported faster but was often restrictive for unusual sample types and manual transfer of data often resulted in transcription errors.

By the mid-1990s the personal computer and software to manage the unwieldy laboratory activities was available. Work flows were simplified and sample status tracked easily. Instrument data was uploaded automatically. Quality control was incorporated with blanks, duplicates and reference materials and the tedious task of producing control charts was automated. The laboratory today is nearly paperless.

WHAT HASN'T CHANGED

There are certain classical methodologies that show little sign of yielding to more productive methods. Fire assay is more so than anything else. It's been around since the mid-16th century and yet continues to be the primary method for precious metals. Improvements have been made to furnace control. The AA and ICP finishing step improved productivity and detection limit. Robotics have improved the material handling process but still are not common. Attempts to automate the entire process using robotics and Arc/Spark spectroscopy have had limited success for platinum group metals but is still problematic for gold analysis. The basic principles Agricola wrote about in the 16th century and the texts written by Bugbee and that from Shepard and Dietrich continue to be used for teaching today (Bugbee 1940; Shepard and Dietrich 1940).

Classical wet methods are still in use for referee analysis of base metal concentrates when ultimate accuracy is required and time is not a factor. While most of the classical methods have been replaced by newer instrumental methods, the classics will never be lost from the texts as one method does not fit all sample types, concentrations and interference control. Change is not easily adopted but occasionally legislation, metallurgical process improvements and/or metal prices dictate change in methodologies.

Microwave ovens had been around since the late 1940s, but it was not until 1986 that microwaves began to make their ways into the laboratories. With all analysis, the trend for faster turnaround continued and microwave digestions promised to deliver just that. Early high pressure vessels were cumbersome to assemble and space was limited to 6–10 samples at a time. Temperature was the limiting factor and cooling times were long. The actual heating in the microwave field is dependent on the type of sample and sample quantity. Rarely will two samples exhibit the same behavior. Only small sample sizes could be used. While the microwave digestion has a place for niche applications and for environmental samples, after 25 years it still has not proven to be a productivity improvement tool in the mining laboratory.

The DC Arc spectrometer, which lost popularity due to the advent of ICP-OES and ICP-MS in the 1980s, is back in 2012 to fill the lost niche of measuring trace impurities in copper, steel, iron and other alloys. Swings in metal prices dictate production rates and quick quality measurement. Detection limits are lower and measurement is more automated but the basic principles are the same.

PITTSBURGH CONFERENCE

In the 1940s and 1950s, Pittsburgh was one of the largest industrial centers in the U.S. Industrial complexes in and around the city produced chemicals and industrial machinery, glass and aluminum, and, most prominently, steel. These industrial concerns employed a large community of chemists, including analytical researchers with a distinct orientation to industrial applications.

In 1950, the Pittsburgh Conference on Analytical Chemistry & Applied Spectroscopy—the first Pittcon—took place. Eight hundred attendees paid $2.00 in registration to take in 56 papers and to visit the booths of 14 exhibitors over three days. Pittcon has since moved away from Pittsburgh in order to accommodate as many as 30,000 scientists and vendors and continues to be the largest trade show for scientific equipment in the world. Manufacturers today still use Pittcon to introduce new analytical laboratory innovations to the world.

THE FUTURE OF THE ANALYTICAL LABORATORY

Fire Assay replacement could be a reality within 10 years. Studies in at least one laboratory have demonstrated the use of gamma activation analysis on gold, silver and 30 other elements. Results could be obtained down to very low levels in samples weighing a few hundred grams with minimal sample preparation in less than a few minutes (Process 2009).

Old technologies will be improved for special applications. An example is the Microwave Plasma-Atomic Emission Spectrometer (MP-AES) released in 2010. Microwave plasma was pushed aside in the 1970s as ICP-OES emerged as the technology winner because of its stable plasma but has reemerged as a hybrid of an ICP-OES and flame AAS. A major benefit is the elimination of costly argon gas and the ability to use it in remote laboratories where argon gas is difficult to obtain.

Process Analytical Technology will become more prominent with sophisticated on-line analysis for operations control. The use of multivariate data analysis and advanced software will correlate fast measurement from tools such as NIR, Raman spectroscopy and other sensors with traditional analytical tests.

REFERENCES

Agricola G., Translated Text of *De Re Metallica.*

Anderson D., 1967, Leco Corporation, Carbon Measurement in Steelmaking: Where Fast Is Not Quick Enough.

Blakemore J.S., 1985, *Solid state physics*, pp. 1–3, Cambridge University Press.

Boss C. and Fredeen K., 1997, *Concepts, Instrumentation and Techniques in Inductively Coupled Plasma Optical Emission Spectrometry*, 2nd Edition.

Bugbee E., 1940, *A Textbook of Fire Assaying*, third edition, 1991 reprint.

Consumer Price Index (estimate) 1800–2012.

French M.,1997, *US Economic History since 1945.*

Leco Corporation 75th Aniversary Video 2011, St. Joseph, MI, 2011.

Lenahan W.C. and Murray-Smith R.de L., 1986, Assay and Analytical Practice in the South African Mining Industry.

Lodge W.R., 1908, Notes on assaying and metallurgical laboratory experiments, second edition revised.

Macleod W. and Walker C., 1903, *Metallurgical Analysis and Assaying.*

McKay A., 1976, *Surprise and Enterprise–Fifty Years of Science for Australia*, CSIRO Publishing.

MIT. 2012. G.R. Harrison Spectroscopy Lab, http://web.mit.edu/spectroscopy/.

Process, 2009, CSIRO Research Newsletter, Oct.

Robinson J.W., Skelly Frame E.M., Frame G., 2005 *Undergraduate Instrumental Analysis*, Sixth Edition.

Scott W., 1907, *Standard Methods of Chemical Analysis—A Manual of Analytical Methods and General Reference for the Analytical Chemist and for the Advanced Student.*

Shepard O.C. and Dietrich W.F., 1940, *Fire Assaying.*

Szabadváry F., 1960, *History of Analytical Chemistry.*

Thomsen V., 2006, *Spectroscopy* 21(10) October.

Ostwald W., 1894 and 1917, *The Scientific Foundations of Analytical Chemistry.*

ABOUT THE AUTHOR

Meg Dietrich has worked as an analytical and process chemist in metallurgical labs for nearly 30 years. She held positions in the Met Lab at Magma Copper Smelter Operations in San Manuel, Arizona; was chief chemist at BHP Minerals Research Lab in Reno, Nevada, was senior process chemist at Itronics Metallurgical, Reno, Nevada; and is currently the chemist director at Newmont Metallurgical Services, Englewood, Colorado. She has focused her career on expanding analytical laboratory capabilities to support metallurgical process development.

Environmental Innovations

The Generation and Treatment of Acid Rock Drainage

Thomas Wildeman, Katherine Vatterrodt, and Linda A. Figueroa
Colorado School of Mines, Golden, Colorado, United States

ABSTRACT: This review of the chemistry of acid rock drainage (ARD) concentrates on what the author thinks are important but not obvious geochemical principles in the generation and treatment of ARD. In ARD generation, the role Fe(III) plays in the oxidation of sulfides, the generation of further acidity, and the release of minor constituents of concern (COCs) is explained. In ARD treatment, three geochemical principles are described. Of most importance, is the concept of mineral acidity and the prominent role iron and aluminum play in the treatment of ARD. Second in importance is the role oxidation and reduction play in the sequence of treatment steps and in the difficulty of removing minor constituents such as arsenic and selenium. Finally, how complex ion formation can frustrate the removal of some metals such as aluminum and zinc is explained. For the removal of trace COCs in ARD, the role of adsorption and co-precipitation onto the oxyhydroxide precipitates is important to the achievement of environmental guidelines.

INTRODUCTION

Acid rock drainage (ARD) is created when water, air, and sulfide minerals (especially pyrite) come in contact and react with the aid of bacteria. Because all of these constituents have been in the environment for eons, ARD was being generated long before there was mining. Consequently, the geochemical principles that apply to the generation and treatment of ARD are well established, and can be found in standard aquatic chemistry texts (Stumm and Morgan 1996, Langmuir 1997). There are also specific reviews of the properties of ARD that can be found on the Internet (GARD Guide, Office of Surface Mining). This review will highlight some of the properties of ARD that the author considers to be important to its generation and treatment but not always recognized.

GENERATION OF ARD

Stumm and Morgan (1996) proposed the reaction sequence for the development of ARD as follows:

$$FeS_2 + \tfrac{7}{2}O_2 + H_2O \rightarrow Fe^{2+} + 2SO_4^{2-} + 2H^+ \quad (1)$$

$$Fe^{2+} + H^+ + \tfrac{1}{4}O_2 \rightarrow Fe^{3+} + \tfrac{1}{2}H_2O \quad (2)$$

Reaction 2 is catalyzed by bacteria. Reaction 1 shows the interaction of air, water, and sulfide minerals, but Reaction 2 is important relative to eventual effects of ARD. Fe(III) is the important product because it is an oxidizing agent for sulfide minerals.

$$PbS + 8Fe^{3+} + 4H_2O \rightarrow Pb^{2+} + SO_4^{=} + 8Fe^{2+} + 8H^+ \quad (3)$$

$$PbS + 2O_2\,(g) \rightarrow Pb^{2+} + SO_4^{=} \quad (4)$$

Table 1. Constituents in mg/L in examples of ARD and suggested aquatic guidelines for U.S. waters

	Quartz Hill	Big Five	Coal	Cu Raffinate*	Zn Raffinate*	Aquatic Stds.†
pH	2.5	2.9	2.9	2.0		6.5–9
Al	60	20	36	2600	32	0.1
Fe	750	50	180	150	38	1.0
Mn	80	30	50	790	303	1.0
Cu	55	2.0		70	6.7	0.01
Zn	150	10		200	2300	0.1
Cd	0.80	0.03		1.5	30	0.005
Pb	0.14	0.01		7.9	1.6	0.05
As	1.5	0.10		0.38		0.01
SO_4	4000	2100	2050	25,000	15,000	250‡

* Raffinates from a copper heap leach extraction and a zinc hydro metallurgical facility.
† Approximate aquatic guidelines for U.S. waters (USEPA 2009).
‡ Secondary drinking water standard.

Compare Reaction 3 with the oxidation of galena by oxygen (Reaction 4) where no hydrogen ions are generated. Furthermore, this creation of Fe(III) and hydrogen ions frustrates schemes to abate ARD by cutting off air and water from the sulfide minerals. The Fe(II) produced in Reaction 3 can be oxidized by Reaction 2, which is catalyzed by bacteria, and an ARD generation cycle is created where Reaction 1 is circumvented.

Table 1 lists the important constituents dissolved in some examples of ARD. Also included in Table 1 are suggested aquatic toxicity or drinking water guidelines for the listed constituents (USEPA 2009). The examples in Table 1 have acidic pHs and high concentrations of iron and aluminum. However (as seen in the Table 1 examples) another characteristic of ARD, especially when it is generated from a metal-mining situation, is the high concentrations of other constituents of concern (COC). These are generated because Fe(III) can oxidize and dissolve other sulfides as seen in Reaction 3, and the hydrogen ions generated can dissolve other minerals such as rhodocrosite.

$$2H^+ + MnCO_3\ (s) \rightarrow Mn^{2+} + H_2O + CO_2\ (g) \tag{5}$$

Wildeman et al. 1974 noted that the high concentrations of Cu, Zn and Pb in ARD generated from the Central City ore deposit were correlated with the abundance of pyrite and not with the highest occurrence of lead, zinc and copper minerals. This concept was expanded by Plumlee et al. 1999 to show that the type of deposit and the pH of the ARD correlate well with the total concentration of heavy metals that are found in the water. Such a correlation is quite useful in providing a prediction of what type of ARD could be generated when a specific type of mineral deposit is developed. The presence of high concentrations of heavy metals also presents a greater treatment problem for ARD from metal-mining operations than ARD from coal-mining operations where the concentrations of heavy metals are quite a bit lower (Wildeman 1990).

ISSUES IN THE TREATMENT OF ARD

An extensive review of the treatment of ARD has been published by the Society for Mining, Metallurgy & Exploration (SME) (Gusek and Figueroa 2009). Active and passive treatments are the two general treatment methods. The practical use of passive treatment is strongly controlled by the level of flow and the amount of mineral acidity in the water (Gusek and Figueroa 2009). The higher these two variables, the less likely it is that passive treatment is possible. In modern mine planning, it is becoming more the case that design for treatment of mining influenced waters (MIW) such as ARD is being developed along with the metallurgical process (Murphey et al. 2009, Bucknam et al. 2012). When this is done, it is easier and probably requires less capital to construct an active water treatment plant along with the metallurgical facilities instead of waiting for the inevitable. Because, this appears to be the trend in design, this review will concentrate on active treatment.

Three important geochemical principles control the treatment of ARD. Acid/base conditions and the removal of mineral acidity is probably the most important principle that controls the treatment of ARD. The oxidation state of the COCs is also quite important in the design of an effective treatment method. Finally the formation of complex ions in the water can be most vexing when the treatment goal is to remove a constituent to a concentration that is below the aquatic toxicity limits. These three geochemical principles are developed below.

Mineral Acidity

Figure 1, adapted from Stumm and Morgan (1996) is familiar to most practitioners of water treatment. It shows that as pH increases, the concentrations of most COCs decrease in solubility. This is one of the reasons that pH limits are included in drinking water standards. In the figure, Fe(III) and Al(III) stand out because they precipitate at pH values that are quite a bit lower than the other metals. Indeed, these two cations are the primary contributors to mineral acidity in ARD.

$$Fe^{3+} + 3H_2O \Leftrightarrow Fe(OH)_3 + 3H^+ \quad \textbf{(6)}$$

$$Al^{3+} + 3H_2O \Leftrightarrow Al(OH)_3 + 3H^+ \quad \textbf{(7)}$$

Reactions 6 and 7 show that when Fe(III) and Al(III) precipitate as oxyhydroxides, hydrogen ions are generated. As pH increases, Fe(III) is first removed at pH between 3 and 4, and then Al(III) is removed between pH 5 and 6 (Langmuir 1997, Wildeman et al. 1997). These two constituents contribute more to mineral acidity than do the concentration of hydrogen ions. This leads to a number of important treatment guidelines.

Because Fe(III) and Al(III) form oxyhydroxides at lower pHs, they must be removed before any of the other COCs can be removed. Any treatment process that selectively attempts to remove copper or zinc and leave the iron and aluminum in the water will be problematic. Also, alkalinity in the form of bicarbonate in the water is not possible until the pH is above 5 (Stumm and Morgan 1996). So creating an environment conducive to aquatic life cannot be possible until Fe and Al are removed. Finally, below a pH of 5, most metal ions in water exist as bare cations (Stumm and Morgan 1996). This form of the dissolved metal species is the most toxic to aquatic organisms (Santore et al. 2001). These reasons make removal of iron and aluminum in ARD as the most important objective in the treatment of ARD.

To remove Fe and Al and eliminate mineral acidity, it is necessary to add a source of alkalinity. Traditional sources of alkalinity are sodium hydroxide, calcium hydroxide (lime), or sodium bicarbonate (Gusek and Figueroa 2009). Currently lime is the reactant of choice because it is the least expensive of the three reactants. Also, calcium addition to the water by using lime raises the hardness of the water and this protects aquatic organisms from heavy metal toxicity (Santore et al. 2001, Pagenkopf 1983).

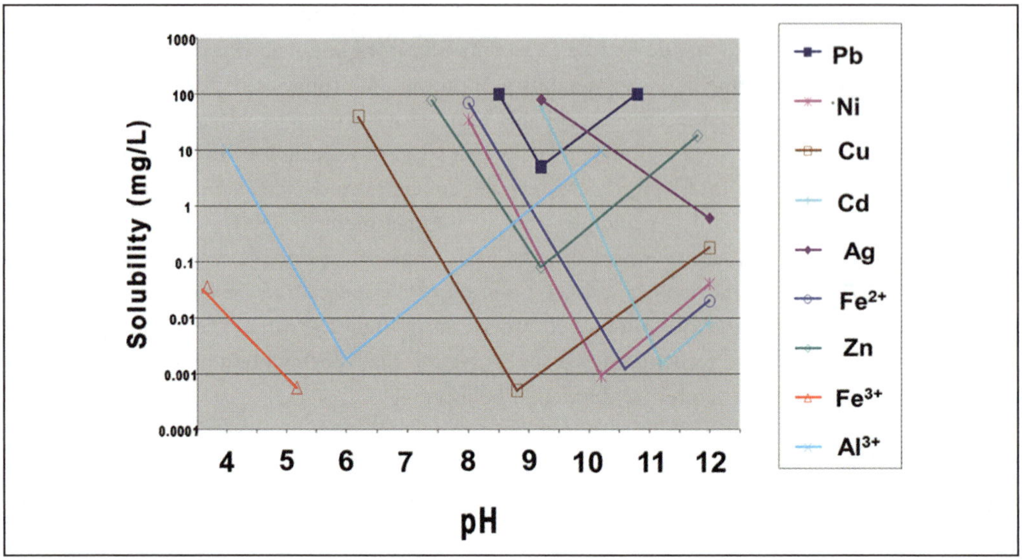

Figure 1. Hydroxide solubility of metals in water

Oxidation and Reduction

A few COCs in ARD are capable of having different oxidation states. Iron and manganese are the best examples. To understand how oxidation state can control the removal of a metal, it is instructive to consider **Figure 2** that was generated by Rose et al. 1979. In this figure, the ionic charge (oxidation state) is plotted against the ionic radius. Broad regions of mobility (solubility) are also defined. Cations with low ionic charge and large ionic radii, such as Na^+ and Ca^{2+}, are generally soluble. Those cations with ionic charges of +3 and 4 are generally insoluble. Species with oxidation states of +5 and +6 are quite small and react with water to form soluble ions; sulfate S(VI), arsenate As(V), and UO_2^{2+} U(VI) are good examples that are shown in **Figure 2**.

In active treatment of ARD the system is exposed to air, and thus it is an oxidizing environment. Iron and manganese are two important metals in ARD with multiple oxidation states. Manganese exists in ARD as Mn^{2+} and most of the iron also is in the +2 oxidation state, and this makes them quite soluble. Oxidation to Fe(III) and Mn(IV) is important to render these metals to have low solubility, and to make the produced sludge much more stable. This is why aeration is an important component of an active ARD treatment program.

Changing the oxidation state of a species in water is not as easy as changing the pH. Oxidation/reduction reactions require moving electrons and most of these reactions are slow. **Figure 3** is a diagram of how long it takes to oxidize half of the concentration of Fe and Mn (half-life) in solution (Stumm and Morgan 1996). Oxidizing Fe^{2+} and Mn^{2+} in solutions where the pH is below 3 takes a very long time. To speed up the reaction, the pH must be raised. For Fe(II), the pH needs to be raised to 7 to make the reaction speed on the order of minutes. So in ARD treatment, if one aerated and then raised the pH, the success would be problematic. Consequently, the order of treatment is to raise the pH and then start aeration.

Figure 3 shows that oxidation of Mn^{2+} to Mn(IV) is more difficult; a pH of over 8 is necessary to have a reasonable reaction speed. Usually, the pH is raised to around 10 to remove Mn as

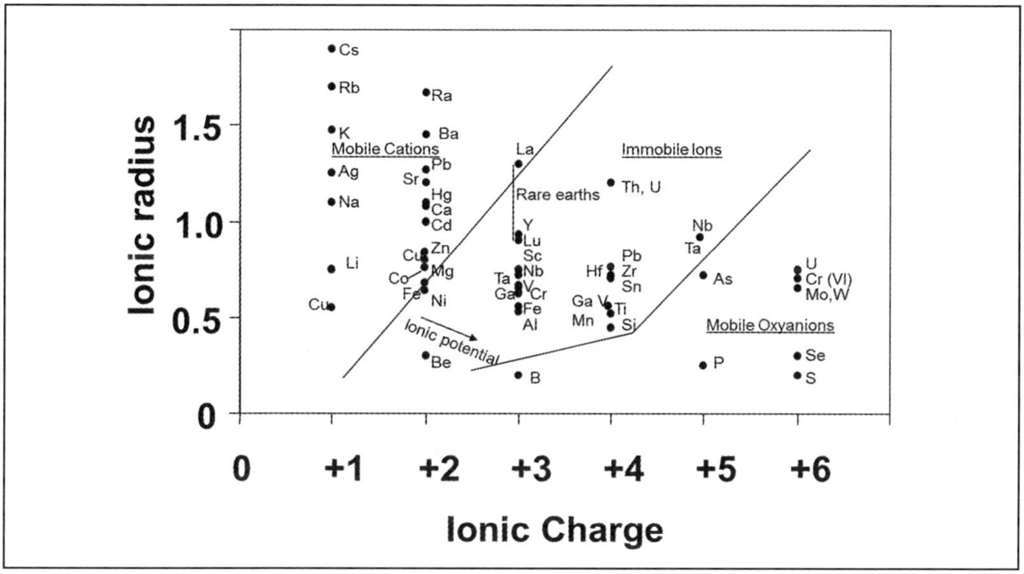

Figure 2. Mobility (solubility) of metals in water (Rose et al. 1979)

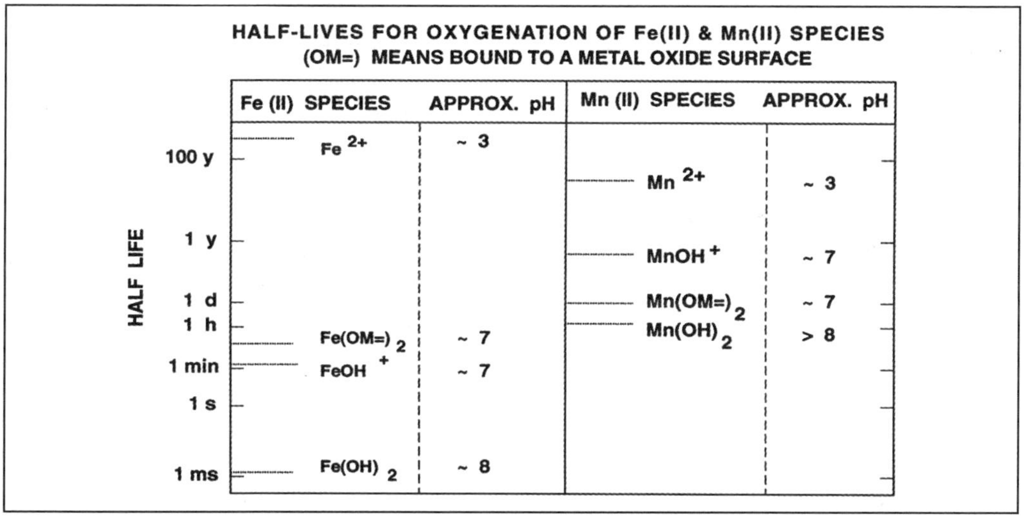

Figure 3. Half-lives for the oxidation of Mn(II) and Fe(II) (Wehrli & Strumm 1989)

$Mn(OH)_2$ and then the sludge is allowed to age and oxidize. Because of the difficulty in removing manganese from ARD, a useful rule of thumb is that if Mn is removed, most other COCs are removed.

Figure 2 contains a wealth of information that can be applied to other possible COCs in ARD. For example, note that As(V) as arsenate, Se(VI) as selenate, and S(VI) as sulfate are all soluble in ARD because they have such high oxidation states and are hydrated to anions. If there is arsenic or selenium in the water and a one stage treatment process is tried, it is very likely that these two species will be in concentrations above the regulatory limits. Of course removing sulfate from water is

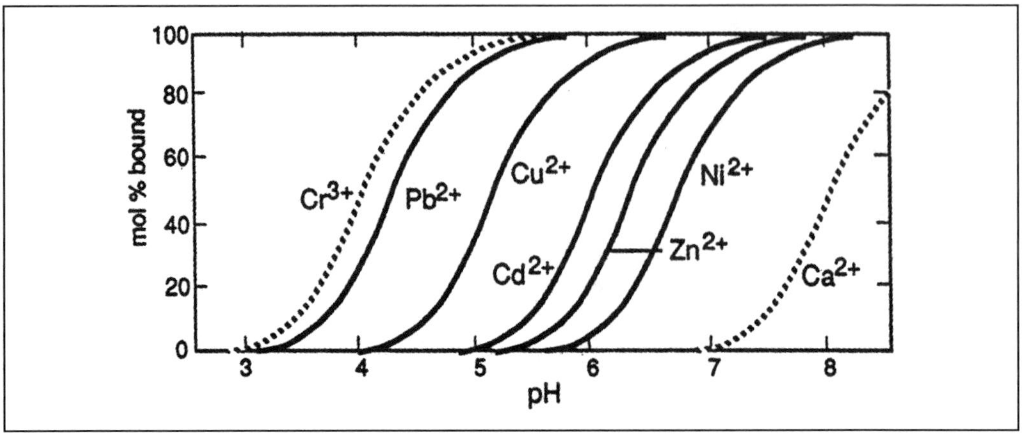

Figure 4. Adsorption of cations versus pH for Al_2O_3 (Stumm and Morgan 1996)

becoming a most vexing problem because many regulatory agencies around the world make sulfate concentrations of 250 or 500 mg/L a mandatory rather than a suggested limit. For sulfate as well as arsenic and selenium, many treatment operations resort to reverse osmosis for removal.

Complex Ion Formation

Figure 1 that shows metal insolubility with pH also reveals increased solubility at high pH values for Al, Cu, Zn, and Pb. In these cases, the metals react with water to form soluble complex ions such as $Al(OH)_4^-$ and $Zn(OH)_3^-$. As seen in Table 1, aluminum is a major constituent in ARD. Figure 1 shows that the minimum solubility of aluminum occurs at around a pH of 7. Depending upon the environmental regulations, a two stage lime treatment process may be necessary for the ARD. If the aquatic toxicity guideline of 0.1 mg/L (shown in Table 1) is used, it is most likely that alkalinity addition to raise the pH to around 7 would be needed followed by clarification to remove Al to a concentration that is below the environmental guideline. Then the supernatant solution would be sent through a second stage to raise the pH to around 10 to remove manganese and other minor COCs.

In treatment of the raffinate from the zinc hydrometallurgical circuit that is shown in Table 1, formation of complex ions by zinc will be an issue. Because zinc is the major metal in the ARD, it will control the chemistry of treatment. This implies that even with ideal pH conditions, achieving removal to below the aquatic limit of 0.1 mg/L will require special attention to co-precipitation reactions that could mitigate the formation of soluble zinc species at high pH and special handling of the $Zn(OH)_2$ sludge to insure excellent flocculation.

ADSORPTION AND CO-PRECIPITATION IN TREATMENT

Fe(III), Al(III), and Mn(IV) oxyhydroxides solids are capable of generating high surface charges that attract other cations and cause their adsorption. Figure 4 from Stumm and Morgan (1996) plots the binding of COCs onto Al_2O_3. It is interesting that compared with hydroxide precipitation (Figure 1), metals are adsorbed at lower pH values and are not released at high pH values. Adsorption is typically defined as molecular attachment of an ion onto the surface of a solid, whereas co-precipitation is inclusion of the ion into the structure of the precipitate. Most studies on adsorption and co-precipitation are conducted under controlled conditions and they are run to equilibrium (Stumm and Morgan 1996, Langmuir 1997, Pantuzzo et al. 2008, Twidwell and Leonhard 2008). Active treatment of ARD has time constraints, solids are imperfectly formed, and COC removal has to meet environmental standards (Miller et al. 2008). Consequently, bench-scale tests on the actual water to be treated are essential. In these actual treatment tests, the terms adsorption and co-precipitation probably do not apply; the terms loosely and tightly bound may be more appropriate.

Table 2 shows the results of a bench-scale treatment test on an actual ARD using hydrated lime. A two stage process was employed with lime addition in tanks 1 and 2 followed by clarifiers. Oxygen was sparged into tanks 1 and 2, and 25 mg/L of Fe(III) along approximately 100 ppm of an anionic flocculent was added to tank 2. Tank 3 was an attempt to add settling time so that the flocculent could be more effective. Tank 1 was maintained at a pH of 7.0 ± 0.5 and tank 2 was held at pH 10.0 ±0.5. The residence time in each tank was approximately 20 minutes, and the residence time in the clarifiers was 70 minutes.

Except for aluminum and thallium, all COCs are removed to concentrations below the environmental guidelines. For aluminum, the guideline is achieved in the first stage where the pH is held at 7.0. However, it appears that some sludge leaked into tank 2 and Al was dissolved. in that tank. Note that according to Figure 1, Zn and Cd should not begin to precipitate as hydroxides until after pH 7.0. However, significant portions of these two COCs are removed in the first stage. This most likely is due to adsorption onto the oxyhydroxide precipitates of Al and Fe(III). $FeCl_3$ was added to tank 2 and it was immediately precipitated as iron oxyhydroxide. This acted not only as a flocculent but also as a center of adsorption for other trace constituents in the water that was delivered to tank 2. Thus, Cu, Cd, and Zn all were reduced to below the environmental guidelines when Fe and Mn oxyhydroxides precipitated in tank 2. Finally it is important to note that arsenic concentrations were reduced in the pH 7.0 stage and then further reduced in the pH 10 stage. Up to pH 7.0, much of the arsenic removal may be attributed to the formation of $FeAsO_4$. However, above a pH of 7 adsorption of arsenic onto oxyhydroxides probably plays a significant role (Pantuzzo et al. 2008).

Table 2. Constituents in mg/L in a bench-scale treatment test using hydrated lime

	Initial ARD	Tank 1	Clarifier 1	Tank 2	Tank 3	Clarifier 2	Aquatic Stds.*
pH	2.8	7.0	7.0	10.0	10.0	10.0	6.5–9
Al	70	0.063	0.025	0.176	0.169	0.160	0.1
Fe	188	0.0056	0.021	0.023	0.010	0.018	1.0
Mn	14.2	8.9	9.3	0.151	0.0089	0.0068	1.0
Cu	49	0.181	0.37	0.0027	0.0022	0.0022	0.01
Zn	18	3.5	4.4	0.0127	0.0090	0.0087	0.1
Cd	0.75	0.34	0.36	0.0006	0.0003	0.0003	0.005
Pb	0.024	<0.0015	<0.0015	<0.0015	<0.0015	<0.0015	0.05
As	0.35	0.0062	0.0069	<0.004	<0.004	0.0050	0.01†
Tl	0.111	0.104	0.110	0.055	0.052	0.048	0.002†

*Approximate aquatic guidelines for U.S. waters (USEPA 2009).
†Primary drinking water standard.

This bench-scale test was one in a series of tests that were conducted to determine how well thallium could be removed from ARD (Vatterrodt et al. 2012). The objective of this specific test was to determine how much thallium can be removed by adsorption onto Fe and Mn oxyhydroxides. In this test, thallium was spiked into the ARD to a concentration of 0.100 mg/L. Vatterrodt et al. 2012 found that Fe and Mn precipitates were the most effective in thallium removal. For this actual ARD, no thallium was removed in the pH 7.0 stage, and approximately 50% of the thallium was removed in the pH 10 stage. In synthetic ARD waters, the removal of thallium was above 50% when Fe and Mn were added to the solution. However, the tests were one stage processes where the pH was increased to 10 by the addition of lime. Clearly, more research is needed to devise a good removal scheme for thallium.

CONCLUSIONS

In the generation of ARD, the role of Fe(III) is important. It oxidizes all sulfides with the generation of hydrogen ions. These hydrogen ions along with the Fe(III) cause the release of other important COCs. In the treatment of ARD, Fe(III) and Al(III) are the primary source of mineral acidity and must be removed before other COCs can be removed and the water made suitable for aquatic organisms. Oxidation of Fe and Mn is also important in treatment. Because these constituents are more rapidly oxidized at pHs above 7, it is important to add alkalinity and then aerate rather than reversing the two procedures. Finally, generation of complex ions at high pH values can frustrate the removal of metals such as Al, Zn, Pb, and Cu. However, adsorption and co-precipitation of metals that tend to form complex ions by oxyhydroxides of Al, Fe, and Mn can promote the removal of these metals at high pH.

REFERENCES

Bucknam, C., Snyder, D., Wildeman, T., Sexsmith, K., and Barazzoul, L. 2012. Oxic and anoxic tailings slurry aging studies for environmental assessment of a proposed mining and processing development with emphasis on cyanide species., Proc. of 9th International Conference on Acid Rock Drainage (ICARD).

GARD Guide. http://www.gardguide.com/index.php/Main_Page.

Gusek, J.J., and Figueroa, L.A. 2009. *Mitigation of Mining Influenced Water*, Volume 2. Littleton, CO: SME.

Langmuir, D. 1997. *Aqueous Environmental Chemistry*. Upper Saddle River, NJ: Prentice-Hall.

McLemore, V.T. 2008. *Basics of Metal Mining Influenced Water*, Volume 1. Littleton, CO: SME.

Miller, A., Figueroa, L., Wildeman, T., 2011. Zinc and nickel removal in simulated limestone treatment of mining influenced water. *Applied Geochemistry* **26**(1): 125–132.

Murphey, K., Bucknam, C., Wildeman, T., Sexsmith, K., Barazzuol, L., and Chapman, J. 2009. Oxic and anoxic tailings slurry aging studies at the Hope Bay Project, Nunavut. In, *Tailings and Mine Waste '09*. David Sego et al. Eds, November 1–4, 2009, Banff, Alberta, Canada.

Office of Surface Mining. http://www.osmre.gov/browse.shtm#azindex.

Pagenkopf, G.K. 1983. Gill surface interaction model for trace metal toxicity to fishes: Role of complexation, pH, and water hardness. *Environ. Sci. Technol.* 17: 342–347.

Pantuzzo, F.I., Ciminelli, V.S.T., and de Brito, W., 2008. New evidences for the role of precipitation and adsorption during Fe(III)-As(V) coprecipitation. In *Hydrometallurgy 2008*. Edited by C.A. Young, P.R. Taylor, C.G. Anderson, and Y. Choi. Littleton, CO: SME.

Plumlee, G.S., Smith, K.S., Montour, M.R., Ficklin, W.H., and Mosier, E.L. 1999. Geologic controls on the composition of natural waters and mine waters. In *The Environmental Geochemistry of Mineral Deposits, Part B.* Case Studies and Research Topics. Edited by L.H. Filipek and G.S. Plumlee. Reviews of Economic Geology, Vol. 6A. Littleton CO: Society of Economic Geologists.

Rose, A.W., Hawkes, H.E., and Webb, J.S., 1979. *Geochemistry in Mineral Exploration*, 2nd Ed. London: Academic Press.

Santore, R.C., Di Toro, D.M., Paquin, P.R., Allen, H.E., and Meyer, J.S. 2001. Biotic ligand model of the acute toxicity of metals. 2. Application to acute copper toxicity in freshwater fish and *Daphnia*. *Environmental Toxicology and Chemistry*, 20, 2397–2402.

Stumm, W. and Morgan, J.J. 1996. *Aquatic Chemistry, 3rd Ed.* New York, NY: Wiley Interscience.

Twidwell, L.G., and Leonhard, J.D., 2008. Removal of cadmium, copper, nickel, and zinc from aqueous solution by adsorption of ferrihydrite and long-term storage stability of the metal loaded product. In *Hydrometallurgy 2008*. Edited by C.A. Young, P.R. Taylor, C.G. Anderson, and Y. Choi. Littleton, CO: SME.

USEPA, 2009. National Recommended Water Quality Criteria. United States Environmental Protection Agency. Office of Water (4304T). Available at http://www.epa.gov/scitech/swguidance/standards/criteria/current/index.cfm.

Vatterrodt, K., Davies, M. Figueroa, L., Wildeman, T., and Bucknam, C., 2012. The effects of aluminum, iron, manganese, and hydrogen peroxide on thallium removal. Proc. of 9th International Conference on Acid Rock Drainage (ICARD).

Wehrli, B., and Stumm, W. 1989. Vanadyl in natural waters: Adsorption and hydrolysis promote oxygenation. *Geochim. Cosmochim. Acta*. 53: 69–74.

Wildeman, T.R. 1991. Drainage from Coal Mines: Chemistry and Environmental Problems. In: *Geology in Coal Resource Utilization*. Edited by D.C. Peters, pp. 499–511, Fairfax, VA: Tech Books.

Wildeman, T.R., Cain, D., and Ramirez, A. 1974. The relation between water chemistry and mineral zonation in the Central City Mining District, Colorado. *Am. Water Res. Assn., Proc.* No. 18, p 219–229.

Wildeman, T.R., and Schmiermund, R., 2004, Mining influenced waters: Their chemistry and methods of treatment. In *Proceedings of the 9th Billings Conference and 2004 National Conference of the. American Society for Surface Mining and Reclamation*, pp. 2001–2013.

Wildeman, T.R., Dinkel, J.W., Smith, R.M., and McCallister, M.L., 1997. Field assessment of Fe(III), Al, and dissolved O_2 for passive treatment options. In: *Proceedings of the Fourth International Conference on Acid Rock Drainage*, Volume IV, pp. 1659–1672. MEND Ottawa, Ontario: MEND Canada.

ABOUT THE AUTHORS

Thomas Wildeman has been teaching, doing research, and consulting for more than 45 years and is currently professor emeritus of chemistry and geochemistry at the Colorado School of Mines. He and Jim Gusek from Golder have been developing passive treatment systems for 25 years. Along the way, they have twice been awarded the Colorado and National award from the Consulting Engineers Council in the environmental engineering division for work on passive treatment. In 2006, Wildeman was awarded Researcher of the Year by the American Society of Mining Reclamation. Recently, he and Charles Bucknam from Newmont have been working on active treatment and tailings slurry assessment.

Katherine Vatterrodt is currently pursuing her master's degree in environmental science and engineering at the Colorado School of Mines. She works closely with Linda Figueroa and Thomas Wildeman the Colorado School of Mines and Charles Bucknam at Newmont Mining to assist in the research and development of new water remediation methods for mining influenced water. Before attending Colorado School of Mines, Vatterrodt attended the University of Minnesota–Twin Cities where she received her B.S. degree in recreation resource management along with minors in Spanish and environmental science, policy, and management. The focus of her degree was sustainable development in Central America, specifically Costa Rica, where she studied for several weeks. Vatterrodt graduated in May 2013 with a developed skill set in land/water use policies, geochemistry, and water treatment for mining influenced waters.

Linda A. Figueroa is an associate professor of civil and environmental engineering and the associate director of the Nuclear Science and Engineering Center at the Colorado School of Mines. She is a registered professional engineer and has more than 27 years of experience in research and design in environmental engineering. Before joining CSM, she held environmental engineering positions with private industry. Her research emphasis since coming to CSM in 1990 is on the microbially facilitated immobilization of metals and radionuclides. She is an author of approximately 70 journal and conference papers in the area of treatment and remediation and is co-editor of *Mitigation of Metal Mining Influenced Water* published by SME in 2009. Her current projects are in the areas of active and passive treatment of metals and radionuclides as well as source control and resource recovery.

A Perspective of Innovations in Tailings Management

Dirk van Zyl
University of British Columbia, Vancouver, British Columbia, Canada

ABSTRACT: This paper presents an overview of innovations in tailings management from the author's perspective. It reviews both technical and procedural/process related advances that represent innovations. During the 20th Century tailings management became a reality and the innovations were driven by higher expectation from stakeholders and shareholders as well as technological advances. It is today possible to reach the goal of protecting human health and safety and the environment at all tailings management facilities. The next area of innovation will be in the broader training and capacity building of future mine materials engineers to serve industry and governments around the world.

INTRODUCTION

During the 20th Century tailings disposal changed from "out the back door and down the hill" to "tailings management." This period has been marked by many tailings failures that impacted communities, the environment and the economic well being of companies. During this period there was also a sea change in the regulatory environment on national and international scales. The expectations from civil society, governments and shareholders were responsible for these improvements in practice.

A recent publication (Minalliance 2012) summarizes 100 innovations in the mining industry. The selection criteria used to list an innovation were: the innovation must improve performance, the innovation must be useful to the Québec industry (the publication was compiled in Québec) and the innovation must currently be in use or soon be used by the industry. This publication focuses on the mine life cycle and the tailings related innovations are all listed in the "remediation" or closure related section. They include: reusing mine tailings, neutralization of tailings, reusing bauxite waste (especially the red mud tailings), and solar panels for tailings sites.

This paper does not set out to provide a state-of-the-art review of tailings management in the early years of the 21st Century, it instead provides a brought review of changes that occurred that were innovative in the opinion of the author, it is therefore a personal perspectives of changes in practice and management culture as well as the incentives from influences external to the industry. These innovations include both technical issues (such as tailings deposition methods) and process/procedural issues (such as regulations, guidance documents, etc.). Due to space limitations the innovations are described in broad terms. A large body of literature is available on tailings management and much of it has been published since the 1970s.

The next section of the paper discusses the use of terminology followed by reviewing an updated paradigm of tailings and overall mine material management, which is beginning to take hold in the industry. This is followed by a review of technologies and their implementation during the last century. Regulatory and corporate governance requirements and guidelines represent an important aspect of present day tailings management and some examples will be provided. Finally, the importance of site-specific solutions to tailings management is emphasized; one size definitely does not fit all.

TERMINOLOGY

It is still accepted practice to use the terms *tailings dams* and *mine waste dumps*. These terms do not provide a clear sense that the industry "manage" the tailings and mine rock materials. Some would argue that the use of specific terminology is steeped in regional practices and preferences, however the mining industry is a global industry where it is appropriate to consider the use of global terminology.

Tailings Management Facility

It is proposed that the term *Tailings Management Facility (TMF)* be adopted to describe the facility where the tailing material is placed, stored and finally closed and reclaimed. The term Tailings Storage Facility is also used in practice but it does not convey a clear sense that the tailings material is managed.

The term *Tailings Dam*, used in South Africa and elsewhere, does not provide any comfort to this author as a number of these "dams" have failed due to water being stored on them, e.g., Merriespruit (Wagener et al. 1998). Most TMFs are not designed as water storage facilities and the term *dam* may provide a false sense of confidence that water can be placed on the facility.

Rock Management Facility*

It is proposed that the term *Rock Management Facility (RMF)* be used to describe the facility where mine rock is managed. A mine waste dump can convey a number of negative images; it could be considered that the mine waste (something that nobody wants) will be dumped (without specific care be taken). It is proposed that RMF or Mine Rock Management Facility (MRMF) be used to emphasize that a management plan is in place for these materials.

UPDATED PARADIGM FOR TAILINGS AND MINE MATERIAL MANAGEMENT

Klohn (1972) expresses the typical view of the time: "Tailings are a waste product of the mining industry. Their disposal adds to the costs of production; consequently, it is desirable to dispose of the tailings as cheaply as possible."

About 20 years ago the author was on the way to a new mining project in the Western United States in the company of John Haan, at that time with Hecla Mining Company. During the discussion John suggested that the mining industry should rethink their paradigm for mining similarly to the paradigm shift that took place in the rental car industry. In the rental car industry the companies realized that they are more in the car selling business than the rental business because the resale value of the cars is an important economic consideration. This shift resulted in the rental car companies buying cars with more features that prospective buyers will find attractive.

The paradigm shift proposed for mining is that mining be considered a large-scale "waste management" activity and that the profit of a mining company is the amount remaining after paying

* It is recognized that this paper does not deal specifically with mine rock management; however, this short section is included to emphasize the importance of terminology in handling the excess materials from the mine and the mill.

for the management of these materials. While this is not universally accepted and practiced, many mining companies understand the importance of mine materials management as well as the costs related to managing these materials. The decisions of managing these materials throughout their whole life cycle are major drivers in selecting mine material management options.

PAST AND PRESENT APPROACHES AND TECHNOLOGIES FOR TAILINGS MANAGEMENT

This section will discuss a number of past and present approaches and technologies for tailings management. While some of these are difficult to recognize as tailings "management" practices as they do not do not necessarily do a good job of "managing" the materials, they are included in the discussion to provide a perspective of how much progress has been made. The list of approaches and technologies are not necessarily exhaustive but are meant to provide the major milestones.

Defunct Approaches

This section provides a short description of tailings disposal practices that are not used anymore but provide insights in the progress made in managing tailings during the last century.

Out the Back Door or Into Rivers

There are many examples in old mining districts around the world where tailings material was discharged directly out the "back door" of the processing facility into the creek or other low spot downstream. In most cases no containment structures were constructed and the tailings came to rest as they dewater and the shear strength increased sufficiently to stop flowing.

There are also a number of projects where the tailings material was directly deposited into streams. This was done in Butte, Montana, as well as the mines along the Coeur d'Alene River, Idaho. Tailings deposition into the Coeur d'Alene River was only banned in 1968, until then multiple mines deposited tailings into the river. Both the Clark Fork and Coeur d'Alene rivers had been listed under Superfund and are in various degrees of progress.

Tailings discharge in river systems also took place in other parts of the world, e.g., Chile and Peru. In both cases large scale deposits formed on the beach where the tailings entered the ocean. At Chañaral in Chile a large copper tailings deposit along the beach changed the beachfront of the town and has also resulted in environmental and human health concerns (e.g., Wisskirchen and Dold 2006). Deposition of tailings into the El Salado River took place from 1926 to 1975.

Tailings Management in the 20th Century

The following sections will provide a high level discussion on some of the technologies and approaches in the 20th Century that the author considers as innovative. Some of these innovations were regional in nature while others had larger global importance, especially those that were introduced towards the end of the 20th Century.

South African Gold Mine Tailings Practices

The South African gold mine tailings practice became well established during the 20th century. Ruhmer (1974) provides an extensive description of the practice followed by the Anglo American Group. It is common to refer to these TMFs as "slimes-dams," a local term that is still used. The majority of South African gold mines are located in fairly flat terrain in semi-arid climate and the practices evolved around that aspect as well as the fact that they were initially not constructed much taller than 100 ft. The general shape of these facilities is rectangular or some other geometry that is determined by land ownership. The TMF is raised using tailings and the wall is constructed through upstream construction. The initial earthworks for the wall construction consist of a series of trenches and small embankments. From the outside a surface water trench sized to intercept surface water runoff from the catchment. A low berm next to it helps in containing the flows. A space of about 40 ft is left to he effluent drain trench which transfers the mill water return from the penstock to the mill water return dam. A low berm is also constructed next to this, which forms the outside wall of the sediment collection "paddy dams" next to the toe of the wall. Sediment that erodes from the wall is collected in these paddy dams that are located along the circumference of the wall. Another low berm is constructed about 50 ft from the previous which forms the inside wall of the "paddy dams" as well as the toe for the wall. Transverse walls are constructed between these outer and inner walls to form the "paddy dams" along the toe. The inner wall trench and berm is located about 60 ft from the toe wall. Trenches on the inside of the toe wall berm and inner wall berm are important features for the start of operations. A series of cross walls are constructed between the toe wall and the inner wall to establish paddocks for deposition of the tailings.

Tailings deposition starts in the paddocks between the toe wall and inner wall and the paddocks are filled to within about 9 in. from the top of the small berms. The berms are then raised with tailings excavated from the paddock and placed on top of them. Up to about 1980 this used to be a very labor intensive operation using hand shovels to construct the raises. Mechanical equipment is now used to accomplish that.

A key part of the deposition scheme is that the paddocks are flooded with slurry tailings and the free water plus fine materials that settles the slowest is decanted to the center area of the TMF. During the night all tailings are discharged into the center part of the TMF and allowed to sediment, water is decanted through a penstock system. The outer paddocks and raised wall is referred to as the "day wall" as it is constructed under very well controlled conditions. Ruhmer (1974) provides very good guidance as to the detailed depositional schemes. The day wall is raised at a higher rate than the internal part of the TMF and therefore provides sufficient freeboard. It is also allowed to dry in the semi-arid climate and therefore provides a stable configuration. The rate of rise of the TMF, based on experience and analytical evaluations, is kept to about 8 ft/year, which allows for complete consolidation of the deposited tailings.

The construction method for gold TMFs in South Africa resulted in stable structures that could be constructed economically while also providing environmental protections. Variations to the techniques were developed but for the state of technology, as well as the imperative of the 20th Century to limit the amount of expenditure on "mine waste," the approach served the industry well.

Slurry Tailings Construction for Improved Seismic Stability

Klohn (1972) provides an excellent description of mine tailings design and construction approaches used in practice at the time. The advantages and disadvantages of upstream, downstream and centerline construction methods are discussed as well as the importance of using cyclones to separate coarse and fine tailings and compacting the sand to improve stability and provide better seepage collection controls. He emphasizes that: "A good tailings dam design must satisfy the basic requirements of safety, pollution control, storage capacity and economy." The liquefaction failure of a number of Chilean tailings impoundments that were constructed using upstream methods (Dobry and Alvarez 1967) changed the accepted practice in Chile. Considerable advances were made during the last 30 years in improving TMF practices in seismically active regions (Barrera et al. 2011). Compaction of the sand embankments is often done to further improve the seismic stability of the TMF.

Slurry Tailings Deposition

Slurry tailings deposition progressed from point discharge of the total slurry stream to carefully planned spigot discharge from a series of discharge points. Single point discharge results in the formation of a large plunge pool and meandering slurry streams on the surface of the impoundment. The segregation of particles can occur along the beach but is poorly controlled under the high flow velocities. Carefully planned spigot discharge distributes the flow more evenly over a larger area of the beach, ideally resulting in laminar flow of the slurry along the beach and much more controlled particle segregation. Ideally coarser particles* settle near the deposition point providing a deposit with higher permeability than the deposited fines further along the beach. This is advantageous for upstream construction as the wall area will be well drained and will consolidate reasonably rapidly allowing upstream construction to proceed.

Slurry tailings can also be deposited in a similar fashion in a TMF with downstream constructed embankments of mine rock. Such construction provides excellent seismic stability of a TMF.

The beach slope of slurry deposited tailings is dependent on many factors including the specific gravity of the solids, the solids content of the slurry, beach flow characteristics of the slurry and the particle size distribution of the tailings. Beach slopes in the order of 0.5% to 1% are typically achieved. The overall beach profile is convex.

Slurry deposition can be successfully done over a range of slurry solid contents. This is important as the solids content of slurry delivered from the mill may vary on a daily basis.

Cyclone separation of coarse and fine particles in a slurry provides excellent embankment construction material for upstream, downstream and centerline construction. The challenge is to produce a consistent product when there is variation in slurry solids content.

Thickened Tailings

The storage efficiency of a TMF can be improved if a steeper beach can be established as well as implementing discharge from a central point instead of from the top of the wall. By thickening the slurry to higher percent solids, say 55% and higher (depending on the rheological behavior of the material), will result in a steeper beach slope. This concept was first proposed by Robinsky (1999) and implemented at the Kidd Creek mine in Ontario. Much refinement took place over a long period to improve the reliability of this deposition technique and it is currently used at a number of TMFs around the world.

* It is more accurate to express this as heavier particles settle near the deposition and lighter particles further down the beach. While this translates into larger and smaller particles if the specific gravity of the particles is uniform, it is clearly not the case when there are differences in the specific gravity of particles.

Filtered Tailings

Filter press and belt filter technology have been widely implemented in various applications, including mineral extraction. In some of these plants the tailings material was filtered for full recovery of metals, or other purposes, and then re-slurried for transport to the TMF. There are significant advantages having filtered tailings instead of slurry tailings. The filtered tailings can be transported by conveyor belt or truck and the filtered tailings stack can be constructed using earthmoving equipment thereby forming a stable deposit that can be reclaimed as it is developed. An early application of this technology in North American mining was at the Greens Creek mine near Juneau, Alaska in the early 1980s.

Paste Tailings

The development of deep thickeners provided the opportunity to thicken a total tailings stream to high solids content. This technology was first used for the production of backfill material in underground mines in the late 1980s. Since then this technology has been expanded to surface tailings deposits and requiring much higher capacity thickening equipment. Paste tailings technology received considerable international attention over the last 15 years resulting in a series of international conferences held annually in various locations (Paste 2013, June 2012, Belo Horizonte, Brazil, is the 16th annual meeting) as well as Guidance Documents (Jewel et al. 2006).

Consistent production of uniform quality paste is difficult when there are significant fluctuations in tailings slurry solids contents. The beach slope of paste tailings is higher than that for slurry tailings as well as thickened tailings and there is no segregation of particles along the beach. Research is still underway on the liquefaction resistance of paste tailings.

GOVERNANCE

Governance, as used here, refers to the legal and regulatory frameworks of a jurisdiction, the policy and guidance in companies, guidance documents issued by Associations for use by their members (and others) as well as the implementation of independent review boards by companies during the design as well as operation of a TMF. While some of these documents contain design requirements they most often are more "process" related, i.e., focused on the design and construction of a TMF that will be protective of human health and safety and the environment. The combined outcome of all these governance aspects is that more consistency and higher reliability in TMF performance is achieved internationally.

Government Regulations and Guidelines

Extensive changes in government regulations and guidelines took place in the 20th Century and played a major part in the innovations in tailings management. Many of these documents are now relying on "good practices" in other jurisdictions and this all help to "level the playing field" and result in higher quality practice. Similarity of regulatory frameworks may not result in similarity of implementation due to the inconsistent nature of regulatory capacity, both financial and professional.

Corporate Policies

Most large corporations have clear policies and guidelines for the design, operation and closure of TMFs. These policies are typically reserved for internal use and they are not public documents.

Consultants working with a specific company are expected to follow these; similarly all operations of a company should follow these, thereby resulting in consistent practices within a company.

Mining Associations

Mining Associations serve their members, both on the legislative and regulatory front as well as the dissemination of technical information. Mining associations have different missions and approaches and few focuses specifically on tailings management. The Mining Association of Canada (MAC) developed a series of guidance documents for the design, operations, and review of TMFs:

- *A Guide to the Management of Tailings Facilities*
- *Developing an Operation, Maintenance and Surveillance Manual for Tailings and Water Management Facilities*
- *A Guide to the Verification of Tailings Facility Management*
- *Towards Sustainable Mining Assessment Protocol—A Tool for Assessing Tailings Management Performance*

Committees consisting if representatives of the member companies as well as experienced consultants compiled these documents. All MAC members are expected to implement these guidance documents. These documents are public and can be used by the international mining community. Not only are the topics and approaches innovative, making these documents freely available on the MAC website is also innovative.

Review Boards

A very significant development during the last 10 to 20 years of the 20th Century was the formation of independent review boards for reviews of TMF designs and operational performance. These review boards are appointed by the owner of the mine to provide direct feedback to them. They review the work of the mine's consultants and mine personnel as it relates to the TMF. Review Boards were originally used in the design and construction of water storage dams during the last 40 years or so of the 20th Century.

Morgenstern (2010) discusses a number of advances that led to the improved safety of tailings management facilities. He introduced review boards as follows: "The appointment of an Independent Tailings Dam Review Board (ITRB) to provide third party advice on design, construction, operation and closure of tailings dams has become increasingly common and is recognized to provide value. World Bank and other Lenders Groups are requiring formation of an ITRB. International Finance Corporation/World Bank guidance and operating principles OP4.01 and OP4.37 establish the requirement to review the development of tailings dam design, construction and initial dam filling. Maintaining an ITRB through operations and closure will depend upon the scale of the facility."

Other

Government and industry joint efforts have also resulted in innovative approaches to tailings management. Two examples are: the Code of Practice for Mine Residue Published by the South African Bureau of Standards in 1998 (SABS 1998) and the Tailings Management Handbook published by the Australian Government (DRET 2007).

The International Commission on Large Dams (ICOLD) has published more than 10 reports on mine tailings since 1982. These publications are compiled by tailings practitioners from around the world that serves as country representatives on the ICOLD committee.

A review of innovations in tailings management is incomplete without acknowledging the role of the internet and the ease of access to large volumes of information from diverse sources. While it is easy to search the internet on any of the topics in this paper a website that presents useful summary information is www.tailings.info. There is also considerable interest in posting conference proceedings on the web for free access, e.g., that of Tailings and Mine Waste 2011 is available at https://circle.ubc.ca/handle/2429/37210 and that for Tailings and Mine Waste 2012 will be posted in the same area in 2013.

SITE-SPECIFIC DESIGN

Development of a tailings management plan for a mine must take into account mine site-specific conditions. These include the physical aspects of the site (location, topography, seismicity, climate, etc.) as well as the regulatory framework of the jurisdiction where the mine is located. There are no "cookie cutter" solutions and designers learn from their previous experience and from each other. It is essential that the design, operations and closure for each TMF is focused on the specific site conditions.

CLOSING REMARKS

This paper provides a personal view of the author on innovations in tailings management during the last Century. Much has changed in how tailings management is approached by industry and regulators. Technologies have changed significantly and so have the governance aspects of tailings management. Both regulatory and industry expectations have resulted in more uniform guidance and approaches in the design, operations and closure of TMFs.

It is today possible to reach the goal of protecting human health and safety and the environment at all TMFs. While experienced individuals are available for the successful design, construction, operations and closure of TMFs many more will be needed in the 21st Century. The next area of innovation will be in the broader training and capacity building of future mine materials engineers to serve industry and governments around the world.

REFERENCES

Barrera, S., Valenzuela, L. and Campaña, J. 2011. Sand Tailings Dams: Design, Construction and Operation, In: *Tailings and Mine Waste '11*: Proceedings of the 15th International Conference on Tailings and Mine Waste. Vancouver, BC. Boca Raton FL: CRC Press. 13 pp. http://hdl.handle.net/2429/38018. Accessed December 2012.

Department of Resources, Energy and Tourism (DRET). 2007. Tailings Management. Leading Practice Sustainable Development Program For the Mining Industry. 88 p. http://www.ret.gov.au/resources/Documents/LPSDP/LPSDP-TailingsHandbook.pdf. Accessed December 2012.

Dobry, R. and Alvarez, L. 1967. Seismic Failures of Chilean Tailings Dams. *J. of the Soil Mechanics and Foundations Division,* Proc. 93. SM6. New York: ASCE. pp. 237–260.

Jewel, R.J. and Fourie, A.B. (Ed.). 2006. Paste and Thickened Tailings—A Guide. 2nd Edition. Australian Centre for Geomechanics. Perth. 256 pp.

Klohn, E.J. 1972. Design and Construction of Tailings Dams, *CIM Bulletin.* April. Québec: CIM. pp. 28–44.

Minalliance. 2012. 100 Innovations in the Mining Industry. http://www.minalliance.ca/pdfs/Minalliance_100_innovations_en.pdf. Accessed December 2012.

Morgenstern, N.R. 2010. Improving the safety of mine waste impoundments. In: *Tailings and Mine Waste '10:* Proceedings of the 14th International Conference on Tailings and Mine Waste. Vail, CO. Boca Raton FL: CRC Press. pp. 3–10.

Robinsky, E.I. 1999. Thickened Tailings Disposal in the Mining Industry, Toronto: E.I. Robinsky Assoc. 210 pp.

Ruhmer, W.T. 1974. Slimes-dam Construction in the Gold Mines of the Anglo American Group. *Journal of the South African Institute of Mining and Metallurgy,* February. Johannesburg: SAIMM. pp. 273–284.

SABS. 1998. *Code of Practice—Mine Residue.* South African Bureau of Standards. SANS 10286 Edition 1. 127 pp.

Wagener, F. von M., Craig, H.J., Blight, G., McPhail, G., Williams, A.A.B. and Strydom, J.H. 1998. The Merriespruit Tailings Dam Failure—A Review, In: *Tailings and Mine Waste '98*: Proceedings of the 5th International Conference on Tailings and Mine Waste. Rotterdam: Balkema. pp. 925–952.

Wisskirchen, C. and Dold, B. 2006. The Marine Shore Porphyry Copper Mine Tailings Deposit at Chañaral, Northern Chile. In: *7th ICARD Conference.* Littleton: SME. pp. 2,480–2,489.

ABOUT THE AUTHOR

Dirk van Zyl is professor and chair of mining and the environment at the Norman B. Keevil Institute of Mining Engineering, University of British Columbia, Vancouver, British Columbia. Van Zyl has more than 35 years of experience in research, teaching, and consulting in tailings and mine waste rock disposal and heap leach design. During the last decade, much of his attention has been focused on mining and sustainable development and mine life cycle management. Van Zyl is a registered professional engineer and has consulted internationally on many mining projects. These projects covered the whole mining life cycle, from exploration to closure and post-closure, in a large range of climatic and geographic environments. Most of this work has been focused on geotechnical and environmental mining engineering aspects to provide solutions for environmental and human health protection.

Innovations in Water and Brine Treatment Using Membrane Technology

Larry A. Lien
Membrane Development Specialists, Solana Beach, California, United States

ABSTRACT: Membrane technology has been selectively applied over the past 20 years to a variety of waste and process waters. With the proper pretreatment and process design these membrane based water treatment plants have been both reliable and robust. The opportunity exists to expand the use of membrane separation technologies directly into process design circuits which will lead to increased efficiencies in metals extraction and simultaneously recover and reuse process waters that ultimately reduce wastes.

INTRODUCTION

Basic membrane technology is about 60 years old. Initially membrane technology was part of the nuclear weapons industry that morphed into nuclear power industry post–World War II. Membranes were developed to physically separate and concentrate uranium compounds and as the membrane technology developed it was discovered these same membranes could remove salts from water. In the 1960s spiral wound concept for reverse osmosis membranes was developed which allowed for high packing density of the membrane in a relatively small volume. Simultaneously a burgeoning nuclear power industry in 1960s gave the promise of inexpensive energy. Since electric energy for pressure driven membrane systems was, and still is, the major operating cost for reverse osmosis membranes, inexpensive electrical power gave great promise to the fledgling industry. Being able to inexpensively desalt seawater using spiral wound reverse osmosis membranes would alleviate water shortage problems around the world.

The first practical reverse osmosis membranes to remove salts from water were cellulose acetate. These membranes were applied widely and many applications were tried that were not practical based on the limitations of cellulose acetate. The cellulose acetate will hydrolyze in both low and high pH. Many of the first membrane applications in mining and metal finishing were to remove metal ions from low pH ARD or plating waste waters, but the membrane life was short and replacement costs high. Furthermore cellulose acetate was vulnerable to bacterial attack, so the membrane technology got reputation of being touchy and delicate.

In the 1970s and early 1980s great strides were made to develop non-cellulose membranes. The resulting products were called thin film composites membranes that were made from chemically stable polymers that would neither hydrolyze nor become food for bacteria. As this technology evolved a group of nylon based polymers were used as the barrier layer to reject salts. In addition to having much higher removal of salts than cellulose, these new thin film nylon membranes operated at ½ to ¼ the pressure. The development thin film composite membranes brought robustness to

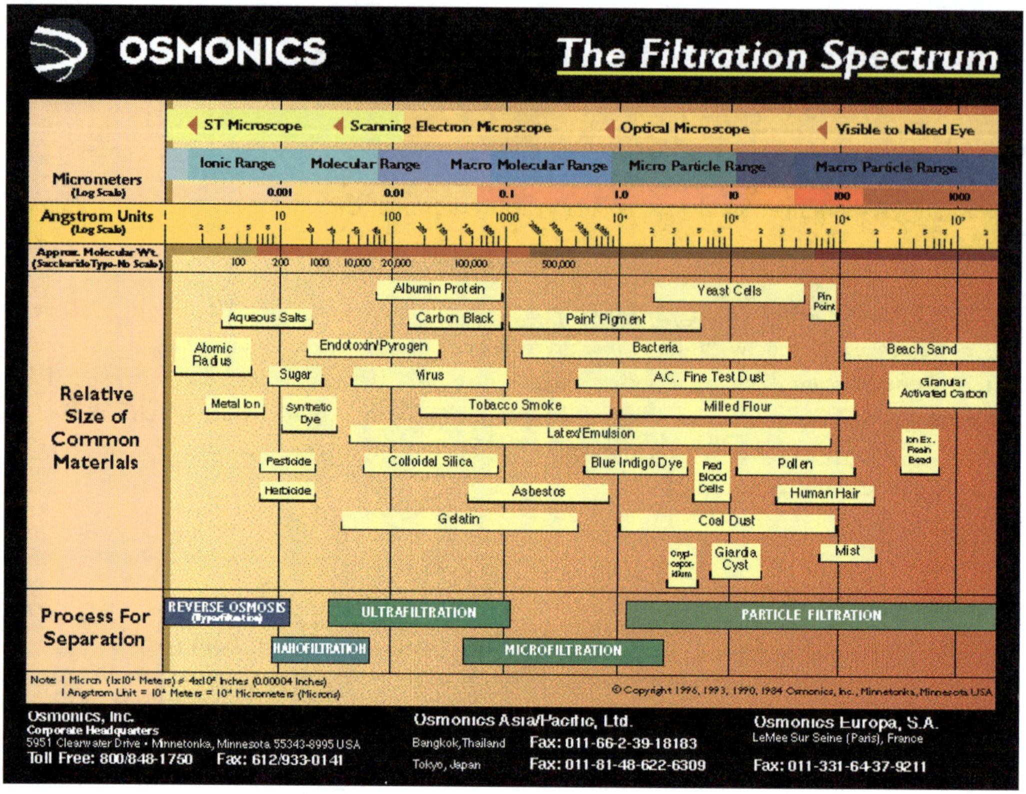

membrane technology that would allow membranes to be used in difficult and demanding mining applications.

As reverse osmosis thin film technology membranes developed the pore size of the membranes increased to impart unique separations. For example reverse osmosis membranes have a pore size of <0.001, but the polymeric layer beneath the thin film composite may have a pore size of 0.02 microns. Hence this substrate layer now has use as to reject proteins from milk whey or color particles from acid solutions. Now there are membranes that can be applied to separate ions, molecules, particles, proteins and bacteria from solutions to allow for concentration or purification processes. To illustrate this membrane separation spectrum a chart below gives the reader a sense of the types of separations possible.

CURRENT APPLICATIONS IN MINING

Using thin film composite membranes as the basis for engineered systems membrane technology has been widely applied to a variety of applications in mining, refining and metal finishing. These engineered membrane systems have been applied to large low pH ARD waste waters, that recovered copper and zinc that otherwise would have been lost to lime precipitation. And high pH cyanide barren concentrating cyanide for reuse and simultaneously producing water that can safely be discharged into the environment. A number of installations are listed below as an example of the breadth of applications and size of plants that can be engineered using modular designs.

Low pH applications:

- Phelps-Dodge (Freeport-McMoRan) 1992 (20 years), 40 m^3/h, pH 1 with an element life is 30–36 months
- Cananea de Mexicana ARD copper recovery 1000 m^3/h, pH 1.2
- Kennecott Copper (Rio Tinto) ARD 250 m^3/h pH 3.0 with copper recovery
- Korea Zinc Refinery low pH waste water 200 m^3/h pH 2.5
- Zijin Copper ARD 250 m^3/h with copper recovery permeate used for floatation at refinery

High pH applications:

- Newmont Yanacocha Barren Leach Solution 3500 m^3/h for cyanide recovery and membrane permeate water discharge to environment pH 10 on line since 2003
- Newmont Waihi Gold barren leach solution post-physical-chemical membrane permeate directly discharged into the environment 250 m^3/h
- Inmet Cobre Las Crucas Zero Liquid Discharge Facility 1000 m^3/h high pH ARD from copper mining permeate from membrane systems directly discharged into the environment
- Barrick Lagunas Norte 800 m^3/h barren leach solution post-physical-chemical treatment for direct discharge into the environment
- Barrick Pierina 110 m^3/h barren leach solution post-physical-chemical treatment for direct discharge into the environment

SYSTEMS DESIGN

These systems have functioned in difficult applications for several reasons. All applications were first pilot tested to develop a process system design, pretreatment and clean in place requirements. If there is a reason why membrane systems are not widely deployed it is because not enough emphasis is placed on two key areas: pretreatment and cleaning. The pretreatment has evolved from sand filters and bag filters to where 50% of the installations listed above use hollow fiber backwashable ultrafiltration 0.02 micron as cost effective pretreatment. This is the ultimate in pretreatment, because all colloids, bacteria and suspended solids are removed and now one can focus on the water chemistry, which in most mining applications is difficult and challenging.

The second important aspect of system design is clean in place system that cleans each stage of the system effectively and efficiently. This means valves and a cleaning system in place to use hot water and cleaning solutions to remove fouling rapidly. Many systems do not deploy semi-automatic cleaning skids, because it is expensive, but for difficult mining waters it is absolutely necessary to be able to clean the systems in place and return to service as rapidly as possible. My experience says there are no short cuts here, one simply needs to deploy and install the proper pretreatment and clean in place to deliver a reliable and robust system.

PROCESS AND ELEMENT DESIGN

Many of the mining applications require high recovery (conversion of feed to permeate) to reduce the amount of waste generated or increase the amount of metal (copper or gold) recovered in the concentrate. No free lunch here, because to increase the recovery, one has to rely on scale inhibitors to prevent precipitation of salts like calcium sulfate (gypsum) and calcium carbonate from forming in the membrane elements and special materials for element construction that also inhibit scale formation and if scale then forms allows for effective clean in place to recover and return the membrane systems to service.

In the past 10 years, cost-effective real-time local and remote expert monitoring of key measured system parameters allows one to understand when processes are out of control. For instance accurate measurement of conductivity for the feed, permeate and concentrate can tell an expert that the antiscalant is not being fed to the system properly or that the cleaning procedure just executed was not completely effective. In the past this information was sent to an expert weeks after the event and in many cases this was too late to clean and recover the membranes. Hence real-time monitoring can benefit everyone: end-users, operators and vendors.

It is inevitable that at some time all systems must be cleaned. As part of the pilot study process and years of experience allows one to select cleaning solutions that are efficient and cost effective. For instance, in systems where calcium sulfate precipitation may be an issue, those high chloride cleaning solutions are made available to the semi-automatic clean in place system to remove calcium sulfate scale. Likewise, the elements for this application would have special materials that inhibit the formation of calcium sulfate scale inside the element.

CONCLUSIONS

Currently there are several large membrane systems are deployed in the mining and refining industries but mostly on the waste water and water balance side of the process. As environmental permitting and sustainability issues face mining companies, governments and NGOs membrane technology can play a vital part of the solution to mining operators. In geographic areas where water balance is an issue, membrane technology can cost effectively produce water that can treat waste waters so they can meet the most stringent limits and safely be discharged into the environment. Conversely, in areas where make-up water supplies are limited, then waste waters can be recovered for reuse by the mine operators. Moreover, some of the waste waters contain recoverable metals otherwise lost, but are recovered as concentrates from the membrane process as the permeate water is reused as process make-up water.

The author believes that many more opportunities exist to improve mining and extraction processes if a membrane system were included in these process designs. For instance, nearly every high pH extraction process could use a high pH membrane element to concentrate metals and simultaneously produce clean caustic solutions for reuse back to the process. Likewise, nearly every low pH extraction process could use a low pH membrane element to concentrate metals from acids and return a purified acid back to the extraction process. Recovery of large volume extraction solutions and concentration of metals could reduce individual process foot print and dramatically reduce the amount of waste caustics and acids that are ultimately sent to waste treatment. Once these waste streams reach a waste treatment facility they longer have process value and are viewed, rightly so, as strictly operating cost. With lower and lower discharge limits for certain toxic metals treating waste water to discharge can become more and more expensive. However, recovering process streams inside the metal extraction and metal production facilities that could reduce extraction costs and minimize or eliminate waste is a real possibility and the author encourages every mining process to be evaluated in this manner.

ABOUT THE AUTHOR

Larry A. Lien has been involved in developing membranes, membrane elements, membrane applications, and equipment for these applications in the mining and metallurgical industries for 30 years. Starting at Desal in 1983, then through the acquisition of Osmonics and GE, until in 2006 establishing Membrane Development Specialists (MDS), Lien has worked on numerous mining applications using membrane technologies from acid rock drainage wastewaters to improvements in the hydrometallurgical processes. He is the owner and general manager of MDS.

Innovations in Chemical and Bacterial Water Treatment

Brian Park, Jay McCloskey, and Larry Twidwell
Montana Tech, Butte, Montana, United States

ABSTRACT: Innovations have occurred in chemical and bacterial water treatment in the mining and metals production industry. In general these have been in response to regulatory drivers, such as implementation of new water quality criteria for discharge, or reductions in water quality criteria for discharge. Some innovations have occurred in response to process improvement efforts, and more recently, innovations have occurred in response to the overall trend toward sustainability in the mining industry. This paper presents the major innovations that have occurred in chemical and bacterial water treatment, with discussion of why they were significant and why they occurred. Discussion on areas of future innovation and future drivers is also presented.

INTRODUCTION

As part of the Innovations Symposium associated with the 2013 SME Annual Meeting, this paper discusses key historical applied innovations that have occurred in chemical and bacterial water treatment in the mining and metals-production industry. There are a wide variety of waters produced in this industry, and there can be multiple uses for treated water.

The mining and metals-production industry produces the following types of waters that may require treatment:

- *Mining:* orebody dewatering; waste rock runoff or seepage; pit lakes; ground water pump-back; acid rock drainage (Note: Acid rock drainage is discussed in a separate paper for this symposium and will not be discussed in-depth in this paper).
- *Milling:* tailings return; unplanned excess water; bleed streams.
- *Heap leach:* rinsing for closure; unplanned excess water.
- *Smelting:* offgas treatment blowdown; acid plant blowdown; bleed streams.
- *Refining:* bleed streams.

These waters can range from fairly dilute groundwaters to highly concentrated process waters. They can vary widely in their characteristics, including pH, oxidation/reduction potential, total dissolved solids, total suspended solids, concentrations of major ions (calcium, magnesium, sodium, potassium, sulfate, chloride, bicarbonate, and possibly others), types and concentrations of contaminants (both metals and non-metals [e.g., cyanide, ammonia, nitrate]), radionuclides, organics, etc.

These waters may have varying desired end-uses including re-use at the mine for dust suppression, revegetation, drilling fluids, etc.; re-use in the mill, smelter, or refinery for process water; or discharge off-site to a natural waterway or aquifer. Any end-use has water quality requirements for the intended use.

Some of these waters may meet or nearly meet discharge requirements derived from water quality criteria, but many others do not. Among the common water quality issues observed historically are the following:

- pH (either high or low)
- Metals contamination (e.g., copper, zinc, lead, cadmium, nickel)
- Highly toxic metals contamination (e.g., arsenic, mercury, selenium)
- Non-metals (e.g., cyanide, nitrate, ammonia, oxygen demand)
- Organics
- Sulfate
- Total dissolved solids

For purposes of this paper, the term "innovative" will refer to technologies that (1) represent a new approach to solving a particular problem; and (2) have become fairly proven and adopted within the industry. Water treatment technologies are continually being developed, modified, combined with other technologies, bench-tested, pilot-tested, etc. Space limitations dictate that this paper take a high-level view.

INNOVATIONS IN CHEMICAL TREATMENT

Chemical treatment involves the usage of chemicals to accomplish a particular treatment goal. For purposes of this paper, this may include the use of physical treatment means (e.g., membranes) at times, since some innovative processes combine aspects of both.

Alkaline Treatment (Particularly Lime and Limestone)

Probably the single most significant chemical water treatment innovation in the mining industry has been alkaline treatment, and treatment using lime or limestone. Many mine waters and smelting and refining process waters and bleed streams are acidic, and sometimes very strongly so. Historically, these waters were simply discharged to nearby waterbodies, causing severe environmental and probably human health effects from acidity and dissolved metals. Simple neutralization with an alkaline reagent, such as limestone, lime, or sodium hydroxide (lye), significantly improved the water quality by removing acidity and greatly reducing most metal concentrations. **Figure 1** presents solubilities vs. pH for metals commonly occurring in mining operations. As seen in the figure, while all the curves are different, a pH of between 9 and 10 provides significant removal of all the metals shown. Lime and limestone are particularly effective, since they are relatively inexpensive, and they additionally may reduce sulfate concentrations by forming gypsum ($CaSO_4 \cdot 2H_2O$), if the initial solution contains high levels of sulfate. This results in an overall reduction in total dissolved solids, in addition to acidity and metals, further lessening the treated water's effect on the environment. **Figure 2** presents a typical lime treatment system utilized throughout the world.

High Density Sludge (HDS) Enhancement of Lime Treatment

Conventional lime treatment of acidic mining and process waters consists of raising the pH of the process stream to approximately 9–10. As described in the previous section, this causes most of the dissolved metals present to precipitate from solution as oxides or hydroxides, and if initial sulfate levels are high enough, this causes a portion of the sulfate to precipitate from solution as gypsum. The precipitated solids are settled and thickened in a thickener, with the thickener underflow either pumped to a disposal or drying pond, or dewatered in some fashion for off-site disposal. In either case, the settled solids are fine, "fluffy," slow to settle, and can be difficult to dewater. In general, the solids are difficult to handle.

The High Density Sludge, or HDS, process was developed as an enhancement to conventional lime treatment, providing a sludge that is much easier to handle. Specifically, HDS sludge typically settles faster and to a smaller volume, with much-improved dewatering characteristics. Figure 3 presents the basic HDS process configuration. The key feature of the HDS approach is to recycle the thickener underflow to a sludge conditioning vessel upstream of the primary reaction vessel, where it is mixed with lime slurry for a short time (e.g., several minutes). Neutralization treatment of the feed stream is then accomplished with this mixture. By following this approach, the newly precipitated metal oxide/hydroxides and gypsum form on existing recycled particles, rather than forming new, very small particles. In this way, larger particles are formed during extended operation, which settle faster and dewater much easier.

A number of variations of the basic approach have been developed, with minor differences in where certain streams are directed and mixed. A good discussion of these may be found in the GARD Guide, Chapter 7: Drainage Treatment (INAP 2010).

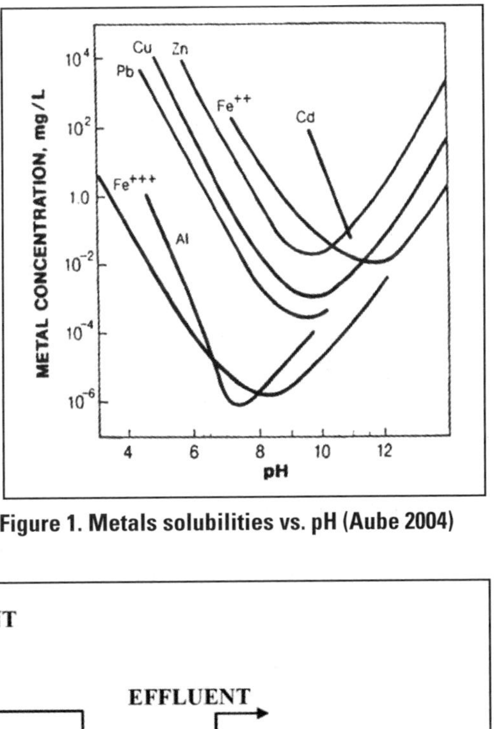

Figure 1. Metals solubilities vs. pH (Aube 2004)

Ferrihydrite Adsorption/Co-Precipitation

As alkaline neutralization became commonly implemented, and study and research progressed, it was observed that iron was a common contaminant in mining and process waters. This is not at all surprising, as iron is one of the most common elements in the earth's crust, and iron minerals are very commonly found in ore deposits. It was also observed that precipitated ferric iron (Fe^{+3})

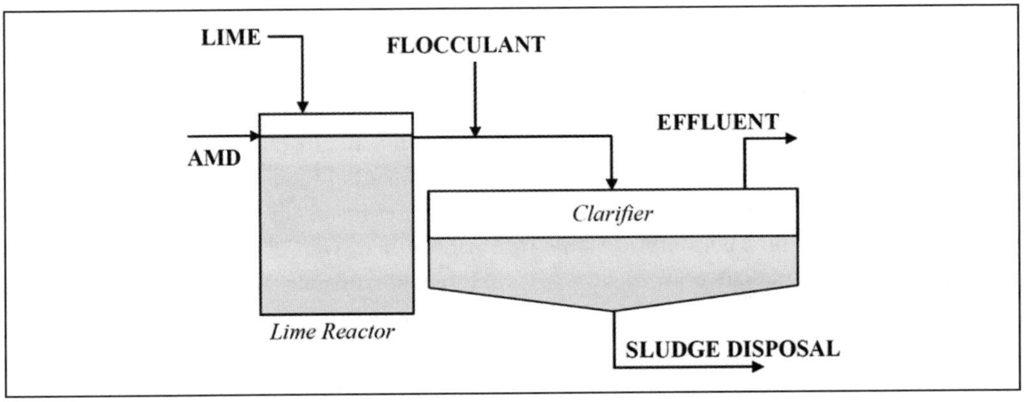

Figure 2. Conventional lime treatment plant (Aube 2004)

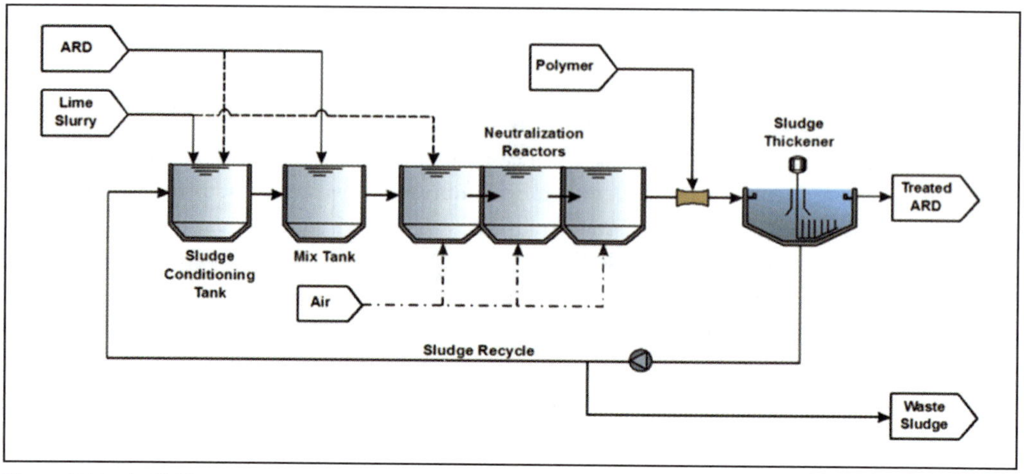

Figure 3. Basic HDS process configuration (INAP 2010)

possessed interesting properties for the adsorption of ionic constituents. Specifically, it was found that ferrihydrite had an ability to adsorb cations such as Cu^{+2}, Zn^{+2}, and Ni^{+2} at a pH above neutral, and an ability to adsorb anions such as arsenate (AsO_4^{-3}), selenite (SeO_3^{-2}), and molybdate (MoO_4^{-2}) at a pH below neutral. The application of these adsorptive properties was a significant innovation, particular for treatment of highly toxic constituents such as arsenic, since it was effective, relatively simple, and relatively inexpensive.

The precipitation chemistry of ferric iron is complex; it depends on the initial solution matrix, and the precipitated iron oxides/hydroxides can form multiple crystalline forms depending on conditions and aging time. However, in the context of water treatment applications it is most commonly referred to as ferrihydrite, with a formula of $5Fe_2O_3{\cdot}9H_2O$ (Paktunc 2010). The reason for the affinity of cations at higher pH and anions at lower pH is the ferrihydrite surface charge. The point of zero charge (pzc) for ferrihydrite in aqueous solution varies depending on solution conditions, but typically is between pH 7 and 9. At pH values below this range, the surface charge is positive, explaining the affinity for negatively-charged anions. Conversely, at pH values above this range, the surface charge is negative, with an affinity for positively-charged cations. Ferrihydrite adsorption treatment has, in particular, been particularly effective for arsenic removal from wastewaters. It was found that, while arsenic could be effectively removed from solution as calcium arsenates when treating with lime, the solid product was unstable when stored in an environment exposed to air (such as a tailings impoundment). The calcium arsenate, being less stable than calcium carbonate, would release arsenate when converting to calcium carbonate. When properly stored, ferrihydrite with adsorbed arsenic is much more stable (Twidwell 2011).

A system utilizing ferrihydrite adsorption to remove cationic metals (such as Ni^{+2} or Pb^{+2}) to low levels would include a conventional lime treatment system, with a subsystem added to dose ferric chloride or ferric sulfate prior to raising the pH to approximately 9 to 10. A system utilizing ferrihydrite adsorption to remove anions (such as arsenate or selenite) would include an additional precipitation and solid/liquid separation step, over and above a conventional lime treatment system. This is because ferrihydrite adsorption for anions requires the much lower pH (on the order of 4 to 6) than that needed for metals precipitation, and the material must be separated from the

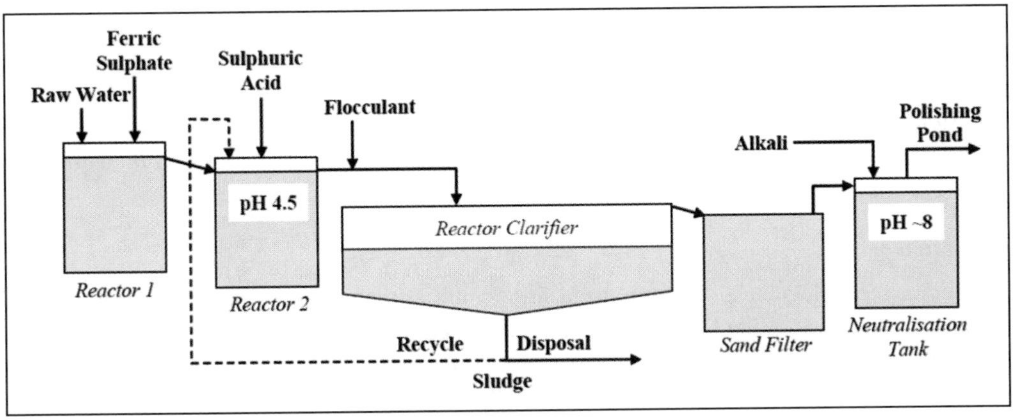

Figure 4. Typical system for combined ferrihydrite adsorption for anions removal and metals precipitation (Aube 2004)

process stream to avoid release of the anions at the elevated pH downstream. Figure 4 presents a typical system for arsenic removal as well as metals precipitation.

Cyanide Detoxification by Oxidation

Cyanide (CN^-) is a highly toxic material used in a number of industries, and used most commonly in the mining and metals production industries for gold leaching. While cyanide alone in water naturally degrades fairly rapidly in a natural environment, cyanide forms complexes of varying strength with a wide variety of metals. These complexes can cause cyanide to linger for long periods. The use of chemical oxidation approaches to oxidize cyanide to less toxic forms was an innovative concept and has become ubiquitous. A wide variety of oxidation approaches are in use today:

- Alkaline chlorination
- Hydrogen peroxide
- SO_2/air
- Caro's acid
- Ozonation

These are in use at mine and mill sites around the world. All have advantages and disadvantages, and the choice for a given location depends on local environmental regulations, site-specific process conditions, local reagent costs, etc.

Metals Removal/Recovery by Sulfide Precipitation

While sulfide precipitation for metals removal or recovery is not a new or novel concept, this has seen increased focus in the last ten years. In particular, the Chemsulphide® process marketed by BioteQ Environmenal Technologies, Inc. (BioteQ) has been successfully applied at a number of different locations and applications around the world. This type of approach is only applicable to metals that form sulfide precipitates, such as copper, zinc, nickel, cobalt, cadmium, etc. Typically, metal sulfide precipitates are much less soluble than oxide/hydroxide precipitates, and separations can be achieved by judicious use of pH and oxidation/reduction conditions, along with sulfide reagent dose. The applications developed by BioteQ have generally been low grade waste streams such as runoff from low grade stockpiles, bleed streams, acid rock drainage, etc. Further, these applications have been in locations where there is a ready market for the metal sulfide precipitates produced, i.e., a copper sulfide at a copper operation, a nickel sulfide at a nickel operation, etc. This type of approach will likely increase in the future as a sustainability enhancement; e.g., used as a front-end step prior to a conventional lime treatment system, or used in combination with other process technologies such as cyanide removal or recycling.

Membranes

Membrane technology, such as reverse osmosis, nanofiltration, ultrafiltration, and microfiltration, was a significant innovation in the overall water treatment industry and is becoming more popular in the mining industry. Mining industry waters can be quite challenging for membranes due to the variety of suspended solids that may be present, and the presence of dissolved salts that may be near or at solubility limits. Pretreatment is generally required to manage these issues. Numerous companies produce antiscalants to inhibit or delay precipitation of dissolved constituents as they are concentrated by the membrane process. In addition, while early membrane technology utilized polymers to fabricate the membranes, more durable membranes made of ceramics are now available.

A potential drawback of membrane treatment is the possibly significant volume of concentrate (or retentate) produced, and what to do with it. In some situations this can be an advantage. Nanofiltration membranes have been used in combination with lime treatment, where the membrane concentrates the salts to higher levels, and lime treatment is used to treat the concentrate. Following solid/liquid separation, this treated concentrate is recombined with the membrane permeate, and lower overall concentrations of metals and sulfate result.

INNOVATIONS IN BACTERIAL TREATMENT

Bacterial treatment involves the use of bacterial activity to accomplish treatment goals. There frequently are cases where chemical treatment of a particular problem is unacceptably slow, but where bacteria are able to overcome significant activation energy barriers to achieve much more rapid treatment. A good example of this is removal of selenium from water.

Selenium Removal

Selenium is extremely difficult to remove from solution if it is present as the selenate form (SeO_4^{-2}). Unfortunately, selenate is the most oxidized form of selenium, and is commonly found in mining and smelting wastewaters. There are essentially no effective, reasonably inexpensive options for removing selenate by chemical treatment, though selenite (SeO_3^{-2}) is very effectively removed by ferrihydrite adsorption. Unfortunately, reduction of selenate to selenite by chemical treatment means is very slow and expensive. While not yet widespread in the mining and metals production industries, bacterial reduction of selenium has become the treatment method of choice in other comparable applications, such as scrubber blowdown from coal-fired power. While already being applied in some U.S. Appalachian coal-mining areas, it is probably just a matter of time until more strict regulatory enforcement of selenium in the mining industry becomes widespread.

There are several bacterial treatment approaches being marketed currently. These include the ABMet system by GE Power & Water, the Fluidized Bed Reactor (FBR) by Envirogen Technologies, and the iBIO® system by Degrement Technologies, and possibly others. All of these systems involve the use of bacteria to reduce selenate to selenite and ultimately to elemental selenium, which is quite insoluble. In doing so, these systems add nutrients and a carbon source to feed the bacteria. These systems generally have beneficial secondary traits inherent in the process, such as nitrate removal by denitrification, and metals removal by precipitation as sulfides as a result of sulfate reduction. These secondary processes are also biologically driven, and typically go hand-in-hand with biological selenium reduction. It should be note that these systems generally require aerobic treatment following the selenium reduction/removal steps, to remove ammonia and biological oxygen demand developed by the upstream steps.

Biological Cyanide Detoxification

Cyanide degradation by biological means has been successfully applied to detoxifying gold mining heap leach pads and tailings solutions. While not yet widely accepted, biological treatment typically requires a much lower operating cost than chemical treatment, and therefore would be advantageous for long-term operation. In this treatment approach, bacteria convert free and metal-complexed cyanides to bicarbonate and ammonia. The metals freed in the process are either adsorbed by the biofilm or are precipitated from solution in subsequent treatment steps. The ammonia produced can be removed by conventional biological means via nitrification to nitrate and denitrification to nitrogen gas. The Homestake Mining Company was a leader in developing this process, having applied it at the Homestake Mine in Lead, South Dakota, and at its Nickel Plate Mine in Canada.

FUTURE INNOVATIONS

In most other subject areas of the SME Innovations Symposium, the primary drivers for innovation are generally production-related, i.e., increasing recoveries, increasing grades, increasing throughput, reducing costs, increasing reliability, etc. In this particular subject area, Chemical and Bacterial Water Treatment, however, the main driver has historically been regulatory discharge requirements. This usually has involved either the implementation of new discharge requirements, or decreases in existing discharge concentrations. It is anticipated that innovations will continue to occur in response to those drivers, though possibly at a slower rate. They will likely arise from research conducted by universities, governments, and major companies, and will probably borrow from other industries with similar specific issues.

It seems likely that, as biological processes become better understood through the use of techniques such as bacterial genomics, more innovations may be seen in the bacterial treatment arena. Further, approaches focused on sustainability will probably see more focus. An example of this would be the development of technologies for recycle of cyanide rather than cyanide degradation/destruction. There are a number of these technologies being developed and applied, such as the AVR process (acidification, volatilization, reneutralization) and the SART process (sulfidization, acidification, recycle, and thickening). Lastly, a trend that is appearing in other industries is the so-called Zero Liquid Discharge (ZLD) approach, which avoids any discharge of water, thereby eliminating risk of discharge violations and reducing or eliminating makeup water requirements. While always site-specific, ZLD generally combines membrane and other technologies to concentrate and remove suspended and dissolved constituents from the site water, allowing recycle of the water and producing solid waste streams for disposal. This approach may make its way into the mining and metals production industry.

SUMMARY

This paper presents the major innovations that have occurred in chemical and bacterial water treatment in the mining and metals production industry. These innovations have generally occurred in response to regulatory drivers, i.e., implementation of new water quality criteria or reductions in existing water quality criteria values. Some innovations have occurred as a result of a desire for process improvement, such as the development of the HDS process improvement over conventional lime treatment. More recently, the mining industry has greatly increased its focus on sustainability. As a result, technologies are more commonly being applied with life-cycle cost in mind. Examples of this would be biological technologies with possibly higher initial costs, but lower long-term operational costs, and application of technologies or combinations of technologies for recovery of metals and recycle of reagents.

As water becomes more valuable over time, these sustainability trends are expected to continue and increase, and become the primary driver for innovation in chemical and bacterial water treatment. It does not seem likely that new regulatory drivers will appear, though there may be increased enforcement in areas of selenium, TDS, nutrients (nitrogen and phosphorus), and anions (sulfate, chloride). The desire to minimize long-term cost and risk will probably be the primary driver.

REFERENCES

Aube, B., 2004. The Science of Treating Acid Mine Drainage and Smelter Effluents. Retrieved September 15, 2012 from: http://www.infomine.com/library/publications/docs/Aube.pdf.

INAP, 2010. "The GARD Guide: The Global Acid Rock Drainage Guide, Chapter 7: Drainage Treatment." Retrieved September 15, 2012 from: http://www.gardguide.com/index.php/Chapter_7#7.5.1_Active_Treatment_Technologies.

Paktunc, D., K. Bruggeman. Solubility of Nanocrystalline Scorodite and Amorphous Ferric Arsenate: Implications for Stabilization of Arsenic in Mine Wastes," Applied Geochem. 25, 2010, 674–683.

Twidwell, L.G., J. McCloskey, Removing Arsenic from Aqueous Solutions and Long-Term Product Storage, JOM, August 2011, pp. 94–100.

ABOUT THE AUTHORS

Brian Park has more than 30 years of industrial experience in environmental aspects of mine operations and remediation, including in situ treatment of pit lakes, active and passive water treatment, waste rock stabilization, and prevention of waste rock oxidation. Park also has experience in hydrometallurgy; metals recycling; bench- and pilot-scale testing along with process development and demonstration; process modeling and simulation; process modification for waste minimization; heat transfer analysis, design, and testing; and project management. He has held positions at General Dynamics/Space Systems Division, MSE Technology Applications, the Center for Advanced Mineral and Metallurgical Processing, and Montana Emergent Technologies, for whom he currently serves as president and senior process engineer. Park has a BS degree in mechanical engineering from Purdue University and an MS degree in metallurgical engineering from Montana Tech, and is a registered professional engineer in four states.

Jay McCloskey has more than 30 years of industrial experience in mining, smelting, environmental remediation, and project management. His career includes positions with American Smelting Refining Company, Montana Resources, MSE Technology Applications Inc., the Center for Advanced Mineral and Metallurgical Processing at Montana Tech, Aspen Ridge Technical Services, and Montana Emergent Technologies (MET). McCloskey has BSc and MSc degrees in metallurgical engineering and mineral process engineering. He currently serves as vice president and senior process engineer for MET. He is or has been a member of numerous professional organizations: MMSA, TMS, SME, IPMI, ITA, REITA, FCHEA, NMA, NWMA, and IDGA. In 2006, he was awarded the TMS–EPD Best Technology Paper Award for his work on *Arsenic Removal from Mine and Process Waters*.

Larry G. Twidwell has a DSc degree and is emeritus professor in the Department of Metallurgical and Materials Engineering at Montana Tech. He has 40 years of experience in teaching and directing research in metallurgical engineering at Montana Tech. Twidwell and his graduate students and co-workers have conducted research emphasizing extractive metallurgy and the application of extractive metallurgical engineering to the treatment of metallurgical wastes, process waters, and by-products. Four books, 58 theses, 95 publications, and 108 national and international presentations have resulted from their research efforts. Twidwell has received several outstanding teacher/researcher awards during his tenure at Montana Tech: TMS-EPD Distinguished Lecturer Award, 2011; TMS-EPD Best Technology Paper Award, 2006; AIME Frank Aplan Award, 2004; Montana Tech Lifetime Distinguished Researcher Award, 2002; AIME Mineral Industry National Education Award, 2001; Montana Academy of Sciences Mershon Award as Outstanding Montana Researcher, 1989; Outstanding Scholar Award, Montana Tech, 1985; and Outstanding Educator Awards, Montana Tech, 1970, 1971, 1973, 1974.

Innovations in Air Pollution Control

Catherine "Carrie" MacDougall, Michael D.S. Blois, and Robert W. Reisinger
CH2M HILL, Las Vegas, Nevada, United States

ABSTRACT: Since the United States enacted Amendments to the Clean Air Act in 1977, mining and smelter operations around the world have been required to develop and implement new technologies to meet air quality regulatory requirements. Hydrometallurgy, acid gas plants, and high-heat bag houses are some of the more common technologies that were developed specifically to meet air quality regulatory requirements. Air quality regulations have continued to become more stringent and new pollutants are becoming regulated. Stricter SO_2 emission standards and toxic air compound regulations are being implemented around the globe, forcing further innovation or possible facility closures. Additionally, Greenhouse Gas regulations are being developed, mandating greater fuel efficiencies for mines and smelters. As the regulations continue to be promulgated, new technologies and methods of operation will need to be developed. Because new technologies are being developed, electronic air quality control resources that are currently being updated on a regular basis have been listed to aid in information transfer.

INTRODUCTION

Air quality concerns and the metals and minerals industry have been linked for thousands of years. As early as 20 AD, strong odors destructive of life were reported (Chandler 1992). Air quality in mines was not the only concern. Before 1250, complaints from Queen Eleanor of England about the objectionable odors from the burning of "sea coale" from Queen Eleanor of England were recorded (Chandler 1992). The smoke from burning of "sea coale" in lime kilns in London became a serious problem as early as 1285. The problem persisted and, in 1306, King Edward I compelled all but smiths to cease burning "sea coale" and return to the use of wood (Chandler 1992). The regulation of the minerals industry and of air quality had begun.

Early efforts for improving air quality in mines began as far back as the 16th century. As shown in Figure 1, great effort was made to improve air quality in the mines. Dust was another concern, and vaulted dust chambers (as depicted in Figure 2) were used. Fumes from the furnaces rose into a wide vaulted chamber. The chimney in the dust chamber had thin iron plates fastened into the walls and the dust adhered to the plates as the fumes ascended through the chimney. In the 19th century, dust in the coals mines was controlled by keeping the coal moist (Chandler 1992).

Since the enactment of the Clean Air Act of 1970 (CAA), several technological innovations have been developed in air pollution control for mining, mineral processing, and smelting operations, and behind each of these innovations is a driver. The role of air pollution regulations as a driver in technological innovations for mines, mineral processing plants and smelters is significant. The regulatory requirement to reduce air emissions was the primary driver for acid gas plants

A—Machine first described. B—This workman, treading with his feet, is compressing the bellows. C—Bellows without nozzles. D—Hole by which heavy vapours or blasts are blown out. E—Conduits. F—Tunnel. G—Second machine described. H—Wooden wheel. I—Its steps. K—Bars. L—Hole in same wheel. M—Pole. N—Third machine described. O—Upright axle. P—Its toothed drum. Q—Horizontal axle. R—Its drum which is made of rundles.

Figure 1. A ventilating device in which bellows are used to move the air (Agricola 1556)

at nonferrous smelters (United States General Accounting Office 1986) and the advancement of hydrometallurgy for several metals (Kovacs-Figueroa 1976, Habashi 2005).

In addition to innovations in technology for mining, mineral processing and smelting practices, add-on pollution control equipment was further developed to meet new, stricter regulatory standards for the mining and smelting industry. Notably, the high-heat bag house and air curtain control systems were developed for smelters.

Changes in industry practices occurring as a result of air quality regulations have been seen in the coal industry in the United States. As part of the 1970 CAA, a sulphur dioxide (SO_2) standard of 1.2 pounds per million British thermal units (lb/MMBtu) was established. This caused a major shift in the coal marketplace, as power plants sought coal that naturally contained less sulphur to avoid the costs of installing scrubbers to collect SO_2 (Michon 1982). Many power producers in the

Figure 2. A dust chamber (D) for collecting particulate from the exhaust before they escaped into the atmosphere through the opening (E) in the chimney (F) (Agricola 1556)

Midwest and Gulf areas of the United States chose to transport low-sulphur coal in lieu of burning local coals with higher concentrations of sulphur.

As more studies are completed on the effects of air pollutants on humans and the environment, regulatory agencies throughout the world are adding new regulations covering more compounds and making existing standards stricter. Notably, the regulation of "toxic" air contaminants and Greenhouse Gas (GHG) pollutants have increased in the last decade and are expected to increase in the foreseeable future. These new regulations have resulted in the closure of some facilities and a demand for more fuel-efficient processes.

A summary of the regulatory framework and advancements in air quality requirements throughout the world that have driven new technologies is presented in the following sections.

Table 1. Comparison of national air quality standards

Emissions Standards by Country	Emissions Standard by Pollutant									
	CO		NO_2		SO_2		O_3		PM_{10}	$PM_{2.5}$
South Korea	10.3 mg/m³ 8-hr 28.6 mg/m³ 1-hr	9 ppm 8-hr 25 ppm 1-hr	56 µg/m³ Annual 110 µg/m³ 24-hr 190 µg/m³ 1-hr	30 ppb Annual 60 ppb 24-hr 100 ppb 1-hr	52 µg/m³ Annual 131 µg/m³ 24-hr 390 µg/m³ 1-hr	20 ppb Annual 50 ppb 24-hr 150 ppb 1-hr	30 µg/m³ 8-hr 50 µg/m³ 1-hr	60 ppb 8-hr 100 ppb 1-hr	50 µg/m³ Annual 100 µg/m³ 24-hr	
U.S.A.	10 mg/m³ 8-hr 40 mg/m³ 1-hr	9 ppm 8-hr 35 ppm 1-hr	100 µg /m³ Annual 188 µg/m³ 1-hr	53 ppb Annual 100 ppb 1-hr	78 µg/m³ Annual 370 µg/m³ 24-hr 200 µg/m³ 1-hr	30 ppb Annual 140 ppb 24-hr 75 ppb 1-hr	150 µg/m³ 8-hr 240 µg/m³ 1-hr	75 ppb 8-hr 120 ppb 1-hr	150 µg/m³ 24-hr	12 µg/m³ Annual 35 µg/m³ 24-hr
U.K.	10 mg/m³ 8-hr	8.7 ppm 8-hr	40 µg/m³ Annual 200 µg/m³ 1-hr	21 ppb Annual 106 ppb 1-hr	350 µg/m³ Annual 125 µg/m³ 24-hr	134 ppb Annual 48 ppb 24-hr	100 µg/m³ 8-hr	51 ppb 8-hr	40 µg/m³ Annual 50 µg/m³ 24-hr	25 µg/m³ Annual
The Netherlands	6 mg/m³ 8-hr 40mg/m³ 1-hr	5.2 ppm 8-hr 35 ppm 1-hr	40 µg/m³ Annual 200 µg/m³ 1-hr 400 µg/m³ 1-hr [3]	21 ppb Annual 106 ppb 1-hr 213 ppb 1-hr [3]	125 µg/m³ 24-hr 350 µg/m³ 1-hr 500 µg/m³ 1-hr [3]	48 ppb 24-hr 134 ppb 1-hr 191 ppb 1-hr [3]	110 µg/m³ 8-hr [1] 120 µg/m³ 8-hr [2] 180 µg/m³ 1-hr	56 ppb 8-hr [1] 61 ppb 8-hr [2] 92 ppb 1-hr	40 µg/m³ Annual 50 µg/m³ 24-hr	

(table continues)

Table 1. Comparison of national air quality standards (continued)

Emissions Standards by Country	Emissions Standard by Pollutant									
	CO		NO_2		SO_2		O_3		PM_{10}	$PM_{2.5}$
Belgium/EU	10,000 µg/m³ 8-hr	8.7 ppm 8-hr	40 µg/m³ Annual	21 ppb Annual	125 µg/m³ 24-hr	48 ppb 24-hr	120 µg/m³ 8-hr	61 ppb 8-hr	40 µg/m³ Annual	25 µg/m³ Annual
			200 µg/m³ 1-hr	106 ppb 1-hr	350 µg/m³ 1-hr	134 ppb 1-hr			50 µg/m³ 24-hr	
WHO	10,000 µg/m³ 8-hr	8.7 ppm 8-hr	40 µg/m³ Annual	21 ppb Annual	20 µg/m³ 24-hr	7.6 ppb 24-hr	100 µg/m³ 8-hr	51 ppb 8-hr	20 µg/m³ Annual	10 µg/m³ Annual
			200 µg/m³ 1-hr	106 ppb 1-hr	500 µg/m³ 10-min	191 ppb 10-min			50 µg/m³ 24-hr	25 µg/m³ 24-hr

Notes:
[1] highest 8-hr Average
[2] highest progressive 8-hr a day, <25 day/yr
[3] 1-hr for the area at least 100 km²
[4] 3-hr average for area at least 100 km²

µg/m³ – micrograms per cubic metre
EU – European Union
hr – hour
km² – square kilometres
ppb – parts per billion
ppm – parts per million
U.K. – United Kingdom
U.S.A. – United States of America

Included are technologies that have been largely driven by air quality requirements, with a description of the technlogies themselves. Trends that could lead to stricter regulations and newly regulated pollutants are summarized, and a brief listing of up-to-date electronic resources is included.

REGULATORY FRAMEWORK

The Air Pollution Control Act of 1955 (P.L. 84-159) was the first federal air pollution law in the United States: it authorized the federal government to conduct and support air pollution research, training, and technical assistance. Much of the research performed as a result of this act addressed the combustion process and the behavior of combustion products in the process and atmosphere. The 1963 CAA (P.L. 88-206) continued the research, training, and technical assistance program of the 1955 act and also initiated a number of new activities. For example, under the CAA, the federal government was given the authority to make grants to state and local air pollution control agencies for the development, establishment, and improvement of air pollution prevention and control programs. The federal government was also given authority to abate interstate air pollution, because many large cities were near the boundaries of states. The CAA also required the federal government to compile and publish air quality criteria, including those for oxides of sulphur and nitrogen, and required federal agencies to build and operate their facilities to minimize the emission of pollutants.

The 1970 CAA (P.L. 91-604) resulted in a major shift in the federal government's role in air pollution control from previous law. Four major regulatory programs affecting stationary sources were initiated: the National Ambient Air Quality Standards (NAAQS), State Implementation Plans (SIPs), New Source Performance Standards (NSPS), and National Emission Standards for Hazardous Air Pollutants (NESHAPs). On December 2, 1970, Congress provided authority for the establishment of United States Environmental Protection Agency (USEPA) and empowered the agency to regulate emissions from sources of air pollutants.

The CAA was amended again in 1977 and 1990 (USEPA 2012a). The 1977 CAAA primarily addressed provisions for the Prevention of Significant Deterioration (PSD) of air quality in areas attaining the NAAQS. The 1977 CAAA also contained requirements pertaining to sources in nonattainment areas for NAAQS and established major permit review requirements to ensure attainment and maintenance of the NAAQS.

The Clean Air Act Amendments of 1990 (1990 CAAA) (P.L. 101-549) further amended clean air legislation by addressing the growing concern of acidic deposition, global warming, depletion of the ozone layer, and emission of toxic air pollutants. The 1990 CAAA required that USEPA promulgate standards based on "maximum achievable control technology" (MACT) and added a comprehensive operating permit program.

In the time since the NAAQS were established under the 1970 CAA, ambient air quality standards, or limits, have been established throughout the world to protect human health and the environment. Table 1 summarizes some of the standards that have been established throughout the world and also presents the World Health Organization (WHO) recommended standards.

Ambient air quality standards lead to stationary source emission reductions through the monitoring of ambient air quality pollutant concentrations. When the USEPA established an SO_2 standard, it required smelters to monitor the ambient air surrounding their facilities to demonstrate they were meeting the standard. Smelters were required to use active controls to reduce SO_2 emissions and not rely on dispersion techniques such as taller stacks, with more air volume.

Table 2. Particulate NAAQS historical summary

Year	Primary/ Secondary	Indicator	Averaging Time	Level	Form
1971	Primary	TSP*	24-hour	260 $\mu g/m^3$	Not to be exceeded more than once per year
			Annual	75 $\mu g/m^3$	Annual average
	Secondary	TSP	24-hour	150 $\mu g/m^3$	Not to be exceeded more than once per year
1987	Primary and Secondary	PM_{10}	24-hour	150 $\mu g/m^3$	Not to be exceeded more than once per year on average over a 3-year period
			Annual	50 $\mu g/m^3$	Annual arithmetic mean, averaged over 3 years
1997	Primary and Secondary	$PM_{2.5}$	24-hour	65 $\mu g/m^3$	98th percentile, averaged over 3 years
			Annual	15.0 $\mu g/m^3$	Annual arithmetic mean, averaged over 3 years
		PM_{10}	24-hour	150 $\mu g/m^3$	Initially promulgated 99th percentile, averaged over 3 years; when 1997 standards for PM_{10} were vacated, the form of 1987 standards remained in place (not to be exceeded more than once per year on average over a 3-year period)
			Annual	50 $\mu g/m^3$	Annual arithmetic mean, averaged over 3 years
2006	Primary and Secondary	$PM_{2.5}$	24-hour	35 $\mu g/m^3$	98th percentile, averaged over 3 years
			Annual	15.0 $\mu g/m^3$	Annual arithmetic mean, averaged over 3 years
		PM_{10}	24-hour	150 $\mu g/m^3$	Not to be exceeded more than once per year on average over a 3-year period
2012	Primary	$PM_{2.5}$	Annual	12.0 $\mu g/m^3$	Annual arithmetic mean, averaged over 3 years
	Secondary		Annual	15.0 $\mu g/m^3$	Annual arithmetic mean, averaged over 3 years
	Primary and Secondary		24-hour	35 $\mu g/m^3$	98th percentile, averaged over 3 years
	Primary and Secondary	PM_{10}	24-hour	150 $\mu g/m^3$	Not to be exceeded more than once per year on average over a 3-year period

Source: USEPA 2012c.
* TSP – Total Suspended Particulate

The USEPA worked with the states to reduce smelter sulphur emissions to meet the air quality standards (USEPA 2012a). The United States General Accounting Office reported that non-ferrous smelters reduced their total SO_2 emissions from 3.5 million tons in 1970 to 0.97 million tons in 1983. This 72 percent reduction was attributed in part to the closure of 10 smelters. The USEPA stated that only 5 of these smelters did not have adequate controls to meet the SO_2 ambient standards. The other smelters were expected to reopen when economic conditions improved. Sixty percent of the emissions were attributed to SO_2 controls on the operating non-ferrous smelters. The two technological advancements leading to the control of SO_2 emissions were in the areas of acid gas plants and hydrometallurgy which will be discussed in greater detail below. The USEPA did not prescribe the technology to meet the ambient air quality standard. The innovations were developed by the industry, but the driver for the technological innovation was the ambient air quality standard.

These standards are continually being updated. Most recently, in 2012, the United States revised the annual average particulate matter $(PM)_{2.5}$ standard downward from 15 $\mu g/m^3$ to 12 $\mu g/m^3$ (USEPA 2012b). Table 2 summarizes a history of the United States NAAQS for particulate.

As seen in Table 2, not only has the standard been set lower over time, the types of particulate being regulated have also changed. When the new NAAQS was set for $PM_{2.5}$, smelters and mines

Table 3. Established BAT for mining and smelting activities

Title of BAT	Date Established
Ferrous Metals Processing Industry	December 2001
Non-Ferrous Metals Industries	December 2001
Smitheries and Foundries Industry	May 2005
Management of Tailings and Waste-Rock in Mining Activities	January 2009
Iron and Steel Production	2010

were required to place a greater emphasis on controls from combustion sources. As a result, the amount of fuel used was more directly related to air regulatory compliance and not only an economic concern for the facility.

As innovations were developed to meet air quality requirements, these innovations were adopted as requirements for implementation as a matter of practice. In addition to establishing ambient air quality standards, several countries, local governments, and the EU have established specific control standards for mines and smelters. Some of these controls are well established and published. For example, the EU has established Best Available Techniques (BAT) for several industries; some of the BAT that apply to mines and smelting operations are listed in Table 3.

Specific technologies are prescribed as BAT. For example, in the BAT for Non-ferrous Metals Industry (European Commission, Integrated Pollution Prevention and Control 2001):

> The production of sulphuric acid in a double contact sulphuric acid plant with a minimum of four passes, or in a single contact plant with gypsum production from the tail gas and using a modern catalyst, are considered to be BAT.

In addition to specific technologies for emissions controls, the BAT also make recommendations to minimize energy use and therefore, reduce air emissions by reducing fuel combustion. Table 4 presents a listing of the types of controls now required for non-ferrous metal facilities throughout the EU and the United Kingdom.

The technological advancements made in the past 40 years that are now considered feasible by air quality regulatory agencies are discussed in more detail in the following sections.

TECHNOLOGICAL ADVANCEMENTS

Metals Innovations

Most innovations developed for the mining and processing of metals though air quality regulations have impacted mining practices for coal as well. Each of the innovations described here were in response to an air quality regulation of some kind: either an ambient air quality standard or an emissions limitation.

Acid Gas Plants

The treatment of SO_2 emissions from smelters was developed in the early 1970s in response to new SO_2 ambient air quality standards. At the Morenci copper smelter in Arizona, as early as 1971, it was shown that the sulphuric acid produced from the smelter offgas could be largely used for the tailings leaching (Kovacs-Figueroa 1976). The practice has been refined over time with multiple

Table 4. BAT for non-ferrous metal facilities

Abatement Technique	Associated Range	Comment
Fabric filter	Dust 1–5 mg/Nm^3 Metals–dependent on dust composition	Depends on characteristics of dust.
Carbon or Bio filter	Total organic C < 20 mg/Nm^3	Phenol < 0.1 mg/Nm^3
Afterburner (including temperature quench for dioxin removal)	Total organic C < 5–15 mg/Nm^3 Dioxin < 0.1–0.5 ng/Nm^3 TEQ PAH (OSPAR 11) < 200 μgC/Nm^3 HCN < 2 mg/Nm^3	Designed for gas volume. Other techniques are available to reduce dioxins further by carbon/lime injection and catalytic reactors/filters.
Optimised combustion conditions	Total organic C < 5–50 mg/Nm^3	
Wet EP Ceramic filter	Dust < 5 mg/Nm^3	Depends on characteristics e.g., dust, moisture or high temperature.
Wet or semi-dry alkaline scrubber	SO_2 < 50–200 mg/Nm^3 Tar < 10 mg/Nm^3 Chlorine < 2 mg/Nm^3	
Alumina scrubber	Dust 1–5 mg/Nm^3 Hydrocarbon < 2 mg/Nm^3 PAH (OSPAR 11) < 200 μgC/Nm^3	
Chlorine recovery	Chlorine < 5 mg/Nm^3	Chlorine is re-used. Possible accidental fugitive releases.
Oxidizing scrubber	NO_x < 100 mg/Nm^3	From use of nitric acid–recovery followed by removal of traces.
Low NO_x burner	< 100 mg/Nm^3	Higher values are associated with oxygen enrichment to reduce energy use. In these cases, gas volume and mass emission are reduced.
Oxy-fuel burner	< 100–300 mg/Nm^3	Including mercury scrubber using Boliden/Norzink process or thiosulphate scrubber–mercury <1 ppm in acid produced.
Sulphuric acid plant	> 99.7% conversion (double contact) > 99.1% conversion (single contact)	
Cooler, EP, lime/carbon adsorption and fabric filter	PAH (OSPAR 11) < 200 μgC/Nm^3 Hydrocarbons (volatile) < 20 mgC/Nm^3 Hydrocarbons (condensed) < 2 mgC/Nm^3	

Source: European Commission, Integrated Pollution Prevention and Control 2001.
μgC/Nm^3 – micrograms carbon per normal cubic metre
EP – electrostatic precipitator
HCN – hydrogen cyanide
mg/Nm^3 – milligrams per normal cubic metre
mgC/Nm^3 – milligrams carbon per normal cubic metre
NO_x – nitrogen oxide
OSPAR – Oslo and Paris
PAH – polycyclic aromatic hydrocarbon
TEQ – toxicity equivalent

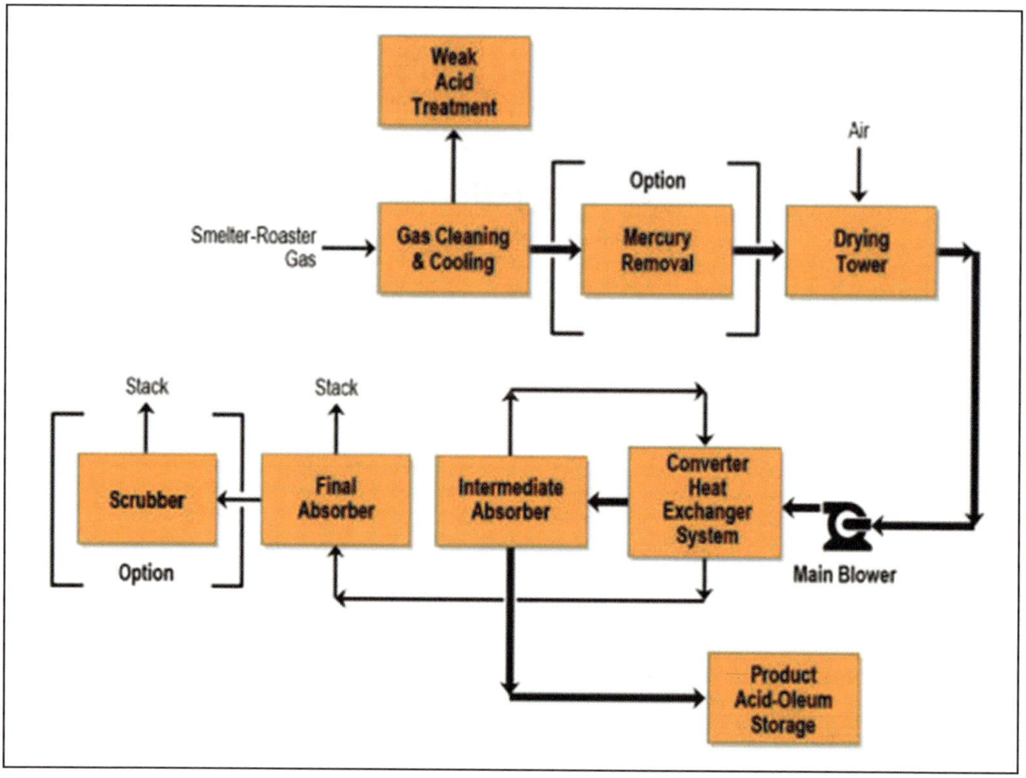

Figure 3. Schematic of acid gas plant (Jacobs Engineering 2013)

pass converter arrays and the use of vanadium catalysts. Use of a sulphuric acid plant to treat smelter effluent gas streams requires that particulate-free gas containing minimum SO_2 concentrations, usually of at least 3 percent SO_2, be maintained. Figure 3 provides a basic diagram of the process. The converter and the heat exchanger system are the heart of the process where SO_2 is converted to SO_3, typically with the use of a catalyst. As listed in Table 5, greater removal efficiency can be achieved by multiple passes through the converter or more catalyst layers in a single pass.

Table 5 lists the concentration of SO_2 in gases from the various refinery processes. As seen from the values in Table 5, significant reductions in SO_2 emissions from smelting can be achieved with the installation and operation of an acid gas plant. Further emission reductions can occur when a scrubber, NO_x control device and a mercury unit (where applicable) are added to the exhaust from the acid gas plant (see Figure 3).

Though commonly used where required in many countries, acid gas plants do require additional energy and additional costs for metal production. In a report sponsored by the EU Directorate-General Enterprise and Industry, it was noted that EU non-ferrous metals industries were at a disadvantage with competitors where environmental policies lead to lower costs of compliance (Ecorys 2011). However, in countries such as the United States, Japan, Canada, and European Free Trade Association vountries, where environmental policies are applied more uniformly, the EU non-ferrous metals industries competed well.

Table 5. Typical off-gas concentrations from smelting processes (USEPA 1986)

Process	SO_2 Concentration (Volume, %)
Multiple hearth roaster	1.5–3
Fluidized bed roaster	10–12
Reverberatory furnace	0.5–1.5
Electric arc furnace	4–8
Flash smelting furnace	10–70
Continuous smelting furnace	5–15
Pierce-Smith converter	4–7
Hoboken converter	8
Single-contact sulphuric acid plant	0.2–0.26
Double-contact sulphuric acid plant	0.05

Figure 4. Converter in copper smelter blowing

Air Curtain Control Systems

Capturing escaping gases from the converters can be difficult. As seen in the Figure 4, when converters are rolled for blowing, capturing the related emissions can be difficult.

Similarly, when converters are charged and skimmed, emissions are difficult to capture. Though not as widely used, air-curtain control systems have also been developed to capture emissions from converters. The basic design of an air curtain is to place baffle walls on either side of the converter aisle. Large volumes of air are moved from one side of the converter aisle to the other, several metres above the height of the converters, at a high velocity that forms a "jet stream" the length of the converter aisle. The air is then collected on the opposite, downwind, side of the converter aisle, where controls are used to remove pollutants before the air is released to the atmosphere. Fugitive emissions from the converters are caught up in the jet stream as they rise above the converters and are vented through the control system.

Various designs of the air-curtain technology have been tried, though success has been limited (Burckle 1984). Contact with cranes and ladles can damage the baffle walls and air blowers. The control uses a great deal of energy. If not properly designed and installed, inadequate fan capacity can lead to dead spots. Heating problems can occur during cold spells because most of the heat is evacuated with the air emissions. If used effectively, approximately 90 percent of the gaseous emissions can be collected and controlled with an air curtain.

Hydrometallurgy

Though hydrometallurgy can be traced back to the first century AD, advances in hydrometallurgy to minimize emissions of air pollutants can be directly linked to the implementation of ambient air quality standards and emission limitations. Most notable is the change in the hydrometallurgical production of zinc in Canada during the 1970s. Pressure leaching of zinc sulphides fully replaced the horizontal retort process based on roasting zinc sulphide (ZnS), reduction of zinc oxide (ZnO) by carbon, and distillation refining of metallic zinc developed in the 18th century (Burckle 1984). The aqueous oxidation of ZnO produces elemental sulphur during the hydrometallurgical process avoiding the need to produce sulphuric acid to control the SO_2 emissions, reducing sulphur emissions by a significant amount. The process also uses less energy than the pyrometallurgical process.

Scrubbers

Sodium (sodium hydroxide) -based wet scrubbers, Figure 5, have been used extensively in flue gas desulphurization (FGD) service. With the addition of acid gas plants at metallurgical facilities, these scrubbers have been added to the end of the acid gas plant to control sulphuric acid mist with caustics. The use of scrubbers at smelters has increased in the past 30 years. SO_2 removal efficiencies of 90 to 95 percent are quite common, and can be as high as 98 to 99 percent.

The scrubbers used for this service have been mostly spray towers: counter-current systems wherein the flue gas flows vertically through a series of spray zones where the recycle scrubbing liquor is sprayed through nozzles to ensure good gas/liquid contact. Multiple spray zones are used depending upon the removal efficiency required. The pH is maintained in the scrubbing solution by the continuous addition of sodium hydroxide. A blow-down stream is usually based on scrubber liquid specific gravity to prevent excessive salt build up. Scrubber liquor is continuously recycled from the spray tower sump to the spray headers by recirculation pumps. A mist eliminator section is provided at the top of the spray tower to remove entrained water droplets. The exhaust can be ducted to a free-standing stack, or an integral stack located on top the spray tower vessel. The materials of construction are generally fibreglass reinforced plastic (FRP) or stainless steel. Generally, the induced draft (ID) fans needed to provide the motive force to move the flue gas through the spray tower is located upstream of the spray tower. Horizontal spray scrubbers have also been used for smaller gas volumes.

Sea Water Scrubber

Sea water scrubber technology is basically a counter-current packed bed or spray tower scrubber that uses sea water to neutralize SO_2, instead of caustic. A particulate control device (electrostatic precipitator or fabric filter) is located upstream of the packed bed absorber.

Figure 5. Example of a wet scrubber

The sea water main supply for the process is split into two streams. One passes through the scrubber for the absorption of SO_2, after which the now acidic sea water flows further by gravity from the scrubber sump to an aeration channel or basin. The scrubber discharge is premixed with the other stream of fresh sea water. Within the aeration system, air is bubbled up through the waste water stream, and converts the SO_2 and sulphur trioxide into harmless sulphate ions. Sulphate ions are a natural and major constituent of sea water. The treated sea water is then discharged into the sea without causing any harm to the marine environment.

Sea water scrubbing is a unique technology that may be very well suited as an FGD technology. The technology is very mature, having been first deployed on ships in 1968, and there are multiple vendors available. Sea water scrubbers have been deployed in Malaysia and Indonesia.

Spray Dryer Absorber

Spray Dryer Absorbers (SDAs), Figure 6, have primarily found widespread usage for FGD service for industrial processes. SDA systems are calcium-based FGDs. The calcium raw material is either "burnt lime" (CaO) or hydrated lime ($Ca(OH)_2$), where the raw material decision is based on local availability and costs. In the case of burnt lime, it is delivered as "pebble lime" to the SDA facility. An onsite hydration system (that is, a lime slaker system) hydrates the lime onsite and mixes it with water to make an approximately 20 to 30 percent by weight slurry. The slurry is pumped to the SDA reactor, where it is atomized into the flue gas. Both rotary atomizers and dual-fluid nozzle atomizers have been used. The finely atomized slurry, once injected into the SDA vessel, begins to dry and simultaneously adsorb SO_2 from the flue gas. Adsorbed SO_2 reacts with the calcium in the droplets to form calcium sulphate and calcium sulphite.

The SDA reactor vessel is a large-diameter down flow (co-current) reactor vessel with 8 to 10 seconds of flue gas residence time. By the time the flue gas has reached the bottom of the SDA vessel, all of the atomized water droplets have evaporated and the flue gas contains dry, finely divided calcium sulphite/calcium sulphate particles. The flue gas and particulates are then ducted to either an electrostatic precipitator or fabric filter for removal. Fabric filters are preferred because additional SO_2 removal occurs as the flue gas passes through the dust cake on the filter bags. The cleaned flue gas (both particulate matter, SO_2, and other acid gases such as hydrochloride [HCl]) is then ducted to induced draft (ID) fans and then to the stack.

Typical SO_2 removal efficiencies of 90 percent are easily achieved. Higher efficiencies (92 to 94 percent) are possible, but excess lime must be used to achieve them.

Figure 6. Exterior of spray dry absorber

Particulate Standards

In addition to the development of ambient air quality standards for SO_2, standards were set for particulate beginning in 1971. As shown in Table 2, the ambient standards for particulate have changed, both in the size of the particles being regulated as well as the ambient concentration considered acceptable. The first standard for TSP was generally considered particles 50 µm in diameter or smaller. The PM_{10} standard set in 1987 limited the regulated particles to those 10 µm in aerodynamic diameter and smaller. Beginning in 2006, the USEPA further regulated the smallest of the PM_{10} particles to those 2.5 µm in diameter and smaller.

While the response to SO_2 ambient standards was largely through process changes, particulate controls have largely been postprocess or through add-on controls. The following sections summarize the many innovations and work practices established since the early 1970s to control particulate emissions.

High Heat Bag Houses

Fabric filters or bag houses, Figures 7 and 8, have been used to control particulate for a long period of time, and were originally developed to collect valuable dusts from non-ferrous processes (Chandler 1992). It was the invention of heat resistant fabrics in the late 1970s that made these controls compatible with metals processing (Neundorfer Particulate Knowledge 2013), as these high heat fabrics are able to withstand temperatures above 350°F. Fabric filter bag houses are able to remove over 99 percent of particulate from a gas stream, though the efficiency will vary with the types of particles, particle size distribution, and bag house design.

Fabric filter collectors remove particles from the gas stream by passing the gas through a tightly woven fabric. Particulate is collected on the fabric by sieving and other mechanisms. The dust cake that forms on the filter from the collected particulate contributes significantly to collection efficiency.

Practical application of fabric filters requires the use of a large fabric area to avoid an unacceptable pressure drop across the fabric. To provide a large fabric area in a small space, the fabric is formed into cylindrical bags. Each bag may be 6 to 9 m long and 130 to 300 mm in diameter; a fabric filter for a 50-MW boiler may have 1,000 separate bags, with a total fabric area approaching 9,300 m^2. Groups of bags are placed in isolable compartments, to allow cleaning of the bags, or to allow replacement of some of the bags without shutting down the entire fabric filter.

Figure 7. Outside of industrial bag house

Fabric filter size for a particular unit is determined by the choice of air-to-cloth ratio, or the ratio of air flow to cloth area, typically expressed in feet per minute (cubic feet per minute of flow divided by square feet of fabric area). The selection of air-to-cloth ratio depends on the particulate loading and characteristics, and the cleaning method used. A

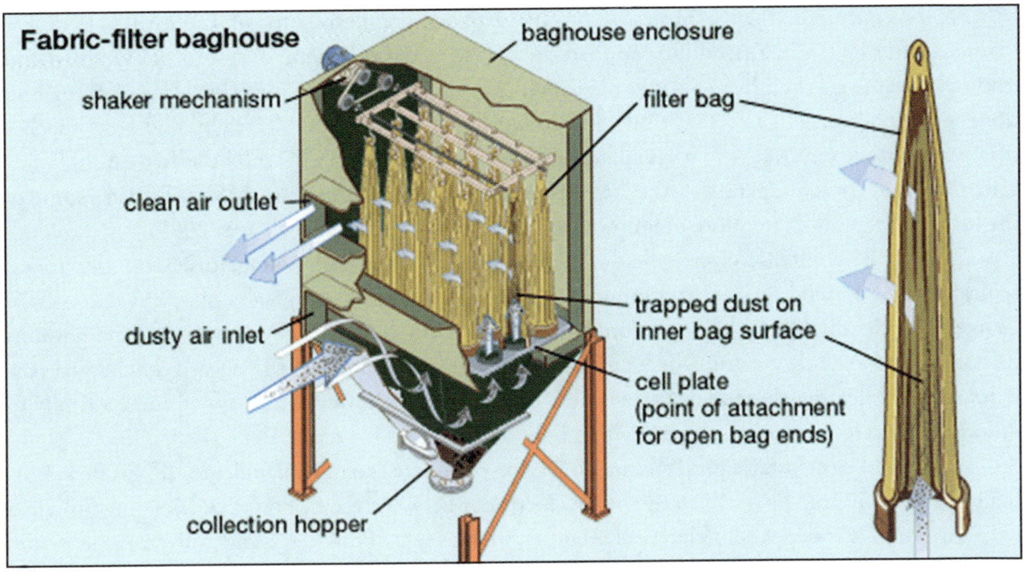

Figure 8. Fabric filter bag house schematic (Encyclopaedia Brittanica 2000)

high particulate loading will require the use of a larger fabric filter in order to avoid forming too heavy a dust cake, resulting in an excessive pressure drop.

Two major fabric-filter types are used: reverse-air fabric filter and pulse-jet fabric filter. In a reverse-air fabric filter, the flue gas flows upward through the insides of vertical bags which open downward. The fly ash collects on the insides of the bags, and the gas flow keeps the bags inflated. To clean the bags, a compartment of the fabric filter is taken offline, and the gas flow in this compartment reverses. This causes the bags to collapse and collected dust to fall from the bags into hoppers. (Shaking or other methods may be necessary to dislodge the dust from the bags.) The cleaning cycle in a reverse-air fabric filter typically lasts about 3 minutes per compartment. Because reverse-air cleaning is gentle, reverse-air fabric filters typically require a low air-to-cloth ratio of 0.01 m/s.

In a pulse-jet fabric filter, dirty air flows from the outside of the bags inward, and the bags are mounted on cages to keep them from collapsing. Dust that collects on the outsides of the bags is removed by a reverse pulse of high-pressure air. This cleaning does not require isolation of the bags from the flue gas flow, and thus may be done online. Because pulse-jet cleaning is harsh, the bags remain relatively clean, so that a higher air-to-cloth ratio of 75 $m^3/m^2/h$ (that is, a smaller fabric filter) may be used in industrial applications.

Fabric filters are good for collecting particles with resistivities either too low or too high for collection with electrostatic precipitators. Fabric filters therefore may be good candidates for collecting fly ash from low-sulphur coals, and possibly for fuel oil combustion (which may have high resistivity) or fly ash containing high unburned carbon levels (which has low resistivity).

Electrostatic Precipitators

Electrostatic precipitators (ESP) use electrical fields to remove particulate from flue gas (Nic et al. 2006). Because precipitators act only on the particulate to be removed and only minimally hinder gas flow, they have very low pressure drops and low energy requirements and operating costs.

Fredrick Cottrell, then a professor of chemistry at the University of California, Berkeley, invented ESPs in 1907. Cottrell first applied the device to the collection of sulphuric acid mist and lead oxide fume emitted from various acid-making and smelting activities (Nic et al. 2006). ESP designs have advanced since that time, and today ESP units use less electricity and have mechanisms for particle removal. The particulate removal efficiency for an ESP can be as high at 99.9 percent. The high costs of operation, the large space needed, and maintenance of the ESP are some of the reasons bag houses are more predominant today (Drummond and Chaves 2009).

In an ESP, an intense electric field is maintained between high-voltage discharge electrodes, typically wires or rigid frames, and grounded collecting electrodes, typically plates. A corona discharge from the discharge electrodes ionizes the gas passing through the precipitator, and gas ions subsequently ionize fly ash (or other) particles. The electric field drives the negatively charged particles to the collecting electrodes. Periodically, the collecting electrodes are rapped mechanically to dislodge collected particulate matter, which falls into hoppers for removal.

In a typical electrostatic precipitator, collecting plates are arranged parallel to the gas flow, normally 230 to 450 mm apart, with discharge electrodes between them, Figure 9. Most precipitators have three to five independent electrical sections, such as sets of discharge and collecting electrodes with independent power supplies in series. Each independent section removes a fraction of the particulate in the gas stream. This arrangement allows the use of higher voltages in the first sections of the precipitator, where there is more particulate to be removed. Lower voltages must be used in the final, cleaner precipitator sections to avoid excessive sparking between the discharge and collecting electrodes. In a precipitator with only one electrical section, the power input would be limited to

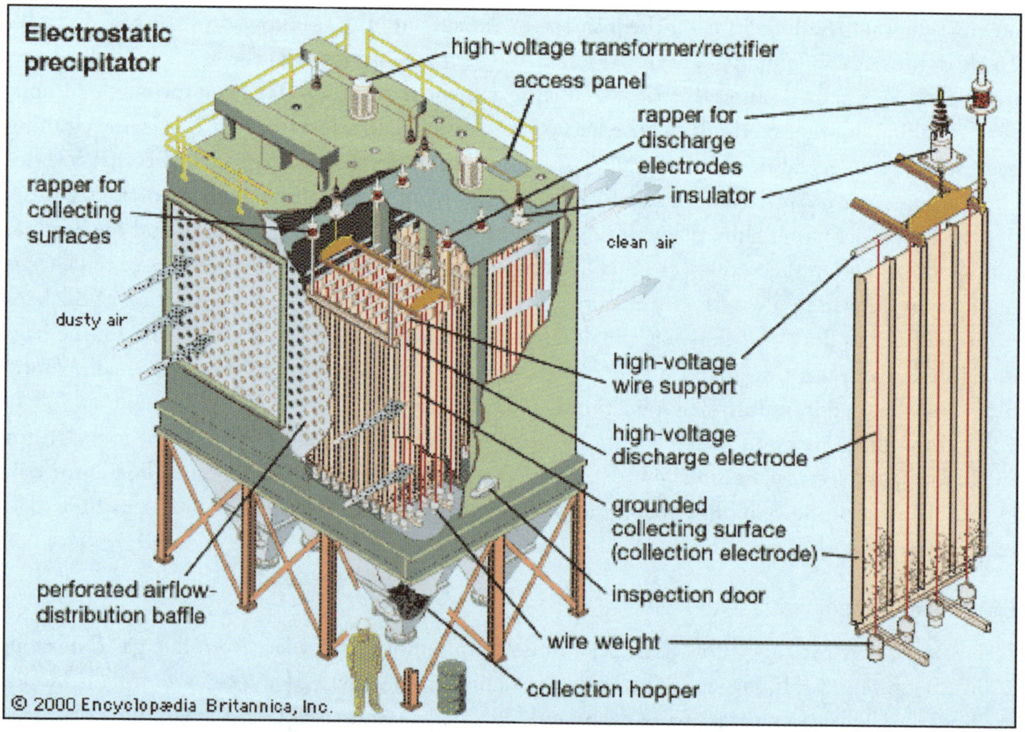

Figure 9. Electrostatic precipitator schematic (Encyclopaedia Britannica 2000)

the input, which would cause sparking at the precipitator exit, thus limiting the performance of the entire precipitator.

Precipitator sectionalisation has the added advantage that particles re-entrained in the flue gas stream by rapping may be collected in downstream sections of the precipitator, thus minimizing net rapping re-entrainment losses.

Factors limiting precipitator performance are flow non-uniformity and re-entrainment. More uniform flow will ensure that there are no high gas velocity, short treatment time paths through the precipitator. Attaining flow uniformity also will minimize "sneakage," or gas flows bypassing the electrical fields.

Re-entrainment of collected particles may occur during rapping. Proper rapper design and timing will minimize rapper re-entrainment. Maintenance of appropriate hopper particulate levels and of flow uniformity will minimize re-entrainment of particulate matter from the hoppers.

Unlike dry electrostatic precipitators, which use rapping to remove particulate from the collecting electrodes, wet electrostatic precipitators use a water spray to remove this particulate. A typical wet configuration has (vertical) cylindrical collecting electrodes, with discharge electrodes located in the center of the cylinders. Wet precipitators are useful in obtaining low opacities through the removal of acid gases and mists in addition to fine particulate. In addition, these devices have no rapping re-entrainment losses, and no back corona.

Wet electrostatic precipitators are substantially more expensive than dry electrostatic precipitators and are typically applied on smaller gas volumetric flow rates or where acid mist and or condensing organic aerosol control are needed.

Particulate Controls for Mines and Raw Material Handling

Fugitive emissions of particulates from mines, run-of-mine ore, and tailings were largely unregulated until the standard for PM_{10} was promulgated in 1987, Table 2. Now these activities are routinely controlled and new innovations are being developed.

Haul Roads. Haul roads are now routinely watered and sometimes a dust suppressant is applied. Common dust suppressants include magnesium chloride, lignin sulphonate, mineral oils, petroleum based oils, and rock; in rare instances, roads are paved. Road paving is very expensive and sufficient paving must be applied to stand up to the heavyweight haul trucks. In areas where rain, snow and ice occur during colder months, paving is not an option. The effectiveness of the different methods depends on the type of vehicles, the native soil content, and the road configuration. Regardless of the type of palliative, it is more difficult to control sections of roadways with slopes and curves, and where vehicles change velocity.

Secondary impacts resulting from the use of dust suppressant were studied by an expert panel sponsored by the USEPA (Piechota et al. 2002), and the recommendations of the expert panel summarized. If there is a potential for runoff from roadways to enter a potable water source, panel recommendations should be reviewed prior to the use of a dust suppressant.

Raw Material Handling. Fugitive emissions from mine ores and coal feeds are commonly controlled today. Conveyors are enclosed and water sprays are employed to minimize particulates. Telescoping chutes and lower drop heights are used to minimize emissions from stockpiling. Crushers use partial enclosures and water sprays. Where full enclosures are possible, bag houses and cyclones are added. If stockpiles are undisturbed, covers or watering is sometimes required. Dust control is not only used for environmental management purposes: in the South African gold mines, for example, it is common for the dust to have a gold grade twice that of the total ore. Therefore, there is a direct economic driver for dust mitigation.

Blasting. More recently, emissions from blasting are being regulated. Particulate controls used include minimizing the blast area and not clearing the area to be blasted of native vegetation and rock, as much as possible. After blasting, applying water to the area as soon as the all-clear signal is provided helps to minimize fugitive emissions.

Tailings Dams. Tailings dams associated with ore processing that have been in place for decades can become quite large. In Globe, Arizona, the United States Geological Survey updated the topographical map to include the tailings from a local smelter when the tailings dam exceeded 500 feet in height. As metal recovery technologies have improved, some operations are mining the tailings for additional metals production; however, fugitive particulate emissions are still an issue for many facitlities.

At the Freeport-McMoran facility in Miami, Arizona, vegetation reduces the amount of fugitive dust. Vegetation establishment has been facilitated through a cattle operation.

The tailings impoundment owned by Kennecott Utah Copper outside of Salt Lake City, Utah, has an air permit. The facility is required to keep the tailings impoundment watered, apply dust palliatives, and, for long-term dust control, establish vegetation on undisturbed areas. The tailings from the concentrator are placed in the impoundment in a slurry, which essentially fully mitigates fugitive emissions during placement.

The coal industry continues to face increasingly tighter air pollution requirements. For example, the Utility Maximum Achievable Control Technology Rule (UMACT), a rule finalized in February 2012, targets emissions of mercury and other air pollutants from power plants. The rule imposed new numerical emission limits for mercury, particulate matter, and HCl at electric generating units that use coal or oil.

INNOVATIONS TO REDUCE AIR POLLUTION IN THE COAL INDUSTRY

The 1990 CAAA constituted a major reauthorization measure that had immense impacts on the production and burning of coal. The 1990 CAAA promoted the use of low sulphur coal and natural gas, as well as innovative technologies to clean high sulphur coal. For example, in Kentucky, where much of the coal production has a higher sulphur content, production of coal decreased from 173.3 million tons in 1990 to 108.8 million tons in 2011 (United States Energy Information Administration 2012).

The efforts of federal government and industry in the United States have resulted in significant technological advances in reducing air pollution related to the coal industry. A major public/private partnership over the past 15 years has been the Clean Coal Technology (CCT) Program established by the U.S. Department of Energy (DOE). Since 1986, public/private partnerships of the CCT Program have engaged in innovative research, development, and demonstration of clean coal technologies that ultimately can be brought to large-scale deployment. Thirty-five pioneering projects in 18 states became part of the original CCT Program, carried out from 1986 through 1993. Advances in technologies included low NO_x burners, selective catalytic reduction (SCR), FGD, fluidized bed combustion (FBC), and integrated gasification combined cycle (IGCC). DOE's Clean Coal Power Initiative (CCPI) is providing government co-financing for new coal technologies that can help utilities cut sulphur, nitrogen, and mercury pollutants. The CCPI includes an effort to integrate advanced technologies in the world's first near-zero-emissions coal plant employing carbon capture technologies. DOE's Power Plant Improvement Initiative was established in October 2000 to further the commercial-scale demonstration of clean coal technologies at existing and new electric generating facilities.

According to DOE's National Energy Technology Laboratory (NETL 2011), the United States has reduced its NO_x emissions by 88 percent and SO_2 emissions by 82 percent from 1970 through 2011 (NETL 2011). Major contributors to these reductions have been the scrubbers, low-NO_x burners, and selective catalytic reduction systems demonstrated through the CCT Program. In addition, mercury controls research conducted as part of the CCT Program has resulted in reducing the cost of mercury emitted by coal power plants by 50 to 70 percent (NETL 2011). A signature technology of mercury research is activated carbon injection (ACI).

TRENDS IN AIR QUALITY REGULATIONS

As shown in Table 1, the WHO is recommending a 24-hour ambient standard for SO_2 of 20 μg/m^3 and a 10-minute standard of 500 μg/m^3. This is significantly lower than any standard adopted by an individual country or area.

The standards documentation note the following (WHO 2005):

> There is still considerable uncertainty as to whether SO_2 is the pollutant responsible for the observed adverse effects or whether it is a surrogate for ultrafine particles or some other correlated substance.

Included in this statement are many of the emerging trends in air quality regulation. SO_2 standards are becoming stricter and being implemented in areas without previous standards, such as Singapore. The country of Singapore has established a goal of meeting the WHO standard by 2020.

The movement toward regulating smaller and smaller particulate (as summarized in Table 2) is continuing. The WHO uses the term "ultrafine particles" to describe particles smaller than 1 mm. The USEPA has asked for more information regarding this smaller particle and while preliminary, the regulation of these smaller particles is being discussed in scientific circles.

A concern of particulate isn't just the size of the particle but the chemistry of the particle. Lead, added to the list of criteria pollutants in 1977, had a maximum allowable level of 1.5 μg/m^3 of ambient air averaged over a calendar quarter and measured at or outside the industry fence line (Witkowsky et al. 1986). On October 15, 2008, USEPA substantially strengthened the NAAQS for lead. The revised standards are 10 times tighter than the previous standards and will improve health protection for at-risk groups, especially children. USEPA has revised the level of the primary (health-based) standard from 1.5 μg/m^3 to 0.15 μg/m^3, measured as TSP. USEPA has revised the secondary (welfare-based) standard to be identical in all respects to the primary standard. Scientific evidence about lead and health has expanded dramatically since USEPA issued the initial standard of 1.5 μg/m^3 in 1978. More than 6,000 new studies on lead health effects, environmental effects, and lead in the air have been published since 1990. Evidence from health studies shows that adverse effects occur at much lower levels of lead in blood than previously though (USEPA 2008).

As the largest integrated lead producer in the Western Hemisphere, The Doe Run Company (Doe Run) has a longstanding interest in technology to improve the performance and efficiency of production. Doe Run is based in St. Louis, Missouri, a state rich in leadbearing ore with a long history of mining and lead metal production.

Doe Run's goals for any process to convert lead concentrates to finished metal include the following: (1) provide a high recovery rate for metal from lead concentrates, greater than 96 percent, (2) achieve Doe Run's highpurity product standards, 99.99 percent pure lead, (3) ensure cost competitiveness, (4) meet and outperform current and future anticipated regulatory standards, and (5) develop a process that delivers on community and societal expectations for lead metal production in the 21st century. Doe Run evaluated commercially available alternatives and concluded they could not meet these goals (Pyatt and Maccagni 2010).

Doe Run has been pursuing a new way to produce lead for almost 20 years. Known as "Flubor," the new technology was patented by Engitec Technologies S.p.A., an Italian engineering company (St. Louis Today 2010). This hydrometallurgical approach involves the dissolution of the lead concentrates in a fluoboric acid solution. The lead is subsequently recovered from the solution using a proprietary electro-winning technology.

It is projected that the hydrometallurgical process will significantly reduce air emissions of regulated gases, particulates, and lead. SO_2 regulations adopted by the Missouri Air Conservation Commission in 2009 will require a phased elimination of SO_2 emissions at the Herculaneum smelter by 2017. The NAAQS adopted in 2008 mandate a 90 percent reduction of lead in ambient air from 1.5 $\mu g/m^3$ to 0.15 $\mu g/m^3$, also by 2017. The new technology achieves a reduction in lead emissions of more than 99 percent and virtually eliminated emissions of SO_2.

Doe Run is an example of a company that realized that the continued downward trend in air emissions regulations would likely make lead smelting an uneconomic technology. As a result, Doe Run has looked to innovations in the hydrometallurgical approach for the recovery of lead.

The newest class of pollutants to be regulated is the GHGs. These pollutants are regulated in different ways throughout the world, or not regulated at all. The EU has implemented a cap-and-trade program, and a similar program proposed in Australia has been stayed. Whether the cap-and-trade program results in true GHG reductions of "carbon leakage" as energy production and GHG-emitting facilities are moved to areas without cap-and-trade requirements remains to be seen (St. Louis Today 2010).

The United States now requires major new sources of GHG emissions to implement Best Available Control Technologies (BACT). Without successful carbon capture and sequestration demonstrations being made, BACT is currently limited to energy efficiency. Several countries have signed onto the Kyoto accord and have implemented voluntary carbon emission reductions. Several other countries have acknowledged the need to reduce GHG emissions but have not implemented any mandatory programs.

CONCLUSIONS

Air quality regulations have driven many innovations in the mining and smelting industries and have spawned new technologies in hydrometallurgy. Regulations have become stricter over time, and the types of air emissions being regulated have increased. The current trends are for lower limits on emissions, smaller particle regulation, greater energy efficiency, and renewable energy resources. Government and industry have made substantial gains to reducing air pollution through technological developments over the past several decades. New innovations must be made to meet the growing demand for clean air and lower GHG emissions.

ON-LINE RESOURCES

Because air quality regulatory innovations are not stagnant, a list of online resources that are routinely updated is provided below. This list is a starting point for finding new innovations as they are developed.

Best Available Techniques: http://eippcb.jrc.es/reference/.

Compilation of Air Pollutant Emission Factors, AP-42: http://www.epa.gov/ttnchie1/ap42/.

Department of Energy, Clean Coal Technology: http://www.fossil.energy.gov/programs/powersystems/cleancoal/.

RACT/BACT/LAER Clearinghouse: http://cfpub.epa.gov/rblc/index.cfm?action=Search.BasicSearch&lang=eg.

Western Regional Air Partnership (WRAP) Fugitive Dust Guidance: http://www.wrapair.org/forums/dejf/fdh/index.html.

REFERENCES

Agricola, A. 1556. *De Re Metallica.* Translation from Latin by H.C. Hoover and L.H. Hoover, Modern publication by Dover Publishers, Inc., New York 1950.

Air Pollution Control Act of 1955 (Public Law 84-159, Chapter 360, 69 Statute 322).

Burckle, John. 1984. *Evaluation of an Air Curtain Flooding System for a Primary Copper Smelter Converter, Vol. 1.* Report prepared for the U.S. Environmental Protection Agency, Industrial Environmental Research Laboratory. February 1984, p. 4.

Chandler, S. 1992. Mankind, Minerals and the Environment. Emerging Process Technologies for a Cleaner Environment. Society for Mining, Metallurgy & Exploration Inc. p. 3–6.

Clean Air Act of 1963 (Public Law 88-206, 77 Statute 392).

Clean Air Act Amendments of 1963 (Public Law 91-604, 81 Statute 486).

Clean Air Act Amendments of 1990 (Public Law 101-549, 42 Statute 7661).

Drummond, M.V., and Chaves, C.A. 2009. "Sustainable Steel Plant on an Environmental Protection Area." SDIMI Conference, Gold Coast, Queensland. July 8, 2009, p. 123–133.

Ecorys. 2011. *Competitiveness of the UE Non-ferrous Metals Industries Final Report.* April 5, 2011, p. 89–100.

Encyclopedia Britannica, Inc. 2000.

European Commission, Integrated Pollution Prevention and Control (IPPC). 2001. Reference Document on Best Available Techniques in the Non Ferrous Metals Industries. December 2001.

Habashi, F. 2005. A Short History of Hydrometallurgy. *Hydrometallurgy, Vol. 79*, p. 15–22.

Jacobs Engineering. 2013. Acid Plants. Reference http://www.jacobs.com/products.aspx?id=6912.

Kovacs-Figueroa, C. 1976. Morenci Tailings Leach Plant. *Transactions of Society of Mining Engineers, AIME, Vol. 260*, p. 170–173.

Michon, A.E. 1982. A New Source of Energy for Texas – Western Coal. *Presentation at the SME-AIME Annual Meeting, Dallas, Texas.* February 14–18, 1982, p. 1–5.

National Energy Technology Laboratory (NETL). 2011. 2011 Accomplishments. U.S. Department of Energy.

Neundorfer Particulate Knowledge. 2013. reference http://neundorfer.com/knowledge_base/baghouse_fabric_filters.aspx.

Nic, M., J. Jirat, and B. Kosata, eds. 2006. "Electrostatic Precipitator." *IUPAC Compendium of Chemical Terminology* (Online ed.). doi:10.1351/goldbook.E02028. ISBN 0-9678550-9-8. http://goldbook.iupac.org/E02028.html.

Piechota, T., van Ee, J., Batista, J., Stave, K., James, D., editor. 2002. Potential Environmenatal Impacts of Dust Suppressants: Avoiding Another Times Beach, An Expert Panel Summary. May 30–31, 2002.

Pyatt, J.L. and Maccagni, M. 2010. Meeting the Needs of Tomorrow: a Breakthrough Technology for Producing Lead Metal. *Lead – Zinc 2010.* Vancouver. October 2010, p. 47–57.

St. Louis Today newspaper. 2010. "Doe Run's New Technology Could End Need for Lead Smelter." March 21, 2010.

United States Energy Information Administration. 2012. Annual Coal Report 2011. November 2012. www.eia.gov.

United States Environmental Protection Agency. 1986. AP-42, Fifth Edition, Volume I, Chapter 12: Metallurgical Industry, Section 3, Primary Copper Smelting, p. 6, reference http://www.epa.gov/ttn/chief/ap42/ch12/index.html.

United States Environmental Protection Agency. 2008. Fact Sheet: Revisions to the Lead Ambient Air Monitoring Requirements. October 2008 http://www.epa.gov/air/lead/pdfs/20081015pbfactsheet.pdf.

United States Environmental Protection Agency. 2012a. History of the Clean Air Act. reference http://epa.gov/oar/caa/caa_history.html.

United States Environmental Protection Agency. 2012b. EPA Revises the National Ambient Air Quality Standards for Particulate Pollution. December 14, 2012. reference http://www.epa.gov/pm/actions.html.

United States Environmental Protection Agency. 2012c. History of the National Ambient Air Quality Standards for Particulate Matter During the Period 1971–2012. reference http://www.epa.gov/ttnnaaqs/standards/pm/s_pm_history.html.

United States General Accounting Office. 1986. "Sulfur Dioxide Emissions from Nonferrous Smelters Have Been Reduced." Report to the Chairman, Subcommittee on Oversight and Investigations, Committee on Energy and Commerce House of Representatives, AIR POLLUTION.

Witkowsky, D.S., R.D. Smith, O.A. Kiehn, D.R. Wilburn, and R.C. Bowyer. 1986. Lead Reduction in Ambient Air: Technical Feasibility and Cost Analysis at Domestic Primary Lead Smelters and Refineries. *SME Fall Meeting, St. Louis, Missouri*. September 1986, p. 1–13.

World Health Organization. 2005. WHO air quality guidelines for particulate matter, ozone, nitrogen dioxide and sulfur dioxide, Global update 2005, Summary of risk assessment.

ABOUT THE AUTHORS

Catherine "Carrie" MacDougall has more than 25 years of experience in air quality studies with mines and smelters since earning her chemistry degree. She has developed emissions measurement techniques for mines and smelters, written numerous articles on emission controls and permitting strategies, and represents mines and smelters on regulatory issues. Her recent work includes fugitive air toxic compound emissions testing from smelters and mine reclamation plans in China and Malaysia. She is known for the "MacDougall Method" for estimating wind-blown particulate emissions from land without vegetation (i.e., tailings), served on the USEPA Dust Palliative Expert Panel, and is a fellow member of the Air & Waste Management Association.

Michael D.S. Blois is vice president of mining, water & environment at CH2M HILL. He graduated from the Royal School of Mines, Imperial College, London, with a degree in mineral technology. He also has a master's degree in business leadership from the University of South Africa. He is a member of SME; a chartered engineer through the Institute of Materials, Minerals and Mining, a professional engineer through the Southern African Institute of Mining & Metallurgy; and a qualified professional member of the Mining and Metallurgical Society of America. He has more than 35 years of international experience related to minerals processing and extractive metallurgy including plant operations, feasibility studies, process design, detailed design, construction, commissioning, and project management. He is currently the vice president of CH2M HILL's Mining, Water and Environment group. He has published five papers relating to mineral processing and metallurgy including "Plant Design, Layout and Economic Considerations" for SME's *Mineral Processing Plant Design Handbook* 2002 Update.

Robert (Bob) W. Reisinger is a certified professional engineer and has degrees in mining engineering and environmental engineering. Bob has more than 30 years of experience in the mining industry focusing on closure/reclamation, engineering feasibility studies and economic analyses, environmental impact assessments, and assessment of regulations related to mining sites throughout the world. He has a strong mining and environmental background with experience in both the private and public sectors.

Index